Setting Up Control Charts Using Exponentially-Weighted Moving Averages

$$\overline{X} = \sum_{t=1}^{k} X_t / k$$

$$s_X = \left[\sum_{t=1}^{k} (X_t - \overline{X})^2 / (k-1) \right]^{1/2}$$

$A_t = rX_t + (1-r)A_{t-1}$ where $A_0 = \overline{X}$

$V_t = rD_t + (1-r)V_{t-1}$ where $D_t = |X_t - A_{t-1}|$ and $V_0 = s_X$

A-Chart
CL: $\overline{X}$

$\text{LCL}_A, \text{UCL}_A: \overline{X} \pm A^* s_X$

V-Chart
CL: $d_2^* s_X$

$\text{LCL}_V, \text{UCL}_V: D_1^* s_X, D_2^* s_X$

r	A^*	D_1^*	D_2^*	d_2^*
0.100	0.688	0.390	1.247	0.819
0.200	1.000	0.197	1.486	0.841
0.250	1.132	0.109	1.597	0.853
0.333	1.342	0.000	1.780	0.874
0.500	1.732	0.000	2.164	0.921

Process Capability Formulas

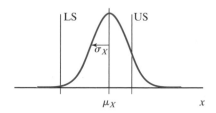

$Z_{LS} = (LS - \mu_X)/\sigma_X$ $Z_{US} = (US - \mu_X)/\sigma_X$

$\%\text{Conforming} = P(Z_{LS} \leq Z \leq Z_{US})$

$C_p = (US - LS)/6\sigma_X$

$C_{pk} = \text{Min}(-Z_{LS}, Z_{US})/3$

Some Useful Formula

Sampling Distribution of $\overline{X}$

$\mu_{\overline{X}} = E[\overline{X}] = \mu_X$

$\sigma_{\overline{X}}^2 = \text{Var}[\overline{X}] = \sigma_X^2/n$

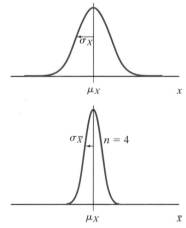

Additive Law of Variances

If $W = A + B + C$, with A, B, and C as independent random variables, then $\sigma_W^2 = \sigma_A^2 + \sigma_B^2 + \sigma_C^2$.

Quadratic Loss Function

$L(x) = k(x - m)^2$, where m is the nominal value.

Expected Loss from a Process

$E[L(X)] = k[(\mu_X - m)^2 + \sigma_X^2]$

Formulas for Setting Up $100(1 - \alpha)\%$ Confidence Intervals for the Mean

Variance Known	Variance Unknown
$\overline{X} \pm z_{1-\alpha/2}\sigma_X/\sqrt{n}$	$\overline{X} \pm t_{\nu, 1-\alpha/2} s_X/\sqrt{n}$

Calculating Probabilities Based on the Normal Distribution

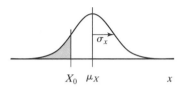

$P(X \leq x_0) = P(Z \leq z_0)$ where $z_0 = (x_0 - \mu_X)/\sigma_X$

Statistical Quality Design and Control

Contemporary Concepts and Methods

Statistical Quality Design and Control

Contemporary Concepts and Methods

SECOND EDITION

Richard E. DeVor
Department of Mechanical Science and Engineering
University of Illinois at Urbana–Champaign

Tsong-how Chang
Department of Industrial and Manufacturing Engineering
University of Wisconsin–Milwaukee

John W. Sutherland
Department of Mechanical Engineering–Engineering Mechanics
Michigan Technological University

PEARSON
Prentice
Hall

Upper Saddle River, New Jersey 07458

Library of Congress Cataloging-in-Publication Data

DeVor, Richard E.
   Statistical quality design and control : contemporary concepts and
methods / Richard E. DeVor, Tsong-how Chang, John W. Sutherland.
     p. cm.
   Includes bibliographical references and index.
   ISBN 0-13-041344-5
   1. Quality control—Statistical methods. 2. Process
control—Statistical methods. I. Chang, Tsong-how II. Sutherland, John W.
(John William), 1958- III. Title.
   TS156.D53 2006
   658.5'62—dc22

                                   2006045791

Vice President and Editorial Director, ECS: *Marcia J. Horton*
Senior Editor: *Holly Stark*
Executive Managing Editor: *Vince O'Brien*
Managing Editor: *David A. George*
Production Editor: *Wendy Kopf*
Director of Creative Services: *Paul Belfanti*
Creative Director: *Jayne Conte*
Cover Designer: *Bruce Kenselaar*
Art Editor: *Greg Dulles*
Manufacturing Buyer: *Lisa McDowell*
Senior Marketing Manager: *Tim Galligan*

Printed in the United States of America

10 9 8 7 6 5 4 3 2 1

ISBN 0-13-041344-5

Pearson Education, Inc., *Upper Saddle River, New Jersey*
Pearson Education Ltd., *London*
Pearson Education Australia Pty. Ltd.,*Sydney*
Pearson Education Singapore, Pte. Ltd.
Pearson Education North Asia Ltd., *Hong Kong*
Pearson Education Canada Inc., *Toronto*
Pearson Educación de Mexico, S.A. de C.V.
Pearson Education—Japan, *Tokyo*
Pearson Education Malaysia, Pte. Ltd.

To

Jearnice

Linda, Alec, Anthony, and Amy

Brenda, Jenny, and Beth

DEDICATION

In October 1992, Professor Shien-Ming Wu departed this life. "Sam," as he was affectionately known, was our mentor and to this day continues to be a constant source of inspiration and guidance. Sam was the consummate teacher and the one most responsible for introducing us to the concepts and methods of statistics and their application to the world of engineering. With great respect and loving memory, we dedicate this second edition to Sam.

Contents

Preface

When the first edition of this book was published in early 1992, the watershed decade of the 1980s had just passed, during which the emphasis on quality in American industry had undergone tremendous change. With the leadership of several large companies, most notably those in the ground vehicle sector, and with the guidance of giants like Deming and Taguchi, a quality renaissance had occurred that left many buoyed by the momentum it had created, but only cautiously optimistic, owing to the experiences of the 1950s and 1960s. However, what companies like Ford and General Motors had done for themselves and their enormous supply bases in the 1980s was carried forward in the decade of the 1990s by companies like Motorola, General Electric, and others, and it soon became clear that this was a movement that was not going away. Domestic and international initiatives such as the National Institute of Standards and Technology's Malcolm Baldridge Award Program and ISO 9000 further codified and solidified "quality" as a quintessential ingredient in the quest for a competitive position in the world marketplace. Additionally, with the globalization of the economy over the last decade, driven by the confluence of a number of factors so aptly documented by Friedman[1] in *The World is Flat*, we have renewed confidence that this will likely not change in the years to come.

New to the Second Edition

This book has also now undergone some changes since its first publication in 1992. However, its core philosophy and principles remain unaltered. We built its message and its lessons on the backs of those whose vision and wisdom gave rise to that philosophy and those principles—Shewhart, Deming, Taguchi, Box, and others—and we hope that still comes out loud and clear. We have attempted to strengthen that element of the text with additional cases—actual case study experiences—that originally helped us to demonstrate that philosophy and those principles at work. At the same time, we have also gone back to the basics to strengthen our core with treatments like Deming's "Red Bead Experiment" and the "funnel experiment." Our treatment of the design of experiments has been expanded in this edition to include comprehensive coverage of simple comparative experiments and experimental optimization using the tools and methods of *Response Surface Methodology*. We have also expanded our treatment of tolerances—both design and analysis—and added more basic

---

[1]Thomas L. Friedman, *The World is Flat: A Brief History of the Twenty-First Century*, Farrar, Straus and Giroux, New York, 2005.

material on sampling distributions and inference via hypothesis testing and confidence interval methods. We have also updated the historical evolution of the quality movement with discussion of topics like "Six Sigma" and certification and award programs.

Computer-Based Workshops

Our use of computer-based workshops to demonstrate the principles and methods at work and provide hands-on experience with the application of the subject matter remains another core element in our presentation. These PC-based workshops were originally designed to allow students to simulate the operation of an actual process, collect data, develop statistical charts, read the statistical signals, diagnose the cause(s) of the process faults, formulate improvement actions, and then witness the impact that those actions have on the performance of the process. In this second edition we have added workshops on the design of experiments and updated the implementation of these workshops through the software support of today. This edition of the textbook now contains a total of five computer workshops. The software in support of these workshops is available for download on the web at http://www.me.mtu.edu/quality_book. Also available at the site are detailed workshop explanations and descriptions of software usage. In addition, the website has suggested student assignments for each workshop. We have also added many new homework exercises at the end of the chapters and updated the *Solutions Manual,* which is available for download from Prentice Hall.

Acknowledgments

No textbook can emerge without both the inspiration and the help of many others. The acknowledgment of others so important to the development of the first edition of this text remains as important to us now as it was then and so we repeat that acknowledgment here.

There are few involved in the teaching and practice of quality design and improvement who have not been influenced by and stand in continued awe of the work of Dr. W. Edwards Deming. His philosophy and teachings are a constant source of knowledge and inspiration and represents a level of ethics and conscience that is to be strived for continually but probably not reached by others.

Our associations and experiences with a large number of companies over the years have all been invaluable. But it is both fair and necessary to say that the opportunities we have had to interact with many within the Ford Motor Company and Caterpillar, Inc. have had a great influence on our thinking. For that we are forever grateful. At Ford, we are particularly grateful for our associations with many people in the then Plastic and Trim Products Division (formerly, the Plastic Products Division, and before that, the Plastics, Paint, and Vinyl Division) over a period of more than eight years. It is impossible to recognize all of those there who contributed in one way or another to this effort, but certainly the names of Jerry French, Robert Deacon, Peter Belaire, and Roger Sadowski quickly come to mind. Many others have contributed to our understanding of the concepts and methods herein and have participated in many of our efforts during this period. Three of those that should be mentioned are Professor Donald Ermer of the University of Wisconsin–Madison, Dr. Mark Lindsay, and Professor Kevin Dooley, now of Arizona State University. We would also like to acknowledge the encouragement and input we have received from several of our colleagues during the preparation of various drafts of this book. In addition to those already mentioned, we should include Professor Herbert Moskowitz of Purdue University and Professor Emanual Sachs of the Massachusetts Institute of Technology.

Dr. Thomas Babin of Motorola, Inc., participated with us in a variety of activities and in stimulating discussions that had a tremendous impact on this book. He is valued colleague and friend, and we most gratefully acknowledge his many contributions to this effort.

The preparation of the original manuscript was a long and humbling experience. The valuable assistance of Theresa Jordan, Mark Tornberg, and Daniel Waldorf in preparing many of the homework problems and solutions is greatly appreciated. The students in our classes over the years have provided invaluable comments and suggestions. The contributions of Amanda Horner at the University of Illinois at Urbana–Champaign in helping us to prepare various drafts and work our way through the many steps required in seeing this project through to completion will always be remembered and greatly appreciated. Becky Mullins of Champaign, IL has, over the years, provided valued support to many of the activities that have affected much of what is now contained herein.

Finally, the authors are most grateful for the support of their institutions, namely, the University of Illinois at Urbana–Champaign, the University of Wisconsin–Milwaukee, and the Michigan Technological University during this project.

There are also others whose contributions to this second edition have been enormous and we wish to single them out now. We are most grateful to Dr. William A. Berezowitz of General Electric Health Care Division for his assistance with Six Sigma material. Dr. Thomas Babin of Motorola's Physical Realization Labs also contributed suggestions in terms of Six Sigma and Baldridge content, and was most helpful in the introduction of a new case study in Chapter 23. A number of graduate students at Michigan Technological University in Houghton, MI have contributed to the second edition. Among those is Vishesh Kumar, whose efforts were crucial in maintaining our momentum and Karl Haapala, who led the charge in developing the computer workshops. Others who assisted in the development of homework exercises were Liu Hsiang Yun, Rajiv Nanjundareddy, Timothy Jenkins, Daniel Adler, and Prasad Shirodkar. Kirsten Holles at MTU was most helpful in keeping things moving. At the University of Illinois at Urbana-Champaign two students, Courtney Crump and Ricky Santoso, provided invaluable assistance in the development of the Solutions Manual. There were also a number of people who reviewed our plans for the second edition, steering the content and giving important feedback on both the original and the new material. These people include Bhaba R. Sarker (Louisiana State University), Val

Hawks (Brigham Young University), Mohamed H. Aboul-Seoud (Rensselaer Polytechnic Institute), and Alan Scheller-Wolf (Carnegie Mellon University).

Remembrance

When we were preparing the first edition of this book, our editor, David Johnstone, was not only our greatest champion and driving force, but someone who became a dear friend as well. His senseless death as a result of a random act of violence served to demonstrate both how fragile life can be and how important it is to take advantage of each moment.

R. E. DeVor
Champaign-Urbana, IL

T. H. Chang
Milwaukee, WI

J. W. Sutherland
Houghton, MI

Fundamental Concepts
and Methods

1

Evolution of Quality Design and Control

1.1 Introduction

The field of quality control underwent major changes over the last two decades of the 20th century. Since its birth as a specific discipline in the 1920s, quality control has taken its place as a central activity in the industrial system. Most industrial organizations, at least through the 1970s, had a "quality control" manager of a "quality control" department which had a function to administer "quality control" as outlined in the "quality control" manual. The American Society for Quality (Control) has been in existence as a major national organization since 1945. Quality control had been taught as a distinct and almost universally present curriculum component in industrial engineering programs at the university level for decades. But the face of quality control experienced a dramatic change in the years between 1980 and 2000. This change has not occurred as the result of a major breakthrough in the technology and science of quality control. Rather, it has occurred out of absolute necessity, as America's position in the world marketplace slowly eroded to alarming levels, beginning in the late 1970s.

The dramatic change in the meaning and application of quality control as a discipline has both a philosophical and an analytical side. This book covers both of these aspects in a comprehensive and balanced way. In this chapter we present a historical perspective of quality control with the hope that an understanding of its evolution will point to the need for emphasis on the philosophy of quality control as well as the analytical techniques for quality control. That these two aspects of the problem are presented and understood together is essential.

The first two chapters of this book are directed toward the philosophical side of quality control. In Chapter 1 we review the history of the field, introduce some of the pioneers and their basic philosophies, and begin to formulate the conceptual framework for quality design and improvement. In Chapter 2 we review the contributions of Deming and Taguchi, discuss the meaning of quality, and comment on the cost-of-quality issue, which has in the past gotten and will continue in the future to get more attention than it probably deserves.

1.2   Quality Revolution of the 1980s

During the 1980s a major revolution took place in the industrial sector in America, as manufacturers strove to regain their competitive position in the world marketplace. As we stand today in the early part of the 21st century and look back on the decade of the 1980s, it is fair to say that the period was probably the most influential in the whole history of quality control. One element of the revolution of the 1980s centered around a renewed emphasis on quality, with an approach aimed at preventing defective materials from being manufactured through improved process monitoring and diagnosis and at designing quality into the product from the very beginning. The concepts and methods of Deming and others have had a profound impact on the way we view quality from the manufacturing/process perspective. The simple but powerful statistical methods for process control developed by Shewhart more than 80 years ago had been successfully revived and applied on a very broad base. In the engineering design arena we looked to the Japanese and found that the methods of Taguchi had been used successfully for more than 40 years to provide a sound basis for improved product/process design. From the total system point of view the concept of *companywide quality control* (CWQC) that has been practiced in Japan for some time began to receive considerable attention. In particular, methods like *quality function deployment* (QFD) began to be practiced as a means to transmit customer needs through the organization both vertically and horizontally.

Major Change in Our View of Quality Control

Central to the quality movement of the 1980s was the tremendous emphasis on the role of statistical thinking and methods. Across the industrial and service sectors in America statistical education and training for employees from the shop floor to the boardroom was unprecedented in terms of both breadth of exposure and depth of conviction. No other issue in management or science and technology had received such attention over the several decades that preceded the 1980s. To many, the concepts and methods were totally new. But what was new about the use of statistical methods for quality and productivity design and improvement? It was not the methods themselves, for they had been with us for quite some time. What was new was the mindset that drove the use of these and other analytical methods toward the never-ending pursuit of quality and productivity design and improvement.

Beginning with the early 1980s, our thinking has undergone considerable transformation. The impetus lies largely with the teachings of men like W. Edwards Deming, who had played a central role in shaping the Japanese approach to manufacturing excellence over the previous 40 years, and who during the 1980s had a similar impact on the American manufacturing community. The renaissance lay not with the influx of advanced technology in manufacturing, although the impact of technological change in manufacturing cannot be dismissed as irrelevant either then or now. The renaissance lay, rather, with the management of manufacturing systems within a structure based on principles that guide how we should view the design and improvement of products and their associated manufacturing systems.

The work of Shewhart in the 1920s led to a sound approach to the scrutiny of process variation and the diagnosis and removal of process faults. However, the statistical approach to the sampling of process output prior to shipping to determine the extent to which it conformed to specifications dominated the quality field from the 1930s through the 1970s. Today, it is recognized that this product control approach to quality assurance contributes little to the enhancement of competitive position. Our recognition that quality and productivity can move together in the right direction only when we attack the process, finding the root causes of process faults and taking action to remove them, is today reshaping the meaning and intent of quality control.

Quality Effort: Comparison of the Approaches of Japanese and U.S. Companies

In the 1980s many U.S. companies began to examine carefully the "quality control element" within their organizations to identify the new directions that had to be cultivated to improve competitive position significantly in the long term. Companies sought to develop an overarching quality philosophy that could be implemented through policy and operating procedures to ensure that customers' needs could be met competitively. One thing that became clearly evident was the need to push the quality issue farther and farther upstream so that it became an integral part of every aspect of the product life cycle (e.g., marketing, product design and engineering, process engineering, and production).

In the decade of the 1980s U.S. companies gradually but steadily built a new quality base, moving to accomplish as quickly as possible what had evolved in Japan over a period of about 40 years. Given the crisis situation in the late 1970s and the way in which the U.S. approach to quality evolved through the 1950s and 1960s, U.S. efforts in quality had for some time been strongly centered in the problem-solving/firefighting arena, with less attention to the other end of the spectrum—product development. Figure 1.1 shows a comparison of the U.S. and Japanese quality effort by activity, as it was viewed then. Many U.S. managers quickly realized that to have "their curve" move to the same position as the Japanese curve over the next decade would require that the quality issue be pushed farther and farther upstream. The recognition of contributions of Genichi Taguchi in this regard was significant. The United States had to learn to execute and learn patience and persistence, virtues that had not necessarily been prominent in the character of American management.

The general acceptance of the concepts and methods of Taguchi within American industry over the decade of the 1980s was impressive, to say the least. Although in some quarters the willingness of industry to embrace the use

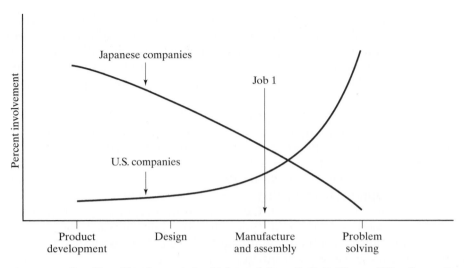

Figure 1.1   Quality effort by activity (Adapted from L. P. Sullivan, "The Seven Stages of Company-Wide Quality Control," *Quality Progress*, **May 1986.)**

of statistical methods for process control and design of experiments had been guarded, those advocating the methods of Taguchi met with considerably more enthusiasm. The "Taguchi movement" had strong and organized support and gathered momentum as reports of success were well documented and disseminated. In particular, the engineering community was quite receptive to the overall approach while being somewhat slower to embrace the more rigorous statistical approaches to process control and design of experiments. Whereas the latter found its roots in, and tended to be promoted by, the statistical community, the methods of Taguchi came from the direction of the engineering community and therefore appeared to be more easily accepted by those in design and manufacturing. Although some viewed the situation as strictly an either/or choice, it turned out that a thoughtful marriage of the two perspectives provides for a powerful approach to quality design and manufacturing.

1.3   Historical Perspective of the Management of Quality

Many of the systems firmly entrenched in the fiber of the management structure of corporate America must be changed. They simply are not consistent with the new philosophy of doing business that has been evolving over the last 20 years. They may be, in fact, the very things that kept us from properly viewing this philosophy years ago. We will not attempt to examine management style and structure comprehensively in this book, but we will look at a few key issues that dictate to a great extent our ability to be successful. The history of quality control goes as far back as history itself, but for our purposes here it is useful to pick up the time line sometime prior to the industrial revolution that began after the Civil War in the United States. Figure 1.2 traces this history forward to today.

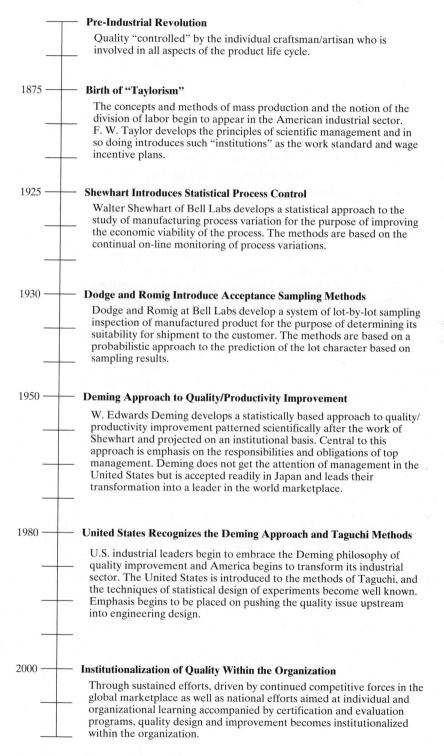

Pre-Industrial Revolution

Quality "controlled" by the individual craftsman/artisan who is involved in all aspects of the product life cycle.

1875   **Birth of "Taylorism"**

The concepts and methods of mass production and the notion of the division of labor begin to appear in the American industrial sector. F. W. Taylor develops the principles of scientific management and in so doing introduces such "institutions" as the work standard and wage incentive plans.

1925   **Shewhart Introduces Statistical Process Control**

Walter Shewhart of Bell Labs develops a statistical approach to the study of manufacturing process variation for the purpose of improving the economic viability of the process. The methods are based on the continual on-line monitoring of process variations.

1930   **Dodge and Romig Introduce Acceptance Sampling Methods**

Dodge and Romig at Bell Labs develop a system of lot-by-lot sampling inspection of manufactured product for the purpose of determining its suitability for shipment to the customer. The methods are based on a probabilistic approach to the prediction of the lot character based on sampling results.

1950   **Deming Approach to Quality/Productivity Improvement**

W. Edwards Deming develops a statistically based approach to quality/ productivity improvement patterned scientifically after the work of Shewhart and projected on an institutional basis. Central to this approach is emphasis on the responsibilities and obligations of top management. Deming does not get the attention of management in the United States but is accepted readily in Japan and leads their transformation into a leader in the world marketplace.

1980   **United States Recognizes the Deming Approach and Taguchi Methods**

U.S. industrial leaders begin to embrace the Deming philosophy of quality improvement and America begins to transform its industrial sector. The United States is introduced to the methods of Taguchi, and the techniques of statistical design of experiments become well known. Emphasis begins to be placed on pushing the quality issue upstream into engineering design.

2000   **Institutionalization of Quality Within the Organization**

Through sustained efforts, driven by continued competitive forces in the global marketplace as well as national efforts aimed at individual and organizational learning accompanied by certification and evaluation programs, quality design and improvement becomes institutionalized within the organization.

Figure 1.2   Historical evolution of quality control

Impact of F. W. Taylor

Much of what we now find at the foundation of management of industrial systems was born of necessity as far back as 120 years ago. As the industrial sector began the transformation from individual craftsman to cadres of workers organized toward the mass production of products, the principles of scientific management and the methods of shop management began to emerge. Great men such as Frederick Winslow Taylor rose to the task and pioneered the field of industrial management.

The most fundamental change came with the need to divide labor into component tasks. The worker no longer had the opportunity to fabricate from "beginning to end," personally monitoring and controlling the quality and productivity of his or her efforts. Now workers became responsible only for a small element of the complete manufacturing cycle, unable to see clearly or have any impact on what happened upstream or downstream from their new world. They saw only their assigned incremental tasks and often had no conception of the final product and its required function.

As a result of the division of labor and the subsequent narrow scope of involvement of a given worker in the total product life cycle, two basic problems arose. One relates to the quality issue. Since the worker seldom saw the final product or had any contact with its field use and since the tasks were quite repetitive and hence boring, it was more difficult for workers to generate ownership and sustain pride of workmanship in what they were doing. As a result the quality of the product suffered. The other basic problem relates to the productivity issue. For some of the same reasons stated above, workers did not always perform the tasks with the speed required to sustain a profitable situation.

Recognizing the problems associated with this new division of labor, Taylor and others realized that the management of workers must be put on a scientific basis. Workers must know what management expects of them in terms of personal performance. Management must be able to control the balance of flow in the system to achieve a steady and productive output. The simplest measurement of performance was time—how long a given task should take. All workers should be directed in some way to work steadily at a relatively fast pace. So the work standard was born and the methods of time study were developed to establish such standards scientifically. Amazing as it may seem, a look at Taylor's 1911 treatise[1] will reveal that the foundations of time study, work standards, and wage incentives, although massaged and embellished over the next 70 years, were still playing the same role philosophically in the world of the 1980s as they had in the infancy of America's industrialization. So entrenched did these methods become that even today they are taken for granted in some quarters as essential elements of the management of manufacturing.

Taylor's efforts in the science of management tended to deal with only one of the two problems created by industrialization, the productivity problem. With the introduction of the work standard, a fundamental change took place

---

[1]F. W. Taylor, *Principles of Scientific Management*, Harper & Brothers, New York, 1911.

that may, in part, be viewed as a driving force in bringing us to the point we are at today. Human beings were no longer responsible for the quality of what they were doing, but rather only the speed at which they were able to do it. Time study specialists observed people at work on an established task using predetermined methods and developed a standard time for that task. Certainly, these studies revealed faults with the division of tasks and the assignment of methods and led to improvements, but the preoccupation was on time—time study, time standards. The fundamental purpose was to invoke incentive pay systems: get people to work harder and pay them more for their efforts.

The establishment of the concept of work standards—a standard time or number of units to be produced per unit time—had a profound influence on the evolution of production management.

We must remember that the concept of the work standard and the methods of the time study were originally created by Taylor to deal with the fundamental issue of wage incentives—the need to make people work harder to drive down the labor cost of manufacture. These principles were, of course, picked up by others and used for other purposes. Gilbreth and others extended time study to the study of motions as the field of work design became formalized. The motivation again was sound: to improve the efficiency of tasks performed by labor and provide a better marriage of worker and machine in the workplace. But work standards were also used as a quantitative basis to negotiate rates of labor and reasonable expectations of workers. Work standards were used by management for internal budgeting purposes and as a basis for many decisions, such as the cost of internal manufacture versus outsourcing. In such instances the work standard is viewed as the accepted benchmark for what we can do at our best and hence will serve to camouflage the presence of improvement opportunity.

Today we recognize the impact of work standards as being completely opposite to that initially intended. As a target or objective, the work standard, in a sense, actually places an upper bound on productivity and stifles the pursuit of improvement beyond the standard. If the standard is being achieved, it would appear that there is no reason or motivation for continually looking for improvement opportunities. The methods introduced later by Shewhart, for example, could not be interpreted properly or adopted broadly, since they were viewed as embracing the issue of quality, not productivity, viewed as a tool to control the quality of the product, not the efficiency of the process.

Shewhart's View of Quality Control

In the 1920s at Bell Labs, Walter Shewhart, a physicist, began to formulate a statistically-based approach to quality control/improvement. It was Shewhart who first recognized the strong relationship between consistency of performance of products and their associated quality in terms of field performance. Furthermore, it was Shewhart who was able to link this to the manufacturing process in terms of its variation in performance. In his 1931 treatise,[2] Shewhart

---

[2] W. A. Shewhart, *Economic Control of Quality of Manufactured Product*, D. Van Nostrand, New York, 1931. Reprinted by the American Society for Quality (Control), Milwaukee, Wis.

speaks of the economic control of manufacturing operations and of the use of statistical methods for detecting a lack of control in processes through the study of the variation pattern of product and process quality characteristics over time. That the process was driven solely by a constant system of forces of variation was deemed necessary by Shewhart to guarantee the economic success of the process.

Shewhart's concept was to detect and eliminate the sources of variation that could not be attributed to the routine operation of the process. Unfortunately, again the presence of the work standard probably made it very hard to adopt this new philosophy. If the work standard was being met, the process must have been operating acceptably; we must look, then, for other ways to deal with the quality issue. The work standard had now become a misinterpreted and abused instrument. It was not being used for the purpose that Taylor had intended; instead, it had become an overall measure of performance of the combination of man and machine—a measure of process performance— perhaps *the* measure of process performance.

Inspection and the Quality Control Function

The second profound influence of Taylorism on manufacturing management was the need to establish an external (to the process) function for the control of product quality. Although Shewhart, and later Deming, promoted the establishment of definitions of quality that embraced the process, quality control was viewed as a screening process separate from manufacturing. It was the job of production to get the numbers out—meet the quotas established through the work standards. Since workers had only to achieve a productivity standard, someone else had to look out for quality.

The quality control function was developed to take care of the business of detecting and containing bad product—that which was outside of tolerances— so that it would not reach the customers' hands. The quality control department became the arch-adversary of the production department, one driving for the numbers, the other driving for product quality despite the numbers through inspection of product. And as the numbers grew, it became difficult to inspect all the parts. There were those who felt further that inspecting all the parts was unnecessary and ineffective. Probabilistic thinking was adopted to develop sampling procedures. A field of scientific research and development emerged, which was richly cultivated for more than 50 years. The underlying philosophy had not changed, only the methods used to invoke that philosophy.

While Shewhart was developing and refining his methods for the economic control of processes, H. F. Dodge and H. G. Romig also at Bell Labs were developing a system for the lot-by-lot inspection of in-process and finished goods.[3] Their "acceptance sampling" systems were founded on a solid probabilistic basis and were designed as a means to pass judgment on large quantities

---

[3]H. F. Dodge and H. G. Romig, *Sampling Inspection Tables: Single and Double Sampling*, 2d ed., Wiley, New York, 1959.

of goods—lots—using relatively small samples. Unfortunately, although these methods were probabilistically sound within their framework of product control, they overwhelmingly emphasized and hence overwhelmingly allocated resources to the "detection and containment of defective material."

One by-product of the Dodge–Romig approach to quality control was the acceptance of the concept of an acceptable quality level (AQL)—a target level of defective material, supposedly based on economic grounds. The most devastating effect of the AQL concept is that it promotes quality improvement up to an acceptable plateau, beyond which further improvements appear to be unjustified economically. Scrap became an accepted part of the business, unfortunate but inevitable. Certainly, scrap reduction wars were waged from time to time, but by and large, scrap became an integral part of the production and business planning activity. Deming clearly pointed out the fallacies in this way of thinking, but for many years (the 1930s through the 1970s) this basic approach to quality control prevailed.

World War II further fueled the fires of sampling inspection as the government, working in the best interests of the war effort, developed formal standards for sampling inspection, and strict adherence to the use of these standards became part of the contractual agreement signed by all suppliers of military goods. The job of managing the war procurement effort to ensure that high-quality goods were supplied to the government was ably handled by a then little-known senator from Missouri, Harry S. Truman. And in the short term such methods of inspection were undoubtedly necessary and effective in guaranteeing that only high-quality products reached field use. One can even argue that the philosophy on which the Military Standard 105D[4] was founded tended to force the supplier to try to do better when material was repeatedly rejected. But "reject and return" is in most cases not a feasible alternative. A rejected lot is merely a signal that someone has to do better. But how to do better?

It would be unfair to suggest that the quality efforts during World War II were solely of a product control—acceptance sampling—nature. During the famous War Production Board 8-day short courses on statistical methods for quality control, extensive teaching of Shewhart's control chart principles took place. But for numerous reasons that will come out as this book evolves, these methods did not gain widespread popularity in the industrial sector. The fundamental reason may be, as Deming pointed out later, that top management was not aware of the overarching concepts that drive Shewhart's approach and hence did not understand the broad-based benefits of the process control approach.

World War II ended, but sampling inspection had only just begun to flourish. Many companies adopted its use both internally and in dealing with suppliers. Military Standard 105D became heavily used and unfortunately abused on both philosophical and technical grounds. So strong was the impact of the military standard for sampling inspection that by the early 1960s an international standard was adopted. If such attention had then been given to the use of

---

[4]Military Standard 105D, *Sampling Procedures and Tables for Inspection by Attributes*, U.S. Government Printing Office, Washington, D.C., 1963.

statistical charting for process improvement, one can only speculate about its impact then or now, for that matter.

We should not dwell on the use of sampling inspection or inspection by any other means, since these are but tools devised to carry out a philosophy that grew out of the establishment of a separate and adversarial function to production—the quality control department. Work standards drove production and the worker, so someone had to take the responsibility for quality. Some quality control departments made use of statistical charting—sometimes to a fetish—diving headlong into the business of making charts. Occasionally, an action might be taken to solve a production problem based on charting information. But in general such work was not effective. How could it be, given the relative roles of production and quality control as defined and supported by management?

Deming: The Need for Change

Yet by and large, through all of this historical evolution, American industry prospered. Since everyone was basically "playing the game" by the same general rules, relative differences in performance tended not to be extreme. But then something happened to change all of this for certain types of industries. A strong competitor in the world marketplace began to emerge—Japan. And the competition was not necessarily playing by the same rules—not driven by the same internal forces that had been guiding the evolution of industrial development in America over a period of more than eight decades.

To respond to this new external force it would be necessary to adopt a new philosophy of production management. To this end, W. Edwards Deming developed his "14 points" to show management how to put quality control on an institutional basis rather than a departmental basis.

This phenomenon of "change of necessity" is not a new one. A bit of study of American industrial history shows that it is but a repetition of the past. At the dawn of the 20th century, Taylor proposed an important thesis that appears worthy of consideration today. In commenting on the practice of the management and labor in some of the important industries of the day, he said: "In these industries, however, although they are keenly competitive, the poor type of shop management does not interfere with dividends, since they are in this respect all equally bad. It would appear, therefore, that as an index to the quality of shop management the earning of dividends is but a poor guide." If the United States was to compete effectively in the world marketplace, management would have to slaughter some of the sacred cows of traditional production management and find new ways of doing business.

In the early 1980s, one of this book's authors attended a meeting at a large plant in one of the U.S. automotive companies. Hourly inspectors were presenting some of their results and experiences in using statistical charting for problem solving on the production floor. A few months prior to this meeting the plant management had eliminated the quality control department as such, folding all hourly inspectors and quality control staff personnel into the production function itself. As one hourly inspector explained what he was doing, he became somewhat emotional and began to expound on the change in attitude that had taken place within himself. "I've worked in this plant for 18 years and I can tell you this: I've done more good for this company in the last two months than in all the time combined prior to that!" He went on to explain that not only had management for the first time taught him some tools and methods that enabled him to do an effective job in finding problems and working with others to solve them, but his whole attitude about his work had changed. He said that he used to watch the clock continually, take long breaks and lunch hours, and look for places to "hide." Recently, he caught himself on several occasions forgetting about his break altogether, and on two occasions he had suddenly realized that his shift had been over for about 45 minutes!

This incident clearly demonstrates that we do not need monetary incentive plans to motivate our work force. Just give the worker some tools to do a better job. Restore pride of workmanship by taking off the handcuffs and removing the barriers that keep people from doing what they really like to do—work together as a team to resolve problems and search for improvement opportunities. In the next chapter, a definitive road map for accomplishing this is described. This road map includes the Deming philosophy for the enhancement of competitive position and Taguchi's approach to quality engineering.

1.4   Quality and the Engineering Design Process

Traditional View of the Engineering Design Process

At least two fundamental issues concerning the engineering design process as we have known it over the years may need more thought. One has to do with the way we do design itself, the other with the relationship between design and manufacturing. By addressing the first issue in a fundamental way, with quality design as the focal point, we may really be addressing both issues at the same time.

Figure 1.3 shows the more traditional engineering design process. In this model, system design embraces the application of the technology of the day to arrive at a product design configuration and to select the design parameter

Figure 1.3   Traditional design process model

nominal values. Prototype construction and testing generally follow. Testing in this context is not to be confused with experimentation. Although experimentation generally involves the comparison of design configurations with purposely varying parameter values, usually in some structured design-of-experiments matrix, testing generally means the placement of numerous prototype design replicates on the life test to determine the functionality and reliability of a specific design configuration, that is, one with fixed design parameter values.

One criticism of the design process model of Fig. 1.3 is that it tends to lead to iteratively looping through the system design and testing phase, often in a somewhat unstructured manner. Furthermore, little concern in this model is directed toward the impact and importance of the manufacturing processes that will be employed to make the product. Once function appears to be served through the design process, others will need to worry about making the product economically.

The relationship of the design process to manufacturing and the design/manufacturing interface, in particular, has received considerable attention over the past several decades. CAD/CAM has dealt primarily with the computerization of the design and manufacturing processes and in particular with the translation of design specifications into manufacturing procedures and activities. Design modeling and analysis has been strengthened and extended. Standardization and rationalization of both design characteristics and manufacturing process characteristics have been advanced and the efficiency of the interactive processes of design and manufacturing has been improved. The computer has greatly facilitated this translation, and to some extent lead times have been reduced.

The concepts of design for manufacturability and design for assembly have been the subject of considerable research and development. Numerous specific models have been proposed and refined and will not be discussed here. All of this work is aimed at overcoming the difficulties precipitated by the traditional over-the-wall design philosophy depicted graphically in Fig. 1.4.

Alternative Model for the Engineering Design Process

The Taguchi approach[5] to quality engineering through design has a number of significant strengths. Taguchi has emphasized the importance of minimizing variation as a means to improve quality and of bringing the mean of the process to the design-mandated target. The idea of designing a product the performance of which is insensitive to environmental conditions, and making this happen at the design stage through the use of design of experiments, has been a cornerstone of the Taguchi methodology for quality engineering.

Some of the strengths of Taguchi's approach are

1. The center of gravity of the overall approach: the engineering design process.
2. Definition of the relative roles of the factors that influence product/process performance.

---

[5]G. Taguchi and Y. Wu, *Introduction to Off-Line Quality Control*, Central Japan Quality Control Association, Meieki Nakamura-Ku Magaya, Japan, 1979.

Figure 1.4   Traditional (over-the-wall) approach to design and manufacturing

3. Robust design: the parameter design concept.
4. The use of the loss function as a means of assessing the economic impact of variation.

We elaborate on each of these points as this book unfolds.

Taguchi methods focus attention on the engineering design process: in particular, the projection of a three-stage design process model of system design, parameter design, and tolerance design. Figure 1.5 provides a diagrammatic representation of Taguchi's design process model. In the initial stage, system design, the available science, technology, and experience bases are used to develop/select the basic design alternative to meet customer needs. A variety of techniques may be useful in specifically mapping the relationship between customer needs and the selection of design configuration and parameters that will meet those needs effectively.

At the parameter design stage, interest focuses on the selection of the specific nominal values for the important design parameters. The overarching selection criterion is to "identify those nominal values that minimize the transmission of variation to the output performance as a result of the presence of noise factors operating in the environment in which the product and/or process is functioning." That is, we select the nominal values that are the most robust/least sensitive to noise variation. It is at the parameter design stage that Taguchi strongly advocates

Customer needs

Robust product/process design

Figure 1.5   Taguchi's three-stage design process model

the use of design-of-experiments methods. The tolerance design stage of the design process concentrates on the selective reduction of tolerances to reduce quality loss at the expense of increasing manufacturing cost. The loss function concept of quality is used to provide a basis for striking the economic trade-off.

Parameter Design: Concept of Robust Design

The concept of robust design is an intuitively pleasing one and is perhaps one of the most significant contributions of Taguchi in terms of advancing the state of the art of the engineering design process. Figure 1.6 clarifies the notion of robust/parameter design through the transfer function model for the product. In Fig. 1.6(a) we see a representation of a product functioning in the field and we note that it is subject to certain noise factors, the variation of which is transmitted through the product design to the output measure of performance. To improve the quality of the product, we need to reduce the functional variation, that is, improve the consistency of the output response. To do this, the parameter design concept is invoked wherein the nominal values of the control factors—those factors that can be manipulated by the designer—are selected that reduce/minimize the transmitted variability due to the noise factors. This result is depicted in Fig. 1.6(b).

An extremely relevant measure of product quality is the ability of the product to function in a sound fashion—according to design intent—over a broad range of conditions. In particular, good quality means that the product function maintains a consistent level of performance in the face of noise factors, sometimes referred to as a *robust* product. In attempting to minimize the effects that noise factors have on product quality, certain countermeasures may be taken. The most important countermeasure is that of design. In the past, engineering design in the United States has emphasized system design and, particularly, tolerance design to achieve quality at the expense of manufacturing cost.

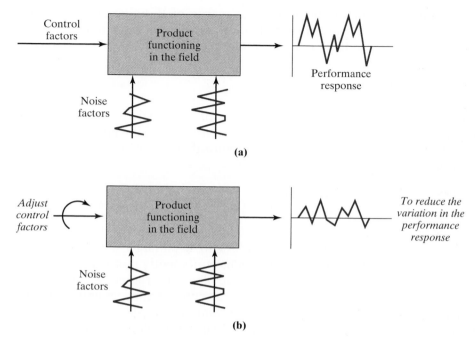

Figure 1.6 Product functioning in the field

Robust design reduces variation—improves quality—without increasing manufacturing costs.

The design model discussed above and the robust design idea should not circumvent or ignore manufacturing but rather should embrace it in an integral way. On the one hand, the influence of the manufacturing process on our ability to be on target with smallest variation can and should be considered at all three stages of the design process, as portrayed by Taguchi. On the other hand, the idea of robustness should be applied to the manufacturing process environment and the noises inherent in this environment that will be transmitted to the resulting product. In fact, it would be useful to apply this concept to both product design parameters and manufacturing process parameters at the same time, in a common setting, during design. In the future, one person will do this, not two, but this person will need both the knowledge and the tools and concepts to make this happen.

1.5 Strategic View of Quality Design and Improvement

We will now begin to lay out a conceptual framework for quality and productivity design and improvement, using the concepts of robust design. In this regard this section begins to provide insight into the following two important issues, which will stand as recurring themes throughout this book:

1. Sources of variation and the consideration of variation as a fundamental measure of product or process performance.

2. The concept of signal-to-noise ratio as a way of thinking about how we measure product or process performance, and how in a general sense we may choose strategically to enhance performance (i.e., increase the signal-to-noise ratio).

On Target with Smallest Variation

Quality is a matter of product function. In Chapter 2 we talk of both traditional and more contemporary definitions of quality. In general, the failure of a product to meet the intended function, as communicated by the customer, can arise from either or both of two basic problems:

1. Failure to achieve the nominal performance required by design.

2. Excessive variation about the intended nominal performance level.

These two issues may be graphically depicted as shown in Fig. 1.7.

What we should really be interested in is to be on target with the smallest variation. The key is whether we can become sufficiently motivated to pursue the smallest variation part of the question. The importance of this issue—smallest variation—lies at the very center of the concept of never-ending quality and productivity improvement. Two important relationships that today are being better understood hold the answer to the question of motivation. They are

1. The productivity/variability relationship.

2. The quality/variability relationship.

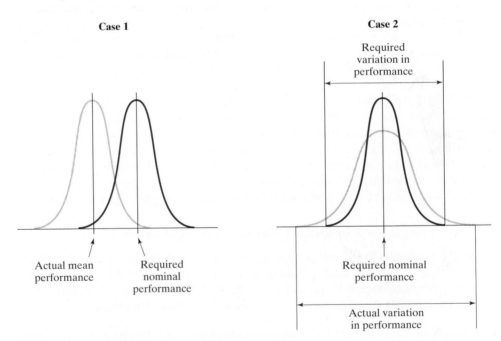

Figure 1.7   **On-target and smallest-variation problems**

The concept "on target with smallest variation" has been promoted in recent years through the methods of statistical process control. W. Edwards Deming and others have demonstrated that sources of variation are sources of *waste* and *inefficiency* and that for every source of variation identified and removed we will experience increases in both quality and productivity. The concept of quality as measured by loss due to functional variation will further reinforce our newfound striving to be on target with smallest variation. The role of experimental design in achieving this goal has been clearly demonstrated by Taguchi and others. But to a great extent, experimental design methods have emphasized the revelation of improvement opportunities on the average, and have therefore been more useful in the "being on target" part of the equation. The following illustration demonstrates this idea and a way we might improve on it in a very simple way.

Suppose that we are attempting to control the part weight of a molded product and we study by a simple experiment how two molding machine parameters influence part weight. We run an experiment and obtain the results shown in Fig. 1.8, where the points on the square correspond to the four test conditions examined. Each weight response is the average of 25 parts molded consecutively, under the given conditions. The results of the experiment seem to indicate that by manipulating control variable 1 we can control part weight. Control variable 2 seems to have little to do with part weight, on average. We can set it (or vary it) however we wish and nothing will happen to change part weight.

Now suppose that someone came in and made a time plot of the individual part weights for the 25 consecutive parts made under each condition. The results could look like those shown in Fig. 1.9. The individual charts in the figure

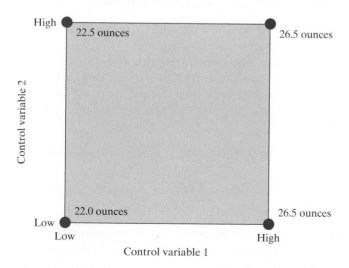

Figure 1.8   Factorial design of experiment for the molded part example

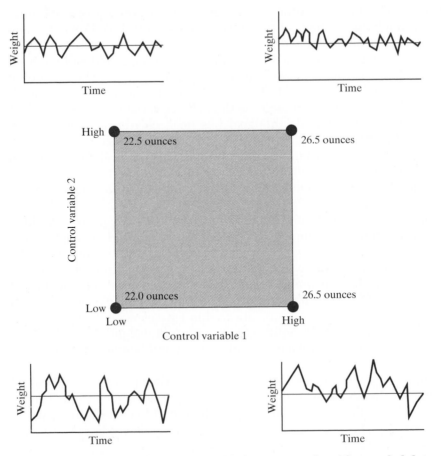

Figure 1.9   Factorial design for the molded part example with sampled data

reveal something important that was not known previously—that although control variable 2 does not affect part weight on the average, it does appear to affect *variation* in part weight from part to part. Such could certainly be measured by the standard deviation as a response of the experiment. The results of the experiment seem to be telling us:

1. To control the average part weight to the target, manipulate control variable 1.

2. To do item 1 with the smallest variation, set control variable 2 at its high level.

In the past it has been less common for us to use variation as an experimental measure of performance and to seek ways to reduce variation in product/process performance through designed experimentation. Later in the book we will see how this can be achieved.

Concept of Signal-to-Noise Ratio as a Measure of Performance

In the preceding example, which dealt with the control of part weight for an injection molding process, improved quality/product performance is obtained by shifting the process mean to the design nominal via control variable 1. In the terminology of the communications field we say that we are altering the signal. Similarly, in this example, improved quality/product performance is obtained by reducing the variation in performance about the design nominal via control variable 2. In the terminology of the communications field we say that we are reducing the noise. Sometimes it is convenient to put these two aspects of the problem together through use of the signal-to-noise ratio.

In the communications field, the signal-to-noise ratio is commonly expressed as a power ratio in decibels or as a voltage or current ratio in decibels. For our purposes at this time, a perhaps more basic and relevant way of examining the signal-to-noise ratio is to use the following definition:

$$\text{S/N} = \frac{\text{average}}{\text{standard deviation}}.$$

Figure 1.10 graphically depicts the signal-to-noise-ratio concept. In the top part of the figure we have a signal that represents the input to a certain process. For the process to be successful, this input must have a value strictly greater than zero.

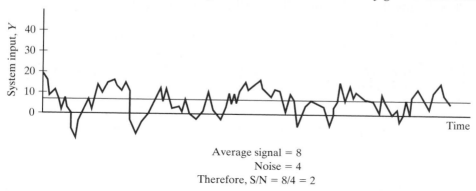

Average signal = 8
Noise = 4
Therefore, S/N = 8/4 = 2

Solutions to the problem

Figure 1.10   **Signal-to-noise-ratio concept**

In Fig. 1.10 it is clear that such is not always the case. There are basically two ways to solve this problem, both depicted in the lower part of Fig. 1.10. To the left in Fig. 1.10 we see that the average signal has been increased, and therefore even with the same amount of variation about the average, the signal is now always above zero. To the right, the problem has been solved by reducing the variation in the input without the need to increase the average value of the signal. Contrast these two strategies for increasing the signal-to-noise ratio in your own mind, thinking of all the problems you may have solved over the years, either personally or professionally. Have you tended to do so by increasing the signal or by reducing the noise?

The following simple examples illustrate how these contrasting strategies for quality design and improvement can be found in many of the design and manufacturing systems now in place.

Using Additional Resources

Signal-to-noise ratio could be a measure of the acceptable number of parts produced. Too often, the rate of production of defective parts is large—the process efficiency or yield is low. One way to increase the yield is to make more parts—increase the signal—either by increasing the production rate of the equipment in use or by adding additional equipment/machines (i.e., more machine time) or by reducing the cycle time. This is illustrated in Fig. 1.11. The signal is increased and therefore the usable process output is increased. Another approach would be to use the same equipment/machines but to improve their efficiency—reduce the noise. Statistical thinking and methods, utilizing techniques such as statistical process control (SPC) and design of experiments, will provide considerable insight into how such improvements can be made.

Figure 1.11   Using additional machines

Parallel Redundancy in Design

The *mean time between failures* (MTBF) of a given product is an indication of quality similar to the signal-to-noise ratio. Let us suppose that the MTBF,

$$\text{MTBF} = \frac{1.0}{\phi},$$

is unacceptably low, where ϕ is the failure rate (failures per unit time). We can increase the signal-to-noise ratio (MTBF) by including parallel redundancy in the product design (i.e., the system now will be successful if *either* component 1 *or* 2 is functioning), as shown in Fig. 1.12. For parallel redundancy and n components with identical failure rates, ϕ:

$$\text{MTBF} = \frac{1 + 1/2 + \cdots + 1/n}{\phi}.$$

Clearly, parallel redundancy will increase the signal-to-noise ratio by increasing the signal with no regard for the level of noise (ϕ).

An alternative strategy receiving considerably more attention today is that of attacking the noise—in this case the component failure rate—to determine what design and/or manufacturing improvements could lead to a reduction in this rate. The techniques of design of experiments and Taguchi's concepts such as robust design are likely to be quite useful in this regard.

Screening Inspection/Product Control

In the field of electronics, signal-to-noise-ratio enhancement may be effectively accomplished by filtering out the noise. For example, many signals of interest have relatively low frequencies, with bandwidths extending only over a narrow range, perhaps a few hundred hertz. In these cases a simple low-pass filter can effectively limit the measurement-system bandpass to that necessary to pass the desired signal frequencies. Such is shown in Fig. 1.13. In quality control, the analogous situation may be the 100% screening inspection of outgoing (producer)

Figure 1.12   Parallel redundancy

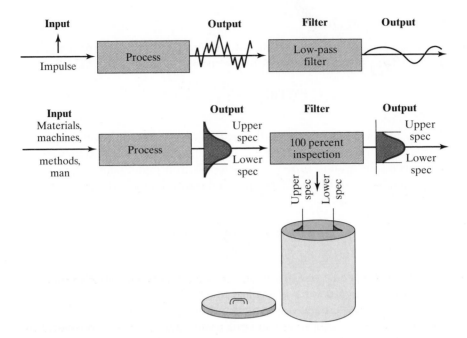

Figure 1.13   Low-pass filter vs. 100% inspection

or incoming (consumer) product. In Fig. 1.13 as with the low-pass filter, 100% inspection does not remove the source of the noise; it merely filters out the effect of the noise on the output the customer sees.

Earlier we saw that one approach to increasing the signal-to-noise ratio of a manufacturing process would be to increase the signal by adding resources (more machines, overtime, etc.). In this last example, increasing the signal-to-noise ratio is accomplished by reducing the noise by filtering or inspection, which is a post-processing activity. Actually, these two countermeasures are generally used together; if the defect rate is high, we make more parts and screen the bad ones out. This is very costly and clearly does not provide opportunities for improvement.

Statistical process control (SPC) is a countermeasure that attempts to attack the noise in the manufacturing process, eliminating sources of waste and inefficiency that produce variation in the function of the product. Statistical process control increases the signal-to-noise ratio by maintaining the target and reducing the noise component. Figures 1.13 and 1.14 contrast inspection (product control) and SPC (process control) as approaches to improving the signal-to-noise ratio. Through process observation, evaluation using such techniques as statistical charting and fault diagnosis to identify the root cause of the problem, sources of variation may be eliminated altogether.

Using More Expensive Raw Materials

In certain situations increasing the signal-to-noise ratio can be accomplished at the product design stage, again using the idea of filtering. Here, at the design

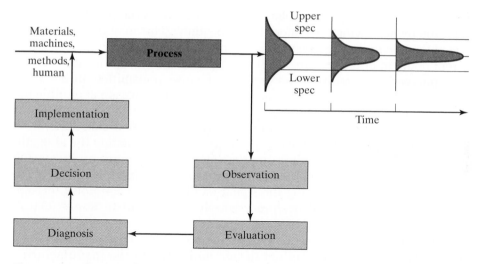

Figure 1.14   Statistical process control as a means to increase signal-to-noise ratio

process, through the use of more consistent raw materials at a higher cost we may reduce the effect of the material noise on product performance. We are, in effect, tightening the performance specifications on the material proper-ties to reduce the amount of noise to be transmitted through the product. It is important to note that the use of more expensive raw materials or purchased parts is a sort of filtering process, initiated at the design stage. Better, more expensive materials are themselves the filter that diminishes the effect of the noise on the signal.

   We hope to show that such filtering at the design stage is an important strat-egy but needs to be invoked in a different fashion. Figure 1.15 illustrates two

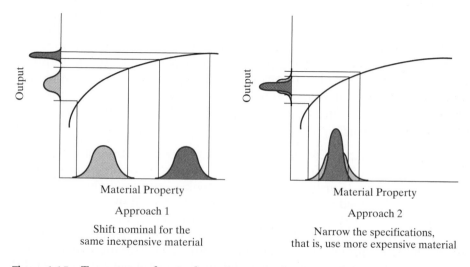

Figure 1.15   Two approaches to the noise transmission problem

contrasting approaches to the problem. Approach 2 solves the problem of noise transmission by narrowing the material specifications. This may greatly increase the cost of the material. Approach 1 takes a much different tack; it tries to locate a different nominal or target value for the cheaper material that for the same tolerance/input variation reduces the amount of noise transmitted through the product/material to the output performance. In approach 1 we are exploiting the nonlinearity in the system. By applying the same raw material, but in a different way, we reduce the transmitted variation and produce a more robust design.

There are many situations we can point toward that illustrate the altogether too common approach to increasing the signal-to-noise ratio by increasing the signal. We have examined several common examples of increasing the signal-to-noise ratio by either increasing the signal or filtering out the noise through postprocessing. The notion of *postprocessing* here is meant to imply that although the noise still "reaches" the signal and may significantly corrupt it, it may be "removed" after the fact by some form of filtering. As we have just seen, some of these situations relate to product design and some to the manufacturing process, but the overriding theme is always the same—a failure to attack the problem at the root cause and eliminate, through improvement actions, the forces that are producing variation and waste. In the foregoing scenarios the approach to solving the problem was oriented more toward treating the symptom than treating the cause. The concepts and methods that follow in this book will put us in a better position to treat the causes.

1.6   Beyond the 1980s—The Institutionalization of Quality in the Organization

Since the watershed decade of the 1980s, quality has become a driving force in the struggle for competitive position in the world marketplace. Organizations and management are clearly more cognizant that quality must be not only a central metric in the evaluation of performance of products and services and the systems that lead to those products and services, but also an essential design criterion. A number of efforts and programs that have evolved since the mid-1980s have institutionalized "quality" in many organizations at unprecedented levels. We will now briefly review some of these mechanisms and driving forces in the institutionalization process.

Certification Programs

In the early 1980s, many U.S. industries were under mounting pressure from foreign competition. In response, beginning in 1980, the Ford Motor Company, on the advice of W. Edwards Deming, undertook a corporatewide, top-down program approach aimed at institutionalizing quality in all of its systems. One aspect that received early attention at Ford was pushing the quality concept down into its vast supplier base. This ultimately led to a well-defined, disciplined system of supplier training, evaluation, and certification. The immediate goal of Ford's "Q1" program was to strengthen and improve supplier quality. The Q1 program embodied the philosophies of Deming (14 Points) and Juran (Quality Trilogy) relative to quality management, quality planning, process control, and process and product quality improvement. The long-term goal was to reduce the number of suppliers overall and at the same time develop strong, longer-term relationships with those suppliers who had committed to the new philosophy of quality design and improvement. Other companies noted Ford's quality improvement and were quick to follow suit, and an extremely broad-based emphasis on quality began to form in the manufacturing sector.

In the latter part of the 1980s, company-specific supplier certification programs drove the establishment of an international standard that today is firmly entrenched in the fabric of our manufacturing and service industry sectors. The International Organization for Standardization in 1987 released a family of standards (ISO 9000) aimed at promoting adherence to customer requirements and compliance with regulations related to product quality and organizational consistency. It provides the organization with a structure for managing its processes and activities. The focus of the ISO 9000 standards is on the quality management system, documentation, and internal standards of operation. ISO 9000 certification by independent auditors may be pursued to confirm that a company's quality system conforms to the standard's principles.

The ISO 9000 family is among ISO's most widely known standards ever, having been implemented by thousands of organizations in over 100 countries. ISO 9000 has become an international reference for quality management requirements in business-to-business dealings. The ISO 9000 family focuses on 'quality management.' In particular, it addresses what the organization needs to do, in the words of the standard, to fulfill:

- the customer's quality requirements, and
- applicable regulatory requirements, while aiming to
- enhance customer satisfaction, and
- achieve continual improvement of its performance in pursuit of these objectives.

ISO 9000, as a family of international quality management standards and guidelines, has an established global reputation as the basis for establishing quality management systems.

The original three standards, ISO 9001, ISO 9002, and ISO 9003, have now been integrated into the new ISO 9001:2000.[6] Those interested are directed to the ISO website, where detailed information on implementation of the standard can be found. Some of the helpful information found there includes the following discussion:

> ISO 9001:2000 specifies requirements for a quality management system for any organization that needs to demonstrate its ability to consistently provide product that meets customer and applicable regulatory requirements and aims to enhance

---

[6]http://www.iso.org

customer satisfaction. ISO 9001:2000 has been organized in a user-friendly format with terms that are easily recognized by all business sectors. The standard is used for certification/registration and contractual purposes by organizations seeking recognition of their quality management system.

The greatest value is obtained when you use the entire family of standards in an integrated manner. It is suggested that, beginning with ISO 9000:2000, you adopt ISO 9001:2000 to achieve a first level of performance. The practices described in ISO 9004:2000 may then be implemented to make your quality management system increasingly effective in achieving your own business goals. ISO 9001:2000 and ISO 9004:2000 have been formatted as a consistent pair of standards to facilitate their use. Using the standards in this way will also enable you to relate them to other management systems (e.g., environmental), many sector specific requirements (such as ISO/TS/16949 in the automotive industry) and will assist in gaining recognition through national award programmes.

The Six Sigma Program

Six Sigma's history can be traced back to the early/mid-1980s at Motorola. This quality program embraces two overarching quality concepts: the Deming Cycle (to be discussed in Chapter 2) and the Taguchi loss function. Originally, the term focused on the metric—a capability so good that virtually no parts are beyond the specifications—but over time, Six Sigma has evolved into a business improvement methodology. Six Sigma should be viewed as both a business philosophy and an organized process and set of tools to identify and analyze problems in a structured way.

The term "Six Sigma" has been used a variety of ways: a Six Sigma company, a Six Sigma process, Six Sigma quality improvement, and so on. As used by practitioners, there are also many facets to this term. At its highest level, Six Sigma represents a vision and a philosophy for quality improvement that is adopted by an individual or a corporation. It embodies a goal that is lofty yet tangible and has been associated with quality greatness at an emotional level. At the next level it represents a systematic methodology following a five-step procedure, derived from the Deming cycle: Define–Measure–Analyze–Improve–Control. Each step is clearly defined with specific deliverables that quantify improvement or success. Statistical and problem-solving tools are employed at each step of a Six Sigma implementation process. Following Taguchi's idea of loss function, quality improvement can be effectively achieved by the reduction of variation. With the surge of interest in Six Sigma across corporate America, management at all levels now understands that variation is one of their top quality issues. Six Sigma brings a way to measure it, analyze it, attack it, and eventually improve the overall performance and hence the predictability of the business.

The methods associated with Six Sigma are now applied across all business issues, including quality. Structured approaches are employed such as DMAIC (Define, Measure, Analyze, Improve, and Control) for problem solving and DMADV (Define, Measure, Analyze, Design, and Verify) for creating new products and processes. Six Sigma involves focusing on the top line and bottom line and using data to make decisions with a significant emphasis on the use of statistical methods. In addition, dedicated personnel resources (with titles derived from the martial arts such as Black Belt and Green Belt) are involved in teams responsible for product/process improvements that directly affect customers. As such, Six Sigma has its own unique form of people-level certification within a company that promotes learning within the organization. The American Society for Quality, several universities, and many consulting firms offer training programs that, when combined with project work, result in certification of individuals at the various belt levels. It may be noted that the establishment of a set of certified "experts" within a company provides both a technical and mentoring support structure to pursue and maintain quality-related initiatives.

The Malcolm Baldrige Award

When William Conway, former CEO of the Nashua Corp. and an early Deming disciple, traveled to Japan in the late 1970s to learn firsthand how the Japanese were so effectively dealing with the quality issue, he found it difficult to set up an initial meeting because "everybody was working on the Deming Prize!" The Deming Prize was established by the Japanese Union of Scientists and Engineers (JUSE) in appreciation of Deming's work in Japan and his generosity in donating his course-notes royalties to JUSE. The late Kenichi Koyanagi, then managing director of JUSE, proposed using this money to fund a prize to commemorate Deming's contribution and to promote the continued development of quality control in Japan. The Deming Prize, especially the Deming Application Prize given to companies, has exerted an immeasurable influence on the development of quality control/management in Japan. More than three decades later in the United States the Malcolm Baldrige Award was established, and like the Deming Prize, it has had a major impact on the evolution and concomitant institutionalization of quality management principles and methods.

The Malcolm Baldrige National Quality Award is managed by the National Institute of Standards and Technology (NIST) and was signed into law in the United States in 1987. Its purpose is to improve quality and productivity within the United States by recognizing the achievements of companies that improve the quality of their goods and services. The seven 2005 Baldrige categories, outlined in Fig. 1.16, are leadership; strategic planning; customer and market focus; measurement, analysis, and knowledge management; human resource focus; process management; and business results. Points are earned within each category based on an assessment of the organization.

With respect to the seven categories, the following topics are of interest:

1. *Leadership.* Vision, values, communication, performance, governance system, social responsibility, legal and ethical behavior, and community support;

2. *Strategic planning.* The process and objectives of strategy development, and the development and deployment of action plans and performance projections (including metrics);

Figure 1.16   Baldrige award model (adapted from NIST, 2005).

3. *Customer and market focus.* Customer and market segmentation and group-ing, customer-requirements determination, customer relationship building, managing complaints, satisfaction measurements, and customer follow-up;

4. *Measurement, analysis, and knowledge management.* Selecting/collecting data and information, maintaining the performance management system, bench-marking processes and performance, data availability and accuracy, best practices sharing, and quality of information;

5. *Human resource focus.* Organization of work, performance management sys-tem, hiring, career progression, succession plans, employee learning and motivation, training delivery approaches, and employee well-being and satisfaction;

6. *Process management.* Value creation processes, key design processes, produc-tion/service/delivery processes, cost of poor quality, and support processes and operational planning;

7. *Business results.* Product and service outcomes, customer-focused results, financial and market results, human resource results, organizational effec-tiveness results, and leadership and social responsibility results.

There are currently five classifications of the award: manufacturing busi-nesses, service businesses, small businesses, education organizations, and health care organizations. Up to three awards may be given in each class each year.

Not only has the Baldrige Award had a direct impact on quality design and improvement in its own right, but it has also spawned the creation of many other quality award programs at both state and local levels. The total impact of this national program has clearly been very significant.

Quality and the Organization

One of the most significant effects of the renewed attention to quality improvement has been the rethinking of where quality fits within an organization. Applications of quality improvement methods can be found in just about every business function, from finance to facilities. As Deming so clearly points out (Point #14 of his 14 Points, to be discussed in the next chapter), quality is the responsibility of top management in the organization. Management must examine carefully each and every system within the organization to determine whether current practice promotes or inhibits the improvement of quality, productivity, and competitive position.

Evidence of the institutionalization of quality within the organization begins to emerge as individuals and groups at all levels and across all functions in the organization begin to seek and exploit opportunities for improvement on a regular basis. This requires the understanding and ongoing practice of "systems thinking"—that is, the ability to make decisions in the best interest of the organization as a whole, not merely some subsystem within the organization. It also requires recognizing that one's customer is not merely the person or organization that is the recipient of the products and services of the company, but also the person and entity within your own organization that depend on your output. In what follows in this book, in terms of both overarching philosophies and tools and methods, we hope to bring these issues to life.

■ CHAPTER SUMMARY

It has now been over 25 years since the beginning of the renewed emphasis on quality in the United States. What many cynics thought would be a year-or-two phenomenon has sustained its momentum for a quarter of a century. The fact that several large companies (Ford, Motorola, General Electric, and others) have demonstrated continued drive and leadership in this effort has probably been instrumental in sustaining the movement. We now have a national quality award; have seen millions of workers, salaried and hourly, trained in the methods of Deming, Taguchi, and others; and have witnessed major advancement in competitive position on the part of many companies. Yet technological growth in the use of the concepts and methods discussed in this book has been hampered by organizational and cultural factors that continue to hold the American industrial enterprise in a straightjacket.

■ EXERCISES

1.1 Explain the change in the view of quality control in the United States between the decades of the 1950s, 1960s, 1970s, and today.

1.2 What changes must take place for the United States to become competitive again in the world marketplace?

1.3 What four major milestone events/circumstances have led to major steps forward in the concepts, methods, and implementation of quality control?

1.4 What was the effect of division of labor on the standard of quality of goods manufactured in the United States? Why did this occur?

1.5 Is the use of work standards and/or wage incentives consistent with our view of quality today? Explain.

1.6 What danger is associated with the use of acceptance sampling for quality control? Why does the type of attitude it promotes hinder the enhancement of competitive position?

1.7 What are some possible effects realized by the general acceptance of the idea of AQL?

1.8 What was Shewhart's basic approach to quality control?

1.9 How did Shewhart's approach to quality control differ from that of Dodge and Romig?

1.10 What is wrong with the phenomenon of "change of necessity"?

1.11 Why has the United States over recent decades become concerned with improving quality? What forces may be at work now that did not exist 30, 40, 50 years ago—or did they?

1.12 Who has responsibility for initiating quality control? What things need to be done?

1.13 Explain why it is important to consider variation about the nominal as a measure of product/process performance; that is, why is it not sufficient or advisable to consider only performance on average?

1.14 How/why do some of the newer concepts/ways of viewing quality encourage manufacturers to reduce the variability about the nominal in their products/processes; that is, what motivates the continual pursuit of variation reduction?

1.15 A major automobile manufacturer advertises that it performs 100% inspection on all processes that produce parts for the brake systems in its cars. It claims that the consumers can feel good about being safe in their cars. What should the consumers feel bad about? Is there any situation in which measuring every part might be useful?

1.16 The percent defective from a process that makes injection-molded plastic keychains has gradually been rising to the point where the process can no longer meet production demands in a single 8-hour shift per day. The shop floor manager sends you a memo recommending that a second shift be added on this machine in order to get the product out the door. As his or her boss, how would you respond to this memo?

1.17 After experimentation with a certain product's performance in the field, it is found that better quality raw materials significantly improve the product's performance in all areas. Based on these results, would you recommend that the more expensive raw material be used for this process? Of what is this an example? How might we better deal with this issue?

1.18 Provide an example in your personal or professional life in which you solved a problem by (a) increasing the "signal," and (b) decreasing the "noise" in the signal-to-noise ratio you might use to characterize the performance of the process. Carefully explain your answer.

1.19 Your boss likes the color scheme you specified for the decoration on a new motorcycle helmet and begins production immediately in order to have the first ones off the line in time for his son's birthday. Although you have had no chance to consider design changes that may prove necessary, he assures you that later modifications will cause minimal downtime on the line. Is your boss considering all the costs involved? How would you prove to him that his decision was a poor one?

1.20 Explain the difference between testing and experimentation. Which do you think is more useful in the engineering design process, and why?

1.21 What is meant by the term "robust" in the context of product design?

1.22 What criterion does Taguchi recommend using when selecting nominal values for the controllable variables in a product/process design?

1.23 Relate the differences between the U.S. and Japanese design life cycles to the differences between the two countries' quality efforts by activity (Fig. 1.1). Which approach would seem to be more costly? Why?

1.24 What do you think is Taguchi's most important contribution in terms of his approach to quality design and improvement relative to the more traditional, Western approach to quality control?

1.25 The purpose of the Baldrige Award program is to improve quality and productivity in companies and organizations to achieve efficiency, competitiveness, and profitability. Hundreds have won the award since its inception. Collect information on at least 10 companies to make an evaluation of their success, particularly in terms of quality and productivity, since receiving the award.

1.26 Describe each of the structural steps of DMADV of a Six Sigma program for product and process design.

1.27 Many different Six Sigma structures have been employed for applications, each with a somewhat different focus on certain aspects of process improvement. Compare two of these structures, DMAIC vs. DMADV, for their specific focuses.

1.28 What are some advantages and disadvantages of a quality certification program in general and of ISO 9000 in particular?

1.29 Do you think a quality certification program such as ISO 9000 ensures continuous improvement? Why or why not?

1.30 Briefly explain the interconnections among the seven steps in the Baldrige Award model presented in this chapter.

2

Conceptual Framework for Quality: Design and Control

2.1 Introduction

In this chapter we focus on the work of two persons whose contributions to quality design and improvement have had a lasting impact on the world we live in today. Although many have contributed in a significant way to the field of quality, Deming and Taguchi have provided not only the tools and methods necessary to make a difference but, more important, the conceptual underpinning necessary to embrace and institutionalize their approaches.

W. Edwards Deming was largely responsible for bringing statistical thinking and methods for quality improvement to Japan after World War II. An eminent statistician in America during the 1930s and 1940s, Deming was at the forefront of the statistical quality control scene during World War II. After the war, his philosophies and teachings fell on the deaf ears of American management, and the quality effort fires that burned so brightly during the war years slowly went out. America was in a period of industrial boom. Every product that could possibly be manufactured could be sold. In Japan, the situation was very different. A devastated industrial base was struggling to rebuild, natural resources were scarce, and a reputation for poor, shoddily manufactured products was rapidly solidifying. The Japanese became aware of Deming's work and invited him to Japan, where he met with and caught the attention of top management. A technically oriented management listened patiently to Deming—and the rest is history.

Shortly after World War II, Genichi Taguchi began to cultivate an understanding of how the methods of design of experiments could be used to improve the productivity of research and development work in Japan. Motivated by the need to improve telephone communications, the Japanese government established the Electrical Communications Laboratory (ECL) in 1949 and Taguchi was asked to join the lab and head up an effort to improve the efficiency of R&D

activities. He quickly recognized that most of the work of the lab involved costly, time-consuming experimentation, and he set out to see how the already present field of design of experiments might be brought to bear on the problem. As Taguchi began to cultivate the application of design of experiments in the R&D work of the ECL, a unique and profound approach to quality design and improvement slowly began to evolve.

2.2   Deming Philosophy of Never-Ending Improvement

When Deming traveled to Japan in 1950 at the invitation of the Japanese Union of Scientists and Engineers, he found a climate that was quite conducive to the promotion of his concepts and methods. On the one hand, Japan appeared to have a solid base of statistical expertise, although its energies had been directed primarily toward mathematical theory and the application of that theory in non-manufacturing environments such as agriculture. On the other hand, Deming found an industrial leadership base in Japan very eager to listen to what he had to say. With the lessons learned from his experience in attempting to promote quality improvement in America, Deming's first order of business was to conduct a series of top-management training seminars in which he laid out what needed to be done to place quality improvement on an institutional basis within any organization. The obligations and responsibilities of management that he spelled out in these seminars came to be known as his *14 Points* for management.

It is perhaps the combination of upper management perspective and a firmly rooted background in mathematical statistics that enabled Deming to sustain his efforts in a leadership position over more than five decades, until his death in 1993 at the age of 93. Deming was able to tell management what they ought to do and then provide them with the rigorous analytical tools and methods to carry out his directions. The teachings of Deming are heavily oriented toward the use of statistical thinking and methods to identify opportunities for quality and productivity improvement. Deming developed a plan for management to follow to enhance competitive position over the long run. This plan, referred to as his 14 Points, management's obligations, is discussed briefly below. It should be noted that in the very spirit of his 14 Points, Deming continued until his passing to refine and improve these tenets. An excellent discussion of Deming's 14 Points is provided by William Scherkenbach.[1] Some of his interpretations are included below.

Deming's 14 Points

1. Create constancy of purpose for the improvement of product or service.

Deming was particularly critical of the shortsighted thinking of American management. Driven by the quarterly report, management has had a tendency to develop strategies that may show profit in the short term but are counterproductive to the long-term well-being of the organization. Central to creation of

---

[1]W. W. Scherkenbach, *The Deming Route to Quality and Productivity: Road Maps and Roadblocks*, Mercury Press/Fairchild Publications, New York, 1987.

constancy of purpose is the commitment on the part of management to the long run, making significant investment in research and education so that innovation can be the hallmark of the business plan. Management must believe that the firm will be in business for a long time and therefore develop a business strategy that is based on long-term thinking. This business strategy must be strongly a function of the customer, and this requires that management understand clearly what the needs and wants of the customer are and be able to translate these into the products, processes, and systems that can address those needs competitively.

2. Adopt the new philosophy.

It is clear that we are in a new economic age created, in part, by the leadership of the Japanese in the world marketplace. Management must be willing to discard the old philosophy of accepting defective products. The thought that quality need only be improved up to an acceptable plateau, after which no further improvement is economically justified, fails to recognize the many benefits of variation reduction and the inseparable relationship between variation reduction and productivity improvement. To adopt the new philosophy, management must recognize its roles and responsibilities and assume a leadership posture. To adopt the new philosophy, management must be willing to understand the meaning and importance of the other 13 Points and be willing to do whatever is necessary to embrace them fully.

For many years the management of quality was based on the premise that from a cost-of-quality point of view, there existed some optimal level of quality that should be strived for. Beyond this level, the cost of further improvement outweighed the return in terms of better quality. Of course, what "optimal level of quality" really means is an optimal level of defective material! The old philosophy is that of product control, and product control as a way of life is directed toward balancing the costs of quality to define the appropriate percent defective as a corporate objective and then maintaining that objective through a quality system built on "detection and containment." Unfortunately, such systems have failed miserably over the last several decades and now must be totally abandoned. The new philosophy is built on the process control way of thinking and is oriented toward never-ending improvement at the process, on the one hand, and pushing the quality issue farther and farther upstream into engineering design, on the other.

3. Cease dependence on mass inspection for quality control.

In adopting the new philosophy, it is essential that management recognize the need to abandon defect detection and containment as a means of controlling quality and emphasize the importance of defect prevention. Figure 2.1 is a graphical depiction of the product control model for quality control. In the product control model, output from the process is subjected to inspection procedures, usually carried out by the quality control department. Defective product is detected and contained, either to be reworked or scrapped. Although such an approach may keep the outgoing stream of product pretty much defect-free, its cost erodes competitive position. More important, such a system never addresses

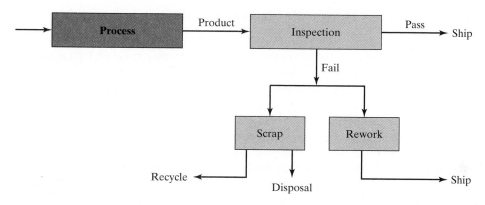

Figure 2.1   Product control model for quality control

the problem at the origin, and hence the root cause of the defective material remains active in the system.

The product control approach to quality control promotes the notion that
quality and productivity are conflicting rather than jointly achievable goals.
Under the mass inspection approach the belief is that if quality is to improve,
productivity must suffer, and vice versa. The goal of production is to get the
numbers out, while the goal of the quality control department is to keep the
flow of shipped material free of defectives.

We can cease our dependence on mass inspection when we adopt the
process control model for quality control. Figure 2.2 depicts this model graphically. The process control model is directed toward the identification and ultimate removal of the underlying causes of the problem. Hence its central focus
is preventing faults rather than containing defective material. As a result, both
quality and productivity will be enhanced.

Competitive position can only be enhanced when the causes of problems can
be identified and appropriate remedial actions taken at the manufacturing process
or upstream during the design process. Therefore, it is clear that the process control way of thinking and of managing manufacturing operations must be adopted.

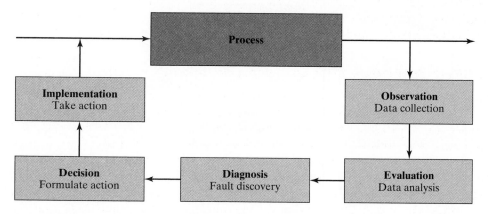

Figure 2.2   Process control model for quality control

4. End the practice of awarding business on the basis of price tag.

For many years the common practice in American industry was to maintain several suppliers for any given commodity. There were probably two basic reasons for taking this approach. First, it provided multiple sources of goods that could alleviate short-term problems when one or more suppliers could not meet customer schedules because of quality problems. In essence, the multiple suppliers became buffers against their own inability to deliver quality goods in a timely fashion. The second main reason was related to price. By maintaining multiple relationships on an ongoing basis, the customer could continually apply pressure to reduce the unit costs. By awarding contracts on this basis there was a feeling that the best deal was always being made and the suppliers were continually being kept in line, pricewise.

Multiple sourcing might be likened to a common design strategy—parallel redundancy—as a means to ensure a continually successful system. This analogy is shown in Fig. 2.3, where the mean time between failures (MTBF) of the assembly process due to supplier failure to deliver is

$$\text{MTBF} = \frac{1 + 1/2 + 1/3}{\phi}$$

for three parallel "components" with equal failure rates ϕ. The assembly will be successful if any of the three suppliers is successful (delivers quality goods on time).

Under this way of doing business, the relationship between the customer and the supplier was often somewhat adversarial, with a certain amount of mistrust present. The job of the quality assurance representative was to visit the supplier on a frequent basis, spot-checking for bad parts in shipments about to go out and subsequently "beating up" the supplier when it appeared that quality was poor. As a result, suppliers would counter with practices for quality control which were centered around mass inspection to ensure that no bad parts were found. There was often a general feeling

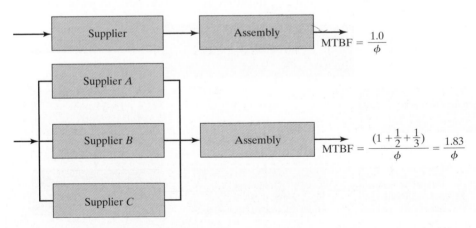

Figure 2.3   Multiple sourcing as parallel redundancy

that suppliers would try to put one over on customers if they could and so, of course, customers would develop equally devious tactics.

Many companies now develop long-term partnerships with their suppliers. The automotive companies, for example, have provided training and implementation assistance in the use of statistical methods to their suppliers. They have developed certification programs such as the Ford Motor Company Q1 program and have rewarded suppliers who have developed and maintained process-control-oriented quality programs with sole sourcing and multiyear contracts.

5. Improve constantly and forever the system of production and service, to improve quality and productivity, and thus constantly decrease costs.

One thing Deming stressed in his discussion of continuous improvement is the "how" of it all. Everyone wants to do better—to improve—but how will we accomplish this? Working harder and having a positive attitude, enthusiasm, and willingness to accept the need to improve are all important ingredients, but we need more. Deming repeatedly asked, **"By what method will we improve?"** What specific concepts and tools will we need to learn and invoke to foster "continuous improvement"?

The process of continuous improvement is built on the Deming cycle, the following iterative four-step procedure:

1. Recognize the opportunity.

2. Test the theory to achieve the opportunity.

3. Observe the test results.

4. Act on the opportunity.

This cycle is clearly implementable through the process control model and its representation as a feedback control loop, as depicted in Fig. 2.2. Figure 2.4 depicts the Deming cycle graphically.

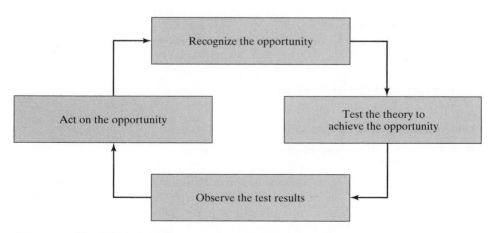

Figure 2.4   The Deming cycle

To recognize the opportunity, it is essential that customer needs and expectations are clearly understood. The opportunity is basically defined by the gap between the customer's needs and one's ability to meet those needs. Hence both must be clearly defined in an operational way. Furthermore, data are needed to estimate this gap and must be conceived of and analyzed based on sound statistical principles. To seize the opportunity, one must formulate a plan, a theory for improvement, and the theory must be tested and proven. The use of statistical principles to design an experiment/develop a data collection plan to test the theory is essential. Of course, to realize the gain, one must act on the opportunity. To do this, it is important to institute change in a manner that will allow one to hold the gain. Too often the action on the opportunity is simply an extension of the experiment; it may provide further proof of the value of the change but be unable to sustain its presence.

To identify customer needs and expectations, we must develop *operational definitions* for them. But what do we mean by "operational definition"? Deming said: "An operational definition puts communicable meaning into a concept." Words like good, on-time, reliable, uniform, unsafe, round, and flat have no communicable meaning unless they are articulated in terms of (1) a criterion, (2) a sampling procedure, and (3) a test/measurement procedure and scale. An operational definition is something that everyone involved can agree upon and communicate to each other in the same precise way today, tomorrow, and the next day.

For example, air pollution is a hazard. But how do we define air pollution? Answer: Air with 100 parts per million of carbon monoxide is polluted air. This sounds pretty precise, but you have to think of the actual measurement process. Where will you take the sample? At the surface of the earth? One hundred feet in the air? When will you take the sample? At any instant in time? In the morning before 6 A.M.? At 5 P.M.? How will you capture the sample? How will you measure the parts per million of CO?

Suppose that we say that an object should be round (this is Deming's example from *Out of the Crisis*, Chapter 9). What do we mean by "round"? The dictionary offers some help: A figure is *round* if it is everywhere in Euclidean two-dimensional space equidistant from a point called the center. This defines the *concept* of round, but it is not an operational definition. The question really is,

- How far can it depart from roundness?
- What criterion will we use for roundness?
- How will we measure roundness?
- What units will we use to express roundness?

We might define *out-of-roundness* as *the average difference of four measurements of the diameter of the object from their average, each taken at intervals of 90°.* The measurements are to be made with a Vernier caliper and rounded to the nearest 0.001 inch. This is an operational definition of roundness. We can do business with this definition.

The act of precipitating continual improvement through the cyclic feedback system described above requires that this cycle become an integral part of the system, that is, be put on an institutional basis throughout the organization. This can happen only when two conditions are present. First, the motivation for seeking continual improvement must be clear to the people working within the system. Each person must come to realize the importance of this approach and must have the tools in hand to achieve this goal. This is a responsibility that management must accept. Second, the management styles and systems, and the operating precedents and technologies of the organization, must be in complete harmony with the concept of never-ending improvement. All such systems must be critically examined by management to ensure that this is the case. If systems are found that appear to inhibit the institutionalization of the new philosophy, they must be changed or eliminated.

6. Institute more thorough, better job-related training.

Training and education are essential to continuous improvement. However, if it is to be the basis for improvement, training must be followed by application. Many inhibitors stand in the way of the effective use of training. Management must change the system so that people are encouraged to use the results of training. Some of the major inhibitors are: "We probably need to get some of our people trained but we're very busy with new product introductions and haven't got the time to spare"; "It's for my people, not for me"; "SPC training won't help us, our problems are different"; "We rely on experience, our people learn on the job." All of these comments have been used again and again. It is regrettable but true that in many organizations when hard times hit one of the first things to be cut is the training budget—it should be the last. If the Deming cycle is to be implemented on a continuous basis, people must be given the tools to make it happen. Statistical thinking and methods are essential.

William Conway, former CEO of the Nashua Corporation, told a story[2] that exemplified the need for giving people the proper tools to do their jobs effectively. He told of an employer who decided to reward one of two employees, let's call them Bill and Jack, with a lavish vacation at company expense. The two were to compete for the prize. The employer first called Bill into his office and described the trip, detailing the many extravagant features of this perk. Needless to say, Bill became very enthusiastic, with great motivation to compete to his best ability to win the trip. The employer then called Jack in and proceeded to get him quite excited and motivated about winning the trip. The day came for the competition and both Bill and Jack were eager to perform. The employer told both that the task they were to complete was to secure six screws into a piece of wood. The employer gave Bill the screws, the piece of wood, and a nice screwdriver to aid in completing the

---

[2]Adapted from a videotape presentation by Conway to a group of Ford executives, Ford Motor Company Radio, TV, and Film Dept., File No. 81V44-1, Dearborn, MI, 1981.

task. He turned to Jack and gave him the screws and the wood but told him he would not have the use of a similar tool. On the count of three, both men enthusiastically began to complete the task. In a matter of a couple of minutes, Bill was done, with little effort. In the meantime, Jack broke three of his fingernails, cut open the palm of his hand, and was unable to get even one screw into the board.

It is clear that both Bill and Jack were highly and about equally motivated to do the work assigned to them. Both wanted to do a good job. However, only one of them was given the tools by his manager to do the job properly. Motivation and desire are clearly not enough; people must be given the proper tools to do their work effectively. In Chapter 12 we will look at a similar case in an actual manufacturing process setting. We will find that the best efforts of an otherwise skilled operator fell short of the mark simply because she was not given the knowledge and tools to do her job properly. It is clear from this account that while attitude and psychology are important to the never-ending pursuit of improvement, they must be clearly separated from the tools required to do the job properly and efficiently. Training and education are essential to providing these tools.

Based on the authors' experience, through many years of teaching in both the university and industry, there are a few guiding principles that may be important to those developing and implementing education and training in the workplace. These are:

1. *Responsibility for the learning process.* More of the responsibility for the learning process should be shifted from the instructor to the student.

2. *Learning in the workplace.* The learning process should be part of the normal workday routine. It should become a normal expectation of the individual to be continually striving to expand his/her capacity to do his/her work.

3. *Role of the facilitator/teacher.* The facilitator/teacher's "contact time" should be shifted away from the formal classroom setting to smaller and more personal group meetings and individual discussions. The process should involve less of the formal lecture/discussion presentation format and more informal discussion.

4. *Application/problem driven.* The learning process should take place in a context-specific application setting familiar to the students involved.

5. *Shared goals and experience.* The learning process should be shared by a small group of individuals who share a common concern or goal. This is consistent with the self-directed work-team concept now practiced by a number of companies. Just as the team plans and executes its work, so should it plan and execute the process by which it learns and applies new technologies.

6. *Reinforcement of the learning process.* The learning process should engender support for and provide reinforcement of the disciplines that enable the learning organization to evolve. Whether explicitly or implicitly present, the technologies of systems thinking, personal mastery, mental models, shared vision, and team learning should be a part of the process.

7. Institute leadership.

For many years Deming referred to this point as: "Improve supervision." But he came to realize that you cannot improve something that does not exist. First you must institute supervision; only then can you improve on it. In this new economic age, a supervisor must be a coach and a teacher, not a watchman. A supervisor must continually seek ways to provide his or her people with the tools to do an effective job and the opportunity to use those tools. Continual improvement requires the "catching someone doing something right" mentality, not "catching someone doing something wrong."

Too many management systems today are oriented toward the management of things, not of people. They are oriented toward the management of the outgoing stream of product, not of the management of the people and machines that make the product. Leadership in management should have as its primary goal to motivate people to work to their maximum levels of performance, but too often the management system being used does nothing more than foster mediocrity. Under the system of *management by objective*, the main focus often becomes the job of setting the objective. Just as the engineering specification can be a focus of negotiation to the detriment of quality and productivity, so can the setting of objectives for individual performance. Fear of failure may cause a subordinate to negotiate with his boss to arrive at a lesser objective for his performance for the year. Once the objective is set, however, the only thing that is important from that point is the objective itself and the employee's performance relative to it.

8. Drive out fear, so that everyone may work effectively for the company.

Deming felt this to be one of the more critical of his 14 Points, since it affects directly so many of the other 13. Fear is an all-pervasive quality that touches every corner of the organization in many different ways. Fear of failure and fear of embarrassment inhibits our ability to capitalize on opportunity. Fear causes us to misinterpret the intentions of others. Fear breeds overreaction, which can only increase variability in the system, thereby increasing waste and inefficiency. To reduce fear requires a fundamental change in the philosophy of management. Implementation of Point 7 is an essential prerequisite to the substantial elimination of fear throughout the organization.

9. Break down barriers between departments.

People in all activities within the organization must work as a team on the implementation of the Deming cycle. To accomplish this, Point 8 is crucial. But to truly invoke continuous improvement on a companywide basis requires that everyone come to a common understanding of the needs of the customer and the present capability of the organization to meet those needs. To implement this point requires that the system foster teamwork. This may mean that management systems that destroy teamwork, such as the personnel performance appraisal system, must be changed. In his book, Scherkenbach singles out performance appraisal systems as potentially one of the most significant inhibitors to quality and productivity design and improvement.

10. Eliminate slogans, exhortations, and targets for the workforce that ask for zero defects and new levels of productivity.

Personal motivation and improved awareness of the needs of the organization will contribute to the reduction of waste and inefficiency. But such are not a substitute for providing training and giving people the tools to do a better job. As we discussed earlier, William Conway, then CEO of the Nashua Corporation, told a story about two employees who were competing for a major company perk. Both were asked to do the same job and both were equally enthusiastic about the reward. Motivation was at a fever pitch. One, however, was given the proper tools by management to do the job while the other was simply told to do the best he could. Clearly, the one who had the proper tools won the perk. Both had the same desire to do a good job but only one was given what was required to make it happen. Everybody wants to do a good job. But management must create an environment that enables people to improve continuously.

11. Eliminate work standards on the factory floor.

Clearly, we have already said much about this point in Chapter 1. Work standards place a cap on productivity improvement and are totally contrary to the concept of never-ending improvement.

12. Remove the barriers that rob employees at all levels in the company of their right to pride of workmanship.

To make this happen, major changes in the way we manage people must take place. Supervisors must emphasize the need for quality, not for getting out the numbers. The systems that evaluate the performance of people on a regular basis must probably be changed. Financial management systems that tend to focus on the short term must change. The use of daily or weekly production performance reports must be abandoned. Such reports encourage the management of things, not people, and often lead to overreaction, which, as we have already said, increases the variation and hence the waste in the system.

13. Institute a vigorous program of education and self-improvement.

People must not be treated as a commodity but rather as an asset. People in the organization must believe that management is willing to reinvest in their future, providing them not only with the opportunity for change, but also with the tools to take advantage of the opportunity.

Education and self-improvement, however, require that both individuals and the organization as a whole overcome a number of inhibitors that commonly stand in the way of this process. In his book *The Fifth Discipline: The Art and Practice of the Learning Organization*, Peter M. Senge of MIT's Sloan School of Management discusses a number of learning disabilities that are characteristic of many organizations. These learning disabilities stand as significant inhibitors to the growth of an organization's ability to learn and expand its capacity to do its work more effectively. Some of them are:

1. **"I am my job."**
   Most people view themselves as a small part of a large system in which they have little control, either of their destiny or the organization's. When people focus only on what they do—their position—they can gain little sense of what the results can be like when all of the positions interact.

2. **"The enemy is out there."**
   Often people within an organization develop the attitude that forces beyond their control account for their inability to succeed in what they do. Often, they view the enemy as the organization itself, or others in another part of the organization, or outside influences such as foreign competition, government regulations, and labor unions. This feeling of lack of control makes it difficult for many to take a proactive stance in dealing with problems.

3. **"Fixation on events."**
   Organizations tend to be geared toward reacting to events, whether they be "a sudden drop in sales" or "the need to become a certified supplier of a major customer." Being event driven entails at least two basic problems. First, we tend to build strategies and programs around events, so that when the event fades in memory, so do the programs. In recent decades this has been a real problem for many firms that have built employee education programs around the desires of their customers. Second, "event-based thinking" tends to make it difficult for us to see the real causes of our problems, which are usually the result of gradually evolving processes.

4. **"The parable of the boiled frog."**
   As the parable goes, if a frog is placed in a pot of boiling water it will immediately scramble to jump out. If a frog is placed in a pot of room-temperature water and the water is slowly heated up, the frog will do nothing. In fact, it probably will find the warmth soothing and relaxing. Eventually, it will become so groggy that it cannot get out of the water even if it wants to. People in organizations tend to react the same way. Our "fixation on events" causes us to recognize only big and often sudden changes. We have not been educated and conditioned to observe slowly evolving trends. For example, from 1962 to 1981—a period of two decades—the Japanese share of the U.S. auto market gradually rose from 4% to 21.3%. Detroit's reaction to this growing problem did not even begin to manifest itself until the 1980s.

5. **"The delusion of learning from experience."**
   An important characteristic of adult learners, from a self-recognition standpoint, is the apparent power and importance of learning from experience. This may be one of the greatest personal learning disabilities of all. The delusion that we have over the years picked up a great deal of valuable knowledge from experience is even more pervasive in organizations than we can imagine. There are several problems with "learning from experience." To begin with, in most complex systems, the real consequences of our actions are too far removed from

our sphere of operation in either time or space to be observed in the first place. Therefore, what we often observe and correlate with our actions and decisions is happenstance rather than causal in nature. The belief that we have learned something important from experience has another inhibiting consequence. It causes us to jump immediately to the answer, circumventing the process of rationally developing the decision and action plan. The delusion of learning from experience is reinforced by the apparent pressures of time and cost to "do it the way we have done it in the past—we *know* that works!"

14. Put everybody in the organization to work to accomplish the transformation.

The transformation is everybody's job. What ingredients, though, must be present within the organization to make this happen? At least four things need to be part of the process.

1. **Top-Down Approach**.
   At some reasonably high level, a commitment by the top manager to work toward this goal as a total system is required, and the commitment of his/her leadership in the effort must be in place. In principle, this person should always be the top in the organization, but the size of the organization has some bearing on this. In a smaller organization, it must be the top person. In a medium-sized organization, it could be someone who reports to the top person. In a large organization, it could be someone responsible for a large portion of the organization. Why is this important? There are at least three answers to this question: (1) Proper institutionalization requires strong intellectual leadership; (2) Proper institutionalization requires change in the system itself; and (3) People look for signals from the top. When they see such signals, they are encouraged that what they are thinking about and want to do is supported. When they do not see these signals, they become jaded in their perspective on the system and often bitter about the extent to which their views are accepted. Grass-roots efforts can work and should not be discouraged, but the process is slow, painful, and frustrating, and must be done with an eye toward winning the top over, not fighting the top.

2. **Concurrent Education and Application Activities**.
   The transformation cannot take place unless education—new knowledge—is available. But knowledge alone is not enough; it must be put to work to effect change. Similarly, problem-solving activities without knowledge of new concepts and tools will likely lead to very little progress and lots of churning and frustration. The education process stimulates thinking and provides new ideas; it is a growing process that helps make change seem more natural. What education is required? Several things must be considered, including: (1) the Deming philosophy; (2) problem-solving tools and methods, e.g., statistical methods; and (3) context-specific education—for example, for an accounting problem, learn more accounting; for a machining problem, learn more about new, emerging machining technologies. Above all, make sure you secure the services of competent people or attend courses conducted by competent people.

3. **Willingness to Look for Inhibitors and Change the System.**
   Often, the transformation is blocked by inhibiting forces in the system that cut across the grain of what one is trying to accomplish. Examples include:

   - Personnel evaluation/reward systems.
   - Standards for work performance.
   - Management-by-objectives approach.
   - Arm's-length labor/management relationship.
   - Purchase by low bid practices.
   - Quality assurance—end-item inspection practices.
   - Managing outcomes—Management by the numbers.

   But identification of the presence of the inhibitor is only the first step, important as it is. We must remove the inhibiting force if real progress is to be made. A common myth is that "I must replace the inhibitor with something else!" But as Deming said on many occasions, "If you have an illness, do you replace it or simply get rid of it?" Removing inhibitors means that fundamental system change must take place, and only management can change the system.

4. **Agreement on a Common Mission and Goals.**
   Deming said, "Put everyone in the organization to work on the transformation." Then everyone must be operating by the same set of guiding principles.

   The 14 Points are clearly the responsibility of management. They, themselves, define the essential elements of the institutionalization of quality and productivity improvement through statistical thinking and methods. Quality and productivity improvement is everyone's job, but only management can create the structure in the organization that enables the 14 Points to be carried out. The 14 Points define the elements of a process—a new way of conducting all aspects of the business. As is obvious from the discussion above, they cannot be deployed selectively; success in carrying out any one point depends on carrying out the others. The 14 Points provide important guidance in terms of what we must do. It is up to management to redesign the system to achieve the 14 Points.

Deming's Seven Deadly Diseases of Management

The 14 Points—management's obligations—just discussed lie at the foundation of Deming's philosophy of management. To lend credence to this philosophy while at the same time pointing out common ills and barriers that management often faces, Deming articulated in his teachings *Seven Deadly Diseases of Management* and some associated obstacles to the institutionalization of his philosophy. We now briefly review these deadly diseases.

1. Lack of constancy of purpose.

Too many companies muddle through their existence with no real sense of purpose and long-term plan that can render them responsive to the voice of the customer. This is generally the product of short-term, often reactive, thinking.

Continual reaction to short-term financial figures, management fads, and perceived changes in market forces saps the energy from the organization and makes the best efforts of well-meaning employees irrelevant. "Overcontrol" applies not only to a particular process but to the organization as a whole, increasing total system waste and inefficiency.

2. Emphasis on short-term profits.

One of the most debilitating forces within many organizations is preoccupation with the "quarterly report" and the price of the stock. Generally it results in actions that might "make the numbers look good" in the short term but irreparably damage the company in the long term. For example, it is not at all uncommon to see the slashing of the R&D budget or the elimination of employee training programs to "improve" profits in the short term.

3. Evaluation of performance, merit rating, and annual reviews of performance.

This deadly disease seems to be deeply rooted within most organizations. Its consequences include destroying teamwork and developing fierce personal rivalry within the organization. To a great extent, performance of individuals is driven by the system forces beyond their control and, too often, by system forces that are badly flawed. Performance evaluation under such circumstances can leave individuals both bitter and beaten, further eroding their contributions to the organization. In addition, personnel evaluation systems too often include many levels of "defined" performance that makes it impossible to determine a rational basis for establishing where a given individual lies on the "performance scale." Thinking of people's performance as being represented by a "bell-shaped" curve, the conclusion can only be that performance is either within the system, perhaps in the wide area of two sigma to either side of the mean, or outside the system—that is, beyond the two-sigma limits and significantly better or worse than average.

4. Mobility of top management.

In his comparisons of U.S. and Japanese management practices, Deming often pointed to the rapid movement of managers in and out of jobs as a tremendous barrier to constancy of purpose and optimization of the total system. Managers often spend more time thinking about *what* their next job will be and *how* to get there than about *what* their current job is and *how* to do it more effectively. Continual reorganization—"the rearranging of the deck chairs"—is a barrier to constancy of purpose and a waste of valuable system resources that could otherwise be directed at understanding what future markets might be and planning for products and services for those markets to ensure jobs and profitability for the long term.

5. Running a company on visible figures alone.

Management by outcomes or results—that is, taking immediate action based on visible figures, such as number of defects, customer complaints, and poor sales figures for a given quarter—can lead to more trouble in the long run than it can ever avoid in the short run. Deming preached that the most important factors governing the long-term success of an organization are often unknown and unknowable. Such tampering is another form of overcontrol and can only increase the variation—waste and inefficiency—of the system.

6. Excessive medical costs.

Little did Deming realize how prophetic his words were when he preached to the auto industry in the decade of the 1980s on the ills of excessive medical costs! This may be the most all-pervasive problem facing scores of companies and the U.S. economy as a whole today. Enough said here.

7. Excessive legal damage awards swelled by lawyers working on contingency fees.

This, too, has become a major socioeconomic issue plaguing the United States today. Deming was always quick to point out that the United States had more lawyers per capita than any other country in the world. Deming felt that tractable solutions to this and the excessive medical costs problem would likely require government intervention.

In addition to these seven deadly diseases, Deming in many of his writings, videos, and class lectures discussed a number of other obstacles to the institutionalization of his management philosophy. Some of these were:

- The supposition that automation, gadgets, and new machinery will transform industry; spending money on these instead of on "brains."
- Searching for examples, i.e., copying existing solutions.
- Excuses such as "our problems are different."
- Reliance on the quality control department.
- Blaming the workforce for problems.
- Quality by inspection—product control, not process control.
- False starts (e.g., wholesale training in statistical methods without the necessary change in company philosophy).
- Meeting customer specifications.
- Inadequate testing of prototypes; insufficient investment in R&D.
- "Anyone that comes to help us must understand all about our business."

2.3   Traditional View of Quality

Shortcomings of Traditional Definitions of Quality

Before we can really understand what type of a strategy needs to be adopted to move in the direction of enhanced competitive position in the world marketplace, we need to ask some basic questions:

- What do we mean by quality products?
- What do we mean by improving quality?

- What do we mean by quality design?
- Quite simply, what do we mean by *quality*?

Over the years many definitions of quality have been offered. A quick look into the first chapter of dozens of books written on the subject over the years provides a variety of definitions of quality:

- Fitness for use/fitness for function.
- The degree to which a specific product satisfies the wants of a specific customer.
- The degree to which a product conforms to design specifications.
- Providing products and services that meet customer expectations over the life of the product or service at a cost that represents customer value.
- The characteristics or attributes that distinguish one item or article from another.
- Conformance to applicable engineering requirements as described in engineering drawings, specifications, and related documents.

While most of the words sound okay, such definitions have not been very useful; that is, they have failed to help us do a better job, to be more competitive. The question is: Why? There may be at least three reasons why such definitions have not served us well:

1. *They are attribute based and qualitative in nature.* The precise quantitative definition of quality has not been established through any of the definitions above. These definitions of quality generally prescribe that something is either in one state or another: good or bad, defective or nondefective, within or outside the specifications. As a result, they promote the improvement of quality only up to an acceptable level and therefore are a serious inhibitor to continuous and never-ending improvement.

2. *They are manufacturing rather than design based.* Many definitions of quality attempt to compare the product on completion of manufacturing to the design specifications that have been imposed on the manufacturing process. As a result, they become the final filter that we attempt to push the product through rather than the criteria on which the product design is based. That is, the application of these definitions of quality is taking place at the end rather than at the beginning of the product development life cycle. Over the years, the most common and widely used measure of quality has been "conformance to specifications." However, specifications are generally a product of engineering design, not a design objective. Too often, specifications are driven by the limitations of the manufacturing process. Sometimes, the specification results from considerable negotiation between design and manufacturing personnel.

3. *They do not clearly establish the proper link between customer wants/ needs/expectations and product function.* It is important to distinguish between customer likes in terms of preferences—color, style, and so on—and

customer needs in terms of the performance of a function. A certain customer may prefer a red automobile over a blue one, but once that issue, which is one of subjective preference, is settled, the only issue that remains is that of performance of the function. Function is judged by such things as product life, power consumption, trouble in the field, and production of harmful effects during use. A product may be sold—gain market—by virtue of its subjective characteristics (style, color, etc.) but may lose reputation—market share—by virtue of its function, its quality.

In today's economic age we simply cannot afford to use concepts and measures of quality that do not relate the achievement of function in the field to the engineering design process but instead administer "quality control" through defect detection and containment after manufacturing. Definitions of quality that promote improvement to an acceptable plateau of performance will inhibit the continual pursuit of never-ending improvement and hence have a weak and perhaps opposing relationship to process performance in terms of efficiency/productivity. Rather, to improve competitive position our view or definition of quality should:

- Provide a quantitative basis to move the quality issue upstream to engineering design—a design criterion.

- Promote focus on the process, not the product in a manufacturing sense—prevention rather than containment.

- Be strongly tied to the issue of productivity and therefore promote continual pursuit of never-ending improvement.

- Quantify/measure the loss imparted to the customer as a result of poor quality rather than the loss imparted to the producer—consumer vs. producer orientation.

Quality and the Engineering Specification

From the above it is clear that the traditional definitions of quality appear to have fallen short in terms of their ability to articulate quality in a way that can foster improvement in competitive position. In particular, the association of quality with *conformance to the engineering specification* puts the measurement of quality on an attribute basis and provides little more than a *shipping criterion*, when it is essential that we articulate quality in a manner that enables it to be used as a *design criterion*. The view of quality as conformance to specifications promotes the product control approach to quality control and hence stands as a significant inhibitor to the adoption of a process control approach to manufacturing and the integration of quality and the design process.

Figure 2.5 provides a graphical depiction of the traditional interpretation of the engineering specification. In Fig. 2.5(a) and (b), two representations of a process as a statistical distribution of measurements are shown relative to a certain bilateral specification. Although the bell-shaped process distribution in (a) might be preferred to the more loaf-shaped process distribution in (b), some people would not make much distinction between the two cases in terms of quality; that is, in both cases virtually all of the process/parts are conforming to the specifications. However, when one begins to contemplate Fig. 2.5(c), one

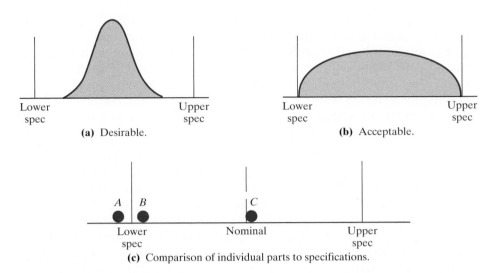

(a) Desirable.

(b) Acceptable.

(c) Comparison of individual parts to specifications.

Figure 2.5   **Classical interpretation of the engineering specification**

begins to realize that the relationship between quality and the engineering specification is not as strong as we may have viewed it in the past. Considering the dots labeled *A, B,* and *C* as representing three different manufactured parts, we generally interpret part *A* to be unacceptable because it is outside the specifications, while parts *B* and *C* are considered acceptable because they are within the specifications. The crucial point in this interpretation is the fact that we generally do not make a distinction between parts *B* and *C* as far as quality is concerned. It does not take too much thought, however, to come to the realization that as far as performance is concerned, it is unlikely that there is very much difference between parts *A* and *B,* although we might well conclude that part *C* should perform considerably better than either *A* or *B.*

If you were a teacher and you were considering the performance of three students on an exam who earned scores of 59, 61, and 95, you would probably view the performance of the two students who got the scores of 59 and 61 as about the same, and much poorer than that of the student who got 95. Even though 60 might be your lower limit for a passing grade, you would probably not view the two students who got grades of 61 and 95 as being equal from a quality standpoint. Why, then, should we think of manufactured product and its potential for quality performance in the field any differently?

2.4   Taguchi's Definition of Quality

As discussed in Chapter 1, Taguchi's approach to quality engineering has a number of significant strengths. In particular, Taguchi has emphasized the importance of minimizing variation as the primary means to improve quality. The ideas of designing products whose performance is insensitive to environmental conditions and making this happen at the design stage through the use of design of experiments have been cornerstones of the Taguchi methodology for quality engineering.

The Loss Function Concept

Taguchi offers a view of quality that directly addresses the concerns raised in the preceding section. He suggests that it is important to think of quality in terms of the "loss imparted to society during product use as a result of functional variation and harmful effects." Harmful effects refer to side effects that are realized during the use of the product and are unrelated to product function. Functional variation refers specifically to the deviation of product performance from that intended by design, that is, deviation of performance from the design target, the nominal.

Taguchi uses a loss function concept to quantify quality as "loss due to functional variation." He argues that the loss to society as a whole is minimized when performance is at the design nominal and as performance deviates from the nominal, loss increases. For many applications a quadratic loss function may be appropriate. Figure 2.6 illustrates the loss function idea. The loss function concept as a way to measure quality clearly suggests that the goal of design and manufacturing is to develop products and processes that perform "on target with smallest variation." (Recall the quality/variability relationship we spoke of in Chapter 1 as a motivation to seek continual reduction of variation.) The loss function places tremendous importance on the reduction of variation to achieve the most consistently performing products and processes. To drive performance toward this overall objective, it is important that we learn more about the way in which the important parameters of the product/process influence performance. Over the years the methods of design of experiments have proved to be an effective way to do this.

Figure 2.7 provides an interpretation of the engineering specification when the loss function concept of quality is superimposed. In this case no loss would be realized as long as the quality characteristic in question lies within the lower and upper specifications. Outside of the specifications the product is

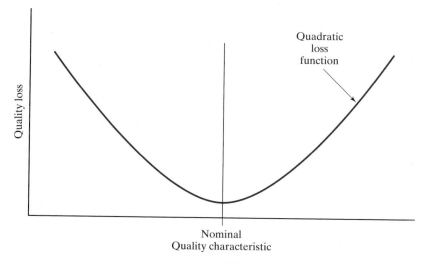

Figure 2.6   Loss function representation of quality

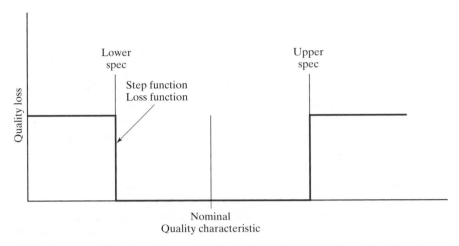

Figure 2.7   Loss function interpretation of the engineering specification

considered to be unacceptable, and therefore a constant loss would be realized. This loss would probably be measured in terms of scrap cost or rework cost.

Variation Reduction and the Loss Function: A Case Study

The loss function view of quality suggests that there exists a clear economic advantage to reducing variation in the performance of a product. The case study described below clearly indicates how such reduced manufacturing imperfection can lead to a reduction in costs and hence an improvement in competitive position.

Several years ago one of the automobile companies performed a study to compare the manufacturing variations evident in certain transmission components for comparable transmissions made in the United States and in Japan. Random samples of transmissions were selected in each case, the transmissions were taken apart, and a number of critical dimensions were measured and recorded. Figure 2.8 is representative of the general findings of the study. In particular, it is seen that in the case of U.S. transmissions, the critical dimensions generally consumed about the middle 75% of the tolerance range specified in each case. Normally, one would conclude based on generally accepted standards that the capabilities associated with the manufacturing processes were therefore well within normal expectations. As the figure shows, however, the same critical dimensions for the Japanese counterpart of this transmission consumed only about the middle 25% of the same tolerance range. The question that begs to be asked is: Why would the Japanese strive to make the parts to such tight tolerances? The answer lies at the foundation of what this book is all about.

There are two very good reasons why the Japanese have worked so hard to reduce variation, as shown in Fig. 2.8. The first of these is seen in Fig. 2.9,

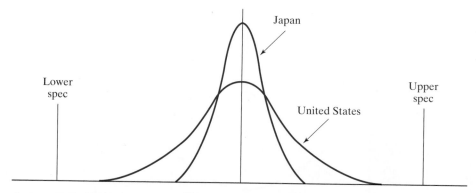

Figure 2.8   Comparison of critical dimensions for U.S. and Japanese transmission components

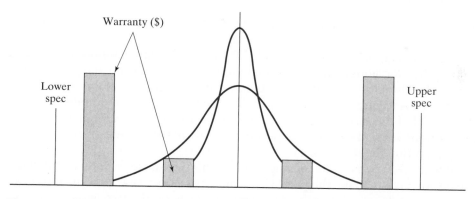

Figure 2.9   Comparison of warranty costs for U.S. and Japanese transmissions

where the shaded bars represent the costs associated with warranty claims for the two different transmissions. These warranty cost bars have been plotted at the extremes of the two distributions of the critical dimensions to depict the relative costs associated with variability. It is clear that for U.S. transmissions, the cost associated with repair and replacement is significantly larger than for their Japanese counterparts.

This comparison of the economic data associated with the two transmissions would suggest that the loss function idea is, in fact, at work, whether we realize it or not. Figure 2.10 illustrates the appearance of Fig. 2.9 if one considers that the relative magnitudes of the warranty costs are simply estimating the loss associated with variation in the critical dimensions under study. The figure strongly suggests that there exists a definite relationship between variability and cost in terms of loss incurred due to functional variation. One could think of the warranty cost values as literally mapping out the loss function.

There is still another side to this variation reduction situation, which suggests an important relationship between variation reduction and reduction in

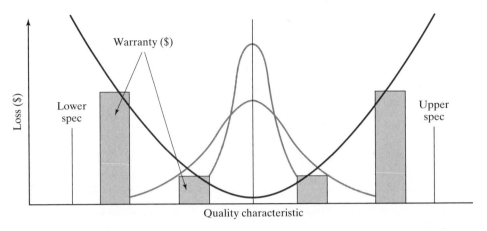

Figure 2.10   Loss function interpretation of the results of the transmission study

waste and inefficiency. Some years ago, one of the authors was sitting in a presentation on statistical thinking and listening to the speaker[3] recount an exchange
between a manager in a U.S. automobile company plant and a representative of a
Japanese firm for which they were making some parts. When the engineering
drawings for the parts arrived at the plant, it became immediately obvious that
something was wrong. There were no tolerances on any of the dimensions. The
manager called the Japanese firm and asked for a new set of drawings with the
specifications indicated. He was promptly told that there were no tolerances
specified for the dimensions.

After a brief pause, the manager asked his Japanese friend: "Well, how
close do we have to make the parts to the nominal dimension?" "How close can
you make them?" was the reply. After some discussion the two agreed upon an
allowable amount of variation. The Japanese company representative indicated
that this negotiated variation level would apply for one year. Furthermore, he
indicated: "After the first year, we will ask you to reduce the amount of variation in the parts you are supplying. We know that to do this you will have to
improve your processes and if you are successful you will reduce some of the
waste and inefficiency in your operations. As a result, your costs will go down.
We want you to share that saving with us and so at the end of the first year
when we ask you to reduce the variation, we will also ask you to reduce the
price of the parts."

The manager was stunned! How could he possibly be expected simultaneously to tighten up on the specification for the parts and reduce the price? "I
did not ask you to tighten up on the specification on the parts," was the reply,
"I asked you to improve the consistency of your processes! If you do this by
finding the root causes associated with the variation in your processes and take
action to remove those problems, the parts will be made closer and closer to

[3]Presented by Howard B. Aaron, Lafayette, Indiana, ASQC chapter seminar on "Quality for Today
and Tomorrow," March 17, 1988.

Figure 2.11   Economic consequences of an improvement action

the nominal value and the efficiency of your processes will improve. We just want you to share your good fortune with us."

2.5   Cost-of-Quality Issue

The loss function concept may precipitate the thinking that since we have put quality (loss) on economic grounds, we should carefully weigh the cost to achieve quality—reduce variability—against the derived gain using the loss function. Figure 2.11 depicts this scenario graphically. The basic question we are asking is: Are such continual improvement efforts always cost-effective? Are we reaching a point of diminishing returns?

Some caution may be needed in the quantitative use of the loss function idea. Because the loss function establishes a rigid economic criterion, it may in fact tend to run somewhat contrary to the concept of the continual pursuit of never-ending improvement. In spirit it may be very attractive and its words quite easy to embrace, but its literal translation may be shortsighted, and generally inhibiting, particularly when the question of the cost of quality arises.

In fact, Taguchi and others have given examples to demonstrate the weighting of costs to precipitate, say, a 50% reduction in product variation against the benefit of obtaining such a reduction in terms of functional variation. Such is also depicted in Fig. 2.11. Some have said that from an economic standpoint, there is a weakness in Shewhart's theory, that if the cost for control is larger than the profit from the reduction of variation, we should not do anything. In other words, it is the duty of a production department to reduce variation while maintaining the required profit margin. Such thinking could be counterproductive and inhibit us from realizing the full and broad-based benefits of the use of statistical thinking and methods for process improvement.

Perhaps the broader question we ought to ask is: What are the costs of not having quality? What are the costs associated with lost opportunity? Philosophically, we can, and perhaps must, argue that attempts to weight the input costs continually against the output gain are counterproductive in the long term (i.e., run contrary to the concept of the continual pursuit of never-ending improvement)—that such thinking will continually inhibit our ability to make the right long-term decision.

What is the traditional view of quality costs? Some have suggested that the following quality cost factors should be considered:

- **Prevention Costs.** Maintaining a quality control system.
- **Appraisal Costs.** Maintaining a quality assurance system.
- **Internal Failures.** Manufacturing losses, scrap, and rework.
- **External Failures.** Warranty, repair, customer, and product service.

Figure 2.12 shows how these factors are often combined to "optimize" the overall quality requirement. The old concept of AQL—acceptable quality level—is derived from such thinking.

When we ask, "What do all these costs have in common?" the answer is sobering. They are all relatively easy to measure, but they are not representative of the factors that enhance competitive position. The real question we need to ask is: "What is the cost of not having quality?" The answer can be seen through the following case study.

A comparative study was done by a manufacturer (Sony) of color television sets made in both the United States and Japan.[4] When many sets were examined

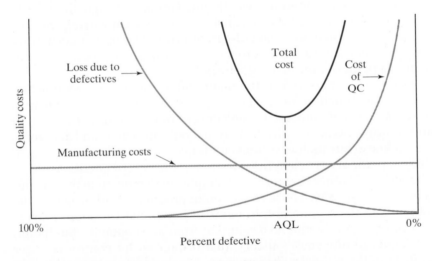

Figure 2.12   Traditional economic view of quality costs

---

[4]*The Asahi*, Japanese language newspaper, April 15, 1979. Also reported by Genichi Taguchi during lectures at AT&T Bell Labs in 1980.

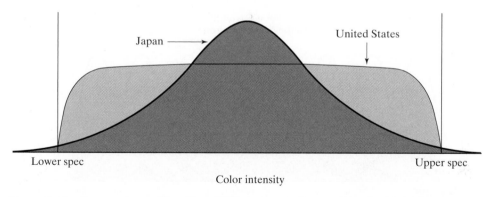

Lower spec   Upper spec

Color intensity

Figure 2.13   Comparison of quality of TV sets manufactured in the United States and in Japan

to observe the color-intensity quality characteristic, the results were as depicted in Fig. 2.13. For the U.S. sets, virtually none could be found whose color intensity was outside the specifications. The variation pattern followed the loaf-shaped appearance in Fig. 2.13. For the Japanese sets, a small percentage were found to have color intensities outside of the specifications, but the variation pattern followed the familiar bell-shaped curve in Fig. 2.13. The U.S. variability pattern is unnatural. It should signal to us that something is radically wrong in the process. And what is wrong is probably costing us a great deal of money.

We need to ask ourselves, "How could a result such as that for the U.S. sets in Fig. 2.13 be made to occur?" Figure 2.14 suggests one answer. We see that the sets originally manufactured out of the specification are adjusted/reworked after inspection to bring the color intensity within the specifications. This is accomplished only at a tremendous cost, the cost associated with not having quality in the manufacturing process.

Manufacturing   Inspection
process output   process output

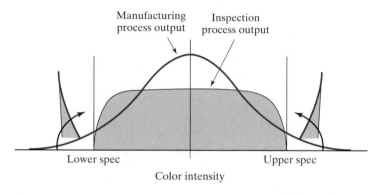

Lower spec   Upper spec

Color intensity

Figure 2.14   Quality "inspected" into the process

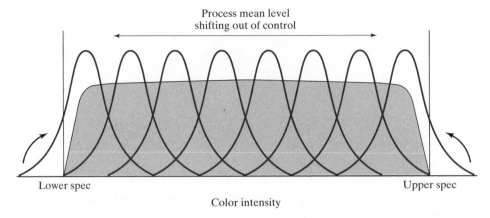

Figure 2.15   Out-of-control process causing quality erosion

Figure 2.15 shows another manifestation of the loaf-shaped process of Fig. 2.13. Although in principle the process in Fig. 2.15 appears quite capable (if stable), its erratic behavior with respect to its mean level causes the color-intensity value to be distributed broadly across the specifications. As in Fig. 2.14, if out-of-specification sets are made, inspection will detect this and they will be reworked. Even if all the sets are within the specifications, the erratic process behavior is likely to cause waste and inefficiency.

Figure 2.16 shows still another way that the loaf-shaped process of Fig. 2.13 can arise. Letting the process drift in this manner to "make full use of the wide specifications" can mask the occurrence of other problems that might otherwise show up on a control chart. Further, from a loss function point of view, the cost of not having quality is clear.

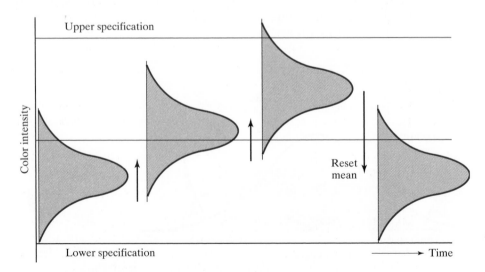

Figure 2.16   Making use of the full range of the specifications

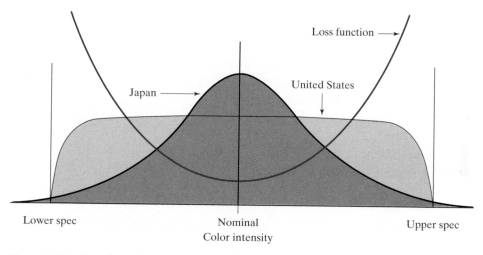

Figure 2.17   Loss function assessment of quality

Yet another cost of not having quality is derived from Taguchi's loss function approach. If the loss function is superimposed on the distribution of color intensity, as in Fig. 2.17, then it is clear that total loss, over all sets, is greater for the U.S.-made sets. Furthermore, although the previous examples of the cost of not having quality were manufacturing-related costs, costs on the user end are included in Taguchi's quality definition. For example, cost incurred resulting from having to rent a set while the broken set is being replaced is a cost of not having quality.

In considering all of the above, we have not even mentioned some of the more significant benefits of the continual pursuit of quality improvement/variation reduction. We can also view these as the lost opportunities associated with:

- **Improved Production Management.** Scheduling/inventory control/MRP.
- **Increased Flexibility/Adaptability.** Adjusting to changing needs/taking advantage of new market opportunities.
- **Enhanced Environment for Breakthrough.** Encouraging larger signal-to-noise ratio/ability to more clearly see trouble: easier to study process improvement actions—experimentation.
- **More Positive Employee Attitudes.** Looking for improvement rather than hiding poor quality; trying to catch someone "doing something right."
- **Identification of Process Flaws.** Searching for hidden waste and inefficiency.

In summary, we can look at this issue of quality costs in a couple of ways. We can spend our time trying to quantify the apparent costs of achieving a certain level of quality, or we can spend our time trying to understand what the cost of not pursuing quality is and then reap the many, many benefits of the continual pursuit of quality improvement.

■ EXERCISES

2.1   What is the purpose of Deming's 14 Points? That is, why did Deming feel the need to develop and teach these tenets?

2.2   What is the most fundamental point of departure of the philosophies of product control and process control?

2.3   How did Deming feel that customer–supplier relations ought to be approached?

2.4   Explain the term "hold the gain" in reference to improvement.

2.5   What sort of leadership did Deming feel is necessary in the realm of manufacturing?

2.6   Define "quality."

2.7   Many definitions of the word "quality" have been used in the past. What are some reasons why many of them do not serve us well?

2.8   What is the problem with the use of conformance to design specifications as a definition of quality?

2.9   How does the loss function concept differ from the classical definition of the engineering specification?

2.10 What is wrong with the loaf-shaped distribution that is characteristic of the output of processes for so many manufacturers?

2.11 When does the pursuit of quality and productivity improvement become economically inadvisable? Explain your answer.

2.12 What is the real cost-of-quality issue?

2.13 Deming stated that it is one of management's responsibilities to provide workers with the proper tools for the job. To what do you think he was referring here? Can you provide several examples, either real or contrived?

2.14 The elimination of work standards has been one of the more difficult of Deming's Points for management to accept. Why do you think that is? How would the elimination of work standards positively affect productivity? Quality?

2.15 Deming's Tenth Point states that work slogans and targets for the workforce should be eliminated. Why? And how will doing so affect quality and productivity?

2.16 Which of Deming's 14 Points show the detrimental effect that weekly performance reports have on overall quality? Explain why these inhibit improvement.

2.17 How will staying with single suppliers benefit a company in the long run? When multiple suppliers are used, where does the emphasis tend to be placed? How can a company be sure that its single-source suppliers will not fail to meet demand?

2.18 Which of Deming's 14 Points addresses the problems associated with the practice of over-the-wall design that was discussed in Chapter 1? Explain.

2.19 What is the major limitation of the loss function concept in terms of encouraging the continual pursuit of improvements in quality?

2.20 How can a customer benefit from improved quality in a supplier's product? How will this improved situation affect inventory management for the customer?

2.21 Discuss briefly each of the Seven Deadly Diseases of Management as articulated by Deming in his teachings.

2.22 Six Sigma may be thought of as a quality program that embraces both the Deming cycle and Taguchi's loss function idea, two of the distinctive quality concepts that emerged in the quality movement during the early 1980s. Elaborate on which part or parts of Six Sigma embrace the Deming cycle and which embrace the loss function.

2.23 Think of one point that could be added to Deming's 14 Points, particularly in view of the high-tech environment and global economy of the 21st century.

2.24 Do you find that any of the Deming's 14 Points need to be rephrased or eliminated for the 21st century?

2.25 "Quality" is a word that generally carries a positive connotation. Why does Taguchi treat it as a loss or cost that generally carries a negative connotation?

2.26 Provide a concrete example to illustrate each of Deming's Seven Deadly Diseases of Management.

2.27 Discuss a situation in your own life where a misunderstanding has arisen due to a lack of an appropriate operational definition. Knowing what you know now, describe a good operational definition for that situation.

2.28 The chapter discusses a number of learning disabilities that have been identified by Senge. Discuss each of these briefly and provide a concrete example for each.

3

Statistical Methods and Probability Concepts for Data Characterization

3.1   Introduction

Throughout our study of statistical methods for quality and productivity improvement, we will be concerned with essentially the same problem: using data obtained from the process to draw conclusions known as inferences about how it is or has been operating. We may be concerned with an injection molding machine that is making truck grilles; or a robot inserting clips into molded instrument panels; or an automatic screw machine making fasteners; or an accounts payable system processing customer invoices; or many other manufacturing processes and service operations. In each case we will find it useful to the improvement of quality and productivity to collect data from the process and model it using statistical concepts and methods. Such will be necessary since the results of the data collection activity will represent only a portion, sometimes quite a small portion, of the entire process.

3.2   Characterizations and Representations of Data

A process engineer is examining an injection molding process that is producing a certain kind of light fixture for automobile headlights. The purpose of this study is to collect data on part weights to help monitor the performance of the molding press as to its state of statistical control and to evaluate its capability in meeting part-weight specifications. Table 3.1 gives 44 samples, each of which is a record of five ($n = 5$) part weights in grams of the light fixtures drawn from the press every hour. The table also lists, in the last two columns, the average weight and the range for each sample.

TABLE 3.1	Light-Fixture Weights (Grams)						

Sample	X_1	X_2	X_3	X_4	X_5	Sample Mean, $\overline{X}$	Sample Range, R
1	167	169	187	160	180	172.6	27
2	186	178	172	153	170	171.8	33
3	155	143	171	170	152	158.2	28
4	195	180	168	167	169	175.8	28
5	168	194	169	183	170	176.8	26
6	165	157	148	161	162	158.6	17
7	167	159	174	173	162	167.0	15
8	171	163	181	165	182	172.4	19
9	145	161	171	164	166	161.4	26
10	156	174	170	177	182	171.8	26
11	161	156	171	187	161	167.2	31
12	159	148	157	162	170	159.2	22
13	167	179	166	197	186	179.0	31
14	179	167	166	181	180	174.6	15
15	157	167	166	162	159	162.2	10
16	178	176	164	165	153	167.2	25
17	169	172	180	176	191	177.6	22
18	167	172	173	163	165	168.0	10
19	159	150	172	157	163	160.2	22
20	165	168	161	169	162	165.0	8
21	182	165	193	172	167	175.8	28
22	164	181	165	166	162	167.6	19
23	154	153	162	171	161	160.2	18
24	170	143	168	170	166	163.4	27
25	181	196	169	166	165	175.4	31
26	167	165	168	166	187	170.6	22
27	159	168	167	163	174	166.2	15
28	170	186	167	161	166	170.0	25
29	159	156	186	158	172	166.2	30
30	172	176	177	161	171	171.4	16
31	186	165	169	193	170	176.6	28
32	159	172	162	169	176	167.6	17
33	192	164	167	176	163	172.4	29
34	168	167	166	189	159	169.8	30
35	173	164	155	174	176	168.4	21
36	160	180	181	169	181	174.2	21
37	167	164	170	147	162	162.0	23
38	179	177	174	175	173	175.6	6
39	170	175	193	169	171	175.6	24
40	170	166	167	162	164	165.8	8
41	176	156	157	157	166	162.4	20
42	153	180	170	167	163	166.6	27
43	178	187	164	163	167	171.8	24
44	166	179	181	178	170	174.8	15
						$\overline{\overline{x}} = 169.02$	$\overline{R} = 21.93$

Measures of Central Tendency and Dispersion

The arithmetic average of the observed values of a sample is defined by

$$\overline{X} = \frac{\sum_{i=1}^{n} X_i}{n}. \tag{3.1}$$

In Table 3.1 the part weights of light fixtures from a molding press are listed five in a row for each hourly sample. The sample mean is calculated using Eq. (3.1). For example,

$$\overline{X}_1 = \frac{167 + 169 + 187 + 160 + 180}{5} = 172.6.$$

Each sample mean is an estimate of the true but unknown process mean, μ_X.

An important measure of the variability in data is the sample variance, which is the average of the squared deviations of the data from their sample mean. The sample variance, denoted by S_X^2, is defined by

$$S_X^2 = \sum_{i=1}^{n} \frac{(X_i - \overline{X})^2}{n-1}. \tag{3.2}$$

In this equation for the calculation of a sample variance, it is noted that the denominator is $(n-1)$ instead of n, which is the actual sample size. This is done because in computing S_X^2 using Eq. (3.2), there are n deviations $(X_i - \overline{X})$, but only $(n-1)$ of them are independent. The positive square root of the sample variance is called the sample standard deviation S_X, which is often more useful because it is expressed in the same units as the data. The sample variance is used as an estimate of the true process variance, which is denoted by σ_X^2. Similarly, the sample standard deviation is an estimate of the process or population standard deviation denoted by σ_X.

Another important measure of the variability is the range, denoted by R, which is the difference between the largest value and the smallest value of the sample. Given the sample measurements from sample 1 in Table 3.1, 167, 169, 187, 160, 180, the range of this sample is

$$R_1 = 187 - 160 = 27.$$

In general,

$$R = X_{\text{largest}} - X_{\text{smallest}}, \tag{3.3}$$

where X_{largest} and X_{smallest} refer to the largest and smallest values in the sample, respectively.

The range R may be used to obtain an estimate of the population standard deviation of the individuals X, σ_X. Under certain assumptions, an exact mathematical relationship between the true mean range and σ_X has been established, namely,

$$\sigma_X = \frac{E(R)}{d_2}, \tag{3.4}$$

where d_2 is a constant that is determined according to the sample size n, and $E(R)$, read "the expected value for the range," is a symbolic representation for the true mean range. In practice, an average of many sample ranges denoted by $\overline{R}$ is used to ensure a good estimate of the population standard deviation.

The average range, $\overline{R}$ (read as "R bar"), of the 44 sample ranges of Table 3.1 is 21.93 grams. The value of d_2 for a sample size of 5 is 2.326 from Table A.2 in the Appendix. An estimate of the standard deviation of individual light-fixture weights of the molding process can be obtained using Eq. (3.4) as follows:

$$\hat{\sigma}_X = \frac{21.93}{2.326} = 9.43 \text{ grams.}$$

As a comparison, the sample standard deviation s_X calculated from the 220 individual weights is 10.20 grams.

Both the range and standard deviation are measures of the amount of variation, or the degree of scatter of data. The efficiency of using the range as a variability measure decreases rapidly as the sample size gets very large. But due to its simplicity and the fact that usually small sample sizes are used in quality control applications, the sample range, R, is commonly used.

When the variabilities of two or more sets of data are compared, particularly if their averages are likely to be different, the use of pairs of the measures, such as $\overline{X}$ and S_X, for comparative analysis of the data sets may not be meaningful. For instance, if the variability of the light-fixture weights is compared with that of the instrument panels, another molded part, the standard deviation of the instrument-panel weights can be several times greater than that of the light fixtures and still be considered tolerable because the instrument panels are many times heavier. A dimensionless measure of the variability, called the *coefficient of variation*, may be used, and is defined by

$$CV = \frac{S_X}{\overline{X}}. \tag{3.5}$$

When the CV's in weights of the light fixtures and the instrument panels are compared, they provide a relative measure of the variability of the same injection molding process in producing two different products. The coefficient of variation can also be used as a measure of quality or productivity of a process over time, as either one or both of the mean and the variance of a process are subject to change.

The measures of central tendency and variability may also be combined differently to indicate the overall performance of a process. One such measure, the *signal-to-noise ratio*, could be expressed by

$$S/N = \frac{\overline{X}}{S_X},$$ (3.6)

which is the reciprocal of the CV. As the name suggests and as pointed out in Chapter 1, it is desirable to seek improvement of a process to produce larger and larger values of S/N.

Dot Diagram

To appreciate the manner in which the data are arising and to gain some insights as to the nature of the process, it will be meaningful to display the data pictorially. The simplest form of graphical representation of data is the dot diagram. Each piece of data is represented as a dot on a graph according to its value. When all data are plotted, the diagram, or the graph, presents an overall distribution of the data, such as the one shown in Fig. 3.1. When a sufficient amount of data is available, say, 25 or more measurements, such a simple graphical representation may be quite insightful. However, such a graph becomes tedious to construct with very large sets of data such as Table 3.1 and somewhat difficult to interpret if a data set contains many different values.

Frequency Histogram

With large data sets of, say, 150 data points or more, we will group the data into cells (or intervals) and make note of the frequency of observations falling within each cell. When we do this grouping, all the observations within a cell

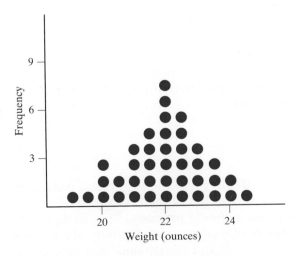

Figure 3.1   **Dot diagram of part weights**

are considered to have the same value, which is the midpoint value of the cell. The following are guidelines for choosing cell boundaries and the number of cells:

1. Over the full range of the data, somewhere between 10 and 20 cells or intervals are chosen, depending on the size of the set.
2. Equal class intervals, or cell widths, are used in general.
3. Although the measured scale may actually be continuous, measurement or reading limitations allow only a finite number of possible readings to be recorded as data. Cell boundaries are chosen to fall midway between two possible readings.

For our example of the light-fixture data, the observed weights range from 143 to 197 grams. All observations are recorded as integers because the scale used for the measurement reads only to the nearest gram. The data are grouped into 19 cells with a common cell-width of 3 grams. The boundaries of the cells will be at xxx.5.

The first step toward constructing a frequency histogram, given that the appropriate class intervals have been chosen, is to tally the data. Table 3.2 is a tally representation for the 220 light-fixture weights, which are grouped into 19 cells. The numerical frequencies f_i for each cell are also listed on the righthand side in

TABLE 3.2	Tally Sheet for the Data of Table 3.1		
Cell Boundaries	Cell Midpoints	Frequency per Cell	f_i
141.5–144.5	143	11	2
144.5–147.5	146	11	2
147.5–150.5	149	111	3
150.5–153.5	152	11111	5
⋮	155	11111 11	7
	158	11111 11111 1111	14
.	161	11111 11111 11111 11111 1	21
.	164	11111 11111 11111 11111 1111	24
.	167	11111 11111 11111 11111 11111 11111 11111 111	38
.	170	11111 11111 11111 11111 11111 11111 1	31
.	173	11111 11111 11111 11	17
.	176	11111 11111 11	12
.	179	11111 11111 1111	14
.	182	11111 11111 1	11
.	185	111	3
.	188	11111 11	7
.	191	11	2
.	194	1111	4
195.5–198.5	197	111	3
		Sum of f_i = 220	

the tally. To convert the tally sheet to a histogram, we allow the number of observations within each cell, the cell frequency, to represent the height of the frequency bars on the ordinate and the difference between adjacent cell boundaries to represent the width of the frequency bars on the abscissa. This is shown in Fig. 3.2.

From the frequency histogram we note three things of importance about the way in which the data arise:

1. The data tend to cluster about a certain value (i.e., more weights arise toward the center of the data).
2. The data are spread out over a broad range (i.e., a wide spectrum of part weights are represented).
3. The histogram has a characteristic shape (i.e., the frequencies fall off rapidly as we move away from the center of the data). Further, the frequencies are approximately symmetric about the center of the data.

The succinct description of these three characteristics of the data—(1) central tendency, (2) dispersion or variability, and (3) distributional shape of frequencies—is central to the successful use of such data as a decision aid.

To characterize a set of data fully, particularly for probability evaluations and statistical analysis, the shape of the population distribution must be established. Usually, a certain probability distribution is assumed to describe the population either based on theoretical grounds or from the appearance of graphical displays of the sample data. To check the validity of the assumption, a number of methods, including graphical representations and statistical analyses, are often employed. We will discuss some of these methods.

Uniqueness of the Population or the Process "In Control"

Given a set of data, one can always calculate from it a number of sample statistics, such as the sample mean and sample range, or graphically represent it in plots such as the dot diagram and histogram. But these characterizations are

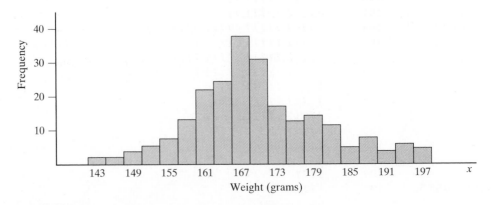

Figure 3.2   Histogram for individual light-fixture weights

meaningless and may provide very misleading information about the product or the process unless all the data are from one unique population or are observations of a process that is subject to a constant system of variability causes.

For a data set consisting of samples collected over time, an effective method to check whether or not every observed sample is from the same unique process is with Shewhart-type control charts. Beginning with Chapter 5, we devote a large portion of this book to the use of such control charts. Control charts are graphical displays of statistics plotted according to the order of their observation. If no nonrandom patterns or extreme points are shown on a control chart, then the sample data are considered to have been observed from a process that was subjected to one common-cause system during the period of observation. Such a process is said to be in statistical control. The control charts (Fig. 3.3) of the averages and ranges of the data on light-fixture weights (Table 3.1) show no values beyond the "control limits" (no extreme points). The uppermost chart in the figure, the $\overline{X}$ chart, considers the behavior of the sample mean light-fixture weights and appears a little suspicious—it seems to demonstrate a somewhat cyclical behavior (a nonrandom pattern). In Chapter 6, we will establish a set of tests or rules for interpreting control charts. The application of these rules to the charts in Fig. 3.3 would reveal that both the average part weight and the weight variability are in a state of statistical control.

3.3   Some Important Concepts of Probability and Probability Distributions

Concepts of Probability

A random experiment is a process that yields any one of several possible outcomes in a trial. The actual occurrence of a particular outcome on any given trial cannot be predicted with certainty. It can only be prescribed in terms of probability. Repeated trials of a random experiment under identical conditions produce a distribution of experimental results. Statistical methods help us to analyze and interpret the results of a random experiment. Common examples of random experiments include receipt of invoices in an accounts payable department, sampling inspection of product quality, and number of incoming telephone calls at a switchboard. In each case, the actual outcomes are driven by chance causes.

Suppose that a random experiment has a total of m different outcomes, $A_i, i = 1, 2, \ldots, m$. If n trials of the experiment resulted in the outcomes $A_1, A_2, \ldots, A_m$, which occurred $n_1, n_2, \ldots, n_m$ times, respectively, one can estimate the probability of outcome A_i by

$$\hat{p}_i = \frac{n_i}{n}, \qquad i = 1, 2, \ldots, m.$$

Since $n_i \le n$,

$$0 \le \hat{p}_i \le 1, \qquad i = 1, 2, \ldots, m \tag{3.7}$$

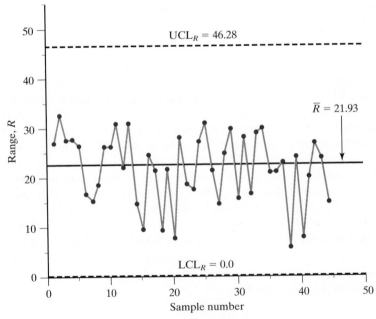

Figure 3.3 $\overline{X}$ and R charts for the light-fixture data

and $n_1 + n_2 + \cdots + n_m = n$, we have

$$\sum_{i=1}^{m} \hat{p}_i = 1. \tag{3.8}$$

Equations (3.7) and (3.8) are two important laws of probability. Simply stated, these two laws mean that a probability can only assume values between 0 and 1, and the total probability, or the sum of the probabilities of all possible outcomes, equals exactly 1.

▶ EXAMPLE 3.1

When a fair die is rolled, there are six possible outcomes with equal probability of occurrence:

$$p_1 = p_2 = p_3 = p_4 = p_5 = p_6 = \frac{1}{6}.$$

This example illustrates the notion of *equally likely outcomes*. Not all random experiments produce equally likely outcomes.

▶ EXAMPLE 3.2

A box contains 10 bolts, 20 nuts, and 50 screws. The probability of drawing a bolt from the box at random is $\frac{10}{80}$ and the probability of drawing a screw is $\frac{50}{80}$. Thus these two outcomes are not equally likely to occur in a drawing.

This example shows that the probability of different outcomes may not be the same. It also serves to illustrate the classical approach in assigning probabilities to experimental outcomes.

In the examples above the probability of each possible outcome can be determined a priori, or before experimentation. In most actual situations, one must estimate probabilities of various outcomes by the *frequency approach*, or a posteriori. In these situations the probabilities can be estimated only after sufficient observations of the outcomes have been made.

As part of a summer promotional effort, a rental car company has developed a fleet of 50 cars with the following composition in terms of color and model.

5	White, full-size	[W, F]
10	Red, full-size	[R, F]
10	Blue, full-size	[B, F]
15	Red, luxury	[R, L]
10	Blue, luxury	[B, L].

Patrons pay a special rate and draw keys from a bowl to determine exactly what type of car they will get. For the sake of illustration here we will assume that keys drawn, one at a time, are always from the full population of 50 cars.

(a) If a key taken at random from the bowl is for a car that is red in color, what is the probability that it will be a luxury car?

The required probability is called a *conditional probability* because a condition—color—has been specified, known, or given.

$$P(\text{luxury, given red}) = P(L|R)$$

$$= \frac{\text{no. red luxury cars}}{\text{no. red cars}}$$

$$= \frac{15}{25}$$

$$= \frac{3}{5}.$$

(b) If a key is taken at random from the bowl, what is the probability that the car is red and that it is a luxury car?

The required probability is called a *joint probability* because the outcome is a compound outcome that possesses two or more characteristics simultaneously. This joint probability may be obtained by

$$P(R \text{ and } L) = P(R, L)$$

$$= \frac{(\text{no. red luxury cars})}{(\text{total no. of cars})}$$

$$= \frac{15}{50}$$

$$= \frac{3}{10}.$$

This joint probability of two characteristics can also be obtained by

$$P(R, L) = P(L|R)P(R)$$

$$= \left(\frac{3}{5}\right)\left(\frac{25}{50}\right)$$

$$= \frac{3}{10}. \tag{3.9}$$

Equation (3.9) is the *multiplication law of probability*. The quantity $P(R)$ is known as the marginal probability of the outcome that a key taken at random from the bowl will be red. If we were going to use Eq. (3.9) to find the probability of drawing a car that is blue and luxury, we would get

$$P(B, L) = P(B|L)P(L) = \left(\frac{10}{25}\right)\left(\frac{25}{50}\right) = \frac{1}{5}.$$

In this case, since $P(B) = \frac{20}{50}$, and $P(B|L) = \frac{10}{25}$, the given condition "luxury model" provides no additional information to the estimation of the probability of getting a blue car, and the events B and L are said to be statistically (or probabilitywise) independent to each other. For independent events, the multiplication law of probability becomes

$$P(B, L) = P(B)\, P(L), \qquad \text{where } B \text{ and } L \text{ are independent.} \qquad (3.10)$$

In words, the joint probability of independent events is the product of the marginal probabilities of the independent events.

(c) If a key is drawn at random from the bowl, what is the probability that it is either for a red car or for a luxury model?

The required probability is given by

$$P(\text{a car is either red, or luxury, or both red and luxury})$$

or

$$P(R + L) = P(R) + P(L) - P(R, L)$$

$$= \frac{25}{50} + \frac{25}{50} - \frac{15}{50}$$

$$= \frac{7}{10}. \qquad (3.11)$$

Equation (3.11) is the *addition law of probability*. To use Eq. (3.11) to calculate the probability of getting a car that is either white or luxury we would write

$$P(W + L) = P(W) + P(L) - P(W, L).$$

Since there are no white cars of the luxury model, the joint probability $P(W, L) = 0$, and we have

$$P(W + L) = P(W) + P(L)$$

$$= \left(\frac{5}{50}\right) + \left(\frac{25}{50}\right)$$

$$= \frac{3}{5}. \qquad (3.12)$$

Equation (3.12) is a special case of the addition law of probability. The fact that the two events, W and L, do not coexist simplifies the calculation of the probability of one of several events that are mutually exclusive.

Random Variables

If a real value is assigned to each of the possible outcomes of an experiment, the set of all assigned values is called a *random variable*, which is different from an ordinary variable (deterministic) in that the values of a random variable can only be determined probabilistically.

▶ EXAMPLE 3.3

The possible outcomes of tossing two balanced coins are

$$(H, H) \ (H, T) \ (T, H) \ (T, T).$$

If one dollar is paid out for each head that turns up in the tossing, we have

X (dollars)	0	1	2
Outcomes	(T, T)	(T, H) or (H, T)	(H, H)

where X is the random variable, which assumes values 0, 1, and 2 with probabilities 0.25, 0.50, and 0.25, respectively.

Random variables that take on only integer values, or isolated and distinct values, are called *discrete random variables*. Many random variables, however, can have an infinite number of values within a small finite interval, such as time, temperature, and many other dimensional measurements. These are called *continuous random variables*.

Probability Function and Cumulative Distribution Function

There is a probability assigned to each possible value of a discrete random variable. The set of all the probabilities associated with the values of a discrete random variable X is called the *probability function* of the random variable, denoted by $p(x)$. For example, in the problem of tossing two coins, the probability function of X (in dollars) is

x	0	1	2
p(x)	0.25	0.50	0.25

The *cumulative distribution function* (cdf) of a random variable X is defined by

$$F(x) = P(X \le x) = P(X \text{ has values up to and including } x)$$

$$= \sum_{t \le x} p(t).$$

For example, the probability of paying out no more than 1 dollar in a tossing of two coins is

$$F(1) = P(X \le 1) = \sum_{t \le 1} p(t) = p(t = 0) + p(t = 1)$$

$$= 0.25 + 0.50 = 0.75.$$

Probability Density Function

If a random variable X is continuous, it is meaningful only to consider the probability of X assuming values within an interval. This requires a different function to describe the probability behavior. A function called the *probability density function* (pdf), $f(x)$, is used for a continuous random variable. A pdf must satisfy the following:

$$f(x) \geq 0, \quad \text{all } x, \quad \text{and} \quad \int_{-\infty}^{+\infty} f(x) \, dx = 1, \tag{3.13}$$

over the interval for X. The corresponding cumulative distribution function (cdf) of a continuous random variable with pdf $f(x)$ is given by

$$F(x) = P(X \leq x) = \int_{-\infty}^{X} f(t) \, dt. \tag{3.14}$$

▶ EXAMPLE 3.4

A random variable X has a pdf $f(x) = 0.5x, 0 \leq X \leq 2$. The probability of X to be no more than 1.5 is

$$F(1.5) = \int_{0}^{1.5} 0.5t \, dt = 0.25t^2 \Big|_{0}^{1.5} = 0.5625.$$

Similarly, $F(2) = 1$ and $F(1) = 0.25$. The probability of X having values between 1 and 1.5 is given by

$$P(1 \leq X \leq 1.5) = F(1.5) - F(1) = 0.5625 - 0.2500 = 0.3125,$$

which is shown as the shaded area in the sketch of $f(x)$ in Fig. 3.4. This probability is also noted on the cdf in Fig. 3.4.

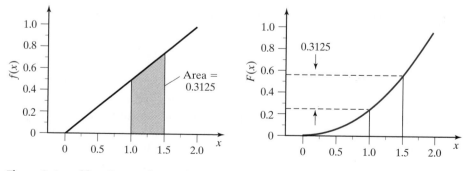

Figure 3.4   pdf and cumulative distribution function

Mathematical Expectation

The concept of *mathematical expectation* or expected value of a random variable is important in probability and statistics. It is noted that any function of a random variable or a function of several random variables is also a random variable. The expectation of a function $g(x)$ of a random variable X is defined as

$$E(g(X)) = \sum_{\text{all } x} g(x)p(x) \qquad \text{for discrete } X, \qquad (3.15)$$

$$= \int_{-\infty}^{+\infty} g(x)f(x)\ dx \qquad \text{for continuous } X. \qquad (3.16)$$

Some important expectations:

1. $g(x) = x$: the expectation of X, $E(X)$, is the true arithmetic mean value of X, usually denoted by μ_X.
2. $g(x) = (x - \mu_X)^2$: the expectation of $(X - \mu_X)^2$ is the true variance of X, which is denoted by σ_X^2.

▶ EXAMPLE 3.5

Referring to the example of the coin-tossing game, the mean and the variance of the random variable X are

$$\mu_X = E(X) = (0)(0.25) + (1)(0.5) + (2)(0.25) = 1$$

$$\sigma_X^2 = \text{Var}(X) = (0-1)^2(0.25) + (1-1)^2(0.5) + (2-1)^2(0.25) = 0.5.$$

Some useful theorems of mathematical expectation include:

1. $E(cX) = cE(X)$, if c is a constant.
2. $E(X \pm Y) = E(X) \pm E(Y)$, where X and Y are two random variables.
3. $\text{Var}(cX) = c^2\ \text{Var}(X)$, if c is a constant.
4. $\text{Var}(X \pm Y) = \text{Var}(X) + \text{Var}(Y)$, if X and Y are independent.

3.4   Normal Distribution

Over 200 years ago it was observed that errors of measurement seemed to follow a definite pattern with respect to their relative frequency of occurrence. Repeated measurements of the length of certain objects seemed to arise in a frequency sense according to a bell-shaped distributional curve, symmetric about the mean, which fell off quite rapidly beyond a distance of about one standard deviation from the mean.

This distribution, known as the *normal probability distribution* or *normal distribution*, does an excellent job of approximating the relative frequencies of many natural and man-made phenomena (e.g., heights or weights of people, dimensions of machined parts, and strengths of steel specimens). Mathematically, the normal distribution is defined by the equation

$$f(x) = \frac{1}{\sigma_X\sqrt{2\pi}}e^{(-1/2)[(x-\mu_X)/\sigma_X]^2},\qquad(3.17)$$

where the mean μ_X and the standard deviation σ_X are the two parameters that define the probability density function $f(x)$ as shown above.

Determining Probabilities Using the Normal Curve

The function $f(x)$ has been scaled so that the total area under the curve over the full range of $X(-\infty$ to $+\infty)$ equals 1.0. Then the proportion of the whole or the probability that an observation will arise over a given range of values is determined by finding the area under the curve over that range; that is, the probability of occurrence between x_1 and x_2 equals the area under curve between x_1 and x_2, as shown in Fig. 3.5.

For the example data of Table 3.1, suppose we want to find the chances of producing a light fixture that has a weight of 150 grams or less. One method to assess this chance or probability based on the data at hand would be to count the number of fixtures that have weights of 150 grams or less, from among the 220 fixtures measured. The corresponding percentage could then be used as an estimate of the probability. Since there are 7 such fixtures of the 220 that were sampled, one would estimate this probability to be 7/220 = 0.032 or approximately 3%.

A more general way to answer the question posed above would be to use the data and the associated frequency histogram of Fig. 3.2 to develop a mathematical model for the frequency distribution of the phenomenon being considered, light-fixture weights. For example, a normal distribution could be proposed and

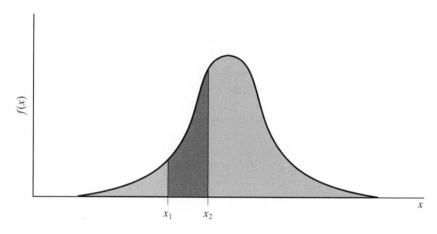

Figure 3.5   Partial areas under the curve as probabilities

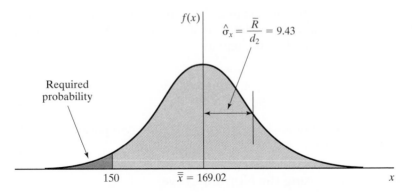

Figure 3.6   Probability of a fixture weighing less than 150 grams

verified as appropriate using analytical or graphical tests for goodness of fit. We will examine such tests later in the chapter. The area under the normal curve for weights less than or equal to 150 grams will be the estimate of the desired probability. This normal distribution will be completely specified given its mean and standard deviation as seen in Eq. (3.17). The mean and standard deviation may be estimated by the sample mean ($\bar{\bar{x}}$ = 169.02 grams) and standard deviation estimate ($\bar{R}/d_2$ = 9.43 grams). This situation is depicted graphically in Fig. 3.6. The shaded area under the normal curve shown in Fig. 3.6 is an estimate of the probability of interest. This estimate is a valid prediction of the future chances of obtaining a light-fixture weight less than or equal to 150 grams, provided the process remains in statistical control at the same mean and standard deviation.

Standard Normal Distribution

Since the mean and standard deviation of the normal distribution can take on many different values from situation to situation, it is convenient to define and work with a standardized normal distribution. Such a standard normal distribution for the random variable Z is defined to have a mean $\mu_Z = 0$ and a standard deviation $\sigma_Z = 1$. Partial areas under this standard normal curve have been calculated and put in a table. It remains for us to establish the relationship between the particular normal distribution we are working with and the standard normal distribution so that we may use it to determine required areas or probabilities. Figure 3.7 illustrates this relationship.

To move from the $f(x)$ normal distribution to the $f(z)$ normal distribution, we use the following transforming equation:

$$Z = \frac{X - \mu_X}{\sigma_X},$$

(3.18)

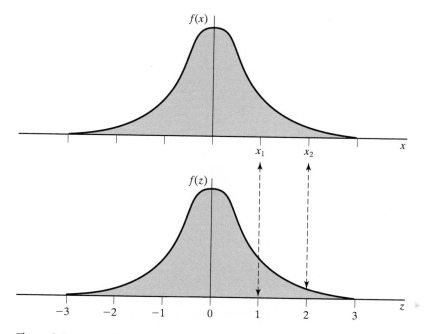

Figure 3.7   **Transformation to the standard normal distribution**

so that if x_1 and x_2 are 1 and 2 standard deviations to the right of μ_X, respectively, then

$$z_1 = \frac{x_1 - \mu_X}{\sigma_X} = 1.0$$

$$z_2 = \frac{x_2 - \mu_X}{\sigma_X} = 2.0.$$

In the following section we illustrate use of the normal probability table to obtain probability estimates.

Using the Standard Normal Table

The diameters of shafts turned on a lathe are considered to have a normal distribution with mean $\mu_X = 1.00$ inch and standard deviation $\sigma_X = 0.01$ inch. Suppose you wish to know the probability that a given shaft will have a diameter between 0.985 and 1.005 inches. We note that the probability $P(0.985 \le X \le 1.005)$ equals the area under the curve from 0.985 to 1.005. Graphically, this is depicted in Fig. 3.8. We may define the equivalent area in the standard normal distribution by determining the number of standard deviations

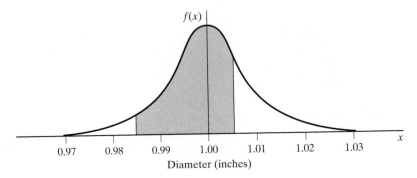

Figure 3.8   Graphical depiction of portion of the process of interest

that the boundaries of the area (0.985 inch and 1.005 inch) are from the mean (1.000 inch):

$$z_1 = \frac{x_1 - \mu_X}{\sigma_X} \qquad z_2 = \frac{x_2 - \mu_X}{\sigma_X}$$

$$= \frac{0.985 - 1.000}{0.01} \qquad = \frac{1.005 - 1.000}{0.01}$$

$$= \frac{-0.015}{0.01} \qquad = \frac{+0.005}{0.01}$$

$$= -1.5 \qquad = +0.5.$$

Then

$$P(0.985 \le X \le 1.005) = P(-1.5 \le Z \le 0.5).$$

We can use the standard normal distribution table (Table A.1 in the Appendix) to find the required probability.

1. The area under the curve from $z = -\infty$ to -1.5 is 0.0668.
2. The area under the curve from $z = -\infty$ to 0.5 is 0.6915.
3. Then the area under the curve from $z = -1.5$ to 0.5 is $0.6915 - 0.0668 = 0.6247$.

The probability that a turned shaft will take a value between 0.985 and 1.005 inches is 0.6247.

3.5   Some Other Useful Graphical Representations of Data

Box-and-Whisker Plot

Another graphical method for displaying data distribution is the box-and-whisker plot. Some commonly used statistics for such a plot are sample quartiles.

The first, second, and third quartiles, denoted by Q_1, Q_2, and Q_3, are, respectively, the values of those data points that are positioned in the data sequence of ascending order such that 25%, 50%, and 75% of the data are equal to or smaller in magnitude. The second quartile is also known as the *median*, which is the value of the data point in the middle of the ordered sequence of the sample data. When the sample size n is odd, the median is the value of the $[(n + 1)/2]$th point in the ordered sequence. If n is even, the median is estimated by the average of the $(n/2)$th and $[(n/2) + 1]$th data points.

To construct a box-and-whisker plot for any given sample size n, we record the smallest value, the largest value, and the three quartiles. Then a box is drawn at the first and third quartiles with the median placed inside the box. Finally, the two whiskers extend out from each side of the box with lengths determined according to the smallest and largest values. As an illustration of the construction of a box-and-whisker plot, we list the 44 sample means of Table 3.1 in ascending order in Table 3.3. From the table we have

$$n = 44$$
$$\bar{x}_{(1)} = 158.2, \text{ the smallest value}$$
$$\bar{x}_{(44)} = 179.0, \text{ the largest value}$$
$$Q_1 = \bar{x}_{(11)} = 165.0, \text{ the first quartile}$$
$$Q_2 = [\bar{x}_{(22)} + \bar{x}_{(23)}]/2 = 169.1, \text{ the median}$$
$$Q_3 = \bar{x}_{(33)} = 174.2, \text{ the third quartile.}$$

A box-and-whisker plot of the sample means of the light-fixture data is shown in Fig. 3.9.

TABLE 3.3	Sample Means in Ascending Order, Light-Fixture Weights				
Order No., i	$\bar{x}_{(i)}$	Order No., i	$\bar{x}_{(i)}$	Order No., i	$\bar{x}_{(i)}$
1	158.2	16	167.0	31	172.4
2	158.6	17	167.2	32	172.6
3	159.2	18	167.2	33	174.2
4	160.2	19	167.6	34	174.6
5	160.2	20	167.6	35	174.8
6	161.4	21	168.0	36	175.4
7	162.0	22	168.4	37	175.6
8	162.2	23	169.8	38	175.6
9	162.4	24	170.0	39	175.8
10	163.4	25	170.6	40	175.8
11	165.0	26	171.4	41	176.6
12	165.8	27	171.8	42	176.8
13	166.2	28	171.8	43	177.6
14	166.2	29	171.8	44	179.0
15	166.6	30	172.4		

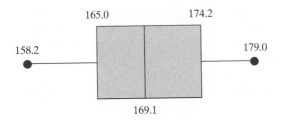

Figure 3.9   **Box-and-whisker plot of sample means of light-fixture weights**

When a set of data is perfectly symmetric about the mean, the median will coincide with the mean and the two whiskers will have the same length. Therefore, a box-and-whisker plot presents the degree of symmetry of the data according to the relative positions of Q_1 and Q_3 to Q_2 and the relative lengths of the whiskers. The box-and-whisker plot of Fig. 3.9 shows some asymmetry in the sample means because the median is not in the center of the box and the lefthand whisker is longer than the one on the righthand side. Overall, the sample means of the light-fixture weights are distributed with heavier clustering on the right but with more skewness to the left tail.

Box-and-whisker plots are often used to compare two or more sets of data. When several data sets are compared for their central tendencies, variabilities, and distribution shapes, all the corresponding boxes and whiskers are plotted, one for each set, to the same scale on one chart. Figure 3.10 shows two such plots of 220 individual light-fixture weights and the 44 sample mean weights for the data of Table 3.1. As shown by the figure, the effects of data grouping ($n = 5$) appear to be that

1. The variability of the data, or the spread, is drastically reduced.

2. The sample means are distributed more symmetrically about the mean, as the median of the sample means is closer to the center of the box.

Box Plots of Percentiles

The method of box plots can be applied to display the sample distribution of data for checking its closeness to an assumed theoretical probability distribution. To do this, one could plot the box based on the percentiles, instead of the quartiles, of the data. The choice of the percentiles can be determined according to the distributional characteristics of the theoretical distribution in question. For instance, if we are interested in checking whether a normal distribution would approximate well the data of light-fixture weights, a box plot of percentiles could be drawn as shown in Fig. 3.11. To better depict the distribution of data for normal fitting, we have used five percentiles, P_{16}, P_{31}, P_{50}, P_{69}, and P_{84} (which correspond to z values at -1, -0.5, 0, 0.5, and 1), together with the smallest and largest values in the data set. If a normal distribution is a reasonably good fit to the data, then according to the

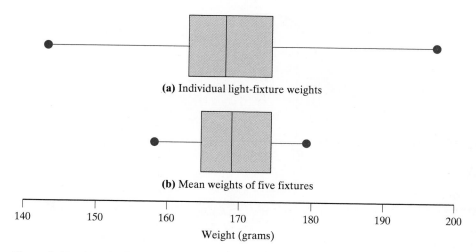

Figure 3.10   Box-and-whisker plots of individual and mean weights of light fixtures

Figure 3.11   Box plots of individual and mean weights of light fixtures

characteristics of a normal distribution illustrated in Fig. 3.12, all the following should hold:

1. P_{50} = mean of the data.
2. The four boxes should have the same width, or $a = b = c = d$, as in Fig. 3.12.
3. The two whiskers should have the same length and be each about three times the width of a box.

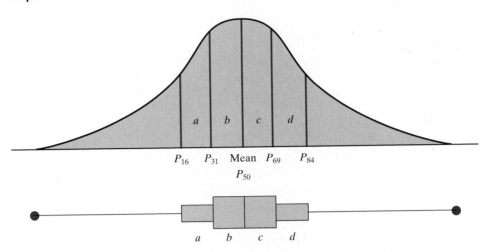

Figure 3.12   Expected box plot of percentiles for normally distributed data

Examining the two plots of Fig. 3.11, we see that the one for the data of individual weights, Fig. 3.11(a), shows somewhat more departure from a normal distribution as the median, P_{50}, is different from the mean, although the other two conditions are not significantly different from their expectations. The other plot, Fig. 3.11(b), appears to be closer to a normal distribution except that the whiskers are too short. For actual data analysis, both of the box plots of Fig. 3.11 would serve as the basis for normal approximation to the data.

Although the box plots, as well as the histograms discussed earlier, are quite effective in graphically displaying data distributions for large samples, they become less adequate when the sample size is small. Another common concern regarding the use of these plots is the question about the uniqueness of the population from which the sample is obtained. Generally, this means that the sampled process must be in a state of statistical control before any valid interpretation or inference can be made from these plots. If a set of data contains a few extreme points or outliers due to special causes or from other populations, we would like to have our graphical method be able to isolate them so that they are not improperly included for parameter estimation or distributional display. One such method is the use of probability plots, one of which is discussed in the following section.

Normal Probability Plots

Probability plots are a graphical means to judge whether or not a set of data may be reasonably characterized by a specified probability distribution. They may also be used to identify outlying data in a sample. *Normal probability plots* are commonly used and are based on the normal probability distribution.

Consider an example in which a sample of $n = 20$ observations is drawn from an unknown distribution. These data are summarized in the second column of Table 3.4. The sample mean and sample standard deviation for this set of data are $\bar{x} = 10.0065$ and $s_X = 0.33427$.

These data may be graphically displayed using a dot diagram such as that illustrated in Fig. 3.13. As is apparent, for such a small set of data it is difficult to tell the shape of the underlying distribution from the dot diagram.

Another way to graphically display the sampled data is with a cumulative probability plot. To construct such a plot, the members of the sample are sorted in ascending order (column 3 in Table 3.4) and each member, x_i, of the sample is assigned a cumulative probability, p_i. Since it is desired to characterize the shape of the underlying distribution based on the sampled data, we must rely on each sample member to tell us something about the distribution. In this case, each sample member represents $\frac{1}{20}$th of the information available about the distribution. Therefore, we can use the smallest X_i value to characterize the lowest 5% of the distribution, the next smallest value to represent the next 5% of the distribution, and so on. Since the smallest X_i value represents the portion of the distribution with cumulative probabilities from 0% to 5%, it may be assigned a cumulative probability value at the midpoint of this interval, that is, $p_i = 2.5\%$. The next smallest X_i value represents the portion of the distribution with cumulative probabilities from 5% to 10%, so it is assigned a cumulative

TABLE 3.4	Calculation Table for the Sampled Data				
Index, i	x_i	x_i Values Sorted in Ascending Order	p_i (%)	z_i	q_i
1	9.63	9.34	2.5	−1.96	−1.99
2	9.86	9.51	7.5	−1.44	−1.49
3	10.20	9.63	12.5	−1.15	−1.13
4	10.48	9.69	17.5	−0.94	−0.95
5	9.82	9.75	22.5	−0.76	−0.77
6	10.07	9.82	27.5	−0.60	−0.56
7	10.39	9.86	32.5	−0.46	−0.44
8	10.03	9.89	37.5	−0.32	−0.35
9	9.34	9.96	42.5	−0.19	−0.14
10	10.26	9.98	47.5	−0.06	−0.08
11	9.89	10.03	52.5	0.06	0.07
12	10.67	10.07	57.5	0.19	0.19
13	9.69	10.13	62.5	0.32	0.37
14	10.15	10.15	67.5	0.46	0.43
15	10.32	10.20	72.5	0.60	0.58
16	9.98	10.26	77.5	0.76	0.76
17	9.51	10.32	82.5	0.94	0.94
18	10.13	10.39	87.5	1.15	1.15
19	9.96	10.48	92.5	1.44	1.42
20	9.75	10.67	97.5	1.96	1.98

Figure 3.13 Dot diagram of data in Table 3.4

probability, p_i, of 7.5%, and so on. In general, the cumulative probability associated with a given x_i may be estimated by

$$p_i = \frac{100(i - 0.5)}{n},$$

where p_i is the estimated cumulative probability in percent. The estimated cumulative probabilities for the sample data are shown in column 4 of Table 3.4. The data, (x_i, p_i) pairs, may then be plotted to form the cumulative probability plot. Figure 3.14(a) shows the cumulative probability plot for the sampled data of this example and Fig. 3.14(b) displays the cumulative distribution function for a normal distribution having the same mean and standard deviation as the sampled data.

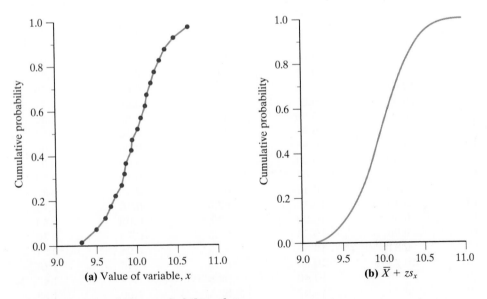

Figure 3.14 Cumulative probability plots

In examining Fig. 3.14(a) and (b), it is evident that the cumulative probability plots of the sampled data and the normal distribution both have "S" shapes. If the S shape associated with the sample is close to the S shape associated with the normal distribution, we may conclude that the sample was drawn from a distribution that is normally, or approximately normally, distributed. Unfortunately, it is not an easy matter to assess the "closeness" of two different S shapes. A normal probability plot solves this problem by using a special scale for the vertical (cumulative probability) axis so that the cumulative distribution function for a normal distribution becomes a line instead of being S shaped. One may visualize the construction of such normal probability paper by imagining a normal cumulative distribution function plotted on graph paper made of thin rubber. The rubber graph paper is stretched along the cumulative probability axis direction in such a way that the normal cumulative distribution function of Fig. 3.14(b) becomes a straight line. The nonlinear scaling of the "stretched" Cartesian scale now defines the scale of the ordinate for normal probability paper.

Commercially available normal probability paper has this specially scaled cumulative probability axis and simplifies the construction of normal probability plots. Figure 3.15 shows a normal probability plot of the sampled data.

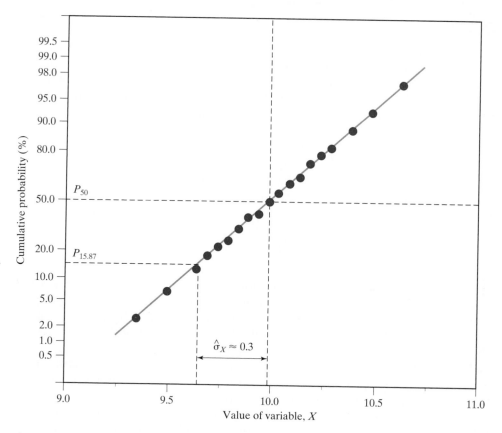

Figure 3.15   Normal probability plot for data in Table 3.4

As is evident, the sampled data in Fig. 3.15 appear to be fairly well described by a straight line. Since the vertical scale for the normal probability plot has been selected so that a normal cumulative distribution function will appear as a line, we may conclude that the distribution from which the sample was drawn is approximately normally distributed.

If all data points appear to be randomly distributed along a line drawn through the data on a normal probability plot, and the line passes on/near the intersection of the mean of X and 50% probability, the fit of the data to the normal distribution is considered good. Conversely, if the points appear to be S-shaped, then the suggestion is that the data are not normally distributed. Sometimes, a line describes a majority of the data but does not characterize a few extreme points. These off-line points are outliers that may have arisen from a different distribution. If the data, part or all, naturally cluster about a straight line that does not go through the mean at 50% probability, then the estimated mean value is probably in error.

We can obtain a graphical estimate, $\hat{\sigma}_X$, of the distribution's standard deviation by finding the difference, measured along the X-axis, between the X values associated with P_{50} and $P_{15.87}$. This estimated standard deviation is shown in Fig. 3.15.

Q-Q Plot

With the help of a normal probability table, the cumulative probabilities, p_i, can be converted into their corresponding standardized normal values, $P(Z \leq z_i) = p_i$. These normal values are shown in the fifth column of Table 3.4.

Provided that the mean and variance of the variable X are known, the sampled data may be standardized using the transformation:

$$q_i = \frac{x_i - \mu_X}{\sigma_X}, \quad \text{for } i = 1, n.$$

Since μ_X and σ_X are generally unknown, we may alternatively use the transformation:

$$q_i = \frac{x_i - \overline{X}}{s_X}.$$

The standardized data values, q_i, are shown in the sixth column of Table 3.4. A plot of the (q_i, z_i) pairs on regular/Cartesian graph paper, called a Q-Q plot, may also be used for judging the normality of a set of data. If the same scale is used for both the q_i and z_i axes, then the line that describes the points (q_i, z_i) should be 45° from the abscissa, provided that the sample is indeed from a normal distribution (and good estimates of the mean and standard deviation have been used to standardize the data). Although a Q-Q plot requires several additional calculations compared to a normal probability plot, for a Q-Q plot there is no need to subjectively place a line through the data points on the graph paper. When it is desired to check the normality of several samples of data, a Q-Q plot also makes the relative assessment of goodness-of-fit among the samples much

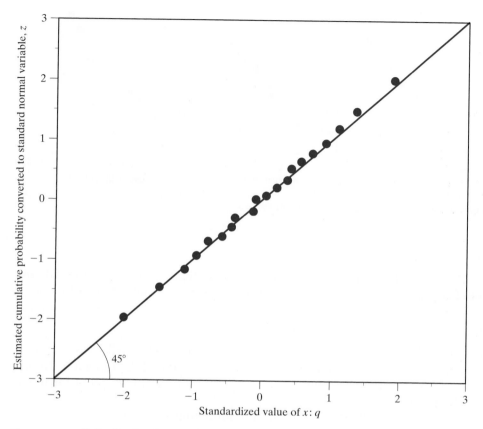

Figure 3.16   Q-Q plot for data in Table 3.4

easier. A Q-Q plot using the same data of Fig. 3.15 is shown in Fig. 3.16, where it is noted that the same scale is used for both the abscissa and ordinate.

Normal probability and Q-Q plots may be constructed using the following steps.

1. Arrange the data, x_i's, in ascending order.
2. Calculate the estimated cumulative probability, p_i, for each member of the sample based on the sample size n.
3. **(a)** For a normal probability plot, plot the points, (x_i, p_i), on normal probability paper. Attempt to fit a line, usually by eye, to the points on the paper so that it is close to all the data points and they are *randomly* distributed about the line.

   (b) For a Q-Q plot determine $\bar{x}$ and s_X for the data and calculate the standardized normal values, z_i, corresponding to the cumulative probabilities, p_i. Calculate the standardized data values, q_i. Plot the points, (q_i, z_i), on regular graph paper. Put a 45° line on the plot.
4. Interpret the plot and draw conclusions.

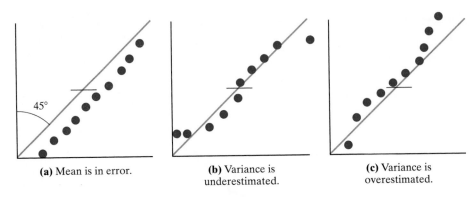

(a) Mean is in error.   (b) Variance is underestimated.   (c) Variance is overestimated.

Figure 3.17   **Poor fits due to estimation errors**

Figures 3.17(a), (b), and (c) illustrate some typical cases of "poor fit" evident from Q-Q plots resulting from estimation errors in the mean and variance.

Normal Plots of the Light-Fixture Data

Suppose that we would like to check whether a normal distribution describes the sample means of the light-fixture data discussed earlier. Table 3.5 summarizes all the calculations for the normal probability and Q-Q plots for the 44 sample means.

The columns in Table 3.5 contain the following:

Column (1): The index, i.

Column (2): The raw data, x_i. The sample mean and sample standard deviation of this data are: $\bar{x} = 169.022$ and $s_X = 5.790$. The sample size is 44.

Column (3): The raw data sorted in ascending order.

Column (4): The cumulative probabilities for a sample of size n, using the equation: $p_i = 100(i - 0.5)/n$.

Example: $p_3 = 100(3 - 0.5)/44 = 5.7\%$.

Column (5): The standard normal values corresponding to p_i. These are obtained by $p_i = P(Z \le z_i)$.

Example, $z_5 = -1.27$ for $p_5 = 10.2\%$.

Column (6): The standardized data values calculated from $q_i = (x_i - \bar{x})/s_X$.

Example, $q_{41} = (176.6 - 169.022)/(5.790) = 1.31$.

The plot of the 44 pairs of (x_i, p_i) data, displayed on normal probability paper, is shown in Fig. 3.18 along with a line drawn through the points. Since the majority of the points are close to the line (except for a few at either end of

TABLE 3.5	Calculation Table for the Light-Fixture Weights				
Index, i	Raw Data, x_i	Raw Data in Ascending Order	Cumulative Probability, p_i (%)	z_i Associated with p_i	Standardized Data, q_i
1	172.6	158.2	1.1	−2.29	−1.87
2	171.8	158.6	3.4	−1.83	−1.80
3	158.2	159.2	5.7	−1.58	−1.70
4	175.8	160.2	8.0	−1.41	−1.52
5	176.8	160.2	10.2	−1.27	−1.52
6	158.6	161.4	12.5	−1.15	−1.32
7	167.0	162.0	14.8	−1.05	−1.21
8	172.4	162.2	17.0	−0.95	−1.18
9	161.4	162.4	19.3	−0.87	−1.14
10	171.8	163.4	21.6	−0.79	−0.97
11	167.2	165.0	23.9	−0.71	−0.69
12	159.2	165.8	26.1	−0.64	−0.56
13	179.0	166.2	28.4	−0.57	−0.49
14	174.6	166.2	30.7	−0.51	−0.49
15	162.2	166.6	33.0	−0.44	−0.42
16	167.2	167.0	35.2	−0.38	−0.35
17	177.6	167.2	37.5	−0.32	−0.31
18	168.0	167.2	39.8	−0.26	−0.31
19	160.2	167.6	42.0	−0.20	−0.25
20	165.0	167.6	44.3	−0.14	−0.25
21	175.8	168.0	46.6	−0.09	−0.18
22	167.6	168.4	48.9	−0.03	−0.11
23	160.2	169.8	51.1	0.03	0.13
24	163.4	170.0	53.4	0.09	0.17
25	175.4	170.6	55.7	0.14	0.27
26	170.6	171.4	58.0	0.20	0.41
27	166.2	171.8	60.2	0.26	0.48
28	170.0	171.8	62.5	0.32	0.48
29	166.2	171.8	64.8	0.38	0.48
30	171.4	172.4	67.0	0.44	0.58
31	176.6	172.4	69.3	0.51	0.58
32	167.6	172.6	71.6	0.57	0.62
33	172.4	174.2	73.9	0.64	0.89
34	169.8	174.6	76.1	0.71	0.96
35	168.4	174.8	78.4	0.79	1.00
36	174.2	175.4	80.7	0.87	1.10
37	162.0	175.6	83.0	0.95	1.14
38	175.6	175.6	85.2	1.05	1.14
39	175.6	175.8	87.5	1.15	1.17
40	165.8	175.8	89.8	1.27	1.17
41	162.4	176.6	92.0	1.41	1.31
42	166.6	176.8	94.3	1.58	1.34
43	171.8	177.6	96.6	1.83	1.48
44	174.8	179.0	98.9	2.29	1.72

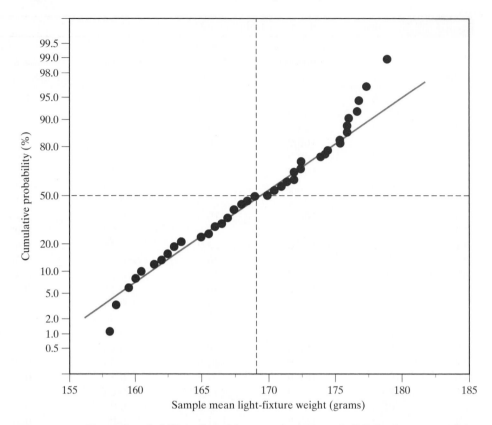

Figure 3.18   **Normal probability plot of the sample means of all light-fixture weights**

the line) and it cuts through the intersection of the mean and the 50% probability, it is concluded that the sample mean light-fixture weights, x_i values,
approximately follow a normal distribution as expected based on the central
limit theorem (to be discussed in the next chapter).

In general, a Q-Q plot requires more calculations than a normal probability plot but may reveal more of the finer disagreements between a normal
distribution and the actual distribution of the data. For instance, the Q-Q
plot of Fig. 3.19 indicates a possible overestimation of the standard deviation of the raw data. Another interpretation of the Q-Q plot of Fig. 3.19 is
that the sample distribution of the data has more spread than would be expected for a normal distribution. A closer examination of both Figs. 3.18 and
3.19 seems to indicate that the data are a bit more uniformly distributed between the mean and about 2 standard deviations than would be expected for
a normal distribution.

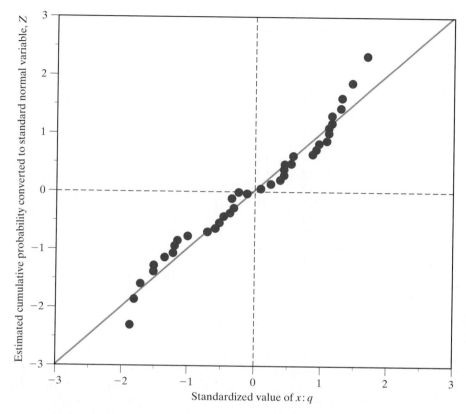

Figure 3.19   Q-Q plot of the sample means of light-fixture weights

▉ EXERCISES

3.1 The scores on a test given to students in a large class are normally distributed with a mean of 57 and a standard deviation of 10. The passing score for the exam is 30. If a student is randomly selected, what is the probability that he or she passed the exam? A score of 75 or greater is needed to obtain an A on the exam. What percentage of the students received an A?

3.2 The data in the accompanying table represent the weight in ounces of molded instrument display panels. The samples were collected at half-hour intervals during a two-shift period. Prepare a tally sheet, like that shown in Table 3.2, of the individual measurements and then prepare a frequency histogram of the data, clearly labeling the cell boundaries. Comment on the shape of the distribution.

Sample	X_1	X_2	X_3	X_4	X_5
1	14	15	13	14	13
2	20	18	14	17	8
3	14	17	14	11	14
4	15	16	11	18	14
5	9	17	18	13	12
6	19	15	14	15	16
7	16	13	14	13	17
8	14	17	9	16	15
9	14	14	12	13	13
10	15	13	17	14	16
11	18	18	16	15	11
12	20	12	13	17	14
13	11	8	9	12	7
14	12	14	16	14	20
15	18	17	12	19	18
16	19	17	16	16	17
17	14	13	15	16	18
18	14	17	12	16	11
19	18	15	16	15	12
20	15	9	12	13	20

3.3 Given the data in the accompanying table, which represent the weld joint strength in ksi from a process known to be in statistical control, make a tally sheet and plot two frequency histograms. For your first histogram use $k = 20$ cells, and for the second use $k = 10$ cells. Clearly show the cell boundaries for each histogram. Comment on the difference between the two histograms.

Sample	X_1	X_2	X_3	X_4
1	23.6	27.2	27.8	24.2
2	23.4	25.2	31.2	23.4
3	29.2	25.8	28.2	29.2
4	26.0	20.8	30.8	29.6
5	29.8	30.8	25.6	24.8
6	25.6	25.8	32.4	33.8
7	33.2	28.2	26.2	31.4
8	29.2	23.2	20.8	26.0
9	23.8	31.4	30.8	28.8
10	32.4	28.6	32.8	25.4
11	24.6	27.2	21.6	24.8
12	31.6	24.4	30.6	28.0
13	30.4	31.2	23.6	20.6
14	28.2	25.6	26.0	33.2
15	15.4	24.4	28.6	27.8
16	30.0	23.8	30.2	28.6
17	22.0	28.2	28.8	23.2

(Continued)

Sample	X_1	X_2	X_3	X_4
18	21.8	23.0	26.6	24.8
19	26.4	29.8	29.4	25.2
20	25.6	21.4	27.6	31.8
21	33.4	26.2	21.2	26.4
22	22.8	28.2	28.4	27.4
23	28.2	28.2	28.2	31.0
24	27.6	29.8	23.0	23.4
25	28.0	30.6	25.8	26.4

3.4 Given the data in the accompanying table, which represent the surface finish in microinches of a cast aluminum part, create a tally sheet of the individual measurements, clearly showing the cell boundaries, and plot a frequency histogram. Also, plot a dot diagram for the individual measurements. What might the shape of the frequency distributions be saying about the process from which the data came?

Sample	X_1	X_2	X_3	X_4	X_5
1	46	46	47	42	41
2	48	44	47	46	52
3	47	45	46	47	42
4	48	37	46	42	45
5	47	43	43	48	46
6	44	45	39	43	49
7	46	41	47	39	45
8	49	50	53	42	45
9	45	39	42	51	46
10	47	41	42	46	46
11	47	45	52	45	45
12	45	44	43	47	46
13	40	43	42	41	46
14	45	43	44	40	46
15	45	39	44	50	47
16	45	39	39	34	41
17	45	37	43	38	47
18	45	42	34	42	34
19	47	43	41	42	44
20	38	35	43	38	43
21	42	40	43	37	36
22	35	36	40	41	39
23	33	44	32	38	33
24	38	44	37	41	43
25	42	38	50	37	47

3.5 You are producing gear blanks that are to have a nominal diameter of 1.060 inches. Samples of size $n = 4$ have been collected once an hour, producing the data shown in the accompanying table. Make a tally sheet with clearly labeled cell boundaries, and then plot a frequency histogram and comment on its shape. What might the shape of the distribution be saying about the process from which the data came?

Sample	X_1	X_2	X_3	X_4
1	1.060	1.063	1.053	1.058
2	1.057	1.053	1.048	1.081
3	1.069	1.077	1.038	1.066
4	1.060	1.050	1.080	1.062
5	1.049	1.048	1.071	1.083
6	1.055	1.064	1.041	1.068
7	1.070	1.080	1.065	1.055
8	1.047	1.040	1.054	1.068
9	1.066	1.025	1.035	1.072
10	1.053	1.064	1.065	1.053
11	1.047	1.050	1.044	1.044
12	1.051	1.084	1.052	1.061
13	1.084	1.058	1.096	1.050
14	1.068	1.035	1.077	1.065
15	1.065	1.074	1.061	1.050
16	1.077	1.064	1.045	1.057
17	1.072	1.062	1.075	1.036
18	1.042	1.094	1.065	1.052
19	1.077	1.052	1.065	1.047
20	1.055	1.036	1.071	1.057
21	1.050	1.066	1.063	1.072
22	1.073	1.059	1.043	1.068
23	1.049	1.048	1.080	1.077
24	1.052	1.049	1.062	1.053
25	1.063	1.074	1.037	1.055
26	1.044	1.083	1.059	1.068
27	1.062	1.046	1.062	1.053
28	1.072	1.057	1.055	1.059
29	1.054	1.076	1.070	1.046
30	1.046	1.065	1.036	1.070

3.6 Determine $\overline{X}$, S_X, and R for each sample in

(a) Exercise 3.2, samples 1 to 5.
(b) Exercise 3.3, samples 6 to 10.
(c) Exercise 3.4, samples 11 to 15.

3.7 You have been told to continue monitoring the resistance of an electronic component with the existing Shewhart control charting scheme using the sampling method used by the person on the previous shift. You see in a log that for the samples he collected, $\overline{R}$ was 0.525 ohms, and the estimate of σ_X that he determined from his data was 0.207 ohms. There is, however, no record of the sample size he used. Based on the information available to you, what sample size do you use to continue testing in the same manner?

3.8 You and a co-worker have been asked to analyze a boring process used to increase the inside diameter of a pipe flange. Two lathes are used in the shop to perform the boring operation, and the two of you take 15 samples of four individual measurements of the bore surface roughness from each machine; the results in microinches are given in the accompanying tables. Use whatever statistical methods you deem appropriate to compare the output of the two machines using the data given. Be sure to include graphical representations of the data.

Machine 1:

Sample	X_1	X_2	X_3	X_4
1	124.4	121.5	126.0	125.0
2	126.8	127.0	125.9	123.6
3	122.6	123.8	122.3	126.4
4	122.5	125.4	123.0	126.4
5	126.5	128.2	124.3	124.8
6	126.8	124.2	126.7	121.6
7	126.3	126.4	126.4	122.1
8	122.1	124.1	124.2	126.0
9	125.9	123.0	124.2	124.3
10	127.6	126.8	127.2	126.2
11	125.9	123.3	125.5	123.6
12	123.4	125.2	125.2	123.5
13	124.9	123.5	128.1	126.1
14	126.3	123.0	126.3	123.3
15	125.3	126.3	124.5	124.4

Machine 2:

Sample	X_1	X_2	X_3	X_4
1	127.7	129.4	128.2	124.7
2	120.7	125.7	123.8	132.5
3	126.1	125.4	118.1	123.8
4	119.9	123.3	120.4	120.8
5	127.6	126.1	124.7	127.4
6	124.3	121.8	125.1	128.3
7	125.4	119.2	123.1	125.6
8	131.2	126.6	129.6	126.0
9	126.7	125.4	129.3	128.7
10	121.2	121.7	117.7	123.1
11	124.1	126.1	128.9	125.7
12	125.3	120.9	128.3	123.9
13	126.6	128.1	122.2	122.1
14	131.9	126.3	121.0	122.5
15	125.2	125.1	127.0	127.9

3.9 A fastener company is supplying washers to a manufacturer of kitchen appliances who has specified that the outside diameter be 0.500 ± 0.025 inch. Fifty samples of $n = 5$ washers show that the process is in good statistical control, the diameters are normally distributed, and μ_X and σ_X for the process are 0.505 and 0.0065 inch, respectively.

   (a) What percent of the washers made by this process are within the specifications? Sketch the appropriate graphs to represent the distribution of the process along with the specifications.

   (b) What would happen if the process variability were reduced by about half (i.e., σ_X changed to 0.003 inch)?

3.10 Assuming that the data in Exercise 3.3 can be approximated by the normal distribution, within what range of weld strength would you expect 95% of the welded joints to fall? Within what range would 99% fall? Provide a graphical representation of your answer.

3.11 For a process that is in good statistical control and for which the individual measurements follow a normal distribution with $\mu_X = 39.87$ and $\sigma_X = 0.56$, what percent of the process is expected to be in the range 39.34 to 40.52?

3.12 For a process that molds polycarbonate face guards for motorcycle helmets, the thickness of the plastic can be assumed to have a normal distribution with $\mu_X = 1.825$ millimeters and $\sigma_X = 0.026$ millimeter. Below what value will 10% of the thicknesses be?

3.13 A process that drills holes in crankshafts that are subsequently ground has been found to be in good statistical control, with the following statistics used to estimate the mean and the variance of the hole diameter:

$$\overline{\overline{x}} = 15.050 \text{ millimeters,}$$

$$\overline{R} = 0.110 \text{ millimeter.}$$

It is also known that 95% of the hole diameters lie within the range 14.945 to 15.155 millimeters. Find the size of the samples employed to produce the ranges R used to estimate the standard deviation of the process. Assume that the hole diameters are normally distributed.

3.14 After taking many samples of size $n = 4$ of the length of a pipe, μ_X and σ_X were determined to be 0.973 and 0.003 meter, respectively. The process is in good statistical control and the individual lengths seem to follow the normal distribution.

   (a) What percent of the pipe lengths would fall outside specification limits of 0.965 ± 0.007 meter?

   (b) What is the effect on the percent conforming to specifications of centering the process?

   (c) What would the effect be if $\mu_X = 0.973$ meter and the process standard deviation were reduced to $\sigma_X = 0.0025$ meter?

Represent each situation above by providing a graphical representation.

3.15 A supplier is producing ball bearings for a major auto producer who has provided a specification on the diameter of 0.250 ± 0.005 inch. Sampling has shown that the distribution of the parts is centered at the nominal, but too many part diameters are outside specifications. What would the standard deviation of the process need to be for 99% of the ball bearings to be within specifications if it were assumed that the diameters follow the normal distribution?

3.16 The data in the accompanying table represent the average thickness for samples of aluminum sheets after going through a rolling process. Draw a box-and-whisker plot and a box plot of the data.

Sample	Average (inches)	Sample	Average (inches)
1	0.043	21	0.051
2	0.048	22	0.046
3	0.048	23	0.050
4	0.052	24	0.050
5	0.048	25	0.052
6	0.052	26	0.052
7	0.048	27	0.055
8	0.059	28	0.047
9	0.043	29	0.055
10	0.054	30	0.055
11	0.050	31	0.054
12	0.061	32	0.047
13	0.045	33	0.053
14	0.053	34	0.050
15	0.051	35	0.050
16	0.053	36	0.054
17	0.052	37	0.056
18	0.055	38	0.053
19	0.045	39	0.052
20	0.046	40	0.048

3.17 The process for forming steel sheets of gage 8, with a nominal thickness of 0.16440 inch, is in question after many complaints have been received. Your manager would like to see a graphic representation of the mean and spread of the thickness of these sheets. You have decided that a box-and-whisker plot would be the best way to represent the data.

(a) Construct the plot using the sample means in the first table. The data have been collected in samples of $n = 5$ at random intervals throughout the day.

Sample	Average (inches)	Sample	Average (inches)
1	0.16425	16	0.16444
2	0.16431	17	0.16447
3	0.16473	18	0.16435
4	0.16459	19	0.16469
5	0.16446	20	0.16420
6	0.16449	21	0.16461
7	0.16433	22	0.16441
8	0.16439	23	0.16449
9	0.16419	24	0.16480
10	0.16454	25	0.16454
11	0.16454	26	0.16419
12	0.16480	27	0.16425
13	0.16513	28	0.16429
14	0.16457	29	0.16439
15	0.16423	30	0.16428

(b) After seeing the data, changes were made to improve the process. The data in the second table show the results of a second series of samples collected as were the data in part (a). Prepare a box-and-whisker plot of these new data to the same scale as the plot in part (a), and compare the two processes.

Sample	Average (inches)	Sample	Average (inches)
1	0.16439	16	0.16427
2	0.16424	17	0.16441
3	0.16449	18	0.16436
4	0.16450	19	0.16435
5	0.16430	20	0.16432
6	0.16439	21	0.16441
7	0.16444	22	0.16422
8	0.16448	23	0.16446
9	0.16435	24	0.16452
10	0.16448	25	0.16446
11	0.16455	26	0.16433
12	0.16435	27	0.16446
13	0.16448	28	0.16453
14	0.16437	29	0.16443
15	0.16440	30	0.16438

3.18 It has been suggested that a process for drawing copper sheets is no longer in control; the data seem to be more uniformly than normally distributed about a nominal value of 0.180 inch. In the accompanying table are averages from samples of $n = 5$ consecutively drawn sheets taken every few minutes throughout the day. Using a box-and-whisker plot, what can be said about the distribution of the averages?

Sample	Average (inches)	Sample	Average (inches)
1	0.183	19	0.192
2	0.189	20	0.164
3	0.183	21	0.175
4	0.185	22	0.174
5	0.179	23	0.178
6	0.174	24	0.190
7	0.177	25	0.185
8	0.181	26	0.179
9	0.189	27	0.175
10	0.168	28	0.175
11	0.181	29	0.192
12	0.177	30	0.190
13	0.182	31	0.166
14	0.190	32	0.173
15	0.196	33	0.197
16	0.183	34	0.177
17	0.206	35	0.170
18	0.181		

3.19 Using the following data obtained from a destructive testing procedure, draw a conclusion as to whether or not the results indicate that the tensile strength values follow the normal distribution. Construct a dot diagram, cumulative probability plot, normal probability plot, and Q-Q plot in doing your analysis. The data have been obtained by subtracting 50,000 psi from the actual measurement and dividing the result by 1000 psi.

−0.980	0.992	1.224	0.921
1.432	1.321	−1.391	1.098
−1.199	−0.987	−0.889	−0.998
1.145	1.042	1.427	

3.20 It is desired to determine if the surface finish obtained in a face milling operation used in machining engine blocks can be approximated by the normal distribution. Use the data in the table to construct a dot diagram, cumulative probability plot, normal probability plot, and Q-Q plot. Draw the appropriate conclusion about the data, given the results of your graphical analysis.

Sample	Surface Roughness (microinches)	Sample	Surface Roughness (microinches)
1	118.4	5	117.7
2	113.4	6	118.5
3	115.0	7	117.3
4	115.4	8	114.4

(Continued)

Sample	Surface Roughness (microinches)	Sample	Surface Roughness (microinches)
9	117.1	13	120.0
10	117.9	14	122.5
11	115.2	15	121.1
12	121.6	16	117.6

3.21 The sample data in the accompanying table represent the heights (cm) of 15-year-old girls, which were drawn from five different classes with 20 girls per class. Make a tally sheet with clearly labeled cell boundaries. Plot a frequency histogram and comment on its shape.

165	160	170	167	165.5
154	154	162	160.5	168.5
163	150.5	156	157	160
158	159	168	153	156.5
167	153	175.5	159.5	167.5
152.5	170.5	170	154	148
162	163.5	164	164	153
158.5	166.5	162.5	154	153.5
154.5	150	152	169.5	154
161	180	153.5	166	168.5
168.5	174	167	149	154
170	165	159.5	159	149
160	165.5	153	153.5	154
167	163	168.5	148.5	156.5
158	157	161.5	164	170.5
176.5	166	162	150	172
164.5	154	148.5	168	157
167.5	151.5	167	157.5	171
169	157	171	168.5	170
161.5	153.5	167.5	153	167.5

3.22 The sample data in the accompanying table are the thicknesses of plastic sheets. Draw a box-and-whisker plot and a box plot of the data.

Sample	Thickness (mm)	Sample	Thickness (mm)
1	0.413	11	0.428
2	0.422	12	0.373
3	0.418	13	0.395
4	0.351	14	0.439
5	0.397	15	0.391
6	0.37	16	0.384
7	0.376	17	0.386
8	0.426	18	0.385
9	0.38	19	0.376
10	0.358	20	0.428

3.23 For a process that is in statistical control and follows a normal distribution with $\mu_x = 2.4$ m and $\sigma_x = 0.12$ m, what percent of the process is expected to be in the range 1.8 to 2.6 m?

3.24 A firm manufacturing automobile brake liners has a liner thickness tolerance of 15.0 ± 0.25 mm. The process is in control and the mean and standard deviation of the liner thicknesses are estimated to be 15.11 and 0.09 mm, respectively. Estimate the fraction of brake liners that are expected to fall outside the specifications. The thicknesses may be assumed to follow a normal distribution.

3.25 The weight specification for packets of noodles is 100 ± 5 grams. Inspection has revealed that 95% of the packets are within the specifications. However, the process is not centered at 100 grams. The packet-filling process is in a state of statistical control and the packet weights are normally distributed with a standard deviation of 1.2 grams. Find the value for process mean so that only one packet in 10,000 falls below the lower specification. For this value of the process mean, what percent of the packets will be above the upper specification?

3.26 Data on the pitch velocity (mph) has been collected for one player during a baseball game. Using the data in the accompanying table, make a tally sheet with clearly labeled cell boundaries and plot a frequency histogram for the data. Comment on its shape. What might the shape of the distribution be saying about the pitcher?

89	92	96	94	107	91	95	99	107	98
92	93	104	101	94	91	95	94	96	94
90	97	97	96	102	98	95	93	101	97
87	94	91	101	101	97	96	94	99	101
88	85	98	94	91	93	95	97	104	94
90	96	93	98	101	88	99	107	91	93
97	102	100	98	97	94	98	98	99	97
95	94	91	99	101	104	97	101	98	104
99	95	94	101	94	96	97	104	100	102

3.27 An engineer is studying the noise level of an operating engine. The noise levels (in dB) may be described with a normal distribution that has a standard deviation of 5.6 dB. If the mean of the distribution is 75 dB, find the probability that a given engine selected at random will have a noise level higher than 85 dB.

4

Sampling Distributions and Statistical Hypothesis Testing

4.1   Introduction

This chapter deals with the principles and methods associated with making inferences using data derived from sampling. It considers four basic thrusts. First, it describes the purpose and the general nature of sampling and discusses the important distinction between analytic and enumerative studies. Next, it introduces the concept of the sampling distribution of sample means. Sampling distributions form the foundation for statistical estimation and inference related to population parameters. The chapter then describes the concept and methods of hypothesis testing. Finally, it introduces the method of confidence-interval estimation of a population mean.

Since samples are usually only small subsets of a population, the sample means calculated from different samples taken from the same population will not be identical to each other due to sampling variations. Therefore the sample mean is a random variable with its own probability distribution, which is called the sampling distribution of $\overline{X}$. It characterizes in a frequency sense how sample means behave. In fact, any statistic calculated from a sample will follow a sampling distribution of its own. We will discuss in this chapter the sampling distribution of $\overline{X}$ where the population variance is either known or can be assumed as known. We will also examine the sampling distribution of $\overline{X}$ when the variance must be estimated from sample data, and introduce the t-distribution in this context. Some other distributions that are useful in different sampling situations, such as χ^2- and F-distributions, will be introduced later in the book when they are needed.

4.2   Purpose and Nature of Sampling

The Notion of Sampling

To find out how the process is behaving in terms of one or more output quality characteristics on the average and in terms of variation, we will observe only part of the output of the process. We will then use these data to infer something about the actual process performance—that is, the performance of all the output that comprises the population. *Population* is a concept that means to imply and include all possible realizations of a process within a certain frame. A population is really defined by a constant system of variation causes characterizable by a succinct set of *parameters* that in an average or collective sense describe the salient features of the variation system.

It is generally neither practical nor necessary to observe the entire population. Rather, we observe only a small subset of it at a time, often doing so periodically. When we do it this way, we are said to be *sampling* the population or process. How much we sample each time, how often and/or when we sample, and how we specifically take each sample are critical issues that must be addressed. But what we do with the data, particularly the information extracted therefrom, will dictate to a great extent the degree to which quality and productivity may be improved.

Various functions of the data may be used to calculate measures, each of which reflects some special feature of the population. These sample measures are called *statistics*, each of which may be used as an estimate of the corresponding population parameter. Together, a set of the parameter estimates is used to characterize the process performance. Because statistics are based on sample data, which represent only part of the population, they are uncertain estimates of the population parameters.

In many instances we need to examine a certain context or environment for the purposes of learning about its makeup or performance. We could be interested in the parts we have on hand, the bills we need to pay, or the character and attributes of our workforce. We may wish to know these things because we would like to evaluate the effects of receiving certain types of orders, or evaluating our cash position, or understanding the impact that an early retirement policy might have. The problems at hand here are really "counting" problems: determining how many of "these" and how many of "those" we have.

We often answer the types of questions posed above by sampling—looking at only a fraction of the whole—because an exhaustive count would be prohibitive in a cost and time sense. These are the types of problems on which statistical sampling methods and the theory of statistical inference thrive. In these situations a frame or population is fixed and we are simply trying to observe certain of its characteristics. We are not asking how the parts got into the inventory, why they are there, or what the chances are that the quality of the next ones to join the group might get better or worse. We only want to know what is there right now.

Enumerative Studies

What we have just described are commonly referred to as *enumerative studies*. W. Edwards Deming frequently pointed out the importance of distinguishing between the two types of statistical studies: enumerative and analytic.[1] Enumerative studies are statistical analyses of sample observations from a fixed frame, the kinds of things we have just described above. The purpose is to describe a clearly defined population on the basis of sample information. A large body of theory has been developed to deal effectively with these issues. Statistical methods, including significance tests, can be effectively employed to make inferences about the population that has been sampled. In general, the inference becomes more precise as sampling error is reduced, which can be achieved by taking larger samples at random. But even in the case where we are dealing with a fixed frame, there is no guarantee that the assumptions required by the statistical tests employed will be valid. We will study many concepts and techniques in this book that will enable us to check out these assumptions or precipitate an environment in which the assumptions are approximately valid. The control chart is one of these techniques. Certain principles of the design of experiments are others.

In a statistical sense, making a control chart of a day's production is an enumerative study. We collect the data, which is a sampling of the day's production, and we make charts for the purpose of evaluating what took place during that day. We are interested in finding out something about what has already taken place. But why? There can be only two reasons. One reason keeps the problem in the realm of the enumerative study, but the other does not. If our purpose is to determine whether or not we have to take some action concerning the parts we made that day, say, 100% sort them, then the study remains an enumerative one. In this case any action that we decide to take will be within the frame we are examining. But if our purpose is to learn how to do better tomorrow with an eye toward actions that will precipitate change, then we no longer have an enumerative study, and statistical analysis has done all it can for us. We now are searching for ways to change the system to improve its performance in the future. To bridge the gap between what we did today and what we hope to do tomorrow we will need much more than statistics; we will need knowledge derived from process experience, engineering practice, and, perhaps, basic science to solve the problem.

Analytic Studies

When sample results are analyzed for the purpose of understanding the causal formation of the population, Deming called these studies *analytic*. In an analytic study, no statistical inference could be made about the population, which is undefined or undergoing changes. This is illustrated in Fig. 4.1(a), where the

---

[1] W. E. Deming, "On the Use of Judgement-Samples," *Reports of Statistical Application Research, JUSE*, Vol. 23, No. 1, March 1976, pp. 25–31.

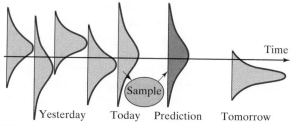

(a) Process not in statistical control: population changes over time.

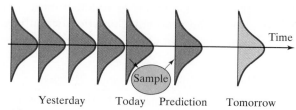

(b) Process in statistical control: one unique causal system.

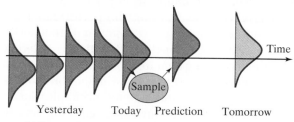

(c) Process in transition: in search of a causal system.

Figure 4.1   Various processes under analytic study

underlying process is changing over time without an identifiable pattern or regularity. The sample information is relevant to the process behavior only at the time the sample is taken. It has no statistical basis to represent either the past or the future of the process. Such sample data can be usefully interpreted by the experimenter only to explain, discover, or predict the dynamics of the process using his or her expert knowledge. Note that in these studies the sample size has very little to do with sampling error, which is undefinable.

Statistical interpretation of the current sample data will be valid for the purpose of explaining or discovering what may have happened in a process within a given frame. But methods such as Shewhart control charts are effective only when the information derived from their patterns, fortified by a statistical interpretation, can be overlaid on our knowledge of the process and its workings to identify possible actions for improvement in the future.

Figure 4.1(b) depicts a process in control. Here control charts are an effective means for analytic studies where the objective is to assess the performance of the process and its capability, and to maintain a process in statistical control. As long as the process remains in a state of statistical control, prediction of future process performance may be inferred statistically from the past data.

Unfortunately, many real-world processes are not stable. For example, in the early development of a product or a process, there is no unique process yet in existence. A stable process is frequently studied by purposeful introduction of changes that will destabilize it to discover opportunities for further improvement. In these situations, usually some knowledge of the process dynamics is available to guide the experimenter in defining a relatively narrow band of uncertainty in his or her studies. In other words, the process as depicted in Fig. 4.1(c), where there is a certain degree of regularity or some inertia of the process dynamics, is more likely to be encountered than the case of Fig. 4.1(a). Again, knowledge derived from process experience, engineering practice, and, perhaps, basic science is required to deal with the problem. Statistically analyzed sample information often provides the only basis for actions for these processes.

Statistical Inference in Analytic Studies

In quality design and control, we are concerned primarily with the development and maintenance of statistical control/process stability and the exploration of improvement opportunities of a process such as the ones in Fig. 4.1(b) and (c). Statistical methods can be used effectively in estimating the performance of a process already in control, or for determining that it was not, within a certain frame. For processes in Fig. 4.1(c), however, the method of statistical inference has to be employed with extreme care in drawing conclusions. To improve the validity of the statistical results in these situations, the experiments or sampling method must be planned ahead following certain fundamental principles such as randomization and classification of sources of variation. After sampling, any statistical inferences derived from the sampling results must be confirmed by additional sample results. This is most effectively done in the environment in which the consequences of the action based on the inference are desired to be felt, which may be quite different from the environment in which the inference was developed. It is with these understandings that statistical methods may be used for making inferences about the process performance. Some of these methods are presented in this chapter. Others are presented elsewhere throughout the book.

4.3   Sampling Distribution of $\overline{X}$

We have seen that the sample mean provides us with an estimate of the population mean, μ_X. If we sampled a population several times, each time calculating an $\overline{X}$, we would find that the $\overline{X}$ values vary from one another, simply due to sampling variation. This variation is, in part, a function of the variation of the items in the population and in part a function of the number of items comprising the average. In other words, how good $\overline{X}$ is as an estimate of μ_X (that is, how close we expect $\overline{X}$ to be to μ_X) depends on the amount of variation inherent in the process/population (σ_X) and the size of the sample, n. We need to understand these relationships more specifically—for instance, how the sample means behave with respect to their own mean, standard deviation, and more important, their own sampling distribution.

To characterize the distribution of the sample mean, $\overline{X}$, for a given sample size n, we need to know the mean and standard deviation of the sample means. These parameters can be estimated from the sample data using the general relationship between the mean and the variance of $\overline{X}$ and, respectively, the mean and variance of X as follows:

$$\mu_{\overline{X}} = E(\overline{X}) = E\left(\sum_{i=1}^{n} \frac{X_i}{n}\right) = \sum_{i=1}^{n} \frac{E(X_i)}{n} = \frac{n\mu_X}{n} = \mu_X \qquad (4.1)$$

$$\sigma_{\overline{X}}^2 = \mathrm{Var}(\overline{X}) = \mathrm{Var}\left(\sum_{i=1}^{n} \frac{X_i}{n}\right) = \sum_{i=1}^{n} \frac{\mathrm{Var}(X_i)}{n^2} = \frac{n\sigma_X^2}{n^2} = \frac{\sigma_X^2}{n}. \qquad (4.2)$$

The relationships above hold true for any random variable, provided that the sampling is done independently from one unique population (i.e., a constant system of common causes, a process in statistical control). Note that the mean of the sample means, denoted by $\mu_{\overline{X}}$, is equal to the mean of the individuals, μ_X, while the variance of the sample means $\sigma_{\overline{X}}^2$ is only $1/n$th of the variance of individuals, σ_X^2. When the sample size is very large, say, equal to the population size, the variance of the sample mean becomes zero, which should not be surprising, because the sample mean is exactly the population mean. Figure 4.2 illustrates the effect of sample size on the variability of sample means. The illustrations show that as the sample size increases, the sample means, $\overline{X}$'s, tend to cluster more about μ_X with reduced spread. This implies that the precision of $\overline{X}$ as an estimate of μ_X increases as sample size increases.

Central Limit Theorem

The illustrations in Fig. 4.2 also show that the sampling distributions of sample means are normal even though the samples may be drawn from a nonnormal population such as a uniform distribution as shown in the figure. A mathematical justification for this relationship is the *central limit theorem*. This theorem states that if $x_1, x_2, \ldots, x_n$ are outcomes of a sample of n independent observations of a random variable X with mean μ_X and variance σ_X^2, the sum of the x's, or equivalently the mean of the x's, will be distributed approximately as a normal distribution with mean μ_X and variance $\sigma_{\overline{X}}^2 = \sigma_X^2/n$. The approximation gets better as n gets larger. In general, the population from which the samples are drawn need not be normally distributed for the sample means to be normal. This constitutes a most remarkable and powerful result.

The following illustration demonstrates the effect of the central limit theorem. Suppose that a particle moves along a line one unit at a time either forward or backward every microsecond with equal probability. Suppose further that each move of the particle is made independent of all previous moves. This is a random variable that can assume two values, -1 and $+1$, each with probability $\frac{1}{2}$. If the particle starts at position 0, it can be at many positions between $-n$

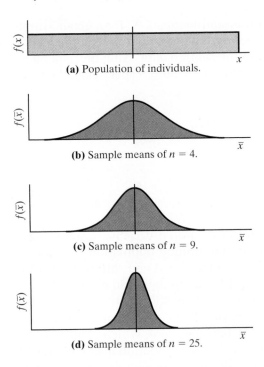

(a) Population of individuals.

(b) Sample means of $n = 4$.

(c) Sample means of $n = 9$.

(d) Sample means of $n = 25$.

Figure 4.2   Sampling distributions of sample means from samples of various sizes n

and $+n$ after n microseconds as depicted in Fig. 4.3. However, the probability that the particle lands at each one of the possible positions will not be the same. Let $X_1, X_2, \ldots, X_n$ be the n independent moves of the random variable X. Each independent move is defined by

$$X_i = \begin{cases} +1 & \text{with probability 0.5, if move forward} \\ -1 & \text{with probability 0.5, if move backward} \end{cases}$$

where $i = 1, 2, \ldots, n$. Then the final position of the particle after n moves is a random variable, the value Y, which is given by

$$Y = X_1 + X_2 + \cdots + X_n, \quad -n \leq Y \leq n.$$

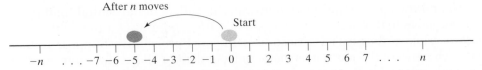

Figure 4.3   Movements of a random variable that can assume a value of -1 or $+1$

Y is described by a probability distribution that will approach a normal distribution with a modestly large n. To find the exact distribution of Y, consider the case of $n = 3$:

Possible Outcomes	Move			x_1	x_2	x_3	Sum	
	1	2	3				y	Probability
1	F	F	F	+1	+1	+1	+3	1/8
2	F	F	B	+1	+1	−1	+1	1/8
3	F	B	F	+1	−1	+1	+1	1/8
4	B	F	F	−1	+1	+1	+1	1/8
5	F	B	B	+1	−1	−1	−1	1/8
6	B	F	B	−1	+1	−1	−1	1/8
7	B	B	F	−1	−1	+1	−1	1/8
8	B	B	B	−1	−1	−1	−3	1/8

In the table above, F and B denote, respectively, forward and backward moves. The final position of the particle, Y, which is the sum of $n = 3$ moves, can be only one of four possible values, with probabilities given as

y	$P(Y) = f(y)$
−3	1/8
−1	3/8
1	3/8
3	1/8

Similarly, a probability distribution can be constructed for any n. Figure 4.4 shows the plots of these distributions for $n = 3$, $n = 9$, and $n = 12$. In each case the resultant sum of just a few moves tends to behave like a normal distribution, although each individual move is either −1 or +1. If the sum of independent variables (of any distribution) approaches a normal distribution, so do the sample means, because they are the sums divided by a constant.

▶ EXAMPLE 4.1

Six batteries taken at random from a large lot were subjected to life tests with the following results:

$$19, 22, 18, 16, 25, \text{ and } 20 \text{ hours.}$$

Find the probability that the sample mean of $\overline{x} = 20$ hours, as observed here, or one less than 20 hours could arise if the true mean of the lot of batteries is 21.5 hours, assuming a known variance of $\sigma_X^2 = 10$ hours² for the individual batteries in the lot.

Solution. Given: $n = 6$, $\mu_X = 21.5$, $\sigma_{\overline{X}}^2 = \frac{10}{6} = (1.291)^2$.

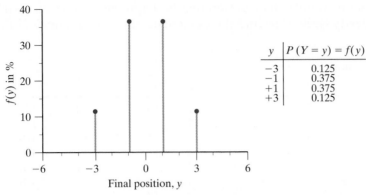

y	$P\,(Y = y) = f(y)$
-3	0.125
-1	0.375
$+1$	0.375
$+3$	0.125

(**a**) Probability distribution of position after
three independent moves ($n = 3$).

y	$f\,(y)$
-9	0.00195
-7	0.01785
-5	0.07031
-3	0.16406
-1	0.24609
1	0.24609
3	0.16406
5	0.07031
7	0.01758
9	0.00195

(**b**) Probability distribution of position after
nine independent moves ($n = 9$).

y	$f\,(y)$
-12	0.00024
-10	0.00293
-8	0.01611
-6	0.05371
-4	0.12085
-2	0.19336
0	0.22559
2	0.19336
4	0.12085
6	0.05731
8	0.01611
10	0.00293
12	0.00024

(**c**) Probability distribution of position after
twelve independent moves ($n = 12$).

Figure 4.4   Central limit effects on sums

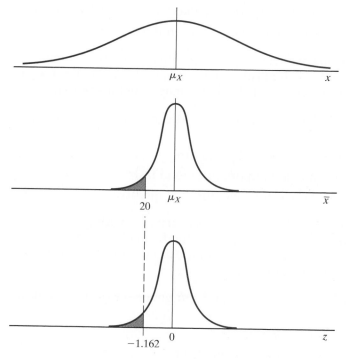

Figure 4.5   Distributions of X, $\overline{X}$, and Z

Using the normal distribution for the sample mean lives, we have

$$Z = \frac{\overline{X} - \mu_X}{\sigma_{\overline{X}}},$$

$$P(\overline{X} \le 20) = P\left(Z \le \frac{20 - 21.5}{1.291}\right)$$

$$= P(Z \le -1.162)$$

$$= 0.123.$$

Figure 4.5 shows the relationships between the distribution of individual battery lives and the distribution of mean lives of samples of size $n = 6$, and between the distribution of $\overline{X}$'s and the standard normal distribution of Z.

4.4   Hypothesis Testing of a Population Mean

Often we are called on to make decisions or draw conclusions on the performance of a process or a product on the basis of sampled data. For example, we might collect data to answer the following questions:

1. Has a new generator design reduced our warranty cost?

2. Is the foam molding process out of statistical control with respect to part weight?

3. Is the concentration of an additive important to the taste of a new food product?

4. Does a new assembly method improve the final product quality?

In making such decisions, we typically form a hypothesis concerning what we believe to be the true state of affairs and then use sampled data to test it. For example, using the scenarios described above, we may hypothesize that:

1. There *is no* difference between the new and old generator designs with respect to the warranty costs.

2. The foam molding process *is* in statistical control.

3. The additive concentration *is not* important to food taste.

4. The new assembly method *does not* affect product quality.

If the data collected to test a hypothesis are inconsistent with it statistically, we must reject the hypothesis. For example, if the data collected for the four scenarios described above are inconsistent with the hypotheses, we must reject them. Therefore, we would conclude that:

1. There *is* a difference between the new and old generator designs with respect to warranty costs.

2. The foam molding process *is not* in control.

3. Additive concentration *is* important to food taste.

4. The new assembly method *does* affect product quality.

In effect, we are rejecting one hypothesis (the one not supported by the data) and accepting another, since the four statements above are also hypotheses. Even if the data collected to test a hypothesis *do* support it, we cannot necessarily accept the hypothesis as the truth. The fact that the data are consistent with the particular hypothesis proposed does not discount the possibility that another hypothesis could be proposed that would also be supported by the data. The best that we can say is that the hypothesis has not been disproven. The hypothesis may or may not be true.

Hypothesis testing is very similar to a court of law. In a court of law we assume that the defendant is innocent (we assume that the hypothesis is true) at the outset of the trial. Testimony is presented (data are collected). The jury, or judge, then decides whether the testimony is strong enough to reject the assumption concerning the innocence of the defendant (data do/do not support the hypothesis). If the assumption of innocence is rejected, we conclude that the defendant is guilty. If the assumption of innocence is not disproven, the court does not conclude that the defendant is innocent; rather, it simply concludes that the defendant is not guilty.

In statistical hypothesis testing we generally formulate two hypotheses. The *null hypothesis*, denoted by H_0, will be rejected or nullified if the sample data do not support it. Any hypothesis that is different from H_0 is called an *alternative hypothesis*, denoted by H_1. Whenever an H_0 is rejected, the stated H_1 will be considered accepted. In a court of law the null hypothesis is that the defendant is innocent (H_0) and the alternative hypothesis is that he is guilty (H_1).

In a court of law, guilt must be proven beyond a reasonable doubt. Although it is difficult to quantify a reasonable doubt, we must do so in a statistical test of hypothesis. In all hypothesis tests, the conclusions must be stated with a probability of error. If a sample result differs markedly from that postulated under the null hypothesis (H_0)—that is, the chance that it could occur if H_0 is true is some probability α or less—one would begin to think that such a large difference could not have been caused by chance variation alone. In this case, the large difference is said to be statistically significant at a significance level α (the probability of error), and the null hypothesis is rejected. If H_0 is rejected on the basis of data analysis, one accepts H_1 as being true. Because the decision "to reject" or "not to reject" an H_0 cannot be made with absolute surety, one needs to specify the probability of error or risk, α. The process of decision making is called statistical because the determination of significance is probabilistic.

Three Ways of Setting Up the Alternative Hypothesis

Statistical hypothesis testing is a method by which one attempts to prove a point, the alternative hypothesis, H_1, by way of rejecting the null hypothesis, H_0. Therefore, it is important to state the alternative hypothesis correctly. The following example illustrates the subjective element in formulating the alternative hypothesis.

In the 1970s when several foundries in Milwaukee, Wisconsin, were closing down or moving out of the state, a newspaper reported that the mean wage rate of local foundry workers was $16 an hour. The management of one of the foundries believed that the wage rate was more than $16, while the foundry union believed that its members actually earned less than that. At the same time, a labor economics professor also wanted to dispute the accuracy of the newspaper's figure. In each case, a sample survey could be taken to test the same null hypothesis:

$$H_0: \mu_X = 16,$$

since all had the same objective, which was to "reject" it (μ represents the true mean value of the wage). But the three parties were to set up three different alternative hypotheses according to their real interests in the event that H_0 was rejected. The three different alternative hypotheses that might be set up are as follows:

Management	Union	Professor
$H_0: \mu_X = 16$	$H_0: \mu_X = 16$	$H_0: \mu_X = 16$
$H_1: \mu_X > 16$	$H_1: \mu_X < 16$	$H_1: \mu_X \neq 16$

The alternative hypotheses above for the management and the union are both called a *one-sided test*; the professor's hypothesis is a *two-sided test*.

Stepwise Approach to Hypothesis Testing

A statistical hypothesis test consists of the following steps:

1. State the null and alternative hypotheses clearly. Define the test statistic used to analyze the situation.
2. Determine a significance level, α, at which the test will be made.
3. Collect the data and calculate the test statistic result.
4. Define the reference distribution for the test statistic.
5. Compare statistically the test statistic and its reference distribution, under H_0. Carry out the necessary analysis of the data.
6. Assess the risk. Weight the economic or otherwise measurable consequences of saying that H_0 is false when it is true, or failing to recognize that it is false when it is false. This may lead to a redefinition of α. Draw appropriate conclusions.

Hypothesis Testing on the Mean

To test the newspaper claim that the mean wage rate of local foundry workers is $16 an hour, a worker is chosen at random by his union, and his wage rate is found to be $11.50. Historical data suggest that the wage rates follow the normal distribution and that the standard deviation of wage rates is $3. Can the union say that the overall mean wage of Milwaukee foundry workers is less than $16 an hour for $\alpha = 0.05$?

STEP 1   The null and alternative hypotheses are H_0: $\mu_X = 16$, H_1: $\mu_X < 16$. The test statistic being used to test H_0 is a single wage rate value, X.

STEP 2   The significance or risk level, α, has been assumed to be 0.05.

STEP 3 $x = 11.50$.

STEP 4   The wage rate values, X's, are known to follow the normal distribution.

STEP 5   Let us compare the specific X value to its corresponding reference distribution under H_0, as shown in Fig. 4.6. Now we must answer the question: Does the observed wage rate value of $11.50 represent a significant departure from the null hypothesis? To answer this question, we can calculate the probability of obtaining a wage rate value less than or equal to $11.50, given that H_0 is true. To perform this probability calculation, we must transform the distribution of wage rates to the standard normal distribution as shown in Fig. 4.7. The Z value associated with the statistic, $x = 11.50$, may now be calculated:

$$z = \frac{x - \mu_X}{\sigma_X} = \frac{11.50 - 16.00}{3} = -1.50.$$

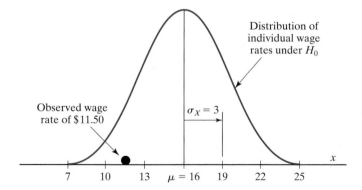

Figure 4.6 Reference distribution for the statistic under H_0

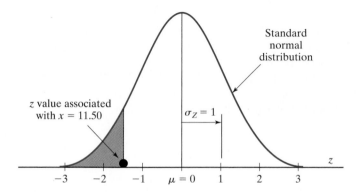

Figure 4.7 Reference distribution for the standardized statistic under H_0

The standard normal distribution table (Table A.1 in the Appendix) indicates that

$$P(Z \leq -1.5) = 0.0668.$$

Therefore,

$$P(X \leq 11.50) = 0.0668.$$

STEP 6 Since the alternative hypothesis is concerned only about the wage rate's being less than \$16, we should compare the calculated probability (0.0668) with $\alpha = 0.05$. Since 0.0668 is greater than 0.05, we cannot reject the null hypothesis. Therefore, we conclude that there is no reason to doubt the newspaper's claim.

Now let us suppose that 25 foundry workers are randomly surveyed by the professor. She finds the average wage rate for this sample of workers to be \$14.50. Having analyzed labor data for years, the professor feels confident in assuming that the standard deviation of wage rates is $3. Can the professor say that the true mean wage is not $16 an hour at an $\alpha = 0.05$?

STEP 1 H_0: $\mu_X = 16$. H_1: $\mu_X \neq 16$. The sample statistic is $\overline{X}$, the sample mean.

STEP 2   Assume that $\alpha = 0.05$.

STEP 3 $\overline{x} = 14.50$.

STEP 4   Regardless of the distribution followed by the individual wage rates, the averages of a number of individual wage rates are approximately normally distributed due to the central limit theorem (Fig. 4.8). The standard deviation of the normally distributed $\overline{X}$'s is given by

$$\sigma_{\overline{X}} = \frac{\sigma_X}{\sqrt{n}} = \frac{3}{\sqrt{25}} = 0.6.$$

STEP 5   We must judge whether we could reasonably expect the observed sample mean to have arisen from the distribution of sample means defined under H_0. To calculate the probability of obtaining an $\overline{X}$ less than or equal to 14.5, we must perform the transformation to the standard normal distribution (Fig. 4.9). The critical values for Z which place $\alpha/2$ $(0.05/2 = 0.025)$ in each tail of the standard normal distribution (we are concerned with both tails of the distribution, since this is a two-sided test) are

$$z_{0.025} = -1.96 \quad \text{and} \quad z_{0.975} = 1.96,$$

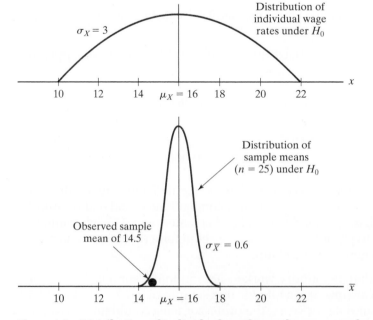

Figure 4.8   **Distribution of individuals and sample means under H_0**

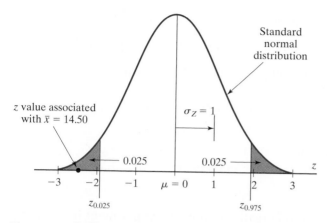

Figure 4.9   Reference distribution for the standardized statistic under H_0

where the subscript on z represents the area under the normal curve to the left of that z value. The calculated z value corresponding to the observed sample mean of 14.50 is then

$$z = \frac{\bar{X} - \mu_{\bar{X}}}{\sigma_{\bar{X}}} = \frac{14.50 - 16}{0.60} = -2.50.$$

STEP 6 H_0 is rejected at $\alpha = 0.05$, since z is less than $z_{0.025}$. Equivalently, H_0 may be rejected, since $P(Z \le -2.5) = 0.0062$ is less than $\alpha/2 = 0.025$. Therefore, the professor has disproved the newspaper's reported wage rate of $16 at a 0.05 significance level.

To carry out a test of the hypothesis as discussed above, the significance level α is interpreted as the probability of rejecting a null hypothesis that is true. This is also referred to as committing a *type I error*. Similarly, a *type II error* is said to have been committed if H_0 is not rejected at a significance level α when in fact it is false. The risk of this type II error, denoted by β, is the probability of failing to reject H_0 given that the true parameter value is at a certain different value, denoted by H_1. Table 4.1 summarizes the errors associated with various conclusions of such a hypothesis test.

TABLE 4.1	Decision Errors in Hypothesis Testing	
Decision	*True H_0*	*False H_0*
H_0 not rejected	No error with probability $1 - \alpha$	Type II error with probability β
H_0 rejected	Type I error with probability α	No error with probability $1 - \beta$

Summary

One can think of the hypothesis testing procedure as proceeding in six basic steps.

1. State the null and alternative hypotheses clearly. Define the test statistic used to analyze the situation.
2. Determine the significance level, α, at which the test will be made.
3. Collect the data and calculate the test statistic result.
4. Define the reference distribution for the test statistic.
5. Compare statistically the test statistic and its reference distribution, under H_0.
6. Assess the risk.

In defining the test statistic one really defines the experiment in a general way. In analyzing the results of the experiment, we need to assess the magnitude of the test statistic of interest, in light of the natural variability inherent in the experiment. The question we are really asking is:

"Could this apparently large (or small) statistic have arisen solely due to chance or is there a real difference between the parameter estimate and the hypothesized parameter value?"

To answer this question we must measure the level of chance variation, usually through genuine test replication, and then determine the probability that we could have observed a result as extreme as or more extreme than the experimental result if only these chance causes were at work. This is the role of statistical inference in making decisions about a certain state of the system or nature. These studies are usually made to provide a basis for action. It is therefore important to clearly state the purpose of the study so that appropriate statistical methods of sampling and data analysis are used to ensure the validity of the results and conclusions.

Of course, as the above suggests, there may be situations in which our measurement of "the level of chance variation" is the only measure that we have of the variability. In other words, there may be instances where we do not know, or there is no assumed value for, σ_X^2. For such hypothesis testing instances, we will be forced to replace σ_X with an estimate for it determined from our sample data, and in so doing, introduce additional uncertainty into the problem.

Hypothesis Testing of Population Mean—Variance Unknown

In our discussion of hypothesis testing to this point, a value for σ_X has been either known or assumed. One might begin to wonder, what happens if σ_X is not known? Or, what are the consequences if σ_X is estimated from sampled data? To begin to answer these questions, consider the following sampling experiment.

A Simple and Illuminating Sampling Experiment

Early on in this chapter, we discussed the distribution of sample means. One excellent way to illustrate the behavior of a statistic as a result of sampling is through simulation. For example, imagine that 2000 samples of size $n = 5$ are taken from a uniformly distributed population between 0 and 10, and the sample mean, $\overline{X}$, is calculated for each sample. We could prepare a histogram for these sample means, but instead let us plot the standardized values for the $\overline{X}$'s, obtained from the following relation: $(\overline{X} - \mu_X)/(\sigma_X/\sqrt{n})$. In this relation, $\mu_X = 5$ and $\sigma_X = 2.887$.[2] The histogram of the standardized $\overline{X}$ values is shown in Fig. 4.10.

We know from our previous discussion of the central limit theorem that the standardized values in the histogram of Fig. 4.10 will be approximately normally distributed. The values in Fig. 4.10 indeed do look normally distributed.

For the case of Fig. 4.10, we were able to standardize the $\overline{X}$ values by subtracting the mean and then dividing by $(\sigma_X/\sqrt{n})$. One might ask, what if the value for σ_X is not known, and we instead divide by $(S_X/\sqrt{n})$, where S_X is the sample standard deviation calculated separately for each simulated sample? Figure 4.11 shows the histogram of the transformed $\overline{X}$ values, where the relation $(\overline{X} - \mu_X)/(S_X/\sqrt{n})$ has been employed. As is evident from the figure, the histogram is bell shaped just like Fig. 4.10; however, the values are more spread out—a consequence associated with the introduction of additional uncertainty by replacing σ_X with S_X. W. S. Gosset first noted this behavior approximately one hundred years ago, while working at a brewery in Ireland; his findings were published under the pen name *Student*,[3] and for this reason, the statistical distribution relating to this histogram is often referred to as Student's t-distribution.

According to the theory, if the samples of size n are drawn from a normal distribution, and the population variance, σ_X^2, is unknown and is estimated by the sample variance, S_X^2, the quantity

$$t_{\text{calc}} = \frac{\overline{X} - \mu_X}{S_{\overline{X}}} = \frac{\overline{X} - \mu_X}{S_X/\sqrt{n}}$$

is said to be distributed according to Student's t-distribution with $\nu = n - 1$ degrees of freedom. The t-distribution is symmetric about its mean of zero and has varying spread, depending on the amount of information (degrees of freedom) that forms the estimate of the variance, S_X^2. The t-distribution has a wider spread than the standard normal distribution due to the uncertainty introduced by replacing the population variance with the sample variance. The spread in the t-distribution decreases as the sample size increases (i.e., the degrees of freedom increase), and in fact it approaches the standard normal distribution as the sample size goes to infinity. Figure 4.12 shows a t-distribution with $\nu = 4$ degrees of freedom along with a standard normal distribution. Table A.3 in the Appendix provides the t values associated with various α-risk levels for several different degrees of freedom, ν. Sketches of t-distributions may be constructed using the t-distribution ordinates given in Table A.4. Note from Fig. 4.12 and Table A.1 that while 2.5% of the area under the normal curve lies beyond 1.96, about 6% of the area under the t-distribution (for $\nu = 4$) lies beyond 1.96.

---

[2]For the uniform distribution, $\sigma_X^2 = (b - a)^2/12$, for X varying on the interval $[a, b]$.
[3]"Student" (W. S. Gosset), "The Probable Error of a Mean," *Biometrika*, Vol. 6, No. 1, 1908, pp. 1–25.

Figure 4.10   Histogram of transformed values for $\overline{X}$ based on σ_X

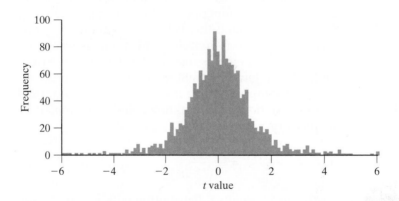

Figure 4.11   Histogram of transformed values for $\overline{X}$ based on S_X

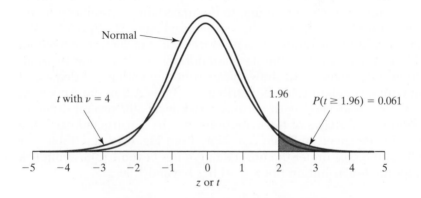

Figure 4.12 t-distribution ($\nu = 4$ degrees of freedom) and standard normal distribution

While the shape of the t-distribution approaches that of the normal (z) distribution as $\nu \to \infty$, as a practical matter, when ν becomes 25 to 30, the two distributions are very close. Strictly speaking, the t-distribution requires the underlying population to be normal. However, the t-distribution is fairly robust to departures from normality of the parent population, so in practice we do not worry too much about the normality assumption. Figure 4.11 seems to confirm this since the simulated samples were taken from a uniform distribution.

To make certain there is no confusion here, the t-distribution arises only in cases where the variance is unknown, regardless of the size of the sample. Remember, it is the quantity $(\overline{X} - \mu_X)/S_{\overline{X}}$ that follows the t-distribution. The sole purpose of the t-distribution is to account for the uncertainty introduced when estimating the population variance with the sample variance.

▶ EXAMPLE 4.2

A foundry wage rate situation similar to that explored previously is under investigation in another city. The local newspaper claims that the mean wage rate is \$17, and the wage rates are known to be normally distributed. Using the hypothesis testing procedure, the steps below are undertaken:

Solution

1. H_0: $\mu_X = 17$ and H_1: $\mu_X \neq 17$; the sample statistic is the sample mean, $\overline{X}$.
2. Assume an α of 0.05.
3. Collect data. A sample of size $n = 6$ hourly wage rates are randomly taken from foundry workers in the city. The sample results (in \$) are (18.02, 21.43, 14.75, 17.16, 20.69, and 18.98), and the sample mean and sample variance are calculated to be

$$\overline{x} = 18.51 \quad \text{and} \quad s_X^2 = 5.94.$$

4. Define the reference distribution for the test statistic. The individual wage rates are known to be normally distributed, and therefore the sample means are also normal. Figure 4.13 displays these distributions under the null hypothesis.

   We would like to evaluate how extreme our sample mean is, given the hypothesized mean. But, since we do not know σ_X, we will use the t-distribution to do this. Therefore, we need to refer our calculated t value to a t-distribution with $n - 1 = \nu = 5$ degrees of freedom. The calculated t value is

$$t_{\text{calc}} = \frac{18.51 - 17}{2.44/\sqrt{6}} = 1.52,$$

   and this value is shown relative to the appropriate t-distribution in Fig. 4.14.

5. We are interested in the probability of obtaining a value more extreme than $\overline{x} = 18.51$ under H_0. Or, $P(\overline{X} \geq 18.51) = P[t_{\nu=5} \geq (t_{\text{calc}} = 1.52)] = 0.0945$. This probability is shown as the shaded area in Fig. 4.14.

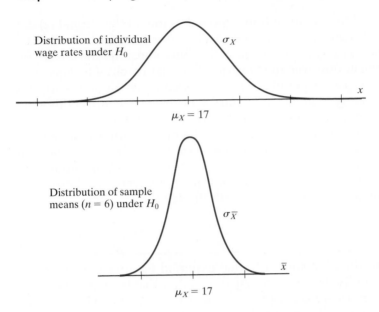

Distribution of individual
wage rates under H_0

σ_X

x

$\mu_X = 17$

Distribution of sample
means ($n = 6$) under H_0

$\sigma_{\bar{X}}$

$\bar{x}$

$\mu_X = 17$

Figure 4.13   Distributions of individuals and sample means under H_0

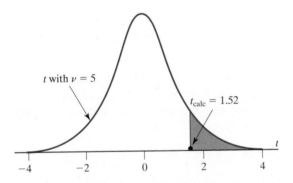

t with $\nu = 5$

$t_{\text{calc}} = 1.52$

t

-4 -2 0 2 4

Figure 4.14   Reference distribution for calculated t value

6. We may not reject H_0, since 0.0945 is greater than our critical risk level of $\alpha/2 = 0.025$. Alternatively, we may compare t_{calc} with the critical value for t corresponding to the selected risk level, $t_{\nu,1-\alpha/2}$, which for our case is $t_{5,0.975} = 2.571$. Since $1.52 \leq 2.571$, we are unable to reject H_0. We have not secured any evidence to disprove the newspaper's claim, and the claim may be true.

4.5   Confidence-Interval Estimation of the Mean

In the previous section, we learned how to set up and test a hypothesis on the mean of a population, based on sample data. In several of the cases, the null hypothesis remained the same—the proposed value for the mean was $16—and the

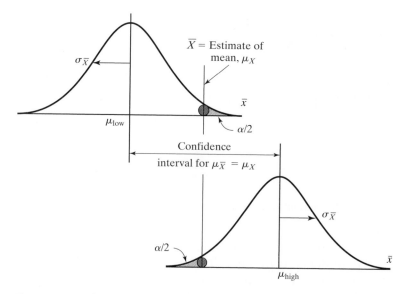

Figure 4.15   Idea of confidence interval for μ

alternative hypothesis changed depending on the point of view of the person set-ting up the test. In examining the section, the reader may have begun to recognize that the value proposed for the null hypothesis was just one of many values that could have been proposed. In fact, the reader may have realized that, given the sample data used for the test, there existed a range of values that could have been proposed for H_0, all of which would have led to failure in rejecting H_0. This, in fact, is the main reason why we always say, "H_0 cannot be rejected" rather than saying, "H_0 is accepted." Figure 4.15 depicts this situation and suggests an alternate way of making an inference about the population mean based on the sample data and our "point estimate" of μ_X, the average of the sample, $\overline{X}$.

Figure 4.15 shows the sampling distribution for the mean positioned at two possible locations that represent the lowest and the highest value that could be proposed for H_0, both of which would not be rejected given the data. In this case, it is assumed that while the mean is unknown, the variance of the popu-lation is known. Clearly, in addition to μ_{low} and μ_{high}, any value in between these values could also be proposed for H_0 and would also lead to the conclu-sion "cannot reject H_0." In this section, we will come to refer to this range of plausible values as a "confidence-interval" estimate of the population mean, μ_X.

Construction of a Confidence Interval for a Population Mean

Suppose a sample of size n is taken from a population with unknown μ_X and "known" σ_X^2. Let the sample mean be $\overline{X}$. According to the central limit theorem, a normal distribution may be used to describe $\overline{X}$. At a selected confidence level $(1 - \alpha)$, an interval can be determined such that it will include the true μ_X with a confidence $(1 - \alpha)$, or it will fail to include the μ_X with a probability α.

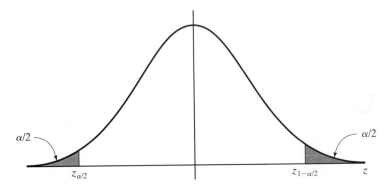

Figure 4.16   **Selection of z's for a symmetrical interval**

Since the normal distribution is symmetric about the mean, the smallest inter-val to produce $1 - \alpha$ confidence is the one that has a probability $\alpha/2$ that the true mean μ_X lies below the lower limit of the interval and a probability $\alpha/2$ that the true mean μ_X lies above the upper limit. Thus, we have

$$z_{\alpha/2} \leq \frac{(\overline{X} - \mu_X)}{\sigma_X/\sqrt{n}} \leq z_{1-\alpha/2} \tag{4.3}$$

where $z_{\alpha/2}$ and $z_{1-\alpha/2}$ are standardized normal values as shown in Figure 4.16.

Equation (4.3) is often written in the following form when calculating a $100(1 - \alpha)\%$ confidence interval for μ_X:

$$\left\{ \overline{X} + z_{\alpha/2}\left(\frac{\sigma_X}{\sqrt{n}}\right) \right\} \leq \mu_X \leq \left\{ \overline{X} + z_{1-\alpha/2}\left(\frac{\sigma_X}{\sqrt{n}}\right) \right\}. \tag{4.4}$$

▶ **EXAMPLE 4.3**

The average height of $n = 100$ male freshmen, selected at random, at a large university is found to be $\overline{x} = 69.45$ inches. The standard deviation, σ_X, is known to be 2.92 inches. Find a 95% confidence interval for the mean height of all male freshmen on campus.

Solution

For a 95% confidence interval ($\alpha = 0.05$) for the mean height of male fresh-men at the university, we use $z_{\alpha/2} = z_{.025} = -1.96$ and $z_{1-\alpha/2} = z_{.975} = 1.96$. Given $n = 100, \overline{x} = 69.45$, and $\sigma_X = 2.92$, we can use Eq. (4.4) to determine the 95% confidence interval for μ_X:

$$\left\{ 69.45 + (-1.96)\left(\frac{2.92}{\sqrt{100}}\right) \right\} \leq \mu_X \leq \left\{ 69.45 + (1.96)\left(\frac{2.92}{\sqrt{100}}\right) \right\}$$

or

$$68.88" \leq \mu_X \leq 70.02".$$

The above interval means that we are 95% confident that the true mean height of all male freshmen on campus is between 68.88 and 70.02 inches. The true mean height could be less than 68.88 or greater than 70.02 but only with a probability of 5%.

▶ EXAMPLE 4.4

The freshmen height problem is re-examined for another university, and this time the average height of $n = 20$ male freshmen, selected at random, is found to be $\overline{X} = 69.80$ inches. For this university, the individual heights are known to be normally distributed. In this case, the standard deviation, σ_X, is not known but has been estimated from the sample to be $s_X = 3.1$ inches. Find a 95% confidence interval for the mean height of all male freshmen on this campus.

Solution

The confidence interval specified by Eq. (4.4) assumes that the value for σ_X is known. When σ_X is replaced with s_X in the equation, we must work with the t-distribution rather than the normal distribution. Making the appropriate adjustments, the confidence interval becomes:

$$\left\{ \overline{X} + t_{\nu,\alpha/2}\left(\frac{s_X}{\sqrt{n}}\right) \right\} \leq \mu_X \leq \left\{ \overline{X} + t_{\nu,1-\alpha/2}\left(\frac{s_X}{\sqrt{n}}\right) \right\} \qquad (4.5)$$

For 19 degrees of freedom and a 95% confidence interval ($\alpha = 0.05$) for the mean height of male freshmen at this university, we use $t_{\nu,\alpha/2} = t_{19,0.025} = -2.093$ and $t_{\nu,1-\alpha/2} = t_{19,0.975} = 2.093$. Given $n = 20$, $\overline{x} = 69.80$, and $s_X = 3.1$, we can use Eq. (4.5) to determine the 95% confidence interval for μ_X:

$$\left\{ 69.80 + (-2.093)\left(\frac{3.1}{\sqrt{20}}\right) \right\} \leq \mu_X \leq \left\{ 69.80 + (2.093)\left(\frac{3.1}{\sqrt{20}}\right) \right\}$$

or

$$68.35" \leq \mu_X \leq 71.25".$$

Hypothesis Testing via Confidence Intervals

To demonstrate the correspondence between the hypothesis testing method and the confidence-interval approach, we now show how significance tests of statistical hypotheses can be carried out equally well by the method of confidence intervals. As discussed previously, a $(1 - \alpha)$ confidence interval is an interval that covers the true parameter value with probability $(1 - \alpha)$. Therefore, if an interval constructed from the sample data fails to include the parameter value under H_0, then the null hypothesis is rejected at a significance level α.

Figure 4.17 **Confidence-interval approach to hypothesis testing for mean hourly wage rate**

We will present this method using the example on average wage rates of Milwaukee foundry workers in the last section, where we have:

$$H_0: \mu_X = \$16$$
$$H_1: \mu_X \neq 16$$

$$\alpha = .05 \quad \text{or} \quad 1 - \alpha = .95$$

and

$$\sigma_X = \sqrt{9} = 3 \text{ is assumed known}, \quad n = 25, \quad \bar{x} = 14.50.$$

The 95% confidence interval for the true population mean is:

$$\bar{X} \pm z_{1-\alpha/2}\left(\frac{\sigma_X}{\sqrt{n}}\right) = 14.50 \pm (1.96)\left(\frac{3}{\sqrt{25}}\right) = 14.50 \pm 1.176$$

$$= [13.32 \leftrightarrow 15.68].$$

Since the hypothesized mean wage is $16.00, which falls outside the 95% confidence interval, the null hypothesis of $\mu_X = \$16$ should be rejected at a significance level of 5%. The same conclusion was obtained in Section 4.4. Figure 4.17 depicts the confidence-interval approach to hypothesis testing for the example.

■ EXERCISES

4.1 Discuss the differences between enumerative and analytic studies with an example for each. Explain their significance.

4.2 What does the central limit theorem state about the distribution of sample means?

4.3 Five veal cutlets selected at random at the local grocery store had the following weights: 3.05, 2.75, 3.3, 3.0, and 2.9 ounces.
   (a) Calculate the sample mean.
   (b) If the true mean of the weight distribution is 3.1 ounces and the standard deviation of the individual weights is 0.2 ounce, find the probability of obtaining a sample mean as small as or smaller than the one calculated in part (a).

4.4 Four measurements of the concentration of inhalable airborne particulate are randomly taken within a production facility: 0.8, 1.6, 1.3, and 1.1 mg/m^3. Previous investigations have revealed that the air quality measurements have a true mean and standard deviation of 1.0 and 0.3 mg/m^3, respectively.

(a) Calculate the sample mean of the particulate measurements.

(b) Find the probability of obtaining a sample mean as large as or larger than the one calculated in part (a).

4.5 Based on extensive testing, a cutting tool manufacturer knows that the distribution of individual tool lives for one of its tool grades has a mean of 60 minutes and a standard deviation of 10 minutes. Furthermore, the distribution is known to be normal.

(a) Prepare a sketch of the distribution of individual tool lives.

(b) A sample of four tools is taken at random from the distribution of individuals and the sample mean is calculated. Prepare a sketch of the distribution associated with the sample mean.

(c) What is the probability of obtaining an individual tool life less than 55 minutes?

(d) What is the probability of a sample mean of four tool lives being less than 55 minutes?

4.6 The resistances of ten randomly selected 100-ohm resistors have been recorded: 79.9, 85.9, 87.0, 96.7, 107.4, 100.8, 115.4, 117.9, 98, and 102.6 ohms. The individual resistances are known to follow a normal distribution with a mean of 100 ohms and a standard deviation of 10 ohms.

(a) Calculate the sample mean resistance associated with the ten resistors.

(b) For a sample size of ten, find the probability of obtaining a sample mean as large as or larger than the one calculated in part (a).

(c) For a sample size of ten, find the probability of obtaining a sample mean with a resistance between 95 and 105 ohms.

4.7 A filling process is used to put green beans in a can. The advertised weight of a can is 16 ounces. The company has established a requirement that when a sample size of $n = 10$ cans is taken from the process, the production line will be shut down if the sample mean is less than 15.9 ounces. The standard deviation of individual weights is 0.5 ounces.

(a) If the process is centered at 16.4 oz., what is the α risk?

(b) Where should the process be centered so that the α risk is 0.005?

4.8 What is hypothesis testing? Use an example to illustrate the hypothesis testing procedure.

4.9 What is the difference between a type I and type II error? Provide an example of each error type.

4.10 A grocery store advertises 3-ounce veal cutlets. Five veal cutlets selected at random had the following weights: 2.95, 2.65, 3.2, 2.9, and 2.8 ounces. The standard deviation of weights is known to be 0.2 ounce. Using the hypothesis testing procedure described in the chapter, evaluate the grocery store's claim. Assume a risk level, α, of 0.05.

4.11 A professor claims that she is so experienced in creating exams, she can guarantee the mean midterm score will be 80 with a standard deviation of 10. Following a midterm exam, four (random) students get together and compare their scores: 68, 87, 59, and 76. Perform a statistical test of hypothesis to evaluate the professor's claim using a risk level, α, of 0.02.

4.12 A bottling company claims that the mean amount of fluid in its beverage bottles is 16 ounces. Past investigations of the bottle filling process have shown the amount of fluid in the bottle to be distributed normally with a standard deviation of 0.5 ounce. Nine bottles are selected at random and the amount of fluid is measured: 15.5, 16.6, 15.5, 16.1, 14.4, 15.7, 15.4, 14.5, and 15.5 ounces. Assess the company's claim at $\alpha = 0.05$.

4.13 The nurse at Stephenson High School claims the expected heart rate of a healthy 16-year-old female student is 80 beats/minute (bpm). Heart rates are known to follow a normal distribution with a standard deviation of 5 bpm. Five 16-year-old students are selected at random and their heart rates are measured as 77, 70, 75, 65, and 62 bpm. Evaluate the nurse's claim. Assume a risk level of $\alpha = 0.02$.

4.14 The Kingwood Tool Company claims that the life of a new grade of cutting tool is 70 minutes on average. Previous studies have shown that the distribution of tool lives is normal with a standard deviation of 8 minutes. A random sample of six cutting tools is tested, producing the following results: 55.0, 70.1, 58.4, 53.2, 62.0, and 67.9 minutes. Assess the company's claim using a statistical test of hypothesis at a risk level of 0.02.

4.15 A turning operation is used to manufacture shafts. The manufacturer claims that the mean diameter of the shafts produced by the process is 20.5 mm. Prior study of the process has revealed that the shaft diameters are normally distributed with a standard deviation of 0.5 mm. A random sample of seven shafts is taken from the process: 20.2, 22.1, 20.5, 21.0, 21.1, 21.1, and 20.9 mm. Using a risk level of 0.05, evaluate the manufacturer's claim.

4.16 A stamping process produces a part that has a feature with a critical thickness specification of 32.00 ± 0.40 mm. It is believed that the stamping process is centered at the nominal value. The distribution of thicknesses is known to follow a normal distribution with a standard deviation of 0.15 mm. A sample of three stamped parts is taken, and their thickness measurements are found to be 32.10, 32.40, and 32.65 mm. Using a risk level of $\alpha = 0.02$, evaluate the belief that the process is centered.

4.17 Evaluate the claim of Exercise 4.10, using a statistical test of hypothesis and assuming that the standard deviation is unknown and must be estimated from the data that were collected.

4.18 Evaluate the claim of Exercise 4.12, using a statistical test of hypothesis and assuming that the standard deviation is unknown and must be estimated from the data that were collected.

4.19 Evaluate the claim of Exercise 4.14, using a statistical test of hypothesis and assuming that the standard deviation is unknown and must be estimated from the data that were collected.

4.20 Evaluate the claim of Exercise 4.15, using a statistical test of hypothesis and assuming that the standard deviation is unknown and must be estimated from the data that were collected.

4.21 Using the data provided in Exercise 4.11, develop a 98% confidence interval for true mean midterm exam score. Use the value for the standard deviation provided in the exercise. Based on this confidence interval, comment on the professor's claim.

4.22 Using the data provided in Exercise 4.12, develop a 95% confidence interval for true mean amount of fluid dispensed into each bottle. Use the value for the standard deviation provided in the exercise. Based on this confidence interval, comment on the bottling company's claim.

4.23 Using the data provided in Exercise 4.13, develop a 90% confidence interval for true mean student heart rate. Use the value for the standard deviation provided in the exercise. Based on this confidence interval, comment on the nurse's claim.

4.24 Using the data provided in Exercise 4.16, develop a 98% confidence interval for true mean thickness. Use the value for the standard deviation provided in the exercise. Based on this confidence interval, comment on the belief related to the centering of the process.

4.25 Using the data provided in Exercise 4.10, and assuming that the standard deviation is unknown and must be estimated from the data, develop a 95% confidence interval for the true mean cutlet weight. Based on this confidence interval, comment on the grocery store's claim.

4.26 Using the data provided in Exercise 4.14, and assuming that the standard deviation is unknown and must be estimated from the data, develop a 98% confidence interval for the true mean tool life. Based on this confidence interval, comment on Kingwood Tool's claim.

4.27 Using the data provided in Exercise 4.15, and assuming that the standard deviation is unknown and must be estimated from the data, develop a 95% confidence interval for the true mean shaft diameter. Based on this confidence interval, comment on the manufacturer's claim.

4.28 Your company has received numerous complaints about the life of some of the hand tools you are manufacturing. In investigating these complaints, management decides to analyze the process used to harden steel wrenches. Several techniques have been suggested for collecting the hardness numbers that will be used. They include:

(a) Measuring the first three wrenches produced each day and assuming that if they are within the specifications, the others that follow will also be satisfactory.

(b) Taking a hardness reading on a wrench drawn from a population at four random times throughout each shift.

(c) Taking samples of four consecutively produced wrenches at approximately half-hour intervals throughout the shift.

(d) Measuring the hardness of each wrench produced.

What benefits and potential problems are associated with each of these methods? Which method would you recommend?

4.29 For Exercises 3.2 and 3.3, prepare dot diagrams for the average values for each sample and compare the dot diagram in each case to the associated histogram of the individual measurements. How do the average-value dot diagrams differ from the individual-measurement histograms? Does this make sense? Be specific and quantitative in your answer.

4.30 Four sample toys have been picked up at random from a toy-making process, and the weights of the toys are 120, 122, 121, and 125 g. If the expected weight for a toy is 122 g, which ones, if any, should we reject at a risk, $\alpha = 0.01$, assuming a normal distribution for the toy weights with standard deviation of 0.5 g?

Process Control and Improvement

5

Conceptual Framework for Statistical Process Control

5.1   Introduction

This may well be the most important chapter in this book from the standpoint of successful implementation of statistical process control. We discuss here five fundamental concepts central to the successful use of statistical process control as a tool for quality and productivity improvement:

1. The origins and nature of *variability.*
2. Process evolution over *time.*
3. Shewhart's concept of *statistical control.*
4. *Managing variability* using control charts.
5. The *process* of statistical process control.

One's ability to discard the old philosophy of inspection for product control and embrace the philosophy of never-ending improvement using process control depends entirely on how well these five concepts are understood. This chapter is devoid of equations, tables of numbers, and other numerical/analytical representations and summaries. Those will follow in great abundance but may well be misinterpreted and misused without a firm understanding of the concepts that follow.

The chapter concludes with a presentation on Deming's famous "Red Bead experiment" that pulls together many of the concepts introduced here.

5.2   Origins and Characteristic Behavior of Variability

Sources of Functional Variation

To reduce functional variation—that is, increase the consistency of product/process performance—it is essential that we identify the basic sources of functional variation so that appropriate countermeasures can be formulated and implemented. Taguchi suggests that variation in product or process function (he refers to this as *functional variation*) arises from three basic sources.

1. *Outer noise.* External sources or factors that are operating in the environment in which the product is functioning and whose variation is transmitted through the design to the output performance of the product. Examples of outer noise factors include such things as temperature, humidity, contaminants, fluctuations in voltage, vibrations from nearby equipment, and variations in human performance. In Taguchi's definition of outer noise, these are sources of noise which are influencing performance as measured during field use under actual operating conditions.

2. *Inner noise.* Internal change in product characteristics such as drift from the nominal over time due to deterioration. Inner noise may be precipitated by such things as mechanical wear and aging.

3. *Variational noise.* Variation in the product parameters from one unit to another as a result of the manufacturing process. For example, the design nominal for a resistor may be 200 ohms, but one manufactured resistor may have a resistance of 202 ohms, another a resistance of 197 ohms. Variational noise is sometimes referred to as manufacturing imperfection.

The significance of recognizing the sources of variation described above becomes evident as one begins to think in terms of the fundamental countermeasures one might invoke to mitigate them. It quickly becomes clear that the forces of outer noise and inner noise can be dealt with upstream effectively only by the engineering design process. Variational noise is a matter of manufacturing imperfection and hence can be dealt with, in part, at the process via techniques such as statistical process control. However, mitigating the forces of variational noise should also be considered to be a product and process design issue as well. In fact, it is likely that variational noise can be dealt with in a more significant and fundamental way if it is thought of as a process and product design problem. Certainly, the concept of design for manufacturability and the emphasis on the simultaneous engineering of products and processes have a bearing on this issue. Since manufacturing processes are themselves subject to external sources of variation, the concept of robust design applies to them as well as to the products they are manufacturing.

Figure 5.1 illustrates how the sources of variation defined above are transmitted to the quality response of the product. At the product design stage the

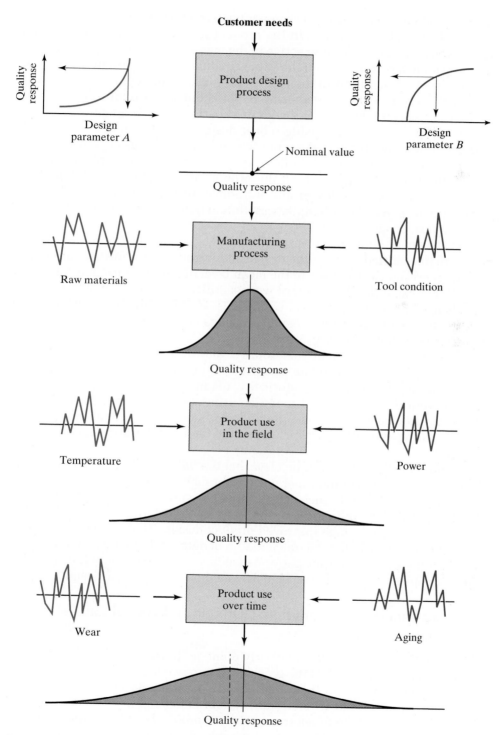

Figure 5.1   **Impact of sources of variation on quality performance**

nominal values of the critical design parameters are selected to produce a pre-specified level of performance in the product in terms of one or more quality responses. Sources of variation active in the manufacturing processes cause the design parameters to vary from the values intended by the design process, and hence introduce variation in the quality response. Although these are sources of variational noise or manufacturing imperfection, the extent to which noise in the process is transmitted to the product performance is a matter to be considered at the design stage, using robust design concepts in concert with the principles of design for manufacturability and simultaneous engineering.

Once the product is put into field use, outer noise sources of variation active in the environment in which the product is being asked to function are transmitted through the design and introduce further functional variation in the quality response. For example, variation in temperature may cause product performance to vary.

As time goes on and the use of the product continues, the forces of inner noise are transmitted through the design and introduce still further variation in the quality response. Changes in the mean level of performance may take place as well. For example, as a result of wear, a critical design parameter may actually change its mean value. Wear could cause a change in size, resistance, viscosity, and so on, thereby causing a change in quality performance, as is depicted at the bottom of Fig. 5.1.

From Fig. 5.1 it is clear that sources of variation enter the picture and have an impact on the ultimate performance of the product during manufacturing and during field use. Historically, the majority of the efforts at quality control were directed toward manufacturing sources of variation. Only recently have design-based concepts and methods emerged that are aimed at understanding and taking action against the forces of variation arising during field use.

Upon reflection, it should be clear that the only way to take action against the forces of outer noise and inner noise is through the design process. Since the root causes of outer and inner noise cannot be economically reduced or eliminated directly, their influence on the functional performance of the product must be dealt with through improved product design. That is, the design of the product must be such that the transmission of outer and inner noise through the design to the product output performance is suppressed as much as possible. This is the primary objective of Taguchi's method of parameter design.

Manufacturing/Process Variation: Common and Special Causes of Variation

Shewhart, Deming, and Juran all clearly point out in their teachings that quality and productivity problems at the process fall into two basic categories. Shewhart described the variation in the process as arising from either *chance causes* or *assignable causes*. Deming referred to *system faults/common causes* and *local faults/special causes*; Juran referred to *chronic problems* and *sporadic problems*. Each provides a lucid description of the overall situation. Assignable causes, local faults, or sporadic problems may be thought of as problems that arise periodically in a somewhat unpredictable fashion and can usually be dealt with effectively at the machine or workstation by the immediate level of supervision or the operator. These are the disturbances to the process that interfere with its "routine" operation and jeopardize the economic viability of the process. It is the primary purpose of statistical process control concepts and methods to identify the presence of these local faults/assignable causes/sporadic problems and then to identify the proper corrective action at the machine.

However, local faults are but a small fraction of the total picture. System faults/chance causes/chronic problems constitute problems with the system itself and are ever present, influencing all of the production until found and removed. System sources of variation are also referred to by Deming as common causes, since they are common to all of the manufacturing output. It is the attack on the system problems that Deming stressed since about 80% to 85% of all the problems we encounter are of this nature, while only 15% to 20% are of a local-fault, sporadic-problem nature.

Table 5.1 summarizes the basic nature of process variation and provides some examples in each case. In the case of common causes/system faults it is clear by their very nature that management must take the responsibility for their removal. Only management can take the actions necessary to improve the training and supervision of the workers; only management can take the responsibility for the redesign of a poor workstation layout; only management can establish new methods or procedures for the process. Clearly, it is essential that we properly differentiate between the occurrence of these two types of problems so that we assign the proper responsibility for solving the problems—taking corrective action.

TABLE 5.1	The Nature of Faults in the Process		
	All processes are subject to two fundamental types of faults or problems.		
Faults	Local faults Special causes Sporadic problems Assignable causes ⋮	vs.	System faults Common causes Chronic problems Chance causes ⋮
Examples	Broken tools Jammed machine Material contamination Machine-setting drift ⋮		Poor supervision Poor training Inappropriate methods Poor workstation design ⋮

Action/
  by Whom
	Correctable locally,
	  at the machine, by the
	  operator or first level
	  of supervision
			Requires a change in the
			  system—only management
			  can specify and implement
			  the change

Summary: Nature of Variation in Product/Process Performance

- It is essential that we understand clearly where variation comes from—the basic sources of variation—so that we know what actions to take and where to assign the responsibility for taking the actions.

- Quality as loss due to functional variation should be observed in product performance during field use. Deviations in parameter values (deviation from the nominal) observed after manufacturing are but one of several sources of functional variation.

- Although some variation can be dealt with effectively through study of the manufacturing process, the engineering design process plays a strong role in determining the extent to which some variation sources, outer noise and inner noise, influence the amount of functional variation—variation in product/process performance—in the field, under working conditions.

5.3   Process Behavior over Time

A succession of parts emanating from a process under statistical control will exhibit variability in the quality characteristic of interest due to a constant set of common causes. These variable measurements tend to collect themselves into a predictable pattern of variation that can easily be described by a few simple statistical measures, such as a mean, a standard deviation, and a frequency distribution, say, the normal distribution. These measures stand as a model that predicts how the process will behave if subject only to a constant set of common causes. Figure 5.2 illustrates how such a statistical model may emerge. Measurements of thickness, for example, may be recorded for a succession of automobile head gaskets produced by the process, and these data may be used to develop a statistical model for the process.

If we can develop a model for the process measurements when subject only to a constant system of common causes, then when a major disturbance affects the process, the ensuing measurements will be seen not to conform to this model; that is, they will stand out clearly in the common-cause variability pattern. This is the case for the situation depicted in Fig. 5.3. For the measured gasket thickness represented by the dot in Fig. 5.3, we ask: Could this deviation of this gasket's thickness from the process mean of 0.045 inch be explained by the forces of common-cause variation? If only common-cause forces are at work, the vast majority, say, 99.73% of all measurements, fall between A and B, so our answer will be no, it does not fit into the common-cause variability pattern!

The statistical description and interpretation above is useful; however, a very large step has been taken in simply taking a set of measurements over a period of time by sampling the process output and collecting them into the graphical summaries of Figs. 5.2 and 5.3. This very large step has ignored a crucial characteristic of the data: the time order of production! To build statistical models to represent product/process characteristics of importance, we must

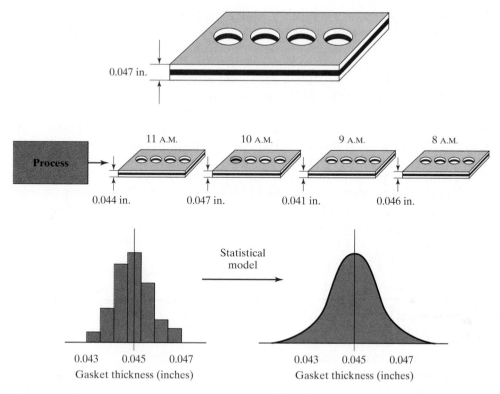

Figure 5.2   Process representation as a statistical distribution of variable measurements

Figure 5.3   Process behavior inconsistent with the common-cause system

first determine if the process data indicate consistency over time (statistical control) and to do this we must consider the process as it evolves over time of production. Far too often we simply take a set of sampled data, calculate an average,
calculate a sample standard deviation, assume a statistical model such as the
normal distribution, and then apply that model for some analytic purpose, such

as assessing the process capability, without checking to see if the sampled data can reasonably be thought of as emanating from a constant system of common causes. Such abuses are commonplace and unfortunately often lead to gross misinterpretations of process performance.

Figure 5.4 depicts the time-order evolution of the process introduced through Fig. 5.2. In Fig. 5.4(a) the process mean and variation stay constant over time and hence the process is said to be stable (i.e., in statistical control). In Fig. 5.4(b), the process mean begins to shift about erratically at 11:00 A.M. At this point the process is said to have gone out of control. Processes under statistical control are driven solely by common causes of variation. Instability, out-of-control behavior, is driven by special causes of variation.

Over time, a process can be subject to several kinds of disturbances (special causes) that can produce a variety of unstable behaviors with respect to either its mean level or level of variability or both. Figures 5.5 and 5.6 graphically portray this concept of process behavior. In Fig. 5.5 only the mean level of the process is exhibiting a changing nature over time.

In Fig. 5.5(a) the mean level of the process is varying from time to time in an erratic fashion. Figure 5.5(b) shows the appearance of a process for which the mean level has experienced an abrupt and sustained shift. In Fig. 5.5(c) the

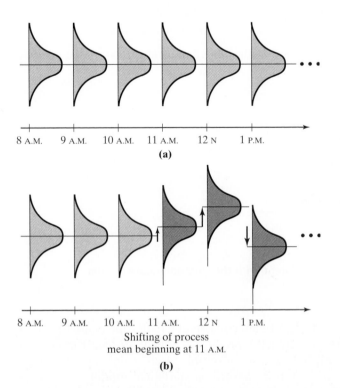

Figure 5.4   Changing process behavior over time

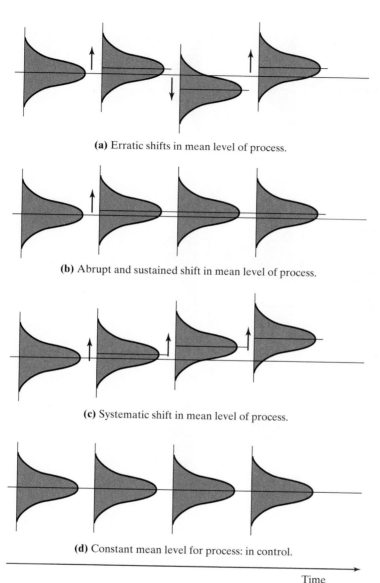

(a) Erratic shifts in mean level of process.

(b) Abrupt and sustained shift in mean level of process.

(c) Systematic shift in mean level of process.

(d) Constant mean level for process: in control.

Time

Figure 5.5   Behavior of mean level of the process over time

process mean is changing in a somewhat systematic or gradual fashion. Finally, in Fig. 5.5(d) the mean level of the process is stable; that is, the process is in control with respect to its mean level. In each case except (d), the indication is that a special cause (or causes) has entered the system and is responsible for the change in the process mean level. It is clear that such changes will influence the

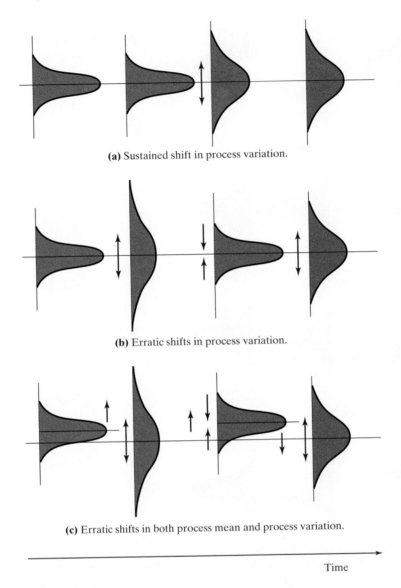

(a) Sustained shift in process variation.

(b) Erratic shifts in process variation.

(c) Erratic shifts in both process mean and process variation.

Time

Figure 5.6   Behavior of the level of variation of the process

quality of the product, particularly when we think in terms of the loss function interpretation of quality. The root cause of the mean shift is also likely to create a loss of productivity.

Figure 5.6 illustrates the appearance of a process for which the variation level has changed over time. This, again, signals in each case the presence of special causes that will have deleterious effects on both the quality and productivity of the process. In Fig. 5.6(a) the process has experienced an abrupt and sustained shift in the variation level. In Fig. 5.6(b) the changes in the amount of process variation are occurring in a more erratic fashion. Figure 5.6(c) exemplifies the worst of all worlds,[1] a process that is unstable in terms of both its mean level and amount of variation. For a process to have any chance to operate in an economical fashion it must at a minimum be in a state of statistical control. From the standpoint of the time behavior of the process this means that the process must have the appearance shown in Fig. 5.5(d).

Summary: Time Evolution of Process Behavior

- Data collected over time from a process may be used to develop a statistical model for the process, as long as the data are collected during a period when the process is subject to a constant system of common causes.

- The most common changes in process performance over time are changes in the process mean level and changes in the amount of variation in the process. These changes can take on a variety of characteristic forms (erratic, systematic, etc.), depending on the physical forces that are driving them. These changes indicate an instability of process performance.

- By tracking process performance over time through periodic sampling, it is possible to detect such changes in performance through the characteristic changes in the process model, namely, changes in the mean and changes in the standard deviation.

5.4   Shewhart's Concept of Statistical Control

Economic Control of Manufacturing Processes

Collecting production performance data and keeping charts of these data is a common practice in industrial operations. The observation of dimensions, strengths, weights, yield, defects, and other quality characteristics can be of great utility not only in documenting performance relative to requirements, but more important, as a tool to aid the identification of ways to improve performance. Regrettably, such data are often examined in a most cursory manner and then filed away, never again to see the light of day. Often, such data are quickly summarized into daily or weekly management reports, and as a result, a great deal of the important information content is lost, *particularly as this information relates to the time history of the process performance*. A basic problem with the use of such production data, even when retained in its raw state, lies with the fact that it is generally not conceived of, modeled, and then interpreted on a statistical basis.

---

[1]During a supplier seminar for a large off-road construction vehicle manufacturer some years ago, one of the authors, using Fig. 5.6, asked the group, "What is the nature of the situation we have in Fig. 5.6(c)?" A loud voice from the back of the room called out, "Chapter 11!"

Many years ago Walter Shewhart at Bell Labs showed how such process data could be developed and interpreted through the use of very simple but profound statistical methods. In his benchmark book on the subject, *Economic Control of Quality of Manufactured Product*,[2] he establishes from the very beginning the overarching philosophy that drives the control chart concept. The first paragraph of the preface of this work clearly sets the foundation for what we know today as statistical process control:

> Broadly speaking, the object of industry is to set up economic ways and means of satisfying human wants and in so doing to reduce everything possible to routines requiring a minimum amount of human effort. Through the use of the scientific method, extended to take account of modern statistical concepts, it has been found possible to set up limits within which the results of routine efforts must lie if they are to be economical. Deviations in the results of a routine process outside such limits indicate that the routine has broken down and will no longer be economical until the cause of trouble is removed.

From Shewhart's statement, several things become immediately obvious:

1. The fundamental focus is on the process: "ways and means of satisfying human wants."

2. The overarching objective is economic operation of the process: "Reduce everything possible to routines requiring a minimum amount of human effort."

3. During normal operation, process behavior falls within predictable limits of variation: "It has been found possible to set up limits within which the results of routine efforts must lie if they are to be economical."

4. Deviations in performance outside these limits signal the presence of problems that are jeopardizing the economic success of the process: "Deviations in the results of a routine process outside such limits indicate that the routine has broken down and will no longer be economical."

5. Improvement in quality and productivity requires that attention be directed at the process to find the root cause of the trouble and remove it: "The routine has broken down and will no longer be economical until the cause of trouble is removed."

Something else is also evident in Shewhart's statement—there is no mention of the product and the conformance of the product to specifications. The total emphasis that Shewhart placed on the process and its economic operation was by many not clearly understood. Hence it was not then, nor is it now, uncommon to hear questions such as the following: "If the product is by-and-large meeting the specifications, why is it important to be concerned about the fact

---

[2]W. A. Shewhart, *Economic Control of Quality of Manufactured Product*, D. Van Nostrand, Princeton, NJ, 1931. Reprinted by the American Society for Quality Control, Milwaukee, WI.

that the process is not in a state of statistical control?" To answer this question, we must understand (1) what we mean by control, and (2) what the implications are of not running a process in a state of control. For these answers we turn again to the teachings of Shewhart:

> A phenomenon will be said to be controlled when, through the use of past experience, we can predict, at least within limits, how the phenomenon may be expected to vary in the future. Here it is understood that prediction within limits means that we can state, at least approximately, the probability that the observed phenomenon will fall within the given limits.

From this definition it is clear that Shewhart is overlaying the statistical characterization of variations on the physical meaning and interpretation of the origins/sources of those variations. This, of course, he made clear in the earlier quote: "It has been found possible to set up limits within which the results of routine efforts must lie if they are to be economical." When a process is not controlled, Shewhart clearly spelled out the consequences:

> Deviations in the results of a routine process outside such limits indicate that the routine has broken down and will no longer be economical until the cause of the trouble is removed.

In short, when a process is not in a controlled state, the productivity or economic success of the process is not guaranteed. We may be making product that meets the specifications but we are not doing so in a manner that will enhance competitive position. In terms of Shewhart's way of thinking, the notion of predictability of the process behavior through the interpretation of the variability pattern in product/process quality characteristics is synonymous with the behavior of a routine process if it is to be economical.

What Shewhart recognized so clearly is simply this. Process instability—the presence of out-of-control behavior—is a signal that forces acting on the process are causing waste and inefficiency (i.e., robbing the process of productivity). To see this fact clearly, we will now consider the important distinction between process capability and process control.

Process Control Versus Process Capability

Perhaps the single most difficult hurdle to clear in coming to a full understanding of Shewhart's teachings is the separation of the notions of the product's conformance to specifications and statistical control of the process. Not only do these issues need to be totally separated, but the value and importance of achieving both of these goals simultaneously and their essential interrelationship must be completely understood. Since both ideas involve the comparison of variation patterns to a set of preestablished limits, the confusion that can arise is considerable. It is essential that we understand that the origin and

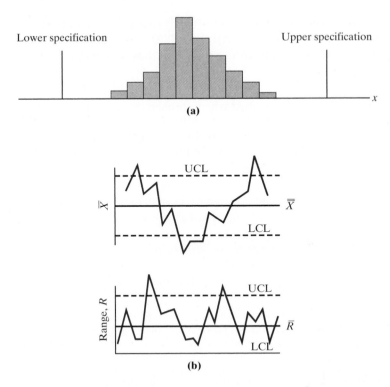

Figure 5.7   Product conformance vs. process control

interpretation of these two sets of limits are totally different. That the quality characteristics of the product fall within the specification limits (derived during engineering design) implies that the customer expectations are being met. That the performance of the process falls also within the limits of variation for routine operation (derived from the observed pattern of variation in the process) implies that the customer expectations are being met in an economic fashion.

Figure 5.7 describes the separate issues of product conformance and control of the process. In Fig. 5.7(a) product quality characteristic measurements are compared to the specifications. It would appear that the product is meeting the specifications. In Fig. 5.7(b) we see the same quality characteristic but this time in the form of Shewhart $\overline{X}$ and R control charts. It would appear that the process variation pattern is not confined to fall within the predictable limits defined by the common-cause system. The process is not in control in terms of either the mean or the amount of variation.[3]

---

[3]During the same supplier seminar referred to in footnote 1, the same author, referring to Fig. 5.7(a), asked, "What can we conclude from the data portrayed in this fashion?" Again, from the back of the room came a loud voice, "Ship 'em!" The respondent probably didn't realize how right he was ... that's about all we can conclude.

But as Shewhart has pointed out so clearly, if we fail to control the process in the statistical sense, the process is not performing in an economical fashion. We are spending more money to make the product than we would if the process were in control. In other words, we are not operating the process in the manner in which it was designed to be run. Fundamental faults in the process are robbing us of productivity. The Shewhart control chart is a tool to alert us to the fact that the process routine has in some way broken down and that continued operation under such conditions will be wasteful.

Summary: Shewhart's Concept of Statistical Control

- "Through the use of the scientific method, extended to take account of modern statistical concepts, it has been found possible to set up limits within which the results of routine efforts must lie if they are to be economical. Deviations in the results of a routine process outside such limits indicate that the routine has broken down and will no longer be economical until the cause of trouble is removed."

- "A phenomenon will be said to be controlled when, through the use of past experience, we can predict, at least within limits, how the phenomenon may be expected to vary in the future. Here it is understood that prediction within limits means that we can state, at least approximately, the probability that the observed phenomenon will fall within the given limits."

- That the quality characteristics of the product fall within the specification limits (derived during engineering design) implies that the customer expectations are being met. That the performance of the process also falls within the statistical limits of variation for routine operation (derived from the observed variation in the process) implies that the customer expectations are being met in an economic fashion.

5.5   Control Chart Approach to the Management of Process Variation

In the first section of this chapter we simply referred to the variation evident in process data as arising from common-cause/chronic problems and from special-cause/sporadic problems. In a subsequent section we saw how Shewhart viewed these sources of variation as arising (1) from the routine operation of the process, and (2) from forces above and beyond the routine operation of the process. We now must be much more specific about the nature of the variability pattern in the data. This is necessary so that we may understand what actions can be taken against the forces of process variation, and more important, who is to be held responsible for formulating and implementing such actions.

Detecting the Presence of Special Causes of Variation

Figure 5.8 illustrates a situation in which a stable process is suddenly subject to erratic shifts in its mean level. As we have just discussed, whatever is causing the process mean to begin to shift after 11:00 A.M. is reducing the process efficiency below what it would be if that force were not present. For example, the shifts in the mean level could be resulting from operator overadjustment of some process setting, thereby not only increasing the variability in the process and the product but also reducing the productivity—the process routine has broken down. It is, therefore, important to be able to detect as soon as possible when such a change in the process occurs. We can then look for the cause of the change and formulate actions that will bring the process back to routine and keep such changes from recurring. This is the job of the Shewhart control chart.

To illustrate the workings of the control chart idea, let's return to the gasket-making process we mentioned briefly above. Suppose that periodically we take small samples of size $n = 5$ gaskets from the production as depicted in the following chart.

	10:00 A.M.		9:30 A.M.		9:00 A.M.	
	0.044 inch		0.046 inch		0.043 inch	
	0.043		0.044		0.044	
Process ...	0.045	...	0.045	...	0.044	...
	0.044		0.045		0.042	
	0.044		0.047		0.047	
	$\overline{X} = 0.0440$		$\overline{X} = 0.0454$		$\overline{X} = 0.0440$	
	$R = 0.0020$		$R = 0.0030$		$R = 0.0050$	

Each small sample tells us how the process is behaving at any one time with respect to its mean level and its variation. The summary statistics, namely

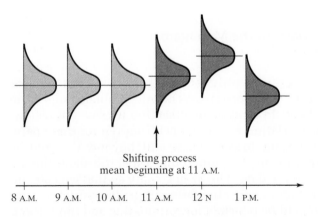

Shifting process
mean beginning at 11 A.M.

8 A.M.   9 A.M.   10 A.M.   11 A.M.   12 N   1 P.M.

Figure 5.8   **Process going out of control**

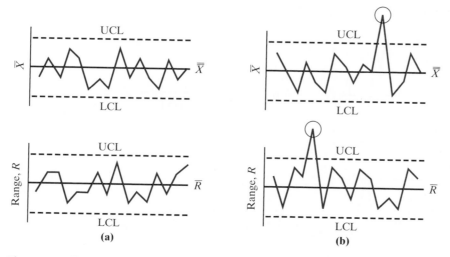

Figure 5.9   Detecting the presence of a special cause

the average and the range of the sample, provide estimates of two process parameters, the mean and the standard deviation. The question is: Is the behavior of the process changing from time to time, sample to sample? If the process is subject only to common-cause variation, the $\overline{X}$'s and R's should vary randomly within certain probabilistic limits of variation as shown in Fig. 5.9(a). Occasionally, the process may experience some real change in its mean level or amount of variation due to some special causes. When this happens, an $\overline{X}$ or R can indicate this by exceeding the natural variability limits, as shown in Fig. 5.9(b).

Shewhart used the statistical model portrayed above to indicate when the results of a stable/routine process were being disturbed by external forces. In this event the deviations would manifest themselves as variations that could not be explained by or expected to arise within the probabilistic limits of variation of the process when under control. He indicated further that such deviations signaled the presence of forces that were breaking down the routine of the process and causing it no longer to be economical. He referred to such deviations as arising from "assignable" causes, as compared to the "chance" causes driving the routine operation of the process. These assignable causes or special causes signal the occurrence of "local faults" in the process that can be dealt with by the operator or other local authority at the process, provided that these people have the knowledge and understanding regarding the root cause of such faults and the associated action(s) required to remove them.

Figure 5.10 shows an example of $\overline{X}$ and R charts being maintained on a particular process. For a period of time the process appears quite unstable, with numerous indications of out-of-control behavior (i.e., the process is not predictable). The implications of the lack of control of the mean level of the process in Fig. 5.10 are serious. First, since the process is not in a controlled

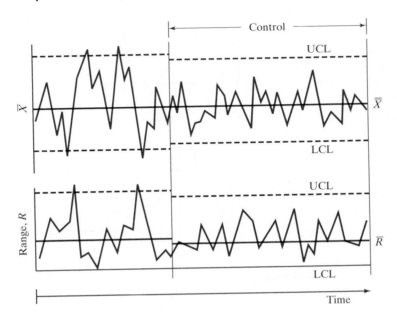

Figure 5.10   **Bringing a process into statistical control**

state, its results are not predictable. The capability of the process is not guaranteed and, in fact, cannot even be assessed reliably. But the equally serious implication of the lack of control or stability in the process is that it signals that the process is not operating in a routine manner and therefore that its economic feasibility is jeopardized. Until the root cause of the problem is found and action against it is taken, the process will continue to perform in a subpar fashion. However, as seen in Fig. 5.10, it appears that after the source of the trouble has been identified and action has been taken to bring the process into control, its pattern of variation is within the "limits of a routine process." The process is now running as it was designed to run, in a stable fashion, subject only to a constant system of common causes of variation. This movement from instability to control is an essential first step toward competitive manufacturing.

Moving from Control to Improvement

The identification of the presence of special causes and their subsequent elimination brings the process into a state of statistical control. But the presence of control only signals the fact that the process is now operating as it was intended to. We really have not improved the process—we simply have it running as it should! Once control is established, those at the process operation level can only be expected to use the charts to maintain control. To improve the process, we must attack the common-cause system. Such requires a more fundamental examination of the process and can be accomplished only through the involvement of management.

Earlier, Shewhart was quoted as indicating that "it has been found possible to set up limits within which the results of routine efforts must lie if they are to be economical." The suggestion here is that there exists a consistent or homogeneous set of forces, inherent to the process/system, which is driving the variation pattern in the process data. These forces are ever present and cause the process to behave in a stable and predictable though variable way. Shewhart referred to this state of affairs as indicating that the process was being driven by the forces of chance.

It was Shewhart's belief that these chance sources were many in number, each contributing only a small amount to the total, but collectively constituting an observable and measurable variation pattern. As a result, it was clear to Shewhart that the laws governing the behavior of random variations applied and hence that limits could be established to predict in probabilistic terms the behavior of these variations. Shewhart further reasoned that if the process operation in terms of its variation could be reduced such that it was to be affected solely by these chance forces, the process operation would be in a sense "doing the best it could." But these "chance" forces are real variation sources so clearly the opportunity for improvement exists and must be seized. The fact that Shewhart's control chart idea has prevailed over more than seven decades attests to the validity and simplistic beauty of the foregoing reasoning.

Over the years, Deming, Juran, and others have built on and refined the theories of Shewhart, and in the process a rich and tremendously useful approach to quality and productivity design and improvement has evolved. The set of chance causes referred to by Shewhart have come to be described as the *common-cause system* and constitute those sources of variation that are part of the system. As such, they are ever present and affect all realizations of the process. Of course, it is possible that the common-cause system variability could be excessive, and as a result, process capability is jeopardized. It then follows that to improve the capability, the system must be changed.

The discussion above leaves no confusion as to who is responsible for the improvement of the capability of processes that are already showing good statistical control. Clearly, management has this responsibility. The operator at the machine is helpless against the forces of common-cause variation. Although the operator may have important input to the fault diagnosis process, he or she cannot unilaterally initiate the action(s) required to change the system—only management can do this. Juran has reinforced this notion by referring to the forces of common-cause variation as describing the level of chronic variation in the process. A chronic condition is clearly one that is ever present as a part of the system and requires the attention of a higher level of authority to treat its root cause.

To illustrate the impact of solving a chronic problem at the process, let us suppose that a certain process has been brought into a state of statistical control. However, an assessment of the process capability has revealed that the process is not very capable; that is, a significant amount of the stable production from this process falls outside the quality characteristic specifications. This state of affairs is illustrated in Fig. 5.11, which also shows the impact that

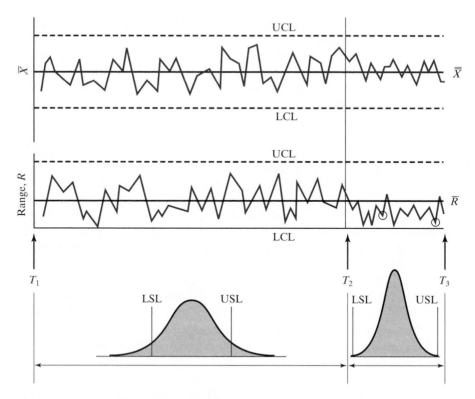

Figure 5.11   Breakthrough by attacking the common-cause system

a significant reduction in the amount of common-cause variation in the process can have on the capability of the process. Figure 5.11 shows that over the time period T_1 to T_2 the process was running in a state of control, but the process output relative to the specifications was not very satisfactory. Something needed to be done—but what and by whom? Now suppose that a team studying the process discovered a problem with the fixturing used and that through re-design, the problem was solved at time T_2. From the appearance of the R chart in Fig. 5.11 one would conclude that this change in the fixturing has caused a major reduction in the amount of common-cause variation in the process. A chronic problem that was compromising the process capability has been solved and the result is clear. As shown in Fig. 5.11, the process capability, as evaluated through process data collected over the time period T_2 to T_3, has been greatly improved. A major "breakthrough" in improving process perform-ance has taken place. It is important to note that a basic change in the process has been made: a change in the fixturing. Such would require a level of author-ity above the operator. Again, while the operator might have been instrumental in identifying the cause of the trouble, he or she would not be in a position to unilaterally formulate and carry out the actions required to make the change in the process. This is the role and responsibility of management.

Control Chart Approach to Identifying Improvement Opportunities

In developing a discipline of continual improvement, it must be clear what we mean by improvement. Bringing a process into a state of statistical control simply means that we have brought the process back to where it should have been to begin with! Figure 5.12 depicts the basic difference between control and improvement. The act of bringing the process into control, as seen in Fig. 5.12, should be viewed as one of simply properly implementing a system as it was conceived of, planned for, and developed during design. At this point it is then possible to begin to assess rationally the present ability of the process to realize the process potential and to seek improvement opportunities. The process may be failing to realize this potential because its implementation is flawed or because the design of the process itself is flawed. In either case, the root cause(s) of the chronic/common-cause problem must be identified and removed at the system level. Such constitutes a breakthrough in performance; that is, an improvement in the process has taken place. The results of such a breakthrough are depicted in Fig. 5.12.

Summary: Management of Process Variation Using Statistical Methods

- Shewhart used the statistical model for common-cause variation, with associated probability limits, to indicate when the results of a stable/ routine process were being disturbed by external forces. In this event the deviations would manifest themselves as variations that could not be explained by the forces of the common-cause system (i.e., expected to arise within the limits of variation of the process, when under control).

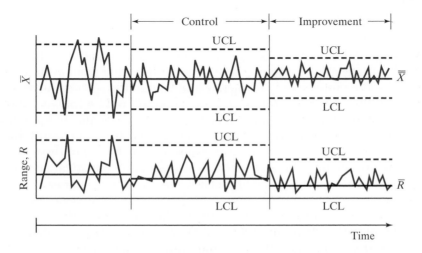

Figure 5.12   Moving from instability to control to improvement

- When the source of the trouble has been identified and action has been taken to bring the process into control—that is, its pattern of variation is within the "limits of a routine process"—the process is now running as it was designed to run, in a stable fashion, subject only to a constant system of common causes of variation.

- From a statistical standpoint, the fact that the process is subject only to the forces of common causes implies that it is in statistical control. A process in statistical control is "doing the best it can under the present system"; to do better, the system must itself be changed.

- Once a process is brought into statistical control it is possible to begin to assess rationally the present ability of the process to realize the potential it was initially intended to have. The process may be failing to realize this potential because its implementation is flawed or because the design of the process itself is flawed. In either case, the root cause(s) of the chronic/common-cause problem must be identified and removed at the system level. Such constitutes a major breakthrough in performance; that is, an improvement in the process has taken place.

5.6   The *Process* of Statistical Process Control

Purposes of Shewhart Control Charts

There are two fundamental purposes that Shewhart control charts can usefully serve.

1. To serve as a means to assist in the identification of both sporadic and chronic faults in the process and to help provide the basis to formulate improvement actions.

2. To serve as a tool to provide a sound economic basis for making a decision "at the machine" as to whether to look for trouble/adjust the process or leave the process alone.

Although these two purposes may sound essentially the same, they are not. The former is strictly an off-line activity while the latter is an on-line activity. On line, the operator can only recognize and take action against the presence of special causes of variation. If the process, as monitored on line, is free of special causes (operating within the limits of routine efforts), the operator is doing the best he or she can. In short, if the process is in statistical control, the operator should do nothing at all to disturb the routine behavior of the process.

On the other hand, the fact that the process is in statistical control in no way guarantees that it is a capable process, meeting the customer expectations as they may be communicated through a specification. However, the operator can certainly not be expected to be responsible for this, although he or she often has some valuable input to the problem. A lack of process capability is basically a product/process design and planning problem. To solve the problem may require the efforts of many different individuals, working together off line to find the basic faults, their root causes, and the actions necessary to remove them. Furthermore, if a process exhibits considerable erratic behavior with little knowledge of the causes of such behavior, the operator will be helpless in his or her efforts to take actions to improve the stability of the process—that is, to remove the forces of special causes.

Perhaps the most fundamental failing in the use of control charts in the past was the failure to recognize them as an important tool to discover off line the presence of improvement opportunities from both the quality and productivity points of view. Those advocating the use of control charts were often too quick to put them at the machine and place total responsibility for their use on the operator. As a result, they were too often viewed as a tool to be used by operators on the shop floor to keep production tracking within the product specifications. If "points" showed that out-of-control conditions were present but the individual parts were within specifications, often no action was taken. Such misuse of the Shewhart control chart concept was commonplace and eventually often led to the abandonment of the methods.

Closing the Feedback Loop at the Process

For the techniques of statistical process control—in particular, Shewhart control charts—to be employed successfully as an off-line problem identification and problem-solving tool, we must keep in mind that we are working with a three-step process:

1. Use statistical signals to find improvement opportunities through identification of the presence of process faults.
2. Use experience, technical expertise, and fault diagnosis methods to find the root cause of the fault that has been identified.
3. Develop an action plan to correct the fault in a manner that will enable any gains that are realized to be held.

This three-step process may be usefully explained through the classical feedback control system perspective shown in Fig. 5.13, which depicts the classical feedback control loop applied to a process. There are five distinct stages in this generic control loop. These five stages facilitate the three-step process outlined above in the following way:

STEP 1   Using statistical signals   (1) Observation
                                      (2) Evaluation
STEP 2   Fault diagnosis   (3) Diagnosis
STEP 3   Action plan   (4) Decision
                                      (5) Implementation

Each of these stages will now be discussed in detail.

Observation. There are two aspects of observation: the physical system aspect of what to observe and how to observe it, and the statistical sampling

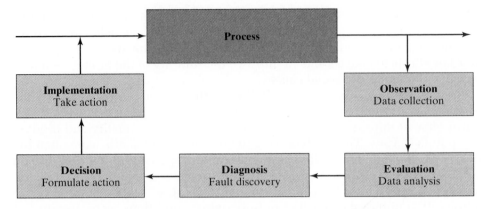

Figure 5.13   Classical control system view of SPC implementation

issue (e.g., how many measurements to take, how often they should be taken, and method of defining the sample). In this book we focus on the latter issue, statistical sampling, in considerable detail in Chapter 7.

Evaluation.   At the evaluation or comparison stage the data are compared to a model that describes how they should look under a predetermined circumstance/state of affairs. The Shewhart control chart is a statistical model that tells us how the data should look/behave if the process is stable/in control—that is, its mean level and level of variability are constant/unchanged over time. This comparison stage provides the basis for step 1 of the three-step SPC implementation process, namely, "reading" the statistical signals, which may tell us that a specific opportunity for quality and productivity improvement exists.

Diagnosis.   Diagnosis is perhaps the most crucial stage in the feedback control loop cycle. Being able to move from comparison to diagnosis marks the difference between making charts and solving problems. The key here is to be able to correlate the statistical signals with the physical state(s) of the process—for example, associate a point on a control chart above the upper control limit with a change in a raw material property. Numerous fault diagnosis aids are available, ranging from the simple but powerful graphical methods of Pareto charts and cause-and-effect diagrams to decision trees for failure modes effects analysis. If the closed-loop control system is to break open, it is likely to happen at the fault diagnosis stage. It is unfortunately all too easy to get caught up in making charts and forget why we are doing it.

Decision.   Once a fault is identified at the root cause level, it is necessary to formulate the appropriate action to eliminate the fault.

Implementation.   Finally, once the appropriate correction has been decided upon, we must define the specific means to make the correction. It is important to study carefully the full implications of the action on all elements of the system so that the action taken may be lasting. *Holding the gains* over the long term requires that the system truly accept the total value of the action.

Summary: *Process* of Statistical Process Control

The fundamental purposes of control charts are

- To serve as a tool to provide a sound economic basis for making a decision "at the machine" as to whether to look for trouble/adjust the process or leave the process alone.
- To serve as a means to assist in the identification of both sporadic and chronic faults in the process and to help provide the basis to formulate improvement actions.

In order to employ the techniques of statistical process control—in particular, Shewhart control charts—successfully as an off-line problem-identification and problem-solving tool, we must keep in mind that we are working with a three-step process:

- Use statistical signals to find improvement opportunities through identification of the presence of process faults.
- Use experience, technical expertise, and fault diagnosis methods to find the root cause of the fault that has been identified.
- Develop an action plan to correct the fault in a manner that will enable any gains that are realized to be held.

5.7   Deming's Famous Red Bead Experiment[4]

Deming often employed simple sampling experiments to demonstrate the workings of the "common-cause system" and the senselessness of blaming workers for causes in the system that are out of their control. The most famous is the "Red Bead experiment" that Deming performed in all of his short courses and seminars, with the help of several members of the class who were chosen as the "Willing Workers," inspectors, and the recorder of the data from inspection. The "script" below describes the experiment and discusses the lessons learned from it. The Red Bead experiment is particularly useful in reinforcing the material described earlier in this chapter. Note: The authors have taken the liberty of using some of Deming's unique language and comments, which made listening to and watching this demonstration so rewarding and enjoyable.

The "Red Beads": The Purpose of the Demonstration

The purpose of this demonstration is to illustrate through mechanical sampling of a mixture of *White Beads*—*good parts* and *Red Beads*—*bad parts* the

---

[4]Adapted from W. Edwards Deming, *The New Economics*, MIT, Cambridge, MA, 1993.

appearance of variation that will arise in performance from worker to worker and from time to time, and—most importantly—to clearly show that when the system is operating in a stable fashion, this variation in performance is driven by forces inherent in the system itself, not by the individual workers.

Lessons to Be Learned from the "Red Beads"

1. *Stability of performance of a system.* You will recognize the appearance and importance of a stable performance system, one in which the results in the future are predictable. Performance will fluctuate from day to day, but in a stable system those fluctuations are predictable—bounded.

2. *Where does the variation come from?* The performance of several individuals working together in a stable system cannot be distinguished from each other. The variation comes entirely from the system itself. In a stable system, people's performance is driven completely by forces in the system that are beyond the control of the individual worker.

3. *Where to look for improvement?* It is futile to ask people to do better, to work harder, and then reward or punish them accordingly, when their performance is solely driven by the system itself, which is already operating in a predictable fashion at its best level of performance. To improve performance of the individual, the system must be changed. As you will come to appreciate, the "Willing Workers" are doing their very best. No more can be expected of them.

4. *Proper management of a system?* You will get a view of a very rigid system in which the workers must follow lock-step procedures to the letter and have no chance to provide input and feedback for the purpose of improving the system.

5. *Treating causes, not symptoms.* Ultimately, it will become clear that the way to improve performance is to keep the "Red Beads" from getting into the box in the first place. We must eliminate the forces that give rise to the production of bad parts.

Props Needed for the Experiment

The experiment requires 4000 beads—3200 White and 800 Red—and a tub of sufficient size and shape to hold them, allowing thorough mixing and easy access to the beads via a paddle with 50 semicircular indentations to "catch and hold" the beads when a sample is drawn.

The "Setup"

- The Company is expanding to accommodate a new customer.
- The new customer needs White Beads—it will take no Red Beads!
- The Company must hire ten new employees to produce White Beads.
- If too many Red Beads are made, the Company will not be able to keep the new customer supplied and be profitable.

New Employees

- *6 Willing Workers.* Job requirements: Must be willing to put forth their very best efforts. Job security depends on performance. Education requirements are minimal. No experience in pouring beads is required!

- *2 Inspectors.* Job requirements: Must be able to tell red from white and to count to 20! No experience required.

- *1 Inspector General.* Same job qualifications as the inspectors.

- *1 Recorder.* Job requirements: Must write legibly and be good in addition and division.

The Training/Apprentice Program

- The training will be for three days. You can ask questions during the training, but once production starts, there is no time for questions. The procedures are fixed and must be followed to the letter! There must be no variation in the production output. No comments; no suggestions; just do your job as you have been instructed.

- Your jobs depend on your performance. Those who do not follow the procedures and who make too many Red Beads will lose their jobs!

- Procedures for dismissal are in place. There are many others here today ready to take your job!

- Work standards are in place here. Everyone must make precisely 50 beads per day, no more, no less.

- Inspectors must count and record their results independent of each other. The Inspector General will make sure that the counts match—we must find all of the Red Beads that we make!

- Our quality control procedures are rigid—no Red Beads must get to the customer! That is why we have such extensive Quality Control procedures—we use 100% inspection to find all of the Red Beads manufactured.

- Does everyone understand what is expected?

Procedures for Each Willing Worker for Bead Manufacture

STEP 1 Prepare the incoming raw material.

- Grasp the tub with both hands, gripping one corner and the opposite wide side.

- Pour from the corner.

- Tilt the tub but do not shake or simply dump the tub over—let gravity do the work! Gravity is cheap!

- Repeat the steps above, following procedures to the letter.

STEP 2   Manufacture the beads.

- Grasp paddle by the handle.
- Insert paddle into beads.
- Agitate paddle briefly.
- Stop agitation, tilt paddle to exactly 44°.
- Raise paddle vertically, paddle remaining at 44° to the horizontal axis.

Does everyone understand the procedure?

STEP 3   Inspect the production.

- Carry work to Inspector No. 1. (Inspector records results on paper, no talking, Inspector 2 must not see Inspector 1's result!)
- Carry work to Inspector No. 2. (Inspector records results on paper, no talking.)
- Inspector General inspects the work of the inspectors.
- Inspector General *loudly* announces the final tally of Red Beads, then announces, "*Dismissed!*"

STEP 4   Record the results.

- Recorder shows the Red Bead tally for each worker on the sheet.
- No recording of results during the apprenticeship.

Quality Training for the Workers

- The company expects good quality at all times.
- We have several slogans and mottoes for our company you should live by:
  "Do it Right the First Time"
  "Take Pride in Your Work"
  "Be a Quality Worker"
  "Increase Productivity"
- You must live up to the quality ethic here!
- We have a rich Quality Tradition here!

Begin the Training

- All six Willing Workers make 50 beads each day for three days—50 beads is the Work Standard!
- The purpose is to learn the procedures and follow them to the letter.
- There must be no variation in the production conditions—Follow Rigid Procedures.

Training Day 1

	Beads Sampled	Red Beads
Worker 1	50	10
Worker 2	50	14
Worker 3	50	10
Worker 4	50	9
Worker 5	50	16
Worker 6	50	12

Remind Worker #5 that no Red Beads are to be produced. All workers must do a better job.

Training Day 2

	Beads Sampled	Red Beads
Worker 1	50	12
Worker 2	50	8
Worker 3	50	7
Worker 4	50	10
Worker 5	50	9
Worker 6	50	8

Overall, the workers have done a better job than on Training Day 1. Nevertheless, there are still too many Red Beads. Review the procedures.

Training Day 3

	Beads Sampled	Red Beads
Worker 1	50	13
Worker 2	50	11
Worker 3	50	8
Worker 4	50	12
Worker 5	50	6
Worker 6	50	7

Worker #5 appears to have made significant improvements during the training days. Other workers need to learn from Worker #5.

It is now time to begin production.

Begin the Production

DAY 1   Each Willing Worker follows the procedures and manufactures precisely 50 beads. The results are inspected and recorded.

	Beads Sampled	Red Beads
Worker 1	50	5
Worker 2	50	9
Worker 3	50	17

(*Continued*)

	Beads Sampled	Red Beads
Worker 4	50	11
Worker 5	50	8
Worker 6	50	10

- The results are somewhat disappointing—too many Red Beads are made.
- One of the workers (Worker #3) does very poorly! There is tremendous variation in the results!
- Workers need to be reminded that they must make only White Beads.
  - We are on the Merit System: We will reward good performance.
  - *The Best Performer will get a merit increase in pay!—Worker #1.*
  - Poor performers will be disciplined!
  - *The Worst Worker will be put on probation!—Worker #3.*

Management Actions

- Merit increases and probations are handed out.
- To improve performance, management has now set a numerical goal: *No more than 3 Red Beads per worker per shift.*

DAY 2   Continued Disappointment!

	Beads Sampled	Red Beads
Worker 1	50	13
Worker 2	50	9
Worker 3	50	7
Worker 4	50	7
Worker 5	50	8
Worker 6	50	9

- Good workers do more poorly today (Worker #1 went from 5 to 13 red beads).
- Some poorer workers improve, but still there are too many Red Beads.
- Management is watching the numbers closely!
- Costs are exceeding revenues.
- Some workers are doing so well—why can't all do better?
- Procedures are rigid—there should be no variation.
- It seems that the merit increase has gone to someone's head! But the worker on probation has done better! Off probation!
- A merit increase is given to today's best workers! But another goes on probation!

DAY 3   Company Announces That Today Is "Zero Defects Day"
Management strongly emphasizes to the workers the need to do better—
try your very best—all jobs depend on you!

	Beads Sampled	Red Beads
Worker 1	50	10
Worker 2	50	8
Worker 3	50	7
Worker 4	50	11
Worker 5	50	11
Worker 6	50	9

- The results continue to be very disappointing.
- Management continues to lose money.
- If results don't improve, the plant will be closed!
- It's all up to the Willing Workers—only you can save your jobs!

DAY 4   Still No Improvement

	Beads Sampled	Red Beads
Worker 1	50	10
Worker 2	50	5
Worker 3	50	10
Worker 4	50	12
Worker 5	50	9
Worker 6	50	6

Management has come up with a way to keep the plant open:

- Some jobs will be saved.
- The three best workers will be retained—surely this will reduce the number of Red Beads manufactured!
- The others are dismissed! They must pick up their checks and leave. It's their own fault—the quality of their work was poor.

Starting with Day 5, the three best workers will now each work two shifts per day! (Workers #2, #5, and #6.)

DAY 5   Continued Disappointment!

	Beads Sampled	Red Beads
Worker 2	50	7
Worker 5	50	14
Worker 6	50	10

(Continued)

	Beads Sampled	Red Beads
Worker 2	50	13
Worker 5	50	8
Worker 6	50	17

- What has happened to the best workers?
- They were so good in the past, now their work has deteriorated.
- The results are no better than when all of the original six Willing Workers were on the job.

Management makes a decision to close the plant!

Summary

Now, let's look at the data. Can they help us understand what has happened here? Let's make a Shewhart control chart based on the data (Fig. 5.14). What is the chart telling us? Recall, there were 4000 beads in the tub: 3200 White and 800 Red. Do the results seem consistent with this? Why or why not?

The Shewhart control chart in Fig. 5.14 shows the Red Bead data for the Willing Workers (Days 1–5). Each worker's output, fraction defective (number of Red Beads/beads sampled), is displayed vs. sample number. The chart shows that the process is in good control—no special causes are evident. The only reason why one worker's output differs from another's is the effect of the common-cause system.

Some Important Questions

- Should we expect that the results will be the same each time—no variation?
- Why do the results vary?
- How do the results vary?
- Did the individual workers influence the result in distinctly different ways?
- How could we improve the results—make fewer Red Beads?

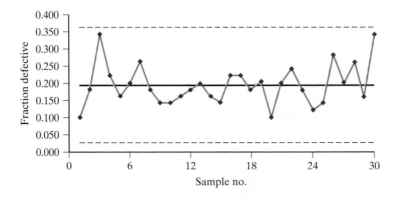

Figure 5.14   Shewhart control chart for data from Red Bead experiment

Lessons of the Red Bead Experiment

1. *The system turned out to be stable, in statistical control, therefore*
   - The results of the Willing Workers are predictable.
   - Production costs are predictable.
   - If production were to be continued, the results would all fall within the established limits of variation; things would be neither better nor worse, regardless of the efforts of the Willing Workers.

2. *All variation among the Willing Workers is due to the system itself.*
There is no evidence here that any worker was performing any better or worse than the others. Their performance was driven solely by the forces at work in the system! We cannot tell the workers apart from each other, so far as their performance is concerned. The workers didn't make the Red Beads—the system did. The workers merely delivered them.

We call this a system driven by common causes of variation. Juran called it *"chronic variation"* because it is inherent in the system itself.

Occasionally, performance may vary due to a "special cause." If a worker filled only half of the paddle, then the number of Red Beads would be affected in a special way. Such behavior is outside the system and is unpredictable. It is likely to fall outside the limits of common-cause variation.

The workers have done the best they could—they have given their best efforts; they can do no better under the present system. To improve performance, the system must be changed. Changing the system means reducing the number of Red Beads in the tub.

3. *Ranking people's performance is wrong.* It is totally ineffective and demoralizing. All we have done is ranked the effect of the system on the people. The performance of the people did not vary by their own hand. Remember, the system made the Red Beads!

We concluded that some workers were performing better than others—this turned out to be totally incorrect. Any other six Willing Workers, following the same procedures, would produce the same results.

4. *We have learned the futility of pay for performance and probation for "poor" workers.* In spite of merit increases, the performance of the Willing Workers did not improve. Their performance, so bad that they all lost their jobs, was governed solely by the system. Rewards and punishment based on performance had no effect on the results.

5. *The entire experiment was a display of bad management.*
   - Procedures were rigid—workers could not make suggestions for improvement. Everyone has a responsibility on the job to improve the system, but the system did not allow this to happen.
   - Management used empty slogans—they did not give the workers tools and methods by which they could improve the output.
   - Management put all of the blame for poor performance on the workers, accepting no responsibility for the poor quality.

- Management should have worked with the supplier of the incoming material to reduce the number of incoming Red Beads. This would change the results—improve performance.

6. *Management used slogans, special quality programs (Zero Defects), and work standards*, all in an attempt to somehow improve the quality—reduce the number of Red Beads in the output. But none of these things did any good.

7. *Knowledge of the proportion of incoming Red Beads does not enable us to predict the proportion of Red Beads in the output.* Our experiment uses mechanical sampling, not random sampling. The results will vary with the particular paddle used, the beads used, the tubs used, the procedures used, and so on. For example, *over the years, Deming obtained different results with different paddles*:

> Paddle 1: Aluminum (1942), $\overline{X} = 11.3$
> Paddle 2: Smaller than Paddle 1, $\overline{X} = 9.6$
> Paddle 3: Made of applewood—bulky, $\overline{X} = 9.2$
> Paddle 4: Made of nylon, $\overline{X} = 9.4$

Deming used the Red Bead experiment for many years in his classes to demonstrate the forces of common-cause variation and to dramatize many of the lessons of his 14 points. He used members of the class as Willing Workers and Inspectors to assist him in acting out the entire scenario. The authors urge others to do the same.

5.8   The Debilitating Effects of Tampering

The Red Bead experiment illustrates the all-too-familiar mistake of confusing common and special causes. A common and debilitating consequence of the failure to recognize and clearly understand the difference between the common-cause system and the presence of special causes of variation is tampering. The best intentions of an operator, often with the encouragement of his/her supervisor, to keep a process "tracking on target" can lead to frequent and unnecessary "adjustments" to the process, and the result is always increased variation, waste, and inefficiency.

Some years ago, one of the authors of this book consulted for a company that manufactured edible casing material for wieners. The complete manufacturing process included a sequence of steps wherein the casing was extruded and then pulled through a series of tanks for further processing. Measurements were regularly taken on a number of characteristics of the solutions in the tanks, and the consultant used these historical data to construct time-ordered plots of the various characteristics. Many of these plots showed a classic oscillatory behavior (the data appeared to be strongly negatively correlated in time), often with large swings above and below the centerline. Becoming suspicious of what the problem might be, the consultant went to the floor to observe the process and spoke with an operator about procedures. One tank in question was a solution for conditioning the casing that required a target pH of 5.5. A plot of the actual pH values taken every two hours looked particularly oscillatory in nature. The operator explained that he was to measure the pH every two hours and make adjustments "as necessary" to keep the pH at 5.5. He indicated that when his measurement was below 5.5, he would make a prescribed addition to the tank to "bring it up." Conversely, if the measurement was above 5.5, he would make a different prescribed addition to "bring it down." His comment was, "It's really hard to keep it right at 5.5, so I have to make adjustments almost every time!"

The problem here is clear—failure to recognize the difference between the common-cause system and the occurrence of a special cause. If left alone, the pH in the tank may well have exhibited a natural/random pattern of variation about a fixed centerline, and the variation in the data would have been smaller than that which resulted from making frequent adjustments. Furthermore, the time spent making the adjustments and the materials expended for this purpose could have been saved. In short, for a process in statistical control, we can achieve better quality and less waste and inefficiency by leaving the process alone.

Deming frequently used a simple experiment in his classes to illustrate the effects of tampering—overadjustment of the process. He attributed the origin of this experiment to his colleague and friend Lloyd S. Nelson. The "funnel experiment" involves positioning a funnel over a cloth on which a target is marked with a "dot." This is depicted in Fig. 5.15(a). A marble is dropped into the funnel, and the position where it comes to rest on the cloth is marked. If the funnel is placed directly above the target and many such drops are made and marked, the data will show a random/natural pattern of variation about the target similar to that in Fig. 5.15(b).

To demonstrate the effects of tampering/overadjustment, a second series of marble drops can be made, only this time the position of the funnel is adjusted for each drop with the hope of reducing the error—the difference between the

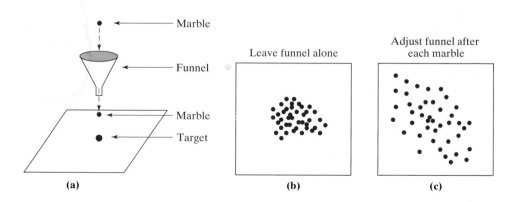

Figure 5.15   Deming's funnel experiment: (a) setup, (b) results with no adjustments, (c) results with adjustments

target and the position of the dropped marble. In the case depicted in Fig. 5.15(c), the funnel is initially positioned above the target and a marble is dropped. The position from the marble to the target is then measured, and the funnel position is adjusted by an equal amount and in the opposite direction from the target. After the next drop, the error is measured again, and the funnel is again adjusted to a new position based on the error from that drop. The drops are continued according to this procedure, with adjustments being made after each drop. The results seen in Fig. 5.15(c) clearly demonstrate that the pattern of variation has increased. Careful scrutiny of the actual data reveals that it is no longer a random/ natural pattern of variation or a stable process. In fact, the process data will tend to increase in variation, oscillating from one side of the target to the other, as the drops are continued. The data obtained through a funnel experiment with the form of tampering precipitated by the adjustment rule used will show a pattern similar to that evident in the data for tank pH described above, but with "exploding" variation!

Scherkenbach has provided a detailed and enlightening account of Deming's funnel experiment.[7] Several different adjustment "rules" are proposed, and in each case the data show a certain characteristic behavior that occurs with that form of tampering. A number of commercial sources provide kits for conducting the funnel experiment in the classroom (e.g., http://www.does.org/masterli/funnel.html).

■ EXERCISES

5.1 Are chance causes/system faults and assignable causes/local faults useful in describing outer noise? Which causes/faults is it the responsibility of management to address?

5.2 Why is it important to consider the time order of production when describing a set of data, particularly if it is being used to help explain the performance of a process of manufacture?

5.3 What is meant by "statistical control" of a process? Why is the achievement of statistical control important?

5.4 An automobile yields consistent performance for several months in use in and around Peoria, Illinois. However, it experiences a considerable drop in mpg when driven for three months at high altitude in Colorado. What is this an example of? In general, how could the performance of this car be improved without restricting it to low altitudes?

5.5 A process is operating under statistical control. Are customer expectations (in terms of conformance to specifications) necessarily being met? Explain.

5.6 A process is manufacturing virtually all of the product within the specifications to satisfy the customer's demand. Is it necessary to monitor this process to see if it is under control? Why or why not?

---

[7]W. Scherkenbach, *Deming's Road to Continual Improvement*, SPC Press, Knoxville, TN, 1991.

5.7 Explain two of the most serious implications of running a process that is not in a state of statistical control.

5.8 A process that is making plastic lids for tennis-ball containers is finally brought into control after some adjustments. Has a fundamental improvement been made? Explain.

5.9 A process manufacturing taillights for a new line of automobiles is seen to be in control, yet it is not meeting customer expectations—too many parts are falling outside the specifications. What should the operator on this line do about the problem? What needs to be done in general?

5.10 Can Shewhart control charts be used off line, away from the shop floor, in a meaningful way? Explain.

5.11 Why is the diagnosis stage in the feedback control loop cycle so important? Why is it so difficult to make the step from evaluation to diagnosis?

5.12 What type(s) of "noise" are affected by parameter design?

5.13 What group(s) within the organization are responsible for reducing the effect that variational noise has on the quality of products that are being manufactured? What role does the operator/first-level supervisor play in solving problems of excessive variational noise?

5.14 Can a process be considered to be running in an economical fashion if it demonstrates a mean that fluctuates out of control periodically, yet still manages to produce virtually all parts within the specifications?

5.15 Are erratic shifts in the mean of a process considered special causes or part of the common-cause variability?

5.16 How could bringing a process into statistical control affect a company's production schedule? What benefits can customers see as a result of doing business with a company whose processes are running in a stable, predictable manner?

5.17 If the time order of data collected was not recorded, would it be possible to detect that special causes may have occurred during the production of products over a shift of operation? Would it be possible to develop a track leading back to the root cause of the local faults? Explain.

5.18 What can be said about the variability pattern in a process that is running as it was designed to run? Explain what is meant by the notion "running as it was designed to run."

5.19 A company initiates statistical charting on a process that has never been charted. The operator is trained in making charts and reading the statistical signals. After charting is begun, the process starts to exhibit erratic shifts in variability at random intervals. The operator cannot find any obvious physical cause for this behavior and becomes very frustrated. Are the control charts serving any useful purpose in this situation? What has been done wrong here? Explain.

5.20 What is meant by the term "holding the gains" in reference to SPC implementation? Why is it so important that this be done?

5.21 A machining cell is often configured and reconfigured for relatively short runs of various machining jobs. How do you apply SPC in this situation, where each run may be too short to provide sufficient data for control charting?

5.22 Using graphical summaries similar to those of Fig. 5.7, sketch each of the four possible scenarios in terms of "product conformance" and "process control" (not conforming/conforming and in control/out of control). What is the conclusion and what action should be taken in each case?

5.23 What do you believe are the three most important lessons of the Red Bead experiment? Why?

5.24 The Red Bead experiment demonstrates a situation in which the workers are expected to produce results to a certain standard, yet clearly the workers have no control over the process result. Describe a case from your own life where you have felt like one of the Red Bead Willing Workers. For this case, what should have been done differently?

5.25 Often well-intentioned workers intervene or tamper with a process. What will generally be the effect of such tampering? Why is this so?

5.26 Describe a situation from your own life, or with which you are familiar, where tampering has led to undesired results.

6

Shewhart Control Charts for Variable Data

6.1   Introduction

In Section 5.6 we presented the complete process of SPC. In this chapter we focus on two important aspects of this process, *evaluation* and *diagnosis*. We first examine the statistical basis for control charts. Then, we focus on the construction of $\overline{X}$ and R charts. Along the way we present some tools that will help greatly in facilitating the fault diagnosis process. In Chapter 7 we will focus on the statistical sampling aspects of *observation*.

Figure 6.1 shows this notion of SPC as a feedback control loop overlaid on the process. A crucial aspect of closing the loop is the ability to convert the statistical signals emanating from the control charts into the root cause(s) of the problem through fault diagnosis. The evaluation stage of the process essentially consists of making control charts from rationally collected data and reading the statistical signals to understand what the charts are telling us about the process behavior in a statistical sense (e.g., behavior of the mean of the process, behavior of the variation of the process). The diagnosis stage is really the key to successful SPC implementation. Once it has been determined that the mean or the range of the process has been influenced by a sporadic/special cause of variation, we must determine the specific physical nature of that cause so that an improvement action can be formulated.

6.2   Statistical Basis for Shewhart Control Charts

Chapter 4 presented the principles and procedures associated with a statistical test of hypothesis. The Shewhart control chart is such a test of hypothesis and therefore also stands much as a jury in a court of law. The information from each sample is judged to determine whether or not it indicates the presence of a special-cause disturbance. Unless the evidence is overwhelmingly in favor of the occurrence of a special-cause disturbance, we enter a verdict: not guilty. In other

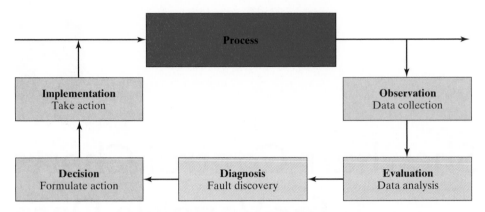

Figure 6.1   Classical control system view of SPC implementation

words, we find no strong reason to believe that forces other than those of common causes are at work. Such an interpretation is, in fact, based on and consistent with the hypothesis testing approach to making statistical inferences about the process based on information contained in the sample.

Two process characteristics are of general interest in the statistical process control scenario: the mean level of the process and the amount of variation in the process. We are particularly interested in determining when these two characteristics/parameters may have changed in magnitude. We have seen that we could make inferences concerning the mean level of a process through sampling and an associated inferential structure that we called hypothesis testing. The null hypothesis was generally our stated belief about the mean level of the process. Given the sampling distribution of the statistic used to estimate the mean level, we could then evaluate a particular sample result to see if its occurrence was consistent with our belief about the process mean.

Figure 6.2(a) graphically depicts the hypothesis testing situation we have just described. If we were to estimate the process mean from the results of a sampling of the process using the arithmetic average $\overline{X}$, the sampling distribution for averages, which is the normal distribution, would provide the probabilistic basis for drawing an inference about the actual process mean. To do all of this, we must have some rational basis for establishing the null hypothesis H_0 in the first place. What do we think the actual process mean is? In the context of the Shewhart control chart method what we do is collect an amount of data (k samples of size n each) that we think is sufficient to allow us to treat the average of all of these data as if it were the actual process mean. Once such a hypothesized mean is established, we can use it as the null hypothesis and conduct the test as illustrated in Fig. 6.2.

In Fig. 6.2(a) we see two sample results ($\overline{X}$ values each from samples of size n), denoted A and B. The reject-H_0 and cannot-reject-H_0 regions are established on the basis of the sample size n, the process standard deviation σ_X, and the α risk. Recall that the latter quantity is the risk we are willing to take in concluding that the null hypothesis is false when, in fact, it is really true. Based on sample result

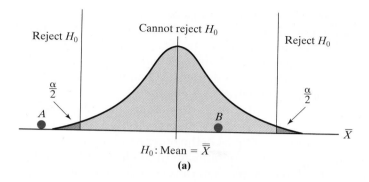

Reject H_0 Cannot reject H_0 Reject H_0

$\frac{\alpha}{2}$

A

B

$\frac{\alpha}{2}$

$\overline{X}$

H_0: Mean = $\overline{\overline{X}}$

(a)

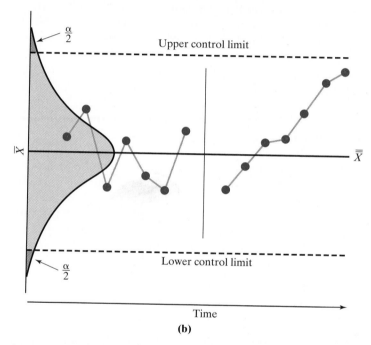

$\frac{\alpha}{2}$

Upper control limit

$\overline{X}$

$\overline{\overline{X}}$

$\frac{\alpha}{2}$

Lower control limit

Time

(b)

Figure 6.2   **Hypothesis testing formulation extended to establish the Shewhart control chart model**

A, we would reject the null hypothesis, while the result B leads to the conclusion that we cannot reject H_0.

It is important to note in the formulation above that even if the sample results A and B were obtained from the consecutive sampling of, say, two batches of raw material, they would be used, in the general hypothesis testing formulation, independent of each other to test the null hypothesis. It was the absence of such serial consideration of sample results that Shewhart recognized when dealing with the results of samples drawn consecutively from an evolving process. His model for the control chart adds this dimension to the problem.

Figure 6.2(b) illustrates how Shewhart extended the hypothesis testing idea to include consideration of the time-evolving nature of process behavior. By turning the classical formulation "on its side," considering the horizontal axis to represent time, and extending the reject-H_0, cannot-reject-H_0 lines, he established a model that enables one to track the behavior of the process mean over time. This is done by examining the results of successive samplings to determine in which region they fall and, of equal importance, the relationship that might or might not be evolving when they are considered relative to each other.

Under the assumption that the sample results arise from a constant system of common causes, they should evolve in time in a random fashion. The sequence to the left in Fig. 6.2(b) may be typical of this. On the other hand, if the sample results show evidence of evolving in a nonrandom pattern, such as the sequence to the right, the suggestion is that forces other than those of common causes are at work and some investigation of their origin is warranted. We will see later that a sequence evolving completely within the cannot-reject-H_0 region (what Shewhart called "within the control limits") but having six or more sample results in a row that are continuing to increase (or decrease) provides us with strong evidence that the mean of the process is shifting upward (downward). In doing this, adding a condition(s) on the evolving $\overline{X}$ values for the rejection of H_0 (adding other criteria), we may be increasing the sensitivity of the overall test, but the effective α-risk is also increased. This is so since 6 $\overline{X}$ values in a row, continually increasing, could occur due to common causes alone. The value 6 was chosen so that this risk value is quite small.

To establish Shewhart control charts for the process mean level and process variability, we begin by defining the sampling procedure to obtain the data. For now we will simply say that we will take "snapshots" of the process at periodic intervals by collecting small samples of consecutively manufactured product. How large these samples should be, how often they should be collected, how many should be collected, and finally, the precise manner in which they should be collected are important questions that must be addressed, and will be in Chapter 7.

6.3   Control Chart for Averages, $\overline{X}$

Let us suppose that we select samples of size $n = 5$ every half-hour from a process manufacturing gasket material blanks. Each sample will contain five consecutively manufactured gasket blanks. Typical data might be as follows, with the measurements being the thickness of the blanks in inches.

	10:00 A.M.		9:30 A.M.		9:00 A.M.	
	0.044 inch		0.046 inch		0.043 inch	
	0.043		0.044		0.044	
Process $\rightarrow$	0.045	$\rightarrow$	0.045	$\rightarrow$	0.044	$\rightarrow$
	0.044		0.045		0.042	
	0.044		0.047		0.047	
	$\overline{X} = 0.0440$		$\overline{X} = 0.0454$		$\overline{X} = 0.0440$	
	$R = 0.0020$		$R = 0.0030$		$R = 0.0050$	

Each small sample tells us how the process is behaving at any one time in terms of the process mean and variability through the sample mean and the range of the sample. The question is: Is this behavior changing from time to time? If the process is subject only to common-cause variation, the $\bar{X}$'s and R's should be randomly distributed within certain probabilistic limits, as illustrated in Chapter 5. Occasionally, the process may experience some real change in its mean level or amount of variation. When this happens, an $\bar{X}$ or R can indicate this when the natural variability limits are exceeded or when a series of either of these values exhibits a nonrandom pattern.

As we saw in Chapter 4, when a series of samples as we defined above are collected from a process in good statistical control, the sample means, $\bar{X}$,

10:00 A.M.	9:30 A.M.	9:00 A.M.
$\bar{X} = 0.0440$	$\bar{X} = 0.0454$	$\bar{X} = 0.0440$

will tend to arrange themselves in a frequency sense according to the normal distribution. This is the central limit theorem effect that was discussed in Chapter 4. This tendency for the sample means obtained from rational sampling of a stable process is illustrated again in Fig. 6.3. We will call the limits shown on the right in Fig. 6.3 the *upper control limit* (UCL) and the *lower control limit* (LCL) for the $\bar{X}$ control chart. For the averages $\bar{X}$ of samples of size n varying about a fixed mean $\bar{\bar{X}}$, these limits are defined by

$$\bar{\bar{X}} \pm \frac{z_{1-\alpha/2}\sigma_X}{\sqrt{n}}.$$

To be precise, $(1 - \alpha)100\%$ of the $\bar{X}$'s should fall within these limits if the process is subject only to common causes of variation. Traditionally, these limits have been established at ± 3 standard deviations from the mean. Hence 99.73% of the $\bar{X}$'s

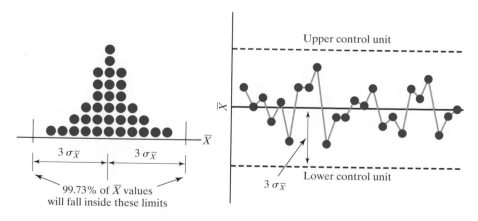

Figure 6.3   **Statistical basis for the control limits on the $\bar{X}$ chart**

should fall within these limits for a process whose mean level is stable—in statistical control.

It is absolutely essential that we keep in mind that these limits are based on the variation we expect to see in averages, not individual measurements. In Chapter 4 the relationship between the distribution of individual measurements and the distribution of averages of several individual measurements was clearly shown. Failure to recognize this important difference is what often leads to an erroneous comparison of control limits and specification limits.

If a process under surveillance by periodic sampling maintains a state of good statistical control, this means that its mean level and level of variability remain constant over time. If such is the case, then, as we have learned, the sample means will have a distribution that follows the normal curve, and therefore nearly all (99.73% to be exact) of the sample means $\overline{X}$ would fall within the band of ± 3 standard deviations of $\overline{X}$ about the established mean of the process, which we estimate generally by $\overline{\overline{X}}$, the average of a number of averages, $\overline{X}$.

Now suppose that a sample mean $\overline{X}$ falls above the upper control limit. Since the chance of this occurring is so small (about $\frac{1}{8}$ of 1%), if the process mean is at the hypothesized mean of the process $\overline{\overline{X}}$, we must assume that some special cause is present that is influencing this $\overline{X}$ to be larger than expected. In particular, we must assume that the process, at least at that point, cannot reasonably have had the mean level that is represented as the null hypothesis, the hypothesized mean used as the centerline of the chart. Rather, it is likely that an upward shift in the mean has occurred, at least at this point in time. This is the hypothesis testing interpretation of the data and the one that Shewhart adopted for the control chart. Figure 6.4 illustrates the presence of such a special cause.

Of course, it is possible that no shift in the mean has occurred and therefore we will mistakenly look for a problem that is not there at all. But we know precisely

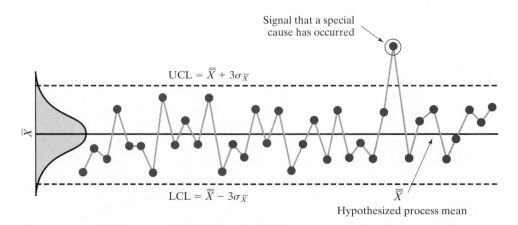

Figure 6.4   Use of probabilistic limits to identify special causes

what the chance is that this error will occur. It is the α risk that we defined earlier, the chance that we will reject the null hypothesis that the process mean is the centerline of the chart when, in fact, it really is.

Control Chart Sensitivity

The fact that all points fall within the 3 sigma limits for $\overline{X}$'s and exhibit a random pattern does not mean that no special causes are present. We must also accept the fact that departures from expected process behavior may not always manifest themselves on the control chart immediately, or for that matter, at all. In other words, the mean has shifted—the null hypothesis is now false—but the data do not show this to be the case. This is the β risk we noted in Table 4.1. Although narrower limits (e.g., 2 sigma limits) would enable us to detect the occurrence of special causes easier and faster, such limits would also increase the chance of false alarms, that is, times when the process is actually in good statistical control but a sample mean $\overline{X}$ falls outside the control limits, and therefore we conclude that it is not in control.

The selection of the appropriate placement of the upper and lower control limits, that is, the selection of the α risk, is an economic issue. The intent would be to fix the limits in such a way as to balance the economic consequences of failing to detect a special cause when it does occur—failing to reject the null hypothesis when it is false—and wrongly identifying the presence of a special cause when it has really not occurred—rejecting the null hypothesis when it is really true. Practice over the years has led to the use of 3 sigma limits as a good balance of these two mistakes. But there is nothing sacred about the 3 sigma limit.

6.4   Control Chart for Ranges

So far we have confined ourselves to a discussion of the detection of shifts in the process mean. However, sporadic disturbances may also cause shifts or changes in the amount of process variability. In Fig. 6.4 we focused only on the process mean behavior, and therefore, have represented only one side of the coin. In so doing, we have violated a fundamental rule—that one should never show an $\overline{X}$ chart without also showing the associated range chart. In fact, it is the range chart that is so essential to the Shewhart control chart concept. If the process variation is not stable over time, then the limits, on the $\overline{X}$ chart calculated from the unstable process, do not reflect only the forces of common-cause variation, and therefore the interpretation of the $\overline{X}$ chart could be quite misleading.

A similar hypothesis testing argument to that above can be put forward for the behavior of a series of ranges of the same samples emanating from a constant system of common causes. The sampling distribution of the ranges then provides the basis for the establishment of the probability limits, which we call the upper and lower control limits when the ranges are plotted in time order of occurrence.

To establish a control chart for the range R, we must be a bit more specific about the distribution of the process (i.e., the quality characteristic of interest X). This is so because the distribution of ranges does not exhibit the same robust behavior as that of the distribution of averages. If we can assume that the individual measurements X are normally distributed, some theory exists that allows us to establish the relationship between the standard deviation of ranges σ_R and the standard deviation of measurements, σ_X.

In general, the distribution of the range may be determined by deriving the distribution of the difference between the largest and smallest values in the sample. By using knowledge of the distribution of the individuals X, such a determination can be made, although the result is much more complicated than that for the $\overline{X}$'s.

For simplicity we will place $+3$ and -3 sigma limits about the average range $\overline{R}$ of a number of sample ranges R to define the upper and lower control limits for R even though the frequency distribution of ranges is not a nice symmetric distribution. Hence the upper and lower control limits for the R chart are given, in principle, by

$$\overline{R} \pm 3\sigma_R.$$

Again, assuming that the quality characteristic X is normally distributed, statistical theory has established the relationship between the standard deviation of the range and the standard deviation of the quality characteristic X:

$$\sigma_R = d_3\sigma_X,$$

where d_3 is a known function of n, the subgroup size. In Chapter 3 the following relationship was provided to relate the true mean range, $E(R)$, and σ_X:

$$\sigma_X = \frac{E(R)}{d_2}.$$

This relation may be used to establish an expression for σ_R as a function of the true mean range:

$$\sigma_R = \frac{d_3 E(R)}{d_2}.$$

Values for d_2 for different sample sizes, n, can be found in Table A.2 in the Appendix. When the true mean range is estimated by $\overline{R}$, we have that

$$\hat{\sigma}_R = \frac{d_3 \overline{R}}{d_2}.$$

Therefore, 3 sigma control limits for the ranges are given by

$$\overline{R} \pm 3\left(\frac{d_3}{d_2}\right)\overline{R}.$$

These control limits reduce to the forms

$$\text{UCL} = D_4\overline{R},$$
$$\text{LCL} = D_3\overline{R},$$

where D_4 and D_3 are known functions of n. Values for D_4 and D_3 are tabulated in Table A.2 and their use will be illustrated later in this chapter. The interpretation of the R chart parallels that of the $\overline{X}$ chart previously discussed—that is, the hypothesis testing approach enhanced by Shewhart to consider the time order of the data/process.

6.5   Setting Up $\overline{X}$ and R Control Charts

Once the statistical basis for Shewhart control charts has been established, it remains to define the necessary elements of the charts mathematically and to establish a standard graphical representation. This section deals with the construction of $\overline{X}$ and R control charts. All of the necessary equations and general procedures for the calculation and graphical representation of the basic elements of the charts are presented in this section.

The first step in setting up $\overline{X}$ and R control charts is the selection of the samples. It is important that all samples are *rational samples*, which are groups of measurements whose variation is attributable only to a constant system of common causes. We must strive to choose our samples in such a way as to minimize the occurrence of special causes within the sample. By so doing, we maximize the opportunity to detect special causes when they occur between samples.

Rational sampling is a crucial concept; an entire chapter (Chapter 7) is devoted to this subject. Sampling from different machines, sampling over extended periods of time, and sampling from products combined from several sources are all nonrational sampling methods and must be avoided. As a good rule of thumb, one should select a minimum of 25, and preferably 40–50 samples/subgroups to provide a solid basis for the initiation of control charts. This serves to provide sufficient data for the estimation of the process mean and process variability with small sampling errors. If the number of subgroups used is too small, say 10 to 20, the sampling errors associated with $\overline{X}$ and $\overline{R}$ may be large and could lead to misguided conclusions about the state of the process. For reasons to be discussed later, the sample/subgroup size should be relatively small; between $n = 3$ and $n = 6$ is common. A sample size of $n = 5$ is very commonly used. Once the process is in a state of statistical control, periodic recalculation of the control limits to update the information is a good practice.

The following simple steps define the procedure necessary to set up $\overline{X}$ and R control charts.

1. For each sample, an average is calculated:

$$\overline{X}_i = \sum_{j=1}^{n} X_{ij}/n, \tag{6.1}$$

where X_{ij} is the jth measurement, $j = 1, \ldots, n$, in the ith sample and where n = sample size.

2. The spread or dispersion within the ith sample is measured by the range R_i:

$$R_i = X_{i,\text{largest}} - X_{i,\text{smallest}}. \tag{6.2}$$

Because of its simplicity and ease of calculation, the range, R, is commonly employed as a measure of within-sample variability.

3. The grand average, $\overline{\overline{X}}$, is the arithmetic average of all the available sample averages. This grand average is an estimate of the process mean μ_X and becomes the centerline of the $\overline{X}$ control chart:

$$\overline{\overline{X}} = \frac{\displaystyle\sum_{i=1}^{k} \overline{X}_i}{k}, \tag{6.3}$$

where k is the number of samples being used to set up the control chart.

4. The average of the sample ranges $\overline{R}$ is an estimate of the mean range, $E(R)$, and becomes the centerline of the R control chart:

$$\overline{R} = \frac{\displaystyle\sum_{i=1}^{k} R_i}{k}. \tag{6.4}$$

5. We are now ready to calculate the trial control limits for the $\overline{X}$ chart. The limits will be in the form

$$\text{grand average} \pm 3 \text{ sigma of sample averages.} \tag{6.5}$$

The grand average $\overline{\overline{X}}$ is the centerline of the $\overline{X}$ chart. We recall from our previous discussion of basic principles in Chapter 4 that

$$\sigma_{\overline{X}} = \frac{\sigma_X}{\sqrt{n}} \tag{6.6}$$

and

$$\sigma_X = \frac{E(R)}{d_2}. \tag{6.7}$$

The standard deviation of sample means is therefore estimated by

$$\hat{\sigma}_{\overline{X}} = \frac{\overline{R}}{d_2 \sqrt{n}}. \tag{6.8}$$

Thus the formulas for the upper and lower limits of the $\overline{X}$ control chart are

$$UCL = \overline{\overline{X}} + \frac{3\overline{R}}{d_2 \sqrt{n}}$$

$$LCL = \overline{\overline{X}} - \frac{3\overline{R}}{d_2 \sqrt{n}}. \tag{6.9}$$

Equation (6.9) can be further simplified. The $3/d_2 \sqrt{n}$ term, which is dependent solely on the sample size n, can be combined into a single constant A_2. This gives us

$$UCL = \overline{\overline{X}} + A_2 \overline{R}$$

$$LCL = \overline{\overline{X}} - A_2 \overline{R}. \tag{6.10}$$

Values for A_2 for different sample sizes n are conveniently tabulated for us and are given in Table A.2 in the Appendix.

6. Note that the distribution for R is not easily determined and is not symmetric about its mean value. However, in the interest of simplicity and ease in establishing the R chart, symmetrical $\pm 3\sigma_R$ limits have been adopted. If the lower limit is less than 0, a lower control limit of 0 is used. As mentioned previously, σ_R can be estimated as a function of $\overline{R}$. Ultimately, the control limit formulas for the R chart take the form

$$UCL = D_4 \overline{R}, \quad LCL = D_3 \overline{R}. \tag{6.11}$$

Values for D_3 and D_4 have been tabulated as a function of the sample size n and are also given in Table A.2 in the Appendix.

6.6   Interpretation of Shewhart Control Chart Patterns

General Patterns Indicating Out-of-Control Conditions

The control chart idea is to monitor the process by periodically drawing small samples from production and estimating the process mean and process variability from the sample by the sample mean $\overline{X}$ and the range R of the sample. Such data serve as the basis for establishing Shewhart $\overline{X}$ and R charts.

But before the charts can be used to identify the presence of special causes on line, the process must be brought into a state of statistical control; that is, the data used to establish the charts and define the control limits must be subject only to a constant system of common causes. Once control is established, the charts may be employed to monitor the future behavior of the process. A rigorous examination of the patterns in the data may be quite fruitful in terms of the identification of quality and productivity improvement opportunities.

Too often, $\overline{X}$ and R control charts are examined only for the presence of extreme points (i.e., sample results that fall outside the control limits). However, the Shewhart control chart model suggests that other conditions should also prevail and the absence of these conditions can signal the presence of special causes—chaotic process disturbances.

Since the distribution of $\overline{X}$ is normal, serious departures from normality can signal the presence of special/assignable causes even if all points are within the control limits. Too many points near the control limits, for example, may signal problems with the process, such as an overcontrol of the process mean level. The occurrence of a trend or recurring cycles in the $\overline{X}$'s can indicate special-cause problems that need attention, such as operator fatigue, excessive tool wear, or systematic adjustment of the process.

Runs of points above or below the centerline on either the $\overline{X}$ or R chart may be indicating shifts in mean level or amount of variability, which may not be large enough to manifest themselves quickly through points outside the control limits. It is essential that we strive to associate with each statistical signal of an out-of-control condition the basic process fault that is producing it. Some useful generic conditions to look for might include:

Trends/Cyclic Behavior. Systematic changes in the process environment, worker fatigue, maintenance schedules, wear conditions, accumulation of waste material, and contamination.

High Proportion of Points Near or Beyond Control Limits. Overcontrol of the process, large differences in incoming raw materials, and/or charting more than one process on a single chart.

Sudden Shifts in Level. New machine/die/tooling, new worker, new batch of raw material, change in measurement system, and/or change in production method.

Figure 6.5 shows $\overline{X}$ and R charts constructed from 50 samples of size $n = 5$ measurements, each drawn in succession from a manufacturing process. Examination of the variability patterns in the charts indicates no reason to suspect that the process is not in good statistical control because:

1. No points exceed the control limits.
2. The points are approximately normally distributed within the limits.
3. The points show no evidence of trends or recurring cycles.
4. The points look quite random with time; that is, no patterns such as runs above or below the centerline are evident.

It is important that we fully appreciate the appearance of a process in good statistical control. We now examine some very specific tests that cover the four items outlined above.

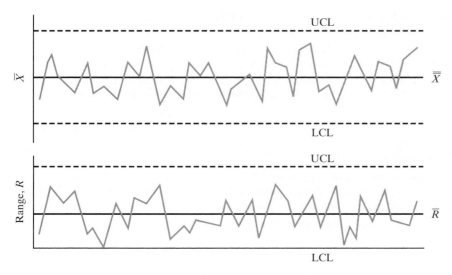

Figure 6.5   **Appearance of a process in good statistical control**

Tests for Control Chart Interpretation

In the discussion above, a number of general control chart patterns have been presented that indicate various types of unnatural process behaviors. Often, these unnatural patterns (e.g., extreme points, trends/cycles, or runs above or below the centerline) are fairly obvious through even the most cursory examination of the charts. However, sometimes special causes produce unnatural patterns that are a bit more subtle, and therefore a more rigorous pattern analysis should generally be conducted.

Several useful tests for the presence of unnatural patterns (signals of special causes) can be performed by dividing the distance between the upper and lower control limits on the $\overline{X}$ chart into zones defined by 1, 2, and 3 sigma boundaries, as shown in Fig. 6.6. Such zones are useful, since the statistical distribution of $\overline{X}$ follows a normal distribution if the process is in control, and therefore we expect

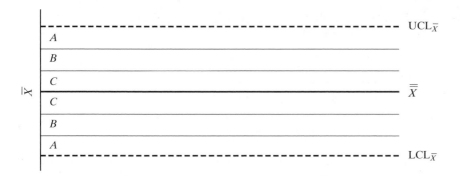

Figure 6.6   **Control chart zones to aid chart interpretation**

that certain proportions of the points fall within plus or minus 1 sigma, between 1 and 2 sigma, and so on.

We will now discuss eight specific tests that may be applied to the interpretation of $\overline{X}$ and R control charts.[1] These tests provide the basis for the statistical signals that indicate to us that the process has undergone a change in its mean level, variability level, or both. Some of the eight tests are based specifically on the zones defined in Fig. 6.6 and apply only to the interpretation of the $\overline{X}$-chart patterns. The other tests apply to both charts. In these cases the zones are not used. Unless specifically identified to the contrary, the tests/rules apply to the consideration of data to one side of the centerline only.

When a sequence of points on a chart violates one of the rules, the last point in the sequence is circled. This signifies that the evidence is now sufficient to suggest that a special cause has occurred. The issue of when that special cause actually occurred is another matter. A logical estimation of the time of occurrence may be the beginning of the sequence in question. Other interpretations are possible.

Test 1: Extreme Points

Test 1 concerns the identification of extreme points—points beyond the control limits—and therefore applies to both the $\overline{X}$ and R control charts. The specific rule is: The existence of a single point beyond a control limit signals the presence of an out-of-control condition. Figure 6.7 shows several examples of the occurrence of a special cause as indicated by this test.

Test 2: Two Out of Three Points in Zone A or Beyond

Test 2 is based on the specific control chart zones and therefore applies only to the $\overline{X}$ chart. The specific rule is: The existence of two of any three successive

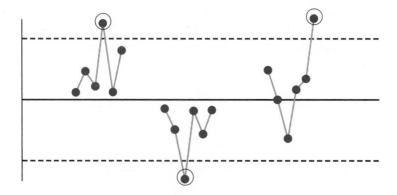

Figure 6.7   **Examples of test 1: extreme points**

---

[1]An excellent article on this subject has been written by L. S. Nelson, "The Shewhart Control Chart: Tests for Special Causes," *Journal of Quality Technology*. Vol. 16, No. 4, 1984, pp. 237–239.

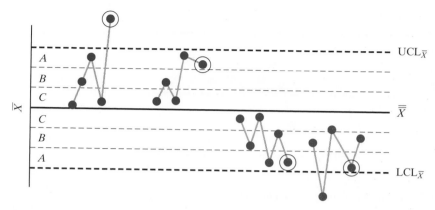

Figure 6.8   Examples of test 2

points in zone A or beyond signals the presence of an out-of-control condition. Figure 6.8 shows several examples of how this test is applied.

Test 3: Four Out of Five Points in Zone B or Beyond

The probabilistic basis for test 3 is the same, in principle, as that of test 2 and hence the test applies only to the interpretation of the $\overline{X}$ chart. The rule states: Four out of five successive points in zone B or beyond signals the presence of an out-of-control condition. Figure 6.9 illustrates the application of this test. As with test 2, this test may help to detect smaller shifts in the process mean which do not give rise to extreme points.

Test 4: Runs Above or Below the Centerline

Test 4 considers long runs (eight or more successive points) either strictly above or strictly below the centerline and therefore applies to both $\overline{X}$ and R charts. The presence of such a run indicates strong evidence that the process mean or

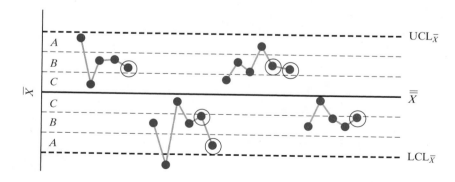

Figure 6.9   Test 3: four-out-of-five rule

Figure 6.10   Runs above or below the centerline test

variability has shifted from the centerline. Figure 6.10 illustrates the run test. As the run continues beyond eight, each additional point is circled, as it represents the endpoint of eight successive points above or below the centerline.

Test 5: Linear Trend Identification

When six successive points on either the $\overline{X}$ or the R chart show a continuing increase or decrease, a systematic trend in the process is signaled. Figure 6.11 shows the occurrence of such trends. Neither the zones nor the centerline come into play for this test.

Test 6: Oscillatory Trend Identification

In the spirit of test 5, when 14 successive points oscillate up and down on either the $\overline{X}$ or R chart, a systematic trend in the process is signaled. Figure 6.12 shows the application of this test. Again neither the chart centerline nor the zones come into play for this test.

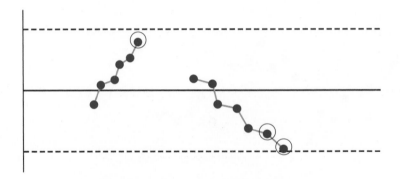

Figure 6.11   Test 5: upward/downward trend test

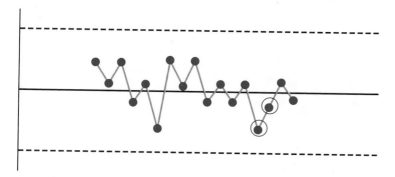

Figure 6.12   Test 6: oscillatory trend test

Test 7: Avoidance of Zone C

When eight successive points occurring on either side of the centerline avoid zone C (Fig. 6.13), an out-of-control condition is signaled. This rule applies only to the $\overline{X}$ chart. This test could be signaling the occurrence of more than one process being charted on a single chart or perhaps overcontrol of the process. It may also be indicating that improper sampling techniques are being used—in particular, process mixing. This will be discussed later.

Test 8: Run in Zone C

When 15 successive points on the $\overline{X}$ chart fall only in zone C, to either side of the centerline, an out-of-control condition is signaled. Such can arise from improper sampling (to be discussed later) or from a change (decrease) in process variability that has not been properly accounted for in the $\overline{X}$-chart control limits. Figure 6.14 illustrates the application of this test.

Using the Tests to Interpret the Charts

The series of tests above are to be applied jointly in interpreting the charts. Several rules may be simultaneously violated for a given $\overline{X}$ or R value, and hence

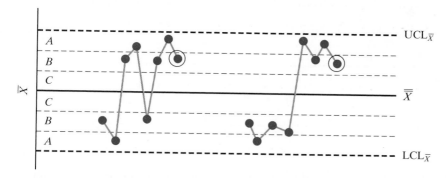

Figure 6.13   Application of test 7: mixing/overcontrol

Figure 6.14   Application of test 8: stratification/reduced variability

that point may be circled more than once. Figure 6.15 illustrates such a situation. In the figure, point A is circled twice because it is the endpoint of a run of eight successive points above the centerline and the endpoint of four of five successive points in zone B or beyond. Another point in this grouping should be circled. Which one, and why? In the second grouping in Fig. 6.15, point B is circled three times because it is the endpoint of:

1. A run of eight successive points below the centerline.
2. Two of three successive points in zone A or beyond.
3. Four of five successive points in zone B or beyond.

Other points in this second grouping should also be circled. Which ones, and why?

The tests discussed above are important to a full interpretation of control chart patterns. These tests may easily be programmed to be carried out by computer. It should be noted that while these tests do improve the sensitivity of the

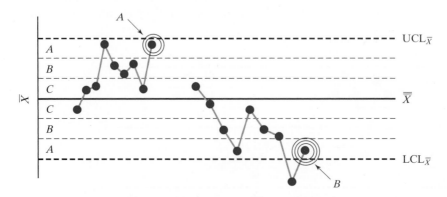

Figure 6.15   Examples of the simultaneous application of more than one test for out-of-control conditions

charts, they increase the α-risk as well. Some work has been done to quantify the specific nature of the α-risks associated with several of these tests.[2]

6.7   Closing the Loop Using the Process of SPC

In previous sections of this chapter we concentrated on only one facet of the complete process of SPC, namely, the evaluation stage (refer to Fig. 6.1). We have established procedures for identifying statistical signals to determine whether changes in the process variability and/or the mean might have taken place. We now present some simple but powerful graphical methods that will help us to take the crucial step from reading statistical signals to diagnosing the cause of the problem (the diagnosis stage of Fig. 6.1).

There are four graphical methods that are particularly useful during the fault diagnosis stage of a quality/productivity improvement study: the flowchart, the Pareto diagram, the cause-and-effect diagram, and the scatter diagram. All of these tools help to focus attention on issues and events relevant to the process at hand and relate effects or symptoms to the factors that are causing them.

Flowcharts

A flowchart is a visual representation of a process. Flowcharts employ a variety of symbols and geometric representations to depict the interrelationships among the various activities within a process, including important precedence relationships. By decomposing the process into its elemental steps or activities, flowcharts can be useful in identifying system flaws, errors, and non-value-added activities. Since one of the principal purposes of the flowchart is to communicate information, it is desirable to use only a few basic symbols that are easily understandable to construct a flowchart. Figure 6.16 shows a simple flowchart example.

A flowchart can be a very useful diagnostic and interpretive tool; it reflects the collective wisdom of the quality improvement team. A flowchart should be constructed to illustrate the manner in which the process is actually performed,

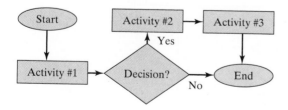

Figure 6.16   **Structure of a simple process flowchart**

[2]G. S. Rahn, S. G. Kapoor, and R. E. DeVor, "Development of Operational Characteristics and Diagnostics for $\overline{X}$-Control Charts," *Proc. of the Sym. on Sensors, Controls, and Quality Issues in Manuf.*, ASME, Dec. 1991, Atlanta, GA.

as opposed to describing the way it is supposed to be executed. The steps within an accounts payable process are shown in the flowchart of Fig. 6.17.[3] This process will be examined in more detail in Chapter 14.

Pareto Diagrams

A Pareto diagram is simply a type of bar chart or histogram that helps us to understand the extent to which a particular phenomenon is occurring relative to the occurrence of others. As such, it helps us to focus on the most frequently occurring problems and prioritize our efforts toward problem solving. This type of diagram is styled after the *Pareto principle*, which says essentially that while the sum total of a certain type of occurrence may arise from quite a number of different sources, it is likely that the vast majority of the occurrences arise from a relatively small subset of these sources. The idea is that if we can identify these few key sources, we can focus our energy on them, thereby hopefully solving most of the problem.

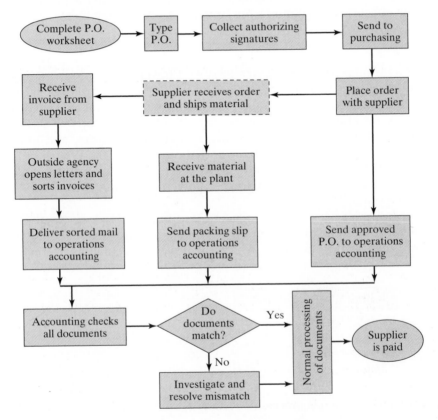

Figure 6.17   Flowchart for an accounts payable process

---

[3]R. J. Deacon, "Quality Improvement in an Accounts Payable System," Independent Study Project Report, Eastern Michigan University, College of Business, Ypsilanti, MI, 1986.

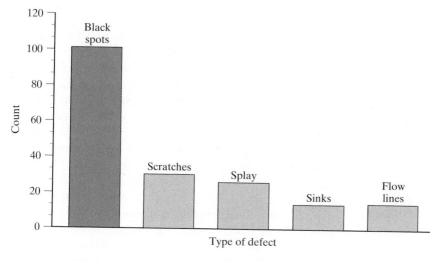

Figure 6.18   Pareto diagram of molding defects

Consider, for example, an injection molding process and the associated defects that can arise on a molded part. These defects could include splay, scratches, flow lines, sinks, and black spots on the surface. We might set up a sampling system whereby we take five successive parts every hour from the production and carefully inspect them for these five types of defects. For each defect we carefully note every incidence of its occurrence. We might be keeping a Shewhart control chart on the number of defects per sample over time. Suppose that the number of defects seems to be fairly stable over time, but the total number is high and we wish to investigate this and perhaps formulate some strategy for the reduction of defects. What should we do first? Which defect or defects should we focus our attention on?

Figure 6.18 is a bar chart of the frequency of occurrence of the various types of defects over a certain period of time. This chart, a Pareto diagram, plots the type of defect along the horizontal axis and the frequency of occurrence of each defect along the vertical axis. The frequencies are plotted in descending order from left to right. The data in Fig. 6.18 are from an actual study conducted in a plant of the Ford Motor Company.[4] This study is presented and discussed in considerable detail in Chapter 14. From Fig. 6.18 it is clear that over one-half of the total defects are of the black-spots type. Since we generally have limited resources to apply to the problem and we would like to make as much improvement as possible and as rapidly as possible, it seems that we might focus our attention on the black-spots issue. As we will see in Chapter 14, this was the course of action chosen, in this case with considerable success.

---

[4] R. J. Deacon, "Reduction of Visual Defects in an Injection Molding Process," *Society of Plastics Engineers, 1983 National Technical Conference Proceedings*, p. 97.

Cause-and-Effect Diagrams

If it is the Pareto diagram that helps us to prioritize our efforts and focus attention on the most pressing problem or symptom, it is the cause-and-effect diagram that helps to lead us to the root cause of the problem. The cause-and-effect diagram is a tool that enables the user to set down systematically a graphical mapping of potential causes for an observed effect or symptom that ultimately leads to the root cause. The cause-and-effect diagram associates with the particular quality concern or symptom the possible factors or causes driving it. This is accomplished in a hierarchical sense, from the broadest of the factor categories to the most specific or lowest-level causal factor under each broad factor.

To construct a cause-and-effect diagram, we first identify the symptom or effect for which we wish to find the root cause. We then develop a small set of broad factors that could be producing the symptom, e.g., the machine, the raw material, the method or process procedures, and the human element in the process. The structure of the cause-and-effect diagram, given these factors, is shown in Fig. 6.19. The cause-and-effect diagram is sometimes referred to as a "fishbone" diagram because of its general appearance. To this coarse "skeleton" we begin to add smaller "bones" by continuing to ask probing questions about the various causes of the problem. Why do we think the machine might be the problem? Perhaps it is a problem with a bearing. What specifically is the problem with the bearing? and so forth. We continue to ask questions until we are at a level of specificity where the root cause can be isolated. This degree of specificity is ultimately needed to allow for the development of an action plan.

Let us return to the injection molding case study. The symptom being focused on is the "black-spot" defect, and attention centers on trying to find the precise cause(s) of black spots. Figure 6.20 shows the fully developed cause-and-effect diagram for the injection molding black-spot defect concern. Many possible causes of the problem have been identified. It remains to focus on one or two causes for which an improvement action(s) can be developed.

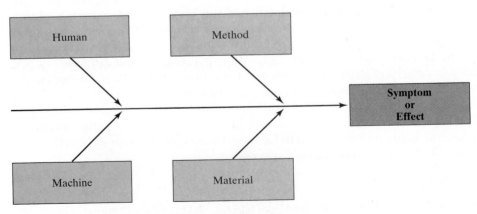

Figure 6.19 **Structure of the cause-and-effect diagram**

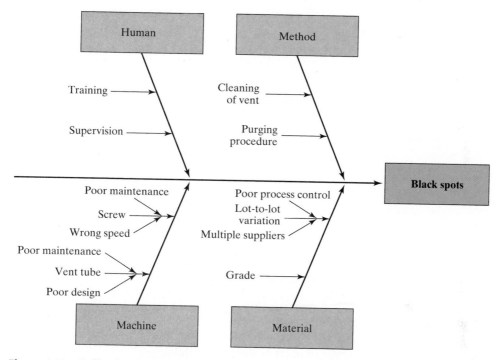

Figure 6.20   Fully developed cause-and-effect diagram for the black-spot symptom

Cause-and-effect diagrams have several distinct advantages:

- They place attention where it ought to be, on looking for solutions to problems at the root cause level rather than on finding ways to treat the symptom.
- They are educational; that is, they summarize the collective wisdom of a group of experts/people who touch the system under study and allow that wisdom to be shared, thereby elevating everyone's understanding of the problem.
- They provide a means to keep the discussion continually focused and converging toward a solution rather than rambling and divergent.
- They provide everyone involved with a quick and clear measure of the level of knowledge currently available on a particular problem. If the diagram does not have very many levels of detail but rather many main factors, this is an indication that either the problem is not well understood or that the right people are not in the room, or both.

Scatter Diagrams

The scatter diagram is a graphical representation aimed at helping to identify correlation that might exist between a quality or productivity performance measure and a factor that might be driving the measure. Figure 6.21 shows one general structure that is commonly used for the scatter diagram. The performance

Figure 6.21   General structure for a scatter diagram

measure variable (Y) is usually plotted on the vertical axis, while the suspected driving variable (X) is usually plotted on the horizontal axis. The point where the axes cross is the average of each of the sets of data.

The basic idea of the scatter diagram is to determine if the particular values of the performance measure seem to depend on the values of the identified independent variables. Although there are a number of more quantitative ways to determine this, the graphical approach is quick, instructive, easy to communicate to others, and generally easy to interpret. Even in the case of the graphical approach, some numbers may be determined—counting the number of occurrences in each quadrant of the plot, for example—to help pin down the presence or absence of correlation. Most often, however, a more qualitative evaluation is made. Figure 6.22 illustrates two hypothetical sets of data plotted on scatter diagrams.

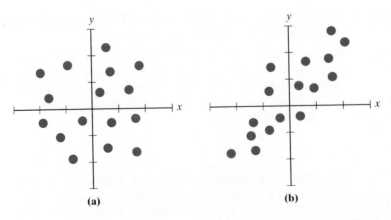

Figure 6.22   Examples of no correlation (a) and positive correlation (b)

In plot (a), the scatter appears quite random and hence we conclude that the two factors seem to have arisen in an unrelated way. In plot (b), however, there appears to be a strong correlation. In particular, the higher the value of X, the higher the value of Y. This is referred to as *positive correlation*. If the plot looked somewhat the same except that the slope of the data was negative, the correlation would be said to be *negative correlation*.

Suppose that $\overline{X}$ and R charts are being kept on the surface finish of a machined part. Further suppose that statistical signals of out-of-control conditions are evident on the $\overline{X}$ chart, while the R chart shows good statistical control. The question is: What is causing the out-of-control conditions on the $\overline{X}$ chart? Suppose that each time a sample was taken for the construction of the charts, observations were also made to record certain key process conditions, such as cutting speed, feed rate, and tool condition. The availability of the latter data provides us with the opportunity to correlate what we see on the charts with what is actually going on in the process.

Suppose, for example, that we think that the out-of-control signals on the chart are driven by changes that might be occurring in the tool wear condition. To examine this hypothesis, we could prepare a scatter diagram that displays $\overline{X}$ values on the vertical axis and the corresponding tool condition on the horizontal axis. These pairs may be plotted in the manner shown in Fig. 6.23. From the figure one might conclude that there is some correlation between the two variables; in particular, the dull tool condition tends to give rise to a smoother surface (i.e., a lower surface roughness value).

The scatter diagram clearly points to the need to collect data in such a way so as not merely to observe the quality characteristic of interest at points in time during production but also to observe the various states of the process which may be having an impact on the quality characteristic. Such information then stands as a basis for us to track down the cause of problems when they occur.

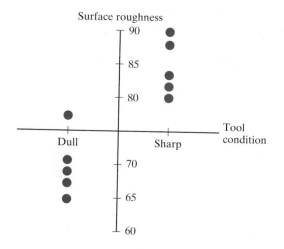

Figure 6.23   **Scatter diagram for surface roughness versus tool condition**

6.8   Cylinder Liner Boring Case Study—Demonstrating the Process of SPC

The following case study demonstrates the application of $\overline{X}$ and R control charts to the study of a manufacturing operation. As will be seen, the case study uses control charts to identify the presence of faults within a cylinder boring operation. It also demonstrates how the simple graphical techniques that have been presented in this chapter may be utilized to find the root causes of the faults and eliminate them. The case study illustrates the complete process of SPC to bring a process into a state of statistical control.

The Situation

As part of recent management change, a supplier to an engine manufacturer has initiated a companywide quality improvement strategy. The company's principal product is a cast iron cylinder liner (or sleeve) that is inserted into the aluminum block produced by the engine manufacturer. Given the reliance of the liner company on this single class of products, it needs to respond quickly to the ever-increasing expectations of the customer. In fact, word has it that the engine manufacturer soon plans to announce new, more stringent specifications for the liner. Given this background, the company decided to investigate the liner production process. A vertically and horizontally integrated study team was formed and asked to identify sources of variation that may pose potential quality problems and negatively impact cost/productivity. The first action taken by the team was to construct a flow chart of the cylinder liner production process (Fig. 6.24).

In examining Fig. 6.24, it is evident that the liners are produced by a centrifugal casting operation and then heat treated. Machining operations are then performed on the liners, i.e., rough turning of outer diameter (OD), turning of external features to manufacturer specifications, and rough boring of liner inner diameter (ID). Each liner is inspected, then an oil film is applied to prevent rusting, and finally the liners are packaged and shipped to the engine manufacturer. It may be noted that once the liners are received at the engine manufacturer, they are inserted as cores in the block casting operation. The engine manufacturer subsequently performs a finish ID boring operation, and a honing operation on the cylinder (liner) wall surface.

The planned new specifications were applied to a subset of the historical liner data to determine which quality characteristics would be affected by the

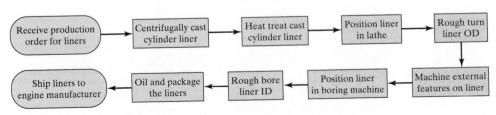

Figure 6.24   **Flow chart of cylinder liner manufacturing process**

Figure 6.25   Pareto chart for cylinder liner defects

more stringent liner specifications. The data analysis revealed instances where liner quality characteristics would fail to meet the new specifications (a defect or nonconformity). For the specific subset of the liner data examined, the defects were grouped by type; the study team then prepared the Pareto chart shown in Fig. 6.25 for the defect data.

An examination of Fig. 6.25 shows that when the new liner specifications are applied to the historical data, a large number of the liner inner diameter (ID) values fall outside the specifications. The Pareto chart reveals that this "defect" occurs most frequently. This revelation was extremely useful to the quality study team; it provided insight into which quality characteristic would be the most sensitive to the planned change in liner specifications. The focus of the study team then shifted to inventorying those factors that could contribute to an out-of-spec liner ID defect. This inventory was captured via the cause-and-effect diagram displayed in Fig. 6.26. The diagram summarizes the collective knowledge relative to potential root causes of the liner ID quality concern. In this case, the construction of the diagram significantly benefited from the involvement of a diverse set of team members. The vertical nature of the team provided differing viewpoints on the system in terms of detail/perspective, and the horizontal character of the team brought together experts from across the liner manufacturing system.

As the team examined and discussed the cause-and-effect diagram of Fig. 6.26, much of the attention centered on the cylinder liner boring process that produces the rough ID. In the figure, four of the categories were associated with the boring operation: tooling, boring machine, machine operator, and fixturing. It was decided to collect new data to assess the stability of the boring process through the construction of control charts. The data collected from the boring process were diameter measurements made to the nearest 1/10,000 of an inch. Samples of size $n = 5$ were taken to obtain some data to initiate $\overline{X}$ and R control charts.

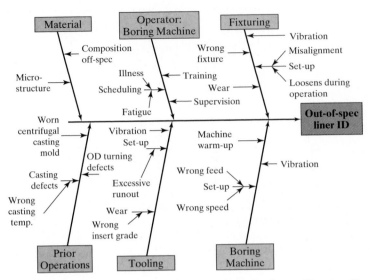

Figure 6.26   **Cause-and-effect diagram for out-of-specification liner ID quality concern**

Initial Control Charts

Forty samples were collected to construct the control charts, as shown in Table 6.1. Samples were collected approximately every half-hour. Each value in the table is the average of three diameter measurements made at various positions inside the liner bore. The measurements are of the form 3.5XXX inch, and the table provides only the last three digits of the recorded value. As is evident from the table, in addition to information on the inner diameter (ID) produced by the boring process, information was also recorded on when the sample was collected from the process. It is important to record data such as the time of production, operator name, material batch, and other measures of the "process state" because such measures are vital to diagnosing the root cause of signals manifested on the charts. In the absence of process state information, it is extremely difficult to "do the detective work" and determine what process changes produced signals evident on the charts. The data in Table 6.1 were used to construct initial control charts.

Determination of Control Limits

For each sample in Table 6.1, the sample mean, $\overline{X}$, and sample range, R, were calculated using Eqs. (6.1) and (6.2). The grand average, $\overline{\overline{X}}$, and average range, $\overline{R}$, were calculated using Eqs. (6.3) and (6.4).

$$\overline{\overline{X}} = 200.25,$$
$$\overline{R} = 7.60.$$

The control limits for the $\overline{X}$ and R charts were determined using Eqs. (6.10) and (6.11). These calculations are summarized below.

	TABLE 6.1	Initial Set of Cylinder Liner Boring Process Data							
Sample Number	Sample Collection Time	X_1	X_2	X_3	X_4	X_5	$\overline{X}$	R
1	8:00 A.M.	205	202	204	207	205	204.6	5
2	8:25 A.M.	202	196	201	198	202	199.8	6
3	8:50 A.M.	201	202	199	197	196	199.0	6
4	9:15 A.M.	205	203	196	201	197	200.4	9
5	9:40 A.M.	199	196	201	200	195	198.2	6
6	10:05 A.M.	203	198	192	217	196	201.2	25
7	10:30 A.M.	202	202	198	203	202	201.4	5
8	10:55 A.M.	197	196	196	200	204	198.6	8
9	11:20 A.M.	199	200	204	196	202	200.2	8
10	11:45 A.M.	202	196	204	195	197	198.8	9
11	1:00 P.M.	205	204	202	208	204	204.6	6
12	1:25 P.M.	200	201	199	200	201	200.2	2
13	1:50 P.M.	205	196	201	197	198	199.4	9
14	2:15 P.M.	202	199	200	198	200	199.8	4
15	2:40 P.M.	200	200	201	205	201	201.4	5
16	3:05 P.M.	201	187	209	202	200	199.8	22
17	3:30 P.M.	202	202	204	198	203	201.8	6
18	3:55 P.M.	201	198	204	201	201	201.0	6
19	4:20 P.M.	207	206	194	197	201	201.0	13
20	4:45 P.M.	200	204	198	199	199	200.0	6
21	8:10 A.M.	203	200	204	199	200	201.2	5
22	8:35 A.M.	196	202	197	201	195	198.2	7
23	9:00 A.M.	197	199	203	200	196	199.0	7
24	9:25 A.M.	202	197	196	199	206	200.0	10
25	9:50 A.M.	202	198	201	199	197	199.4	5
26	10:15 A.M.	203	197	199	197	201	199.4	6
27	10:40 A.M.	204	203	199	199	197	200.4	7
28	11:05 A.M.	198	201	201	195	200	199.0	6
29	11:30 A.M.	201	196	197	204	200	199.6	8
30	11:55 A.M.	203	207	201	195	201	201.4	12
31	1:05 P.M.	206	206	199	200	203	202.8	7
32	1:30 P.M.	197	194	199	200	199	197.8	6
33	1:55 P.M.	200	201	200	197	200	199.6	4
34	2:20 P.M.	199	198	201	202	201	200.2	4
35	2:45 P.M.	200	204	197	197	199	199.4	7
36	3:10 P.M.	195	203	202	210	197	201.4	15
37	3:35 P.M.	205	199	202	201	200	201.4	6
38	4:00 P.M.	193	200	202	196	199	198.0	9
39	4:25 P.M.	195	199	199	198	198	197.8	4
40	4:50 P.M.	201	204	203	202	204	202.8	3

For the $\overline{X}$ chart,

$$\text{UCL} = \overline{\overline{X}} + A_2\overline{R} = 200.25 + (0.577)(7.60),$$

$$\text{LCL} = \overline{\overline{X}} - A_2\overline{R} = 200.25 + (0.577)(7.60),$$

$$\text{UCL} = 204.64, \qquad \text{LCL} = 195.86.$$

For the R chart,

$$\text{UCL} = D_4\overline{R} = (2.115)(7.60),$$

$$\text{LCL} = D_3\overline{R} = (0)(7.60),$$

$$\text{UCL} = 16.074, \qquad \text{LCL} = 0.0.$$

The appropriate values for A_2, D_3, and D_4 are obtained from Table A.2.

With the calculations above completed, we may proceed to construct the $\overline{X}$ and R charts, following a few simple rules.

1. The individual $\overline{X}$ and R values are plotted as solid dots that are connected from sample to sample. This will be helpful in revealing the patterns in the data.
2. Always plot the R chart directly below the $\overline{X}$ chart using the same horizontal axis scale. This makes it easy to compare $\overline{X}$ and R results for individual samples.
3. Use a heavy solid line to denote the centerline of each chart.
4. Use a heavy dashed line to denote the control limits.
5. Write the specific numerical values for the control limits on the charts as well as for the centerlines.
6. Circle any points that extend beyond the control limits and any points that identify a nonrandom pattern (i.e., points that violate any of the tests previously discussed). Points should be circled for each rule that is violated.

It is essential to start with the R chart. This is necessary since the limits of the $\overline{X}$ chart depend on the magnitude of the common-cause variation measured by $\overline{R}$ [recall Eq. (6.10)]. If, initially, some points on the R chart exceed the upper control limit, or other indications of special causes are present, the limits in the $\overline{X}$ chart will be inflated (or perhaps deflated) and may therefore project misleading information about the behavior of the process mean. Figure 6.27 shows the $\overline{X}$ and R charts based on the trial limits determined above.

Interpretation of the Initial Charts

Initially, as we examine the R chart, we see that two points exceed the upper control limit. From this, it may be concluded that there are special causes producing an increase in the process variability, at least at those points. The study team closely examined these points—samples 6 and 16—to see if they could identify a

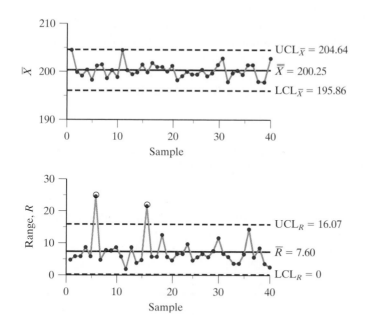

Figure 6.27   Initial control charts for cylinder liner boring process

reason (special cause) for these statistical signals. To focus discussions related to potential special cause(s), the previously constructed cause-and-effect diagram was employed. Discussions relative to the fishbone diagram suggested that the periodic replacement of the regular operator with a substitute was a strong candidate for the root cause of the R chart signals. The study team found the data that they had collected on the process state to be especially useful in checking this hypothesis. The records show that a substitute operator filled in for the regular operator during the midmorning and midafternoon breaks (i.e., 10:00–10:15 A.M. and 3:00–3:15 P.M.). An examination of the control chart data reveals that samples 6, 16, and 36 were produced while the substitute operator was in charge. It may be noted that only samples 6 and 16 manifested themselves as out-of-control conditions on the R chart. Nevertheless, if the substitute operator is indeed the root cause of the increased process variability, then in fact all three samples are subject to the same special cause. This information supports the hypothesis that the substitute operator is the root cause of the R chart signals.

The diagnosis step in the *Process of SPC* has identified the substitute operator as a strong candidate for the source of the R chart signals. Attention now turns to the formulation of a corrective action and its implementation. Discussion with line supervisors and regular and substitute operators uncovered the fact that the substitute operator had received virtually no training in the operation of the cylinder boring machine. The substitute was inexperienced and had not been instructed on how to apply coolant, position the tool, etc. As a corrective action, the study team worked to ensure that the substitute was provided with training on proper machine operation. Of course, only by collecting additional samples from the process can it truly be determined (1) that the substitute was the true root cause of the R chart signals, and (2) that the corrective action is effective in removing special causes from the process.

Control Charts after Addressing Operator Special Cause

Once the substitute operator had received training regarding proper operation of the boring machine, the study team collected an additional 40 samples from the liner boring process, as shown in Table 6.2. Data on the process conditions under which each sample was collected were again recorded.

Determination of Control Limits

The centerlines and the control limit calculations for the charts based on the data in Table 6.2 are

$$\overline{\overline{X}} = 200.24,$$
$$\overline{R} = 6.55.$$

For the $\overline{X}$ chart,

$$\text{UCL} = \overline{\overline{X}} + A_2\overline{R} = 200.24 + (0.577)(6.55),$$
$$\text{LCL} = \overline{\overline{X}} - A_2\overline{R} = 200.24 - (0.577)(6.55),$$
$$\text{UCL} = 204.02, \qquad \text{LCL} = 196.46.$$

For the R chart,

$$\text{UCL} = D_4\overline{R} = (2.115)(6.55),$$
$$\text{LCL} = D_3\overline{R} = (0)(6.55),$$
$$\text{UCL} = 13.85, \qquad \text{LCL} = 0.0.$$

A quick examination of the centerlines based on the new data reveals that the value for $\overline{\overline{X}}$ is virtually unchanged, and that the value for $\overline{R}$ has decreased by approximately 15%. This suggests that the substitute-versus-regular-operator issue did indeed influence the level of process variability. Figure 6.28 shows the $\overline{X}$ and R charts based on the data in Table 6.2 and the control limits determined above.

Interpretation of the Control Charts

Again, we first examine the R chart of Fig. 6.28. No statistical signals are evident in the chart (none of the tests discussed previously have been violated). This confirms that the decisions made by the study team related to the substitute operator were correct. Clearly, efforts to better train the substitute operator were effective. The variability associated with the liner boring process is now in good statistical control. With the variability of the process

TABLE 6.2	Cylinder Liner Boring Data After Operator Special Cause Removed							
Sample Number	Sample Collection Time	X_1	X_2	X_3	X_4	X_5	$\bar{X}$	R
41	8:10 A.M.	204	201	206	207	204	204.4	6
42	8:35 A.M.	200	204	203	205	202	202.8	5
43	9:00 A.M.	204	197	204	203	202	202.0	7
44	9:25 A.M.	202	203	194	195	200	198.8	9
45	9:50 A.M.	202	199	200	205	206	202.4	7
46	10:15 A.M.	197	199	201	199	198	198.8	4
47	10:40 A.M.	199	197	196	202	198	198.4	6
48	11:05 A.M.	192	200	198	201	200	198.2	9
49	11:30 A.M.	197	204	206	199	200	201.2	9
50	11:55 A.M.	197	199	197	198	197	197.6	2
51	1:05 P.M.	204	200	208	206	205	204.6	8
52	1:30 P.M.	203	200	200	203	202	201.6	3
53	1:55 P.M.	201	202	201	202	206	202.4	5
54	2:20 P.M.	200	194	197	201	202	198.8	8
55	2:45 P.M.	191	196	201	203	198	197.8	12
56	3:10 P.M.	198	203	199	199	197	199.2	6
57	3:35 P.M.	199	199	197	201	201	199.4	4
58	4:00 P.M.	198	194	198	201	194	197.0	7
59	4:25 P.M.	199	195	203	200	201	199.6	8
60	4:50 P.M.	196	198	197	201	196	197.6	5
61	8:15 A.M.	203	202	204	206	204	203.8	4
62	8:40 A.M.	208	204	199	200	205	203.2	9
63	9:05 A.M.	202	201	201	201	198	200.6	4
64	9:30 A.M.	193	203	202	197	196	198.2	10
65	9:55 A.M.	199	201	205	198	201	200.8	7
66	10:20 A.M.	202	199	201	201	204	201.4	5
67	10:45 A.M.	200	201	200	201	202	200.8	2
68	11:10 A.M.	197	197	200	198	201	198.6	4
69	11:35 A.M.	205	198	201	200	196	200.0	9
70	11:59 A.M.	199	201	195	198	197	198.0	6
71	1:00 P.M.	207	207	203	204	205	205.2	4
72	1:25 P.M.	197	202	208	201	206	202.8	11
73	1:50 P.M.	199	205	205	199	200	201.6	6
74	2:15 P.M.	194	200	199	206	201	200.0	12
75	2:40 P.M.	197	198	201	196	201	198.6	5
76	3:05 P.M.	200	197	201	200	197	199.0	4
77	3:30 P.M.	201	198	195	201	201	199.2	6
78	3:55 P.M.	198	199	203	202	199	200.2	5
79	4:20 P.M.	201	199	193	197	196	197.2	8
80	4:45 P.M.	192	199	199	203	196	197.8	11

Figure 6.28   Control charts for boring process after operator special cause removed

stable, attention may now turn to an examination of the $\overline{X}$ chart. It is evident that there are a number of statistical signals on the $\overline{X}$ chart. The samples that represent signals and the tests for control chart interpretation that have been violated are shown in Table 6.3.

TABLE 6.3	$\overline{X}$ Chart Test Violations for Fig. 6.28
Sample	*Test(s) Violated*
41	Point above the upper control limit
42	2 out of 3 points in zone A or beyond above the mean
45	4 out of 5 Points in zone B or beyond above the mean
48	Run of 8 Points that avoid zone C
	4 out of 5 Points in zone B or beyond below the mean
50	4 out of 5 Points in zone B or beyond below the mean
51	Point above the upper control limit
60	2 out of 3 points in zone A or beyond below the mean
62	2 out of 3 points in zone A or beyond above the mean
71	Point above the upper control limit
72	2 out of 3 Points in zone A or beyond above the mean

The signals on the $\overline{X}$ chart indicate the presence of a special cause(s) in the process. The root cause(s) for the signals evident on the chart must now be identified. The study team must do the detective work and diagnose the cause(s) for the behavior reflected on the chart. With this in mind, the study team returned to the cause-and-effect diagram of Fig. 6.26. During discussion of the diagram, it was suggested that machine warm-up might be the cause of the behavior. In other words, parts produced early in the day (shortly after the machine is turned on at 8 A.M.) and right after lunch (the machine is shut down for the 12–1 P.M. lunch hour) may deviate from the desired size. An examination of the $\overline{X}$ chart does reveal that $\overline{X}$ values produced during the 8–9 A.M. and 1–2 P.M. time periods are generally larger than many of the other $\overline{X}$ values.

To ascertain the extent to which the warm-up is related to the liner ID, the study team prepared a scatter plot as shown in Fig. 6.29. The $\overline{X}$ values for samples 41–80 were plotted as a function of the elapsed time since machine start-up. An examination of figure showed that the liner inner diameter (ID) appears to reduce in size as the machine warms up. The warm-up period seems to last for about 60 to 90 minutes, after which the liner ID stabilizes at a value near 200.

With some evidence that machine warm-up affects the liner ID, the study team brainstormed strategies to avoid/manage the warm-up effect. The first strategy suggested was to change the operating parameters during the warm-up period of the machine. The second strategy was to have the machine go through an accelerated warm-up procedure and not produce any parts during this period. The final suggested approach was to leave the machine on over the lunch hour, and to have a worker arrive at the facility early to start up the machine in the morning. This final strategy was selected for implementation owing to its simplicity and cost effectiveness. Of course, only by collecting additional samples from the process can it be determined whether the warm-up issue is the true root cause, and whether the proposed corrective action will eliminate the instability in the process mean behavior displayed on the $\overline{X}$ chart.

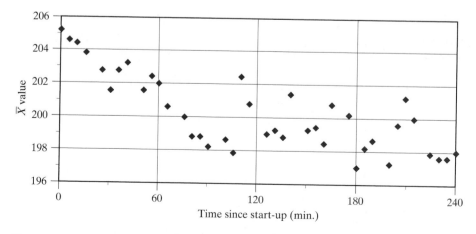

Figure 6.29   Scatter plot for $\overline{X}$ vs. time since start-up

Control Charts After Addressing Warm-Up Special Cause

Following the implementation of the boring machine operation scheme designed to address the machine warm-up issue, 40 additional samples were collected from the process. These data are shown in Table 6.4.

TABLE 6.4	Cylinder Liner Boring Data After Warm-Up Special Cause Removed							
Sample Number	Sample Collection Time	X_1	X_2	X_3	X_4	X_5	$\bar{X}$	R
81	8:05 A.M.	203	200	200	198	201	200.4	5
82	8:30 A.M.	195	203	203	193	199	198.6	10
83	8:55 A.M.	206	200	198	199	202	201.0	8
84	9:20 A.M.	197	202	200	199	202	200.0	5
85	9:45 A.M.	205	194	203	198	201	200.2	11
86	10:10 A.M.	207	202	204	198	204	203.0	9
87	10:35 A.M.	199	198	197	199	199	198.4	2
88	11:00 A.M.	201	206	202	203	198	202.0	8
89	11:25 A.M.	197	202	200	197	200	199.2	5
90	11:50 A.M.	197	199	197	202	193	197.6	9
91	1:00 P.M.	203	194	202	200	200	199.8	9
92	1:25 P.M.	201	202	201	197	202	200.6	5
93	1:50 P.M.	204	195	208	202	202	202.2	13
94	2:15 P.M.	199	202	196	201	195	198.6	7
95	2:40 P.M.	200	201	201	197	198	199.4	4
96	3:05 P.M.	200	200	203	201	201	201.0	3
97	3:30 P.M.	196	201	199	198	197	198.2	5
98	3:55 P.M.	197	202	198	196	197	198.0	6
99	4:20 P.M.	205	193	202	200	196	199.2	12
100	4:45 P.M.	203	200	199	204	202	201.6	5
101	8:00 A.M.	201	205	201	202	199	201.6	6
102	8:25 A.M.	200	203	200	200	203	201.2	3
103	8:50 A.M.	201	203	199	200	201	200.8	4
104	9:15 A.M.	196	201	200	199	204	200.0	8
105	9:40 A.M.	199	199	201	203	201	200.6	4
106	10:05 A.M.	202	201	200	203	201	201.4	3
107	10:30 A.M.	197	201	199	196	199	198.4	5
108	10:55 A.M.	195	200	205	202	195	199.4	10
109	11:20 A.M.	201	202	200	195	201	199.8	7
110	11:45 A.M.	196	200	200	196	196	197.6	4
111	1:15 P.M.	206	200	197	196	198	199.4	10
112	1:40 P.M.	202	198	205	201	196	200.4	9
113	2:05 P.M.	200	198	198	200	197	198.6	3
114	2:30 P.M.	199	201	204	200	205	201.8	6
115	2:55 P.M.	201	201	198	197	198	199.0	4
116	3:20 P.M.	199	206	197	198	201	200.2	9
117	3:45 P.M.	201	197	191	200	201	198.0	10
118	4:10 P.M.	202	198	199	200	204	200.6	6
119	4:35 P.M.	201	203	204	197	202	201.4	7
120	5:00 P.M.	200	204	199	203	199	201.0	5

Determination of Control Limits

The centerlines and the control limit calculations for the charts based on the data in Table 6.4 are

$$\overline{\overline{X}} = 200.005,$$
$$\overline{R} = 6.6.$$

For the $\overline{X}$ chart,

$$\text{UCL} = \overline{\overline{X}} + A_2\overline{R} = 200.005 + (0.577)(6.6),$$
$$\text{LCL} = \overline{\overline{X}} - A_2\overline{R} = 200.005 - (0.577)(6.6),$$
$$\text{UCL} = 203.81, \quad \text{LCL} = 196.20.$$

For the R chart,

$$\text{UCL} = D_4\overline{R} = (2.115)(6.60),$$
$$\text{LCL} = D_3\overline{R} = (0)(6.60),$$
$$\text{UCL} = 13.96, \quad \text{LCL} = 0.0.$$

Figure 6.30 shows the $\overline{X}$ and R charts based on the data in Table 6.4 and the centerlines and control limits determined above.

Figure 6.30   Control charts for liner boring process after warm-up cause removed

Interpretation of the Control Charts

Again we, first examine the R chart of Fig. 6.30 and discern no statistical signals. This provides further evidence that the action taken to train the substitute operator was effective in removing a special cause. Attention then shifts to the $\overline{X}$ chart in the figure—no statistical signals are apparent. This confirms the effectiveness of the study team's corrective action on how to deal with machine warm-up. Both charts demonstrate good statistical control, and the limits may be extended so that the study team can continue monitoring the process.

Summary

This case study points to the importance of maintaining both $\overline{X}$ and R control charts and the significance of first focusing attention on the R chart and establishing its stability. Initially, no points fell outside the $\overline{X}$-chart control limits and we might have inferred that the process mean exhibited good statistical control from the outset. However, the fact that the R chart was initially not in control has caused the limits on the $\overline{X}$ chart to be somewhat wider because of the presence of a special cause that inflates the process variability. Once this special cause of variability was removed, the limits on the $\overline{X}$ chart became narrower and many $\overline{X}$ values showed up as statistical signals with the new limits. A special cause was present in the $\overline{X}$ data but initially was not recognizable because of the excess variability as seen in the R chart.

It is important to note that the root causes identified for the out-of-control conditions on both the $\overline{X}$ and R charts are conjecture until verified through continued process monitoring. It is extremely important that this verification take place, and clearly, the $\overline{X}$ and R charts are the appropriate tool to effect this verification.

6.9   A Note on Computer Software

Since we began preparing the first edition of this text a decade and a half ago, much has transpired relative to the subject matter and its application in practice. We have sought to capture some of this change in this second edition. Perhaps nothing has undergone more development and change than computer software to facilitate the analysis and interpretation of data—for both SPC and DOE applications.

We have always advocated that students, as they first become immersed in the material covered herein, work out all of the necessary calculations by hand. This includes proper plotting of the data, calculation of centerlines and control limits for control charts, normal probability plots of estimated effects of two-level factorial experiments, and two-way diagrams for two-factor interaction effect interpretation. This is not to suggest that competent and useful software for these purposes does not exist—it does. But, students need to see firsthand how all of these tools come to be, and they need to develop the ability and the awareness necessary to question the output of software that does not seem correct or appropriate, for incompetent and flawed software is readily available as well.

In the end, students will naturally gravitate toward programs that they know can help them get the "right" answers. Today's students are a computer-savvy group, and most of the number crunching described herein can be readily accomplished with spreadsheet software (e.g, MS Excel), with which most students are familiar. We frequently encourage students to write their own programs and set up their own spreadsheets, based on the equations and procedures provided in the text. Results may be checked with simple test problems or compared to the solutions of classmates to make sure that things have been done correctly. But in the end, the best guidance is to follow the advice of the instructor, who may wish to use one or more of the statistical packages that provide rigorous analyses, standardized plots, etc. We do not advocate the use of any particular package over another—there are many available that will work just fine. MINITAB (Release 14 in 2005, http://www.minitab.com/products/minitab/14/default.aspx), for example, may be available at some institutions and has comprehensive programs for SPC and DOE. Design-Expert (Release 6, 2005, http://www.statease.com/dx6descr.html) is a popular program for design-of-experiments analysis.

■ EXERCISES

6.1 What is the purpose of Shewhart control charts? What information can be gained from these charts? How can this information be used to improve the quality and productivity of a manufacturing process?

6.2 Given the table of coating weights shown here for samples of size $n = 4$:
  (a) Calculate the $\bar{X}$'s and R's for samples 2 to 8.
  (b) Calculate the $\bar{\bar{X}}$ and $\bar{R}$ values based on all eight samples.
  (c) Determine the appropriate constants that would be used with these data to construct $\bar{X}$ and R control charts, and calculate the control limits.

Sample	X_1	X_2	X_3	X_4	$\bar{X}$	Range
1	18.5	21.2	19.4	16.5	18.90	4.7
2	17.9	19.0	20.3	21.2		
3	19.6	19.8	20.4	20.5		
4	22.2	21.5	20.8	20.3		
5	19.1	20.6	20.8	21.6		
6	22.8	22.2	23.2	23.0		
7	19.0	20.5	20.3	19.2		
8	20.7	21.0	20.5	19.1		

6.3 Control charts are to be kept on the thickness measurements for a process that rolls 10-gage copper sheets. The current specification on the sheets is 0.1340 ± 0.0010 inch. After collecting 30 rational samples of size $n = 4$ measurements at approximately half-hour intervals, the data were used to determine $\Sigma \, \bar{X}_i = 4.014$ inches and $\Sigma \, R_i = 0.027$ inch, $i = 1, 30$. It is desired to set up $\bar{X}$ and R control charts for this process. Use the information above to determine centerlines and upper and lower control limits for the process.

6.4 Averages and ranges of samples of size $n = 5$ are shown in the accompanying table. The data pertain to the depth-to-shoulder measurement of glove boxes manufactured by an American car company. The measurements are in inches.

Sample	$\bar{X}$	Range	Sample	$\bar{X}$	Range
1	0.4402	0.015	16	0.4362	0.015
2	0.4390	0.018	17	0.4380	0.019
3	0.4448	0.018	18	0.4350	0.008
4	0.4432	0.006	19	0.4378	0.011
5	0.4428	0.008	20	0.4384	0.009
6	0.4382	0.010	21	0.4392	0.006
7	0.4358	0.011	22	0.4378	0.008
8	0.4440	0.019	23	0.4362	0.016
9	0.4366	0.010	24	0.4348	0.019
10	0.4368	0.011	25	0.4338	0.015
11	0.4360	0.011	26	0.4366	0.014
12	0.4402	0.007	27	0.4346	0.019
13	0.4332	0.008	28	0.4374	0.015
14	0.4356	0.017	29	0.4339	0.024
15	0.4314	0.010	30	0.4368	0.014

From the first 25 samples, set up and interpret $\bar{X}$ and R control charts. Indicate whether or not control is evident. If the chart shows the process to be in a state of control, plot the next five samples and comment on the results.

6.5 Given the set of data for $\bar{X}$ and R shown in the lower table, plot their values on control charts with the centerlines and control limits shown in the upper table.

	$\bar{X}$ Chart	R Chart
LCL	9.74	0
CL	10.66	1.59
UCL	11.58	3.35

Sample	$\bar{X}$	Range	Sample	$\bar{X}$	Range
1	10.44	1.8	11	10.56	1.5
2	10.46	1.4	12	9.96	2.4
3	9.98	0.7	13	10.44	2.8
4	9.82	2.6	14	10.96	1.0
5	10.90	2.4	15	11.14	1.5
6	10.36	1.1	16	10.04	0.8
7	11.32	0.8	17	11.44	2.1
8	10.86	1.2	18	11.84	1.0
9	10.58	0.5	19	11.14	2.7
10	9.52	1.9	20	11.44	1.6

(a) Do the charts show good statistical control? Why or why not?

(b) What sample/subgroup size was used?

6.6 (a) Using Table A.2 in the Appendix, find the values for A_2, D_3, and D_4 for samples of size $n = 2$, $n = 5$, $n = 10$. How were the values for A_2 originally determined?

(b) If $\bar{R} = 0.034$ and samples of size $n = 5$ have been collected, what is the corresponding estimate of $\sigma_{\bar{X}}$?

(c) For a certain control chart application the upper control limit on the R chart is 1.840 and the lower control limit on the $\bar{X}$ chart is 4.120. The centerline for the $\bar{X}$ chart is $\bar{\bar{X}} = 4.625$. What subgroup size was used for this application?

6.7 An initial set of data has been collected for the construction of control charts to monitor the manufacturing of a spacer with three critical quality characteristics/dimensions: the inside diameter, the outside diameter, and the thickness. Thirty-five rational samples of size $n = 6$ have been co-llected for each dimension. For the inside diameter characteristic, $\Sigma \bar{X} = 121.423$ centimeters and $\Sigma R = 2.714$ centimeters. Set up Shewhart control charts for these data using proper conventions and clearly labeling all relevant entities on the charts.

6.8 Charts are also to be constructed for the thickness quality characteristic in Exercise 6.7. The quality control engineer left for vacation before finishing the charts. You are given her notes and find that all you know is that the upper control limit for the R chart is 0.0031 centimeter and the upper control limit for the $\bar{X}$ chart is 0.5012 centimeter. The raw data page is missing, but from charts on the other characteristics you learn that the sample size was $n = 6$. Finish the construction of the charts, again clearly showing the centerlines, control limits, and zone lines and labeling their values.

6.9 A new operator has been working on a foam molding machine that is producing automobile instrument panels. $\bar{X}$ and R control charts are being used to monitor foam thickness. Samples of size $n = 4$ have been used, but the new operator inadvertently has begun to use a sample size of $n = 5$.

(a) What impact does this have on the ongoing charting process?

(b) What could be the consequences if the operator had inadvertently changed to a sample size of $n = 3$?

6.10 Engineers at two sister plants have decided that they would like to compare the variation in performance of two types/manufacturers of end mills used in a machining operation common to both plants. The end mills are procured under the same specifications but have been purchased from different suppliers. Both machines perform the same task, a slotting operation, and the quality characteristic in question is the surface finish of the slot walls. Shewhart control charts have been kept at each plant to monitor the performance, and the engineers have decided to use these to determine which end mill provides for the more consistent surface finish. However, the subgroup sizes used to create the charts are different. The $\overline{X}$ chart for end mill A, established using samples of size $n = 4$, has an $\overline{\overline{X}}$ value of 85.2 microinches and an upper control limit of 100.3 microinches. End mill B's $\overline{X}$ chart was established using samples of size $n = 5$ and has an $\overline{\overline{X}}$ value of 84.9 microinches and an upper control limit of 98.5 microinches. Both the $\overline{X}$ and R charts have been shown to be in good statistical control over an extended period of time. Which end mill appears to be producing the most consistent/least variable surface finish?

6.11 Suppose that a co-worker has been using Shewhart control charts to analyze a process but has only been using the upper and lower control limits to determine out-of-control conditions in the form of extreme points, without any consideration of the zone rules, runs, and other statistical signals for the $\overline{X}$ chart. When questioned about this, he responds that we only need to worry about the extreme points, since these are the ones that will indicate problems with lack of conformance of the parts to the specifications. How would you respond to his argument?

6.12 Why is it so important to examine the R chart and interpret its patterns before looking at the $\overline{X}$ chart? Be specific in your discussion.

6.13 Match each of the $\overline{X}$- and R-chart pairs (i) to (vi) shown here with one of the time-varying process behaviors (1) to (10), which could have led to the data on the charts. Clearly identify the location and nature of each of the statistical signals that occur. The time of the process changes and the signals they might produce do not necessarily match precisely across the charts, but the general nature of the patterns can be matched.

6.14 Consider charts (i) to (vi), which represent the $\overline{X}$ chart for a certain process. Some of the charts have instances of special-cause variation, while others may show good statistical control.

(a) For each chart, identify (circle) the points where statistical signals show that special-cause variation is indeed occurring. Indicate the exact rule that applies.

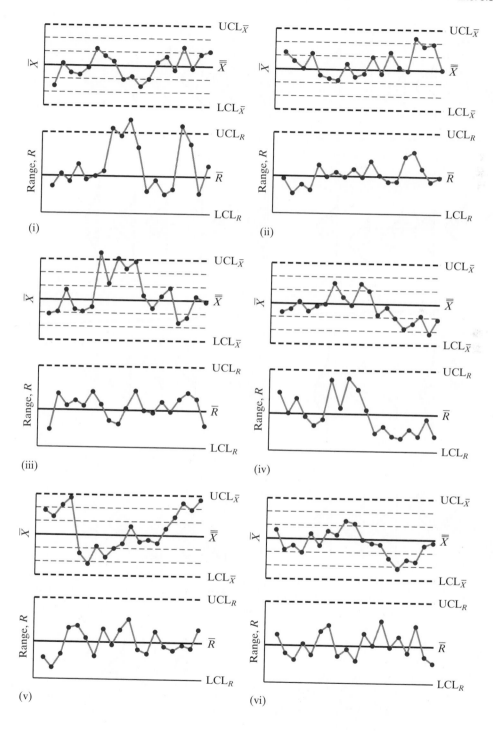

(i)

(ii)

(iii)

(iv)

(v)

(vi)

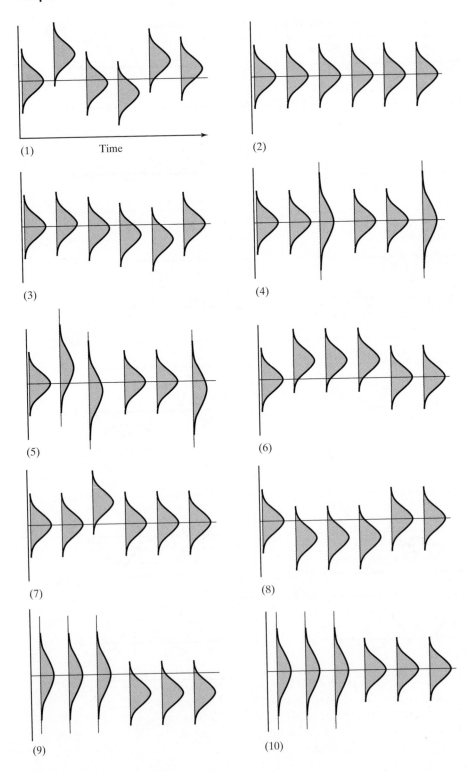

(1)   Time

(2)

(3)

(4)

(5)

(6)

(7)

(8)

(9)

(10)

(b) Identify from the list below [(1) to (9)] the process behavior or sampling fault that is responsible for the appearance of a special cause on the chart.

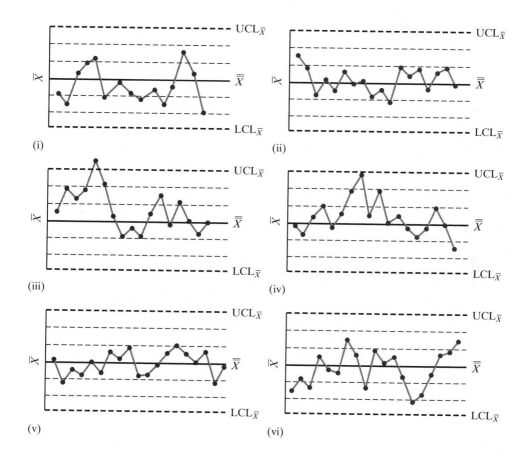

1. Upward shift in the mean.
2. Downward shift in the mean.
3. Upward shift in the variance.
4. Downward shift in the variance.
5. Gradual shift in the mean.
6. Improper sampling.
7. Mixing/overcontrol.
8. Common-cause system only.
9. Trend.

6.15 Use the sketch of the $\overline{X}$ chart shown here as a model in demonstrating the general appearance of the types of process behaviors (a) to (e). On each chart include the occurrence of at least two statistical signals, which could be indicating each type of behavior. Plot 16 points on each chart, clearly identifying the location and type of signals/rule violations.

(a) An abrupt and sustained upward shift in the mean.
(b) Gradual downward shift in the process mean.
(c) Overcontrol of the process.

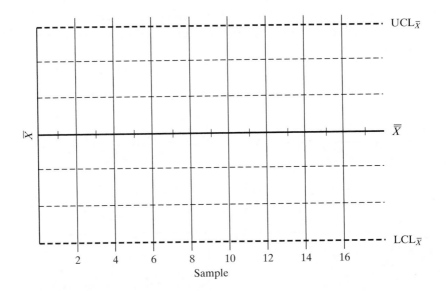

(d) Plotting of two processes on a single chart.

(e) An abrupt but short-lived downward shift in the process mean.

6.16 Engineers at a manufacturing plant have just been to a course to learn the basics of SPC. They have decided to set up their first charting application on a CNC turning center where a motor shaft is being machined. The quality characteristic that they wish to monitor is the outside diameter, which has a nominal value of 2.125 inches and a tolerance of ±0.001 inch. Recently, however, they have been feeling pressure from the company to which they are supplying the parts to improve the consistency of the part dimension in question, and the engineers feel that methods of SPC may be able to assist them in reducing variability. Initially, they collect data on the outside diameter in subgroups/samples of five consecutive parts at intervals of approximately $\frac{1}{2}$ hour, with the results shown in the table. The measurements were made in inches to the nearest ten-thousandth of an inch (e.g., 2.1248 and 2.1259 inches). The data are coded using the last two digits.

Sample	X_1	X_2	X_3	X_4	X_5
1	48	59	42	40	32
2	48	52	51	50	52
3	50	49	60	44	48
4	50	41	48	43	41
5	51	49	60	60	47
6	44	47	54	45	50
7	51	57	53	48	45
8	46	63	63	41	41

(Continued)

Sample	X_1	X_2	X_3	X_4	X_5
9	54	53	53	46	55
10	54	46	55	58	62
11	57	56	53	37	42
12	51	49	67	50	48
13	42	47	48	39	61
14	43	50	44	38	40
15	41	54	59	53	37
16	60	45	52	46	55
17	48	50	45	40	63
18	46	39	43	61	41
19	52	32	54	45	50
20	62	50	47	57	47
21	58	52	57	54	63
22	50	56	44	44	38
23	47	53	42	40	46
24	56	55	44	58	49
25	39	54	54	44	63
26	40	43	59	36	46
27	48	48	34	43	62
28	36	45	43	32	38
29	51	48	24	55	54
30	43	50	40	53	41
31	43	43	49	31	47
32	29	42	43	42	43
33	36	43	43	49	48
34	48	42	43	39	46
35	45	51	43	39	29

(a) Find $\overline{X}$ and R for each sample.

(b) Use all 35 samples to compute the centerlines and control limits for $\overline{X}$ and R charts.

(c) Construct the charts, plotting all of the $\overline{X}$ and R values and including all appropriate information on the charts.

(d) Apply all the appropriate rules to interpret the charts, indicating every incidence of an out-of-control condition, and identifying the rule that is violated in each case.

(e) What can be said about the stability of the process?

6.17 The preliminary results of the investigation in Exercise 6.16 indicated that the cutting tools being used from one of the suppliers did not have the proper edge preparation, causing a more rapid tool wear condition. Once this was determined and verified, the $\overline{X}$ chart seemed to exhibit good statistical control. To continue the charting it was proposed to use control limits determined from the first 25 samples. Twenty-five more samples were collected, again at roughly half-hour intervals, with the resulting data shown in the table. These data are presented in a coded form as in Exercise 6.16, using the last two digits of the actual measurement.

Sample	X_1	X_2	X_3	X_4	X_5	$\overline{X}$	Range
36	50	54	51	45	55	51.0	10
37	54	46	40	53	51	48.8	14
38	60	55	44	47	57	52.6	16
39	46	61	51	56	41	51.0	20
40	39	57	54	56	63	53.8	24
41	52	39	45	48	56	48.0	17
42	49	51	49	54	51	50.8	5
43	51	46	44	50	39	46.0	12
44	54	51	49	48	48	50.0	6
45	54	54	57	50	45	52.0	12
46	54	57	49	47	41	49.6	16
47	46	45	48	51	46	47.2	6
48	49	49	55	53	41	49.4	14
49	45	51	44	55	46	48.2	11
50	54	54	51	47	49	51.0	7
51	43	51	56	52	48	50.0	13
52	51	47	47	56	48	49.8	9
53	50	47	55	59	48	51.8	12
54	48	60	45	49	38	48.0	22
55	54	47	49	50	52	50.4	7
56	41	46	46	48	55	47.2	14
57	52	57	47	58	60	54.8	13
58	53	53	55	52	53	53.2	3
59	55	48	52	56	50	52.2	8
60	48	49	43	54	46	48.0	11

(a) Determine centerlines and control limits for the $\overline{X}$ and R charts using the data from the first 25 samples from Exercise 6.16. Construct the charts, plotting the first 25 sample results, $\overline{X}$, and R. Verify that the charts to this point show good statistical control.

Note: Read part (b) before you start plotting for part (a).

(b) Extend the limits and plot the next 25 sample results (samples 36 to 60).

(c) Evaluate the charts using all applicable rules.

(d) Comment on what the charts are telling us about the process.

6.18 Our engineer friends from Exercises 6.16 and 6.17 have decided to use their newly acquired quality improvement skills on another problem that has developed. Much rework time has become necessary to correct the depth of a keyway that is being milled into the shaft previously studied. Many of the keyways are being found to be too shallow, causing the key not to fit properly during assembly. A sketch of the part is shown. Measurements are collected in samples of size $n = 4$ consecutive parts at approximately 30-minute intervals over the course of three shifts. The slot depth measurements were made in inches to the nearest ten-thousandth of an inch (e.g., 0.1501 and 0.1494 inch). The data are coded using the last three digits.

Depth

Sample	X_1	X_2	X_3	X_4	$\overline{X}$	Range
1	501	501	494	496	498.00	7
2	505	493	509	497	501.00	16
3	497	486	503	500	496.50	17
4	507	507	503	494	502.75	13
5	488	496	496	503	495.75	15
6	505	505	505	488	500.75	16
7	507	496	506	486	498.75	7
8	490	501	494	505	497.50	11
9	490	495	489	505	494.75	15
10	490	493	497	491	492.75	19
11	492	489	485	496	490.50	11
12	496	503	511	499	502.25	15
13	510	507	491	504	503.00	19
14	503	492	496	497	497.00	11
15	503	492	496	497	497.00	10
16	493	500	500	494	496.75	7
17	499	494	512	504	502.25	16
18	505	492	505	493	498.75	13
19	501	505	497	497	500.00	8
20	498	507	496	506	501.75	11
21	502	500	501	501	501.00	2
22	496	492	504	505	499.25	13
23	503	489	501	503	499.00	14
24	490	498	494	509	497.75	19
25	493	502	493	499	496.75	9
26	498	503	490	486	494.25	17
27	500	507	489	491	496.75	18
28	497	502	499	506	501.00	9
29	491	504	506	502	500.75	15
30	502	493	500	501	499.00	9
31	499	504	506	495	501.00	11
32	503	497	502	493	498.75	10
33	493	508	494	496	497.75	15
34	496	495	509	496	499.00	13
35	503	500	508	507	504.50	8

(a) Use the data in the accompanying table to construct $\overline{X}$ and R charts for the process.

(b) Interpret the charts using the appropriate rules. Indicate each incidence of an out-of-control condition.

(c) What are the charts telling us about the behavior of the process in terms of its mean and/or standard deviation?

6.19 A study has been initiated on a filling process. The data in the accompanying table represent the net weights in pounds (above 20 pounds) for 20-pound bags of dry dog food. The samples of $n = 5$ represent five consecutive bags produced by filling head 2 on a four-head filling machine. The samples were collected at 30-minute intervals. Ultimately, the goal is to fill the bags as close to 20 pounds as possible without going under this nominal weight. Regulations allow only 0.1% of the bags to be below this value. The first step in this study was to examine the filling process from a statistical control point of view.

Sample	Day	Time	X_1	X_2	X_3	X_4	X_5	$\overline{X}$	Range
1	1	7:30	0.92	1.01	0.95	1.04	0.90	0.964	0.14
2	1	8:00	1.15	1.02	0.98	0.94	0.99	1.016	0.21
3	1	8:30	0.94	0.91	1.00	1.05	0.95	0.970	0.14
4	1	9:00	1.11	0.94	0.89	1.11	1.00	1.010	0.22
5	1	9:30	0.95	0.97	0.97	0.98	0.86	0.946	0.12
6	1	10:00	1.02	0.89	0.97	0.95	0.97	0.960	0.13
7	1	10:30	1.18	0.84	0.95	1.39	1.03	1.078	0.55
8	1	11:30	0.94	1.15	1.07	0.99	1.03	1.036	0.21
9	1	12:00	1.03	1.20	1.00	1.10	1.09	1.084	0.20
10	1	12:30	0.98	0.82	0.98	1.02	1.13	0.986	0.31
11	1	1:00	0.98	0.95	0.97	1.04	0.89	0.966	0.15
12	1	1:30	1.10	1.12	1.01	1.12	1.04	1.078	0.11
13	1	2:00	1.10	0.94	0.88	0.92	0.91	0.950	0.22
14	1	2:30	1.01	0.99	1.11	0.96	1.05	1.024	0.15
15	1	3:00	1.17	1.30	1.21	0.69	0.82	1.038	0.61
16	2	7:30	0.97	1.03	1.09	1.04	0.94	1.014	0.15
17	2	8:00	0.92	0.88	0.83	0.94	0.87	0.888	0.11
18	2	8:30	0.99	1.00	0.95	1.00	0.90	0.968	0.10
19	2	9:00	0.88	1.09	1.05	1.05	1.01	1.016	0.21
20	2	9:30	0.87	1.08	0.99	0.97	1.04	0.990	0.21
21	2	10:00	1.08	0.99	1.18	1.02	1.07	1.068	0.19
22	2	10:30	0.60	1.28	0.97	0.84	1.01	0.940	0.68
23	2	11:30	0.89	0.99	1.02	0.95	0.99	0.968	0.13
24	2	12:00	1.01	0.90	0.97	1.09	1.13	1.020	0.23
25	2	12:30	0.95	1.01	1.09	1.10	1.10	1.050	0.15
26	2	1:00	1.10	0.96	1.02	1.03	1.01	1.024	0.14
27	2	1:30	0.92	1.05	1.03	0.99	1.08	1.014	0.16
28	2	2:00	1.00	0.87	1.00	1.05	0.97	0.978	0.18
29	2	2:30	0.96	1.03	1.03	1.11	1.05	1.036	0.15
30	2	3:00	1.15	0.84	1.02	1.18	1.05	1.048	0.34

(Continued)

Sample	Day	Time	X_1	X_2	X_3	X_4	X_5	$\overline{X}$	Range
31	3	7:30	0.91	0.85	0.89	0.82	0.95	0.884	0.13
32	3	8:00	0.95	0.92	0.95	0.84	0.92	0.916	0.11
33	3	8:30	0.98	0.98	1.01	1.12	1.19	1.056	0.21
34	3	9:00	0.89	0.90	1.05	1.05	0.87	0.952	0.18
35	3	9:30	1.16	0.96	0.96	1.06	1.00	1.028	0.20
36	3	10:00	1.10	0.87	0.95	1.05	1.14	1.022	0.27
37	3	10:30	0.83	0.75	1.04	1.25	0.77	0.928	0.50
38	3	11:30	0.98	1.02	1.06	0.87	1.00	0.986	0.19
39	3	12:00	0.95	0.88	0.97	1.01	0.85	0.932	0.16
40	3	12:30	1.04	0.95	1.00	1.14	1.06	1.038	0.19
41	3	1:00	0.98	0.96	1.04	1.09	1.05	1.024	0.13
42	3	1:30	0.80	0.99	0.98	1.03	0.89	0.938	0.23
43	3	2:00	1.04	1.00	0.87	1.02	0.91	0.968	0.17
44	3	2:30	0.94	1.02	0.92	1.00	1.01	0.978	0.10
45	3	3:00	0.71	1.36	1.10	1.24	0.81	1.044	0.65

(a) Use the data in the table to establish $\overline{X}$ and R charts for the process.

(b) Interpret the charts. If statistical signals/out-of-control conditions are present, assume that the physical cause was identified. Remove the out-of-control subgroups and recalculate the centerlines and control limits.

(c) Continue until the charts show good statistical control. Comment.

After these steps were taken, a group was formed to study the process further with hopes of reducing the variation in the process. One of the

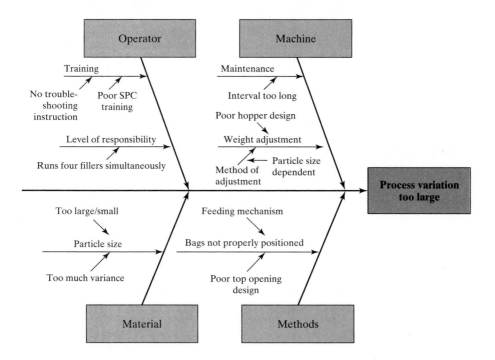

first steps taken was to create a cause-and-effect diagram, as shown. Using this as a guide, it was determined that the most likely cause of variation centered around the fact that adjustments were made assuming a much smaller particle size than actually was present in the dog food. Once this common-cause/chronic problem was addressed, the charting was continued, with the data in the accompanying table being collected.

Sample	Day	Time	X_1	X_2	X_3	X_4	X_5	$\bar{X}$	Range
46	4	7:30	0.97	0.99	1.05	0.98	0.96	0.990	0.09
47	4	8:00	1.02	0.99	1.00	1.00	1.00	1.002	0.03
48	4	8:30	0.96	1.05	0.96	1.02	0.97	0.992	0.09
49	4	9:00	1.05	1.00	1.00	0.98	0.99	1.004	0.07
50	4	9:30	0.98	1.00	0.95	0.97	0.97	0.974	0.05
51	4	10:00	0.98	0.97	0.98	1.04	0.99	0.992	0.07
52	4	10:30	0.99	1.05	1.03	0.99	0.99	1.010	0.06
53	4	11:30	0.95	0.94	0.99	1.03	1.00	0.982	0.09
54	4	12:00	0.97	1.01	1.01	1.01	1.02	1.004	0.05
55	4	12:30	1.02	0.99	0.97	0.99	1.02	0.998	0.05
56	4	1:00	0.99	1.00	1.01	1.05	1.02	1.014	0.06
57	4	1:30	0.99	0.99	1.00	0.98	1.01	0.994	0.03
58	4	2:00	1.02	1.00	1.01	0.99	0.97	0.998	0.05
59	4	2:30	0.99	0.94	0.98	0.99	0.95	0.970	0.05
60	4	3:00	1.04	1.00	1.01	0.98	0.98	1.002	0.06

(d) Use the charts established in part (c) to continue monitoring this process for day 4. Plot the results of the 15 subgroups above. Comment on the continuing process behavior.

6.20 The manufacturer of exercise weights makes a full range of dumbbells. Recently a new set of molds has been developed for the casting of 6-pound iron dumbbells. However, many of the dumbbells made have been falling outside the specifications of 96 + 1.0 ounce, −0.0 ounce.

(a) Use the data in the accompanying table (32 samples of size $n = 4$), collected from one of the new molds roughly at hourly intervals, to establish centerlines and control limits for $\bar{X}$ and R charts. The data are in ounces above 90 ounces.

(b) Plot the values of $\bar{X}$ and R on the control charts. Leave room to plot an additional 12 samples.

(c) What are the charts telling us about the process in terms of the mean and range/variation?

Sample	X_1	X_2	X_3	X_4	$\bar{X}$	Range
1	5.3	4.7	5.1	4.9	5.000	0.6
2	4.7	4.9	5.0	4.9	4.875	0.3
3	5.1	4.5	5.2	5.2	5.000	0.7
4	5.1	5.4	4.6	4.9	5.000	0.8
5	4.9	4.7	4.9	5.2	4.925	0.5

(Continued)

Sample	X_1	X_2	X_3	X_4	$\bar{X}$	Range
6	5.0	4.7	4.8	4.8	4.825	0.3
7	5.0	4.8	4.7	4.2	4.675	0.8
8	4.6	4.8	5.4	4.7	4.875	0.8
9	5.3	4.7	4.9	4.8	4.925	0.6
10	4.8	5.0	5.4	5.2	5.100	0.6
11	4.7	5.3	4.5	5.2	4.925	0.8
12	5.4	4.8	5.6	4.6	5.100	1.0
13	4.6	5.4	4.8	4.8	4.900	0.8
14	5.4	5.6	5.6	5.3	5.475	0.3
15	5.0	4.8	4.5	4.3	4.650	0.7
16	5.3	5.6	4.8	5.3	5.250	0.8
17	5.3	4.7	5.1	4.9	5.000	0.6
18	5.3	6.0	5.2	4.9	5.350	1.1
19	5.1	4.4	5.1	4.5	4.775	0.7
20	5.6	4.3	5.2	5.5	5.150	1.3
21	5.6	5.2	4.7	4.7	5.050	0.9
22	5.8	4.8	4.6	4.6	4.950	1.2
23	5.4	5.1	5.0	4.9	5.100	0.5
24	5.6	5.7	4.9	4.8	5.250	0.9
25	4.6	4.8	4.6	5.1	4.775	0.5
26	4.3	4.7	4.8	5.4	4.800	1.1
27	4.9	4.9	5.6	4.8	5.050	0.8
28	4.9	4.7	4.8	4.9	4.825	0.2
29	5.0	5.2	4.7	5.1	5.000	0.5
30	4.9	4.9	4.9	5.3	5.000	0.4
31	5.2	5.1	5.3	4.8	5.100	0.5
32	4.8	5.6	5.3	4.6	5.075	1.0

A team was established to determine the causes associated with having too many of the dumbbells outside the specifications. It became apparent that the consistent occurrence of blows, which produce voids in the casting, was a common-cause/chronic problem. The cause for the blows was found and removed, and the process was continued, with the following data being collected.

Sample	X_1	X_2	X_3	X_4	$\bar{X}$	Range
33	5.0	4.9	5.0	4.8	4.925	0.2
34	5.3	5.2	5.1	5.1	5.175	0.2
35	5.1	5.0	4.9	4.9	4.975	0.2
36	5.2	5.0	5.1	5.1	5.100	0.2
37	4.9	5.0	5.2	5.1	5.050	0.3
38	5.1	4.9	5.1	4.9	5.000	0.2
39	5.2	5.1	4.8	5.2	5.075	0.4
40	4.9	4.9	4.9	5.0	4.925	0.1

(Continued)

Sample	X_1	X_2	X_3	X_4	$\bar{X}$	Range
41	5.0	5.1	4.8	5.1	5.000	0.3
42	5.1	5.0	5.2	4.9	5.050	0.3
43	4.9	4.9	5.0	5.2	5.000	0.3
44	5.2	4.9	5.1	5.0	5.050	0.3

(d) Plot these data on the already existing charts and continue to evaluate the process. Can it be said that the process is continuing in a state of statistical control? Does the team have cause for concern or celebration?

(e) Comment on the advisability of continuing with these charts. What might be the next step(s) to take in the charting of this process?

6.21 Data are being collected from a metal stamping process where four presses are simultaneously making small metal clips that will be used in the assembly of automobile instrument panels. The quality characteristic that is important for the clips is an inclination angle that will affect the amount of clamping pressure exerted by the clip. The specifications placed on the angle are 15.0 ± 2.5 degrees. Statistical charting has been initiated on this process, and the averages and ranges of 40 samples are given in the accompanying table, along with the associated $\bar{X}$ and R charts. Read the statistical signals and hypothesize as to what may be occurring in this process. Use data/numerical facts to confirm your suspicion.

Sample	$\bar{X}$	Range	Sample	$\bar{X}$	Range
1	14.09	10.71	21	15.42	5.98
2	15.81	14.32	22	15.16	9.16
3	13.93	10.46	23	15.21	10.14
4	13.49	7.38	24	15.59	5.32
5	15.29	4.51	25	15.05	10.87
6	16.25	8.49	26	13.97	9.72
7	14.69	5.13	27	15.58	3.95
8	14.88	8.43	28	15.16	9.16
9	15.54	8.86	29	15.62	8.99
10	15.56	8.62	30	13.80	6.48
11	15.56	9.85	31	14.48	7.97
12	14.34	5.69	32	15.67	1.69
13	16.35	5.81	33	14.15	8.99
14	13.48	8.38	34	15.14	8.74
15	14.38	13.84	35	15.27	4.73
16	14.79	3.63	36	13.86	6.45
17	12.29	5.32	37	13.71	10.99
18	15.35	3.25	38	16.10	12.27
19	17.29	10.92	39	14.28	6.14
20	15.20	12.30	40	13.99	4.56

6.22 Explain the Pareto principle as it might apply to studying visual defects in a quality control application.

6.23 A certain grade of polystyrene is being used in an injection molding process used to manufacture a series of toys for children. A study of the process has revealed the defect-frequency information shown in the accompanying table. Construct a Pareto diagram for the defect information. Comment on the result.

Quality Characteristic	Number of Defects	Quality Characteristic	Number of Defects
Scratches	24	Voids	96
Flow lines	189	Short shots	36
Flash	17	Splay	8

6.24 In packaging a board game, many different parts must be included in specific quantities. Recently it has been noticed that occasionally some of the playing pieces, six different-colored plastic parts, have been missing from the completed packages. Sampling over the past week has shown the frequency of missing pieces as given in the table.

Color	Frequency
Black	9
Blue	45
Green	5
Red	8
White	50
Yellow	149

Create a Pareto diagram for the missing pieces.

6.25 Construct a cause-and-effect diagram to show possible causes as to why the crystal glasses shipped from a distribution center by an overnight mail carrier might have arrived at the final destination with some amount of damage.

6.26 An insurance claims agent is preparing a report regarding an accident caused by a car spinning out during a rainstorm. Construct a cause-and-effect diagram outlining possible causes for the accident.

6.27 Prepare a flowchart to describe the process that you use to get ready for school/work in the morning.

6.28 Prepare a flowchart to describe the process used to prepare and host a dinner at your residence for three of your friends.

6.29 Some of the steps (not necessarily shown in chronological order) associated with securing a professional position following college include:

- Prepare résumé
- Travel to company and interview with company personnel
- Begin new job
- Make company contacts at campus job-fair
- Identify and contact off-campus employers
- Negotiate with company regarding job offer
- Conduct research on company
- Secure letters of recommendation from references
- Prepare cover letter
- Prepare and send out thank-you letter(s)
- If needed, interview with the company over the phone
- If contacted by company, arrange onsite interview

Prepare a flowchart that describes the process for finding a job—note that this may require that you identify additional key steps not shown in the list.

6.30 The manager of Jim's Gym is interested in why people have dropped their membership in the gym. A survey was conducted of former members to find out why they let their membership lapse. The results of the survey are shown in the accompanying table. Construct a Pareto chart for the survey results and comment on the chart.

Reason Cited for Membership Lapse	Frequency
Gym too messy	8
Inadequate parking	19
Inconvenient gym hours	33
Lack of adequate facilities	12
Membership fee too high	91
Rooms too cold	6
Too crowded	14
Unsafe conditions	3

6.31 Briefly describe why Shewhart control charts may be viewed as a statistical test of hypothesis.

6.32 This chapter has described eight tests or rules for interpreting the $\overline{X}$ control chart. Prepare a sketch for each test to illustrate a violation of the rule.

6.33 Control charts have been constructed for a process used to stamp sheet metal parts. The specific quality characteristic of interest is the surface finish of the part. While the R chart shows stability in the process variation, the $\overline{X}$ chart reveals a number of statistical signals. In order to identify the underlying root cause of the statistical signals, it is decided to construct scatter plots for the $\overline{X}$ values as a function of several process variables, i.e., operator, die temperature (°F), and the cumulative number of hits on the die. The accompanying table displays the levels of the process variables and the $\overline{X}$ values for each sample. Prepare the scatter plots and comment on what the scatter plots seem to suggest regarding which process variable merits further investigation.

Sample	Operator	Die Temp.	Die Hits	$\overline{X}$	Sample	Operator	Die Temp.	Die Hits	$\overline{X}$
1	Ken	212	2342	28	12	Pete	211	6533	27
2	Ken	193	2716	36	13	Ken	194	6926	32
3	Ken	195	3089	32	14	Ken	200	7311	47
4	Ken	186	3487	17	15	Pete	208	7681	36
5	Pete	211	3867	21	16	Pete	191	8049	33
6	Pete	220	4267	36	17	Pete	194	8413	38
7	Pete	187	4644	33	18	Pete	204	8784	57
8	Pete	188	5042	31	19	Ken	215	9163	30
9	Ken	201	5416	38	20	Ken	198	9558	27
10	Ken	197	5797	27	21	Ken	204	9953	37
11	Pete	189	6160	24	22	Ken	196	10338	47

(Continued)

Sample	Operator	Die Temp.	Die Hits	$\bar{X}$	Sample	Operator	Die Temp.	Die Hits	$\bar{X}$
23	Pete	192	10727	40	32	Pete	187	14231	40
24	Pete	192	11123	48	33	Pete	201	14600	39
25	Ken	193	11523	30	34	Pete	188	14978	58
26	Pete	201	11901	43	35	Pete	202	15363	43
27	Ken	207	12294	45	36	Pete	209	15726	37
28	Ken	197	12692	50	37	Ken	189	16107	45
29	Ken	181	13067	48	38	Pete	195	16491	42
30	Pete	215	13451	47	39	Pete	203	16887	61
31	Pete	195	13847	39	40	Ken	201	17271	63

6.34 Your score on an exam in one of your classes was not as good as you had hoped. Construct a cause-and-effect diagram to show possible causes.

6.35 A circuit board manufacturing line is under investigation because one particular solder joint connection has demonstrated failures in the field. It has been decided to construct control charts to study the process. Forty samples of size $n = 4$ circuit boards have been collected from the process and for each board the solder joint of interest is assessed; a composite score is determined for the joint that reflects such characteristics as joint coverage, finish, and deposited solder geometry. The composite score data for the solder joints is shown in the accompanying table.

Sample	X_1	X_2	X_3	X_4	Sample	X_1	X_2	X_3	X_4
1	16.4	22.2	16.7	19.7	21	17.6	16.4	17.3	17.7
2	16.2	19.6	19.7	19.6	22	19.1	18.1	15.3	15.2
3	15.7	17.2	21.4	22.1	23	17.9	17.0	20.4	19.1
4	17.4	15.1	20.3	16.8	24	19.3	13.2	20.7	20.4
5	17.0	16.2	21.5	20.4	25	17.6	14.3	17.4	15.9
6	16.8	17.6	16.0	18.8	26	19.0	12.8	15.8	15.9
7	18.9	18.1	20.3	18.6	27	16.6	21.5	18.0	16.0
8	16.2	15.5	18.7	20.2	28	19.6	18.7	16.3	17.3
9	16.0	15.4	14.0	19.4	29	16.2	19.0	19.3	19.8
10	19.8	15.7	16.8	20.3	30	21.4	22.7	19.3	18.8
11	21.4	19.1	14.8	16.6	31	21.0	18.7	18.1	16.9
12	17.8	22.8	20.8	19.8	32	17.1	18.9	22.0	17.2
13	20.2	15.7	17.5	15.5	33	13.2	17.8	18.8	17.4
14	15.4	20.2	18.1	17.5	34	19.7	21.6	17.4	16.8
15	16.0	18.8	15.9	21.5	35	17.3	16.0	21.5	18.3
16	14.9	17.8	16.3	17.6	36	19.7	19.8	17.9	21.5
17	16.7	18.4	16.5	19.9	37	17.2	18.7	18.4	19.6
18	20.9	19.4	21.4	18.2	38	16.1	15.3	19.5	16.5
19	17.6	16.8	20.3	18.4	39	19.3	18.8	18.5	18.6
20	16.1	18.9	16.2	18.6	40	18.8	14.7	18.2	17.7

(a) Find $\overline{X}$ and R for each sample.

(b) Use all 40 samples to compute the centerlines and control limits for the $\overline{X}$ and R charts.

(c) Construct the charts, plotting all of the $\overline{X}$ and R values and including all appropriate information on the charts.

(d) Apply all the appropriate rules to interpret the charts, indicating every incidence of an out-of-control condition, and identifying the rule that is violated in each case.

(e) What can be said about the stability of the process?

6.36 One step in the production of fine wood furniture is the application of a polyurethane finish. The operator employs a hand-held spray applicator to dispense the polyurethane coating. The quality characteristic of interest is the thickness of the coating (specifications of 0.030 ± 0.005 inch). For the purpose of constructing $\overline{X}$ and R control charts to understand whether the finishing process is in control, 30 samples of size $n = 5$ chairs have been collected, with one sample collected in the morning and one in the afternoon over a period of fifteen working days. An examination of the charts reveals several instances of out-of-control conditions on the R chart; i.e., the process variability is not stable. In order to identify the underlying root cause of the statistical signals on the R chart, it is decided to construct scatter plots for the range values as a function of several process variables, i.e., operator, polyurethane batch, and type of chair. The accompanying table displays the levels of the process variables and the range, R, for each sample. Prepare the scatter plots and comment on what the scatter plots seem to be suggesting regarding which process variable merits further investigation.

Sample	Operator	Batch	Chair Type	Range	Sample	Operator	Batch	Chair Type	Range
1	Simon	A	Rocking	0.0116	16	Simon	E	French	0.0211
2	Simon	A	Rocking	0.0116	17	Simon	F	French	0.0118
3	Simon	A	English	0.0156	18	Simon	F	English	0.0213
4	Simon	B	English	0.0273	19	Simon	F	French	0.0319
5	Simon	B	English	0.0189	20	Simon	G	French	0.0157
6	Simon	B	French	0.0180	21	Simon	G	Rocking	0.0148
7	Vicki	B	French	0.0097	22	Simon	G	Rocking	0.0208
8	Vicki	C	English	0.0089	23	Vicki	H	French	0.0065
9	Simon	C	English	0.0243	24	Vicki	H	French	0.0150
10	Simon	C	Rocking	0.0190	25	Vicki	H	Rocking	0.0111
11	Vicki	D	Rocking	0.0050	26	Vicki	H	English	0.0101
12	Vicki	D	Rocking	0.0033	27	Vicki	H	English	0.0054
13	Vicki	D	Rocking	0.0035	28	Vicki	I	English	0.0063
14	Vicki	D	English	0.0056	29	Simon	I	Rocking	0.0112
15	Simon	E	English	0.0139	30	Simon	I	Rocking	0.0272

6.37 Consider the $\overline{X}$ and R charts shown in the figure. For each chart, identify (circle) the points where statistical signals show that special-cause variation is occurring. Indicate the exact rule that applies. In some cases, multiple rules may apply; for these cases, circle the points multiple times.

6.38 The surface finish of lenses produced by a manufacturer is required to be less than 4 microinches. Recently some customers have complained that the lens finish is exceeding the agreed-upon specification. It is decided to establish control charts to examine the diamond turning operation used to produce the lenses. For the purpose of control chart construction, 32 samples of size $n = 3$ are collected from the process, with the surface finish used as the quality characteristic of interest. These data are shown in the accompanying table.

Sample	X_1	X_2	X_3	Sample	X_1	X_2	X_3
1	3.2	3.6	3.2	5	3.4	4.1	2.9
2	2.5	3.7	4.2	6	3.5	2.7	2.7
3	4.1	3.3	3.7	7	3.2	3.5	2.9
4	3.4	2.8	2.8	8	3.3	3.7	4.1

(Continued)

Sample	X_1	X_2	X_3	Sample	X_1	X_2	X_3
9	3.6	3.5	3.3	21	2.4	3.0	2.8
10	3.3	3.5	4.2	22	3.5	3.2	3.0
11	3.8	2.6	3.8	23	3.1	3.9	3.3
12	2.8	2.8	3.5	24	3.3	3.2	3.6
13	2.5	2.1	4.6	25	3.2	2.2	3.3
14	2.7	2.7	4.3	26	3.0	3.4	3.0
15	3.3	2.8	3.6	27	3.2	4.0	1.4
16	3.2	3.9	3.6	28	3.1	3.4	3.4
17	4.1	3.7	2.9	29	2.9	2.8	2.6
18	3.2	2.3	2.9	30	1.8	3.1	2.9
19	2.6	3.4	3.0	31	3.2	3.1	3.0
20	3.3	4.2	3.4	32	3.6	4.0	3.7

(a) Find $\overline{X}$ and R for each sample.

(b) Use all 32 samples to compute the centerlines and control limits for the $\overline{X}$ and R charts.

(c) Construct the charts, plotting all of the $\overline{X}$ and R values and including all appropriate information on the charts. When creating the charts, insure that there is space to plot a total of 60 samples.

(d) Apply all the appropriate rules to interpret the charts. Indicate every incidence of an out-of-control condition, and identify the rule that is violated in each case.

(e) What can be said about the stability of the process?

Based on the analysis of the control charts, an investigation was conducted and it was learned that the lathe had not received maintenance for several months. Following the implementation of the maintenance procedures, an additional 28 samples of three lenses each were collected and their surface finish measured. These new data are shown in the next table.

Sample	X_1	X_2	X_3	Sample	X_1	X_2	X_3
33	2.8	3.9	3.6	47	3.6	3.7	3.0
34	2.3	2.4	2.1	48	4.2	3.9	3.5
35	3.9	3.8	2.8	49	3.4	2.2	2.8
36	3.4	2.4	3.3	50	2.6	2.3	2.5
37	2.6	3.1	3.2	51	2.8	2.4	2.3
38	2.4	3.3	3.7	52	3.4	3.0	2.9
39	3.3	2.4	3.1	53	3.4	2.7	3.1
40	2.6	2.7	3.3	54	3.8	3.5	3.3
41	3.0	3.0	3.7	55	2.7	3.1	3.6
42	2.6	3.1	3.0	56	2.9	3.6	3.4
43	3.7	3.4	4.0	57	3.1	2.6	2.7
44	2.7	3.0	3.0	58	2.9	3.4	3.4
45	3.4	3.0	3.8	59	2.7	2.9	3.3
46	3.5	3.4	3.9	60	2.2	2.3	3.5

(f) Using the data from subgroups 33 through 60, calculate the new $\overline{X}$ and R values and the new control limits. Then plot these data on the new control charts to confirm that the implementation of the maintenance procedures was effective in stabilizing the process variability.

(g) Is the process now in a state of statistical control? What might be the next step(s) in terms of process study?

Importance
of Rational Sampling

7.1   Introduction

We now discuss perhaps the issue most crucial to the successful use of Shewhart control charts: the design and collection of the samples or subgroups. We discuss the concept of rational sampling; talk about sample size, sampling frequency, and sample collection methods; and review some classic misapplications of sampling. Also, we present a number of practical examples of subgroup definition and selection to aid the reader in understanding and implementing this most central aspect of the control chart concept.

7.2   Concept of Rational Sampling

Rational subgroups or samples are collections of individual measurements, whose variation is attributable only to a constant system of common causes. In the development and continuing use of control charts, subgroups or samples should be chosen in a way that provides the maximum opportunity for measurements within each subgroup to be alike (subject only to the forces of common-cause variation) and the maximum chance for the subgroups to differ from one another if special causes arise between subgroups. Figure 7.1 graphically depicts the notion of a rational sample. Within the sample or subgroup only common-cause variation should be present. Special causes/sporadic problems should arise between the selection of one rational sample and another.

Sample Size Considerations

The size of the rational sample is governed by the following considerations:

- *Subgroups should be subject to common-cause variation.* The sample size should be small if we wish to achieve this objective. If the sample is allowed to become too large, variation may be introduced within the subgroup due

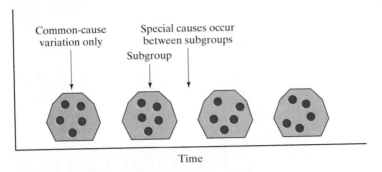

Figure 7.1   **Graphical depiction of the rational subgroup**

to the occurrence of special causes, such as shifts in the process mean level. If subgroups are large enough to allow this to happen often, primarily because the amount of the process captured within a single subgroup is large, the sensitivity of the control chart will be eroded.

- *Subgroups should ensure the presence of a normal distribution for the sample means.* The construction and interpretation of the $\overline{X}$ chart are based on the assumption that the $\overline{X}$ values follow the normal distribution. In general, the larger the sample size, the better the actual distribution of $\overline{X}$ is approximated by the normal curve. In practice, sample sizes of 4 or more generally ensure a good approximation to normality.

- *Subgroups should ensure good sensitivity to the detection of special/assignable causes.* The larger the sample size, the more likely we are to detect a shift of a given magnitude.

- *Subgroups should be small enough to be economically appealing from a collection and measurement standpoint.*

When all of the above are taken into consideration, a sample/subgroup size of 4 to 6 is likely to emerge. Five is the most commonly used number, due to the relative ease of further computation, although this is less important today, since much of the calculation work is done by computers.

Methods of Subgroup Selection

The most obvious basis of selecting subgroups is the time order of production. One method of selecting subgroups by production time is to sample parts all produced at approximately the same time (i.e., consecutively)—for example,

> 5 consecutive measurements at 9:00 A.M.
>
> 5 consecutive measurements at 9:45 A.M.
>
> 5 consecutive measurements at 10:30 A.M., etc.

Here the interval between samples is 45 minutes, but we stress that the selection of the sample generally occurs over a much smaller time period within this interval. Taking measurements consecutively, rather than distributed over the interval, permits the minimum chance for other than common-cause variation within subgroups and the maximum chance for special-cause variation arising between the subgroups to be detected. This method is a more sensitive measure of shifts in the process average and variability.

Rational samples or subgroups should be as homogeneous as possible. By this we mean that we wish to obtain a sample where all variation in measurements within the sample is due entirely to one unique common-cause system. To help accomplish this, we wish for our samples to:

1. *Be comprised of measurements that come at approximately the same time,* so that we do not introduce special-cause variation due to process changes. This suggests that samples should be selected so that the individual observations are consecutive rather than distributed in a production/time-order sense. More will be said on this later.

2. *Have the time for sampling selected randomly* instead of at exactly the same time every day, thus minimizing the effects of occurrences such as shift change, tool wear, and tool change. This may mean that if samples are collected, on average, about once an hour, varying the actual time by ± 15 minutes might be appropriate.

Sampling Frequency

The question of how frequently samples should be collected needs careful thought. In a good number of the applications of $\overline{X}$ and R charts, samples are selected too infrequently to be of much use in identifying and solving problems. Some considerations in sample frequency determination are

1. *General nature of process stability.* If the process under study has not been scrutinized by control charts before and appears to exhibit somewhat erratic behavior, samples should be taken quite frequently to increase the opportunity to identify improvement opportunities quickly. In some cases, particularly where the application of the process (i.e., the rate of generation of the quality characteristic in question) is low, all possible process results may need to be observed to obtain enough data to make good charts. In this case individual measurements would be grouped into samples of consecutive observations. As the process exhibits less and less erratic behavior, the sample interval can be lengthened.

2. *Consideration of frequency of process events.* It is important to identify and consider the frequency with which things are happening in the process. This might include, for example, ambient condition fluctuations, raw material changes, process adjustments such as tool changes or wheel dressings, and so on. If the opportunity for special causes to occur over a 15-minute period is good, sampling twice a shift may be of very little value.

3. *Cost of sampling.* Although it is dangerous to overemphasize the cost of sampling in the short term, clearly it cannot be neglected. If destructive testing is required to observe the quality characteristic of interest, frequent sampling

is difficult unless the problem is so extreme that economic survival of the process is in serious question anyway. In the past, the cost of sampling was routinely calculated and used as a measure of the overall cost of quality. As discussed in Chapter 2, this approach to quality control is ill conceived and should be abandoned. What is often not recognized is that the real waste of money due to sampling arises when samples are collected too infrequently, since then the information/value content of the data is so low. There is considerable evidence, as witnessed by these authors on many occasions, to support the conjecture that many companies waste large amounts of money by sampling so infrequently that the resulting charts have no value.

The bottom line in the sampling frequency question is simply that there is no Santa Claus in the world of quality design and improvement. If you wish to use data to find problems and identify improvement opportunities, you have to collect them. Unless processes are very mature in an SPC sense, exhibiting long-term stability and excellent capability, sampling infrequently will not be very useful. On the other hand, intense sampling over a short period of time may lead to considerable insight on the workings of the process.

7.3   Merits of Consecutive Versus Distributed Sampling

The issue of consecutive vs. distributed subgroup selection requires some special thought. If a process is subject to abrupt shifts in mean level or variability and those shifts are sustained, consecutive sampling may be preferred, since the shift will be more easily detected from one sample result to another. Figure 7.2 illustrates how the process mean might be experiencing abrupt and sustained shifts over time relative to the consecutive subgroup selection. Figure 7.3 shows what the associated $\overline{X}$ and R charts might look like. However, if a process is subject to frequent abrupt but short-lived shifts in mean level, a distributed sample may be preferred, since it provides a better opportunity for detection of a mean level shift through the R chart. Figure 7.4 depicts this type of process behavior and the associated distributed subgroup selection method. Figure 7.5 illustrates how such behavior might be manifested on the $\overline{X}$ and R charts.

If a process is subject to more gradual changes in mean level, this may be detected more readily through distributed sample selection, since a greater opportunity for a large sample range and increased sample mean is present. Figure 7.6 shows a gradual shift in the process mean relative to distributed sample selection. Figure 7.7 shows the associated $\overline{X}$ and R control chart patterns that might arise. Here again, a shift in the process mean may be detected on the R chart by forcing the special cause to occur within the subgroup.

In general, greater attention tends to be focused on the $\overline{X}$ chart and on watching for changes in process mean level. Consecutive sampling generally provides greater sensitivity to the detection of mean level shifts, since the subgroup variability (and hence the $\overline{X}$-chart control limits) is more likely a reflection of only common-cause variation. Distributed sampling should probably be

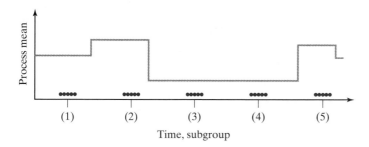

Figure 7.2   Abrupt and sustained shifts in the process mean and the use of consecutive subgroup selection

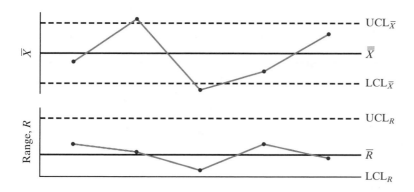

Figure 7.3   Appearance of $\overline{X}$ and R charts for conditions of the nature of Fig. 7.2

reserved for those cases where one believes that certain very well known types of mean shifts are likely to occur.

7.4   Some Common Pitfalls in Subgroup Selection

In many situations it is inviting to combine the output of several parallel and assumed to be identical machines into a single sample to be used in maintaining a single control chart for the process. Two variations of this approach can be particularly troublesome: stratification and mixing.

Stratification of the Sample

In the stratification of subgroups, each of several machines, spindles, filling heads, and so on contributes in the same representative way to each sample—for example, one measurement each from four parallel machines—yielding a

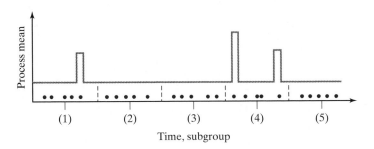

Figure 7.4   Abrupt and short-lived shifts in the process mean and use of distributed subgroup selection

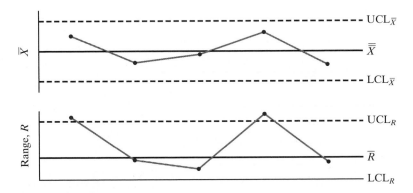

Figure 7.5   Appearance of $\overline{X}$ and R charts for conditions of Fig. 7.4

subgroup of $n = 4$, as seen in Fig. 7.8. In this case there will be a tremendous opportunity for any true differences among the machines to occur within subgroups.

When serious problems do arise, say for one particular machine, they will be very hard to detect because of the use of stratified samples. This sampling problem can be detected on the control charts, however, because of the unusual nature of the $\overline{X}$-chart pattern (recall the previous pattern analysis), and therefore rectified, provided that we understand the concepts of rational sampling. The R charts developed from such data will usually show good control (unless the differences in the process/machine means are very large), while the corresponding $\overline{X}$ control chart will tend to show very wide limits relative to the plotted $\overline{X}$'s and hence their control will appear almost too good. The wide limits result from the fact that the variability within subgroups is likely to be subject to

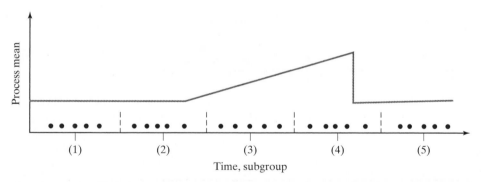

Figure 7.6   Gradual shift in the process mean and the use of distributed subgroup selection

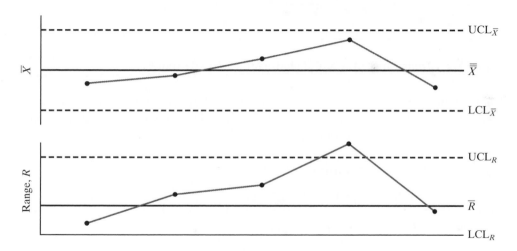

Figure 7.7 $\overline{X}$ and R chart patterns for conditions of the nature of Fig. 7.6

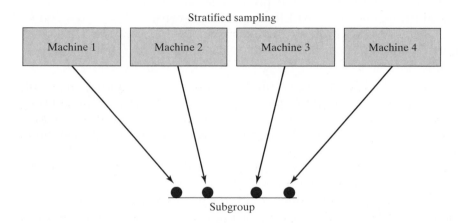

Figure 7.8   Graphical depiction of stratified sample selection

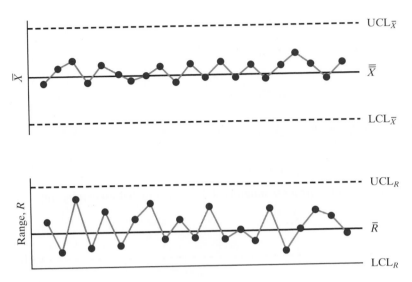

Figure 7.9   **Appearance of $\overline{X}$ and R charts for stratified samples**

more than merely common causes (e.g., real differences in the process means). This is illustrated in Fig. 7.9. Test 8 of the rules for control chart interpretation is specifically directed to this type of sampling problem.

How Stratification of Subgroups Can Occur

Stratification occurs when the subgroup is comprised of representative items that are drawn from similar but distinctly different processes. Stratification occurs primarily for one of three reasons.

1. The process yields a product quality characteristic in a way that gives rise naturally to multiple occurrence of the item. Examples of this would include multiple cavity molds that produce several parts simultaneously, a multihead filling machine that fills several bags, cans, bottles, and so on simultaneously, or a multiple-orifice extrusion die. In each case the parts being made are the same, but they arise from distinctly different applications of the general process.

2. There is a conscious effort to distribute the "points of sampling" in an orderly fashion in time or space. Examples of this type of stratification are sampling uniformly across the width of a sheet as it exits a roll mill, taking several measurements around the circumference of an extruded rubber hose, measuring a dimension on a forging several times over a period of time during cooling, and taking measurements of several different bearing journals after a grinding operation on a crankshaft.

3. An effort is made to shortcut the SPC application by taking a representative sample across several like machines, or lines, selecting one item from each as they proceed roughly in unison. One example here might be drawing one

item each from four parallel injection molding presses, all making the same part. Another example might be taking one item each from three CNC turning centers, all making the same flange.

In each of the situations mentioned above, the common denominator is the fact that the items which comprise the subgroup are different for some known and predictable reason, because of an assignable cause. As such, the principle of rational sampling is violated in each case because the variation within the subgroup is not comprised solely of common causes.

In a roll mill application, sheets exiting the rollers will have a predictable variation in thickness from end to end due to the deflection of the rollers and the associated crowning effect as seen in Fig. 7.10. If four measurements of the sheet thickness are taken across the sheet width, as indicated in the figure, to comprise the subgroup, the known differences in thickness become part of the within-subgroup variation. The subgroup variation will be large and the associated $\overline{X}$ and R charts will have little sensitivity to detect other special causes/sporadic problems when they occur.

Figure 7.11 shows a schematic of a crankshaft with several bearing journals that undergo a final grinding operation. Due to the nature of the fixturing, the deflections during grinding will be different for different journal locations. Because of grinding-wheel differences due to the wheel redress cycle, different

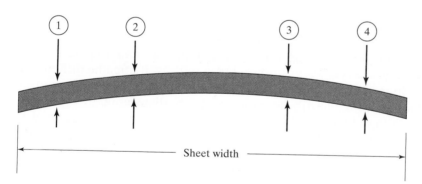

Figure 7.10   Sampling across a sheet of nonuniform thickness

Figure 7.11   Schematic of a crankshaft journal grinding sampling situation

grinding forces will prevail for different journals. If a subgroup of $n = 4$ is formed by measuring each of the four journal locations as indicated in the figure, these predictable/assignable differences will become part of the within-subgroup variation, again seriously inflating the subgroup ranges.

In each of the physical situations described above, each of the individual strata (a mold cavity, a filling head, a sheet thickness location, a single journal) is a separate process and should be treated as such in developing statistical charts. In each case, more would probably be learned that would contribute to improvement even if only one of the strata was sampled and charted.

Mixing the Production from Several Machines

Often it is inviting to combine the output of several parallel machines/lines into a single stream of well-mixed product, which is then sampled for the purposes of maintaining control charts. This is illustrated in Fig. 7.12. Often mixing may be inadvertent or occur totally unknown to those sampling the process. We can think of mixing in much the same way that we think about stratification, in the sense that it tends to arise when processes are present that have multiple applications in close proximity. Many of the examples cited above for stratification can also easily give rise to mixing. In mixing, the basic difference is that instead of having a representative sample with all processes being equally represented in each sample, the composition of the sample arises through a random sampling of the aggregation of items across all of the processes.

When mixing such as that depicted in Fig. 7.12 occurs, a new distribution of the quality characteristic in question is created such that the control charts often appear in good statistical control. Hence not only will mixing mask problems with individual processes/machines, but it will itself often be difficult to detect. If anything, mixing may tend to produce an appearance in the control charts (particularly the $\overline{X}$ chart) where the points are distributed more toward the control limits than they might otherwise be. This may be more apparent if one (or perhaps two) of the machines have means quite different from the others.

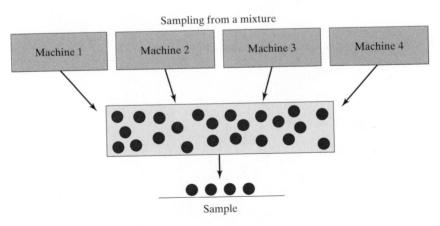

Figure 7.12   Graphical depiction of sampling from a mixture

7.5   Examples of the Stratification Phenomenon

▶ EXAMPLE 7.1 FILLING PROCESS

As an illustration of how stratification during sampling can lead to an unnaturally quiet pattern on the $\overline{X}$ chart, consider the following case. A sample of size
$n = 4$ was taken, one observation from each of four different filling heads on
the same liquid filling machine. However, unknown to the quality analyst
recording the data, the individual outputs (mean levels) of the four filling heads
were somewhat different, as indicated by the four distributions in Fig. 7.13.

For this filling-head example, given that one measurement comes from
each head to comprise the sample, we expect the ranges of the samples to be
quite large (i.e., $\overline{R} \approx 4.0$). Then our estimate of the process standard deviation
will be

$$\hat{\sigma}_X = \frac{\overline{R}}{d_2} = \frac{4}{2} = 2 \text{ (for } n = 4, d_2 \text{ is approximately 2)},$$

whereas if we consider only a single filling head, as seen in Fig. 7.13, the $6\sigma_X$
spread is 1.2 ounces, so

$$\sigma_X = \frac{1.2}{6} = 0.2 \text{ ounce.}$$

Comparing the actual $\sigma_X = 0.2$ with the estimated value resulting from ranges
of stratified samples $\hat{\sigma}_X = 2.0$, we find that they differ by a factor of 10.

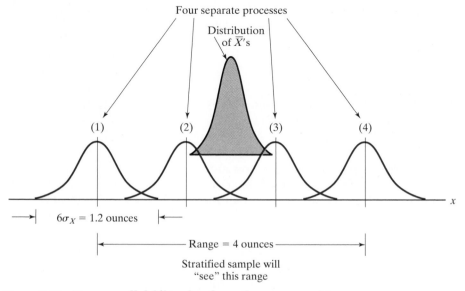

Figure 7.13   **Four parallel filling heads on the same machine**

Thus, based on the $\bar{R}$ from the stratified sample, the control limits on the $\bar{X}$ chart will be unnaturally wide, and the sample values will hug the chart centerline. There will be an absence of points beyond $\pm 1\sigma_{\bar{x}}$. In fact, the distribution of the averages will be centered approximately at the middle of filling heads 2 and 3, as shown in Fig. 7.13.

Another indication of the representative but inappropriate sample will be a relatively flat histogram of all the individual values, which have come from a very specific form of combined populations—the stratified sample. Other examples where stratified sampling can occur include:

- Multispindle screw machine.
- Multicavity mold.
- Grinding multiple journals—crankshaft.

Stratification is, of course, avoided by taking rational subgroups from each filling head or each machine position and setting up $\bar{X}$ and R charts for each process. Remember, control charts are applicable to one and only one process at a time, and there is no gain in taking a representative rather than a rational sample. In fact, one can create a serious misinterpretation of the process behavior through representative sampling.

▶ EXAMPLE 7.2   PROCESS MANUFACTURING RESISTORS

A consultant was visiting a plant that was having some quality problems and was, as a matter of course, reviewing each process running in the plant. In one instance control charts were being maintained for a process making electrical resistors, and the consultant was given the charts to study. The process had already been deemed to be showing excellent long-term stability, and the sample measurements were being routinely collected and charted on the floor. The quality characteristic being measured is the resistance, the nominal/target value being 100 ohms, with a specification of ± 2 ohms. The most recent data used to establish the control charts are shown in Table 7.1 (all measurements are in ohms). The consultant was asked to study this process because, although it showed good statistical control, the process capability study indicated that the process variation was quite large ($\sigma_X = 2$) and as many as one-third of the resistors were out of the specifications. For some time the resistors were being 100% inspected and many bad ones were found. The quality team assigned to improve the process was looking for common causes of variation but with no success. Once the data had been collected, the limits for the control charts were calculated and the charts were constructed, as shown in Fig. 7.14.

$$\bar{\bar{X}} = \sum \frac{\bar{X}_i}{45} = 100.000, \quad \bar{R} = \sum \frac{R_i}{45} = 4.013,$$

$$\text{UCL}_{\bar{X}} = \bar{\bar{X}} + A_2\bar{R}$$

$$= 100.000 + (0.73)(4.013)$$

$$= 102.929,$$

TABLE 7.1	Electrical Resistor Data					
Sample	X_1	X_2	X_3	X_4	$\bar{X}$	R
1	98.18	99.85	100.11	101.46	99.90	3.28
2	97.15	100.15	101.53	101.89	100.18	4.74
3	98.24	98.28	100.56	101.35	99.61	3.11
4	97.32	98.68	101.20	102.22	99.85	4.90
5	97.82	98.16	101.46	102.08	99.88	4.26
6	98.33	98.93	100.99	102.12	100.09	3.79
7	98.21	99.19	100.75	102.41	100.14	4.20
8	97.75	98.76	101.39	102.33	100.06	4.58
9	97.59	98.71	100.00	101.61	99.48	4.02
10	98.02	98.90	101.64	101.74	100.07	3.72
11	98.83	99.49	101.89	100.85	100.27	3.06
12	98.89	99.42	100.26	102.78	100.34	3.89
13	98.01	99.04	101.46	101.95	100.12	3.94
14	98.53	98.72	101.36	102.21	100.20	3.68
15	98.06	98.76	101.07	101.91	99.95	3.85
16	97.86	98.51	101.57	102.21	100.04	4.35
17	97.95	98.89	100.14	102.84	99.95	4.89
18	98.12	99.07	101.56	101.45	100.05	3.44
19	97.88	98.56	101.06	101.19	99.67	3.31
20	97.83	98.81	100.99	102.50	100.03	4.67
21	98.99	100.22	101.14	102.11	100.61	3.12
22	98.49	99.82	100.39	101.80	100.13	3.31
23	97.58	99.15	100.99	102.17	99.97	4.59
24	98.51	99.43	100.92	102.72	100.40	4.21
25	98.27	99.04	101.09	101.93	100.08	3.66
26	97.35	98.57	100.78	102.21	99.72	4.86
27	98.10	99.99	101.33	101.78	100.30	3.68
28	97.12	98.50	101.56	102.70	99.97	5.58
29	97.21	99.66	100.47	102.36	99.93	5.15
30	98.24	99.24	100.38	102.50	100.09	4.26
31	97.99	99.18	101.67	102.07	100.23	4.08
32	98.08	99.06	101.21	101.93	100.07	3.85
33	97.89	98.71	100.98	101.31	99.72	3.42
34	97.72	98.19	101.46	101.43	99.70	3.71
35	97.03	99.23	100.85	102.53	99.91	5.50
36	98.03	98.39	100.93	100.99	99.58	2.96
37	97.72	99.30	101.54	101.57	100.03	3.85
38	97.76	99.00	100.96	100.86	99.64	3.20
39	98.36	99.17	101.28	102.55	100.34	4.19
40	97.89	99.10	100.50	101.71	99.80	3.82
41	97.84	98.46	101.43	101.77	99.87	3.93
42	98.28	98.82	101.00	102.45	100.14	4.17
43	97.82	99.92	100.75	101.66	100.04	3.84
44	98.92	98.37	99.99	102.74	100.00	3.82
45	98.16	98.11	101.13	102.25	99.91	4.14

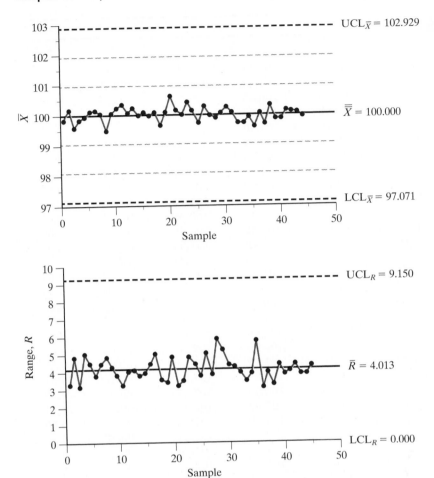

Figure 7.14   Control charts for electrical resistor data

$$\begin{aligned}
\mathrm{LCL}_{\overline{X}} &= \overline{\overline{X}} - A_2\overline{R} \\
&= 100.000 - (0.73)(4.013) \\
&= 97.071,
\end{aligned}$$

$$\begin{aligned}
\mathrm{UCL}_R &= D_4\overline{R} \\
&= (2.28)(4.013) \\
&= 9.150,
\end{aligned}$$

$$\begin{aligned}
\mathrm{LCL}_R &= D_3\overline{R} \\
&= (0)(4.013) \\
&= 0.000.
\end{aligned}$$

The zones referred to in Chapter 6 as A, B, and C on the $\overline{X}$ chart can be found by dividing the distance between the centerline and the upper/lower control limit into three equal widths, the one closest to the centerline being zone C, the middle one zone B, and the farthest outlying one zone A.

Apparently, neither the process operators nor the supervisors had seen problems with this process. The R chart in Fig. 7.14 showed good control, and so did the $\overline{X}$ chart. The $\overline{X}$ chart in Fig. 7.14 shows remarkably good control; it looks almost "too good." All the points are within 1 sigma of the centerline, so immediately it was apparent to the consultant that there was something unusual about the data. The $\overline{X}$ values should arise in a distributional pattern that follows a normal distribution. This means that about one $\overline{X}$ in three should fall beyond ±1 sigma from the centerline.

The management of the company was surprised when the consultant decided to explore further the operation of this process; the process seemed to them to be in very good control. The consultant was quick to point out that the unusual pattern on the $\overline{X}$ chart indicated that something was amiss. The manner in which the points on the $\overline{X}$ chart seem to hug the centerline, combined with the fact that the R chart also appears to have an unusually high percentage of points within the middle third of the chart, points to a possible problem with the sampling procedure. The consultant decided to find out exactly how the samples on this line were being collected.

It turned out that the line was set up so that each machine manufacturing the resistors would produce a stream of parts that would travel to another machine responsible for encasing the parts with an insular coating. This machine was set up to receive one part from each of the four machine streams, coat them, and then eject the four parts simultaneously from the mold. At this point it was convenient to define and physically collect a sample of these four resistors before the parts had a chance to move along a conveyor belt to be color coded for identification purposes farther down the line. A schematic of the process is shown in Fig. 7.15.

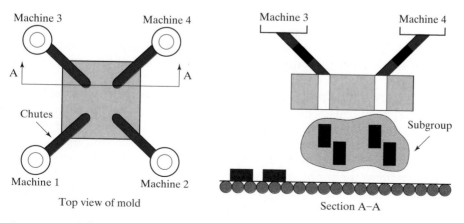

Figure 7.15   **Schematic of the resistor coating process**

It is clear that a fundamental sampling error known as *stratification* is taking place. Four machines are contributing equally to the composition of the sample, which is then observed for control chart purposes. Because the machines are producing the same product does not necessarily mean that they are identical products. The four machines are in actuality four different processes, so combining the product in this way (i.e., stratifying the data) creates a tremendous opportunity for special causes to occur within subgroups rather than between them. As a result, the within-subgroup range could be quite large, reflecting differences in the mean output of each of the four machines. This means that the ranges R will be inflated by such differences, the limits on the $\overline{X}$ chart will be very inflated/wide, and as a result, it becomes very difficult to identify any problems. The process seems to be in control, but the chart is actually very insensitive to changes in the mean level(s) of the individual machines.

The solution is to make separate control charts for each machine that is producing resistors, so that common causes of variation can be properly observed and any special causes properly identified. In this way, variation and inefficiency in the process can be reduced, making the process stable, more economical, and more capable. Once the error in the sampling method was pointed out, the sampling point was moved upstream, so that sampling could be performed before the parts reached the molding process. Under this system, rational samples could easily be taken from each/any of the four machines, so that four independent sets of control charts could be maintained. Under this new system of sampling, it was discovered that although it appeared previously that the process was well centered at the nominal of 100 ohms, none of the four machines was actually centered at the nominal. The new control charts for machine 1, shown in Fig. 7.16, demonstrate this phenomenon. Although the $\overline{X}$ and R charts show good control, the mean for machine 1 lies at 98.017, or approximately 2 ohms away from the nominal value of 100 ohms. As indicated in Fig. 7.17, the mean of machine 1 is actually about 8 standard deviations from the nominal.

Each of the other three processes shows a similar problem with centering. When plotted as distributions on a common graph (see Fig. 7.18), it is easy to see the wide range of mean values that the processes encompass. Although the means of the four processes when averaged produce a grand average at the nominal, obviously each individual process varies from that by a very significant amount.

It is now clear that the problem was not a common-cause problem at all but one in which special causes (machine-to-machine differences in mean values) were responsible for the inability of the process, overall, to be capable. By centering all four machines at the nominal with no further reduction in common-cause variation, each machine by itself is 16 sigma capable. This case study clearly demonstrates the problems associated with poor control chart sampling techniques and the waste of time and effort that would result from the misleading information coming from the control charts.

Figure 7.16   Control charts for machine 1

Figure 7.17   Model of machine 1 process

Figure 7.18   **Distributions of all four machines**

■ EXERCISES

7.1 **(a)** When defining and selecting samples/subgroups, do you want special causes to be occurring within the samples/subgroups or between
them? Why?

   (b) What possible dangers are involved with taking samples at precisely
the same time every day?

7.2 Samples are being taken at 1-hour intervals from an assembly line (600
units per hour). To make up each sample of size $n = 4$, the four units are
taken off the line, one by one, at approximately 15-minute intervals (to
produce one sample per hour).

   (a) Comment on the advisability of this type of sampling to detect:

   (i) Abrupt and short-lived shifts in the mean.

   (ii) Gradual, trendlike shifts in the mean.

   (iii) Abrupt and sustained shifts in the mean.

   (b) What sampling scheme would be best for the detection of each type
of condition in part (a)?

7.3 **(a)** If you suspect that frequent raw material batch changes are behind
special-cause variation in a certain process, would consecutive or
distributed sampling be best used in collecting the samples? Why?

   (b) If you suspect that accelerated tool wear is behind special-cause variation, would consecutive or distributed sampling be best used in collecting the samples? Why?

   (c) If you suspect that very short-lived but somewhat frequent power
surges are behind special-cause variation, would consecutive or distributed sampling be best used in selecting the samples? Why?

7.4 For an injection molding process, a four-cavity mold is being used for a
certain part. It has been proposed to develop $\overline{X}$ and R charts for part
weight, where the subgroup/sample is composed of the four parts from
a single shot (i.e., one part from each cavity). Comment on the appropriateness of this method of sampling. What impact does it have on the
ability of the charts to detect changes in the process? Alternatively, a
proposal has been made that after each shot/cycle the four parts be broken off the sprues and runners and put into a rack. Samples would then

be randomly drawn from this rack. Again, comment on the appropriate-ness of this method of sampling and explain the impact it would have on the charts.

7.5 A process for the manufacture of vinyl for use in automobile instrument panels employs a calender-type process to roll-form the vinyl to a pre-specified thickness. The continuous sheets are measured for their thick-ness several times a shift. Each time this is done, three measurements are taken across the width of the sheet: a left-side, center, and right-side measurement. The quality control engineer decides that she will use these data, samples of $n = 3$, to make Shewhart control charts for this process. Comment on the advisability of her approach. Can you suggest an alternative method of charting this process?

7.6 Two identical molding machines producing the same part are being charted separately and both show good stability. The characteristic being monitored is part weight. Both processes are centered at the nominal, and both are 3 sigma capable. Seeing these facts, the quality control engineer decides to start taking two parts from each line for every sample ($n = 4$) and only use one set of charts. This way, she feels, she cuts the sampling cost in half while continuing to monitor both machines. Comment on the validity of her reasoning, and explain the impact of this change on the overall scheme of things. How would you modify the sampling process to reduce costs but maintain surveillance of the two machines?

7.7 A process that is making plastic parts for CD cases has never been moni-tored with SPC charts but has lately been unable to meet production demands, so a decision is made to begin charting the process. The line produces 650 parts per hour, from 8:00 A.M. to 4:00 P.M. each day. Rec-ommend an appropriate sampling scheme, including sample size, time/frequency of sampling, and whether consecutive or distributed sam-pling should be used. Justify your scheme.

7.8 A process that is producing approximately 10% defects is being reviewed as a candidate for SPC charting. After some consideration, the quality control engineer concludes that it is cheaper to rework the defective parts than to submit the process to the necessary destructive testing to study the characteristic in question properly. If this person were your boss, how would you convince him to do the testing, or would you agree with his decision?

7.9 Why is it so important that sampling schemes be carefully chosen? Give an example of how an unwise selection of a sampling scheme can affect the resulting usefulness of the control charts made.

7.10 Assuming that the R charts associated with each of the $\overline{X}$ charts (i) to (iv) show good control:
   (a) Which of the $\overline{X}$ charts might indicate that a problem with stratifica-tion might exist? Explain.
   (b) Which of the $\overline{X}$ charts may be examples of mixing the production from several machines? Explain.

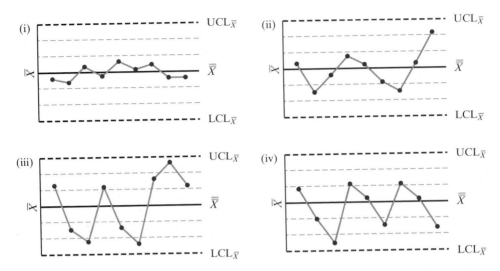

7.11 In this chapter, the practice of taking several measurements around the circumference of an extruded rubber hose is given as an example of stratifying the data. Explain why this is stratification and propose an alternative sampling method.

7.12 At one station in a chrome-plating process for plastic automobile grilles, the parts, hung on 25 individual hooks, are immersed simultaneously for 20 seconds, after which they are removed and held for another 40 seconds before moving to the next station. The level of contaminants on the surface of the part at this point in the process is the quality characteristic of interest here. To measure this, the part must be removed to the lab for testing. Total part cycle time for plating is 20 minutes. To minimize lost production and keep a balanced flow, one part is taken from each successive rack of parts and five consecutive parts are then grouped to form a subgroup for charting. Comment on this scheme. Do you have any suggestions?

7.13 To save money, the quality control engineer at a plant decides to use a sample of size $n = 2$ on a line producing 500 parts per hour. What could you tell him about the advisability of using such a small sample size? What effect will this approach have on the control charts and the information they will yield?

7.14 What is the purpose of collecting all of the observations in a single sample consecutively rather than randomly over the sampling interval?

7.15 What would be some valid reasons for changing from a sample size of $n = 6$ to one of $n = 4$?

7.16 Discuss the considerations/trade-offs involved in selecting the sample size n that should be used for the purpose of constructing control charts for a certain process.

7.17 What considerations come into play when determining the sampling frequency to be used in collecting data from a process for control charting purposes? Explain.

7.18 A certain process involves punching five holes in sheets of metal. The holes have the same diameter specification. The die contains five punches (i.e., all of the holes are punched in one stroke of the die). It is decided to make control charts for this process to monitor hole diameter. One sheet is taken about every 10 minutes, and the five hole diameters on that sheet are measured to form a subgroup for the chart. Comment on this strategy. Do you have any suggestions?

7.19 A process that fills bottles (throughput = 300 bottles/hour) with a certain expensive chemical uses a two-station filling machine. The first station quickly fills the bottles to nearly the proper level by weight, while at the second precision station the bottle is weighed and then carefully topped off to the proper final weight. Owing to concerns about the throughput of this filling process, charting is begun on the precision filling station. Comment on this situation and recommend how the filling process should be charted to work most effectively toward increasing the throughput of the process.

7.20 What is "rational sampling"?

7.21 What are some important factors that must be considered in determining a sample size that will result in rational samples?

7.22 What is the main difference between consecutive sampling and distributed sampling? For which potential types of special causes should each of these types of sampling be employed?

7.23 The following table displays 40 samples of data that have been collected for the construction of $\overline{X}$ and R charts.

(a) Calculate the $\overline{X}$ and R values for each sample. Compute the center-

Sample	X_1	X_2	X_3	X_4	Sample	X_1	X_2	X_3	X_4
1	81.9	91.0	101.2	114.5	21	91.3	101.4	100.9	105.1
2	90.2	89.5	102.7	108.8	22	86.4	93.5	99.1	113.7
3	84.7	96.3	104.5	121.0	23	79.8	95.7	109.3	118.2
4	83.9	97.4	101.9	117.3	24	89.4	101.2	104.2	104.3
5	95.1	93.5	93.8	111.5	25	90.7	98.4	97.8	118.8
6	80.9	96.2	109.2	110.4	26	86.9	90.9	101.2	124.4
7	79.3	98.1	100.7	110.9	27	79.8	98.1	95.8	111.1
8	86.1	93.0	106.3	115.3	28	90.3	93.4	103.6	123.6
9	79.3	86.0	98.5	114.3	29	80.8	103.5	100.9	114.6
10	89.8	97.1	101.2	121.0	30	97.8	95.7	106.9	121.5
11	88.4	94.2	100.8	123.7	31	82.1	92.8	104.3	121.1
12	83.6	93.4	97.4	107.9	32	81.5	91.1	103.5	117.7
13	75.8	78.9	104.4	116.2	33	82.2	98.5	97.6	109.1
14	86.2	97.7	101.9	110.6	34	84.2	94.0	98.6	116.6
15	96.9	98.6	101.4	105.3	35	82.8	98.5	110.3	117.1
16	83.1	95.3	109.9	113.4	36	87.8	88.5	107.7	116.2
17	76.6	95.2	104.6	112.3	37	95.4	97.5	103.8	113.6
18	88.0	95.6	116.8	120.6	38	78.7	91.8	100.6	118.7
19	86.4	97.6	98.5	115.7	39	90.4	105.6	103.4	116.7
20	76.5	90.2	109.6	107.9	40	82.2	93.2	95.5	115.0

lines and control limits for the $\overline{X}$ and R charts.

(b) Construct the $\overline{X}$ and R charts and plot all of the $\overline{X}$ and R values; be sure to display all appropriate information on the charts.

(c) Apply all the appropriate rules to interpret the charts, indicating every instance of an out-of-control condition and identifying the rule that is violated in each case.

(d) Based on the appearance of the charts, what conclusion(s) can you draw about the process?

7.24 The following table displays 40 samples of data that have been collected from a process in the same production facility as in Exercise 7.23.

(a) Find $\overline{X}$ and R for each sample.

Sample	X_1	X_2	X_3	X_4	Sample	X_1	X_2	X_3	X_4
1	44.6	48.0	50.0	57.3	21	43.6	46.9	53.4	55.0
2	45.8	48.2	52.4	59.9	22	48.1	50.4	47.1	58.9
3	46.7	46.8	49.7	54.2	23	43.2	49.3	50.4	55.7
4	42.2	47.0	52.5	53.9	24	46.8	47.3	52.8	56.7
5	45.9	47.6	52.5	58.7	25	44.5	48.2	52.5	56.7
6	48.9	50.1	52.0	53.0	26	41.0	47.0	50.1	57.3
7	45.6	45.6	50.1	57.4	27	47.0	49.7	52.7	55.7
8	43.8	45.9	54.1	56.4	28	40.2	50.3	48.4	57.1
9	46.7	50.5	55.0	55.8	29	42.5	53.2	52.5	56.6
10	40.9	48.9	51.0	57.9	30	44.1	47.5	50.9	54.1
11	45.9	48.1	50.0	55.2	31	43.9	51.5	48.8	60.9
12	44.2	44.7	48.1	58.0	32	45.2	48.7	50.1	56.6
13	46.0	49.5	48.4	57.1	33	42.0	49.3	53.3	56.4
14	45.2	46.6	51.2	55.0	34	42.8	49.8	52.5	56.1
15	45.0	50.4	49.7	58.0	35	44.7	45.7	49.3	58.0
16	45.6	48.2	55.3	53.6	36	48.0	45.6	52.5	54.4
17	43.8	51.1	52.4	55.8	37	46.9	47.1	54.7	55.5
18	43.6	50.0	51.9	55.9	38	46.9	46.1	51.3	55.2
19	43.1	44.9	50.9	54.7	39	44.2	50.4	55.0	55.5
20	44.6	46.2	50.9	59.3	40	44.3	49.7	54.0	54.5

(b) Compute the centerlines and control limits for the $\overline{X}$ and R charts.

(c) Construct the charts, plotting all of the $\overline{X}$ and R values and including all appropriate information on the charts.

(d) Apply all the appropriate rules to interpret the charts, indicating every incidence of an out-of-control condition and identifying the rule that is violated in each case.

(e) What do the charts appear to suggest about the state of the process? Suppose the sample data in the accompanying table were actually

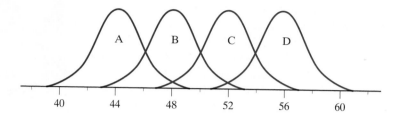

collected from four production lines (A, B, C, and D), which may be described by the four distributions shown in the accompanying figure. The output from the production lines flows into a common bin, from which the samples shown in the table were taken.

(f) Based on this new information and the appearance of the control charts, draw a conclusion about the current situation. What action should be taken at this point?

8

Interpretation of $\overline{X}$ and R Control Charts: Use of Sampling Experiments

8.1   Introduction

This chapter focuses on the interpretation of Shewhart control charts for variable data, in particular, $\overline{X}$ and R charts. To achieve this, the chapter presents the use of sampling experiments to aid in the study of control chart patterns. The results of two sampling experiments are discussed in considerable detail. In the first sampling experiment disturbances are introduced to the data of an initially stable process and chart interpretation rules are invoked, thereby signaling the occurrence of these disturbances. In the second sampling experiment the consequences of one form of nonrational sampling, stratification, are examined in some detail.

In Chapter 5 we discussed the nature of variation and, in particular, noted the distinction between the occurrence of special causes of variation and the constant system of common causes that otherwise drive the performance of the process. A process was said to be in control when it was determined to be free of special causes, that is, driven only by the common-cause system. We have seen that when the performance of a process is driven only by such a causal system, data obtained through rational sampling will behave according to certain probabilistic laws. These laws describe how certain functions of the data, such as averages or ranges, will behave when plotted over time or in a frequency sense. These laws allow us to predict in a probability sense the limits of variation in the data. With such a well-defined statistical basis for the behavior of data emanating from a common-cause system, it is possible to determine the precise probability of occurrence of certain characteristic patterns as they arise over time.

8.2   Analysis of Control Chart Patterns: A Sampling Experiment

In this section a sampling experiment is conducted and studied to illustrate the nature of the forces of common-cause variation and the manner in which control charts indicate the presence of special causes/sporadic problems. The sampling experiment illustrates the application of the control chart pattern rules/tests discussed in Chapter 6 toward the identification of the purposeful introduction of changes in the mean and/or the standard deviation of the process.

Value of Sampling Experiments

The use of sampling experiments as a means to illustrate the workings of common-cause variation and the manner in which special causes of variation manifest themselves on control charts is not new. Working and Olds, in their famous World War II War Production Board eight-day courses on quality control using statistical methods, made extensive use of sampling experiments.[1] Previously, Shewhart had introduced such experiments as a vivid way to demonstrate his principles.

In the days of Working and Olds such experiments were conducted using chips and beads, which were either numbered in proportion to the frequencies of various normal distributions or color coded to provide for various proportions of defective and nondefective parts. Deming's Red Bead experiment, described in Chapter 5, is such an example. Burr[2] provides a tabulation of the relative frequencies to be used for numbering chips to obtain normal distributions with varying means and standard deviations. These are given in Table 8.1. Each set of chips may then be placed in a bag or other convenient container, thoroughly mixed, and samples of various sizes drawn to simulate sampling from a physical process. These authors have used such sampling experiments on countless occasions and have found them tremendously useful. Of course, other normal distributions (different μ_X and σ_X values) may be employed. Today, the use of computers makes it possible to extend and enrich considerably the concept of sampling experiments. In this section we examine the results of one such sampling experiment, conducted using the old technology of bags of chips.

Initial Sampling Results

Table 8.2 gives the results of taking 25 random samples of $n = 5$ observations each from a bag of chips numbered according to one of the normal distributions in Table 8.1 (which bag was used is unknown to the reader at this point). The samples were collected by drawing five chips from the bag, recording the number on each chip, returning the chips to the bag, mixing the chips, drawing

---

[1]H. Working and E. G. Olds, *Manual for an Introduction to Statistical Methods of Quality Control in Industry: Outline of a Course of Lectures and Exercises*, Office of Production Research and Development, War Production Board, Washington, D.C., 1944.

[2]I. W. Burr, *Statistical Quality Control Methods*, Marcel Dekker, New York, 1976.

TABLE 8.1	Frequency Distributions for Sampling Experiments (Frequencies Approximately Normal)			
X	A	B	C	D
+11				1
+10			1	1
+9			1	1
+8			1	3
+7		1	3	5
+6		3	5	8
+5	1	10	8	12
+4	3	23	12	16
+3	10	39	16	20
+2	23	48	20	22
+1	39	39	22	23
0	48	23	23	22
−1	39	10	22	20
−2	23	3	20	16
−3	10	1	16	12
−4	3		12	8
−5	1		8	5
−6			5	3
−7			3	1
−8			1	1
−9			1	1
−10			1	
Number of chips	200	200	201	201
Mean, μ_X	0.0	2.0	0.0	1.0
Standard deviation, σ_X	1.715	1.715	3.470	3.470

another five chips, and so on. For each sample the average $\overline{X}$ and the range R were calculated according to the formulas discussed previously.

Determination of Trial Control Limits

By using the results of the 25 samples in Table 8.2, centerlines and trial control limits were determined as shown below for the construction of $\overline{X}$ and R control charts.

R chart:

$$\text{Centerline:} \quad \overline{R} = \sum_{i=1}^{25} \frac{R_i}{25}$$

$$= \frac{88}{25}$$

$$= 3.52.$$

TABLE 8.2	Results of the First 25 Samples for the Sampling Experiment						
Sample	X_1	X_2	X_3	X_4	X_5	$\bar{X}$	R
1	3	−1	−1	0	−2	−0.2	5
2	−1	1	1	1	−2	0.0	3
3	3	2	1	−2	0	0.8	5
4	0	0	1	1	−1	0.2	2
5	−1	−3	1	1	2	0.0	5
6	0	0	0	−1	−2	−0.6	2
7	0	0	0	1	1	0.4	1
8	0	2	−2	−2	−3	−1.0	5
9	3	−3	−2	−1	−1	−0.8	6
10	0	−1	2	−2	0	−0.2	4
11	2	2	2	2	−1	1.4	3
12	−1	−1	2	−3	−1	−0.8	5
13	0	0	1	0	2	0.6	2
14	0	−2	−1	1	2	0.0	4
15	1	−3	−1	−1	−1	−1.0	4
16	2	−1	−2	−1	2	0.0	4
17	2	2	−1	−1	3	1.0	4
18	0	3	2	−2	−2	0.2	5
19	0	0	−1	−1	−1	−0.6	1
20	−2	−1	0	−2	0	−1.0	2
21	−1	−1	0	2	3	0.6	4
22	0	1	−1	−1	1	0.0	2
23	1	−1	0	−2	1	−0.2	3
24	0	−2	−1	−1	−1	−1.0	2
25	3	2	−1	−1	−2	0.4	5

$$\text{UCL}_R: \quad D_4\bar{R} = (2.11)(3.52)$$
$$= 7.427.$$
$$\text{LCL}_R: \quad D_3\bar{R} = 0(3.52)$$
$$= 0.0.$$

$\bar{X}$ chart:

$$\text{Centerline: } \bar{\bar{X}} = \sum_{i=1}^{25} \frac{\bar{X}_i}{25}$$
$$= -0.072.$$
$$\text{UCL}_{\bar{X}}: \quad = \bar{\bar{X}} + A_2\bar{R}$$
$$= 1.970.$$
$$\text{LCL}_{\bar{X}}: \quad = \bar{\bar{X}} - A_2\bar{R}$$
$$= -2.114.$$

By using the plotting conventions discussed previously, $\bar{X}$ and R charts were constructed and are shown in Fig. 8.1.

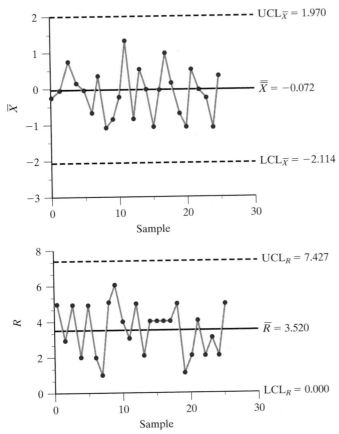

Figure 8.1 $\overline{X}$ and R charts based on the 25 sample results of Table 8.2

Interpretation of the Initial Control Charts

To interpret the patterns in the control charts, we always begin with the R chart. This is important, since if the R chart does not exhibit good statistical control, the $\overline{X}$-chart control limits are not appropriate, and therefore the $\overline{X}$ chart should be interpreted with caution. Clearly, if the R chart is not in control, the $\overline{R}$ value is influenced by the presence of special causes, thereby being either inflated or deflated from the value it would take if only common-cause variation sources were present. As a result, the control limits on the $\overline{X}$ chart will be either wider or narrower than they would otherwise be.

The results of examining the pattern of R values on the R chart in Fig. 8.1 are as follows.

- No extreme points are evident.
- There is a reasonable distribution of points within limits. Roughly two-thirds of the R values should fall in the middle one-third of the chart.
- No trends or cycles are evident.
- No runs of eight or more are evident.

We conclude that there is no reason to believe the process is not stable with respect to its variability.

Since the R chart appears to be in statistical control, we may continue with the interpretation of the patterns in the $\overline{X}$ chart. As discussed previously, it is useful to establish the zones as shown in Fig. 6.6. With these zones in place, numerous tests, as outlined in Chapter 6, can be applied to the pattern of $\overline{X}$'s. The distribution of $\overline{X}$ values by zones follows.

$$\text{Zone } C: \tfrac{16}{25} = 64\% \quad \text{should be } 68.26\%.$$
$$\text{Zone } B: \tfrac{8}{25} = 32\% \quad \text{should be } 27.19\%.$$
$$\text{Zone } A: \tfrac{1}{25} = 4\% \quad \text{should be } 4.28\%.$$

This distribution of points within the limits appears to follow closely what is expected based on the normal distribution. In addition, the analysis indicates that:

- No extreme points are evident.
- No zone rule tests are violated.
- No trends or cycles or other such nonrandom patterns are evident.
- No runs of eight or more above or below the centerline are evident.

We can therefore conclude that there is no reason to believe the process is not stable with respect to its mean value. Since both charts give the appearance of a process in good statistical control, we can extend the centerlines and control limits and continue to sample, using these charts as the model for the way the data should continue to behave if only common causes of variation are at work.

Continuation of the Sampling Experiment

With $\overline{X}$ and R charts in place showing good statistical control, the sampling experiment was continued by selecting an additional five samples of $n = 5$ observations each in the manner discussed previously. Table 8.3 summarizes these sampling results. Note that while such charts should, in practice, be continued on an individual sample-to-sample basis as each new sample result becomes available, we have selected the samples in groups of five for the purpose of the presentation here. The $\overline{X}$ and R values for these additional five samples were calculated and plotted as a continuation of the original $\overline{X}$ and R charts, as shown in Fig. 8.2.

TABLE 8.3	Results of Samples 26 to 30						
Sample	X_1	X_2	X_3	X_4	X_5	$\overline{X}$	R
26	0	1	−1	3	1	0.8	4
27	0	0	−2	−1	2	−0.2	4
28	−1	1	−1	−3	4	0.0	7
29	0	2	−1	2	1	0.8	3
30	0	−1	0	1	2	0.4	3

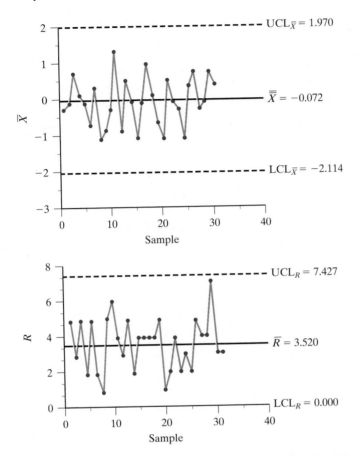

Figure 8.2   **Continuation of $\overline{X}$ and R charts for samples 26 to 30**

In our continuing examination of the evolving patterns on the $\overline{X}$ and R charts, it is important that we adopt a sample-by-sample approach to the interpretation of the charts; that is, we must interpret the charts as if the samples were being taken one at a time, each $\overline{X}$ and R pair being interpreted in turn as it arises. This is what we would do if the charts were evolving at the process, with samples taken every hour or half-hour, and so on.

As we, in turn, examine the results of samples 26 to 30, it becomes clear that no unusual patterns are evident. This conclusion comes not from a cursory "eyeballing" of the data but rather by methodically applying each of the tests for chart interpretation that we discussed in Chapter 6. This is a bit tedious but must be done. Graphical and analytical aids may be developed to assist in this examination. It is clear that the computerization of these rules would be very helpful. Our conclusion in interpreting the patterns in samples 26 to 30 is that there is no reason to suspect that anything has changed from what we observed in the first 25 samples. In short, the process appears to be continuing in a state of control or stability—at least we have no evidence to the contrary.

TABLE 8.4	Results of Samples 31 to 35						
Sample	X_1	X_2	X_3	X_4	X_5	$\bar{X}$	R
31	−1	1	0	−1	−4	−1.0	5
32	−3	−2	−3	−4	0	−2.4	4
33	−3	−2	−2	−2	−1	−2.0	2
34	1	−3	−1	−2	−5	−2.0	6
35	0	−3	−4	−3	−1	−2.2	4

Table 8.4 shows the results of taking another five samples. The $\bar{X}$ and R values for samples 31 to 35 were plotted as a continuation of the previous control charts. The results of the continued plotting of the charts are shown in Fig. 8.3.

Applying all the appropriate tests to the interpretation of the evolving patterns in the R chart leaves us no reason to suspect that the process is not continuing in a state of statistical control as far as its variation is concerned. However, examination of the $\bar{X}$ chart indicates that an unusual pattern of variation in the

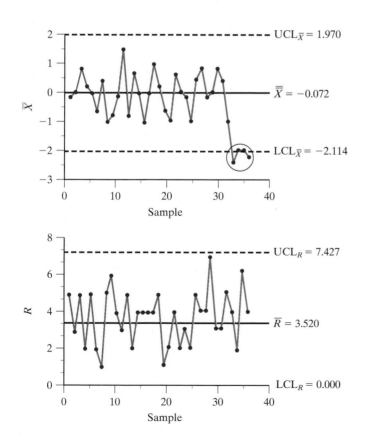

Figure 8.3   Continuation of the $\bar{X}$ and R charts for samples 31 to 35

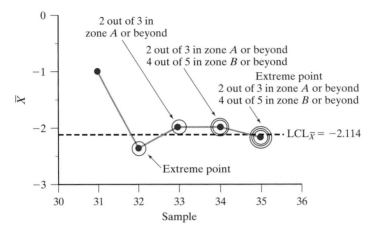

Figure 8.4   Interpretation of samples 31 to 35 on the $\overline{X}$ chart

$\overline{X}$'s begins to arise at sample 32. Here, again, it is important that we carefully apply all of the appropriate tests, sample by sample.

Figure 8.4 focuses on that portion of the $\overline{X}$ chart circled in Fig. 8.3 and shows the results of the proper application of the tests relevant to the $\overline{X}$ chart. The presence of the extreme point at sample 32 is clear evidence that the mean of the process must have shifted downward, at least at that time. This shift appears to be sustained through sample 35. Sample 33 violates the "two-out-of-three samples in zone A or beyond" test. For samples 34 and 35, two and three of the rules are violated, respectively.

Table 8.5 provides that data for the next five samples of this sampling experiment. The $\overline{X}$ and R values are plotted as a continuation of the control charts and are shown in Fig. 8.5. Continuing with our point-by-point examination of the evolving patterns in the $\overline{X}$ and R charts for samples 36 to 40, the following observations can be made:

- The R chart continues to exhibit good statistical control. There is no evidence to suggest that the process variability level has changed.
- The $\overline{X}$ values for samples 36 to 40 all violate more than one rule.
- There continues to be strong evidence of a sustained downward mean shift over samples 32 to 40. It would appear that the process mean has shifted to a level around -2.0.

TABLE 8.5	Results of Samples 36 to 40						
Sample	X_1	X_2	X_3	X_4	X_5	$\overline{X}$	R
36	-4	-4	0	-1	-1	-2.0	4
37	0	-2	-4	-1	-1	-1.6	4
38	-3	0	0	-3	-1	-1.4	3
39	-2	-2	-1	-4	-4	-2.6	3
40	-1	-2	0	-3	-3	-1.8	3

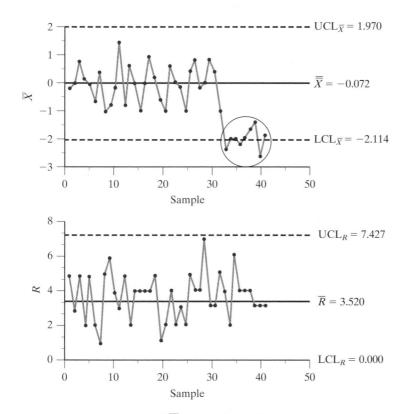

Figure 8.5   Continuation of $\overline{X}$ and R charts for samples 36 to 40

Figure 8.6 shows the section of the $\overline{X}$ chart that includes the results of samples 36 to 40. Each time a rule is violated, a point is circled. From the figure it is clear that each of the $\overline{X}$ values violates more than one of the rules we have been applying. As an exercise the reader should review exactly which of the rules are violated for each of the $\overline{X}$ values. For example, in the case of sample 40 four rules are violated. Which ones are they?

Table 8.6 gives the results of the sampling experiment for samples 41 to 45. The $\overline{X}$ and R values are plotted as a continuation of the existing $\overline{X}$ and R charts. These updated charts are shown in Fig. 8.7.

Upon continuing with our point-by-point interpretation of the evolving patterns in the $\overline{X}$ and R charts, we make the following observations for samples 41 to 45:

- On the R chart, samples 41, 42, 44, and 45 are all extreme values (above UCL_R). Therefore, beginning with sample 41 there is a clear indication of an upward shift in process variability over samples 41, 42, 44, and 45.
- Samples 41, 42, and 43 have $\overline{X}$ values that continue to violate at least one of the zone rules. Under the conditions of a stable R chart this would be

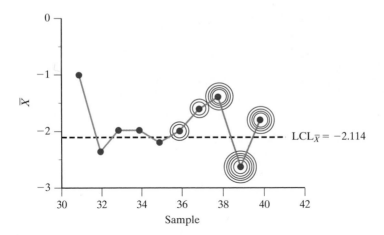

Figure 8.6   **Summary of the rule violations for the $\overline{X}$ value from samples 36 to 40**

TABLE 8.6	Results of Samples 41 to 45						
Sample	X_1	X_2	X_3	X_4	X_5	$\overline{X}$	R
41	0	2	4	−5	−6	−1.0	10
42	−3	−1	−6	3	1	−1.2	9
43	2	4	−1	2	0	1.4	5
44	6	1	−4	3	−3	0.6	10
45	4	−2	−5	−1	−2	−1.2	9

indicating that there is reason to believe that the shift in the process mean is sustained through sample 42. However, we must be careful at this point, since the R chart has shown out-of-control conditions at samples 41 and 42—in particular, an increase in variability. Such would naturally cause the $\overline{X}$ values to exhibit increased variation. Hence we no longer may feel confident that the conditions on the $\overline{X}$ chart at samples 41 and 42 are signaling a shift downward in the process mean. Sample 43 has an $\overline{X}$ value that is now above the original chart centerline but violates the "8 points in a row avoiding zone C" rule. Samples 44 and 45 have $\overline{X}$ values that do not violate any of the rules.

- Note that the variation in the $\overline{X}$ values for samples 41 to 45 appears to be somewhat larger than that evident in the $\overline{X}$'s up to this point. This should not be too surprising, since an upward shift of the process variation as indicated on the range chart should manifest itself on the $\overline{X}$ chart as well. This is so since the variation in $\overline{X}$'s is a function of the variation in the individual observations, σ_X.

In summary, samples 41 to 45 are indicating that the variation in the process has experienced an upward shift. The only exception is the range value for sample 43, which does not indicate that a shift in the variation has taken

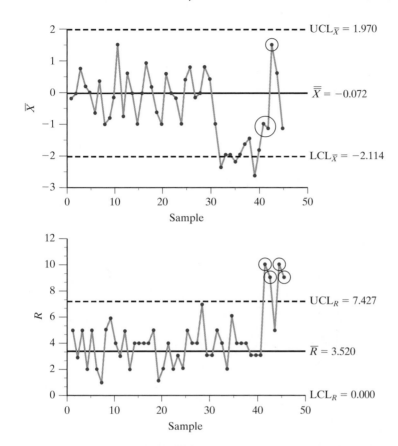

Figure 8.7   Continuation of $\overline{X}$ and R charts for samples 41 to 45

place. Further, it appears that the shift in the process mean that was evident from 32 to 40 is no longer present (from sample 43 to sample 45 no signals are present that would indicate a sustained downward shift). As indicated above, caution must be exercised in drawing conclusions about samples 41 and 42 because of the increase in variation suggested by the R chart.

Table 8.7 gives the results of the last five samples for this sampling experiment. The associated $\overline{X}$ and R values are plotted as a continuation of the previous $\overline{X}$ and R charts. These updated charts are shown in Fig. 8.8.

TABLE 8.7	Results of Samples 46 to 50						
Sample	X_1	X_2	X_3	X_4	X_5	$\overline{X}$	R
46	2	0	−3	6	−5	0.0	11
47	2	−1	1	−2	−1	−0.2	4
48	−3	2	0	−1	2	0.0	5
49	−1	3	3	−1	0	0.8	4
50	0	2	2	−1	0	0.6	3

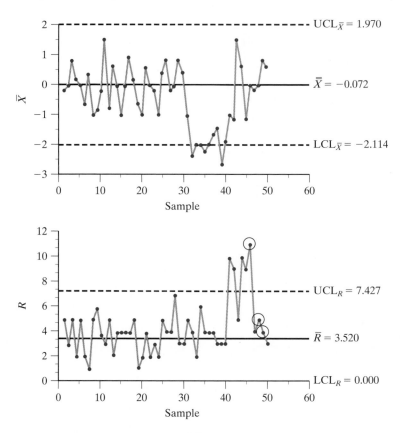

Figure 8.8   Continuation of the $\overline{X}$ and R charts for samples 46 to 50

Interpretation of the evolving patterns in the $\overline{X}$ and R charts yields the following observations:

- On the R chart sample 46 is an extreme point, while samples 48 and 49 violate the run rule. It should therefore be concluded that the evidence continues to indicate a sustained shift in the process variation. The R value for sample 50 does not violate any of the appropriate rules, and therefore one should conclude that there is no longer any statistical evidence to support the existence of a shift in the variation level of the process.
- The $\overline{X}$ values continue to show a pattern not unlike the pattern evident in the $\overline{X}$'s for the first 30 samples. There is no evidence to suggest that the process mean is anything other than the centerline of the $\overline{X}$ chart.

Revelation of the True Nature of Changes in the Process

It is now time for us to reveal the true nature of the changes that were made to the process mean level and the amount of process variation as the sampling experiment evolved. It is clear from the statistical signals emanating from the $\overline{X}$ and R charts that the distributions from which the samples were drawn were changed from time to time. The statistical signals, in fact, suggested where it appeared that these changes had taken place and what their specific nature was. The question is: How good a job have the charts really done? A summary of the nature of the statistical distributions used to generate each of the 50 sample results follows.

- Samples 1–25: $N(\mu_X = 0, \sigma_X^2 = 1.715^2)*$
- Samples 26–30: $N(\mu_X = 0, \sigma_X^2 = 1.715^2)$
- Samples 31–40: $N(\mu_X = -2, \sigma_X^2 = 1.715^2)$
- Samples 41–46: $N(\mu_X = 0, \sigma_X^2 = 3.47^2)$
- Samples 47–50: $N(\mu_X = 0, \sigma_X^2 = 1.715^2)$

Figure 8.9 graphically depicts the nature of the evolving process in terms of the statistical distributions outlined above. In comparing the statistical signals emanating from the charts with Fig. 8.9, it would appear that the charts have done an excellent job of signaling the presence of shifts in the process mean and/or the amount of process variation. For example, the mean of the process was actually shifted downward at sample 31, and this fact showed up on the $\overline{X}$ chart at sample 32. The variation of the process was shifted upward at sample 41,

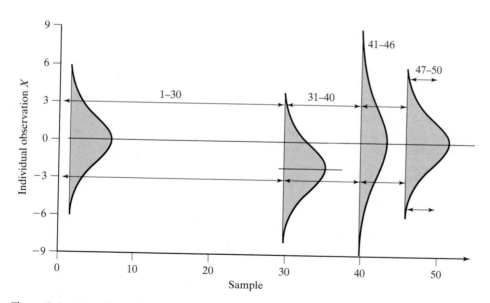

Figure 8.9 Graphical depiction of the true nature of the process over the sampling experiment

---

$^{*}N(\mu_X = 0, \sigma_X^2 = 1.715^2)$ denotes a normal distribution with a mean of zero and a standard deviation of 1.715.

and this fact was immediately detected at sample 41 when an R value went above the upper control limit on the R chart. The process mean level was actually shifted back to the original value of 0.0 at sample 41. This fact was revealed by the $\overline{X}$ chart at sample 43, although for samples 41 and 42 we were cautious in saying that the process mean continued to be shifted downward because of the shift in variation detected at samples 41 and 42. Similarly, the process variation was returned to its original level at sample 47. This fact was signaled by the R chart at sample 50.

Sampling experiments of the type we have just reviewed in detail are an excellent tool for learning purposes. In addition to revealing how shifts in the process mean and/or variation can be signaled through the use of $\overline{X}$ and R control charts, they serve as a means to illustrate inappropriate sampling methods such as stratification and mixing and can be used to simulate other common problems, such as overcontrol of the process. We will now make use of this sampling experiment idea to illustrate the problem of stratification of the subgroup.

8.3   Illustration of Stratification of Subgroups

Simulation of the Stratification Phenomenon

A sampling experiment using several different statistical distributions can be most useful in demonstrating the consequences of poor sampling methods such as stratification. In this section we conduct such a sampling experiment to demonstrate the effect that stratification has on the appearance of the associated $\overline{X}$ and R control charts and to demonstrate the loss of chart sensitivity due to stratification of the sample.

Figure 8.10 shows four normal distributions, each with the same standard deviation, but with means of $\mu_X = 2, 5, 5,$ and 7. To simulate a stratified subgrouping, samples of size $n = 4$ were drawn, with each sample consisting of one measurement from each of the four distributions shown in Fig. 8.10. Table 8.8 gives the data for the first 50 of a total of 100 subgroups simulated in this manner and includes the subgroup average and range values. The data in Table 8.8

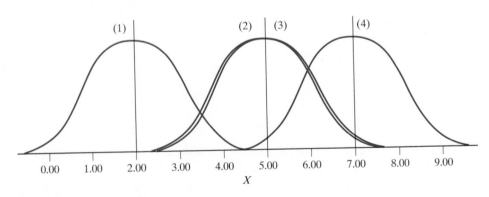

Figure 8.10   **Normal distributions used to simulate stratification**

TABLE 8.8	Data for Stratification Example					
Sample	X_1	X_2	X_3	X_4	$\bar{X}$	R
1	6.88	3.32	5.85	3.55	4.900	3.560
2	8.77	0.72	6.08	6.20	5.443	8.050
3	5.84	0.72	5.36	2.50	3.605	5.120
4	7.10	3.77	5.92	1.52	4.578	5.580
5	6.07	2.47	3.10	2.73	3.592	3.600
6	7.00	0.06	6.54	3.12	4.180	6.940
7	7.79	2.24	4.19	4.59	4.702	5.550
8	7.61	3.68	2.60	6.51	5.100	5.010
9	5.21	5.63	5.38	4.57	5.198	1.060
10	7.84	2.29	8.04	5.76	5.983	5.750
11	10.44	0.75	2.59	4.39	4.543	9.690
12	6.76	2.53	1.79	3.17	3.562	4.970
13	7.73	2.81	3.25	5.80	4.898	4.920
14	6.48	3.28	6.21	4.29	5.065	3.200
15	6.24	0.00	5.18	5.45	4.218	6.240
16	4.53	3.11	2.88	6.03	4.138	3.150
17	6.42	0.48	6.04	7.73	5.168	7.250
18	12.09	3.45	6.98	4.76	6.820	8.640
19	7.13	3.47	5.83	3.91	5.085	3.660
20	2.73	2.92	3.29	3.93	3.217	1.200
21	6.28	2.45	5.84	5.63	5.050	3.830
22	4.68	1.32	2.77	5.86	3.658	4.540
23	7.59	5.10	5.10	3.57	5.340	4.020
24	8.64	2.76	3.52	5.92	5.210	5.880
25	9.54	2.60	4.78	5.95	5.718	6.940
26	7.91	1.27	5.71	1.99	4.220	6.640
27	6.28	0.74	4.94	3.86	3.955	5.540
28	6.60	2.39	5.71	4.87	4.892	4.210
29	6.45	0.00	7.14	4.32	4.478	6.450
30	4.26	5.27	4.46	8.57	5.640	4.310
31	6.64	4.05	1.55	5.34	4.395	5.090
32	7.99	0.45	7.36	3.90	4.925	7.540
33	5.17	1.82	3.78	8.93	4.925	7.110
34	5.53	3.68	7.77	4.95	5.483	4.090
35	5.52	0.80	5.18	7.27	4.693	6.470
36	7.18	1.04	3.92	7.65	4.948	6.610
37	8.34	1.92	5.75	3.81	4.955	6.420
38	1.83	2.84	4.74	4.37	3.445	2.910
39	5.95	2.37	5.46	4.39	4.543	3.580
40	7.11	2.13	6.84	4.41	5.122	4.980
41	7.12	0.70	5.81	6.23	4.965	6.420
42	6.95	1.11	9.27	5.23	5.640	8.160
43	5.02	3.38	0.70	7.67	4.193	6.970
44	5.37	3.45	5.26	6.36	5.110	2.910
45	4.69	4.94	5.80	7.32	5.688	2.630

(Continued)

TABLE 8.8	(Continued)					
Sample	X_1	X_2	X_3	X_4	$\overline{X}$	R
46	9.06	0.00	2.84	5.49	4.349	9.060
47	7.81	4.49	3.08	5.67	5.262	4.730
48	8.51	1.73	7.66	6.61	6.128	6.780
49	6.18	0.05	4.24	4.87	3.835	6.130
50	4.57	3.28	4.77	8.48	5.275	5.200

plus the additional 50 subgroups collected were used to set up $\overline{X}$ and R charts. The calculations are as follows:

$$\overline{\overline{X}} = \frac{\Sigma \overline{X}_i}{100}, \quad i = 1, 100,$$

$$= \frac{481.5}{100}$$

$$= 4.815.$$

$$\overline{R} = \frac{\Sigma R_i}{100}, \quad i = 1, 100,$$

$$= \frac{515.5}{100}$$

$$= 5.155.$$

$$\text{UCL}_{\overline{X}} = \overline{\overline{X}} + A_2\overline{R}$$

$$= 4.815 + (0.73)(5.155)$$

$$= 8.578.$$

$$\text{LCL}_{\overline{X}} = \overline{\overline{X}} - A_2\overline{R}$$

$$= 4.815 - (0.73)(5.155)$$

$$= 1.052.$$

$$\text{UCL}_R = D_4\overline{R}$$

$$= (2.28)(5.155)$$

$$= 11.753.$$

$$\text{LCL}_R = D_3\overline{R}$$

$$= (0)(5.155)$$

$$= 0.0.$$

Figure 8.11 shows the $\overline{X}$ and R charts for the stratified subgroup data of Table 8.8. The R chart looks typical of data arising from a constant system of common causes. No special-cause signals are evident, and the overall appearance of the chart does not reveal anything that would suggest stratification had taken place. Examination of the $\overline{X}$ chart, however, reveals a somewhat unusual pattern of variability. The $\overline{X}$ values seem to cluster quite closely about $\overline{\overline{X}}$; in fact,

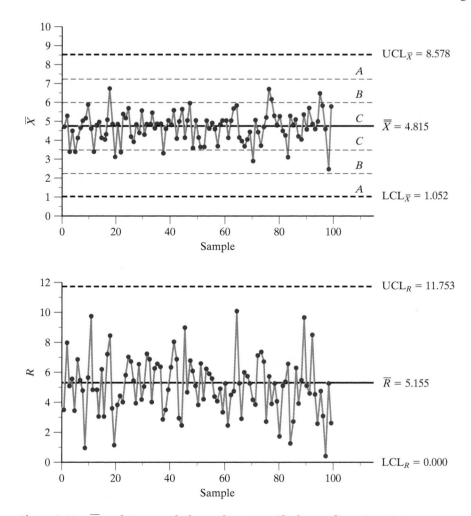

Figure 8.11 $\overline{X}$ and R control charts for a stratified sampling situation

86% of the $\overline{X}$ values are within ± 1 standard deviation of $\overline{\overline{X}}$. Normal distribution theory suggests that we should expect a much smaller percentage: 68%. Furthermore, runs of 15 or more consecutive $\overline{X}$ values within 1 standard deviation of the chart centerline occur twice. These signals indicate that something is wrong—and that something is the sampling method itself.

If samples were drawn separately from each distribution in Fig. 8.10, we would expect the ranges to average about 3.4; that is, $\overline{R}$ should be approximately equal to 3.4 (given that the theoretical standard deviation = 1.715 for each individual distribution, $d_2 = 2.06$ for samples of $n = 4$, and $E(R) = \sigma_X d_2$). However, the observed range is about 50% higher than this expected value. Clearly, this is occurring because we are sampling from several different distributions to make the subgroups, and these distributions have different means. These differences in means are inflating the within-subgroup variation. Fortunately, this characteristic behavior shows up in the $\bar{X}$ chart so that the improper sampling, namely, stratification, is signaled.

Effect of Stratification on Chart Sensitivity

The results of stratification of the sample, as seen in Fig. 8.11, clearly indicate that the special causes due to differences in the process means (refer to Fig. 8.10) that occur within the subgroup significantly inflate the within-subgroup variation. As a result, the limits on the $\bar{X}$ chart seem very wide, and hence one would conclude that the sensitivity of the chart to detect change in process mean level would be greatly reduced. This loss of chart sensitivity can be demonstrated through the use of sampling experiments such as those already conducted in this chapter.

Table 8.9 provides the data for another 50 samples of size $n = 4$ drawn in a stratified way from the four distributions of Fig. 8.10. The data were used to calculate centerlines and control limits for $\bar{X}$ and R control charts. These calculations are given below. The charts are shown in Fig. 8.12.

$$\bar{\bar{X}} = \frac{\Sigma \bar{X}_i}{50}, \quad i = 1, 50,$$

$$= 4.7876.$$

$$\bar{R} = \frac{\Sigma R_i}{50}, \quad i = 1, 50,$$

$$= 5.9201.$$

$$\mathrm{UCL}_{\bar{X}} = \bar{\bar{X}} + A_2 \bar{R}$$

$$= 4.7876 + (0.73)(5.2901)$$

$$= 9.1093.$$

$$\mathrm{LCL}_{\bar{X}} = \bar{\bar{X}} - A_2 \bar{R}$$

$$= 4.7876 - (0.73)(5.2901)$$

$$= 0.4659.$$

$$\mathrm{UCL}_R = D_4 \bar{R}$$

$$= (2.28)(5.9201)$$

$$= 13.4978.$$

$$\mathrm{LCL}_R = D_3 \bar{R}$$

$$= (0)(5.9201)$$

$$= 0.0.$$

In Fig. 8.12 we see the same characteristic pattern in the $\bar{X}$ chart that was evident before as a result of subgroup stratification. Almost 90% of the $\bar{X}$ values fall within the middle one-third of the chart. Furthermore, a run of 18 in a

TABLE 8.9	Data for a Stratification Sensitivity Example					
Sample	X_1	X_2	X_3	X_4	$\bar{X}$	R
1	−1.119	8.274	4.858	7.253	4.817	9.400
2	2.694	2.087	4.261	6.866	3.977	4.780
3	0.749	5.444	5.624	5.477	4.323	4.875
4	3.608	8.089	6.711	7.031	6.360	4.481
5	0.258	8.187	4.149	5.148	4.435	7.930
6	3.366	3.299	5.869	4.697	4.308	2.570
7	2.444	2.890	4.869	8.591	4.699	6.147
8	0.282	3.069	4.916	5.161	3.357	4.879
9	3.090	0.399	6.168	5.354	3.753	5.769
10	0.172	8.028	5.488	7.030	5.180	7.856
11	4.159	6.191	2.902	8.671	5.481	5.769
12	0.318	6.882	4.493	9.546	5.310	9.228
13	−0.484	6.642	8.355	5.745	5.064	8.840
14	2.165	3.288	8.148	5.457	4.765	5.983
15	−0.010	8.611	5.293	10.483	6.094	10.494
16	2.265	5.702	6.388	9.930	6.071	7.665
17	−0.503	4.789	6.770	6.146	4.301	7.273
18	3.977	6.757	4.356	5.932	5.256	2.780
19	1.772	5.639	3.531	6.382	4.331	4.601
20	2.107	2.668	4.063	5.357	3.549	3.250
21	1.427	7.736	5.227	6.179	5.142	6.309
22	1.580	2.175	2.842	10.678	4.319	9.098
23	4.376	4.784	8.072	8.694	6.482	4.317
24	1.076	3.764	5.922	9.986	5.187	8.909
25	2.589	4.114	5.833	3.102	3.909	3.244
26	0.091	4.942	2.641	8.053	3.932	7.963
27	4.139	3.490	6.982	7.704	5.579	4.214
28	0.166	3.706	4.584	5.763	3.555	5.597
29	4.308	5.285	4.831	5.593	5.004	1.285
30	2.982	2.832	4.859	4.936	3.902	2.104
31	1.353	1.591	4.309	9.350	4.151	7.997
32	−0.032	3.708	4.615	6.687	3.745	6.718
33	1.153	7.319	4.395	5.018	4.471	6.166
34	1.281	5.643	3.202	8.619	4.686	7.338
35	2.758	5.794	5.294	6.497	5.086	3.739
36	3.163	5.341	7.753	6.166	5.606	4.589
37	−0.625	3.442	5.817	5.162	3.449	6.443
38	2.637	4.516	6.399	6.058	4.903	3.762
39	0.121	2.281	6.907	6.000	3.827	6.787
40	3.727	4.387	6.093	9.402	5.902	5.676
41	3.448	7.121	5.361	6.781	5.678	3.673
42	4.581	8.303	4.821	6.879	6.146	3.721
43	−0.704	1.043	6.987	5.297	3.155	7.691
44	2.704	5.949	7.170	7.320	5.786	4.616
45	−0.029	3.410	6.048	8.099	4.382	8.127

(Continued)

TABLE 8.9	(Continued)					
Sample	X_1	X_2	X_3	X_4	$\overline{X}$	R
46	4.991	5.533	6.182	9.333	6.510	4.342
47	1.457	5.828	5.394	6.183	4.715	4.726
48	0.076	5.028	3.231	8.947	4.320	8.871
49	2.663	6.572	5.535	5.311	5.020	3.909
50	1.920	3.443	4.823	11.426	5.403	9.505

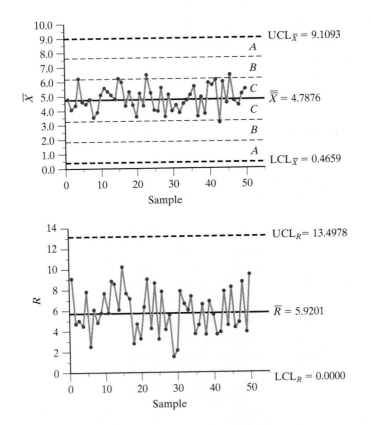

Figure 8.12 $\overline{X}$ and R charts for data of Table 8.9

row within zone C starts at subgroup 5, and another such run of 19 in a row starts at subgroup 24.

Table 8.10 provides the data for 25 additional subgroups chosen in the same fashion as the previous 50 subgroups, with the exception that one of the processes (distributions) has sustained a mean shift of one unit (from 5.00 to 4.00;

Sample	X_1	X_2	X_3	X_4	$\overline{X}$	R
TABLE 8.10	Additional 25 Subgroups for Stratification Sensitivity Example					
51	2.935	4.335	5.180	10.002	5.613	7.067
52	2.560	0.626	6.940	7.368	4.373	6.743
53	1.715	5.106	5.252	9.019	5.273	7.303
54	1.410	4.888	6.784	8.758	5.460	7.349
55	3.022	3.272	5.776	7.514	4.896	4.492
56	4.846	4.234	4.720	6.802	5.150	2.568
57	0.828	4.776	0.931	7.534	3.517	6.707
58	1.374	3.542	7.054	5.652	4.405	5.680
59	1.146	5.580	4.159	5.962	4.212	4.816
60	0.063	3.778	5.731	7.148	4.180	7.085
61	1.096	5.837	7.685	4.574	4.798	6.589
62	−0.590	3.633	6.676	7.676	4.349	8.266
63	7.335	2.977	6.634	5.659	5.651	4.358
64	3.520	3.979	3.186	4.492	3.794	1.306
65	0.847	6.205	5.373	6.471	4.724	5.624
66	0.599	7.179	5.936	6.203	4.979	6.579
67	2.197	6.010	3.436	11.114	5.689	8.917
68	2.058	3.275	4.738	7.541	4.403	5.483
69	3.445	7.595	6.185	6.734	5.990	4.150
70	6.716	1.521	5.529	5.481	4.812	5.195
71	3.175	0.715	4.426	4.960	3.319	4.245
72	1.713	3.613	1.759	6.240	3.331	4.526
73	3.838	6.725	5.326	10.457	6.587	6.619
74	0.117	3.668	5.064	7.183	4.008	7.066
75	0.617	3.582	3.062	4.248	2.877	3.632

again, refer to Fig. 8.10). The limits and centerlines for the $\overline{X}$ and R charts in Fig. 8.12 were extended, and the results of these additional 25 subgroups were plotted on the charts. The updated charts are seen in Fig. 8.13.

In Fig. 8.13 the $\overline{X}$ chart provides no signals that would indicate that the mean of one of the processes had shifted from 5.00 to 4.00. Yet such is clearly a shift away from the chart centerline of 4.7876. In fact, a run of $\overline{X}$ values within zone C that started at subgroup 47 is sustained all the way through subgroup 71, which is an indication that stratification may be occurring. The chart gives no indication that a special cause indicating a shift in the mean actually occurred at sample 51.

To illustrate the result of proper sample selection on chart sensitivity by contrast, still another sampling experiment has been done. In this sampling experiment, 50 subgroups of size $n = 4$ were drawn from a single process (distribution) with a mean of 4.75 (approximately the same mean as the data in Fig. 8.12 that were collected in a stratified way). Table 8.11 provides these new data. $\overline{X}$

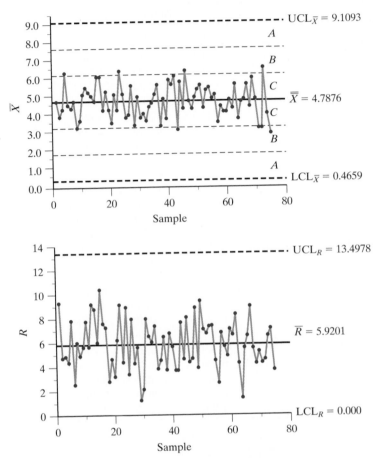

Figure 8.13 $\overline{X}$ and R charts extended from Fig. 8.12 using the data of Table 8.10

TABLE 8.11	50 Subgroups from a Single Distribution/Process				
Sample	X_1	X_2	X_3	X_4	$\overline{X}$	R
1	4.784	3.307	5.980	3.418	4.372	2.673
2	4.891	5.749	3.787	4.314	4.685	1.962
3	5.422	7.241	6.528	7.444	6.659	2.022
4	5.766	3.535	5.313	8.630	5.811	5.094
5	3.460	7.186	3.454	3.473	4.393	3.732
6	3.974	3.053	6.292	6.390	4.927	3.337
7	2.279	4.156	3.654	4.070	3.540	1.877
8	3.950	3.739	4.997	4.902	4.397	1.258
9	2.694	5.347	7.315	6.473	5.457	4.620
10	1.976	5.158	7.685	4.148	4.742	5.709
11	4.279	3.430	2.963	4.984	3.914	2.021
12	6.520	5.643	2.142	6.595	5.225	4.452
13	0.877	3.849	3.205	2.186	2.529	2.972

(Continued)

Sample	X_1	X_2	X_3	X_4	$\bar{X}$	R
TABLE 8.11	*(Continued)*					
14	4.935	3.769	3.803	4.590	4.274	1.166
15	4.839	4.448	4.247	5.725	4.815	1.478
16	4.056	2.644	7.048	3.792	4.385	4.044
17	4.342	4.498	4.695	5.328	4.716	0.987
18	3.874	3.735	6.415	2.927	4.238	3.488
19	5.232	5.915	1.274	1.792	3.553	4.641
20	5.253	6.182	5.349	4.266	5.262	1.917
21	3.285	4.017	6.629	8.246	5.544	4.961
22	3.861	6.273	7.191	4.339	5.416	3.329
23	3.799	7.678	4.199	2.427	4.526	5.251
24	4.867	5.053	4.338	7.405	5.416	3.067
25	6.142	6.501	4.877	5.321	5.710	1.624
26	7.901	2.826	0.116	3.213	3.514	7.785
27	3.751	4.587	3.537	4.545	4.105	1.051
28	7.467	4.433	2.255	2.057	4.053	5.411
29	4.880	3.987	5.475	6.754	5.274	2.767
30	3.485	3.460	3.912	4.899	3.939	1.439
31	7.689	4.258	5.857	6.518	6.081	3.430
32	6.011	3.792	5.382	4.220	4.851	2.219
33	4.057	4.850	7.186	7.181	5.818	3.129
34	7.592	4.418	7.402	4.952	6.024	3.443
35	1.496	4.991	5.725	3.356	3.892	4.229
36	2.933	6.691	2.642	1.751	3.504	4.940
37	7.371	1.092	7.974	4.770	5.302	6.882
38	5.416	1.886	5.213	8.230	5.186	6.344
39	4.653	2.907	8.119	3.151	4.707	5.212
40	2.010	4.013	6.614	5.340	4.494	4.604
41	6.872	4.704	6.172	7.031	6.195	2.328
42	5.990	7.138	4.251	2.819	5.050	4.319
43	3.651	4.407	8.336	6.314	5.677	4.685
44	5.554	5.012	6.868	3.844	5.320	3.024
45	4.041	4.330	4.296	7.405	5.018	3.363
46	2.174	5.822	3.613	8.362	4.993	6.188
47	4.963	1.426	6.440	3.423	4.063	5.014
48	3.536	1.945	5.673	7.296	4.613	5.351
49	4.321	6.945	4.897	4.356	5.130	2.624
50	2.362	3.634	3.715	4.814	3.631	2.452

and R control charts were developed and are shown in Fig. 8.14. Calculations for the centerlines and control limits are given below.

$$\bar{\bar{X}} = \frac{\Sigma \bar{X}_i}{50}, \quad i = 1, 50,$$

$$= 4.7788.$$

$$\bar{R} = \frac{\Sigma R_i}{50}, \quad i = 1, 50,$$

$$= 3.6055.$$

$$\mathrm{UCL}_{\bar{X}} = \bar{\bar{X}} + A_2\bar{R}$$

$$= 4.7788 + (0.73)(3.6055)$$

$$= 7.4108.$$

$$\mathrm{LCL}_{\bar{X}} = \bar{\bar{X}} - A_2\bar{R}$$

$$= 4.7788 - (0.73)(3.6055)$$

$$= 2.1468.$$

$$\mathrm{UCL}_R = D_4\bar{R}$$

$$= (2.28)(3.6055)$$

$$= 8.2205.$$

$$\mathrm{LCL}_R = D_3\bar{R}$$

$$= (0)(3.6055)$$

$$= 0.0.$$

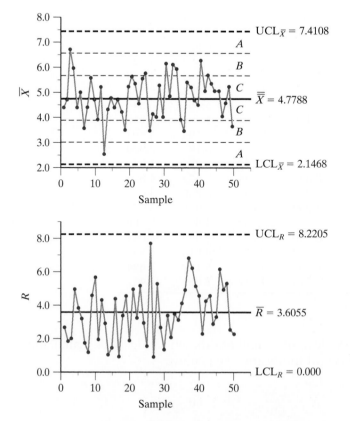

Figure 8.14 $\bar{X}$ and R control charts for data of Table 8.11

In Fig. 8.14 it is seen that the charts show good statistical control with no rule violations or unusual patterns of variation. It should also be noted that a 32% reduction in the within-subgroup variation has led to $\overline{X}$ and R charts with much narrower limits than previously was the case with the stratified samples (refer to Fig. 8.13).

An additional 25 subgroups were then selected from a process (distribution) with a mean of 4.00, which would constitute a mean shift of 0.75 from the previous process with a mean of 4.75. These 25 samples were taken from the same distribution used to introduce a shift in the mean within the stratified subgroup—the 25 samples of Table 8.10. The data for these 25 subgroups are given in Table 8.12, and their $\overline{X}$ and R values are plotted as a continuation of the charts indicated in Fig. 8.14. Figure 8.15 shows charts with extended limits and the additional 25 subgroup results plotted.

The R chart in Fig. 8.15 continues to show good statistical control, as we would expect, since the process variation has remained constant. The $\overline{X}$ chart, however, provides several signals indicating that the mean has shifted. There is a run below the centerline that starts at subgroup 50, which is 8 long. There is also a signal at subgroup 64, where two of three successive $\overline{X}$ values are 2 or

TABLE 8.12	Additional 25 Subgroups Selected from a Process with a Mean of 4.00					
Sample	X_1	X_2	X_3	X_4	$\overline{X}$	R
51	2.242	1.735	5.015	1.535	2.632	3.480
52	4.716	0.375	5.118	6.797	4.251	6.422
53	5.564	4.391	3.000	3.781	4.184	2.563
54	3.006	5.879	4.047	3.931	4.216	2.873
55	3.071	1.888	2.705	3.776	2.860	1.888
56	2.119	5.397	3.309	1.928	3.188	3.469
57	3.528	4.775	6.212	1.252	3.941	4.960
58	4.729	3.695	7.341	4.944	5.177	3.646
59	3.040	5.801	6.811	2.677	4.582	4.133
60	5.644	3.025	4.313	4.718	4.425	2.619
61	3.150	5.879	7.466	1.922	4.604	5.544
62	5.004	5.920	2.943	3.837	4.426	2.977
63	4.522	3.013	0.261	2.277	2.518	4.261
64	−0.282	2.216	3.724	5.948	2.901	6.230
65	5.440	2.651	2.473	5.430	3.998	2.967
66	3.644	4.064	5.519	7.478	5.176	3.834
67	5.127	4.296	0.476	4.582	3.620	4.652
68	3.488	1.404	5.015	6.778	4.171	5.374
69	6.562	4.215	3.418	2.989	4.296	3.573
70	4.544	5.273	4.984	6.797	5.399	2.253
71	5.512	3.636	4.731	2.094	3.993	3.417
72	4.221	0.957	4.762	5.964	3.976	5.006
73	2.984	3.821	2.924	4.456	3.546	1.532
74	2.230	4.922	4.862	3.850	3.966	2.692
75	3.893	3.966	4.609	2.221	3.672	2.388

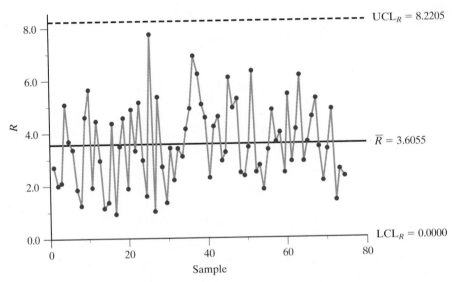

Figure 8.15 $\overline{X}$ **and R charts extended from Fig. 8.14 using the data from Table 8.12**

more standard deviations below the centerline. The indications are clear that the mean has shifted downward.

The sampling experiments above demonstrate the loss of chart sensitivity that occurs when samples are stratified by sampling across several different processes. Process mean differences become part of the within-subgroup variation, thereby inflating the range and widening the control limits of the $\overline{X}$ chart. The comparison of the $\overline{X}$ charts of Figs. 8.14 and 8.15 clearly show the sensitivity issue for the same process shifts in mean level.

SPC WORKSHOP #1: PROCESS SIMULATION, PROSIM

Introduction

In this workshop, a sampling experiment will be performed with the aid of PROSIM, an MS Excel–based Visual Basic program that generates samples of data from a fictional process. $\overline{X}$ and R control charts can then be constructed from the samples to assess the process behavior history.

Learning Objective

The objective of this computer workshop is to provide students with practice in identifying statistical signals on control charts, and in the use of these signals to deduce the underlying process behavior.

Description of the Computer-Based Process Simulator, PROSIM

PROSIM randomly selects a process history from among those stored in a process history library. A process history is a chronological description of the distributions/behaviors used to generate the sample data. PROSIM draws 125 subgroups of size $n = 5$ using the information associated with the selected process history. The succession of samples within a history may have one of the following normal distributions associated with it: $N(0, 2^2)$, $N(2, 2^2)$, $N(-2, 2^2)$, $N(0, 1^2)$, $N(0, 4^2)$, $N(1, 1^2)$, and $N(-1, 1^2)$. Alternatively, a group of samples may be linked to an inappropriate sampling method (stratification or mixing) or the process fault of overcontrol. There are three sampling rules of PROSIM:

1. At least the first 25 samples of size $n = 5$ are taken from the distribution $N(0, 2^2)$.

2. Shifts in the mean and/or variation are sustained for at least five consecutive subgroups.

3. Stratification, mixing, and overcontrol are sustained for at least 20 consecutive subgroups.

As an example, a process history could have been generated as follows:

Samples 1–28:	$N(0, 2^2)$	Samples 78–84:	$N(0, 4^2)$
Samples 29–37:	$N(-2, 2^2)$	Samples 85–109:	Overcontrol
Samples 38–45:	$N(2, 2^2)$	Samples 110–118:	$N(0, 4^2)$
Samples 46–69:	Stratification	Samples 119–125:	$N(0, 2^2)$
Samples 70–77:	$N(0, 1^2)$		

Once PROSIM has generated the sample data inside the Excel spreadsheet, control charts may be constructed for the sample data using Excel, or the data may be exported to another file for subsequent control chart construction via third-party software such as MiniTab. The charts may then be interpreted manually for statistical signals. Ultimately, the student should be able to draw conclusions about the specific distributions/behaviors that were used and the time period over which they were present.

On the Web

PROSIM is available for download on the Web at http://www.me.mtu.edu/quality_book/. Also available at the site is a more detailed workshop explanation and description of PROSIM usage. The website also has a suggested student assignment in connection with the PROSIM workshop.

9

Some Control Chart Methods for Individual Measurements

9.1   Introduction

In previous chapters dealing with control charts for variable measurements, the notion of the rational subgroup was central to data collection, chart construction, and chart interpretation. The ability to observe the forces of common-cause variation alone is critical to establishing the model for $\overline{X}$ and R control charts and providing a sensitive indication of the presence of special causes when they arise.

In certain situations the notion of taking several measurements to form a rational sample of size greater than one simply does not make sense, because only one measurement is available or meaningful at each time that samples are to be taken. For example, certain process characteristics, such as oven temperature, number or dollar value of accounts receivable, suspended air particulate count, and machine downtime, will not vary in close proximity in time or space during the time that such sampling normally occurs, because either the characteristic is determined only at discrete and infrequent points in time (dollar amount of accounts receivable) or the medium being sampled is quite homogeneous in nature (oven temperature). The apparently different values from multiple observations of such processes at each sampling are results of reading and measurement system errors rather than reflections of true process variability. In these situations, special control charts may be used. Presented in this chapter are two types of control charts to be used for process control with individual measurements, namely X and R_m charts and exponentially weighted moving average (EWMA) charts.

9.2 X and R_m Control Charts

The X and moving range charts are perhaps the simplest types of control charts that can be used for the study of individual measurements. The construction of X and R_m charts is similar to that of $\overline{X}$ and R charts except that now X is the value of the individual measurement and R_m, the moving range, is the range of a group of n consecutive individual measurements combined artificially to form a subgroup of size n. Most often the moving range is calculated from two or three consecutive individual measurements. The moving ranges are calculated as shown below for the case of using three consecutive measurements to form the artificial samples of size $n = 3$.

Since the moving range R_m is calculated primarily for the purpose of estimating the common-cause variability of the process, the artificial samples that are formed from successive measurements must be very small in size—taken as close together as possible in time, space, and so on—to minimize the chance of including data in the subgroup that come from out-of-control conditions. If special causes occur within the subgroup, the estimate of the common-cause variability will be inflated and the sensitivity of the chart will be eroded. It is noted that X and R_m are not independent of each other and successive sample R_m values come from subgroups that are overlapping. Considerable caution and good judgment must be exercised in the use of these charts and in the selection of the artificial subgroup, especially since the nature of the circumstances that lead to the consideration of these charts is such that the time between sample observations is often large. This means that the opportunity for special causes to contaminate the subgroup and produce inflated estimates of the level of common-cause variation is larger than it is for the $\overline{X}$ and R chart situation.

X and R_m charts are perhaps the most misused (often, abused) of all of the charts in common use today. The reason is very simple—they seem to suggest that few data need be taken to make a valid control chart. Often, individual measurements are taken and used in cases where natural subgroups can and should be used. The result is (1) a control chart that often has inflated within-subgroup variability because too much time has transpired from one measurement to another, and (2) a control chart constructed with a small number of points and hence subject to large sampling errors. Using such charts may be worse than using no charts at all.

Construction of X and R_m Control Charts: A Step-by-Step Procedure

Let us assume that k consecutive individual observations of a process have been obtained and that they are at least approximately normally distributed. In general, one may set up X and R_m control charts to analyze the process for any appropriate artificial sample size n. However, as noted above, the sample size should be small ($n = 2$ or 3) to minimize the opportunity for special causes to arise within the artificial samples and hence inflate the moving ranges. The following step-by-step procedure provides a basis for the construction of X and R_m charts.

1. Let the k consecutive individual measurements be denoted by $X_1, X_2, \ldots, X_k$, where k should be 25 or more (preferably 40 to 50). Calculate from the data a sample mean by

$$\overline{X} = \frac{\sum\limits_{i=1}^{k} X_i}{k}.$$

This sample mean will serve as the centerline for the X chart.

2. Compute sample moving ranges for an artificial sample of size n starting with R_{m1}, which is the difference between the largest and the smallest value in the first artificial sample $(X_1, X_2, \ldots, X_n)$. Repeat this computation for each succeeding moving sample of artificial size n as follows:

Compute R_{mi} from $(X_i, \ldots, X_{i+n-1})$ for $i = 1, 2, \ldots, k - n + 1$.

3. Calculate an average sample moving range $\overline{R}_m$ from the $(k - n + 1)$ sample moving ranges,

$$\overline{R}_m = \frac{\sum\limits_{i=1}^{k-n+1} R_{mi}}{k - n + 1}.$$

This average moving range will serve as the centerline for the R_m chart.

4. Plot the X's and R_m's using the standard control chart conventions described previously.

5. Use $\overline{R}_m$ as an estimate of the true process range. Calculate the control limits for the R_m chart using $\overline{R}_m$ and the constants D_3 and D_4 in Table A.2 associated with the sample size n.

6. Divide $\overline{R}_m$ by d_2 (value corresponding to sample size n in Table A.2) for estimating the process standard deviation. Determine control limits for the X control chart using $\overline{X} \pm 3\hat{\sigma}_X$. It may be noted that the A_2 value is not used in calculating the control limits for the X chart, because the sample size for the X chart is always 1.

In examining X and R_m control charts, note that since the artificial subgroups are overlapping, the interpretations of X and R_m charts are not independent. Sometimes only the X chart is analyzed for out-of-control signals, while the R_m chart is checked for proper construction of the X chart.

9.3   Case Study: White Millbase Dispersion Process

An automotive paint plant was studying the batch processing of white millbase used in the manufacture of topcoats. The basic process begins by charging a sandgrinder premix tank with resin and pigment along with other additives. The premix is agitated until a homogeneous slurry is obtained and then a small portion of the slurry is pumped through the sandgrinder. The grinder output is evaluated to check for fineness and gloss via an ash test that measures pigment content. A batch may require adjustments by adding pigment or resin to achieve acceptable gloss.

The plant was experiencing a situation where too many batches required two or three adjustments or required additional processing time to achieve an acceptable level of gloss. Each time an adjustment was made, about four hours would elapse before the resulting gloss could be evaluated. No quantitative method of calculating pigment and resin additions was available, so a rule of thumb was used. Sometimes after numerous adjustments the premix tank was so full that further batch adjustments were impossible.

To identify potential areas of quality and productivity improvement, it was decided to begin studying the process using X and R_m control charts. The quality characteristic selected for study in this case was the weight per gallon of the batch. The rationale behind the selection of this quality characteristic was that if a batch was overcharged with pigment (high density), the weight of the batch tended to be large, while if it was overcharged with resin (low density), the weight tended to be small. The weights (pounds per gallon) of 30 consecutive batches from the millbase process were collected and are displayed in Table 9.1.

To determine the R_m values, an artificial sample size of $n = 2$ was selected. The R_m values are therefore given by

$$R_{mi} = |X_{i+1} - X_i|, \quad i = 1, \ldots, 29,$$

that is,

$$R_{m1} = |X_2 - X_1|, \quad R_{m2} = |X_3 - X_2|, \quad \text{and so on.}$$

Table 9.1 displays the 29 calculated moving range (R_m) values that were determined from the data using an artificial sample size of $n = 2$.

Construction of the X and R_m control charts may be accomplished using the step-by-step procedure outlined previously. For the X chart, the centerline is

$$\overline{X} = \frac{\sum\limits_{i=1}^{30} X_i}{30} = \frac{429.33}{30}$$
$$= 14.311.$$

TABLE 9.1	Millbase Data for Samples/Batches 1 to 30 (pounds per gallon)		
Batch	X	R_m	Moving Range Calculations
1	14.56		
2	13.88	0.68	$= R_{m1} = 14.56 - 13.88 = 0.68$
3	13.98	0.10	$= R_{m2} = 13.98 - 13.88 = 0.10$
4	14.50	0.52	$= R_{m3} = 14.50 - 13.98 = 0.52$
5	14.22	0.28	
6	14.36	0.14	
7	14.46	0.10	
8	14.32	0.14	
9	14.27	0.05	
10	14.20	0.07	
11	14.35	0.15	
12	13.84	0.51	
13	14.43	0.59	
14	14.72	0.29	.
15	14.75	0.03	.
16	14.27	0.48	.
17	14.23	0.04	
18	14.60	0.37	
19	14.25	0.35	
20	14.12	0.13	
21	14.13	0.01	
22	14.47	0.34	
23	14.73	0.26	
24	14.02	0.71	
25	14.11	0.09	
26	14.48	0.37	
27	14.36	0.12	
28	13.70	0.66	
29	14.36	0.66	
30	14.66	0.30	$= R_{m29} = 14.66 - 14.36 = 0.30$

For the R_m chart, the centerline is

$$\overline{R}_m = \frac{\sum_{i=1}^{29} R_{mi}}{29} = \frac{8.54}{29}$$

$$= 0.294.$$

In the calculation of averages above, $\overline{X}$ is an average of all 30 individual measurements, and $\overline{R}_m$ is an average of $30 - 1 = 29$ R_m values, because there are only 29 moving ranges for a subgroup size of $n = 2$. If the artificial samples were of size $n = 3$, there would have been only $30 - 2 = 28$ moving ranges.

Once the $\overline{X}$ and $\overline{R}_m$ values are calculated as the centerline values of the X and R_m charts, respectively, the calculation of upper and lower control limits for the R_m chart can proceed in the same way as for a regular R chart, using the artificial sample size n to determine D_3 and D_4 values. However, the upper and lower control limits for the X chart should always be based on a sample size of 1 [i.e., setting the limits at $\pm 3\hat{\sigma}_X$ or $\pm 3(\bar{R}_m/d_2)$]. These calculations are shown below for the millbase data.

1. For the R_m chart: For $n = 2$, $D_3 = 0$ and $D_4 = 3.27$,

$$\text{UCL}_{R_m} = D_4\bar{R}_m = (3.27)(0.294) = 0.963,$$

$$\text{LCL}_{R_m} = D_3\bar{R}_m = (0.0)(0.294) = 0.0.$$

2. For the X chart: To set the control limits for the X chart, we need an estimate of the standard deviation of X. This estimate is obtained from the relationship between the range and the standard deviation, $\hat{\sigma}_X = \bar{R}_m/d_2$. The constant d_2 is found from Table A.2 using $n = 2$. From the table, $d_2 = 1.128$ and therefore

$$\hat{\sigma}_X = \frac{0.294}{1.128} = 0.261.$$

The X-chart control limits are, therefore,

$$\text{UCL}_X = \bar{X} + 3\hat{\sigma}_X = 14.311 + (3)(0.261) = 15.094,$$

$$\text{LCL}_X = \bar{X} - 3\hat{\sigma}_X = 14.311 - (3)(0.261) = 13.528.$$

The resulting X and R_m charts are shown in Fig. 9.1.

When the individual X values follow a normal distribution, the patterns on the X chart are analyzed in the same manner as for a Shewhart $\bar{X}$ chart. A sample X is circled as an out-of-control signal if it falls outside a control limit, if it is the endpoint of a sequence that fails any of the tests discussed in Chapter 6, or if it simply indicates a nonrandom sequence. Generally, the zone rule tests should be used with caution on X charts since, unlike $\bar{X}$ charts, the distribution of the individuals may have considerable departure from the normal distribution.

To check if the white millbase data can be assumed to have a normal distribution, a normal probability plot was constructed from the data in Table 9.1 and is shown in Fig. 9.2. This set of data, with the exception of one or two points, appears to be well described by the indicated line, and thus we may conclude that the data follow a normal distribution. Therefore, in examining the X chart of Fig. 9.1, in addition to looking for points outside the limits and nonrandom patterns, we may also apply the zone rule tests that are based on the normal distribution. It should be noted here, however, that if the X chart as initially viewed showed considerable lack of control, the normal probability plot of the individual measurements would probably show the data to be nonnormal, limiting additional chart interpretation via the zone rules. Since

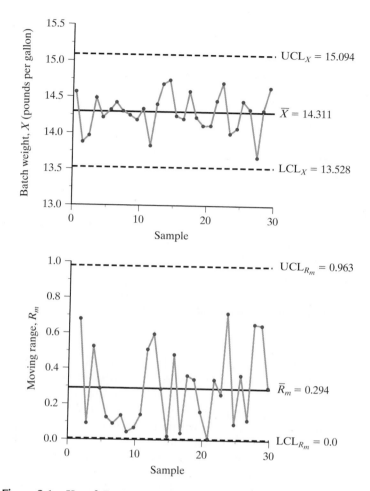

Figure 9.1 X and R_m control charts for the millbase process, batches 1 to 30

the charts of Fig. 9.1 are fairly well behaved, the normal plot is helpful here in learning the extent to which we could use the zone rules for fine-tuning the chart interpretation.

In examining the R_m chart in Fig. 9.1, it may be seen that there are no points beyond the limits. The chart appears to show no other statistical signals. An examination of the X chart shows no signals.

Change in the Millbase Dispersion Processing System

Since the process appeared to be in good control, the engineers involved in the millbase study decided to review the entire system and its operating procedures for possible areas of improvement (i.e., look for system faults). A cause-and-effect diagram was made to help focus on a few areas for more detailed

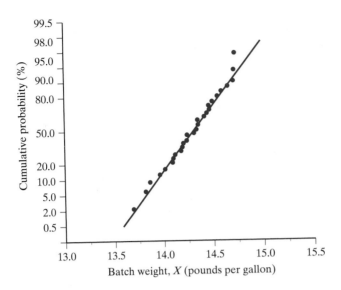

Figure 9.2   Normal probability plot for the white millbase data

study. An analysis of the production data indicated the need for a better method of determining the pigment concentration in the millbase. Several ash tests were performed to ascertain the balance between resin and pigment levels. These tests also verified the expected correlation between the weight per gallon and the ash test reading. Through curve fitting to the test results, a batch adjustment chart was developed using weight per gallon to determine the amount of resin or pigment to be added to bring the millbase to the target level of ash content. More important, a theoretical target batch weight was determined. The adjustment chart indicated that a weight of 13.77 pounds per gallon should produce the target ash content. The weights for 27 new batches using the batch adjustment chart (along with batch 30 from Table 9.1) are shown in Table 9.2. Note that while some time had actually passed between batch 30 and the batch identified as 31, for the purpose of charting the process the data are considered here as if they were consecutive.

To demonstrate the effect of using the batch adjustment chart on the process mean and variability, the X and R_m values were plotted on a continuation of the charts displayed in Fig. 9.1. These new charts that display the data for all 57 batches, with control limits based only on the first 30 batches, are shown in Fig. 9.3. As is readily apparent from an examination of Fig. 9.3, there are a number of statistical signals on the X and R_m charts for batches 31 to 57, indicating that both the process mean and variability have been reduced.

The statistical signals indicate that the batch adjustment chart not only has reduced the average weight, but has also improved the batch-to-batch consistency. The use of the batch adjustment chart has in fact changed the system. Since the R_m chart does show out-of-control conditions past sample 31, the

TABLE 9.2	Millbase Data for Samples/Batches 30 to 57 (pounds per gallon)		
Batch	X	R_m	Moving Range Calculations
30	14.66		Copied from Table 9.1
31	14.04	0.62	$= R_{m30} = 14.66 - 14.04 = 0.62$
32	13.94	0.10	$= R_{m31} = 14.04 - 13.94 = 0.10$
33	13.82	0.12	$= R_{m32} = 13.94 - 13.82 = 0.12$
34	14.11	0.29	$= R_{m33} = 14.11 - 13.82 = 0.29$
35	13.86	0.25	
36	13.62	0.24	
37	13.66	0.04	
38	13.85	0.19	
39	13.67	0.18	
40	13.80	0.13	
41	13.84	0.04	
42	13.98	0.14	
43	13.40	0.58	
44	13.60	0.20	
45	13.80	0.20	
46	13.66	0.14	
47	13.93	0.27	
48	13.45	0.48	
49	13.90	0.45	
50	13.83	0.07	
51	13.64	0.19	
52	13.62	0.02	
53	13.97	0.35	
54	13.80	0.17	
55	13.70	0.10	
56	13.71	0.01	
57	13.67	0.04	$= R_{m56} = 13.71 - 13.67 = 0.04$

limits on the X chart are clearly wrong from that point. Actually, it was the X chart that showed out-of-control conditions first, so the resulting interpretation here is quite sound.

To develop control charts for the new millbase processing system, we may use the data for samples/batches 31 to 57 in Table 9.2.

1. For the R_m chart: The centerline is

$$\bar{R}_m = \frac{\sum_{i=31}^{56} R_{mi}}{26} = \frac{4.99}{26}$$
$$= 0.192.$$

It may be noted that the first R_m value in Table 9.2 (R_{m30}) is not used in calculating the average moving range since it was calculated from two individual

Figure 9.3 X and R_m control charts for the millbase process, batches 1 to 57

measurements that are now believed to come from different processes (X_{30} and X_{31}). The control limits for the R_m chart are ($D_3 = 0$ and $D_4 = 3.27$ for $n = 2$)

$$\text{UCL}_{R_m} = D_4 \bar{R}_m = (3.27)(0.192) = 0.628,$$
$$\text{LCL}_{R_m} = D_3 \bar{R}_m = (0.0)(0.192) = 0.0.$$

2. For the X chart: The centerline is

$$\bar{X} = \frac{\sum\limits_{i=31}^{57} X_i}{27} = \frac{371.87}{27}$$

$$= 13.773.$$

The estimated process standard deviation is ($d_2 = 1.128$ for $n = 2$)

$$\hat{\sigma}_X = \frac{\overline{R}_m}{d_2} = \frac{0.192}{1.128}$$

$$= 0.170.$$

The control limits are therefore

$$\text{UCL}_X = \overline{X} + 3\hat{\sigma}_X = 13.773 + (3)(0.170) = 14.283,$$
$$\text{LCL}_X = \overline{X} - 3\hat{\sigma}_X = 13.773 - (3)(0.170) = 13.263.$$

The X and R_m charts for the millbase process using the batch adjustment chart are shown in Fig. 9.4. In examining the X and R_m charts of Fig. 9.4 no statistical signals are evident. In addition to the analysis of trends and patterns on

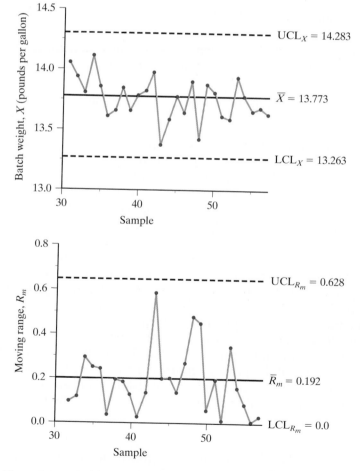

Figure 9.4 X and R_m charts for the millbase data, batches 31 to 57

the X chart, one should also see whether the fluctuations of sample X values are becoming narrower or wider. These narrowing or widening trends may indicate a change in process variability. The result of examining the charts of Fig. 9.4 is to conclude that there is no reason to believe that the process is not in a state of statistical control, at least over the period of the 27 batches examined.

The charts in Figs. 9.3 and 9.4 demonstrate a definite improvement in terms of both the mean and variability level of the new millbase dispersion processing system. Some of the observable effects from this study are listed briefly below.

- The average pigment level was lowered by over 2%, amounting to an annual savings of tens of thousands of dollars in material alone.
- The first gloss check met specifications on 18 consecutive batches, reducing the processing time by hundreds of hours. This served to prove that the theoretical batch weight was correct and it was important to maintain the process average at this level.
- In terms of average sample range, the average batch-to-batch fluctuation of pigment content was reduced by one-third.
- A solid basis was established from which to search for other improvements, including upgrading the silo weighing system and the shop floor measurement procedures.
- A similar approach could be applied to other large-volume products in the plant.

Following the implementation of the batch adjustment chart, the millbase dispersion process was continually monitored with X and R_m control charts. Figure 9.5 shows 29 new batches of data (samples 58 to 86) taken from the millbase dispersion process approximately 6 months after the study described above was performed. These data have been plotted relative to the limits and data of batches 31 to 57. An examination of Fig. 9.5 shows that in both the X and R_m charts there appears to be less variability in the newly collected data. This is most evident in the first 10 new batch weights that were obtained. In fact, there is a run of 10 points beneath the centerline on the moving range chart. Although no special cause(s) was found to explain this statistical signal, it does appear as if the process variability has been even further reduced, perhaps as a result of the added care and attention that are being given to the process.

A final set of X and R_m control charts was constructed, based only on the newly collected data. These charts are shown in Fig. 9.6. The plots of X and R_m of Fig. 9.7 represent the history of the process improvement during this study. Figure 9.8 shows a summary of the improvement in consistency and stability of the weight per gallon of the white millbase dispersion process over the period.

Summary

In summary, a control chart for individuals is less sensitive than an $\overline{X}$ chart, but in those cases where rational subgroups are not possible, the X and R_m charts can

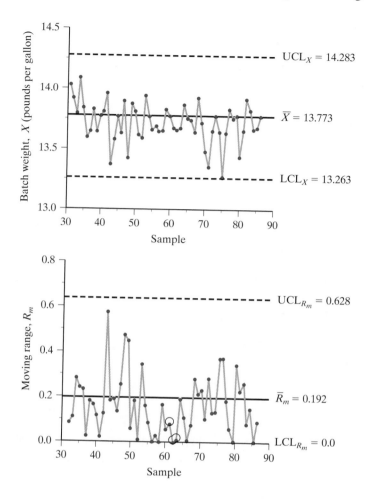

Figure 9.5 X and R_m **control charts for millbase data, batches 31 to 86**

serve to determine whether a process is statistically in control or not, and can lead to important quality and productivity improvement opportunities by proper and careful implementation. The millbase case study clearly demonstrates this fact.

9.4   Exponentially Weighted Moving Average Control Charts

As pointed out in the preceding section, X and R_m charts may not be very sensitive for detecting shifts in the process mean level or amount of process variation, particularly when the shift in the mean or in the variability is relatively small. Another weakness of X and R_m charts is that these two charts are not independent of each other. In this section we introduce the use of exponentially weighted moving averages as the sample means and of exponentially weighted moving deviations as the sample standard deviations for the tracking

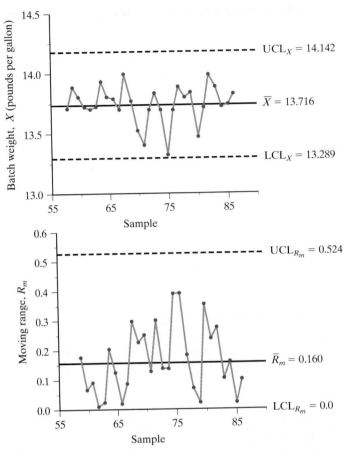

Figure 9.6 X and R_m control charts for the millbase dispersion process, batches 58 to 86

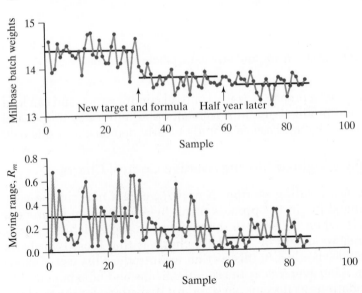

Figure 9.7 Improvement progress of white millbase dispersion process

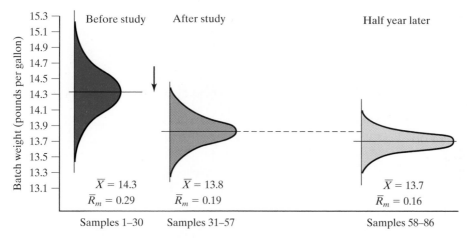

Figure 9.8   Consistency improvement in the millbase dispersion process over a period of about half a year

of process performance. The control charts for these moving statistics are known as *exponentially weighted moving average* (EWMA) control charts.

An exponentially weighted moving average is a moving average of past data where each data point is assigned a weight. These weights decrease in an exponentially decaying fashion from the present into the remote past. Thus the moving average tends to be a reflection of the more recent process performance, because most of the weighting is allocated to the most recently collected data. The amount of decrease of the weights is an exponential function of the weighting factor, r, which can assume values between 0 and 1:

$$W_{t-i} = r(1 - r)^i,$$

where W_{t-i} is the weight associated with the observation X_{t-i}, X_t being the most recently observed data point. When a very small value of r is used, the moving average at any time t carries with it a great amount of inertia from the past; hence it is relatively insensitive to short-lived changes in the process. For control chart applications where fast response to process shifts is desired, a relatively large weighting factor, say $r = 0.2$ to $r = 0.5$, may be used. To see how the weights gradually decrease as a function of r, Fig. 9.9 shows the decaying behavior for some commonly used values of r.

In selecting r, the following relationship between the weighting factor r and the sample size n for Shewhart control charts for sample means is often used:[1]

$$r = \frac{2}{n + 1}.$$

[1]A. L. Sweet, "Control Charts Using Coupled Exponentially Weighted Moving Averages," *Transactions of the IIE*, Vol. 18, No. 1, 1986, pp. 26–33.

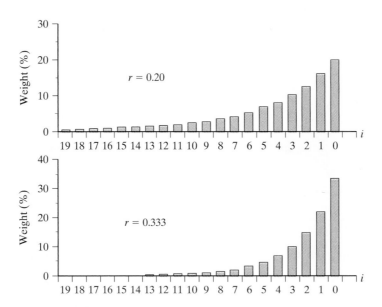

Figure 9.9   **Exponentially decaying weights for $r = 0.2$ and $r = 0.333$**

It is seen from the relationship above that if $r = 1$ is selected for the EWMA, all the weight is given to the current single observation, which is equivalent to the Shewhart control chart for individual measurements.

Construction of EWMA Control Charts: A Step-by-Step Procedure

Let X_t represent the individual measurement of a process observed at time t, when the process mean is μ_X and the process standard deviation is σ_X. At each sampling time t, the value of X_t is assumed to vary according to a normal distribution with mean μ_X and standard deviation σ_X as long as the process is in a state of statistical control. If a shift in the mean occurs, the exponentially weighted moving averages will gradually (depending on r) move to the new mean of the process, while the exponentially weighted moving standard deviations will remain unchanged. On the other hand, if there is a shift in the process variability, the exponentially weighted moving standard deviations will gradually move to the new level, while the exponentially weighted moving averages still vary about the process mean μ_X Similar to the behavior of $\overline{X}$ and R charts, when the process variability increases as a result of a special cause, there is an increasing probability of having out-of-control signals on the control chart for the mean as well.

The following step-by-step procedure leads to the construction of EWMA control charts:

1. Collect at least $k = 25$ samples of individual measurements $X_1, X_2, \ldots, X_k$ (preferably, k should be 40 to 50).

2. Calculate estimates of the process mean and standard deviation by

$$\overline{X} = \sum_{t=1}^{k} \frac{X_t}{k}, \qquad \text{sample mean,}$$

$$s_X = \left[\sum_{t=1}^{k} \frac{(X_t - \overline{X})^2}{k-1} \right]^{1/2}, \quad \text{sample standard deviation.}$$

3. Compute exponentially weighted moving averages, A_t, and exponentially weighted moving standard deviations, V_t, as follows:

$$A_t = rX_t + (1-r)A_{t-1}, \quad \text{where } A_0 = \overline{X},$$

$$V_t = rD_t + (1-r)V_{t-1}, \quad \text{where } D_t = |X_t - A_{t-1}| \text{ and } V_0 = s_X.$$

4. Calculate control limits and centerlines for A_t's and V_t's by

$$\text{UCL}_A = \overline{X} + A^* s_X, \qquad \text{centerline} = \overline{X},$$

$$\text{LCL}_A = \overline{X} - A^* s_X,$$

$$\text{UCL}_V = D_2^* s_X, \qquad \text{centerline} = d_2^* s_X$$

$$\text{LCL}_V = D_1^* s_X,$$

where the constants A^*, d_2^*, D_1^*, and D_2^* are as listed in Table 9.3.

TABLE 9.3	Constants for EWMA Control Charts				
Weighting Factor, r	Equivalent Sample Size, n	For Means: A^*	For Standard Deviations		
			D_1^*	D_2^*	d_2^*
0.050	39	0.480	0.514	1.102	0.808
0.100	19	0.688	0.390	1.247	0.819
0.200	9	1.000	0.197	1.486	0.841
0.250	7	1.132	0.109	1.597	0.853
0.286	6	1.225	0.048	1.676	0.862
0.333	5	1.342	0.000	1.780	0.874
0.400	4	1.500	0.000	1.930	0.892
0.500	3	1.732	0.000	2.164	0.921
0.667	2	2.121	0.000	2.596	0.977
0.800	—	2.449	0.000	2.990	1.030
0.900	—	2.714	0.000	3.321	1.076
1.000	1	3.000			

Source: Adapted from A. L. Sweet, "Control Charts Using Coupled Exponentially Weighted Moving Averages," *Transactions of the IIE*, Vol. 18, No. 1, 1986, pp. 26–33.

5. Plot all k A_t values on the A chart and all k V_t values on the V chart and interpret the charts to determine if the process is in statistical control in terms of both the variability and the mean.

Millbase Dispersion Process Revisited

Let us reconsider again the first 30 samples/batches collected from the millbase dispersion process to illustrate the application of EWMA control charts. First, we calculate the average and standard deviation of the data:

$$\overline{X} = 14.311 \text{ pounds/gallon,}$$

which is the previously calculated $\overline{X}$ for the first 30 batches, and

$$s_X = \left[\sum_{t=1}^{30} \frac{(X_t - 14.311)^2}{30 - 1} \right]^{1/2}$$

$$= 0.266 \text{ pound/gallon.}$$

After choosing a weighting factor of $r = 0.333$, we obtain the following constants from Table 9.3:

$$A^* = 1.342, \quad D_1^* = 0, \quad D_2^* = 1.780, \quad d_2^* = 0.874.$$

The centerlines and control limits are calculated below.

1. For the chart of moving averages, A_t:

$$\begin{aligned}
\text{CL}_A &= \overline{X} \\
&= 14.311, \\
\text{UCL}_A &= \overline{X} + A^* s_X \\
&= 14.311 + (1.342)(0.266) \\
&= 14.668, \\
\text{LCL}_A &= \overline{X} - A^* s_X \\
&= 14.311 - (1.342)(0.266) \\
&= 13.954.
\end{aligned}$$

2. For the chart of moving deviations, V_t:

$$\begin{aligned}
\text{CL}_V &= d_2^* s_X \\
&= (0.874)(0.266) \\
&= 0.233,
\end{aligned}$$

$$\text{UCL}_V = D_2^* s_X$$
$$= (1.780)(0.266)$$
$$= 0.474,$$
$$\text{LCL}_V = D_1^* s_X$$
$$= (0.0)(0.266)$$
$$= 0.0.$$

To start the series of moving averages and moving deviations, we need some initial values for the A's and V's, which can be estimated by $A_0 = \overline{X} = 14.311$ and $V_0 = s_X = 0.266$, as mentioned previously. Now, using the first data point, $X_1 = 14.56$, the value of the first moving average is given by

$$A_1 = rX_1 + (1 - r)(A_0)$$
$$= (0.333)(14.56) + (1 - 0.333)(14.311)$$
$$= 14.394.$$

Since $D_1 = |X_1 - A_0| = |14.56 - 14.311| = 0.249$, the value of the first moving deviation is given by

$$V_1 = rD_1 + (1 - r)(V_0)$$
$$= (0.333)(0.249) + (1 - 0.333)(0.266)$$
$$= 0.261.$$

The moving averages and moving deviations for the first 30 samples of the millbase study are listed in Table 9.4.

Figure 9.10 presents the A chart and V chart for the data given in Table 9.4. Since the control limits of the A chart are constructed using the sample standard deviation of the original data, it is important first to check to determine if the V chart shows stability of the process variation. An examination of the V chart shows a run of 8 points beneath the centerline. It may be recalled that we did not see a run of 8 points in our study of the first 30 samples using the R_m chart. However, there is a basic difference between the Shewhart control charts and the EWMA control charts in their method of construction and interpretation. The Shewhart charts plot independent sample data points, each of which is interpreted according to the probability law of the sampling distribution of the statistic in question. Inferences about possible shifts in the process parameters are made indirectly through the distribution patterns of the data on a Shewhart chart. An exponentially weighted moving statistic is directly an estimate of the corresponding process parameter. Therefore, a series of EWMA data on the chart tends to move slowly to the new level following a shift in the process, or will vary about the centerline with small fluctuations when the process is in control. In general, it is rare to have data points fall outside the control limits on an EWMA chart, so the detection of shifts is based primarily on trend lines of

TABLE 9.4	Exponentially Weighted Moving Statistics for the Millbase Data (pounds per gallon)			
Batch or Sample	Raw Data, X_t	Moving Average, A_t	Absolute Deviation, D_t	Moving Deviation, V_t
1	14.56	14.394	0.249	0.261
2	13.88	14.223	0.514	0.345
3	13.98	14.142	0.243	0.311
4	14.50	14.261	0.358	0.327
5	14.22	14.247	0.041	0.232
6	14.36	14.285	0.113	0.192
7	14.46	14.343	0.175	0.186
8	14.32	14.335	0.023	0.132
9	14.27	14.314	0.065	0.110
10	14.20	14.276	0.114	0.111
11	14.35	14.301	0.074	0.099
12	13.84	14.147	0.461	0.219
13	14.43	14.241	0.283	0.240
14	14.72	14.401	0.479	0.320
15	14.75	14.517	0.349	0.330
16	14.27	14.435	0.247	0.302
17	14.23	14.367	0.205	0.270
18	14.60	14.444	0.233	0.258
19	14.25	14.380	0.194	0.237
20	14.12	14.293	0.260	0.244
21	14.13	14.239	0.163	0.217
22	14.47	14.316	0.231	0.222
23	14.73	14.454	0.414	0.286
24	14.02	14.309	0.434	0.335
25	14.11	14.243	0.199	0.290
26	14.48	14.322	0.237	0.272
27	14.36	14.335	0.038	0.194
28	13.70	14.123	0.635	0.341
29	14.36	14.202	0.237	0.306
30	14.66	14.355	0.458	0.357

the data. Additionally, it is very common to see runs above and below the chart centerlines, because the use of the weighting procedure described previously actually introduces inertia/correlation into the moving averages and deviations.

A run of 8 points beneath the centerline on the V chart does not necessarily constitute a statistical signal in this case. Apart from the run, the moving deviations chart displays no obvious trendlike patterns. An examination of the A chart shows no systematic tendencies, and thus we may conclude that the process mean is stable.

Fundamental Process Improvement

As described previously, at this point in the millbase dispersion process improvement case study a batch adjustment procedure was developed and implemented.

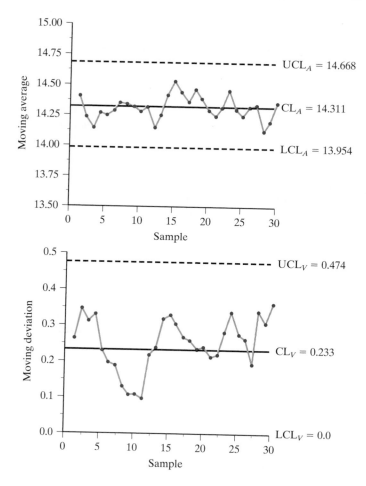

Figure 9.10   Exponentially weighted moving average and moving deviations control charts for the white millbase data, samples 1 to 30

Twenty-seven additional samples were collected to judge the effectiveness of the new procedure. Table 9.5 lists the moving averages and deviations for these 27 additional batches. The moving averages and deviations for these samples are shown in Fig. 9.11. The control limits displayed in Fig. 9.11 are based on the first 30 samples taken from the millbase process. As is evident from an examination of Fig. 9.11, the mean and variability levels are quite different for the samples generated with the new batch adjustment procedure. The effectiveness of the new procedure is clearly demonstrated on these control charts as on the X, R_m charts.

Since implementation of the new batch adjustment procedure effectively created a new processing system, the samples collected under the operation of the new system may be used to create a new set of A and V charts. Using the data from samples 31 to 57, centerlines and control limits for the new charts

Batch or Sample	Raw Data, X_t	Moving Average, A_t	Absolute Deviation, D_t	Moving Deviation, V_t
31	14.04	14.250	0.315	0.343
32	13.94	14.147	0.310	0.332
33	13.82	14.038	0.327	0.330
34	14.11	14.062	0.072	0.244
35	13.86	13.995	0.202	0.230
36	13.62	13.870	0.375	0.278
37	13.66	13.800	0.210	0.255
38	13.85	13.817	0.050	0.187
39	13.67	13.768	0.147	0.174
40	13.80	13.779	0.032	0.127
41	13.84	13.799	0.061	0.105
42	13.98	13.859	0.181	0.130
43	13.40	13.706	0.459	0.240
44	13.60	13.671	0.106	0.195
45	13.80	13.714	0.129	0.173
46	13.66	13.696	0.054	0.134
47	13.93	13.774	0.234	0.167
48	13.45	13.666	0.324	0.219
49	13.90	13.744	0.234	0.224
50	13.83	13.773	0.086	0.178
51	13.64	13.728	0.133	0.163
52	13.62	13.692	0.108	0.145
53	13.97	13.785	0.278	0.189
54	13.80	13.790	0.015	0.131
55	13.70	13.760	0.090	0.117
56	13.71	13.743	0.050	0.095
57	13.67	13.719	0.073	0.088

TABLE 9.5 Exponentially Weighted Moving Statistics for the Millbase Data (Samples 31 to 57: $A_{30} = 14.355$, $V_{30} = 0.357$)

may be calculated. First, we calculate the average and standard deviation of the data using samples 31 to 57:

$$\overline{X} = \sum_{t=31}^{57} X_t/27 = 13.777 \text{ pounds/gallon},$$

$$s_X = \left[\sum_{t=31}^{57} \frac{(X_t - 13.777)^2}{27 - 1} \right]^{1/2} = 0.172 \text{ pound/gallon}.$$

The centerlines and control limits are calculated below.

1. For the chart of moving averages A_t:

$$\text{CL}_A = \overline{X}$$

$$= 13.777,$$

Figure 9.11   **Exponentially weighted moving average and moving deviation control charts for the white millbase data (limits based on samples 1 to 30)**

$$UCL_A = \overline{X} + A^* s_X$$
$$= 13.777 + (1.342)(0.172)$$
$$= 14.008,$$
$$LCL_A = \overline{X} - A^* s_X$$
$$= 13.777 - (1.342)(0.172)$$
$$= 13.546.$$

2. For the chart of moving deviations, V_t:

$$CL_V = d_2^* s_X$$
$$= (0.874)(0.172)$$

$$= 0.151,$$

$$\text{UCL}_V = D_2^* s_X$$

$$= (1.780)(0.172)$$

$$= 0.307,$$

$$\text{LCL}_V = D_1^* s_X$$

$$= (0.0)(0.172)$$

$$= 0.0.$$

The moving averages and deviations for samples 31 to 57 of the millbase study are listed in Table 9.6. To start the series of moving averages and moving deviations, we need some initial values for the A's and V's, which can be estimated

TABLE 9.6	Exponentially Weighted Moving Statistics for the Millbase Data (Samples 31 to 57)			
Batch or Sample	Raw Data, X_t	Moving Average, A_t	Absolute Deviation, D_t	Moving Deviation, V_t
31	14.04	13.865	0.263	0.203
32	13.94	13.890	0.075	0.160
33	13.82	13.866	0.070	0.130
34	14.11	13.948	0.244	0.168
35	13.86	13.918	0.088	0.141
36	13.62	13.819	0.298	0.194
37	13.66	13.766	0.159	0.182
38	13.85	13.794	0.084	0.149
39	13.67	13.753	0.124	0.141
40	13.80	13.768	0.047	0.110
41	13.84	13.792	0.072	0.097
42	13.98	13.855	0.188	0.127
43	13.40	13.703	0.455	0.236
44	13.60	13.669	0.103	0.192
45	13.80	13.713	0.131	0.172
46	13.66	13.695	0.053	0.132
47	13.93	13.773	0.235	0.166
48	13.45	13.666	0.323	0.219
49	13.90	13.744	0.234	0.224
50	13.83	13.772	0.086	0.178
51	13.64	13.728	0.132	0.163
52	13.62	13.692	0.108	0.145
53	13.97	13.785	0.278	0.189
54	13.80	13.790	0.015	0.131
55	13.70	13.760	0.090	0.117
56	13.71	13.743	0.050	0.095
57	13.67	13.719	0.073	0.088

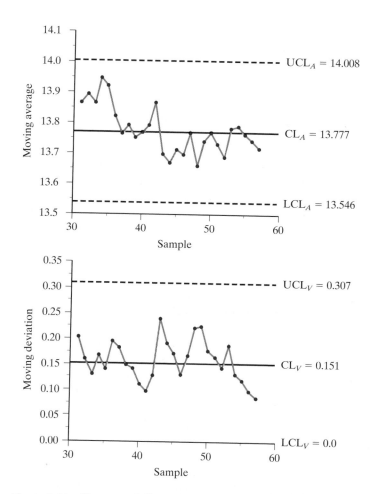

Figure 9.12 **Exponentially weighted moving average and moving deviation control charts for the white millbase data (limits based on samples 31 to 57)**

by $A_{30} = \overline{X} = 13.777$ and $V_{30} = s_X = 0.172$, as mentioned previously. The control charts associated with the data in Table 9.6 are displayed in Fig. 9.12. The V chart appears to show no statistical signals, suggesting that the process variability is constant. The moving averages on the A chart, however, seem to indicate a downward trend in the mean from the fifth batch on, although there are no out-of-control points. This slowly developing trend was not detected from the X and R_m charts shown in Fig. 9.4.

■ EXERCISES

9.1 For the data given in the accompanying table, calculate the moving ranges, R_m, based on a sample size of $n = 2$, $n = 3$, and $n = 4$.

Sample	X_i	Sample	X_i
1	11	6	9
2	13	7	8
3	9	8	12
4	14	9	14
5	10	10	10

9.2 Describe the circumstances under which it is difficult or does not make sense to use $\overline{X}$ and R charts to study a process, and therefore an investigator must use control charts for individual measurements. Give several examples where control charts for individual measurements should be used instead of $\overline{X}$ and R charts. What concerns or cautions should be exercised in using X and R_m charts as compared with $\overline{X}$ and R charts?

9.3 Use the data from Exercise 6.16 to construct X and R_m charts by using only the first measurement (X_1) from each subgroup.
  (a) Compute all the sample moving ranges using $n = 3$.
  (b) Calculate $\overline{X}$ and $\overline{R}_m$.
  (c) Calculate an estimate for the process standard deviation.
  (d) Calculate control limits for the X and R_m charts.
  (e) Construct X and R_m charts for these data and plot the data.
  (f) Evaluate these charts and comment on their appearance. Identify all statistical signals that indicate out-of-control conditions.

9.4 Charting can be continued from Exercise 9.3 using data from Exercise 6.17, again using only the first measurement (X_1) in each subgroup listed there.
  (a) Compute all the sample moving ranges using $n = 3$.
  (b) Continue charting using the control limits and centerlines from the first 35 points. Comment on the appearance of the charts.
  (c) Construct new X and R_m charts based only on the new data. Comment on the appearance of these charts and contrast them to the charts made in part (b). Show all calculations, and explain the reasoning behind the comments you make.

9.5 Consider the third measurement (X_3) from each subgroup listed in Exercise 6.16. Compute all the moving ranges, centerlines, and control limits, and construct X and R_m control charts for the following cases:
  (a) Moving ranges based on $n = 2$.
  (b) Moving ranges based on $n = 3$.
  (c) Moving ranges based on $n = 4$.
  (d) Compare the three sets of control charts constructed above. Comment on the differences between the charts. Be specific.

9.6 In this chapter it is stated that the individual observations used to make X and R_m charts need to follow a normal distribution if zone rules are to be used in analyzing the X chart. Explain why this is so.

9.7 A chemical process that has been operating for some time has produced the data listed in the accompanying table, which are pH measurements of a chemical reactant taken once every hour.

Sample	X_i	Sample	X_i
1	4.7	23	5.1
2	4.9	24	5.7
3	4.5	25	4.8
4	5.4	26	4.7
5	4.7	27	4.9
6	4.7	28	4.7
7	4.8	29	5.2
8	4.8	30	4.9
9	4.7	31	5.1
10	5.0	32	5.6
11	5.3	33	4.9
12	4.8	34	5.2
13	5.4	35	5.0
14	5.6	36	5.0
15	4.8	37	5.0
16	5.6	38	4.9
17	4.7	39	5.1
18	6.0	40	4.9
19	4.4	41	5.1
20	4.3	42	5.0
21	5.2	43	4.9
22	4.8	44	4.9

(a) Construct a dot diagram for the individual measurements. Comment on its general appearance and what this might mean.
(b) Compute all the sample moving ranges using $n = 2$.
(c) Calculate $\overline{X}$ and $\overline{R}_m$.
(d) Calculate an estimate for the process standard deviation.
(e) Calculate control limits for the X and R_m charts.
(f) Construct X and R_m charts for these data and plot all points.
(g) Evaluate these charts and comment on their appearance. Identify all statistical signals that indicate out-of-control conditions.

9.8 Forty consecutive samples of size $n = 1$ were collected from an assembly process during an 8-hour shift. This process had not been studied previously by charting, and some problems with poor quality had been noted. The process measurements, the time (in minutes) required to assemble a product using a bench fixture, are shown in the accompanying table.

Sample	X_i	Sample	X_i
1	2.09	21	3.42
2	3.81	22	3.16
3	1.93	23	3.21
4	1.49	24	3.59
5	3.29	25	3.05
6	4.25	26	1.97
7	2.69	27	3.58
8	2.88	28	3.16
9	3.54	29	3.62
10	3.56	30	1.80
11	3.56	31	2.48
12	2.34	32	3.67
13	4.35	33	2.15
14	1.48	34	3.14
15	2.38	35	3.27
16	2.79	36	1.86
17	1.29	37	1.71
18	3.35	38	4.10
19	5.29	39	2.28
20	3.20	40	1.99

(a) Is it appropriate to use X and R_m charts to study this process? Why or why not?

(b) If you were studying this process, what type of sampling scheme would you develop? Be specific. Give the rationale for your choice of sample size, sampling frequency, number of samples, and so on.

9.9 X and R_m charts are being constructed for a chemical distillation process. Samples of size $n = 2$ have been used to obtain the moving range values. Based on 50 samples, $\overline{X}$ and $\overline{R}_m$ have been calculated to be 10.0 and 3.0, respectively. What are the control limits for the X and R_m charts?

9.10 The temperature (in degrees Celsius, °C) within a refrigeration unit has been recorded every 2 hours. These temperature data are shown in the accompanying table.

(a) Construct a normal probability plot using the individual measurements and comment on its appearance.

(b) Using $n = 2$, calculate the moving ranges. Calculate $\overline{X}$, $\overline{R}_m$, and an estimate for the process standard deviation.

(c) Calculate control limits for the X and R_m charts, and construct X and R_m charts for these data using all 30 observations.

(d) Evaluate these charts and comment on their appearance. Identify all statistical signals that indicate out-of-control conditions.

Sample	X_i	Sample	X_i
1	0.6	16	-1.3
2	-2.0	17	-4.7
3	-3.3	18	0.0
4	-9.6	19	-4.2
5	-5.2	20	-2.1
6	-5.3	21	-1.8
7	-0.6	22	1.6
8	-4.1	23	-4.7
9	-9.4	24	-2.3
10	-0.4	25	3.9
11	-4.7	26	0.4
12	0.4	27	0.1
13	-1.6	28	-0.2
14	-5.3	29	-0.8
15	-0.4	30	-4.3

9.11 Following the analysis of the data described in Exercise 9.10, an improvement action was performed on the temperature control device of the refrigeration unit. To assess the effect of the improvement action, more data were collected and are shown in the accompanying table.
(a) Calculate moving ranges for each new measurement (use $n = 2$).
(b) Continue charting using the control limits and centerlines for the first 30 measurements. Comment on the appearance of the charts.

Sample	X_i	Sample	X_i
31	0.8	46	-3.1
32	-4.4	47	0.0
33	-1.8	48	-1.7
34	-1.3	49	-1.2
35	-3.4	50	-1.0
36	1.0	51	-3.3
37	0.7	52	-2.6
38	0.0	53	-2.2
39	-0.6	54	-4.5
40	-1.3	55	-1.4
41	-0.3	56	-2.1
42	-2.7	57	-1.7
43	-7.2	58	-0.9
44	-5.4	59	-3.7
45	-2.0	60	-2.6

(c) Construct new X and R_m charts based on the new data only. Comment on the appearance of these charts and contrast them to the charts made in part (b). Show all calculations, and explain the reasoning behind the comments you make.

9.12 It is desired to study the chemical concentration of an acid bath via X and R_m control charts. Thirty individual concentration values have been collected and moving range values have been calculated based on a sample size of 2. An examination of the data reveals that

$$\Sigma X_i = 596.1 \quad \text{and} \quad \Sigma R_{mi} = 85.8.$$

(a) Calculate the centerlines and control limits for the X and R_m charts.
(b) Ten additional concentration levels have been collected as shown in the accompanying table. Calculate the missing R_m values.

Sample	31	32	33	34	35	36	37	38	39	40
X_i	21	24	23	26	24	26	18	21	19	20
R_{mi}	1.8									

The normal probability plot shown here has been prepared based on the first 30 concentration levels. X and R_m control charts were constructed for the first 30 points and showed no extreme points or runs.

(c) Comment on the appearance of the normal probability plot. Will the appearance of the plot in this case have any impact on the further interpretation of the X and R_m control charts? Why or why not?

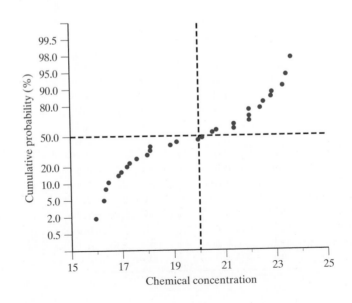

9.13 The moisture content of corn kept in a bulk storage facility is under study. Daily measurements of the moisture content have been recorded for 40 consecutive days and are shown in the accompanying table. Compute all the moving ranges, centerlines, and control limits, and construct X and R_m control charts for the following cases:

(a) Moving ranges based on $n = 2$.

(b) Moving ranges based on $n = 3$.

(c) Moving ranges based on $n = 4$.

(d) Compare the three sets of control charts constructed above. Comment on the differences between the charts. Be specific.

Sample	X_i (% moisture)	Sample	X_i (% moisture)
1	14.3	21	17.0
2	18.7	22	15.2
3	15.6	23	14.4
4	14.9	24	15.4
5	10.0	25	17.7
6	16.2	26	14.2
7	17.5	27	17.6
8	16.4	28	13.4
9	13.8	29	14.5
10	13.3	30	14.8
11	13.5	31	12.2
12	19.1	32	16.5
13	16.8	33	10.4
14	14.7	34	14.3
15	15.3	35	12.9
16	15.6	36	12.8
17	14.1	37	16.0
18	14.6	38	12.6
19	12.8	39	14.7
20	16.5	40	10.9

9.14 Use the data shown in the accompanying table to construct EWMA charts for the process that generated it. Use $r = 0.200$.

(a) Calculate estimates of the process mean and standard deviation.

(b) Calculate A_t and V_t for each observation.

(c) Calculate control limits and centerlines for both charts.

(d) Construct an A chart and a V chart for the process and plot all the points. Interpret the charts and make comments on any patterns you see. What conclusions can you draw about the behavior of the process? Be thorough in your explanation.

Sample	X_i	Sample	X_i
1	1.5	21	1.6
2	0.5	22	1.7
3	0.8	23	1.3
4	−0.8	24	1.2
5	1.1	25	−0.1
6	0.7	26	−0.7
7	1.3	27	−0.5
8	2.0	28	−0.6
9	−0.5	29	0.1
10	1.6	30	−0.9
11	1.0	31	−2.8
12	0.4	32	0.2
13	0.7	33	−1.4
14	0.8	34	0.8
15	1.6	35	−0.4
16	0.7	36	−1.8
17	1.2	37	−0.6
18	0.1	38	−2.1
19	0.3	39	−0.6
20	1.6	40	−1.5

9.15 Consider the data in Exercise 9.14.
   (a) Construct X and R_m charts using $n = 2$. Interpret the charts.
   (b) Construct EWMA charts using $r = 0.66$. Interpret the charts.
   (c) Comment on the relative appearance of the two sets of charts constructed in parts (a) and (b). Are there significant differences? If so, how can these be accounted for or explained?

9.16 The data shown in the accompanying table were collected from an automated chemical process. The measurements represent the deviation of the concentration of a certain chemical (%) from the target value for that factor.
   (a) Calculate estimates of the process mean and standard deviation.
   (b) Calculate A_t and V_t for each observation using r values of 0.2, 0.333, and 0.4.
   (c) Construct A and V charts for r values of 0.2, 0.333, and 0.4. Interpret all three sets of charts.
   (d) Considering the three sets of charts developed in part (c), what effect(s) does changing r appear to have on the charts? What factors might influence the selection of a value for r?

Sample	X_i	Sample	X_i
1	−0.01	21	0.75
2	−0.23	22	−1.07
3	−0.80	23	−0.71
4	−0.17	24	0.23
5	−1.00	25	1.06
6	−1.20	26	−2.65
7	0.33	27	−1.17
8	0.89	28	−2.61
9	−1.18	29	−1.85
10	−1.64	30	−1.68
11	2.07	31	−1.94
12	−0.77	32	−1.83
13	0.83	33	−2.50
14	0.82	34	−0.56
15	0.03	35	−2.58
16	−0.90	36	−2.46
17	1.57	37	−1.84
18	−0.16	38	−2.11
19	−0.52	39	−2.14
20	−0.23	40	−1.03

9.17 A wire drawing process with large variability is being studied in hopes of discovering opportunities for improvement. The data shown in the accompanying table are the measured diameters of the wire (in 0.001 inch), and were collected every 5 minutes for a period of approximately 3.5 hours.

Sample	X_i	Sample	X_i
1	110.5	15	106.3
2	108.4	16	107.4
3	112.0	17	109.3
4	109.5	18	108.7
5	112.0	19	109.2
6	109.4	20	108.1
7	110.2	21	111.9
8	109.9	22	109.0
9	112.1	23	114.1
10	107.7	24	109.9
11	109.3	25	114.2
12	112.6	26	110.4
13	110.7	27	110.3
14	109.3	28	110.4

(Continued)

Sample	X_i	Sample	X_i
29	113.3	35	109.5
30	109.6	36	109.1
31	110.6	37	109.5
32	110.0	38	110.1
33	112.2	39	111.0
34	111.5	40	109.1

(a) Calculate estimates of the process mean and standard deviation.
(b) Calculate A_t and V_t for each observation using $r = 0.333$.
(c) Calculate control limits and centerlines for both charts.
(d) Construct an A chart and a V chart for the process and plot all points. Interpret the charts and make comments on any patterns you see. What conclusions can you draw about the behavior of the process? Be thorough in your explanation.

9.18 A change was made to the wire drawing process described in Exercise 9.17 in an attempt to improve its behavior, and the data collection activity was continued. The data are shown in the accompanying table.

Sample	X_i	Sample	X_i
1	110.6	16	108.3
2	110.0	17	109.8
3	112.1	18	110.4
4	111.5	19	110.6
5	109.6	20	109.0
6	109.2	21	109.4
7	109.5	22	109.1
8	110.0	23	111.2
9	110.9	24	110.6
10	109.2	25	108.6
11	110.0	26	110.8
12	110.8	27	108.4
13	111.4	28	108.8
14	110.8	29	110.1
15	109.6	30	111.4

(a) Using the control limits and centerlines developed in Exercise 9.17, plot the newly collected data.
(b) Comment on the behavior of the new data. Interpret the extended charts and make appropriate comments.
(c) Construct charts based only on the new data. Comment on the appearance of these new charts and compare them to the charts you made in part (b).

9.19 The temperature of a chemical reactant has been determined to be a crucial factor in obtaining satisfactory yield from a chemical process. The nominal value for this temperature is supposed to be 150°C. The temperature of the reactant is recorded every hour for 40 consecutive hours. These temperature measurements, recorded as the deviation from the desired value of 150°C, are shown in the accompanying table.

Sample	X_i	Sample	X_i
1	−4.53	21	−4.87
2	7.52	22	5.12
3	−3.05	23	−2.73
4	3.31	24	−2.11
5	13.27	25	−2.29
6	−0.68	26	5.45
7	8.13	27	0.92
8	−0.35	28	4.26
9	5.74	29	0.03
10	3.92	30	5.14
11	2.43	31	−3.91
12	−0.91	32	3.68
13	−0.67	33	11.49
14	5.76	34	5.45
15	2.44	35	3.93
16	−0.96	36	10.80
17	8.72	37	−1.28
18	13.29	38	10.52
19	2.41	39	0.91
20	−1.82	40	−1.84

(a) Construct X and R_m charts using $n = 3$. Interpret the charts.

(b) Construct EWMA charts using $r = 0.5$. Interpret the charts.

(c) Comment on the relative appearance of the two sets of charts constructed in parts (a) and (b). Are there significant differences? If so, how can these be accounted for or explained?

9.20 A seamless tubing extrusion process that is producing a large number of "wall breakouts" is being scrutinized in hopes of improving its performance. The tubing is extruded and cut into 10-foot lengths. One wall thickness measurement is made for each of 40 consecutive lengths, with the data shown in the accompanying table. It is felt that the process may be prone to exhibit abrupt but short-lived shifts in the mean level.

(a) Calculate estimates of the process mean and standard deviation.

(b) Select an appropriate value for r based on the nature of the process. Explain the rationale for your selection. Calculate A_t and V_t for each observation.

Sample	X_i	Sample	X_i
1	11.7	21	10.7
2	11.1	22	12.3
3	11.4	23	9.9
4	10.7	24	10.7
5	9.4	25	10.8
6	11.0	26	10.6
7	10.2	27	11.4
8	11.0	28	9.9
9	10.9	29	11.0
10	11.4	30	11.7
11	11.9	31	10.8
12	10.9	32	11.4
13	11.7	33	12.0
14	12.4	34	9.8
15	10.3	35	11.4
16	11.4	36	10.5
17	12.7	37	10.4
18	11.6	38	11.2
19	10.4	39	10.4
20	11.3	40	11.2

(c) Calculate control limits and centerlines for both charts.

(d) Construct an A chart and a V chart for the process and plot all points. Interpret the charts and make comments on any patterns you see. What conclusions can you draw about the behavior of the process? Be thorough in your explanation.

9.21 After a change is made to the extrusion process from Exercise 9.20, more data are collected, as shown in the accompanying table.

Sample	X_i	Sample	X_i
41	13.2	56	12.6
42	13.0	57	13.6
43	12.3	58	14.9
44	14.0	59	14.7
45	13.7	60	12.0
46	13.5	61	12.5
47	12.2	62	12.7
48	13.8	63	12.7
49	13.9	64	13.1
50	13.4	65	12.5
51	12.1	66	13.7
52	13.7	67	14.3
53	12.8	68	12.4
54	13.0	69	14.4
55	12.9	70	14.1

(a) Using the control limits and centerlines developed in Exercise 9.20, plot the newly collected data.

(b) Comment on the behavior of the new data. Interpret the extended charts and make appropriate comments.

(c) Construct charts based only on the new data. Comment on the appearance of these new charts and compare them to the charts you made in part (b).

(d) Does it appear that the change made to the process will affect the "wall breakout" problem? Why or why not? Be thorough in your explanation.

9.22 Discuss the similarities and differences between Shewhart-type control charts for individual measurements and EWMA charts for individual measurements in terms of:

(a) The assumption about the relationship among individual points on the charts.

(b) The general manner(s) in which special causes due to shifts in the mean show up on the charts.

(c) The relative appearance of X and R_m and EWMA charts for the same data where frequent, short-lived shifts in the mean are occurring.

9.23 The manager of the injection molding department has asked you to prepare an EWMA chart for a maintenance operation—weekly mold changes. The table of data here shows the number of mold changes for the last 45 weeks.

Sample	X_i	Sample	X_i	Sample	X_i
1	10	16	9	31	11
2	10	17	12	32	9
3	13	18	11	33	9
4	9	19	9	34	9
5	8	20	8	35	10
6	9	21	13	36	10
7	10	22	12	37	9
8	9	23	12	38	10
9	10	24	12	39	10
10	11	25	12	40	10
11	8	26	11	41	8
12	10	27	12	42	9
13	9	28	11	43	10
14	10	29	12	44	9
15	9	30	11	45	10

Construct EWMA charts using $r = 0.33$. Identify any statistical signals on the charts or other trendlike patterns. What can you report to the manager about the mold-changing process?

9.24 Construct X and R_m charts for the data on weekly mold changes of Exercise 9.23, using an artificial sample size of 3 for R_m calculations. What differences do you find between these charts and the EWMA charts of Exercise 9.23? Explain any similarities and/or differences between the two types of charts for individual measurements from their respective bases of chart construction.

9.25 In the construction of X and R_m charts, we need to decide on the artificial sample size, n, for the purpose of calculating the moving ranges. Briefly describe the issues that one must consider in selecting a value for n.

9.26 In the construction of EWMA charts, we need to decide on the value of r. Briefly describe the issues that one must consider in selecting a value for r.

9.27 Explain why the interpretation of the EWMA charts is largely based on identifying trends in the moving average and moving deviation values as opposed to simply applying the rules for $\overline{X}$ and R chart interpretation.

9.28 X and R_m control charts are said to be less sensitive to the detection of small, sustained shifts in the mean compared with the EWMA charts. On the other hand, the EWMA charts are less sensitive to the detection of small, short-lived shifts in the mean compared with the X and R_m charts. Explain.

9.29 Can the interpretation of X and R_m charts and the interpretation of EWMA charts for the same set of data be entirely different? Explain why or why not.

10

Process Capability
Assessment

10.1 Introduction

In this chapter we present both traditional and more modern views of process capability and how it is assessed. The presentation stresses the relationship between the stability of a process and its capability. We emphasize the clear distinction between the engineering specification and the statistical control limits in terms of use and interpretation. By using the traditional engineering specifications to view process capability, this chapter illustrates the consequences of excessive common-cause variation in terms of the manner and extent to which a process produces parts that meet design intent. The loss function approach to the articulation of process capability is presented. In this regard, the consequences of failure to ensure statistical control of the process, as well as poor process centering, are demonstrated in a quantitative fashion. Finally, given the pervasive application of the C_{pk} capability index within industry, a cautionary tale is presented regarding its use—the dilemma of C_{pk}. A case study is used to illustrate the sampling distribution of C_{pk} when it is calculated based on run-off data. Hypothesis testing and confidence-interval approaches for dealing with this situation are critically examined.

10.2 Process Capability Versus Process Control

There are two separate but vitally important issues that we must address when considering the performance of a process: the ability of the process to produce parts that conform to engineering specifications, and the ability of the process to maintain a state of good statistical control. Since most process capability assessments are made on statistical grounds, these two process characteristics are linked in the sense that it is both inappropriate and statistically invalid to assess process capability with respect to conformance to specifications without being reasonably assured of having good statistical control. Although control certainly does not imply conformance, it is a necessary prerequisite to the proper assessment of conformance.

In a statistical sense, conformance to specifications involves the process as a whole, and therefore we are focusing our attention on the distribution of individual measurements. In dealing with statistical control we employ summary statistics from each sample, for example, $\overline{X}$ and R, and therefore are dealing with the distributions of these statistics, not individual measurements. Because of the distinction between populations and samples, it is crucial not to compare or confuse specifications associated with quality characteristics and control limits. In fact, we should never place tolerance/specification limits on an $\overline{X}$ control chart. This is so because the $\overline{X}$ control chart is based on the variation in the sample means, $\overline{X}$, while it is the individual measurements in the sample that should be compared to specifications. By placing specifications on the control chart, we may get the mistaken impression that good conformance exists when, in fact, it does not. Furthermore, it is philosophically wrong to place specifications on a control chart, since such would suggest that the purpose of the chart is to track a quality characteristic relative to specifications.

A process may produce a large number of pieces that do not meet the specified tolerances, even though the process itself is in a state of good statistical control (i.e., all the points on the $\overline{X}$ and R charts are within the 3 sigma limits and vary in a random manner). This may be because the process is not centered properly; in other words, the actual mean value of the parts being produced may be significantly different from the specified nominal value of the part. If this is the case, an adjustment of the machine to move the mean closer to the nominal value may solve the problem. Another possible reason for lack of conformance to specifications is that a statistically stable process may be producing parts with an unacceptably high level of common-cause variation, although it may be centered at the nominal value.

In summary, if a process is in statistical control but not capable of meeting the tolerances, the problem may be one of the following:

- The process is off-center from the nominal.
- The process variability is too large relative to the tolerances.
- The process is off-center and has large variation.

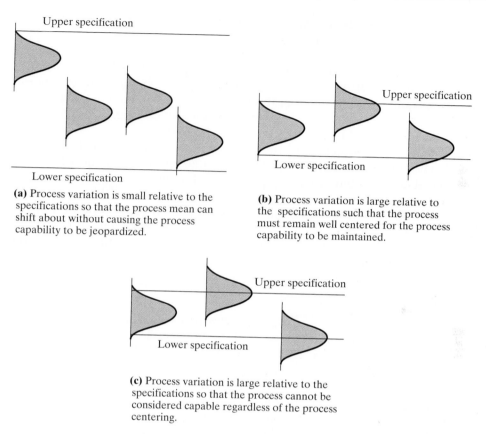

(a) Process variation is small relative to the specifications so that the process mean can shift about without causing the process capability to be jeopardized.

(b) Process variation is large relative to the specifications such that the process must remain well centered for the process capability to be maintained.

(c) Process variation is large relative to the specifications so that the process cannot be considered capable regardless of the process centering.

Figure 10.1   Relationship between process variability and product specifications

Figure 10.1 illustrates the possible relationships that could exist between the process variation and the specifications. Of particular importance may be attention to the leftmost case in the figure. Here the process variation is small relative to the specifications. In such a situation it is easy to forget about the issue of statistical control since the specifications are so wide. In fact, statistically based modified control chart limits have even been suggested to take advantage of this happy state of affairs. Such are probably inadvisable, since they mask the presence of improvement opportunities as a result of sporadic behaviors that rob the process of productivity as well as quality. Again, it is essential that we separate the issues of control and conformance.

10.3   Statistical Assessment of Process Capability: A Case Study Revisited

Recall the machined cylinder liner bore example of Chapter 6. Measurements of the inside diameter of the bore liner were made after the machining process. The new specifications for the liner inner diameter are 3.5199 ± 0.0004 inches. Many liners would be rejected based on 100% inspection using a go/no-go gage because they would fail to meet the new specification.

Following the completion of the investigation using $\overline{X}$ and R charts described in Chapter 6, the capability of the stabilized process may now be examined. The data collected from the process following the identification and removal of the special causes is shown in Table 10.1. These data were taken from the liner boring process at a rate of about one sample every half-hour. The table provides 40 samples of $n = 5$ each. Recall that these data are in coded form, with the last three digits of the diameter measurement shown in the table.

Checking for Statistical Control

The first step in the process capability assessment should be to examine the state of statistical control (or lack of it) of the process. To do this, we will construct $\overline{X}$ and R control charts. This, of course, requires that the data be collected in a manner (samples collected over a period of time) that is consistent with control charting, as was the case here. Recall that in Section 6.8 the process was thoroughly studied and, after the removal of special causes, the process showed good statistical control. The necessary calculations are repeated as follows.

We first estimate the mean of the process by $\overline{\overline{X}}$:

$$\overline{\overline{X}} = \sum_{i=81}^{i=120} \frac{\overline{X}_i}{40}$$

$$= 200.005.$$

We will also need to estimate the range of the process:

$$\overline{R} = \sum_{i=81}^{i=120} \frac{R_i}{40}$$

$$= 6.6.$$

The control limits for $\overline{X}$ and R charts are (using the constants for $n = 5$)

$$\text{UCL}_{\overline{X}}: \ \overline{\overline{X}} + A_2\overline{R} = 200.005 + (0.577)(6.6)$$

$$= 203.81,$$

$$\text{LCL}_{\overline{X}}: \ \overline{\overline{X}} - A_2\overline{R} = 200.005 - (0.577)(6.6)$$

$$= 196.20,$$

TABLE 10.1	Cylinder Liner Boring Process Data—In Control							
Sample Number	Sample Collection Time	1	2	3	4	5	$\overline{X}$	R
81	8:05 A.M.	203	200	200	198	201	200.4	5
82	8:30 A.M.	195	203	203	193	199	198.6	10
83	8:55 A.M.	206	200	198	199	202	201.0	8
84	9:20 A.M.	197	202	200	199	202	200.0	5
85	9:45 A.M.	205	194	203	198	201	200.2	11
86	10:10 A.M.	207	202	204	198	204	203.0	9
87	10:35 A.M.	199	198	197	199	199	198.4	2
88	11:00 A.M.	201	206	202	203	198	202.0	8
89	11:25 A.M.	197	202	200	197	200	199.2	5
90	11:50 A.M.	197	199	197	202	193	197.6	9
91	1:00 P.M.	203	194	202	200	200	199.8	9
92	1:25 P.M.	201	202	201	197	202	200.6	5
93	1:50 P.M.	204	195	208	202	202	202.2	13
94	2:15 P.M.	199	202	196	201	195	198.6	7
95	2:40 P.M.	200	201	201	197	198	199.4	4
96	3:05 P.M.	200	200	203	201	201	201.0	3
97	3:30 P.M.	196	201	199	198	197	198.2	5
98	3:55 P.M.	197	202	198	196	197	198.0	6
99	4:20 P.M.	205	193	202	200	196	199.2	12
100	4:45 P.M.	203	200	199	204	202	201.6	5
101	8:00 A.M.	201	205	201	202	199	201.6	6
102	8:25 A.M.	200	203	200	200	203	201.2	3
103	8:50 A.M.	201	203	199	200	201	200.8	4
104	9:15 A.M.	196	201	200	199	204	200.0	8
105	9:40 A.M.	199	199	201	203	201	200.6	4
106	10:05 A.M.	202	201	200	203	201	201.4	3
107	10:30 A.M.	197	201	199	196	199	198.4	5
108	10:55 A.M.	195	200	205	202	195	199.4	10
109	11:20 A.M.	201	202	200	195	201	199.8	7
110	11:45 A.M.	196	200	200	196	196	197.6	4
111	1:15 P.M.	206	200	197	196	198	199.4	10
112	1:40 P.M.	202	198	205	201	196	200.4	9
113	2:05 P.M.	200	198	198	200	197	198.6	3
114	2:30 P.M.	199	201	204	200	205	201.8	6
115	2:55 P.M.	201	201	198	197	198	199.0	4
116	3:20 P.M.	199	206	197	198	201	200.2	9
117	3:45 P.M.	201	197	191	200	201	198.0	10
118	4:10 P.M.	202	198	199	200	204	200.6	6
119	4:35 P.M.	201	203	204	197	202	201.4	7
120	5:00 P.M.	200	204	199	203	199	201.0	5

$$\text{UCL}_R: \quad D_4\overline{R} = (2.115)(6.6)$$
$$= 13.96,$$
$$\text{LCL}_R: \quad D_3\overline{R} = (0)(6.6)$$
$$= 0.0.$$

Now we can construct $\overline{X}$ and R control charts. These were previously shown in Fig. 6.30 and are repeated here in Fig. 10.2.

Both the $\overline{X}$ and R charts seem to indicate good statistical control with no points exceeding the 3 sigma limits, a reasonably normal distribution of points between the limits, and no discernible trends, cycles, zone rule violations, and so on. The good control of the R chart indicates that our estimate of the process variation,

$$\hat{\sigma}_X = \frac{\overline{R}}{d_2}$$
$$= \frac{6.60}{2.326}$$
$$= 2.8375,$$

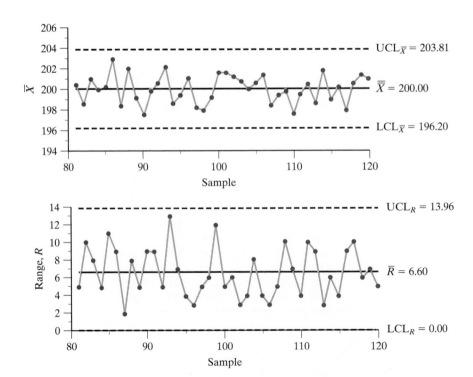

Figure 10.2   Control charts for process capability case study data

is a valid estimate of common-cause variation. With a stable process demonstrated through the $\overline{X}$ and R charts, we may now proceed to evaluate the process with respect to its conformance to specifications.

Statistical Assessment of Process Capability

To get a clear picture of the statistical nature of the data from an individual measurements point of view, a frequency histogram was plotted. This is shown in Fig. 10.3. The histogram seems to exhibit the shape of the normal distribution, but the mean appears to be a little higher than the nominal value of 199. To further clarify the nature of the data, a normal probability plot was also made as shown in Fig 10.4. The normal plot also indicates that it is reasonable to use a normal distribution to describe the individual data.

Using what we've learned about the distribution of the process and our estimates of the process mean $\overline{\overline{X}}$ and process standard deviation $\hat{\sigma}_X$, we can sketch the process as a normal distribution, as shown in Fig. 10.5. The specifications are also shown in the figure. The shaded area under the normal curve represents the probability of obtaining a bore that does not meet the specifications.

To compute the probability of a bore falling outside of the specifications, we calculate Z and use the standard normal curve table (since n is quite large)

$$Z = \frac{X - \overline{\overline{X}}}{\hat{\sigma}_X},$$

where X is the value of either the lower or upper specification, $\overline{\overline{X}}$ is our estimate for the process mean, and $\hat{\sigma}_X$ is our estimate of the process standard deviation.

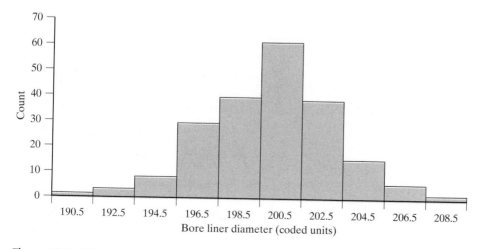

Figure 10.3   Histogram of case study data

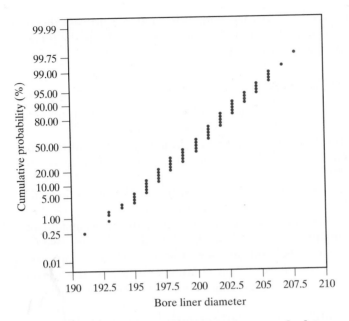

Figure 10.4   Normal probability plot of case study data

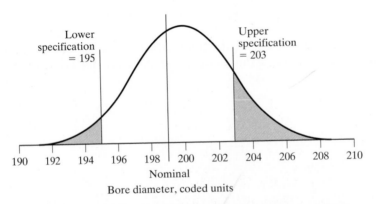

Figure 10.5   Normal distribution model for process capability case study data

To find the probability of a measurement below the lower specification limit, LSL = 195, with $\overline{\overline{X}} = 200.00$ and $\hat{\sigma}_X = 2.8375$,

$$Z_L = \frac{195 - 200}{2.8375}$$

$$= -1.762.$$

Looking this value up in the normal curve table (Table A.1), we find that the probability of $X \leq 195$ is 0.0390. This means we have a 3.90% chance of an individual bore liner measurement falling below the specified lower specification limit.

To find the probability of an individual bore liner measurement falling above the upper specification limit, USL = 203, with $\overline{\overline{X}} = 200.00$ and $\hat{\sigma}_X = 2.8375$,

$$Z_U = \frac{203 - 200}{2.8375}$$
$$= 1.057.$$

The value in the normal curve table for $Z = 1.057$ is 0.8547, but this includes all the area to the left of the USL. Thus, the probability of a bore liner being above the upper specification limit is equal to $1.0 - 0.8547 = 0.1453$. The percentage of product that does not meet the specification is therefore 3.90% + 14.53% = 18.43%

Now we might ask: Would centering the process at our nominal value of 199 help? To check, we construct a new normal curve, centered at 199. The new probability of a point below the lower specification is found by computing Z using the nominal value as the process mean:

$$Z_L = \frac{195 - 199}{2.8375}$$
$$= -1.409.$$

Looking up the area for this Z value in the normal curve table, we get a value of 0.0794, which is the probability of getting a value below the lower specification limit. The probability of getting a value above the upper specification is found by

$$Z_U = \frac{203 - 199}{2.8375}$$
$$= 1.409,$$

which from the table gives an area of 0.9206. Thus, the probability of a bore liner being above the upper specification limit is equal to $1.0 - 0.9206 = 0.0794$. The total new probability of a bore liner not meeting specifications is the sum of these, or $0.0794 + 0.0794 = 0.1588$.

We can see that recentering the process will not help very much. The process is in control—no special causes of variability were indicated. However, over 18% of the part measurements fall outside the specifications (and even with centering still 16%). Possible remedies may include:

1. Sort by 100% inspection.
2. Widen the tolerances (e.g., 3.5199 ± 0.0009).

3. Use a more precise process to reduce process variation.

4. Use statistical methods to identify variation reduction opportunities for the existing process.

Too often, strategy 2 is used (or at least urged). Clearly, we suggest that stronger consideration be given to strategy 4.

Summary: Comparison of Specification Limits and Control Limits

It is important to differentiate clearly between specification limits and control limits. In particular, we must note that specification limits or tolerances for a product quality characteristic are

- Characteristic of the part/item in question.
- Based on functional considerations.
- Related to/compared with an individual part measurement.
- Used to establish a part's conformability to design intent.

Control limits on a control chart are

- Characteristic of the process in question.
- Based on the process mean and variability.
- Dependent on sampling parameters, namely, sample size and the α-risk.
- Used to identify the presence/absence of special-cause variation in the process.

Control limits and specification limits must never be compared numerically and should not appear together on the same graph. Tolerances are limits on individual measurements and as such may be compared against the process as a whole as represented by a statistical distribution, as was done in the previous example to assess overall process capability. Figure 10.6, based on the data of the previous process capability study, depicts the difference between specification and control limits, and clearly indicates the danger of putting specification limits on the control chart. The open dots represent the individual measurements from which the sample means ($\overline{X}$'s) shown in the figure were determined. It is clear that while the $\overline{X}$'s are well within the control limits, many (11 out of 50) individual measurements fall on or outside the specifications.

10.4   Some Common Indices of Process Capability

Process Capability Index, C_p

It is common to measure process capability in units of process standard deviations. In particular, it is common to look at the relationship between the process

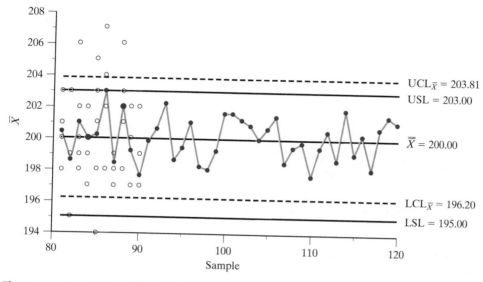

Figure 10.6   **Comparison of individual measurements and averages**

standard deviation and the range between the upper and lower specification limits. This measure is known as C_p:

$$C_p = \frac{\text{USL} - \text{LSL}}{6\sigma_X} \qquad (10.1)$$

The minimum acceptable value for C_p is considered to be 1.

Process Capability Index, C_{pk}

Many companies use the capability index, C_{pk}, which relates the process mean to the nominal value of the specification. For bilateral specifications, C_{pk} is defined in the following manner. First, we determine the relationship between the process mean μ_X and the specification limits in units of standard deviations:

$$Z_{\text{USL}} = \frac{\text{USL} - \mu_X}{\sigma_X}, \qquad Z_{\text{LSL}} = \frac{\text{LSL} - \mu_X}{\sigma_X}. \qquad (10.2)$$

Then we select the minimum of these two values:

$$Z_{\min} = \min[Z_{\text{USL}}, -Z_{\text{LSL}}]. \qquad (10.3)$$

We find the C_{pk} index by dividing this minimum value by 3:

$$C_{pk} = \frac{Z_{\min}}{3}. \qquad (10.4)$$

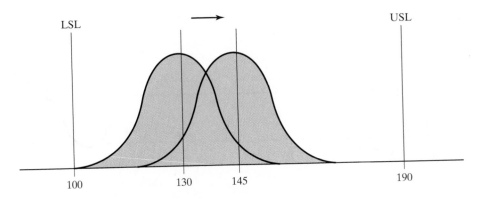

Figure 10.7   **Motivation via C_{pk} to be on target**

Commonly, for the process capability to be considered acceptable, C_{pk} should be ≥ 1.00.

The idea of using the minimum Z value is to promote adherence of the process mean to the nominal value. Figure 10.7 demonstrates this apparent motivation for being on target. Consider initially that the process mean is located at 130, which is somewhat away from the nominal of 145, with a standard deviation of $\sigma_X = 10$. Then

$$Z_{USL} = \frac{190 - 130}{10}$$

$$= 6,$$

$$Z_{LSL} = \frac{100 - 130}{10}$$

$$= -3,$$

$$Z_{min} = \min[6, -(-3)]$$

$$= 3,$$

$$C_{pk} = \frac{3}{3}$$

$$= 1.00.$$

If the process mean were centered at the nominal of 145, then

$$Z_{USL} = \frac{190 - 145}{10}$$

$$= 4.5,$$

$$Z_{LSL} = \frac{100 - 145}{10}$$

$$= -4.5,$$

$$Z_{\min} = \min[4.5, -(-4.5)]$$
$$= 4.5,$$
$$C_{pk} = \frac{4.5}{3}$$
$$= 1.50.$$

Therefore, by centering the process, the value of C_{pk} is increased by 50%.

The foregoing apparent motivation via C_{pk} to adhere to the nominal value must be viewed, however, with some caution. It is possible, with the use of C_{pk}, to use reduction in variation as a means to shift the process mean away from the nominal and still get an increase in C_{pk}, or at least not experience a decrease in C_{pk}. Returning to our previous example, suppose that the process is running at the nominal with a standard deviation of 15. C_{pk} under this circumstance is equal to 1.00. However, by reducing the variation to a standard deviation of 10, the process mean can be shifted well away from the nominal and C_{pk} is not reduced. Figure 10.8 illustrates this problem. The parameter values for each case are shown below.

> *Case 1:* $\mu_X = 145$ and $\sigma_X = 15$, so $C_{pk} = 1.00$.
> *Case 2:* $\mu_X = 130$ and $\sigma_X = 10$, so $C_{pk} = 1.00$.

Therefore, in this case, there is no penalty for purposely moving the process mean away from the target. Other measures of process capability do exist to avoid this problem.

10.5   Loss Function Approach to Capability Assessment

The process capability study discussed earlier in this chapter measured the overall conformance to specifications, which may be useful for comparisons

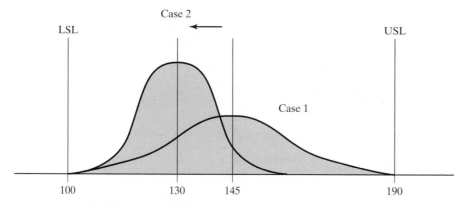

Figure 10.8   Potential problem with C_{pk}

between different processes. To pursue the notion of being on target with smallest variation as introduced in Chapter 1, and perhaps to begin to direct our thinking toward a more global and economic interpretation of the engineering specification, the concept of the loss function advocated by Taguchi and used by economists for decades will now be examined. To see how this concept is applied in the evaluation of process capability relative to the engineering specifications, we will use some illustrations from an article by Peter T. Jessup of the Ford Motor Company.[1]

Defining a Loss Function

The quality characteristic of interest in the Jessup case study is the lever effort on a heater control. Very low efforts would cause customer complaints of flimsy feel or rattle, while very high efforts would also cause customer complaints of stiffness. And in each case, the more extreme the condition, the more likely it would prompt a customer complaint. The total complaint rate is the sum of these two individual kinds of complaints. Between the extremes the net complaints from the two conditions drop to a minimum. This is illustrated in Fig. 10.9.

By choosing the point with the minimum loss on the net loss curve as the target and setting it to zero, we establish a continuous loss function of the quality characteristic as shown in Fig. 10.10. In general, a loss function may be defined by evaluating the expected losses for several values of the quality characteristic. Figure 10.11 shows calculations for one such point.

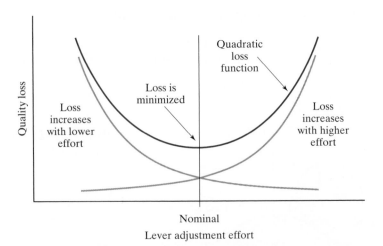

Figure 10.9   Composition of a net loss function

[1]P. T. Jessup, "The Value of Continuing Improvement," *Proc. IEEE International Communications Conference, ICC 85*, 1985.

Figure 10.10   **Loss function representation of quality**

Figure 10.11   **Construction of the loss function**

The calculations for the loss incurred at two particular values for the quality characteristic are as follows.

1. Identify a most desirable or target value:

$$x = 0, \text{ at the target;} \quad L(x = 0) = 0.$$

2. Expected cost of a particular failure type (in the field):

$$x = 1.5, \text{ and}$$

at $x = 1.5$, the cost of a failure = $45; the chance that this failure occurs = 5%; the expected loss = $L(1.5) = 0.05 \times \$45 = \$2.25$.

If it is assumed that a quadratic equation closely approximates the true loss function, then the loss function can be derived from the expected losses estimated from any two points, such as the loss at the target and the loss at one other point. The quadratic loss function can be written as

$$L(x) = k(x - m)^2, \tag{10.5}$$

where $L(x)$ is the loss when the quality characteristic is x units away from the origin, m is the target, and k is the loss coefficient. The loss function above can be rewritten as

$$k = \frac{L(x)}{(x - m)^2}.$$

Using the data evaluated earlier, we have

$$k = \frac{L(1.5)}{(1.5 - 0)^2} = \frac{\$2.25}{2.25} = \$1.0.$$

Loss Function Interpretation of the Engineering Specification

It is possible to use the continuous loss function to give an economic interpretation to a specification limit. This is illustrated in Fig. 10.12. With reference to the figure, one has a choice of either

1. Letting the part pass on to the customer with an expected loss as determined by the loss function, or

2. Reworking or replacing the part at a certain cost.

Figure 10.12   Loss function interpretation of an engineering specification

To minimize the total costs, the alternative with the lower cost should be chosen. The point of indifference between the two alternatives determines the specification limit.

Suppose, in the example here, the replacement/rework cost is $1.00. Since the loss coefficient $k = \$1.00$, then the points of indifference—the lower and upper specifications—would be ±1.0. In general, the points of indifference can be determind by solving

$$L(x) = C,$$

where C is the cost of a failure. Thus

$$k(x - m)^2 = C$$

or

$$x = m \pm \sqrt{\frac{C}{k}}. \tag{10.6}$$

Evaluation of Expected Process Loss

Although a loss function can be used for the economic establishment of specifications, it can be put to an even more illuminating task—evaluating the aggregate expected loss from a process, based on the distribution of actual performance of the quality characteristic in question. Hence the loss function constitutes a new way of assessing the capability of a process. This approach to assessing process capability takes into account the process distribution relative to the specifications in terms of the economic loss to the user due to deviation from the nominal. This average (expected) realized loss is found by weighting the loss function by the probability density function (statistical distribution) of the actual process output. In general this can be written as

$$E[L(X)] = \int_{R_X} L(t)\, f(t)\, dt \tag{10.7}$$

where

$L(x)$ is the loss function for the variable X,
$f(x)$ is the pdf for the variable X,
R_x is the domain of X.

For a quadratic loss function [Eq. (10.5)], the expected loss is simply the loss coefficient k multiplied by the average of the squared deviations of the individual measurements from the target. The squared deviations can be decomposed into two elements, the process variance plus the square of the offset of the process mean from the target, the process bias:

Expected loss per piece $= k(\text{process variance} + \text{process mean offset}^2)$

$$E[L(X)] = k[\sigma_X^2 + (\mu_X - m)^2]. \tag{10.8}$$

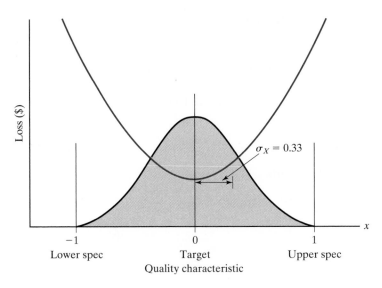

Lower spec   Target   Upper spec
                  Quality characteristic

Figure 10.13   **Evaluation of quality loss of a process that is normally distributed**

To illustrate this use of the loss function, consider a situation defined by the following parameters: $m = 0$, $k = \$1.0$, LSL $= -1.0$, USL $= 1.0$, X is normally distributed with mean 0, and $\sigma_X = 0.33$, as shown in Fig. 10.13. The expected loss is therefore given by

$$E[L(X)] = \$1.00[0.33^2 + (0 - 0)^2]$$
$$= \$0.11.$$

The process in the example above would typically be considered capable in a traditional sense because the specification limits constitute a spread of 6 sigma. Applying our derived quadratic loss function, however, we find that this $1.00 part (the replacement cost at manufacturing) carries with it an additional loss of about $0.11. This is a hidden loss above and beyond direct costs of materials, labor, and processing incurred so far. This incremental cost is experienced downstream by the customers, in terms of additional costs they will incur due to decreased utility of the product.

Now suppose that, instead, X is distributed uniformly between -1 and 1. As shown in Fig. 10.14, the standard deviation of the process would then be $\sigma_X = 0.58$.[2] Consider that the other data are the same as in the previous example. The expected or average loss per part is now

$$E[L(X)] = \$1.00[0.58^2 + (0 - 0)^2]$$
$$= \$0.33.$$

---

[2]For the uniform distribution, $\sigma_X^2 = [(b - a)^2]/12$ for X varying on the interval $[a, b]$.

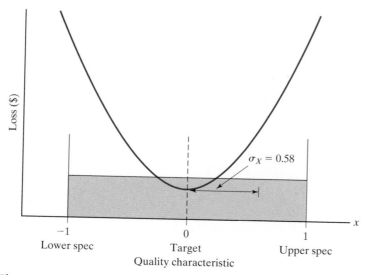

Figure 10.14   Evaluation of the quality loss for a process that follows a uniform distribution

The effect of a change in the process distribution is to increase the loss three-fold. The important question is: What would make the process follow a uniform distribution? The likely answer is that the process is exhibiting instability—shifting of its mean value, for example.

Figure 10.15 shows the effects of reducing the variation of a normally distributed process that remains centered on the target. The process has a standard deviation of 0.10, which constitutes a reduction to about one-third of the previous value. The associated expected loss,

$$E[L(X)] = \$1.00[0.10^2 + (0 - 0)^2]$$

$$= \$0.01,$$

is reduced by a factor of 10. Traditional manufacturing wisdom values such better-than-necessary performance primarily because it can absorb some shocks from special causes of variation without generating nonconforming output. This example, however, shows that additional benefits can be gained from reduced variation even when the process is stable and centered on the target. The loss function as a measure of process capability takes into account the loss incurred by the customer as a result of reduced utility in field use. As such, it clearly demonstrates the presence and importance of this incremental cost from the standpoint of the producer as well. Such loss incurred by the customer is directly transmitted back to the producer in terms of warranty/replacement cost, lost business cost, and so on.

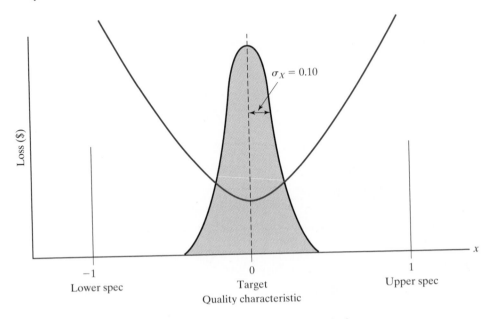

Figure 10.15   Loss reduction due to reduced process variation

Process Not Centered on the Target

When a process shows a great deal more than just marginal capability, one seemingly economically attractive way of capitalizing on this potential is to seek short-term benefit by, for instance, running the process toward the side of the specifications that minimizes, perhaps, material usage or cycle time. Figure 10.16 shows the results of such an approach, taking the preceding example ($\sigma_X = 0.10$) and locating the process mean as far to the low side as possible, consistent with remaining 3 standard deviations away from the specification limit. The loss under this process operating condition is now

$$E[L(X)] = \$1.00[0.10^2 + (-0.7 - 0)^2]$$
$$= \$0.50.$$

This process, which previously generated only $0.01 in hidden losses when centered on the target, now causes $0.50 in customer losses when run close to the lower specification limit. Very few pieces fail to conform to specifications, and such a process would meet the explicit requirements of the bilateral specification. The loss arises because the output is consistently mediocre in a design intent sense—and the customer (and therefore the producer as well) suffers. If running the process at the shifted mean saved $0.10 in direct (manufacturing) costs, there would still be $0.40 in net waste, forever lost, as a result

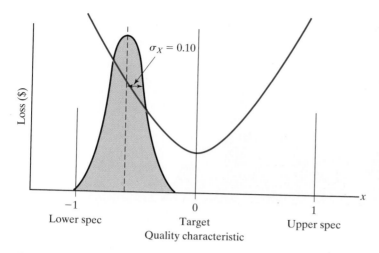

Figure 10.16   Quality loss consequences of shifting the process mean toward the lower specification

of lost utility of the product during field use. Taguchi describes this operating strategy as "more immoral than the actions of a thief."

Processes with Linear Drift

Another way that we often attempt to take advantage of small process variation relative to the specifications is to reduce reset or tooling costs by allowing the process to drift across the specification range. A tool-change discipline that involves setting the process 3 standard deviations away from one specification limit, then letting the tool wear continue until the process is 3 standard deviations away from the other specification limit seems rational. However, these apparent savings are not achieved without risk of hidden customer loss.

The normally distributed process with a standard deviation of 0.10, discussed previously, may be allowed to drift to its maximum extent, with its centers moving from −0.7 to +0.7. Figure 10.17 shows the process distributions at these extreme centers, as well as the resulting product quality characteristic distribution of a complete drift cycle. As can be seen from the figure, this net distribution is no longer bell shaped but more loaf shaped.

The total hidden loss in this example is now $0.33, coming mainly because the lengthy tool change interval causes marginally acceptable products to be produced near the beginning and end of each cycle. The loss calculation results from the fact that what we essentially have as a result of this operating policy is roughly equivalent to the case of Fig. 10.14 (i.e., the quality characteristic follows a uniform distribution). If the tool-change interval was reduced by 30%, for instance, with set and change points at −0.5 and +0.5, the hidden loss would be reduced to almost one-fourth of the loss calculated above. It is clear

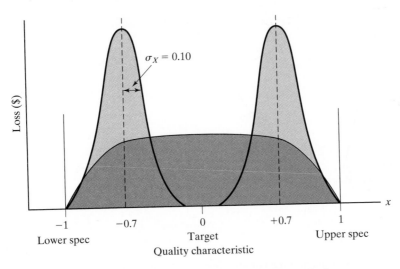

Figure 10.17 Consequences of a process with linear drift

that tool-change policy should really balance the economies of longer change intervals against the downstream/customer-incurred losses to achieve the lowest total cost.

It is also apparent that when a process with very small variations is allowed to meander within the relatively wide specifications, rather than being maintained in statistical control near the target, this lack of shop discipline is not cost-free, but comes at the expense of the customer.

10.6 The Dilemma of C_{pk}: A Case Study

Introduction

In the preceding treatment of the capability index C_{pk} we have simply considered its value as a number, paying no attention to the fact that, like $\overline{X}$'s and R's, this number is derived from a sample and therefore is subject to sampling variability. In short, when estimates for the process mean ($\hat{\mu}_X = \overline{\overline{X}}$) and process standard deviation ($\hat{\sigma}_X = s_X$) are used in place of the true parameter values, μ_X and σ_X, the calculated value for the C_{pk} index is subject to sampling variation just like any other statistic. Uncertainties in our estimates for the process mean and variation translate into uncertainty in the C_{pk} value, and this fact must be remembered when making decisions about the status of a process and the possible actions that might be taken as a result of its evaluation in a given setting. Yet, in most quarters, values for C_{pk} are calculated from a set of process measurements and then interpreted literally, often leading to erroneous conclusions about the process that can end up costing thousands or even millions of dollars.

Consider a situation where a supplier of automotive parts is to demonstrate that a process is capable at a level of $C_{pk} = 1.33$ or greater. The supplier carefully develops a control chart for the process. As we will soon see, it would not be uncommon for the data from that control chart based on 25 subgroups of $n = 5$ items, and showing good statistical control for the process, to produce a C_{pk} of, say, 1.2, when the true process capability is actually 1.5. In such a case, the supplier might lose the job or might be forced to take actions to "improve" the process, when no such actions are truly required. This is a scary scenario, to say the least, but it probably occurs dozens of times each day. What makes matters worse is the fact that such critical decisions are often made using much smaller samples, perhaps as few as 30 individual measurements.

The following case study depicts the dilemma faced by a machine tool builder that was required by its customer to "run off" a machine before shipping it and document that it meets or exceeds the customer's requirements in terms of C_{pk}. The machine in question was a flexline for machining automotive engine blocks, and the total cost of the line was in excess of twenty million dollars. For every day that the line could not be shipped because "it does not meet the customer's C_{pk} requirement" the interest cost alone was many thousands of dollars. And this does not take into consideration the cost of actions taken to make the line capable that, in truth, might be totally unnecessary.

Definition of the Process Capability Index C_{pk}

Many companies use the C_{pk} capability index. As we have discussed, one reason for using this index rather than the index C_p is that C_{pk} penalizes deviations of the process mean from the nominal value of the specification. Just to review the basic definitions, for bilateral specifications, C_{pk} is defined in the following manner. First, we determine the distance of the specification limits from the process mean in terms of standard deviation units:

$$Z_{\text{USL}} = \frac{(\text{USL} - \mu_X)}{\sigma_X}, \quad Z_{\text{LSL}} = \frac{(\text{LSL} - \mu_X)}{\sigma_X}.$$

Then, we calculate $Z_{\min}$ and C_{pk} using the relations:

$$Z_{\min} = \min[Z_{\text{USL}}, -Z_{\text{LSL}}] \quad \text{and} \quad C_{pk} = Z_{\min}/3.$$

In the definition of C_{pk} above, the true values for the process mean and standard deviation (μ_X and σ_X) have been used, but, as discussed previously, in practice we must estimate these true values with data collected from the process.

$$\hat{\mu}_X = \overline{\overline{X}} = \frac{\sum\limits_{i=1}^{N} X_i}{N}$$

and

$$\hat{\sigma}_X = s_X = \sqrt{\dfrac{\displaystyle\sum_{i=1}^{N}(X_i - \overline{\overline{X}})^2}{N-1}},$$

where N is the total number of observations collected from the process (e.g., for 30 subgroups of size 5 collected from a process, $N = 30 \cdot 5 = 150$). When the estimates for the process mean and standard deviation are used in place of the true values for μ_X and σ_X, the resulting calculated value $(\hat{C}_{pk})$ represents an estimate of the true capability index, C_{pk}. This is where the trouble begins.

Machine Run-Off Requirements

According to the customer's specification document, the requirements of machine capability in terms of C_{pk} are defined as follows:

> The purpose of the demonstration run-off at the machine tool builder's location is to ensure that each machine is repeatable and can be targeted to achieve its required tolerance of 1.33 C_{pk} when delivered to the customer's facility.

The interpretation of this requirement is that the "true" or long-term C_{pk} capability of the machine should be 1.33. The meaning of the words "true" or "long-term" is that if a large amount of data is available (say from a month's operation of the machine), the estimates of the process mean and standard deviation would be virtually without error and the ensuing calculated C_{pk} value would be very, very close to its true or long-term value, required to be at least 1.33. When C_{pk} is calculated from fewer data, say from a 120-piece sample or even less, the sampling error associated with the estimate, $\hat{C}_{pk}$, must be considered if informed decisions concerning the capability of the machine are to be made.

Evaluating C_{pk} from Run-Off Data

It is important to recognize that when we calculate C_{pk} based upon estimates of the process mean and standard deviation, it is a statistic just like the average, the range, and the sample standard deviation. All of these quantities are estimates of corresponding parameters (fixed constants). Therefore, a value for C_{pk} determined from data—a run-off result—is subject to error or what we have called sampling variation; as a result, we will refer to the calculated value as $\hat{C}_{pk}$. Recall that we spent a good deal of time in Chapter 4 discussing the concept of the sampling distribution, and now we must consider the sampling distribution of $\hat{C}_{pk}$.

If we took several 40-piece samples, for example, we would not expect to get the same value for $\hat{C}_{pk}$ each time. Rather, we would get some range of values. Precisely what degree of scatter we would see in the $\hat{C}_{pk}$ values is governed by statistical theory, which tells us exactly what to expect. This theory is a bit complicated in this case but the precise mathematical definitions are not important here. Rather, it is the conceptual and altogether too sobering view of sampling error related to $\hat{C}_{pk}$ that we need to come to appreciate through this case study. The sampling distributions that are shown below, while approximate, are consistent with this theory.

The curve in Fig. 10.18 describes the result that we would get if we took a very, very large number of 40-piece samples from a machine that was built so that it has a true—long-run—C_{pk} capability of 1.33. The results here are a little unsettling. They show, for example, that for any given 40-piece sample we could get a calculated value for $\hat{C}_{pk}$ as low as 1.0 or less, or as high as 1.8 or more. In fact, as shown by the shaded area in Fig. 10.19, there is only about a 50% chance that a single 40-piece sample would produce a $\hat{C}_{pk}$ value between 1.2 and 1.46. Furthermore, due to the sampling behavior of $\hat{C}_{pk}$, there is less than a 45% chance of getting a $\hat{C}_{pk}$ value ≥ 1.33 *even if the true C_{pk} is 1.33.*

Often, to provide some insurance that the long-term capability requirement will be met, a customer may call for a more stringent "short-term" process capability requirement. In this particular case study, the customer required that a single 120-part run-off produce a C_{pk} of 1.47 or greater to ensure a long-term capability of at least 1.33. As shown in Fig. 10.20, the consequences of this "short-term" requirement are even more unsettling. The figure illustrates the $\hat{C}_{pk}$ distribution for 120-piece samples, obtained through the required run-off assessment of "short-term" capability. The figure shows that the chance of meeting or exceeding the $C_{pk} = 1.47$ requirement on any given run-off sample pieces is 1 in 7 (about 15%). Said another way, the machine could actually be $C_{pk} = 1.33$ capable, but 6 out of 7 run-offs would fail to show this; that is, they would fail to meet the $C_{pk} = 1.47$ requirement. Clearly, some other approach

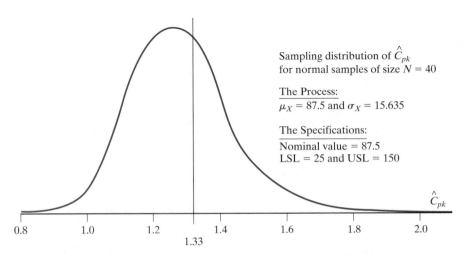

Figure 10.18 Sampling distribution of $\hat{C}_{pk}$ for a sample of size $N = 40$

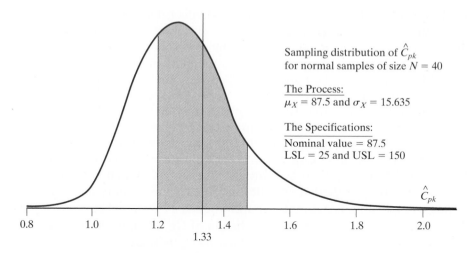

Figure 10.19   Sampling distribution of $\hat{C}_{pk}$ showing a region where 50% of the $\hat{C}_{pk}$ values will arise

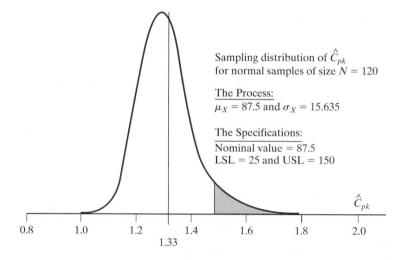

Figure 10.20   Sampling distribution of $\hat{C}_{pk}$ for a sample of size $N = 120$

needs to be developed to demonstrate conclusively that the capability requirements of the customer are satisfied.

Hypothesis Testing Formulation to Test for Adherence to a C_{pk} Requirement

When data are used to test for the adherence of process performance to some standard/customer expectation, it is not uncommon to use a statistical hypothesis testing formulation to compare the sample/data result to the standard/requirement. We have discussed the hypothesis testing idea in detail in Chapter 4. Using the context of the current case study, such a formulation can be established from either the customer or the machine tool builder point of view, but the results—and the resulting interpretation—can be much different.

Customer Point of View. If I am the customer, I would like the machine tool builder to prove to me that the capability requirement is exceeded. Assuming a null hypothesis that "the capability is exactly met," then the alternative hypothesis becomes "the capability is exceeded." For this set of hypotheses, the burden of proof is on the builder to disprove the null hypothesis, that is, prove to the customer that the capability is exceeded. Figure 10.21 shows this situation. Under this formulation, to conclude that the requirement of $C_{pk} = 1.33$ is exceeded at, say, the 10% significance level, one would have to obtain a $\hat{C}_{pk}$ value from the 40-piece run-off sample of about 1.6 or greater. The 10% significance level here really means that if the requirement is exactly met, that is, $C_{pk} = 1.33$, then the chance of getting a sample result of 1.6 or greater is 1 in 10. This point of view certainly favors the customer at the expense of the machine tool builder.

Machine Tool Builder Point of View. If I am the machine tool builder, I would want to put the burden of responsibility on the customer to prove to me that I have not met the capability requirement. In terms of the hypothesis testing formulation, I would want the null hypothesis to be that the capability of the machine is exactly $C_{pk} = 1.33$ and the alternative hypothesis to be that the capability is less than $C_{pk} = 1.33$. Figure 10.22 shows this formulation. Now, taking the 40-piece run-off sample, one will not be able to reject the claim that the C_{pk} value is 1.33, at the 10% significance level, unless the calculated value is about 1.1 or less. If the long-term machine capability is, in fact, $C_{pk} = 1.33$,

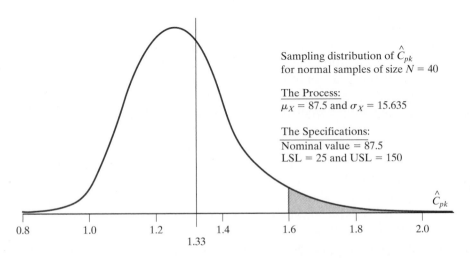

Sampling distribution of $\hat{C}_{pk}$
for normal samples of size $N = 40$

The Process:
$\mu_X = 87.5$ and $\sigma_X = 15.635$

The Specifications:
Nominal value $= 87.5$
LSL $= 25$ and USL $= 150$

$\hat{C}_{pk}$

0.8   1.0   1.2   1.4   1.6   1.8   2.0

1.33

Figure 10.21   Hypothesis testing formulation for meeting a C_{pk} requirement: customer point of view

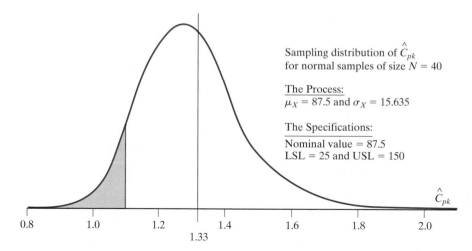

Sampling distribution of $\hat{C}_{pk}$
for normal samples of size $N = 40$

The Process:
$\mu_X = 87.5$ and $\sigma_X = 15.635$

The Specifications:
Nominal value = 87.5
LSL = 25 and USL = 150

$\hat{C}_{pk}$

0.8   1.0   1.2   1.4   1.6   1.8   2.0
                      1.33

Figure 10.22   Hypothesis testing formulation for meeting a C_{pk} requirement: machine tool builder point of view

then there is a 90% chance that the machine tool builder's claim cannot be rejected. In other words, the machine should be judged acceptable.

It is unlikely that either of the above scenarios could be mutually agreed upon. One is very tough on the builder, while the customer would probably view the other as a bit risky. What, then, is the solution to the problem?

SPC—An Insurance Policy for the Machine Tool Builder

One thing is for certain—if I am a machine tool builder engaged in proving machine capability via run-off activities involving C_{pk} calculations, I want to be very sure that the data collected are from a process that is in statistical control. Earlier in this chapter, we emphasized the fact that no attempt at assessing process capability should be made unless the process is proved to be in statistical control. This case study helps to reinforce this requirement. Why? Quite simply, because if the process mean and/or variation are unstable during the run-off, then the ensuing C_{pk} calculation is going to underestimate the true C_{pk} capability of the machine if control/repeatability was present. But why?

Being Off Target. If the process mean is initially at the nominal specification but changes during the run-off, then those changes will be included in the C_{pk} calculation in a way that will cause $\hat{C}_{pk}$ to be smaller than it would have been if the machine had stayed on target. C_{pk} has been specifically crafted to encourage operation at the nominal value. Any departure from this, all other things constant, causes $\hat{C}_{pk}$ to be reduced. The problem is compounded if the standard deviation of the process is estimated using the sample standard deviation of all the data, because shifts in the process mean will inflate the estimate of the process variation, again causing $\hat{C}_{pk}$ to be reduced. On the other hand, estimation of the standard deviation based on $\overline{R}/d_2$ will tend to be less susceptible to fluctuations in the process mean. This is so since $\bar{R}$ is obtained from averages of individual sample ranges, and the ranges will tend to be less influenced by process mean shifts if rational sampling is employed.

A Change in Process Variation. If the process variation experiences some instability in the form of increases in variation, at least for some portion of the run-off, then again the calculated C_{pk} value will be reduced. Of course, if the process variation is reduced during the test, then $\hat{C}_{pk}$ will be increased accordingly.

There is only one way to be confident in the stability of the process during the run-off, and that is to apply SPC principles, namely control charting to the data. If instabilities in the mean level or amount of variation are detected, then no attempt should be made to calculate C_{pk}. The special cause/local fault should be identified and eliminated, and then the run-off should be repeated. If the process is in control, but the $\hat{C}_{pk}$ value is unacceptable, then we know we have a fundamental system problem that needs attention.

Even if as few as 40 or 50 individual parts were taken, some SPC-like plotting and analysis of the data could reveal an obvious problem with the run-off. Of course, it would be better for one's own protection to conduct a run-off using at least 25 subgroups of size 4 or 5 (a total sample of $N = 100{-}125$ parts). Figure 10.23 shows the distribution of $\hat{C}_{pk}$ for several different total sample sizes, and it is evident that the distribution is very sensitive to the number of parts used to calculate C_{pk}.

An Alternative Statistical Formulation: Confidence Interval for C_{pk}

While the hypothesis testing formulation described above may be useful to understand both the varying points of view on the problem and the consequences of variability in $\hat{C}_{pk}$, it does not appear to provide a balanced and meaningful solution to the problem. The confidence-interval approach, however, may be much more useful in this regard. Recall that we discussed this method of

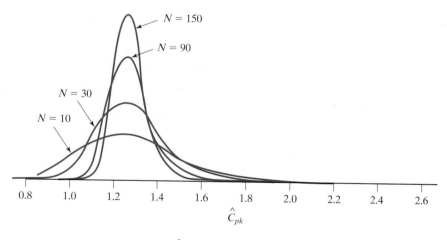

Figure 10.23   Distribution of $\hat{C}_{pk}$ for varying sample sizes

interval estimation in Chapter 4. This approach tends to provide a more comprehensive view of the true state of affairs, based on the data. The confidence-interval approach to estimating the C_{pk} capability of the machine involves forming an interval or range of values about the estimate of C_{pk} we get from the data (from the run-off of the case study) within which we can say that the true value lies with high confidence.

In developing a confidence interval for C_{pk} it is generally assumed that the process data are normally distributed. An approximate $100(1 - \alpha)\%$ confidence interval for the true C_{pk} may then be constructed by also assuming that $\hat{C}_{pk}$ is normally distributed. Under these assumptions, the confidence interval for C_{pk} may be expressed as[3]

$$\hat{C}_{pk} \pm z_{1-(\alpha/2)}\sqrt{\frac{1}{9n} + \frac{\hat{C}_{pk}^2}{2(N - 1)}}.$$

To illustrate this approach, consider a situation in which a 40-piece sample from a run-off test produces a C_{pk} estimate of $\hat{C}_{pk} = 1.22$. For a 95% confidence interval, the value of α is 0.05, and $z_{1-(\alpha/2)} = 1.96$. Using these values, the 95% confidence interval for the true value is

$$1.22 \pm 1.96 \sqrt{\frac{1}{9(40)} + \frac{(1.22)^2}{2(40 - 1)}} \quad \text{or} \quad 1.22 \pm (1.96)(0.148) \quad \text{or} \quad 1.22 \pm 0.290,$$

$$[0.93 \leftrightarrow 1.51].$$

Therefore, we may say that we are 95% confident that the true machine capability is somewhere between 0.93 and 1.51. Since the value we are targeting, $C_{pk} = 1.33$, lies on this interval, the interpretation often used is that 1.33 is a plausible value for the true capability in light of the data, and there is no reason to go further.

The size of the confidence interval is, of course, the thing of real interest. If the interval is narrow, we feel that we have pinned down the true value fairly well. If the interval above seems a little wide, we have not pinned down the C_{pk} value very well. Three things control the size of the interval:

1. The size of the interval depends on how much data you use to calculate $\hat{C}_{pk}$. The more data you have, the smaller the interval. For example, if the value of $\hat{C}_{pk}$ had come from a 120-piece run-off, the 95% confidence interval would be $1.22 \pm (1.96)(0.085)$. Hence, the interval would run from 1.05 to 1.39.

---

[3] R. N. Rodriguez, "Recent Developments in Process Capability Analysis," *J. of Quality Technology*, Vol. 24, No. 4, 1992, pp. 176–187.

2. The size of the interval depends on how much confidence you want to have that the true value for C_{pk} lies on the interval. For example, a 99% confidence interval under the same circumstances as above for a sample of $N = 120$ would be $1.22 \pm (2.575)(0.085)$, or from 1.00 to 1.44.

3. The size of the interval also depends, of course, on the magnitude of the variation in the data.

The confidence-interval approach tends to leave us with a little more "confidence" in our results and our ability to make an informed decision as to what to do. For example, for the case we examined above with the 120-piece run-off, we are 95% sure based on our run-off data that the long-term C_{pk} for the machine is at least 1.05 but no greater than 1.39.

■ EXERCISES

10.1 In your own words, explain the difference(s) between process capability and statistical control. In particular:
  (a) What important relationship exists between these two concepts?
  (b) What are the basic goals/objectives of each?
  (c) Can a process be in control but not capable? Can a process be capable but not in control? Explain.

10.2 Consider a process that continues to produce many parts that are not meeting the required specifications. However, the $\overline{X}$ and R control charts being maintained for the process show it to be in a state of statistical control. Is this possible? Explain why this state of affairs may be occurring. What needs to be done to improve the process capability?

10.3 What steps are involved in performing a process capability study? Be specific in outlining the data collection, analysis, and evaluation procedures.

10.4 Control charts are to be constructed for a process turning the outside diameter of a cylinder. The diameter has specifications of 4.500 ± 0.050 centimeters.
  (a) If $\Sigma \overline{X} = 157.85175$ centimeters and $\Sigma R = 2.18750$ centimeters for 35 samples of size $n = 5$, calculate the centerlines and control limits for the control charts.
  (b) Assume that the charts show that the process is in good statistical control. What percent of the parts produced will be within tolerances if this process continues under a constant system of common causes and can be approximated by the normal distribution?
  (c) What percent would be within the specifications if the process were to be centered at the prescribed nominal?
  (d) Graphically show the situations in parts (b) and (c) in terms of the distributions of the individual measurements.

10.5 Two different sizes of a certain type of fastener are being manufactured by a cold heading process. The nominal outside diameters of the fastener heads, with associated specifications, are

$$\text{Fastener } A:\ \ 0.600 \pm 0.020 \text{ centimeter,}$$

$$\text{Fastener } B:\ \ 0.800 \pm 0.020 \text{ centimeter.}$$

SPC studies show the processes to be in good statistical control. The individual measurements follow the normal distribution. After collecting 50 samples of size $n = 4$:

$$\text{Fastener } A:\ \ \Sigma \bar{X} = 30.310,$$
$$\Sigma R = 0.654.$$
$$\text{Fastener } B:\ \ \Sigma \bar{X} = 40.011,$$
$$\Sigma R = 1.034.$$

(a) For each fastener dimension of interest, find the percent conforming.
(b) If the process mean for fastener A could be brought to the nominal, what would be the percent conforming?
(c) If the variation in the process for fastener B could be reduced by a factor of one-third, what would be the percent conforming?

10.6 In Exercise 6.17 you were given data collected from a process (which machined the outer diameter of motor shafts) after it was brought into statistical control. For that exercise, the nominal value was 2.1250 inches with a tolerance of ± 0.0010 inch. The data provided there were coded to be the last two digits of the measurements. Use those data to determine the percent conforming to specifications. If you assume that the data are normally distributed, verify your assumption with the appropriate data analysis.

10.7 One of your employees has come to you with the data shown in the accompanying table and with associated "control charts" to show that improvements have been made on a metal casting process after an initial study of its capability. The data are based on samples of size $n = 5$. The specifications for the quality characteristic of interest are 3.50 ± 0.10 millimeters. Based on the charts, he tells you that the process is now capable and that further improvements are no longer necessary.
(a) As an informed manager, what are your opinions regarding the process, the approach used to study it, and the conclusion?
(b) If you feel that the situation should have been handled differently, what would you do? Use the data to develop your presentation and conclusions.

Sample	$\bar{X}$	Range	Sample	$\bar{X}$	Range
1	3.49	0.13	26	3.55	0.07
2	3.49	0.13	27	3.55	0.07
3	3.49	0.16	28	3.54	0.13
4	3.50	0.12	29	3.53	0.13
5	3.50	0.16	30	3.54	0.08
6	3.53	0.05	31	3.53	0.07
7	3.49	0.08	32	3.57	0.16
8	3.53	0.21	33	3.50	0.15
9	3.46	0.09	34	3.51	0.14
10	3.45	0.07	35	3.53	0.09
11	3.48	0.12	36	3.51	0.18
12	3.52	0.16	37	3.50	0.13
13	3.47	0.22	38	3.46	0.13
14	3.50	0.09	39	3.54	0.08
15	3.51	0.09	40	3.51	0.11
16	3.50	0.20	41	3.51	0.10
17	3.52	0.17	42	3.53	0.09
18	3.49	0.07	43	3.57	0.09
19	3.50	0.14	44	3.57	0.12
20	3.49	0.08	45	3.53	0.12
21	3.49	0.18	46	3.53	0.08
22	3.49	0.03	47	3.53	0.08
23	3.53	0.11	48	3.48	0.15
24	3.52	0.12	49	3.49	0.08
25	3.51	0.14	50	3.48	0.09

10.8 You need to make some decisions regarding the procurement of bolts from three different suppliers. You have requested that the shank diameter (above the thread) have specifications of 1.500 ± 0.009 inches. The SPC studies done by the suppliers have indicated that their processes are behaving consistently in statistical control with the following process parameters. The individual measurements follow the normal distribution.

$$Supplier\ 1: \ \overline{\overline{X}} = 1.500 \text{ inches}, \ \sigma_X = 0.003 \text{ inch},$$

$$Supplier\ 2: \ \overline{\overline{X}} = 1.500 \text{ inches}, \ \sigma_X = 0.0022 \text{ inch},$$

$$Supplier\ 3: \ \overline{\overline{X}} = 1.4950 \text{ inches}, \ \sigma_X = 0.0015 \text{ inch}.$$

Which supplier would you purchase from? Why? Explain your logic and show calculations and graphical evidence to back it up.

10.9 The specifications of the thickness of a low-alloy steel gear blank are 0.3000 ± 0.0015 centimeter.

  (a) If it is desired that the value of C_{pk} be at least 1.25, what would be the maximum σ_X of a centered process?

  (b) If the specification is changed to 0.3000 ± 0.0010 centimeter, what would be the maximum σ_X for a centered process to have $C_{pk} \geq 1.25$?

10.10 Consider again the steel gear blank in Exercise 10.9. The specifications are 0.3000 ± 0.0015 centimeter.

  (a) If it is desired for the process to have a C_p value of at least 1.25, what is the maximum σ_X if $\overline{\overline{X}} = 0.301$ centimeter? What if $\overline{\overline{X}} = 0.298$ centimeter?

  (b) What does this say about using C_p as a measure of capability?

10.11 The maker of grandfather clocks needs to be very careful about the weight of the swinging pendulum, which is essential to the time-keeping precision. It is desirable for 99.9% of the pendulums to have a weight of 1.0000 ± 0.0002 kilogram.

  (a) If the process is in statistical control, centered, and has a normal distribution, what value of σ_X would be required to meet the criterion?

  (b) If the process is centered, what values of C_p and C_{pk} would the value of σ_X found in part (a) produce?

10.12 A die-casting process is being used by a supplier to fabricate a cast housing that will be supplied to a major client, who will perform further machining operations on the pieces. It has been determined that monitoring part weight is a good indication of quality process performance. The casting should have a nominal weight of 775 grams and a tolerance of ± 20 grams. Use the data given in the accompanying table (coded as the last two digits, such that $79 = 779$ grams, etc.) to perform a capability study. Determine the percent conforming to specifications, C_p, and C_{pk}. If you assume that the data are normally distributed, verify your assumption with the appropriate data analysis.

Sample	X_1	X_2	X_3	X_4	X_5
1	79	82	71	71	73
2	75	63	75	79	74
3	72	82	70	90	75
4	79	76	73	80	79
5	71	63	79	76	87
6	81	67	72	83	71
7	88	77	70	65	59
8	80	77	78	86	80
9	65	72	75	84	85
10	72	73	75	75	65
11	73	74	72	76	77
12	73	70	83	78	85
13	84	74	78	79	76
14	88	70	79	88	78
15	64	71	65	78	82
16	86	73	71	78	71
17	84	67	71	59	87
18	76	81	66	74	73
19	75	69	83	79	71
20	79	75	73	75	74
21	79	86	85	87	84
22	76	78	88	75	89
23	81	88	84	70	82
24	81	74	72	78	60
25	84	65	65	77	79
26	85	72	76	72	78
27	83	72	72	81	66
28	74	78	74	82	93
29	77	73	74	71	83
30	59	71	85	74	80
31	82	73	76	74	85
32	66	74	79	81	73
33	77	73	77	86	74
34	89	83	71	75	82
35	82	78	66	69	67
36	73	82	67	73	70
37	83	69	77	77	61
38	82	82	80	70	72
39	82	70	70	74	75
40	77	69	80	68	79

10.13 To determine the present status of a piston-turning operation where the surface roughness is an important characteristic, it has been decided that a process capability study should be performed. The surface roughness has specification of 105 ± 10 microinches. Use the data shown in the accompanying table, collected in samples of size $n = 5$, to perform this study.

Sample	$\bar{X}$	Range	Sample	$\bar{X}$	Range
1	108.2	13	16	103.6	8
2	105.8	27	17	115.4	24
3	104.0	24	18	103.2	34
4	104.6	16	19	108.0	71
5	106.6	22	20	108.2	57
6	111.4	32	21	100.6	26
7	104.6	12	22	98.2	73
8	102.2	23	23	96.4	47
9	111.0	22	24	101.4	26
10	113.4	18	25	105.6	49
11	102.2	23	26	91.4	68
12	108.0	26	27	116.6	52
13	111.2	22	28	113.6	53
14	103.2	31	29	112.4	42
15	100.4	6	30	99.6	37

10.14 A decision needs to be made regarding the possible purchase of some new casting equipment. To aid in this process, a capability study is being performed to assess the performance of the current equipment. An important cast part that will consume much of the run time of the equipment is to be used for the study. The width of this part at a crucial spot should be 3.000 ± 0.025 centimeters, and rational sampling has produced the data listed in the accompanying table. Perform the capability study on this process using the samples in the table and comment. The samples, of size $n = 5$, were collected every half-hour over two continuous shifts.

Sample	$\bar{X}$	Range	Sample	$\bar{X}$	Range
1	2.991	0.038	17	3.005	0.039
2	3.004	0.027	18	3.008	0.054
3	2.997	0.049	19	3.001	0.009
4	3.004	0.033	20	3.000	0.062
5	2.999	0.043	21	2.993	0.034
6	3.000	0.070	22	2.982	0.012
7	2.988	0.044	23	2.990	0.028
8	2.986	0.050	24	3.003	0.091
9	2.991	0.064	25	3.007	0.049
10	2.994	0.048	26	2.998	0.052
11	3.017	0.070	27	3.012	0.077
12	2.987	0.051	28	2.996	0.040
13	3.013	0.060	29	2.997	0.018
14	2.993	0.019	30	3.004	0.020
15	3.009	0.044	31	2.987	0.063
16	2.990	0.043	32	3.002	0.028

10.15 A company is making 10-ohm resistors to be used in electric thermo-
stats. The resistors have specification of 10.00 ± 0.10 ohms. It has been
determined that if the resistor's actual resistance is 10.15 ohms, a failure
of the thermostat will occur 15% of the time with a cost of $45. A quad-
ratic loss function is assumed to approximate the loss.
  (a) Use these data to find the value of k in the loss function.
  (b) Sampling performed on resistors shows that $\overline{\overline{X}} = 10.01$ ohms and
       that σ_X (approximated by $\overline{R}/d_2$) is 0.025 ohm. Use this information
       to find the average loss per part made.
  (c) What is the average loss per part if the process in part (b) is centered?
  (d) What is the average loss per part if the process in part (b) is centered
       and the variance is reduced by a factor of 2?
  (e) Assuming a process centered on the nominal value, if the replace-
       ment/manufacturing cost of these resistors is $4.32, what should the
       tolerance on the resistors be?

10.16 The company from Exercise 10.15 has been required to have a minimum
C_{pk} value of 1.3 for the resistors. Due to the way the process that makes
the resistors works, it is less expensive by $0.05 per resistor to manufac-
ture the resistors at a nominal of 9.95 ohms instead of at 10.00 ohms.
Through adjustments it has been possible to reduce the standard deviation to half of its original value, to 0.0125, and it has been deemed appropriate to adjust the mean of the process to 9.95 ohms.
  (a) What happens to the C_{pk} value if this change in the variation and
       adjustment to the mean occurs?
  (b) What happens to the average loss per piece if this process mean
       shift is implemented?
  ✓ **(c)** Evaluate the decision to make the proposed process change.

10.17 Two parts in an assembly are in constant contact and have frequent sliding movements relative to each other. Because of this, it is necessary to
have a certain surface roughness so that the pieces are able to move easily
upon application of the proper lubricant. Therefore, in the machining
operation that produces these parts, the surface roughness needs to be
between 110 and 140 microinches, with a nominal of 125 microinches.
  (a) If the loss for parts made with a roughness of 106 microinches is
       $3.29, find the value of k for the loss function. Assume a quadratic
       loss function.
  (b) What would be the loss for a part with a surface roughness of 150
       microinches?
  (c) If the process is in control, approximately normally distributed, and
       centered at the nominal with a C_{pk} value of 1.2, what is the average
       loss incurred per assembly?

10.18 Consider that the data collected during an SPC study indicate that a normal distribution is occurring in the measurement of the depth of a slot
machined in an aluminum fixture. The slot should have a depth of
2.750 ± 0.010 centimeters. It is known that for these pieces if the depth

is 2.765 centimeters, a cost of $100 is incurred by the user to replace the entire fixture due to improper fit.

   (a) Use the information above to find the value for k, assuming a quadratic loss function.

   (b) If the process machining the slot is centered at 2.752 with a standard deviation of $\sigma_X = 0.003$ centimeters, calculate the average loss per part. What effect does centering the process have on the loss?

10.19 There are some important differences in the way the various capability measures respond to changes in the process as it is represented by a statistical distribution. Consider a process producing a product with an important quality characteristic that has a nominal value of 74, an upper specification of 82, and a lower specification of 66. For the following process situations, graphically depict each of the three cases, assuming the process to be in control, and find the values for C_p, C_{pk}, and expected $L(x)$. For the loss function, leave k as a constant.

   (a) $\overline{\overline{X}} = 74$, $\hat{\sigma}_X = 2.67$.
   (b) $\overline{\overline{X}} = 78$, $\hat{\sigma}_X = 2.67$.
   (c) $\overline{\overline{X}} = 82$, $\hat{\sigma}_X = 2.67$.
   (d) Comment on the trends that occur for each index.

10.20 In Chapter 2 we saw the example of the loaf-shaped distribution that occurred in the color-intensity quality characteristic of TV sets produced in the United States. In the television example, it was noted that the quality characteristic of the TV sets produced by the Japanese company followed a normal distribution, which nearly stayed completely within the specifications. Compare these two situations by finding the expected $L(x)$ values for the process approximated by the uniform distribution and by the normal distribution. Both processes are centered at the nominal and have a $C_{pk} = 1$.

10.21 A certain process is being studied to determine its stability and process capability. $\overline{X}$ and R control charts have been developed based on 30 subgroups of 5 measurements each. The charts indicate that the process shows good statistical control. To meet the customer's needs and expectations, the process should also have a C_{pk} value of 1.5 or more. The continuing analysis beyond the control chart study has shown that the process capability, based on this data, is $\hat{C}_{pk} = 1.32$. Is it possible that the process actually does meet the customer expectations? Justify your answer with both the logic and a properly scaled sketch that conveys your argument.

10.22 For an injection molding process the part in question has an outside diameter with the following specification:

$$0.45 \pm 0.01 \text{ centimeter.}$$

Control charts based on 25 subgroups of $n = 5$ measurements each show good statistical control. The centerlines of the $\overline{X}$ and R charts are found to be 0.452 centimeter and 0.0025 centimeter, respectively.

(a) Calculate the $\hat{C}_{pk}$ value for this process, based on the available data.

(b) Calculate a 95% confidence interval for the true C_{pk} of the process.

(c) If the required C_{pk} for this situation is 1.33, comment on the process capability, based on your analysis.

10.23 A machine tool builder is to deliver a special-purpose machine to the customer only after demonstrating that a critical quality characteristic on the part it is to produce has a demonstrated C_{pk} of 1.5 or better. To provide this proof, the builder produces 35 of the parts and measures the critical dimension. The data are then used to determine the $\hat{C}_{pk}$ value for the process, which turns out to be 1.38. This $\hat{C}_{pk}$ value is lower than the target value, and the builder hypothesizes that the part fixturing is the problem; with this in mind, the builder proceeds to redesign and re-build the fixture, and then repeats the 35-piece run. This time the $\hat{C}_{pk}$ is 1.34. The chief engineer on the project is about to have the machine's spindle changed when a newly hired engineer, Mike, makes some suggestions based on what he learned in college studying from this book. Play the role of Mike and provide a detailed critique of the work done to date, pointing out **all** of the possible errors in thinking and analysis.

10.24 Two successive run-offs (done on the same day within a couple of hours of one another) on the same machine produce $\hat{C}_{pk}$ values of 1.11 and 1.77. The run-offs are based on a 120-piece sample in each case.

(a) Assuming that the data showed the processes to be in good statistical control, is such a result plausible? Demonstrate your answer with facts.

(b) What are the prevailing circumstances that might make such a result perfectly plausible?

11

The Design and Analysis of Tolerances

11.1   Introduction

The importance of tolerance specification in the traditional approach to Six Sigma quality and process capability has been discussed respectively in Chapters 1 and 10. Six Sigma programs are for the quality improvement of a process targeted toward a numerical goal against a given tolerance specification. Process capability studies are conducted to estimate a performance measure of a process, again, against a given tolerance specification. In both applications, therefore, how the tolerance is "given" or determined plays a fundamentally important role in the pursuit of a Six Sigma improvement program and in the validity of a calculated C_{pk}. In this chapter, we will reiterate the importance of and actually apply the ideas of the loss function in the analysis and design of tolerances. Once more, we will re-emphasize statistical thinking and statistical process control in dealing with tolerances.

Through illustrations, we will show in Section 11.2 the value of statistical treatment and economic evaluation of tolerances in their design and analysis for enhanced competitiveness. A review of the statistical definition of the conventional Six Sigma interpretation is presented in Section 11.3 to demonstrate the importance of basic SPC concepts in the understanding of quality programs that are aimed at satisfying given tolerances. For economic analysis in setting tolerances, some loss-function-based cost models of unilateral and bilateral tolerance specifications are presented with examples in Section 11.4. The methods of statistical assessment of an assembly from given component tolerances and of assignment of component tolerances to meet a given assembly specification are discussed in Section 11.5. The relationship between the statistically determined specifications and the achievement of their intent at the process via a statistically stable manufacturing environment is discussed in Section 11.6.

11.2   Tolerancing and Competitiveness

Tolerances can be defined as being concerned either with physical and chemical properties, including size, weight, hardness, and composition of a part, or with the geometric characteristics, including dimension, shape, position, and surface finish of some part features. Since it is not possible to produce many parts each to have exactly the same nominal value of a feature, deviations from the design nominal are inevitable and hence allowed or tolerated. When a part deviates too much from the nominal, it fails to perform the intended function. To guard against possible functional failures, design engineers usually determine a maximum allowable deviation known as the tolerance, with upper and/or lower limits specified for each quality feature.

Suppose the design nominal is the best, unique functional value for a quality characteristic. Then any part with a quality that deviates from the nominal may not be expected to produce the optimum function, even if its quality is within the specification limits. Any such deviations, therefore, will surely cause additional losses, over and above the expected loss of those products that are on target or nominal, to the customer and/or producer. Yet it is common to find in practice that:

- Engineers work overtime trying to make as many parts as possible fall within the specification limits whatever their deviations might be;
- Management relies on inspection schemes to screen out parts that are unacceptable so that the rest may be shipped to the market; and
- Customers have suffered, but learned to live with high-cost, all-too-often intolerably bad products because there are no other choices.

Tolerances in Practice

Let us examine some of the common practices in tolerancing with a view toward their quality and productivity implications. The following are taken, respectively, from a manufacturing textbook[1] and an engineering handbook:[2]

1. "Some basic rules are ...: Specify the *broadest* tolerance and *coarsest* surface finish that still satisfies functional requirements."

2. "For tolerance control ... in manufacturing, two of the common criteria are: *Maximum* possible tolerance is assigned to each cut in the process, and every tolerance assigned is *equal to* and preferably *larger than* the process capability."

---

[1] J. A. Schey, *Introduction to Manufacturing Processes*, 2d ed., McGraw-Hill, New York, 1987.
[2] *Tool and Manufacturing Engineers Handbook*, 4th ed., Vol. 1, Chapter 2, Society of Manufacturing Engineers, Dearborn, MI, 1983.

The above seem to reflect the operating philosophy behind actual workings concerning tolerances: Every effort should be made to set the tolerances as large as possible.

Tolerance practice, then, is very much like the driving habits of some of us. The government, by law, puts up a speed limit sign (design specification), and some drivers almost always push to that limit and sometimes go beyond it by a few miles/hour (working tolerances) if no patrol cars are in sight (inspection and control). In the case of manufacturing, however, such practice manifests a really serious misconception about product quality, as illustrated in Fig. 11.1 for a bilateral specification. Parts a and b have about the same amount of deviation from the nominal M, while part c is right on the target. The strategy of assigning the largest possible tolerance must have been based on the notion that part b is drastically different from part a, but about the same as part c. In terms of actual losses due to functional deviations, it implies that all the parts that are within the specification limits incur the same amount of costs, while those falling outside the limits will suffer some considerably higher costs. In other words, the loss function is a step function of deviations, where the steps occur precisely at the lower and upper specification limits, LS and US. As shown in Chapter 10, the loss function, $L(x)$, is more likely continuous, quadratic, and centered at the nominal value where the loss is minimum.

The significance here is that the quadratic loss function promotes the idea of assigning the minimum possible tolerances and of searching continuously for ways to reduce the tolerances. What a contrast!

Quantifying the Quality Loss

In the final analysis, the question to be asked is: Which approach makes more sense in the real world? If all the products are treated to be indifferent from each other as long as they fall within the specification limits, we have shown in the previous chapter that their quality is likely to be distributed with a somewhat uniform frequency, as $f_2(x)$ in Fig. 11.2. On the other hand, a working

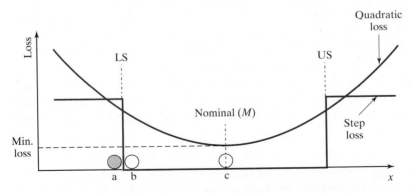

Figure 11.1   Functional loss due to deviations from the nominal

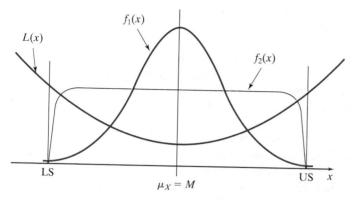

Figure 11.2   **The effects of quality distributions on functional loss**

principle toward tolerance minimization tends to yield a quality distribution very close to the bell-shaped curve, $f_1(x)$. Obviously, $f_1(x)$ will have more happy customers than $f_2(x)$. The expected loss of a centered process, $E[L(X)] = k\sigma_X^2$, as presented in Chapter 10, shows that the average loss is proportional only to the process variance σ_X^2. Let $f_1(x)$ be approximated by a normal distribution with a tolerance interval $(US - LS) = \pm 4\sigma_X$ from the nominal $(C_{pk} = 1.33)$ and $f_2(x)$ be a uniform distribution within the interval (LS, US) as illustrated in Fig. 11.2. Then,

$\sigma_X^2 = (US - LS)^2/64$ is the required variance for a normally distributed X, and
$\sigma_X^2 = (US - LS)^2/12$ is the required variance for a uniformly distributed X.

Thus, the average loss per unit by adopting the approach of maximum tolerance (uniform distribution) is about five times higher than that by adopting the minimization approach (normal distribution). In other words, in terms of the operational costs of the products in use, to pursue the idea of tolerance minimization could make the customers five times happier.

   A more general situation is depicted in Fig. 11.3(a), where the actual process mean is different from the designed nominal, and the nonsymmetrical specification limits are placed such that some rejects are produced as shown in Fig. 11.3(b), in addition to the quality loss of Fig. 11.3(c). Usually the first attempt in reducing the total quality costs (of rejects and losses) for a given specification is to minimize the rejection cost by bringing the process mean from "off target" to "on target" and then to seek ways to reduce process variability.

Designing Tolerances

In our discussions up to now, we have assumed that the given specification of the nominal and tolerance interval is methodically designed according to some optimality criteria. The practice, however, is likely quite different from the optimum, particularly in setting tolerance limits. While the nominal value may

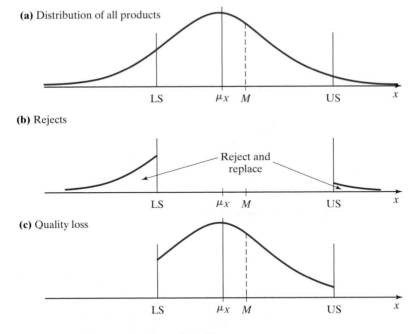

Figure 11.3   Decomposition of quality costs

be designed through careful engineering analysis to be optimum or nearly so, the tolerance interval is often assigned somewhat arbitrarily and after the nominal has been determined. This sequential approach sounds quite logical. Unfortunately, there are at least two problems with it. The first problem exists in the way we actually deal with tolerances on the production floor. Whenever manufacturing has difficulty meeting the specifications, it is usually only the size of the tolerance interval that is reconsidered, with the specified nominal remaining unchanged. If the tolerance interval is changed because of the reconsideration, then the optimality of the designed nominal could become immediately questionable. For quality and productivity considerations, both the nominal and its associated tolerance interval should be designed simultaneously, instead of sequentially. The second issue is addressed below.

Effects of the Nominal in Tolerance Design

The sequential approach to tolerance design would be appropriate if the variability of a quality response were unrelated to the control and/or signal variables. In reality, however, they are often related. In other words, different parameter values of an input variable may produce different amounts of variation in the quality response. As first introduced in Chapter 1, Fig. 11.4 shows the effects of noise transmission from input variable X to response Y due to a nonlinear relationship between them. Suppose M_a and M_b are two possible choices for the input nominal. If their corresponding mean responses a and b

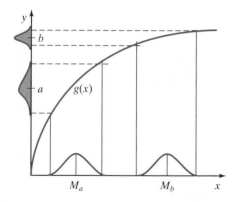

Figure 11.4   The effect of design nominal on response variability

are within the acceptable range, M_b would be a better choice for the nominal of X, because it produces a smaller variance for the response. Of course, the illustration of Fig. 11.4 may not be applicable or even valid if either of the two response means, a or b, is not feasible. Nevertheless, the possible trade-off between M_a and M_b should be evaluated in tolerance design whenever possible. This brings up the second problem of the sequential approach. It precludes the opportunity to explore the interrelationships between the nominal value and its effect on output response variation.

Parameter Interaction Effects in Tolerance Design

Often interaction effects exist among the design parameters or control variables that influence a performance response. A simple case of the parameter interaction analysis in tolerancing is sketched in Fig. 11.5. The two design

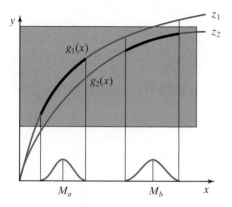

Figure 11.5   Interaction effect on choice of design nominal

parameters, or variables, X and Z have an interaction effect on the response Y; in other words, the nonlinear effect of variable X on the response is dependent on the level for variable Z. Let $g_1(x)$ and $g_2(x)$ be the response functions at two different values of the variable Z, z_1 and z_2, respectively. These two response functions are different due to the dependency between X and Z. Suppose, as before, that M_a and M_b are two feasible nominal values for X under consideration. Figure 11.5 shows that of the four possible settings for the nominal values of X and Z, only two are feasible, as they produce a tolerance range that falls completely within the feasible range (shaded rectangle) of the response. These are (M_a, z_1) and (M_b, z_2). The choice between them will depend on whether or not a larger Y with small variation is better than a smaller Y with larger variation.

This example demonstrates that in engineering tolerancing, not only should the tolerance be designed together with each of the nominal values, but also with all related quality features jointly. Following the sequential approach of selecting the nominal first and then the tolerance, the response variability due to possible nonlinear functional relationships and interaction effects would never be learned. In the case where all the important design parameters are continuous variables, the analysis for tolerance and nominal design may be carried out by the technique of response surface methodology, which will be presented in Chapter 23.

11.3   Statistical Interpretation of Six Sigma Quality

The popular Six Sigma program applies the concepts and methods of statistics to quality improvement. In applications, however, its interpretation of the capability of a process with respect to a given tolerance specification has raised some issues of concern. The first and most important concern is in its treating of the process mean as a variable instead of a constant. We will examine in this section the full potential of correctly interpreted Six Sigma quality following the fundamental notion of statistical process control we have discussed in the previous chapters. To understand this we need to review the conventional definition of Six Sigma.

Conventional Definition of Six Sigma Quality

The Six Sigma quality as it is defined for applications in industry is depicted in Fig. 11.6. A centered process should have its mean equal to the specified nominal value, M, which is exactly 6 standard deviations from both of the tolerance limits, LS or US. The actual process mean, μ_X, however, is allowed to vary from M by a maximum of 1.5 standard deviations in either direction. Such a definition of Six Sigma quality allows the actual process mean to vary and not to be controlled so long it is within $\pm 1.5\sigma_X$ from M. Adopting this definition as an operational philosophy not only directly contradicts Deming's first point, "Create constancy of purpose for improvement," but also makes the application of the concepts and methods of SPC difficult, if not impossible.

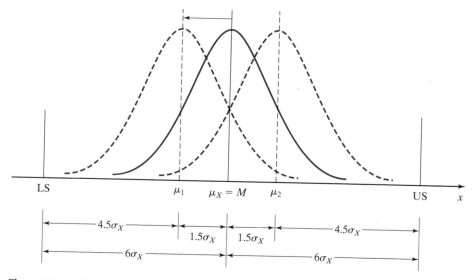

Figure 11.6   Conventional definition of a Six Sigma process

C_{pk} of a Six Sigma Process

Referring to Fig. 11.6, it is obvious that the true C_{pk} of a perfect 6 sigma process is a variable, not a constant, that can assume values between $1.5 = 4.5\sigma_X/3\sigma_X$ and $2 = 6\sigma_X/3\sigma_X$. Note that this variation in C_{pk} is not the one due to sampling errors discussed in Chapter 10, but depends on the operating policies in actual production practice. If the practice is to allow the process mean to vary either systematically or randomly between μ_1 and μ_2, there cannot be any validity of a calculated C_{pk} from the data of a sample run-off. Furthermore, extending the same product control idea of setting tolerance limits for individual part quality to the setting of another allowance interval for process mean not only complicates the statistical methods required for the process control and performance estimation, but also causes severe economic consequences. We will demonstrate by loss function analysis the potential reduction in expected quality loss if the process mean is held constant and on target at the nominal.

Probability Distributions of a Six Sigma Process

Whenever the process mean is allowed to shift or to be set within $\pm1.5\sigma_X$ from the nominal, the resultant distribution of the individual product quality, X, can be of many different shapes. The following are two possible distributions, assuming a normal distribution for the underlying process with a constant variance and a process mean that can vary within limits.

1. If a process is to run continuously with a mean that is set at a fixed point within $\pm1.5\sigma_X$ from the nominal, such as the one shown in Fig. 11.7, the resultant

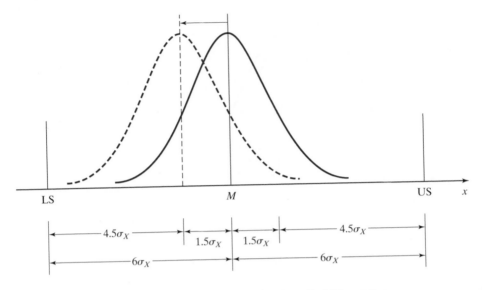

Figure 11.7   Process mean is fixed within an interval of $M \pm 1.5\sigma_X$

distribution of X will also be normal. The true value for C_{pk} will be between 1.5 and 2.

2. If the mean of a process is allowed to vary either gradually and/or systematically from one extreme to the other within the interval of $M \pm 1.5\sigma_X$, the resultant distribution of X may look like the one illustrated in Fig. 11.8. Now there is no way to estimate the C_{pk}. Any calculated estimate of C_{pk} from a sample run-off would surely misrepresent the true process capability.

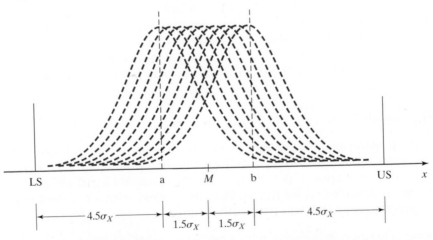

Figure 11.8   The resultant distribution of a process whose mean changes at a constant rate between "a" and "b"

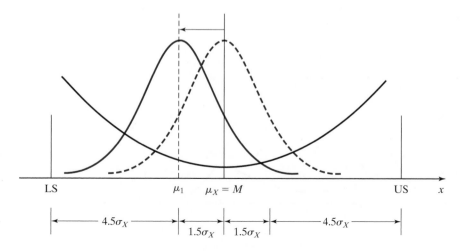

Figure 11.9   The process mean is set at $M - 1.5\sigma_X$

The Effect on Quality Loss

It may be of interest to evaluate the two cases above by their costs of quality loss.

1. When the process mean is kept constant at $M - 1.5\sigma_X$, as shown in Fig. 11.9, the expected loss per unit is given by

$$E[L(X)] = k\{\sigma_X^2 + (\mu_X - M)^2\} = k\{\sigma_X^2 + (1.5\sigma_X)^2\} = 3.25k\sigma_X^2,$$

where k is the quality loss coefficient of the process.

2. Figure 11.10 shows the mixture distribution of Fig. 11.8 with an added quadratic loss function. The estimation of the expected loss in this case is a bit more complicated. When the process mean is allowed to change continuously from $\mu_1 = M - 1.5\sigma_X$ to $\mu_2 = M + 1.5\sigma_X$, however, the expected quality loss per unit can be shown to be approximately

$$E[L(X)] = 1.75k\sigma_X^2.$$

By following Deming's principle of "constancy of purpose" in SPC, the process should always be run with $\mu_X = M$ and $\sigma_X = (US - LS)/12$ shown in Fig. 11.11; then the expected quality loss is

$$E[L(X)] = k\sigma_X^2.$$

When this minimum loss is compared with either of the possible scenarios where the process mean is allowed to vary within limits, the average quality loss can be reduced by over $100\%(1.75 - 1)/1.75 = 40\%$ to nearly $100\%(3.25 - 1)/3.25 = 70\%$ simply by applying SPC in practice.

$$E[L(X)] = 1.75k\sigma_X^2$$

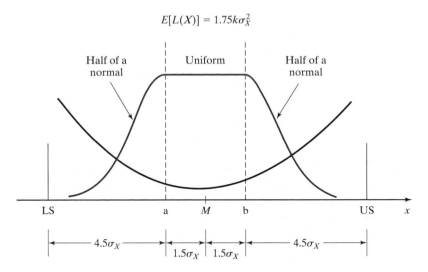

Figure 11.10   Quality loss of a mixture distribution

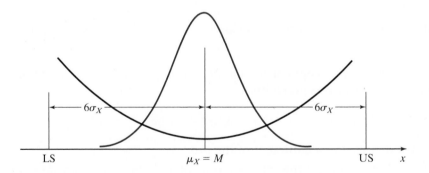

Figure 11.11   A centered Six Sigma process distribution

Discussion

In this section we have shown that the C_{pk} value is reduced and the expected quality loss is increased once a process is allowed to vary its mean even within limits. All we need to do to further improve the process capability as well as its estimation accuracy and to achieve substantial economic gains is to interpret the Six Sigma correctly and follow the simple SPC principles in practice. This seems to support the merit of the first of Deming's 14 points—constancy of purpose. Note, too, that the conventional definition of Six Sigma seems to be based on a belief that in actual manufacturing it is easier to hold the variance constant than to center the process. In reality, the opposite is generally true.

Another point to be noted is that the common definition of Six Sigma is inherently inefficient economically. By allowing the process mean to vary within $\pm 1.5\sigma_X$ from the nominal, it effectively uses an interval of $9\sigma_X$ within a tolerance interval of $12\sigma_X$. What this could mean in applications is that a much more expensive, high-precision machine might have to be assigned unnecessarily to a low-precision job at a higher cost.

In conclusion, the fundamental concepts in SPC should always be followed in quality improvement programs, including Six Sigma quality programs, in order to ensure the attainment of the ultimate economic return.

11.4   Tolerance Design and Analysis

Introduction

Ideally, we would like to see every manufacturing process operated with a variance so small that there would be no need to set tolerances at all. In reality, however, engineering tolerance concerns will remain a part of routine engineering and management. As we have discussed in Section 11.2, such tolerance concerns are best dealt with by design. Knowledge of tolerances, established based on function and derived on economic grounds, is an important input to manufacturing process planning to ensure that the processes chosen and the conditions under which those processes are run are both capable and competitive. Since tolerances have immediate cost consequences as we implement Six Sigma programs and evaluate process capabilities, we will present a cost model for tolerance analysis and design in this section. The primary objective is to demonstrate that tolerances may be rationally developed based on the economics of the situation and that Taguchi's loss function concept, introduced in Chapter 10, can form a basis for such tolerance design and analysis.

Tolerance Models for a Normally Distributed Process

Suppose a quality characteristic X of a process is a normal random variable with mean μ_X and variance σ_X^2. Since quality loss applies only to products that are in field use, the population distribution of these products, denoted by Y, is a truncated normal distribution, as shown in Fig. 11.12. In saying this, we are assuming that only products that meet the specifications are allowed to pass into field use. In general, the actual process mean of X may be different from the nominal M, as illustrated in Fig. 11.12(a). A two-sided tolerance with a nonsymmetrical, truncated normal density function (representing that output of the process that meets functional requirements) is shown in Fig. 11.12(b) together with the loss function $L(x)$. In order to estimate the mean and variance of the truncated normal distribution $f(y)$ for quality loss analysis, we will find the probability density function of Y as follows.

The cumulative probability function of Y is

$$\begin{aligned} F(y) &= P[Y \le y] \\ &= P[X \le x, \mathrm{LS} \le X \le \mathrm{US}] \\ &= P[X \le x]/P[\mathrm{LS} \le X \le \mathrm{US}] \end{aligned}$$

(a) Normal pdf

(b) Truncated normal pdf

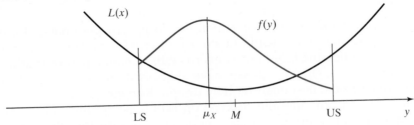

Figure 11.12   **Two-sided truncated normal distribution for quality loss estimation**

Then the probability density function of Y is given by

$$f(y) = F'(y) = \left(\frac{W}{\sigma_X\sqrt{2\pi}}\right)e^{-((y-\mu_X)^2/2\sigma_X^2)}, \tag{11.1}$$

where $W = 1/P[\text{LS} \le X \le \text{US}] = 1/\{\Phi(z_u) - \Phi(z_l)\}$, $z_u = (\text{US} - \mu_X)/\sigma_X$, and $z_l = (\text{LS} - \mu_X)/\sigma_X$ are the standardized normal values of US and LS, respectively, $\Phi(z_u)$ is the normal probability of $(Z_x \le z_u)$ or $(X \le \text{US})$, and $\Phi(z_l)$ can be defined similarly.

On letting $z = z_y = (y - \mu_X)/\sigma_X$, we obtain the pdf of the standardized Y by

$$f(z) = \left(\frac{W}{\sqrt{2\pi}}\right)e^{-z^2/2}. \tag{11.2}$$

From the above, we obtain the mean and variance of Y as follows:

$$\mu_Y = E[Y] = \mu_X + W\sigma_X[\phi(z_l) - \phi(z_u)] \tag{11.3}$$

and

$$\sigma_Y^2 = \text{Var}[Y] = \sigma_X^2[1 + W\{z_l\phi(z_l) - z_u\phi(z_u)\} - W^2\{\phi(z_l) - \phi(z_u)\}^2], \tag{11.4}$$

where $\phi(z)$ is the normal probability density at z standard deviations from μ_X, where the density is given by

$$\phi(z) = \frac{1}{\sqrt{2\pi}}e^{-z^2/2}.$$

Equations (11.3) and (11.4) can be used to find the mean and variance of a truncated normal random variable Y from the original normal random variable X (with mean μ_X and variance σ_X^2) for a pair of specified tolerance limits LS and US. In applications, the lower and upper tolerance limits may be nonsymmetrical when they are set at different distances from the nominal, or where only one tolerance limit, either the lower or the upper, is required. Furthermore, the production process itself may have a mean that is different from the nominal value, or off-centered. For the purposes of illustrating the principles involved here, we will consider only the case of a "symmetrical tolerance specification of a centered process." This is the case that is encountered most often in practice. It is also the simplest case for applications, because $\mu_X = M$ and $z_u = -z_l$. Since $\phi(z_l) = \phi(z_u)$ in this situation, Eqs. (11.3) and (11.4) are reduced to the following for the mean and variance of the truncated variable Y:

$$\mu_Y = E[Y] = \mu_X \quad \text{and}$$

$$\sigma_Y^2 = \text{Var}[Y] = \sigma_X^2[1 + 2Wz_l\phi(z_l)]. \tag{11.5}$$

It is seen from Eq. (11.5) that the variance of Y is smaller than the variance of X, since z_l is negative in general.

The Effects of Tolerances on Variance Reduction

From the above analysis of the truncated normal random variable Y, we notice that the variance of the products in field use is reduced due to tolerance specifications, which is expected. However, the reduction will be appreciably significant only when the two-sided tolerance limits are set relatively close to the nominal. This is because $\phi(z)$ approaches zero much faster than z approaches ∞. As a result, $\text{Var}[Y]$ is almost as large as σ_X^2 when the absolute value of z is very large.

Figure 11.13 shows the change in variance reduction as a function of z for the two-sided tolerance situation. It is seen from the curve in Fig. 11.13 that the variance of Y is not significantly reduced from the variance of X for $z > 3$.

In summary, when the quality characteristic X of the process is normally distributed and C_{pk} is required to be, say, one or greater, the variance of Y will not be much different from the variance of X.

Expected Quality Loss of Various Loss Functions

The loss function proposed by Taguchi for the characterization of product quality is a quadratic function of the deviations of the product quality from the target quality (nominal), where the loss is a minimum. In general, if X has a continuous probability density function $f(x)$ with an arbitrary loss function $L(x)$, the expected quality loss is

$$E[L(x)] = \int_{R_X} L(x)\, f(x)\, dx. \tag{11.6}$$

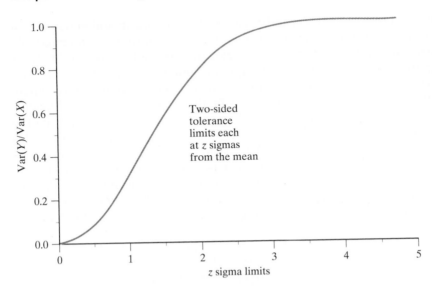

Figure 11.13   Variance ratio as a measure of variance reduction of a truncated normal distribution within the tolerance

Equation (11.6) provides an average measure of the quality loss for all x values including those falling outside the tolerance interval. In tolerance analysis, however, we will be concerned with only those x values that are within the tolerance interval in question. In other words, we will be interested in estimating $E[L(Y)]$ where Y has values within the tolerance interval, as we discussed in the last section.

Due to differences inherent in the quality preference or requirements of X, different criteria for the loss function $L(x)$ may be employed for appropriate applications. There are, in general, three types of loss functions:

1. *Nominal Is Best.* This is the more common type that applies to a two-sided tolerance interval with a nominal M set at the target value m, $M = m$. In this situation, there is a concern of the quality deviation from the nominal in either direction. Some examples of a two-sided tolerance are: the clearance between two parts should be neither too large nor too small; the thickness of a soft gasket cannot be too thick or too thin; and the force required to close a car door must not be too heavy or too light.

   The appropriate loss function for the "Nominal Is Best" quality characteristics is the one we have discussed so far, namely

$$L(x) = k(x - m)^2.$$

Its expectation is given by

$$E[L(X)] = kE[(X - m)^2] = k\{E[X^2] - 2mE[X] + m^2\}$$

$$= k\{E[\mu_X^2 + \sigma_X^2] - 2m\mu_X + m^2\} = k\{(\mu_X - m)^2 + \sigma_X^2\}. \quad (11.7)$$

2. *Smaller Is Better.* The objective here is to reduce the value of X, or maintain it at a low level. Sometimes it is ideal to have X equal to zero. Therefore, for applications, only an upper tolerance limit is to be determined. Examples: wear, deterioration, degradation, shrinkage, noise, harmful effects, impurity, and failure rate.

Since the ideal target is 0 for the "Smaller Is Better" quality characteristics, an appropriate loss function is given by

$$L(x) = kx^2 \quad \text{for } x \geq 0. \tag{11.8}$$

The corresponding expected quality loss is

$$kE[X^2] = k\{\mu_X^2 + \sigma_X^2\}. \tag{11.9}$$

In this case, we would like to have the process mean μ_X be very small. To reduce the average quality loss, we would also try to reduce the process variation.

3. *Larger Is Better.* Here, it is preferred to increase the value of X. Ideally we would like it to be infinite, 100%, or perfect. Examples: chemical yield, tensile strength, system reliability, fuel efficiency, and product life. For such "Larger Is Better" quality characteristics, only one lower tolerance limit is needed in order to ensure that X will not be too small to cause functional failures. An appropriate choice of the loss function is given by

$$L(x) = k(1/x^2) \quad \text{for } x > 0 \tag{11.10}$$

The expectation of the loss of Eq. (11.10) is

$$E[L(X)] = kE[(1/X^2)]$$

Applying a Taylor's series expansion, it can be shown that an approximation of the expectation of $L(X)$ is

$$k\{(1/\mu_X^2) + (1 + 3\sigma_X^2/\mu_X^2)\}. \tag{11.11}$$

Obviously, we would like to have a very large process mean μ_X and a small σ_X^2 in order to reduce the average quality loss in this case.

Total Quality Cost

Consider the total cost of quality to consist of two cost components: cost of rejection and cost of quality loss, as illustrated in Fig. 11.3 in Section 11.2. The expected total quality cost be expressed by

$$E[C_T] = E[L(X)] + qC_R, \tag{11.12}$$

where C_R = rejection cost/unit, q = probability of rejection = $\Phi(z_l) + \{1 - \Phi(z_u)\}$ for a normally distributed X, and $E[L(X)]$ = quality loss/unit.

The general formula of the expected total quality cost for a normally distributed X can be shown to be

$$E[C_T] = k\{\sigma_X^2 + (\mu_X - m)^2 + 2W\sigma_X(\mu_X - m)[\phi(z_l) - \phi(z_u)]$$
$$+ W\sigma_X^2[z_l\phi(z_l) - z_u\phi(z_u)]\} + qC_R. \qquad (11.13)$$

Equation (11.13) may be used to determine the tolerance limits that are optimum in the sense of minimum expected total quality cost. For the case of a centered process with symmetrical tolerances, Eq. (11.13) reduces to:

$$E[C_T] = k\sigma_X^2\{1 + 2z_l\phi(z_l)/[1 - 2\Phi(z_l)]\} + 2\Phi(z_l)C_R. \qquad (11.14)$$

Example of Cost Analysis of Tolerancing for a Normally Distributed Process

We now provide an example to illustrate the use of the total cost formulas on tolerance analysis of a quality characteristic X. A statistically controlled process produces parts that are normally distributed with a mean, $\mu_X = 10$, and a standard deviation, $\sigma_X = 0.5$. It has been estimated that a part with the quality dimension of 8.5 will have a loss of $11.25 to the customer. As a company policy, all parts are inspected. All rejected parts can be repaired at a cost of $2.00 each. Find the average total quality cost per unit if the tolerance limits are set at 8.5 and 11.5.

Assumptions: The given tolerance limits are symmetrical about the nominal at $m = 10$, and there is no quality loss for those parts whose dimension is right on the target. Using Taguchi's quality loss formula, we have

$$k = L(x)/(x - m)^2 = \$11.25/(8.5 - 10)^2 = \$5.$$

The standardized lower tolerance limit is: $z = (8.5 - 10)/0.5 = -3$.

From the normal pdf and the normal area table in Appendix A.1, we obtain $\phi(-3) = 0.0044$ and $\Phi(-3) = 0.00135$. On substituting into Eq. (11.14), we obtain

$$E[C_T] = (\$5)(0.5)^2\{1 + (2)(-3)(0.0044)/[1 - (2)(0.00135)]\}$$
$$+ (2)(0.00135)(\$2) = \$1.2223.$$

Discussion of the Example: Why were the tolerance limits in the example set at ± 3 sigma? Can we find a different tolerance interval that is more economical in terms of average total quality cost? The answer, in general, is yes. For example, if an interval of ± 2 sigma was used, i.e., $z = -2$, it would have an average cost of $1.058/unit.

The problem is to determine the optimum values of z such that $E[C_T]$ is minimized. A numerical optimization procedure may be employed to find such solutions. Alternatively, one may calculate the $E[C_T]$ for each of several selected values of z to obtain an approximate solution such as the ones shown below.

z_I	$E(C_T)$
-3.00	$\$1.222$
-2.00	$\$1.058$
-1.33	$\$0.946$
-1.00	$\$0.998$
-0.50	$\$1.335$

From the above results, one would select the following solution:

$$z^* = 1.33 \quad \text{and} \quad E[C_T] = \$0.946,$$

which reduces the expected total quality cost by $100\%[(1.222 - 0.946)/1.222] = 22.59\%$.

The above shows that it is cheaper to set the tolerance at $\pm 1.33\sigma_X$ rather than at $\pm 3\sigma_X$ from the target. In other words, the minimum cost symmetrical tolerance limits should be set at:

$$10 \pm (1.33)(0.5) = 10 \pm 0.67 = (9.33, 10.67).$$

By selecting the above optimum tolerance for this centered process, the expected total quality cost would consist of:

Expected quality loss $= \$0.579$, or 61% of the total cost.

Expected repair/scrap cost $= \$0.367$, or 39% of the total cost.

The cost breakdown shows a relatively high cost of rejection. Improvement opportunities seem to lie in the reduction of the process variance in order to significantly reduce the rejection rate. With a reduced process variance, the expected quality loss would further reduce the expected total quality cost.

For the data of the example, we have also plotted a total cost curve as a function of z shown in Fig. 11.14. This cost curve exhibits a unique minimum at $z = -1.3$, which agrees very well with the $z = -1.33$ we obtained earlier. Sometimes an examination of a complete cost curve would be useful for choosing alternative solutions.

11.5   Statistical Assessment and Assignment of Tolerances

Suppose that a process is found to be in good statistical control, but a subsequent process capability study indicates that the process does not meet the customer needs in terms of conformance to the specifications. One of two possible problems is evident. Either the process variation is too large or the specifications

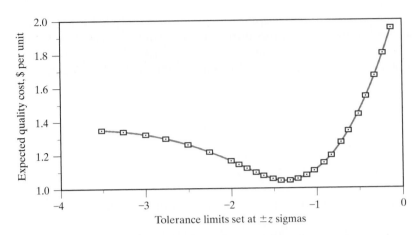

Figure 11.14   Total quality cost curve for Example 11.1 (*Note*: The total quality cost curve also includes an inspection cost of $0.10 per unit.)

have been set too narrow. Our emphasis in this book seems to favor strong attention to the first possibility—that the process variation is too large and that efforts ought to be made to find and remove the major sources of common-cause variation. It is possible, however, that the specifications have been determined in an improper fashion, and as we will soon see, a common improper approach to the determination of specifications can lead to narrower specification limits than are otherwise required when the statistical methods are used.

Previously, we have emphasized the importance of pushing the quality issue upstream, into engineering design. We have briefly introduced the tolerance design concept based on the loss function, which addresses the economics of tolerance assignment based on loss due to functional variation. This approach also assumes that the process may be represented by a statistical distribution, and therefore that the process is in statistical control.

Statistical Determination of Assembly Tolerances: An Example

Consider the three-part assembly shown in Fig. 11.15. The dimension in question is the length of the assembly and is equal to the sum of the three component lengths. We will assume that the processes producing each of the three components are in statistical control, the processes are centered at the nominal value, and the process capability for each is about $\pm 4\sigma_X$; that is, the specifications for each part are $8\sigma_X$ wide.

In examining Fig. 11.15, we might ask: What tolerance should be assigned to the assembly dimension based on the tolerances that have been assigned to each of the individual part dimensions as specified in the figure? An altogether too common answer is that the assembly tolerance is simply equal to the sum of the individual part tolerances. This would result in an assembly tolerance of 0.010 inch + 0.002 inch + 0.005 inch = 0.017 inch. However, under

Figure 11.15 Tolerance determination for a three-part assembly

the conditions of the preceding paragraph, one would find that the actual assemblies would come out much better than this, that is, would have a smaller natural spread. It is important that we understand exactly why this is so.

Two important forces are at work that govern this problem:

1. Statistical distribution of component dimensions.
2. Random assembly.

When in statistical control, the individual processes produce parts that vary over some range due to common-cause variation. This variation tends to arise in a frequency sense according to a statistical distribution and therefore is confined to some range of occurrence by probabilistic limits. If we assumed a normal distribution for each part's individual measurements, the state of affairs would be as shown in Fig. 11.16, with the process capabilities as specified above.

Now, considering random assembly—that is, the parts are drawn randomly from processes 1, 2, and 3, respectively—the assembly deviation from nominal will behave in a predictable fashion based on precise probability laws. Because of this statistical way of thinking about the dimensions of the individual parts emanating from their respective processes, the chances of randomly selecting three parts, all in the far-left tail of their respective distributions (at the 4 sigma distance from the mean), is very, very small. This is depicted graphically in Fig. 11.17.

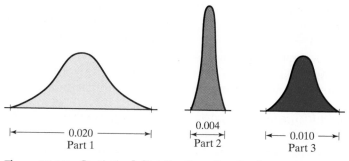

Figure 11.16 Statistical distributions for the dimensions of each part

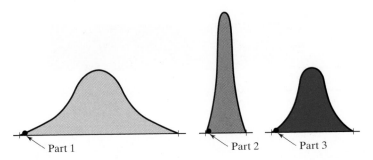

Part 1   Part 2   Part 3

Figure 11.17   Simultaneous selection of parts from the extreme left of their respective distributions

The probability of getting any one part measurement 4 or more standard deviations below the nominal is about 0.00003. The probability of getting all three parts simultaneously at -4 sigma or less under random assembly is $(0.00003)^3 = 0.000000000000027$! It is clear from this that using simple additivity of the individual part tolerances is an unrealistic way to combine them to obtain the assembly tolerance. Then how shall we proceed more realistically? Suppose that for the three parts in question the specifications were set at $\pm 4\sigma_{process}$; then

Part 1 tolerance: ± 0.010 inch implies that $\sigma_1 = 0.0025$.

Part 2 tolerance: ± 0.002 inch implies that $\sigma_2 = 0.0005$.

Part 3 tolerance: ± 0.005 inch implies that $\sigma_3 = 0.00125$.

Now there is a third force at work here, the *additive law of variances*:

$$\sigma^2_{assembly} = \sigma^2_1 + \sigma^2_2 + \sigma^2_3. \tag{11.15}$$

Using the additive law of variances to combine the tolerances produces the following result in terms of the standard deviation of the assembly:

$$\sigma_{assembly} = \sqrt{(0.0025 \text{ inch})^2 + (0.0005 \text{ inch})^2 + (0.00125 \text{ inch})^2}$$
$$= 0.00284 \text{ inch}.$$

If the assembly tolerance is also set at $\pm 4\sigma_{assembly}$, virtually all assemblies will fall within $\pm 4(0.00284 \text{ inch}) = \pm 0.01136$ inch of the nominal assembly dimension, not ± 0.017 inch as obtained by adding part tolerances. In fact, 99.73% of all assemblies will fall within ± 0.0085 inch of the nominal assembly dimension, *one-half* of the value obtained by adding part tolerances. In summary:

- *By adding tolerances for individual parts:* assembly tolerance $= \pm 0.017$ inch.
- *By combining tolerances statistically:* assembly tolerance $= \pm 0.01136$ inch.

Statistical Assignment of Component Tolerances

In practice the problem above is not the one we generally face; rather, it is the reverse problem. Based on functional grounds, it is likely that the assembly tolerance is specified first. The question that we then face is: How do we partition the assembly tolerance; that is, how do we assign tolerances back to the individual parts? To answer this question, we consider another example.

Consider a certain assembly made up of three identical parts as shown in Fig. 11.18. The assembly tolerance is ±0.009 inch. How do we distribute this tolerance among the individual parts that comprise the assembly? By invoking additivity of tolerances, we would obtain the following answer:

Tolerance of the assembly = tolerance of part 1 + tolerance of part 2
+ tolerance of part 3
±0.009 inch = ±0.003 inch + ±0.003 inch + ±0.003 inch.

But we already know that this is not the correct way to combine the individual part tolerances. We now understand that this places tolerance requirements on the parts that are much tighter than they need to be to guarantee the required assembly tolerance. From what we have seen already, it would appear that the individual part tolerances could exceed ±0.003 inch and still the assembly tolerance could be ±0.009 inch.

To determine precisely what the individual part tolerance should be, let us again invoke:

1. Random assembly.
2. Statistical (and in this case normal) distribution of part dimensions.
3. The additive law of variances.

Further, let us assume:

4. The processes making the parts will be ±4σ_{part} capable.
5. The part standard deviations are equal (made by same or similar processes).

Figure 11.18   Part tolerance assignment for an assembly

With the assumptions above we can easily determine the individual part tolerances:

$$4\sigma_{\text{assembly}} = 0.009 \text{ inch.}$$

$$\sigma_{\text{assembly}} = 0.00225 \text{ inch.}$$

Furthermore,

$$\sigma_{\text{assembly}} = \sqrt{\sigma_1^2 + \sigma_2^2 + \sigma_3^2}$$

and we will assume that

$$\sigma_1^2 = \sigma_2^2 = \sigma_3^2 = \sigma_{\text{part}}^2.$$

Therefore,

$$\sigma_{\text{assembly}} = \sqrt{3\sigma_{\text{part}}^2}.$$

But $\sigma_{\text{assembly}} = 0.00225$ inch, as calculated above, so

$$\sigma_{\text{part}} = 0.0013 \text{ inch.}$$

Now if the individual part tolerances are set at $\pm 4\sigma_{\text{part}}$, then a ± 0.0052-inch tolerance for each part ensures that virtually all of the assemblies will fall within ± 0.0090 inch. This is in contrast to a much tighter ± 0.003-inch tolerance as would be specified by using simple addition of part tolerances.

In summary:

> *Simple division:*   part tolerance = ± 0.0030 inch.
> *Statistical division:*   part tolerance = ± 0.0052 inch.

An important implication of the statistical tolerance assignment situation that we saw above is that by inappropriately setting the part tolerance using additivity/division, you could end up with a situation in which you could be scrapping/reworking perfectly good parts. This fact is seen in Fig. 11.19, using the numbers of the previous example.

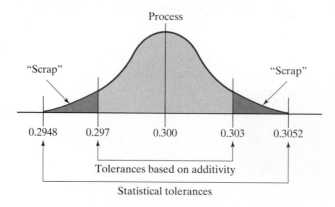

Figure 11.19   Consequences of improper tolerance assignment

It is important that in the above we have the proper view of what we really mean by σ_{part} and setting tolerances, for example, $\pm 4\sigma_{part}$. This notion has meaning only when we think of σ_{part} as a process-generated variation. But the tolerance is based on part function requirements. What the design engineer is doing when he or she sets the tolerance is setting the capability requirement for the process. This, of course, requires that the process is in control. Conformance to specifications cannot be interpreted otherwise.

Importance of Process Centering in Maintaining Capability: Individual Component Versus Assembly Tolerances

Suppose we are concerned with a part that is really a three-piece construction using three identical parts, as shown in Fig. 11.20. Based on design intent the assembly nominal is 0.390 inch and the tolerance is set at ± 0.050 inch. Given the functional needs above, the tolerances for the individual parts are to be set. Using the additive law of variances, assuming random assembly and assuming that the assembly process is to be ± 3 sigma capable, we can determine the required part tolerance.

$$\sigma_{assembly} = \frac{0.050 \text{ inch}}{3}$$

$$= 0.017 \text{ inch.}$$

But as we have seen previously,

$$\sigma_{assembly} = \sqrt{\sigma_1^2 + \sigma_2^2 + \sigma_3^2} = \sqrt{3\sigma_{part}^2}$$

since the three parts are identical. Therefore,

$$0.017 \text{ inch} = \sqrt{3\sigma_{part}^2}$$

$$\sigma_{part} = 0.010.$$

0.390 inch

0.130 inch

Figure 11.20   Three-part assembly

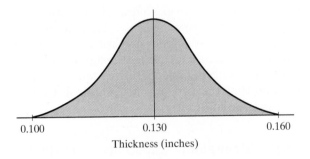

0.100   0.130   0.160

Thickness (inches)

Figure 11.21   **Process for the component parts**

It seems logical, therefore, to set the individual part requirements at 0.130 ± 0.030 inch (lower specification $= 0.100$ inch, upper specification $= 0.160$ inch), as shown in Fig. 11.21.

   Suppose that a stable process for the individual part is established as seen in Fig. 11.22 and that through a series of process improvements, the process common-cause variation has been halved. Now σ_{process} or $\sigma_{\text{part}} = 0.005$ inch. Let us further suppose that as a result of this process improvement there is a change in the operating policy for the process. Since the process variability has been reduced, it is decided to adjust the process mean for "economic" reasons, so that the process is now operating as seen to the right in Fig. 11.22.

   Now it is important that we examine how this change in the component part process operation will affect the product, which is the assembly. Using the additive law of variances,

$$\sigma^2_{\text{assembly}} = 0.005^2 \text{ inch}^2 + 0.005^2 \text{ inch}^2 + 0.005^2 \text{ inch}^2$$
$$= 0.000075 \text{ inch}^2.$$

Therefore,

$$\sigma_{\text{assembly}} = 0.0087 \text{ inch.}$$

0.100   0.130   0.160   0.100 0.115   0.160

Figure 11.22   **Change in the component part process operating policy**

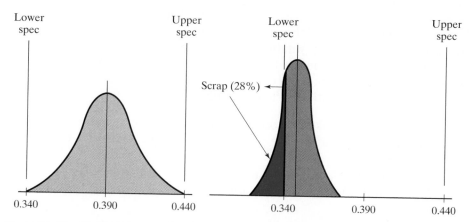

Figure 11.23   Consequences of a change in the component part processing policy on the assembly conformance to specification

The actual assembly mean value will be $(3)(0.115 \text{ inch}) = 0.345$ inch because of the change in operating policy for the process. For the assemblies, 99.73% will now be within 0.345 ± 0.026 inch.

Recall that based on design intent, assemblies must be 0.390 ± 0.050 inch. But the result now seems unbelievable. The average of the actual assemblies is now within 0.005 inch of the lower specification limit. Figure 11.23 shows the consequences of the change in component-part processing policy. As a result of the change in the individual component process mean, *nearly one of every three assemblies will be out of the specification.* It is clear from this example that merely operating within the bilateral specification is not sufficient to guarantee that assemblies of individual components will meet the specifications developed by design based on function. The relationship between design intent and its realization through the process is a strong one and must not be ignored. In the next section this fact is reemphasized.

11.6   Statistical Process Control and the Statistical Tolerance Model

As we continue to examine many of the systems that we must deal with in both the product/process design arena and in manufacturing, we will find that the concept and achievement of statistical process control is an ever-present central issue. An important issue discussed throughout this book is the relationship between the design process and the manufacturing process. Both parties must realize together the common goals they share and the importance of developing a joint strategy to meet those goals.

The issue of setting tolerances and, in particular, the statistical assignment and assessment of tolerances is an excellent example of the need for design and manufacturing to understand what each other is doing and why. The design process can go unfulfilled if the manufacturing process is not operated in a manner totally consistent with design intent. To appreciate the marriage of thinking that must exist between the design and manufacturing worlds, we will now re-examine some of the basic assumptions of setting tolerances that we discussed earlier and their relationship to the manufacturing process. What we are about to see clearly points to the importance of statistical process control relative to the issue of process capability.

The key concepts in statistical tolerancing are

1. The use of a statistical distribution to represent the design characteristic and hence the process output for the product in question relative to the design specifications.
2. The notion of random assembly (i.e., random part selection from these part process distributions when more than one part is being considered in an assembly).
3. The additive law of variances as a means to determine the relationship between the variability in individual parts and that for the assembly.

To assume that the parts can be represented by a statistical distribution of measurements (and for the assumption to hold in reality), the processes must be in a state of statistical control.

Importance of Statistical Control to Tolerance Fulfillment

To illustrate the importance of statistical process control in achieving design intent in a tolerancing problem, we will consider the following example. Figure 11.24 shows drawings for two simple parts, a plate with a hole and a pin that will ultimately be assembled to a third part but must pass through the hole in the plate. For the assembly, it is necessary for its function to be served that the clearance between the plate hole and the pin be at least 0.015 inch but no more than 0.055 inch.

To achieve the design requirement stated above, the nominal values and tolerances for the plate hole and pin were statistically derived and are given in Fig. 11.25. In arriving at these tolerances, it was assumed that:

1. The parts would be manufactured by processes that behaved according to the normal distribution, with the same process variation.

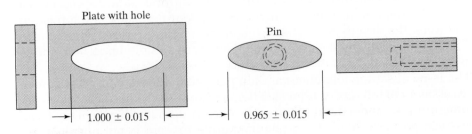

Figure 11.24   Two parts in an assembly

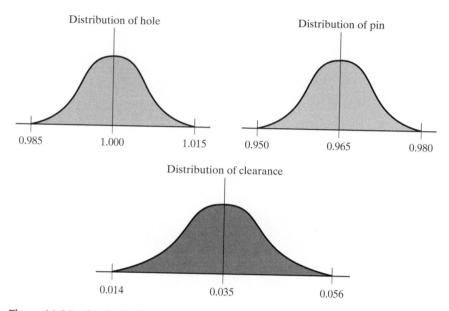

Figure 11.25   Statistical basis for satisfying design intent for the hole/pin clearance

2. The process capabilities would be at least ±3 sigma, the processes would be centered at the nominal values given in Fig. 11.24, *and the processes would be maintained in a state of statistical control.*

3. Random assembly would prevail.

If these assumptions are met, the processes for the two parts, and hence the clearance associated with assembled parts, would be as shown in Fig. 11.25, and therefore design intent would be met at the ±3 sigma level.

Let us suppose now that despite the assumptions made and the tolerances thus derived, the processes used to manufacture the pin and plate hole were not maintained in good statistical control. As a result, when examined, the parts actually more nearly follow a uniform/rectangular distribution within the specifications, as shown in Fig. 11.26. Such could have arisen as a result of sorting or rework of a more variable process(es), in which case the results are doubly distressing—poor-fitting assemblies and increased cost to the system due to sorting and rework.

Figure 11.26 shows the distribution of the clearance if the hole and pin dimensions follow the uniform distribution within the specifications. The additive law of variances has been used to derive the variation in the clearance distribution, but this time assuming the uniform distribution for the individual part processes. The distribution of the clearance will not be uniform, but it will not be near-normal either. It is important to note that now many assemblies will fit quite tightly and may later bind if foreign matter, for example,

Figure 11.26 Clearance implications of poor process control of plate hole and pin dimension

gets into the gap, while others will fit together with much larger clearance than desired. The overall range within which we expect the assemblies to lie is now 0.060 with more uniform probability of occurrence.

The problem here is not a design problem. The plate hole and pin tolerances have been derived using sound statistical methods. However, if the processes are not in good statistical control and hence are not capable of meeting the assumptions made during design, poor-quality assemblies will follow. It should be noted that the altogether too common process appearance of a uniform distribution of measurements within the specifications can arise in several different ways:

1. It can arise from processes that have good potential, in terms of variation, but are not kept in good statistical control.

2. It can arise from unstable and/or large variation processes which require sorting/rework to meet the specifications.

3. It can arise from processes that are allowed purposely to vary over the full range of the specifications to take advantage locally of wide specifications relative to the process variation.

In all three cases mentioned above, additional costs will be incurred and product quality will be eroded. This was the message of the final section in Chapter 2. Clearly, statistical process control is crucial to the issue of setting tolerances in engineering design. Taguchi's loss function model, which is an essential element in tolerance design, assumes similarly that quality characteristics can be represented by a statistical distribution of measurements. Again, for this assumption to be met at the process and hence in the ultimate product in the field, the manufacturing processes must be maintained in a state of statistical control.

▮ EXERCISES

11.1 Since the mid-1980s, two of the most widely adopted quality improvement concepts/programs have been C_{pk} and Six Sigma. Both are implemented with an objective to better conform to a given tolerance specification, the validity or even the appropriateness of which are seldom questioned/challenged in practice. Propose explanation(s) as to why this practice has been followed. Use whatever resources are available to you to answer the question, including your personal experience, quality improvement practitioners, engineers and managers in industry, and professors in engineering and management.

11.2 Answer each of the following questions with an explanation and/or illustration.
  (a) What are the major difficulties in actually applying the loss function concept in practice?
  (b) What is the most fundamental flaw in the conventional definition of Six Sigma quality?
  (c) What are some of the consequences likely to result from the implementation of the conventional Six Sigma program?
  (d) What are some reasons for, or benefits of, a nominal value and tolerance limits being designed/selected simultaneously, instead of sequentially?
  (e) What are some concerns you may have about the idea of tolerance minimization?

11.3 Suppose the data on individual pipe lengths of Exercise 3.15 cannot be adequately described by a normal distribution, as assumed in the analysis there. Actually, the individual pipe-length data behave more closely according to a uniform distribution. The mean and standard deviation of the pipe lengths, X, have been estimated to be 0.973 meter and 0.003 meter, respectively. Under an assumed normal distribution, it has also determined that about 37% of the pipes are expected to fall outside the specifications.
  (a) What percent of the pipes would be expected to fall outside the specification limits of 0.965 ± 0.007 meter, if indeed the pipe lengths are distributed as a uniform distribution with a mean of 0.973 meter? [Recall that the variance of a uniform random variable is equal to $(b - a)^2/12$, where a and b are the limits of the range of the variable.]
  (b) What would be the effect on percent of pipes that fall outside the specification if the same uniform distribution were centered at 0.965 meter?
  (c) If the process is centered at 0.965 meter and assumed to follow a normal distribution, calculations show that about 2% of the pipes are expected to fall outside the specification limits. Observe any

differences between this result and the one you obtained in part (b). Based on your observations, which distribution would you like the process to follow? Why?

11.4 Since the process that produces the pipe lengths of Exercise 11.3 can be recentered at the design nominal of 0.965 meter, the improvement effort now shifts to variance reduction. As a goal for the variance-reduction study, determine a target value for the standard deviation in accordance with the definition of Six Sigma quality for a reduced tolerance interval of 0.965 ± 0.005 meter. Suppose the process is presently running successfully at the targeted mean and standard deviation.

(a) If, later on, the process becomes off-centered at a mean of 0.96625 meter, estimate the percent of lengths that now conform to the new tolerance, assuming the standard deviation remains unchanged.

(b) If, later on, the process remains centered at the nominal value, but the standard deviation increases by 25%, estimate the percent conforming.

(c) If, later on, the process is off-center at a mean of 0.96625 with a standard deviation increased by 25% from the target, estimate the percent conforming to the new tolerance.

(d) Comment on the relative effects on conformance of a process in control but with constant shifts in (1) the mean, (2) the variance, and (3) both the mean and variance.

11.5 Applying the method of loss functions, calculate the expected loss of the newly improved pipe production process of Exercise 11.4, using a loss coefficient of $k = \$0.75/\text{meter}^2$. Compare this with the expected loss of each of the three process situations of parts (a) through (c) in Exercise 11.4. Rank these three process situations in terms of the economics of their expected quality loss.

11.6 Suppose that the process mean of Exercise 11.3 is allowed to vary randomly between 0.96375 and 0.96625 while the new targeted standard deviation remains unchanged.

(a) Estimate the percent of pipes conforming to the specification of 0.965 ± 0.005 meter. [*Hint*: The resultant distribution is a composite of a uniform and two half-normal distributions (see Section 11.3) that can be approximated by considering only the uniform distribution in the center for the purpose of conformance estimation.]

(b) Approximate the expected loss per unit for the portion of the process that falls inside the tolerance (0.96375, 0.96625) using a loss coefficient of $k = \$0.75/\text{meter}^2$.

11.7 Based on your analyses in Exercises 11.5 and 11.6, do the following:

(a) Discuss the expected effectiveness in an actual application of the loss function approach and the Six Sigma program for continuous improvement.

(b) Do you think a Six Sigma program is better than the use of loss function in quality improvement? Why?

(c) Do you think Six Sigma and loss function are mutually exclusive, or can they be applied together in quality improvement? Why?

11.8 In practical applications of either Six Sigma or the loss function for quality improvement, as you may have noticed, neither method makes any direct reference to the idea of statistical process control.

(a) Do you think SPC is important for either or both methods in their applications? In what way?

(b) Comment on the assumptions in terms of their practicality that are made or implied in each of these two methods for quality improvement.

11.9 A die casting process is being used to fabricate a cast housing for a major client. The part weight of these housings is a critical quality concern that must satisfy the customer-specified tolerance of 770 ± 25 grams. A random sample of 40 housings has been taken from the process that has been running in control. The table below displays the measured weights of the sampled housings.

775	773	778	774	775	775	772	780	776	772
774	778	778	781	772	776	774	774	775	775
784	781	781	773	774	777	775	780	776	774
778	775	777	780	772	773	773	777	774	775

For each housing that weighs 745 grams and is put into field use, the customer will expect a loss of $250. If a part is rejected at the plant because it is either too light or too heavy to meet the specification, it is scrapped at a cost of $15.

(a) Estimate the probability that a housing produced by this process will weigh within the specified tolerance interval, assuming that the part weights follow a normal distribution with mean and standard deviation estimated from the sample data.

(b) Find the expected total quality cost per part produced under the current operating conditions and that meet the customer's requirement. (Assume a normal distribution and a quadratic loss function centered at the specified nominal.)

11.10 Referring to the casting process of Exercise 11.9, do you find the customer's current tolerance specification optimum in terms of expected total unit quality cost? If yes, provide a basis for your answer. If no, recommend a different tolerance specification that is based on a minimum expected quality cost.

11.11 In pursuit of continuous improvement, you are asked to propose a quality project to further reduce the expected total unit cost of the casting process in Exercise 11.9. There are in general three possible strategies:

(1) centering the process, (2) reducing process variability, and (3) centering the process and reducing its variability. Obviously, project cost and time requirements can be quite different among the three strategies. It seems logical to first estimate the expected return, such as the expected savings in total quality cost, of each approach to aid in the selection of one of the three strategies. Based on past experience with a similar casting process, it is thought feasible to reduce the standard deviation of this new process to about 10 grams. Make an estimate of the expected total quality cost per unit for each of the three strategies, based on the available data and assumptions given in Exercise 11.9.

11.12 Referring to Exercise 11.9, suppose your client's real concern with respect to housing weight is only when the weight is too low. In other words, heavy castings are not considered to be a problem. A two-sided tolerance has been specified for the current process only because of tradition or convention. Mathematically, one can set the upper tolerance limit as large as infinity to apply the general formula, Eq. (11.13), for the calculation of $E(C_T)$. Determine the lower one-sided tolerance limit based on an approximate minimum expected total quality cost by setting $\mu_x = m$ and $z_u = \infty$ in Eq. (11.13).

11.13 Using a vapor deposition technique, a new process for coating a high-speed steel tool with titanium nitride (TiN) as a refractory compound is being investigated. The design calls for the average TiN thickness to be 1.5 microns. Engineers are concerned with the possibility that the process may produce tool coatings that are too thin, say, less than 1 micron. After a prototype process has been run for some time with evidence of being in statistical control, a random sample of 60 tools coated on this process has been collected and measured. The measurements are listed in the table below.

1.518	1.459	1.259	1.291	1.474	1.399	1.399	1.646	1.413	1.528
1.361	1.263	1.451	1.435	1.550	1.434	1.241	1.527	1.466	1.529
1.535	1.402	1.575	1.278	1.515	1.547	1.518	1.170	1.391	1.596
1.485	1.341	1.504	1.452	1.484	1.458	1.587	1.535	1.673	1.488
1.366	1.520	1.432	1.582	1.191	1.500	1.577	1.505	1.349	1.318
1.330	1.539	1.432	1.382	1.522	1.459	1.559	1.028	1.458	1.527

(a) Plot a histogram of the coating thickness data, observe its likely distribution, and calculate the sample mean and sample variance to be used as estimates of their respective true mean and true variance. Do you think a normal distribution would describe the data well?

(b) Calculate an approximate expected quality loss per tool according to the loss function, $L(x) = k/x^2$, where the loss coefficient, k, has been estimated to be $4.80 micron^2. Also, make a sketch of the loss function to compare its shape with a quadratic loss function using the same loss coefficient value.

(c) Would you consider analyzing this problem as one with a two-sided tolerance using a corresponding quadratic loss function developed in (b) above? Justify your answer.

11.14 (a) Applying the usual method for C_{pk} estimation, calculate the C_{pk} for the coating process of Exercise 11.13. Comment on the validity of this capability measure in general as it is widely used in practice.

(b) Suppose you have been assigned to validate a C_{pk} value submitted by a supplier. To ensure a thorough review, prepare a list of questions to ask, pertinent data to review, and methods to show how the C_{pk} was determined.

(c) If a Six Sigma program is in place at the supplier's company, what is some of the most important information you would want to have provided by the supplier?

(d) If you are asked to determine a C_{pk} for a process in your own company, describe the procedure that you will follow in order to produce a valid C_{pk}.

11.15 The engineers at the lab have completed a pilot run of a new design of a light fixture to be produced on an injection molding press. Their analysis of the part shrinkage data associated with the new design shows an average of 1.3% and a standard deviation of 0.33%. Accordingly, they have concluded that the new fixture satisfies very well the tolerance specification, which states that no fixture should shrink more than 2.5%. In fact, they have also estimated the rejection rate for the new process at about 150 per million. You are requested to review the engineers' analyses and conclusions. The test data used by the engineers is shown in the table below.

1.18	1.20	0.96	1.73	1.17	1.33	1.19	0.86	1.22	1.05
1.36	2.11	1.45	1.03	0.97	1.11	1.31	1.87	1.49	1.12
1.09	1.62	0.99	1.32	1.00	0.87	2.43	1.47	1.18	1.30
0.93	1.44	1.06	1.28	1.61	1.08	1.15	1.26	1.78	1.33
1.26	0.95	1.82	1.30						

Suggest a method of analysis if you find that you do not agree with the assumptions, the method of analysis, and/or the results.

11.16 In your review of the report on shrinkage associated with the new light-fixture design in Exercise 11.15, you may have plotted a histogram of the data that shows a strong skewness to the right. (Create such a plot if you did not do it in your review.) It seems, therefore, highly unlikely that the engineers' test data can be approximated adequately by a normal distribution. Since the methods we have presented so far are based on an assumed normal distribution, any results and conclusions derived from these methods are invalid theoretically and may be misleading if the data distribution is not normal. A relatively simple approach to resolve this situation is to transform the original data with the hope that the transformed data will follow a normal distribution at least approximately. For positively skewed distributions, one commonly employed method is to take a log transformation of the data, e.g., $y_i = \ln x_i$, $i = 1, \ldots, n$, where x_i is the shrinkage associated with the ith fixture and y_i is the natural log of x_i. To decide on whether a normal distribution may be applied to describe the y's, use the usual methods such as histograms and normal probability plots of the transformed data. If a normal distribution appears to fit adequately, then all the needed analysis can be carried out on the y's. After the analysis is complete, take antilog transformations of the results and/or conclusions to return to the original units/dimension. Carry out an analysis of $\ln x_i$ for your review of the fixture shrinkage study of Exercise 11.15.

11.17 The amount of a chemical ingredient, x, in the preparation of a quenching fluid, was found to have an effect on the hardness, y, of a heat treated steel plate. Based on the results of a series of tests, the following prediction equation was found to fit the data well:

$$\hat{y} = 634.5 + 33.3x - 1.92x^2.$$

The experimental range for the ingredient in all the tests was 2 to 20 grams. The tests produced hardness readings that varied substantially, but all the hardnesses remained within the acceptable range. It is critical to have a uniform hardness between plates, since four plates will be stacked together to form an assembly. Given the fact that the ingredient influences the hardness, it is important to learn about how the actual amount of the ingredient X varies in quenching fluid preparation. For this analysis, three nominal x values within the experimental range were selected $(x_1 = 3, x_2 = 11,$ and $x_3 = 19$ grams). For each nominal value, five quenching-fluid mixtures were prepared, with the actual amount of the ingredient, X, varying somewhat from mix to mix. A sample standard deviation of the actual X content in the fluid was calculated for each nominal x value, and these three values were found not to differ significantly from one another. Accordingly, they were pooled to better estimate the common population standard deviation (σ_X). This estimate turned out to be $\hat{\sigma}_X = 0.23$ gram.

(a) Plot the prediction equation above and graphically depict a distribution of X corresponding to each of the nominal X values.

(b) Display an approximate distribution of y for each of the nominal x values.

(c) Examine these distributions of y to select a nominal x value that produces the smallest variation in y.

11.18 In Section 11.2, some discussions were presented on two points related to the practice of tolerancing: tolerance maximization vs. minimization and tolerance design. State your opinion on these two topics from either or both conceptual and practical perspectives. Also, critique the idea of economical design of tolerances based on the criterion of a loss function.

11.19 Four steel plates, as discussed in Exercise 11.17, are to be stacked together in an assembly. The height of the stack is the sum of the thicknesses of the individual plates, each of which has the same nominal value. Because each plate thickness is a random variable, the height of a stack of four plates is a random variable, too. There is a specified tolerance of ± 0.004 inch for the stack. The inner two plates must have less variation in thickness for better functioning during field use. Suppose the inner plates should have a standard deviation in thickness that is 50% of that of the outer plates. Design for a C_{pk} of 1.333 for both individual plates and the stack height. Determine the required maximum standard deviation for the outer plates and for the inner plates.

11.20 An assembly, composed of three parts, has a total nominal length of 1.500 inches with a tolerance of ± 0.050 inch. The three parts are essentially identical in terms of the processes manufacturing each. The processes making the parts (and the assembly) should have a C_p value of at least 1.5.

(a) What should be the standard deviation for the processes making the individual parts to meet the capability criterion above, and what would the individual tolerances be for these parts?

(b) If the tolerances had been determined incorrectly by simple division of the assembly tolerance, what must the individual process values for σ_X be to meet the $C_p = 1.5$ requirement?

(c) If the processes for making the individual parts were such that the average was at 0.5002 inch and the standard deviation was 0.0094 inch, what percent would have been incorrectly rejected if simple division had been used to set the tolerances?

11.21 A hollow cylinder is made up of two semicylindrical parts as shown in the figure. The assembly is held together by an epoxy layer that must be between 0.002 and 0.005 inch in thickness. The two cylindrical halves are made from the same extrusion. If the outer diameter of the assembly is to be 2.000 inches with tolerances of ±0.010 inch, and a ±5 sigma assembly is required, what would be the required nominal value and tolerance for the dimension *A* of the individual parts?

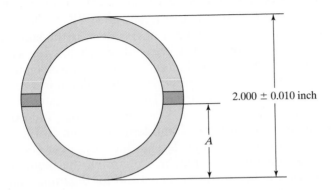

2.000 ± 0.010 inch

A

11.22 An assembly is being produced, as shown in the figure, that is comprised of two aluminum plates with a hard rubber insulation in between. The aluminum plates have a nominal thickness value of 0.09375 inch and the insulation is to be 0.0625 inch thick. Because of the resilience of the hard rubber material, it is deemed necessary that the standard deviation of the process making this material be one-half that of the process making the aluminum. If the overall assembly has specification of 0.2500 ± 0.0040 inch, what are the required specifications on the individual parts? Assume a ±4 sigma assembly.

0.09375 inch

0.0625 inch

0.09375 inch

11.23 A woodworking company is shipping flat, plain doors to another firm, which will be using them for cabinets. The thickness of these doors is 0.75 inch with a tolerance of ±0.03 inch. When the doors are shipped they are put in crates, with each crate holding 10 doors. A piece of cardboard is placed on either side of each door. The cardboard is 0.050 inch thick, with a tolerance of ±0.005 inch. Statistical charting has shown that the process for making the doors is ±4 sigma capable, and the process for making the cardboard is ±2 sigma capable. The shipper wants to be 99.5% certain that 10 doors will fit into each crate, with the assembly at or below the cover level. What should the minimum depth of the crates be?

12

Statistical Thinking for Process Study: A Case Study

12.1 Introduction

The concept of bringing a process into a state of statistical control and then maintaining that state of control is often difficult to appreciate completely because of the tremendous emphasis we have historically placed on the engineering specification as a measure of quality. As a result, the primary focus of production is to maintain the existence of a process that is producing parts that conform to specifications without a great deal of attention given to the notion of achieving and maintaining a stable, consistently performing process. Central to the product control mentality is the notion that the engineering specification is an appropriate measure of quality. If it is, then of course we have put quality on an attribute basis, categorizing finished parts as either good or bad, defective or nondefective, which makes little sense from a functional point of view.

It is important to recognize that the specification is truly an inhibitor to the promotion of the concepts and methods of statistical control of processes. Until rejected as a measure of quality, the specification will continue to stand between us and our ability to make full use of and receive the broad-based benefits of statistical process control.

To convince yourself that the problem posed above is a real one, try the following: Ask someone who is aware of but not familiar with the concept of SPC this simple question: "If you had a process that was making parts which were consistently meeting the design specifications, would you think it necessary or even advantageous to chart this process and bring it into a state of statistical control?" Chances are you may quickly get the response, "No." This alone should be sufficient to convince you that there is a real problem with fully embracing SPC when the concept of the engineering specification is tied so closely with our apparent understanding of the meaning of quality.

Much of what we have discussed in the first eleven chapters has been directed toward overcoming the misconceptions cited above. We have been introduced to the concept of quality as loss due to functional variation and have come to recognize the poor correlation that actually exists, from the customer's point of view, between quality and conformance to the engineering specification, particularly when the specification is misused by manufacturing. This has helped us to see why the pursuit and maintenance of a process in good statistical control is important and how this can affect our ability to identify opportunities for the continual reduction of variability. Furthermore, Shewhart and later Deming have so clearly pointed out to us the important relationship between variation reduction and increase in productivity—variation as an indication of the presence of waste and inefficiency. We have seen that there are still other forces driving the importance of statistical process control. We reexamine all these issues in this chapter as we review in detail an actual case study of the use of statistical thinking and methods for quality and productivity design and improvement.

12.2   Overview of the Case Study

This case study chronicles the activities of a company that, over a period of months, learned and applied the techniques of statistical process control to one of its key processing operations. The study clearly demonstrates the power of and the broad-based benefits to be derived from the use of statistical thinking and methods for quality and productivity design and improvement. The story is real and the results are dramatic. This case study will call on virtually all the concepts and methods of data analysis and interpretation that we have discussed in the first eleven chapters. It points out many of the misconceptions that we have examined as well.

The Dilemma.   A certain manufacturer of gaskets for automotive and off-road vehicle applications awoke one morning to find that a major customer had significantly tightened the bilateral specification on the overall thickness of a hard gasket used in automotive engines. Although the current specification was by and large being met, the product did not come close to meeting the new specification. The first reaction of the gasket manufacturer was to negotiate with the customer to obtain relaxation of the new specification. When efforts in this regard did not bear fruit, the relationship became somewhat strained. At first the gasket manufacturer thought that if he waited long enough, the automotive company would eventually be forced to loosen the requirements and purchase the existing product. However, as time went on it became clear that some positive steps would have to be taken to improve the quality. But what should be done? And by whom?

Figure 12.1   The product: a hard gasket of three-piece construction for automotive applications

The Product.   Figure 12.1 shows the product in question, a hard gasket. A hard gasket is a three-piece construction comprised of two outer layers of soft gasket material and an inner layer consisting of a perforated piece of sheet metal. These three pieces are assembled, and some blanking and punching operations follow, after which metal rings are installed around the inside of the cylinder bore clearance holes and the entire outside periphery of the gasket. The quality characteristic of interest in this case study is the assembly thickness.

The Process.   An initial study by the staff engineers revealed that the variation of the thickness of soft gasket material—the two outer layers of the hard gasket—was large and undoubtedly responsible for a large portion of the total variability in the final product. Figure 12.2 shows the roll mill process, which fabricates the sheets of soft gasket material from which the two outer layers of the hard gasket are made. To manufacture a sheet of soft gasket material, an operator adds raw material, in a soft pellet-like form, to the gap, called the knip, between the two rolls. The larger roll rotates about its axis with no lateral movement, while the smaller roll rotates and moves back and forth laterally to change the size of the roll gap. As the operator adds more and more material to the gap, the sheet, with the application of a starter material, is formed around the larger roll. When the smaller roll reaches a preset destination (final gap/sheet thickness) a bell rings and a red light goes on, telling the operator to stop adding raw material. The operator stops the rolls and cuts the sheet horizontally along the larger roll so that it may be pulled off the roll. The finished sheet, called a pull, is pulled onto a table where the operator checks its thickness with a micrometer. As we will see later, the operator can adjust the final roll gap if he or she feels that the sheets are coming out too thick or too thin relative to the prescribed nominal value.

Figure 12.2   The process: roll mill for the manufacture of soft gasket material

12.3   Product Versus Process Control: Management Strategy

While the staff engineers were studying the problem to formulate an appropriate action plan, something had to be done to make the necessary production requirements for hard gaskets within the specifications. A management decision was made to intensify product inspection and, in particular, to grade each piece of material according to thickness so that the wide variation in thickness could be balanced out at the assembly process. Figure 12.3 shows the selective assembly process that was put in place to provide a solution to the problem. Extra inspectors were used to grade each piece of soft gasket material. The sheets of the same thickness were shipped in separate bundles on pallets to a sister plant for assembly. Thick and thin sheets were selected to make a hard gasket to balance out the highs and the lows. The process worked pretty well and there was some discussion about making it permanent. However, some felt it was too costly and did not get at the root cause of the problem.

Meanwhile, the staff engineers in the company were continuing to study the problem more carefully and concluded that the existing roll mill process

Figure 12.3   Selective assembly process used to assemble hard gaskets that meet the specifications

equipment for making the soft gasket sheets simply was not capable of meeting the new specifications. This conclusion was reached as a result of the examination of production data and scrap logs over the past several months. They had researched some new equipment that had a track record for very good sheet-to-sheet consistency and had decided to write a project to replace the existing roll mills with this new equipment. To help the approval process their boss asked them to get some data on the poor process capability of the existing equipment. They therefore set out to conduct a "statistical process capability study." (The reader is encouraged to carefully scrutinize the decisions/actions of the engineers—perhaps some mistakes were made.)

Initial Process Capability Study

The engineers, confident that the equipment was not capable, selected what they thought was the best operator and the best roll mill (the plant has several roll mill lines) and took careful measurements of the thickness of each sheet made on an 8-hour shift. During that particular shift a total of 72 sheets/pulls were made. This was considered quite acceptable, since the work standard for the process is 70 sheets per shift. The measurements of the sheet thickness (in the order of manufacture) for the 72 sheets made on that shift are given in Table 12.1. The engineers set out to use these data to conduct a statistical process capability study.

Calling upon a statistical methods course that one of the engineers had in college, the group decided to construct a statistical distribution from the data and use it to estimate the percentage of the measurements/process that fell

TABLE 12.1 | Measurements of Sheet Thickness

Sheet	Thickness (in.)	Sheet	Thickness (in.)	Sheet	Thickness (in.)	Sheet	Thickness (in.)
1	0.0440	19	0.0472	37	0.0459	55	0.0425
2	0.0446	20	0.0477	38	0.0468	56	0.0442
3	0.0437	21	0.0452	39	0.0452	57	0.0432
4	0.0438	22	0.0457	40	0.0456	58	0.0429
5	0.0425	23	0.0459	41	0.0471	59	0.0447
6	0.0443	24	0.0472	42	0.0450	60	0.0450
7	0.0453	25	0.0464	43	0.0472	61	0.0443
8	0.0428	26	0.0457	44	0.0465	62	0.0441
9	0.0433	27	0.0447	45	0.0461	63	0.0450
10	0.0451	28	0.0451	46	0.0462	64	0.0443
11	0.0441	29	0.0447	47	0.0463	65	0.0423
12	0.0434	30	0.0457	48	0.0471	66	0.0447
13	0.0459	31	0.0456	49	0.0427	67	0.0429
14	0.0466	32	0.0455	50	0.0437	68	0.0427
15	0.0476	33	0.0445	51	0.0445	69	0.0464
16	0.0449	34	0.0448	52	0.0431	70	0.0448
17	0.0471	35	0.0423	53	0.0448	71	0.0451
18	0.0451	36	0.0442	54	0.0429	72	0.0428

within the specifications. The engineers first estimated the mean and standard deviation of the 72 sheet thicknesses, using the arithmetic average and the sample standard deviation as defined in Chapter 3. These calculations are

$$\overline{X} = \frac{\Sigma X_i}{72}$$

$$= 0.0449,$$

$$s_X = \left[\frac{\Sigma(X_i - \overline{X})^2}{72 - 1}\right]^{1/2}$$

$$= 0.0014.$$

At this point the engineers usually would simply assume that the normal/bell-shaped curve would adequately represent the frequency distribution of the data. But to be more rigorous about their analysis, they decided to construct a frequency histogram of the data to verify this. The following procedure, as outlined in Chapter 3, was followed to construct the frequency histogram:

1. The range of the data was determined by subtracting the smallest value from the largest (i.e., $0.0477 - 0.0423 = 0.0054$).

2. It was decided to use 12 cells of width 0.0005, with midpoints at 0.0XX5 and 0.0XX0.

3. Therefore, the boundaries were chosen such that the first cell went from 0.04225 to 0.04275, the second from 0.04275 to 0.04325, and so on. This ensured that each thickness measurement would fall within a specific cell.

4. Next, a tally sheet of the points was made:

Cell Midpoint	Frequency	Total
0.0425	11111 1	6
0.0430	11111 11	7
0.0435	1111	4
0.0440	11111 1	6
0.0445	11111 11111	10
0.0450	11111 11111 111	13
0.0455	11111 11	7
0.0460	11111	5
0.0465	11111	5
0.0470	11111 11	7
0.0475	11	2

5. Finally, the histogram was plotted, as shown in Fig. 12.4. Also shown in the figure are the product nominal value and the upper and lower specification values. The dark shaded part of the histogram represents the amount of the product that lies outside the specifications.

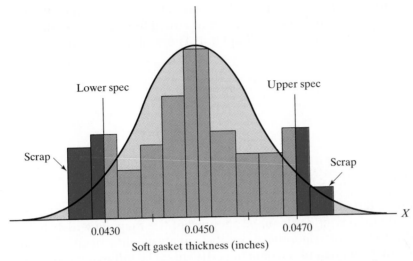

Figure 12.4   **Histogram of data from initial process capability study**

It is apparent from Fig. 12.4 that a large proportion of the process does not meet the individual sheet specification. To quantify the precise nature of the problem, the engineers assumed that the data could be represented by the normal distribution curve as seen overlaid in Fig. 12.4. Despite the somewhat irregular appearance of the frequency histogram when compared to the normal distribution, they contended that the normal distribution was a reasonably good approximation for the process and proceeded to determine the area under the curve that fell inside the specifications. Figure 12.5 shows the process, as represented by the normal distribution, with the product specifications overlaid on the model.

To determine the process capability in terms of percent conforming to the specification, the following calculations were made:

$$P(\text{a part within specs}) = P(0.04300 \le X \le 0.0470).$$

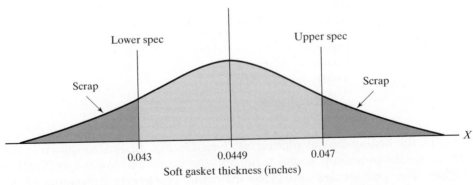

Figure 12.5   **Statistical model for process data for capability evaluation**

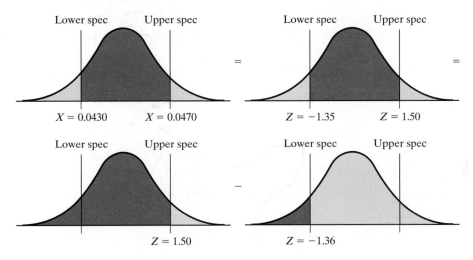

Figure 12.6   Graphical representation of the Z transformation and probability calculation

If we use the standard normal or Z-distribution, the required probabilities can be determined as shown in Chapter 3 and as shown graphically in Fig. 12.6 with σ_X estimated by s_X to be 0.0014 inch, and μ_X estimated by $\overline{X}$ to be 0.0449 inch:

$$Z = \frac{X - \mu_X}{\sigma_X},$$

$$Z_{\text{LSL}} = \frac{0.0430 - 0.0449}{0.0014}$$

$$= -1.36,$$

$$Z_{\text{USL}} = \frac{0.0470 - 0.0449}{0.0014}$$

$$= 1.50.$$

The Z table in the Appendix (Table A.1) can then be used to find the following:

$$
\begin{aligned}
P(\text{a part within specs}) &= P(Z \le Z_{\text{USL}}) - P(Z \le Z_{\text{LSL}})] \\
&= P(Z \le 1.50) - P(Z \le -1.36)] \\
&= 0.9332 - 0.0869 \\
&= 0.8463.
\end{aligned}
$$

Therefore, the process capability in terms of percent conforming to the specifications is estimated to be 85%. Such a capability is clearly unacceptable.

In terms of the process capability index C_p, the process capability was found to be $C_p = 0.477$, which by all measures is totally unacceptable.

$$C_p = \frac{\text{USL}_X - \text{LSL}_X}{6\sigma_X}$$

$$= \frac{0.0470 - 0.0430}{0.0084}$$

$$= 0.477.$$

Calculations were also performed to find the value of the index C_{pk}:

$$Z_{\text{USL}} = \frac{\text{USL} - \overline{X}}{\sigma_X} \qquad\qquad Z_{\text{LSL}} = \frac{\text{LSL} - \overline{X}}{\sigma_X}$$

$$= \frac{0.0470 - 0.0449}{0.0014} \qquad\qquad = \frac{0.0430 - 0.0449}{0.0014}$$

$$= 1.5, \qquad\qquad\qquad = -1.36,$$

$$Z_{\min} = \min[Z_{\text{USL}}, -Z_{\text{LSL}}]$$

$$= \min[1.5, 1.36]$$

$$= 1.36,$$

$$C_{pk} = \frac{Z_{\min}}{3}$$

$$= \frac{1.36}{3}$$

$$= 0.45.$$

This value for C_{pk} is also far from the acceptable value of at least 1.00. Clearly, the results of this "process capability study" show that by any measure the process capability is very poor indeed. This was precisely the answer the engineers were looking for. Armed with their statistical process capability study results, the engineers went to their boss to seek the go-ahead to finalize their report, which would then be presented to the division top management to obtain approval to procure the new roll mill equipment. Their boss was pleased with their work and asked them to prepare the project documentation and presentation materials.

Review of Progress to Date

It is worthwhile to pause for a moment and reflect on what has transpired to this point. In particular, we may wish to examine some fundamental problems with the way this situation has been handled.

1. Product Control Way of Thinking

It is important to note that in all regards the actions of management and the engineering staff have been dictated by the product control way of thinking. Management devised a procedure to attempt to work with the existing level of quality by screening, sorting, and selective assembly. This strategy is consistent with the "detect and contain" approach to quality control. Engineering focused on the notion of process capability purely from the "conformance to specifications" point of view. Their approach left them with a yes/no answer to the question, "Is the process capable?" but with little direction in terms of where to look for improvement other than the drastic action of replacing the existing equipment. At no time to this point was there any attempt to consider what improvements might be made to the existing process to make it more capable.

2. Tunnel Vision of the Engineers

Initially, the engineers rather hastily concluded that the major source of variation in gasket sheet thickness was the equipment—the roll mill. This is only one of several possible sources of variation, which could include the raw materials (properties and preparation), the methods (procedures for set-up, machine operation, etc.), the machine condition (operating settings, maintenance condition), and the operator (training, supervision, technique). With little concern for these issues and no initial data to provide direction, the engineers quickly focused on inherent equipment capability as the problem. It is often the case that we focus quickly on the equipment and propose solutions that call for the purchase of new machines. Deming has, of course, provided us with some insight in this regard. He has cautioned us that *there is too much talk of the need for new machinery and automation—most people have not learned to use what they have.*

3. Selective Assembly Process

The mere thought that such an approach might be an acceptable solution to the problem fails to recognize the fundamental fallacies of product control as a way of doing business. Such an approach can never cause both quality and productivity to move together in the desired direction and therefore will not enhance competitive position. Such can only be accomplished through process control—attacking the process to find the root causes of the problem.

4. Statistical Control/Process Capability Relationship

The engineers have treated the notion of process capability in a classical but totally inappropriate fashion. When they took the data from an entire shift and assumed that in a frequency sense it could be represented by a statistical distribution—in this case the normal distribution—they did not realize that the implicit assumption being made was that the process was in a state of statistical control. This must be the case if the process is to be represented over time by a single statistical distribution, which has a fixed mean and a constant level of variation. To estimate the mean, standard deviation, assume the

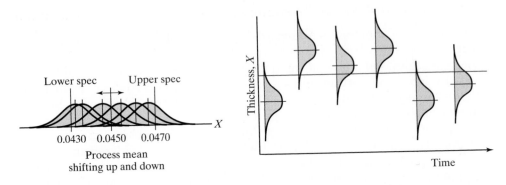

Figure 12.7   **Time behavior of a process that is changing mean level in a sporadic fashion**

normal distribution, and then use this model for analytical purposes, the assumption we must make is that the data have arisen from a constant system of common causes. The engineers, however, did not examine the process data over time and therefore could neither confirm nor deny the process condition in terms of statistical control.

Figure 12.7 illustrates the appearance of a process, represented as a statistical distribution of measurements, when its mean level is changing over time. The right half of the figure shows the time-oriented progression of the process behavior and illustrates the inability of the process to operate in a stable fashion in terms of its mean value. The left part of the figure shows what this means in terms of the process appearance over time when represented by a statistical distribution of measurements relative to the specifications. It is clear from Fig. 12.7 that no single process (distribution) exists over this period of time—rather, many different processes are present. Which one should we use to assess the process capability? Hopefully, the inadvisability of "pushing" all of these processes together by compressing Fig. 12.7 along the time dimension and thereby creating the illusion of the existence of a single, stable process is clear. Furthermore, not only may the results of the capability study by the engineers be totally misleading, but also, their analysis has little diagnostic value—the process is incapable, but why? What are the major sources of variation contributing to the lack of capability relative to the specifications? If a process capability study is to be a meaningful and valid representation of the ability of the process to manufacture product consistently at a certain level of conformance, it must be preceded by statistical charting which confirms that the process is in a state of statistical control.

12.4   Finding Improvement Opportunities Through Charting

A Different Approach to the Problem

While the study of the staff engineers was taking place, the people in the plant were being introduced to Deming's concepts and the philosophies and methods of statistical charting for fault diagnosis and process improvement.

The plant manager decided to form a team to study the problem. In addition to himself, he included the production manager, the quality manager, a process engineer, a product engineer, a production supervisor, and an operator. He arranged for a consultant to meet with the group about once every two weeks to review their progress and make suggestions.

The group began to meet frequently to plan a course of action. As a first step they made a detailed process flowchart, carefully defining the complete sequence of processing steps from raw material handling, the blending of the gasket material for rolling, to the roll mill operation, and on to the cutting/sizing operation, and finally, the blanking operation. At each step along the way they discussed the possible factors that could be contributing to the variation in the final product.

After studying and discussing the complete process flow they concluded that raw material blending and the roll mill process were the two most likely areas to seek and find improvement opportunities. Since the raw material blending operation involves several steps and is somewhat complicated, they decided to focus their attention first on the roll mill process. It was felt that this process was simpler and easier to obtain data from and was therefore more manageable for their first attempt at tackling a process problem using statistical thinking and methods.

The next step for the group was to begin to do statistical charting of the roll mill process using $\overline{X}$ and R control charts. In employing this technique, they hoped to learn more about the nature of the variability pattern of the process data, so that they could identify the root causes of some of the major variation sources. Actions could then be formulated to remove these sources of variability, thereby making the process operate more consistently.

Method of Process Operation

Before we proceed to examine the first set of statistical charts that the group developed, it is important that we understand a bit more about how the process is actually operated. Such information is essential to the fault diagnosis process. Statistical process control chart interpretation should be thought of as a two-step process. First, we read the statistical signals to determine what they are telling us about the behavior of the process mean and process variability. Then we use all available data about the operating conditions of the process; the events that may have taken place at the process during the time the data were collected; and our general knowledge about, and our experience with, the process to determine what is causing the process mean and/or variation to behave in the manner that has been detected. It is our ability to embrace this second step that determines whether we are simply making charts to hang on the wall or whether we are using charts to find improvement opportunities.

Investigation revealed that the operator runs the process in the following way. After each sheet is made, the operator measures the thickness with a micrometer. The thickness values for three consecutive sheets are averaged and that average is plotted on a piece of graph paper that at the start of the shift has only a solid horizontal line drawn on it to indicate the target gage thickness value for the particular soft gasket sheet the operator is making. Periodically, the operator is to review these evolving data and make a decision as to whether or not the process mean—the sheet thickness—needs to be adjusted. This can be accomplished by stopping the machine, loosening some clamps on the small roll, and jogging the small roll laterally in or out by a few thousandths of an inch—whatever the operator feels is needed. The clamps are tightened, the gap is checked with a taper gage, and if it is adjusted properly, the operator begins to make sheets again. Typically, this process takes about 10 to 15 minutes. The questions of when to make such adjustments and how much to change the roll gap for each adjustment are completely at the operator's discretion, based on the evolving sheet measurement averages.

Figure 12.8 shows a series of plots that detail the history of one particular shift over which the operator made several process adjustments based on the data. These data actually come from the shift that the staff engineers used to collect the data for their process capability study. Figure 12.8(a) shows the process data after the first 12 sheets have been made—four averages of three successive sheet thicknesses. At this point the operator judged that the data were telling her that the process was running below the target, so she stopped the process and made an adjustment to increase the final roll gap slightly. She then proceeded to make more sheets. Figure 12.8(b) shows the state of the process somewhat later. Now it appeared to the operator that the sheets were coming out too thick, so she stopped and made another adjustment. As shown in Fig. 12.8(c), the process seemed to run well for a while, but then an average somewhat below the target led her to believe that another adjustment was necessary. Figure 12.8(d) and (e) show points in time where other adjustments were made. Figure 12.8(f) shows the complete history of the shift. A total of 24 × 3 or 72 sheets were made during this shift. When asked, the operator indicated that the history of this shift was quite typical of what happened on a day-to-day basis.

Statistical Charting of the Process

The roll mill operator had actually been collecting data on sheet thickness in a manner such that the data could be used to construct $\overline{X}$ and R control charts. In fact, it was suspected by those in the plant that the peculiar procedure of averaging three successive sheet thickness values and plotting that average on the graph paper, as witnessed in Fig. 12.8, might actually be the remnants of a once-existing SPC charting procedure for this process.

The improvement team in the plant decided to view the process in a series of samples/subgroups of three consecutive measurements. This would enable them to assess the process centering and amount of variation, using an average and a range for each sample at many points in time throughout the day. As an initial charting effort they decided to take all of the sheets from one entire shift and, in fact, were able to use the same data that the staff engineers had collected for their "process capability study." These data are shown again in Table 12.2, but this time they are tabulated in the subgroups that were used. The values of $\overline{X}$ and R for each subgroup are also summarized in Table 12.2. A sample calculation is shown below.

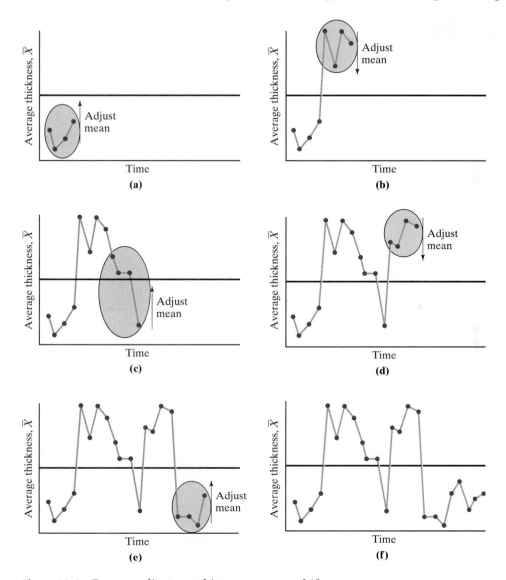

Figure 12.8   Process adjustment history over one shift

Calculations for subgroup 1:

$$\overline{X}_1 = \frac{X_{11} + X_{12} + X_{13}}{3}$$
$$= \frac{0.0440 + 0.0446 + 0.0437}{3}$$
$$= 0.0441 \text{ inch},$$
$$R_1 = X_{1\,max} - X_{1\,min}$$
$$= 0.0446 - 0.0437$$
$$= 0.0009 \text{ inch}.$$

TABLE 12.2	Original Data Grouped into Subgroups of Size $n = 3$ (inches)				
Subgroup	X_1	X_2	X_3	$\bar{X}$	R
1	0.0440	0.0446	0.0437	0.0441	0.0009
2	0.0438	0.0425	0.0443	0.0435	0.0018
3	0.0453	0.0428	0.0433	0.0438	0.0025
4	0.0451	0.0441	0.0434	0.0442	0.0017
5	0.0459	0.0466	0.0476	0.0467	0.0017
6	0.0449	0.0471	0.0451	0.0457	0.0022
7	0.0472	0.0477	0.0452	0.0467	0.0025
8	0.0457	0.0459	0.0472	0.0463	0.0015
9	0.0464	0.0457	0.0447	0.0456	0.0017
10	0.0451	0.0447	0.0457	0.0452	0.0010
11	0.0456	0.0455	0.0445	0.0452	0.0011
12	0.0448	0.0423	0.0442	0.0438	0.0025
13	0.0459	0.0468	0.0452	0.0460	0.0016
14	0.0456	0.0471	0.0450	0.0459	0.0021
15	0.0472	0.0465	0.0461	0.0466	0.0011
16	0.0462	0.0463	0.0471	0.0465	0.0009
17	0.0427	0.0437	0.0445	0.0436	0.0018
18	0.0431	0.0448	0.0429	0.0436	0.0019
19	0.0425	0.0442	0.0432	0.0433	0.0017
20	0.0429	0.0447	0.0450	0.0442	0.0021
21	0.0443	0.0441	0.0450	0.0445	0.0009
22	0.0443	0.0423	0.0447	0.0438	0.0024
23	0.0429	0.0427	0.0464	0.0440	0.0037
24	0.0448	0.0451	0.0428	0.0442	0.0023

Using the data above, the plant group proceeded to determine the center-lines and the upper and lower control limits for the $\bar{X}$ and R control charts. These calculations are provided below.

Centerlines for the $\bar{X}$ and R charts:

$$\bar{\bar{X}} = \frac{\Sigma \bar{X}_i}{24}, \quad i = 1, 24,$$

$$= 0.0449 \text{ inch,}$$

$$\bar{R} = \frac{\Sigma R_i}{24}, \quad i = 1, 24,$$

$$= 0.0018 \text{ inch.}$$

Control limits for the $\bar{X}$ and R charts:

$$\text{UCL}_{\bar{X}} = \bar{\bar{X}} + A_2 \bar{R}$$

$$= 0.0449 + (1.02)(0.0018)$$

$$= 0.0467 \text{ inch,}$$

$$\text{LCL}_{\bar{X}} = \bar{\bar{X}} - A_2 \bar{R}$$

$$= 0.0449 - (1.02)(0.0018)$$
$$= 0.0431 \text{ inch},$$
$$\text{UCL}_R = D_4 \bar{R}$$
$$= (2.57)(0.0018)$$
$$= 0.0046 \text{ inch},$$
$$\text{LCL}_R = D_3 \bar{R}$$
$$= (0)(0.0018)$$
$$= 0.$$

Recall that the values of A_2, D_3, and D_4 are given in Table A.2.

Figure 12.9 shows the $\bar{X}$ and R charts for the roll mill process. Examination of the R chart, which must show good statistical control before the $\bar{X}$ chart can be fully interpreted, indicates no reason to suspect that the process is not in control with respect to its variability. Examining the general patterns and applying the tests discussed in Chapter 6, the R chart does not contain any statistical signals

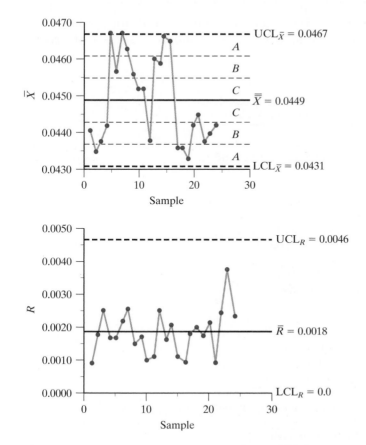

Figure 12.9 $\bar{X}$ and R control charts for the roll mill process

indicating that the process variation is unstable (i.e., there is no indication of any out-of-control conditions on the R chart).

Interpretation of the $\overline{X}$ chart in Fig. 12.9 shows a much different story. Two of the $\overline{X}$ values are extreme points (at or outside of the control limits), in many cases the $\overline{X}$ values violate zone rules, and in one case a run below the center-line occurs. The summary in Table 12.3 identifies each point that constitutes a signal of an out-of-control condition and indicates which test(s)/rule(s) are being violated. Figure 12.10 shows the control charts with all the points appropriately circled that violate rules.

Overcontrol of the Process

In Fig. 12.10 the statistical signals are clear, but what is causing the process mean to shift up and down in a somewhat sporadic fashion? A review of the operating procedures for the process begins to shed some light on this question. A review of Fig. 12.8 and its associated discussion indicates that the statistical signals in Fig. 12.10 correlate quite well with the points in time at which operator adjustments were made. It appears that the operator may have been overcontrolling the process. Why is this occurring? What are the consequences of overcontrol? What can be done to avoid this mistake? The answers to these questions lie at the very foundation of statistical process control.

TABLE 12.3	Rule Violations for Fig. 12.10, $\overline{X}$ Chart
Subgroup	*Rule(s) Violated*
4	1. Four out of five in zone B or beyond
5	1. Extreme point
7	1. Extreme point
	2. Two out of three in zone A or beyond
8	1. Two out of three in zone A or beyond
	2. Four out of five in zone B or beyond
	3. Eight in a row outside zone C
9	1. Eight in a row outside zone C
	2. Four out of five in zone B or beyond
16	1. Two out of three in zone A or beyond
	2. Four out of five in zone B or beyond
18	1. Two out of three in zone A or beyond
19	1. Two out of three in zone A or beyond
	2. Eight in a row outside zone C
20	1. Four out of five in zone B or beyond
	2. Eight in a row outside zone C
22	1. Four out of five in zone B or beyond
23	1. Four out of five in zone B or beyond
24	1. Four out of five in zone B or beyond
	2. Run of eight below the centerline

Figure 12.10 $\overline{X}$ and R control charts of Fig. 12.9 showing out-of-control points

The operator is overadjusting the process because she cannot distinguish between the forces of common-cause variation and special-cause variation (i.e., between chronic variability sources and sporadic variability sources). In fact, she is reacting to patterns in the data that can be explained as coming from the common-cause system as if they were indicating the presence of sporadic disturbances in the process. The consequences of this mistake are devastating. Each time an adjustment is made when it is not warranted, a source of variation is introduced that would otherwise not be there. As a result, quality loss is increased. Furthermore, each time the operator stops to make an unnecessary adjustment, productivity is lost.

It should also be noted that even though the operator is charting the process of making the sheets over time, the mentality is still one of product control. Her primary concern is to keep the sheet thickness as close to the nominal as possible—conformance to the specification. The concept of sporadic and chronic sources of variation is not present. Further, the concept that the nature of the variability pattern could reveal opportunities for productivity improvement is not present; rather, quite the contrary. The entire operation, from management strategy for inspection to the operator's method for making sheets, revolves around a product control way of thinking about quality.

Although we cannot totally avoid making the costly mistake of overcontrol, we can minimize its occurrence by using statistical charting to help distinguish between the presence of only common causes of variation and the occurrence of special causes. If the operator were using charts such as those in Fig. 12.9 instead of the ad hoc graphing of gasket thickness averages as depicted in Fig. 12.8, she would be able to avoid the mistake of process overadjustment.

Figure 12.11 is similar to Fig. 12.4 and shows the histogram of the individual measurements for the sheet thickness values obtained by the staff engineers during their process capability study. A more careful examination of these data indicated some reason to be suspicious from the very beginning. The high percentage of measurements in the tails of the histogram seems to be suggesting that more than one process might actually be represented in the data (i.e., the overcontrol of the process might be recognized by the trained eye, so to speak). The three different normal curves superimposed over the histogram in Fig. 12.11 could well be a plausible result that could easily arise from the type of overcontrol situation which is now believed to be present.

Productivity/Variability Relationship

As emphasized in Chapter 2, W. Edwards Deming pointed out to the world, at every conceivable opportunity, that sources of variability are sources of waste and inefficiency. Each time we identify and eliminate a source of variation,

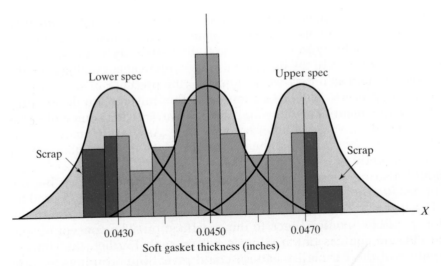

Figure 12.11   Consequences of overcontrol on process performance

productivity must improve/increase. At the same time, Deming noted that elimination of a source of variation means that quality must improve as well. This is absolutely consistent with Taguchi's definition of quality and the loss function.

The importance of the role of statistical process control in improving productivity can be easily illustrated via the problem of overcontrol of a process. Overcontrol arises when operators or other process personnel or electronic/automatic controllers make adjustments to the process in the interest of causing a change in performance when no such adjustments are called for; that is, the process is stable if left alone. Overcontrol clearly affects the process in a negative way in two regards. First, it increases the process variability so that quality is eroded. Second, it decreases productivity, since when the process is being adjusted, no product is being made. The result is a costly mistake from both a quality and a productivity point of view. We have seen this clearly in the gasket-making case study. Overcontrol of a process cannot be totally eliminated, but the frequency of making this mistake can be greatly reduced if statistical signals from control charts are used as a basis for deciding whether to adjust a process or leave it alone. Statistical process control provides an informed basis for this decision-making activity.

Process Improvement Action

As a first process improvement action it was decided by the group to train the operator in the concepts of SPC charting: in the simple calculation of $\overline{X}$ and R, the plotting of these values, and the interpretation of $\overline{X}$ and R control charts. Procedures were established by the production manager and the quality control manager for charting, and the operator was given this new tool to help her to decide when to adjust the process and, more important, when not to adjust the process.

Since the initial control charts developed for the roll mill process showed good statistical control for the variability (the R chart was in control) and since the $\overline{X}$ chart was centered very close to the target value for the process, it was decided to initiate control at the machine simply by extending the centerlines and control limits for these charts. The operator followed the same general procedures as before except that she stopped to adjust the process mean level only if the $\overline{X}$ chart showed that the mean had shifted. The measurements that were obtained for the first shift run in this manner are summarized in Table 12.4.

Figure 12.12 shows the $\overline{X}$ and R control charts for the shift for which the process was run under SPC-based operating guidance. Remember that over this shift, the operator was to make adjustments to the roll gap—process mean— only when the $\overline{X}$ chart gave a signal that the mean had shifted. It turned out that no adjustments were required. The charts in Fig. 12.12 show good statistical control, indicating that if left alone, the process appears to operate in a stable fashion. As can be imagined, the consequences of all of this are quite dramatic.

TABLE 12.4	Process Measurements (inches) for a Shift Run Under SPC Guidance				
Subgroup	X_1	X_2	X_3	$\overline{X}$	R
1	0.0445	0.0455	0.0457	0.0452	0.0012
2	0.0435	0.0453	0.0450	0.0446	0.0018
3	0.0438	0.0459	0.0428	0.0442	0.0031
4	0.0449	0.0449	0.0467	0.0455	0.0018
5	0.0433	0.0461	0.0451	0.0448	0.0028
6	0.0455	0.0454	0.0461	0.0457	0.0007
7	0.0455	0.0458	0.0445	0.0453	0.0013
8	0.0445	0.0451	0.0436	0.0444	0.0015
9	0.0443	0.0450	0.0441	0.0445	0.0009
10	0.0449	0.0448	0.0467	0.0455	0.0019
11	0.0465	0.0449	0.0448	0.0454	0.0017
12	0.0461	0.0439	0.0452	0.0451	0.0022
13	0.0443	0.0434	0.0454	0.0444	0.0020
14	0.0456	0.0459	0.0452	0.0456	0.0007
15	0.0447	0.0442	0.0457	0.0449	0.0015
16	0.0454	0.0445	0.0451	0.0450	0.0009
17	0.0445	0.0471	0.0465	0.0460	0.0026
18	0.0438	0.0445	0.0472	0.0452	0.0034
19	0.0453	0.0444	0.0451	0.0449	0.0009
20	0.0455	0.0435	0.0443	0.0444	0.0020
21	0.0440	0.0438	0.0444	0.0441	0.0006
22	0.0444	0.0450	0.0467	0.0454	0.0023
23	0.0445	0.0447	0.0461	0.0451	0.0016
24	0.0450	0.0463	0.0456	0.0456	0.0013
25	0.0442	0.0451	0.0450	0.0448	0.0009
26	0.0450	0.0471	0.0451	0.0457	0.0021
27	0.0457	0.0432	0.0449	0.0446	0.0025
28	0.0460	0.0453	0.0452	0.0455	0.0008

12.5   Process Capability Versus Process Control

Reassessment of Process Capability

Since the process now exhibits good stability, it is valid and meaningful to conduct a process capability study. To determine if the process could be modeled by the normal distribution, two graphical techniques were employed, the histogram and the normal probability plot. First, a frequency histogram of the data in Table 12.4 was constructed. The steps used for the first set of data were followed once again.

1. The range of the data was determined by subtracting the smallest value from the largest: $0.0472 - 0.0428 = 0.0044$.

2. It was decided to use nine cells of width 0.0005, with midpoints at 0.0XX5 and 0.0XX0.

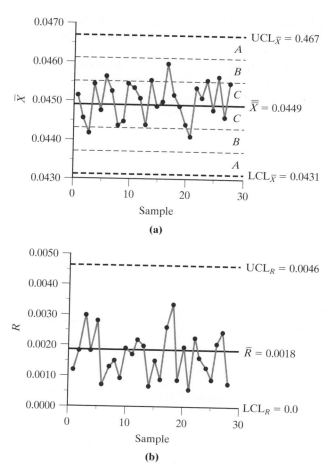

Figure 12.12 $\overline{X}$ **and** R **control charts for the roll mill process under SPC process operation**

3. Therefore, the boundaries were chosen such that the first cell goes from 0.04275 to 0.04325, the second from 0.04325 to 0.04375, and so on.

4. A tally sheet of the points was made.

Cell Midpoint	Frequency	Total
0.0430	11	2
0.0435	11111	5
0.0440	11111 111	8
0.0445	11111 11111 11111	15
0.0450	11111 11111 11111 11111 11	22
0.0455	11111 11111 11111	15
0.0460	11111 111	8
0.0465	11111 1	6
0.0470	111	3

5. The histogram itself was constructed, as shown in Fig. 12.13. This histogram has very much the appearance of the bell-shaped normal distribution.

To investigate further the possible normality of the data, a normal probability plot was made of the measurement points in the first 10 subgroups. Table 12.5 was constructed following the guidelines established in Chapter 3. The plot is

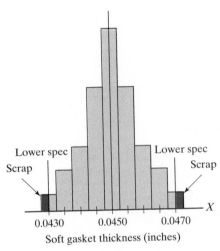

Figure 12.13   Histogram of the gasket measurements in Table 12.4

TABLE 12.5	Data for Normal Probability Plot of Fig. 12.14				
Order	Measurement	Estimated Cumulative Probability (%)	Order	Measurement	Estimated Cumulative Probability (%)
1	0.0428	2	16	0.0450	52
2	0.0433	5	17	0.0451	55
3	0.0435	8	18	0.0451	58
4	0.0436	12	19	0.0453	62
5	0.0437	15	20	0.0454	65
6	0.0441	18	21	0.0455	68
7	0.0443	22	22	0.0455	72
8	0.0445	25	23	0.0455	75
9	0.0445	28	24	0.0457	78
10	0.0445	32	25	0.0458	82
11	0.0448	35	26	0.0459	85
12	0.0449	38	27	0.0461	88
13	0.0449	42	28	0.0461	92
14	0.0449	45	29	0.0467	95
15	0.0450	48	30	0.0468	98

shown in Fig. 12.14. Both the frequency histogram and the normal probability plot seem to indicate clearly that the individual gasket thickness measurements closely follow a normal distribution curve.

Figure 12.15 is a graphical depiction of the process capability for the stabilized process and emphasizes the fact that over the period of time that the process was sampled, it has been established that the process was in statistical control. It appears that when examined relative to the specifications the process variation is greatly reduced. About 95% of the process is now operating within the specifications.

To establish the improved process capability more precisely, the normal distribution was employed as a model for the process. The calculations are as follows, using the data from Table 12.4:

$$\overline{\overline{X}} = \frac{\Sigma \overline{X}_i}{28}, \quad i = 1, 28,$$

$$= 0.04505,$$

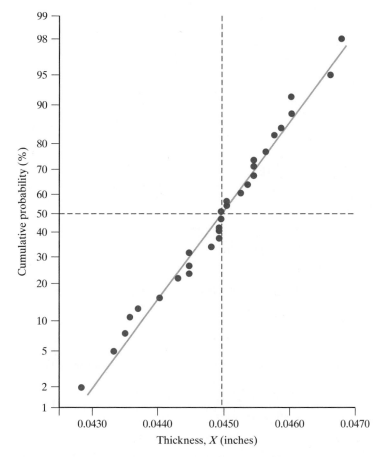

Figure 12.14   Normal probability plot based on the data of Table 12.5

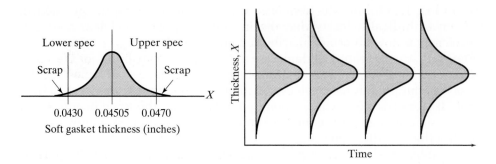

Figure 12.15   Process capability analysis based on the process data of Table 12.4

$$\overline{R} = \frac{\Sigma R_i}{28}, \quad i = 1, 28,$$

$$= 0.00168,$$

$$P(\text{within specs}) = P(0.0430 \le X \le 0.0470)$$

$$= [P(X \le 0.0470) - P(X \le 0.0430)],$$

$$Z = \frac{X - \mu_X}{\sigma_X},$$

$$\hat{\sigma}_X = \frac{\overline{R}}{d_2}.$$

For $n = 3$, $d_2 = 1.693$,

$$\hat{\sigma}_X = \frac{0.00168}{1.693}$$

$$= 0.00099,$$

$$Z_{\text{LSL}} = \frac{0.0430 - 0.04505}{0.00099}$$

$$= -2.07,$$

$$Z_{\text{USL}} = \frac{0.0470 - 0.04505}{0.00103}$$

$$= 1.97.$$

The Z table (Table A.1) can then be used to find the required probability:

$$P(\text{within specs}) = [P(Z \le Z_{\text{USL}}) - P(Z \le Z_{\text{LSL}})]$$

$$= [P(Z \le 1.97) - P(Z \le -2.07)]$$

$$= 0.9756 - 0.0192$$

$$= 0.9564.$$

Compared to the previous determination of the process capability (which, strictly speaking, was not a valid use of the normal distribution), it appears that the process capability has been greatly improved. Previously, by a simple count of the actual number of sheets that were nonconforming (in Table 12.1), the process capability might have been crudely estimated to be about 73%; at least this is the percentage of those 72 sheets that were within the specifications. Now that the process has been stabilized by removing the overcontrol operating fault, the process capability is 95%.

In terms of the C_p index, the process capability is now equal to 0.673, as calculated below. While the process capability is still not acceptable by usual standards ($C_p \geq 1$ is generally required), as indicated above, scrap has been considerably reduced.

$$
\begin{aligned}
C_p &= \frac{\mathrm{USL}_X - \mathrm{LSL}_X}{6\sigma_X} \\
&= \frac{\mathrm{USL}_X - \mathrm{LSL}_X}{6\bar{R}/d_2} \\
&= \frac{0.0470 - 0.0430}{6(0.00168/1.693)} \\
&= 0.673.
\end{aligned}
$$

Calculations were also performed to find the value of C_{pk}:

$$
\begin{aligned}
Z_{\mathrm{USL}} &= \frac{\mathrm{USL} - \bar{\bar{X}}}{\sigma_X} \\
&= \frac{\mathrm{USL} - \bar{\bar{X}}}{\bar{R}/d_2} \\
&= \frac{0.0470 - 0.04505}{0.00168/1.693} \\
&= 1.97, \\
Z_{\mathrm{LSL}} &= \frac{\mathrm{LSL} - \bar{\bar{X}}}{\sigma_X} \\
&= \frac{\mathrm{LSL} - \bar{\bar{X}}}{\bar{R}/d_2} \\
&= \frac{0.0430 - 0.04505}{0.00168/1.693} \\
&= -2.07, \\
Z_{\min} &= \min[Z_{\mathrm{USL}}, -Z_{\mathrm{LSL}}] \\
&= \min[1.97, 2.07] \\
&= 1.97, \\
C_{pk} &= \frac{Z_{\min}}{3}
\end{aligned}
$$

$$= \frac{1.97}{3}$$

$$= 0.66.$$

But improved product quality—reduced variability—is only one-half of the good news of Fig. 12.15. Since no process roll gap adjustments were made, productivity must also have improved. A review of the production data reveals that 84 sheets were made for this shift, compared with the 72 sheets previously manufactured under the original method of process operation. This constitutes a 17% increase in productivity! It is now becoming clearer that when we seek improvements at the process through SPC concepts and methods, we can simultaneously realize improvements in both quality and productivity.

Quality/Variability Relationship

Since one major source of functional variation, as discussed in Chapter 5, is manufacturing imperfection, the methods of statistical process control should stand as an important tool for quality improvement. In this case study operator overcontrol has been identified as a major source of manufacturing imperfection/variability. This source of variation has not only led to a serious loss in productivity but also greatly contributed to the poor process capability.

As outlined in Chapter 2 and discussed in Chapter 10, the loss function approach to defining quality provides a clear motivation for seeking continually to reduce functional variation in product/process performance. Figure 12.16 depicts the change in the process that has been precipitated by the removal of the overcontrol source of variation. Since the loss is a function of the process standard deviation, as was demonstrated in Chapter 10, removal of every source of variation must lead accordingly to a reduction in loss due to functional variation and therefore an improvement in quality.

Holding the Gains

It was made clear by the discussion in Chapter 10 that the loss function provides a quantitative means to assess product/process performance relative to

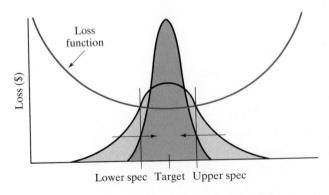

Figure 12.16   Loss function view of quality/variability relationship

the nominal value. This assessment, however, requires that we be able to represent the process in terms of a statistical distribution—for example, the normal distribution. However, such a representation is valid only if the process is in a state of statistical control. If we do not, therefore, strive to keep processes in statistical control, the loss function analysis used for design purposes will go unachieved at the production stage. In general, this will mean that the actual loss encountered by the customer in the field will be greater than what we believe it to be based on the design process, a truly regrettable state of affairs.

The argument above can also be applied to "holding the gains" precipitated by control charting of the process. In the case study at hand both quality and productivity gains have been realized in stabilizing the process through removal of the overcontrol source of variation. However, failure to maintain this state of statistical control can lead to the loss of all that has been gained! This loss, generated at the process, can be illustrated by overlaying the statistical process control and loss function concepts.

Suppose that Fig. 12.17 represents what we can achieve if the process potential is consistently realized; that is, the process, as shown in the figure, is in statistical control with the indicated level of variability. Now if the process is not kept in a state of statistical control (e.g., its mean value is allowed to shift about as depicted in Fig. 12.18), the customer would experience an increased loss due to functional variation. This increase in loss, as measured by the loss function, results from the fact that over the long term the lack of control/stability may give rise to a distribution of measurements of the appearance of Fig. 12.19, a more flat, loaf-shaped appearance.

We could actually calculate the increase in loss as a result of this lack of statistical control of the process by assuming, for example, that the process of Fig. 12.19 can be approximated by a uniform distribution with endpoints equal to the lower and upper specification limits. Then, using the formula for the calculation of the standard deviation of the uniform distribution, we could calculate the loss and compare this value to that when the process is performing consistently (in control) at the nominal according to its position and statistical

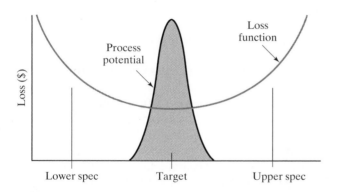

Figure 12.17   Loss if process is in a state of statistical control centered at the nominal

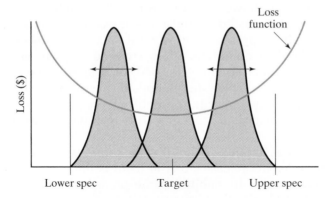

Figure 12.18   Behavior of process if statistical control is not maintained

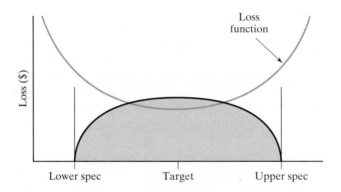

Figure 12.19   Loss if statistical control is not maintained

distribution as shown in Fig. 12.17. It is clear that statistical control is an important concept with significant economic implications from both the producer and consumer points of view.

12.6   Seeking Further Improvement Through Design of Experiments

Encouraged by the marked improvement in process performance, the group in the plant continued to hold their meetings and continued with the charting activity in the hope of finding still other improvement opportunities. At one of the meetings the process engineer suggested that two machine conditions, roll speed and roll temperature (cooling lines in the large roll), might have better settings (i.e., ones that would improve sheet consistency). To investigate this possibility it was decided to run a simple two-level factorial design experiment. This experiment is depicted graphically in Fig. 12.20. A total of $2^2 = 4$ tests are required (graphically, the corners of the square in Fig. 12.20).

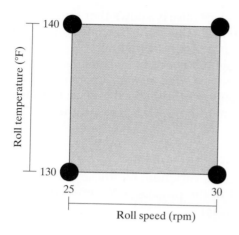

Figure 12.20   Two-level factorial design to study roll mill process

It was decided to run the four tests of the experimental design over a period of one shift, each test being comprised of about 2 hours of production. During the tests, the $\overline{X}$ and R charts would be continued using the previously established values for $\overline{\overline{X}}$, $\overline{R}$, and the control limits, and the R chart would be examined to see if changes in roll speed and/or roll temperature were causing a change in the sheet-to-sheet consistency—that is, any change in the process variability.

Figure 12.21 shows a portion of the data plotted on the R chart maintained during the time that the experiment was being conducted. Examination of Fig. 12.21 indicates that changes in roll speed and/or roll temperature do not appear to be influencing the amount of variation in the process. In fact, if the 2-hour segment boundaries are removed (the dashed vertical lines in Fig. 12.21)

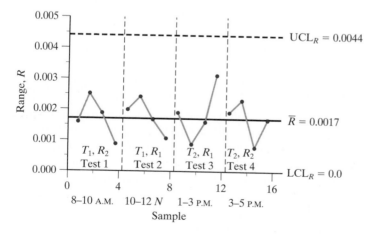

Figure 12.21   Range chart for the process over the period of the factorial experiment

and the range values are "connected," the chart clearly shows that the process variation is in good statistical control. It appears, therefore, that changes in roll speed and/or roll temperature do not influence process variation.

The last sentence of the preceding paragraph needs to be considered more carefully. The truth is that we cannot draw this kind of conclusion about the results of the experiment. The fact that no effects of roll speed and/or roll temperature can be detected from the experiment (Fig. 12.21) can result from either of two distinct possibilities:

1. These variables truly have no effect on process variation.
2. These variables have some effect on process variation, but these effects cannot be detected because the noise level of the process is too great (i.e., the true effects are masked/hidden by the noise).

Too often, possibility 2 is really the factor driving the experimental design results, and therefore either we will end up mistakenly overlooking improvement opportunities that are there but hidden by the noise in the process, or we will conclude that the design-of-experiment techniques themselves are flawed or do not apply to what we are doing. In either case the results are regrettable.

A Major Breakthrough

Disappointed with the results of the factorial experiment but still encouraged by the progress to date, the group in the plant continued to meet to discuss possible improvement actions. One Friday, discussion turned to the amount of material that was added to the roll gap to make each sheet. The operator stated that she had been told years ago simply to add the material one handful after another, evenly distributing it along the roll gap until the bell rang and the light went on, indicating that the smaller roll had reached its final destination (i.e., the sheet was now completed). Clearly, each sheet was not receiving the same amount of material, by weight, for the specified target moisture content. Yet each sheet was constrained by the roll diameter to be the same length and by stops on the rolls to be of the same width. How then, could we expect to control closely the sheet thickness and the sheet density, which was another quality characteristic under study?

To solve the problem above, which is clearly a common-cause/chronic variability problem, it was suggested that a fixed amount of material, by weight, be metered into each sheet's construction. A long discussion about how this would be done ensued and a lot of disagreement followed. A primary concern seemed to be not whether the procedure being used now was a problem but, rather, how this metering problem might be accomplished. Certain members of the group thought that to meter the material into the roll gap more carefully would be too costly in terms of retrofitting a metering device to the roll mill. The quality control and production managers, however, felt that logistics were getting in the way of common sense and reason, so they decided to prepare for an experiment over the weekend.

On Saturday morning they took it upon themselves to measure many sheets of recently manufactured gasket material to determine the proper sheet weight for the nominal thickness. They then took raw material blended the day before that was ready for use on Monday and carefully measured out portions just enough for one sheet, based on the weight/thickness relationship they had established, and placed the metered raw material in plastic bags. They prepared enough of these bags of material for one shift of production. On Monday morning they asked the operator to run all day using the material from each bag to make just one sheet, all to be added for the construction of each sheet, no more, no less. The $\overline{X}$ and R charts were continued. The results are shown in Fig. 12.22 and are dramatic. From the figure it is clear that the effect of metering a precise amount of material into every sheet has been to cause a significant reduction in the common-cause variability in the process. From the R chart we can see that the common-cause variability has been cut almost in half. It should be clear that this result will have a tremendous positive impact on both sheet quality and process productivity.

Figure 12.23 is a graphical representation of the results of a process capability study done on data for which the sheet material metering was in effect. It is obvious that the process capability has been greatly improved in terms of the capability indices C_p and C_{pk}. The process now has a capability of $C_p = 1.253$. Calculations for C_p and C_{pk} are as follows:

$$C_p = \frac{\text{USL}_X - \text{LSL}_X}{6\sigma_X} = \frac{\text{USL}_X - \text{LSL}_X}{6\overline{R}/d_2} = \frac{0.0470 - 0.0430}{6(0.0009/1.693)} = 1.253,$$

$$Z_{\text{USL}} = \frac{\text{USL} - \overline{\overline{X}}}{\sigma_X} = \frac{\text{USL} - \overline{\overline{X}}}{\overline{R}/d_2} = \frac{0.0470 - 0.0449}{0.0009/1.693} = 3.95,$$

$$Z_{\text{LSL}} = \frac{\text{LSL} - \overline{\overline{X}}}{\sigma_X} = \frac{\text{LSL} - \overline{\overline{X}}}{\overline{R}/d_2} = \frac{0.0430 - 0.0449}{0.0009/1.693} = -3.57,$$

$$Z_{\min} = \min[Z_{\text{USL}}, -Z_{\text{LSL}}] = \min[3.95, 3.57] = 3.57,$$

$$C_{pk} = \frac{Z_{\min}}{3} = \frac{3.57}{3} = 1.19.$$

In terms of productivity it may at first be more difficult to understand how material metering might lead to a significant improvement. However, when the operator was questioned about this, she indicated that in the past she had often stopped the machine during the construction of a sheet to check the progress in terms of thickness to determine whether she should add a bit more or less material from that point to the end of the sheet construction. In the case of material metering, since the amount of material to be added was predetermined, such process interruptions were not necessary. As a result, she was not surprised that she was able to make about 90 sheets per shift. This was now a 25% increase in productivity from the outset of the study.

Once the clear advantage of material metering had been proved through the data, interest within the project group centered on finding an efficient

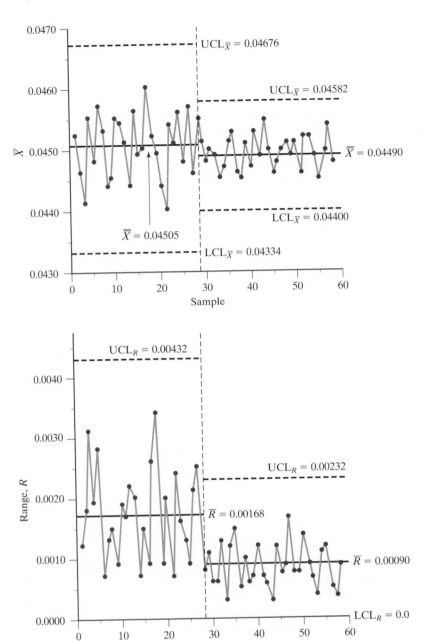

Figure 12.22   Continuation of $\overline{X}$ and R charts during sheet construction under raw material metering

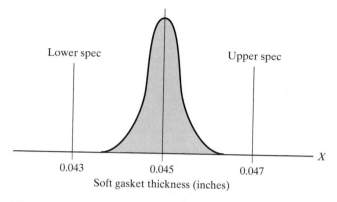

Lower spec

Upper spec

0.043   0.045   0.047

X

Soft gasket thickness (inches)

Figure 12.23   **Process capability after material metering was put into effect**

engineering solution to the material metering problem. A very simple and cost-effective solution was identified rather quickly. Scales were placed under the track that supported the raw material tote pans at the machine. The digital readout for the scale could be set to the prescribed weight and the operator merely had to remove material from the tote pan until the digital readout went to zero. At that point she would know that she had added the correct amount of material to the sheet. This engineering solution was implemented, and some time later the return on investment was calculated to be a mere 13 weeks in terms of both quality and productivity improvement.

It should also be noted that since the raw material metering problem was common to all of the roll mill lines in the plant, the foregoing solution was applied to each line. This is an important aspect of the SPC approach to improvement, that some of the problems discovered through the charting of one machine/process are common to all similar processes. Therefore, solutions may be usefully applied across all machines.

Experimental Design Revisited

It is clear from Fig. 12.22 that the amount of variability of the process has been greatly reduced (i.e., the signal-to-noise ratio has been increased). With this in mind and still with the firm conviction that either roll speed or roll temperature or both might have a real impact on process consistency, the group decided to rerun the same two-level factorial experiment that was conducted previously (see Fig. 12.20). The $\overline{X}$ and R charts employed at this point were based on data obtained from process monitoring after the metering of the raw material was installed as an operating method. Hence with the significant reduction in variability, the charts should be a much more sensitive indicator of process changes when they occur.

Figure 12.24 shows the results of this second running of the factorial experiment. At first the results are quite surprising since now the range chart shows clearly that roll speed appears to have a strong influence on process

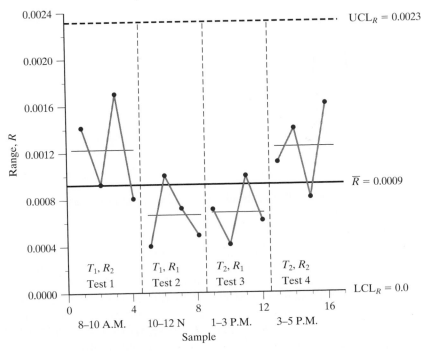

Figure 12.24   Results of the repeated factorial experiment after process material metering was initiated

consistency. In particular, the lower level of roll speed seems to give rise to a significant reduction in process variability. It is important in Fig. 12.24 to note that the vertical scale is approximately half that of Fig. 12.21, which shows the results when the experiments were run earlier. It is now apparent that the excessive noise in the system (due to the variation in the amount of material added to each sheet) was masking the roll speed effect.

Based on the results of the second experiment, the roll speed was reduced, and a significant improvement in the process capability resulted. Over the next several weeks the group continued to study the process and evaluate other potential improvement actions. At one point the operator noted that excessive backlash in the gears often made it difficult to set the roll gap when adjustments were required or when she changed from one gage sheet to another. When the maintenance department made an inspection of the gearing, they found excessive wear and replaced several gears. This greatly increased the probability of arriving at the proper roll gap through a single adjustment. Of course, this led to a further reduction in product variability and a further increase in process productivity. Figure 12.25 graphically depicts the process capability after the roll speed change and the maintenance actions. It is clear that the process is becoming extremely consistent.

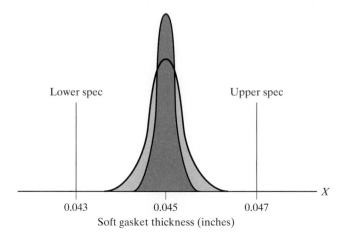

Lower spec

Upper spec

0.043   0.045   0.047

Soft gasket thickness (inches)

Figure 12.25   Process capability after additional improvement actions

Statistical Process Control/Design of Experiments Relationship

From the result above it is clear that the power of design of experiments can be greatly enhanced if the environment in which the experiments are conducted has been affected through variation reduction methods such as statistical process control. Although statistical control of the process is not necessarily a prerequisite to drawing valid conclusions from the results of designed experiment, it can greatly enhance the ability to detect the presence of real variable effects. As we will see later, there are certain fundamental design countermeasures that we can take to ensure experimental validity. SPC is an important countermeasure that we must consider whenever possible to improve experiment sensitivity. Furthermore, in cases where statistical control of a process has been established, subsequent experimentation and associated improvement actions taken are more likely to be realized in future operation of the process. Under statistical control the future is more predictable!

It should also be clear from the above that design of experiments stands as a powerful technique to reveal basic opportunities for improvement during an SPC study. In Chapter 5 we discussed the feedback control loop interpretation of the process control concept. It was indicated then that stage 3 in this loop—fault diagnosis—was critical and that techniques such as design of experiments could be useful for this purpose.

A Postscript

Several months later the consultant who worked with the people in the plant returned to review the current status of their statistical implementation program. He was greeted by the plant manager, who reported enthusiastically that over the past six months they had saved over $500,000 in material costs alone. When the consultant inquired as to exactly how this was accomplished, the

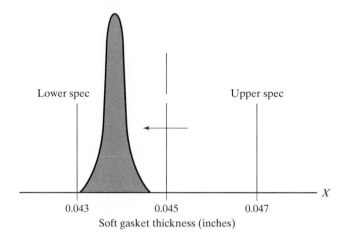

Figure 12.26   Process recentered to reduce material requirements

plant manager replied that since the process capability had been increased to over 12 sigma, it was possible to recenter the process very close to the lower specification, thereby greatly reducing the amount of material required to make a sheet. At the same time he could satisfy the customers' defined product quality requirement in terms of the bilateral specification. Figure 12.26 graphically depicts the action of the plant manager.

Hopefully, by now your understanding of the loss function definition of quality and the notions of statistical tolerance design makes clear the serious implications of such actions. One way to think of this action at the process is that it is equivalent to making a design change in the product, changing the nominal value for the gasket material thickness to a value quite near the lower specification limit. Is it reasonable to think that this can be done without having any effect on product quality?

■ CHAPTER SUMMARY

Through the case study above a number of extremely important concepts and methods central to reducing variability—increasing quality and productivity—have been discussed. What is perhaps most important to note is that by using the same machines, same people, and same materials, but in a much more effective fashion, the problem that this company originally faced has been solved. By applying simple but powerful techniques of statistical charting and fault diagnosis, the company avoided perhaps needlessly spending a large amount of money on new equipment that has been shown to be totally unnecessary. Such results are being experienced by many others as they overlay statistical thinking and methods on their engineering knowledge and experience.

■ EXERCISES

12.1 Construct a cause-and-effect (fishbone) diagram for the gasket sheet roll mill process. Include all aspects of the process that were mentioned in the chapter as well as factors you think may also have existed that were affecting the final product thickness.

12.2 In Section 12.2, the dilemma facing the gasket manufacturer is outlined. The gasket manufacturer realized that "some positive steps would have to be taken to improve the quality [of the gaskets]." What do you think he had in mind when he said that? Would his ideas differ from your concept of quality and quality improvement? Explain.

12.3 Describe the benefits associated with including the process operator in the team studying the problem with the gasket assembly. In general, what should the composition of such teams be like? Why?

12.4 C_p and C_{pk} values are routinely calculated as measures of the quality of a process. Explain how the widespread use of these indices may inhibit the promotion of the concepts and methods of statistical control of processes.

12.5 Explain why the selective assembly process used by the company for the gasket assembly process was not a good solution to their quality problems, although it did allow them to meet the product specifications imposed on them by the customer.

12.6 What types of costs will the company incur to implement the proposed changes in the roll mill process? Comment on the significance of these costs compared with other alternatives that have been suggested for solving the problem.

12.7 What was the effect of metering the raw material in the roll mill process? Would you consider this change in the process as precipitating a major breakthrough or contributing more to the increased stability of the process? Explain.

12.8 How might regular/preventive maintenance on the roll mill machines affect the process as a whole? How would you justify this added cost? Explain.

12.9 At the end of the chapter it was discovered that the process has been recentered below the nominal in order to save on raw material costs. List the reasons why this is a bad practice. Can the savings in excess of $500,000 that they have realized justify the decision to make this change in the process centering? Why or why not? What important concept was missing here?

12.10 From reading about the incident described in Exercise 12.9, describe the notion of quality that might have existed here. How does this idea fit with the notions of Taguchi related to quality?

12.11 Address the issue raised in the question posed in the last sentence of the chapter (just prior to the Summary).

12.12 Discuss the relationship between process capability and process control. Initially, the staff engineers did a capability study to "prove" that the process needed radical change (i.e., new roll mill equipment). Discuss the dangers inherent in their approach.

12.13 It could be argued that the people in the plant were very lucky. They set out to reduce variation through statistical charting, but each time they did, the productivity of the process went up as well. Explain their good fortune.

12.14 Initially, the operator of the roll mill made adjustments in the gap thickness each time she felt, based on her measurements, that such was warranted. What impact did this approach to maintaining the product quality have on the product and the process?

12.15 Why did the initial design of experiments conducted on the roll speed and roll temperature fail to show any effects? What is this saying about the DOE/SPC relationship?

12.16 In this case study, major sources of variation were discovered and acted upon. Categorize these into special-cause and common-cause variation sources. Explain why each is categorized as it is.

12.17 Make a list of the key concepts of quality design and improvement stressed in this book and addressed specifically by this case study. Indicate how each concept was reinforced through this case study.

12.18 Where in the case study is the concept of increasing the signal-to-noise ratio to enhance process performance clearly demonstrated?

12.19 Who do you think played the most important role as a participant in this case study? Why? What is this telling us?

12.20 Sometimes, frustrated by the apparent complexity of implementing a solution to a problem, we tend to attempt to draw attention away from the problem or downplay its importance. Did that happen in this case study? Where?

12.21 In considering the roll mill process investigation, briefly summarize the changes that were made to bring the process into a state of statistical control and then reduce its level of variability.

12.22 At the beginning of the case study, the engineers wanted to purchase a new roll mill. If this had been done, what do you think would have been the outcome in terms of process behavior?

12.23 What does the roll mill process case study suggest about the relationship between sources of variation and quality? Between sources of variation and productivity?

12.24 In the postscript to the case study, the process has been recentered near the lower specification limit. How would you convince the plant manager to shift the process mean back to the nominal value?

■ SPC WORKSHOP #2: TURNING SIMULATION, TURNSIM

Introduction

In this workshop, a sampling experiment will be performed with the aid of TURNSIM, an MS Excel–based Visual Basic program. TURNSIM simulates a single-point lathe-turning operation that is being used to manufacture a bronze bushing. The TURNSIM program is used to "collect data" from the process, which may then be analyzed with $\overline{X}$ and R control charts. Assuming the process is in control, TURNSIM data may also be used to perform a process capability analysis.

Learning Objective

The objective of this computer workshop is to provide students with practice in (1) identifying statistical signals on control charts, (2) relating these signals to physical conditions in a process, and (3) taking actions to eliminate special causes to improve the stability and capability of a process.

Background

A bronze bushing is being machined on its outside-diameter surface to meet functional requirements on its diameter dimension and surface roughness. Inspection has shown that many of the bushings produced by the process have a surface roughness (R_a) outside the 70 ± 15 microinches specification. Changes have been made to the process with no noticeable improvement.

   A list of ten factors that vary from time to time during production has been identified. It is believed that the variation of some of these factors may be causing the poor product conformance to the functional requirements. It has been suggested that these factors actually represent a list of potential special causes. It happens that only three of the ten candidate special causes are truly active in the process simulation program (i.e., produce unusual patterns of variability on the $\overline{X}$ and R control charts, for example, by causing shifts in the process mean and/or variance). The other seven factors, as they vary, have no special-cause impact on the process. It is desired to find out which of the three factors are, in fact, the special causes.

Description of the Turning Process Simulator, TURNSIM

The TURNSIM program simulates the collection of data from the turning process that produces the bronze bushings. The program collects k subgroups (k is specified by the user) of size $n = 5$, suitable for the construction of $\overline{X}$ and R control charts. For each sample, TURNSIM generates surface roughness measurements and levels for each of the ten candidate special causes.

Once TURNSIM has generated the subgroup information inside the Excel spreadsheet, control charts may be constructed for the surface roughness data. The levels (process conditions) for the factors should then be correlated with the patterns on the charts to identify a special cause. It may be noted that the three active special-cause factors produce variations in the surface roughness that can be explained physically based on common knowledge of the lathe-turning process. Once a potential special cause has been identified, the user may re-enter the TURNSIM program, indicate the special cause that has been found, and collect additional data. If the specified special cause is a bona fide special cause, then TURNSIM will "disarm" it as a special-cause factor when additional data are collected. On the other hand, if the user specifies a factor that is not one of the true special causes, nothing will happen to change the future data collected. Of course the process should continue to be sampled to observe the effect (if any) of corrective actions taken inside TURNSIM. Each special cause must be identified and verified separately, one at a time. Additionally, the corrective actions taken to remove the special causes and bring the process into a state of statistical control must be specified.

Once all three special causes have been identified and verified, both through physical reasoning and through the evidence provided by the control charts, a process capability study may be performed (percent conforming, C_p, and C_{pk}) given the part specifications and assuming that the individual measurements are normally distributed. If the process is not capable (e.g., the level of common-cause variability is too large), some corrective actions that might reduce the common-cause variability or otherwise improve the process capability should be offered.

On the Web

TURNSIM is available for download on the Web at http://www.me.mtu.edu/quality_book/. Also available at the site is a more detailed workshop explanation and description of TURNSIM usage. The website also has a suggested student assignment in connection with the TURNSIM workshop.

13

Shewhart Control Charts for Attribute Data

13.1   Data Characterization by Attributes

Observation by Attributes

To this point, the quality assessment criteria for manufactured goods have been of the variable measurement type (e.g., diameter in inches, filled weight in ounces). However, some quality characteristics are more logically defined in a "presence of or absence of" sense. Such situations might include surface flaws on a sheet metal panel, cracks in drawn wire, color inconsistencies on a painted surface, voids, flash or spray on an injection molded part, or wrinkles on a sheet of vinyl. Such nonconformities or defects are often observed visually or according to other sensory criteria and cause a part to simply be defined as a defective part. In these cases we refer to quality assessment as being made by attributes.

Many quality characteristics that could be established by measurements (variables) are often not assessed as such in the interest of economy. During an inspection procedure a go/no-go gage may be employed to determine whether or not a variable characteristic falls within the part specification or not. Parts that fail such a test are simply labeled "defective." The opportunity for making variable measurements may be great in a given plant, but the construction and continuing use of $\overline{X}$ and R charts is more time consuming and expensive. Attribute measurements may be usefully employed to identify the presence of problems that can then be attacked by the use of the more sensitive and informative $\overline{X}$ and R charts.

Some Important Definitions

Before developing control charts for attribute data it is necessary to define clearly some important terms. The following definitions are required in working with attribute data.

Defect.   A fault or nonconformity that causes an article or an item to fail to meet specification requirements is called a defect. Each instance of an article's lack of conformity to specification is a defect.

Defective.   An item or article with one or more defects is a defective item.

Number of Defectives.   In a sample of n items, d is the number of defective items in the sample.

Number of Defects.   In a sample of n items, c is the number of defects in the sample. An item may be subject to many different types of defects. As an item is inspected, every instance of each type of defect is noted. The sum of all defects across the sample is c.

Fraction Defective.   The ratio of the number of defective items in the sample, d, to the total number of items in the sample, n. The sample fraction defective p is therefore equal to d/n.

Operational Definitions

The most difficult aspect of quality characterization by attributes is the precise determination of what constitutes the presence of a particular defect. One reason is that many attribute defects are visual in nature and therefore require a certain degree of judgment; another is the inability to discard the product control mentality. For example, a scratch that is barely observable by the naked eye may not be considered a defect in a product control sense. However, the physical mechanism within the process causing scratches may well be the same for both such a small scratch and much larger ones. Remember, it is the purpose of the control chart to signal the presence of faults in the process so that they may be corrected. Furthermore, human variation is generally considerably larger in attribute characterization, as is pointed out in Fig. 13.1. It is therefore important that precise and quantitative operational definitions be set down for all to observe uniformly when attribute quality characterization is being used. The length or depth of a scratch may be specified, or the diameter of a surface blemish, or the length of a flow line on a molded part.

The issue of the product control vs. process control way of thinking about defects is a crucial one. From a product control point of view, scratches on an auto grille should be counted as defects only if they appear on visual surfaces

Figure 13.1   **Human variability in attribute quality characterization**

that would directly influence part function. From a process control point of view, however, scratches on an auto grille should be counted as defects regardless of where they appear, since the mechanism creating scratches does not know if it is a visual surface or not. Furthermore, by counting all scratches, the sensitivity of the statistical charting instrument used to identify the presence of defects and lead to their diagnosis will be increased considerably. From a process control perspective we may actually wish to be "harder" on the part than if we are only worried about whether to ship or not! The issue is one of chart sensitivity and fault identification and diagnosis.

A major problem with the product control way of thinking about part inspection is that when attribute quality characterization is being used, not all defects are observed and noted. Often, the first detected occurrence of a defect immediately causes the part to be scrapped. Such data are recorded in scrap logs, which then present a biased view of what the problem may really be. One inspector may be keen on scratch defects on a molded part and therefore will seem to see these first. Another may think splay is more critical, so his "data" tend to reflect this type of defect more frequently. The net result is that such data may then mislead those who may be trying to use them for process control and improvement purposes.

Figure 13.2 provides an example of the occurrence of multiple defects on a part. It is essential from a process control standpoint to observe and note carefully each occurrence of each type of defect. Figure 13.3 shows a typical sample result and the careful observation of each occurrence of each type of defect. In Fig. 13.3 the four basic measures used in attribute quality characterization are defined for the sample in question. Some other examples of quality characterization by attributes might be:

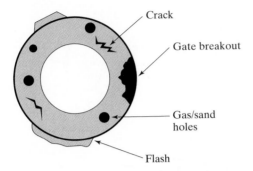

Figure 13.2   Example of defect identification in an attribute quality characterization situation (engine valve seat blank)

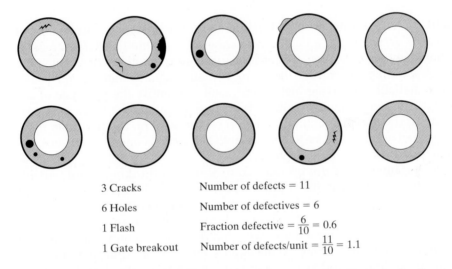

3 Cracks	Number of defects = 11
6 Holes	Number of defectives = 6
1 Flash	Fraction defective = $\frac{6}{10} = 0.6$
1 Gate breakout	Number of defects/unit = $\frac{11}{10} = 1.1$

Figure 13.3   Example of sample result for attribute quality characterization

1. *Final inspection of an assembled engine.* Defects:
   - Missing oil dipstick
   - Paint runs
   - Improper ID tag
   - Dyno test (pass/fail)
   - Crankcase bolts missing

2. *Automobile radiator grille after injection molding process.* Defects:

- Splay
- Oil/grease
- Flash
- Scratches/scuffs
- Black spots

- Voids
- Sinks
- Flow lines
- Burns
- Short shots

3. *Accounts payable process.* Defects:

- No purchase order
- Units conversion issue
- Vendor number missing
- Missing approval signature
- No packing slip
- Material not received

13.2   Control Chart for Fraction Defective

Consider an injection molding machine producing a molded part, an instrument panel for an automobile, at a steady pace. Suppose that the measure of quality conformance of interest is the occurrence of four types of defects: flash, splay, voids, and short shots. If a part has even one occurrence of any of these defects, it is considered to be nonconforming (i.e., a defective part). Recently the scrap rate of this process has been quite high, and it is decided to employ the statistical control chart idea to attempt to determine the source(s) of the problem.

Samples of 10 consecutive panels were examined every 45 minutes. Each incidence of each type of defect was noted. Also, the number of panels in each sample with at least one defect present was noted (i.e., the number of defectives, d). For the purpose of getting an overall impression of the performance of the process, the data for each shift, 10 samples of 10 panels each, were collected into a "shift sample" of $n = 100$ panels. A fraction defective control chart, a p chart, is to be used to study the process. The data in Table 13.1 show the number of panels in each shift sample of $n = 100$ that contain at least one defect and the associated fraction defective data for 30 consecutive shifts.

The process characteristic of interest here is the true process fraction defective μ_p, often denoted as p', so that each sample result is converted to a fraction defective p as shown in Table 13.1,

$$p = \frac{d}{n}.$$

The data (fraction defective p) are plotted for 30 successive samples of the size $n = 100$. This plot is shown in Fig. 13.4, where a centerline has been established on the chart by averaging all 30 p values.

TABLE 13.1	Fraction Defective Data for Shift Samples for Instrument Panel Molding Example	
Sample	*Number of Defectives*	*Fraction Defective*
1	7	0.07
2	8	0.08
3	6	0.06
4	8	0.08
5	6	0.06
6	8	0.08
7	3	0.03
8	5	0.05
9	9	0.09
10	7	0.07
11	7	0.07
12	9	0.09
13	8	0.08
14	7	0.07
15	8	0.08
16	10	0.10
17	10	0.10
18	5	0.05
19	12	0.12
20	11	0.11
21	8	0.08
22	10	0.10
23	4	0.04
24	10	0.10
25	7	0.07
26	7	0.07
27	9	0.09
28	8	0.08
29	10	0.10
30	10	0.10
	237	

The individual values for the sample fraction defective, p, vary considerably, and it is difficult at this point to determine from the plot if the variation about the average fraction defective $\bar{p}$ is due solely to the forces of common causes. Several of the p values are quite low, perhaps indicating that some sporadic behavior is evident in the chart. To help us resolve this problem, the Shewhart control chart idea previously used for variable observations ($\bar{X}$ and R) can be applied. But to develop the Shewhart control chart for the attribute quality characteristic observed as the fraction defective in a sample or subgroup, we will need to know more about how defective items arise, in a frequency sense, when subgroups of several such items are collected.

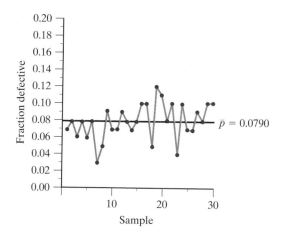

Figure 13.4   Fraction defective values for samples of $n = 100$ instrument panels collected over 30 consecutive shifts

Binomial Sampling Situation

Suppose that we have an hour's production of 1000 parts, of which 50 are defective and 950 are nondefective. If we randomly pick one part from these 1000 parts, there will be two possible outcomes: We get either a defective or a nondefective. Suppose that we record the outcome as "N" for nondefective and "D" for defective. We mix the 1000 parts thoroughly after replacing the first part and randomly pick another part from the lot, recording the outcome as "D" or "N." Remember, we are putting back the part after we have sampled. If we repeat the sampling five times, at the end we will have recorded the result of five trials or outcomes. We could think of this as a sample of $n = 5$ parts, some of them being defective, some of them not.

Now, the problem we are interested in is determining the respective probabilities of getting either 0, 1, 2, 3, 4, or 5 defectives out of five trials or outcomes. In other words, out of five outcomes, we may get no defectives, 1 defective, 2 defectives, 3 defectives, 4 defectives, or even all 5 defectives, and we are interested in knowing the probabilities associated with each of these possible outcomes. In this case we know that there are 50 defective and 950 nondefective parts in the 1000 parts. Therefore, the true fraction defective for these parts is

$$p' = \frac{50}{1000} = 0.05,$$

which may be interpreted as the probability of getting a defective from a single trial. Then, the probability of getting a nondefective on each independent trial is

$$1 - p' = \frac{950}{1000} = 0.95.$$

Note that total probability $[p' + (1 - p') = 0.05 + 0.95]$ is 1.0.

Let us denote the sample size by n. Since we took five random trials, $n = 5$. Now let us try to compute the probability of getting a specific sample result, namely, the combination of 1 defective and 4 nondefectives—that is, $P(1D, 4N)$, in a sample of $n = 5$. If the five draws are independent—that is, the result for one draw does not depend on the specific result of another draw—the probability of obtaining 1 defective and 4 nondefectives is simply the product of the individual trial outcome probabilities,

$$P(1D, 4N) = (p')(1 - p')(1 - p')(1 - p')(1 - p')$$
$$= (0.05)(0.95)(0.95)(0.95)(0.95) = (0.05)^1(0.95)^4 = 0.041.$$

The probability above relates to one particular occurrence of 1 defective and 4 nondefective parts. But we can obtain 1 defective and 4 nondefectives in the following five possible ways, taking the order of the trials into account:

NNNND, NNNDN, NNDNN, NDNNN, DNNNN.

In other words, we could have gotten the defective on either the first draw, the second, or the third, fourth, or fifth draws. For this situation it is easy to enumerate the number of ways we can get 1 defective and 4 nondefectives in a sample of $n = 5$. But if the sample size is $n = 20$, how many ways can you obtain, for example, 4 defectives and 16 nondefectives? The answer is 4845. Clearly, we need a general formula to determine the number of *combinations* we could obtain. This is given by

$$\binom{n}{d} = \frac{n!}{d!\,(n - d)!},$$

where $n!$ is read as "n factorial" (e.g., $5! = 5 \cdot 4 \cdot 3 \cdot 2 \cdot 1 = 120$). In our case, $n = 5$ and $d = 1$ (defective). Thus

$$\binom{5}{1} = \frac{5!}{1!\,(5 - 1)!} = \frac{5!}{1!\,4!} = 5.$$

We saw above that $P(1D, 4N) = (0.05)^1(0.95)^4$ is the probability of only one combination. Since we can get 1 defective and 4 nondefectives in $\binom{5}{1} = 5$ ways, the total probability of getting 1 defective and 4 nondefectives, which we now denote simply as $P(d = 1)$, will be

$$P(d = 1) = \binom{5}{1}(0.05)^1(0.95)^4 = (5)(0.041) = 0.205. \qquad (13.1)$$

We note here that in the sample above the true fraction defective, p', in the lot of 1000 parts is $50/1000 = 0.05$, but the sample result we obtained (1 defective and 4 nondefectives) constitutes a sample fraction defective of $p = 1/5 = 0.20$. We note further that this result will occur about 20% [i.e., $P(d = 1) = 0.205$] of the time! That is, the sample will produce a fraction defective four times the true lot (population) fraction defective for 20% of the samples so chosen. Clearly the forces of chance (common-cause) variation are at work and produce uncertainty in our sample result as it reflects the true situation at hand.

A number of other results could have occurred when we drew the sample (five successive trials). We could have obtained

d	$\binom{n}{d}$	$P(d)$
0D + 5N (0 defective, 5 nondefective)	0	1	0.7737809
1D + 4N (1 defective, 4 nondefective)	1	5	0.2036266
2D + 3N (2 defective, 3 nondefective)	2	10	0.0214344
3D + 2N (3 defective, 2 nondefective)	3	10	0.0011281
4D + 1N (4 defective, 1 nondefective)	4	5	0.0000297
5D + 0N (5 defective, 0 nondefective)	5	1	0.0000003

$$\text{Total} = 1.0000000$$

It is noted that the total probability of all the possible results adds up to 1.0.

In general, we calculate the above probabilities by the following mathematical function, called the binomial probability function,

$$P(d) = \binom{n}{d}(p')^d(1 - p')^{n-d} \qquad (13.2)$$

for $d = 0, 1, 2, \ldots, n$, and where d corresponds to the number of defectives, n corresponds to the sample size, $(n - d)$ to the number of nondefectives, and p' and $(1 - p')$ are, respectively, the probabilities associated with the occurrence of a defective and a nondefective on a single trial. The sum of the probabilities of all the possible results from $d = 0$ to $d = n$ will then be given by

$$\sum_{d=0}^{n} P(d) = \sum_{d=0}^{n}\binom{n}{d}(p')^d(1 - p')^{n-d} = 1. \qquad (13.3)$$

The lefthand-side term of Eq. (13.2) is commonly written as

$$b(d; n, p') = \binom{n}{d}(p')^d(1 - p')^{n-d}, \qquad (13.4)$$

where the lowercase letter b term on the lefthand side means the probability of getting exactly d defectives for a given n and p'. To obtain the probability of x or fewer defectives, we may use the equation

$$B(x; n, p') = \sum_{d=0}^{x} \binom{n}{d}(p')^d(1 - p')^{n-d}, \tag{13.5}$$

where the B term on the lefthand side means the cumulative probability of getting up to and including x defectives.

Equations (13.4) and (13.5) are called the *probability function* and *cumulative probability function*, respectively, of the binomial distribution. The binomial distribution is a discrete distribution and serves as a very useful tool in dealing with statistical problems involving repeated trials where the characteristic observed will have either one of two possible outcomes. We present its applications after we discuss its important properties.

Assumptions of the Binomial Distribution.   In the development of the binomial distribution (probability law) above, we have made several important assumptions about the sampling situation. These assumptions have, in fact, led us to the mathematical formulation that we refer to as the binomial probability model. These assumptions are summarized below:

1. *Only two outcomes are possible for each trial.* Parts can be classified as either defective or nondefective. Each item sampled must be either good or bad; it can be nothing else. The probabilities associated with a defective and a nondefective must therefore add to 1.0.

2. *Sampling with replacement or from a stable process.* By replacing the sampled part after each trial we have ensured that the true fraction defective p' will remain constant from trial to trial. In control chart applications, the true fraction defective p' will remain constant if the sampled process is stable, or in statistical control.

3. *Independence of successive trials.* We have assumed that the outcome of a particular trial is in no way influenced by the outcomes of other trials. Hence we have been able to determine the joint probability of the outcome of the group of trials (total sample) as the product of the probabilities associated with the occurrence of each individual trial.

Properties of Binomial Distribution.   Several important properties of the binomial distribution are now discussed.

1. The binomial distribution has two parameters, n and p'. If we look at Eqs. (13.4) and (13.5), we find that we need only specify n and p' to find the various probabilities.

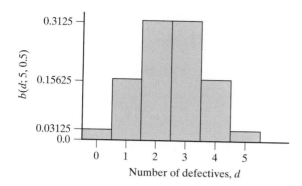

Figure 13.5   Symmetrical binomial distribution

2. An interesting property of the binomial distribution is illustrated in Figs. 13.5, 13.6, and 13.7 for the case $n = 5$ and for $p' = 0.5$, $p' = 0.20$, $p' = 0.80$, respectively. When $p' = 0.50$, Eq. (13.4) becomes

$$b(d; n, p') = \binom{n}{d}(0.5)^d(1 - 0.5)^{n-d} = \binom{n}{d}(0.50)^n$$

and the shape of the binomial distribution is symmetrical, as shown in Fig. 13.5. Note that if p' is less than 0.50, it is more likely that d will be small rather than large compared to n. This is shown in Fig. 13.6 for $n = 5$ and $p' = 0.20$. A probability distribution that has a shape like that of Fig. 13.6 is called *positively skewed*, where the long tail is on the right side. Similarly, if p' is more than 0.50, it is more likely that d will be large rather than small compared to n. This is shown in Fig. 13.7 for $n = 5$ and $p' = 0.80$. A probability

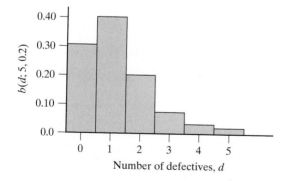

Figure 13.6   Positively skewed binomial distribution

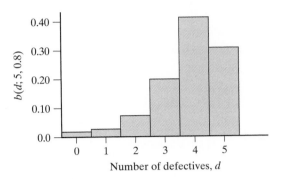

Figure 13.7   Negatively skewed binomial distribution

distribution that has a shape like that of Fig. 13.7 is called *negatively skewed*, where the long tail is on the left side.

3. Mean and variance:
   a. Mean of d, $\mu_d = np'$.
   b. Variance of d, $\sigma_d^2 = np'(1 - p')$.

Examples of the Use of the Binomial Probability Model.   The binomial distribution can be applied in cases where there are only two possible outcomes for each trial: conforming and nonconforming, success and failure, defective and nondefective, and so on.

▶ EXAMPLE 13.1

Suppose we want to find the probability that exactly 9 out of 10 gears are bad, given the probability $p' = 0.80$ that any one of these gears will be bad.

$$b(9; 10, 0.80) = \binom{10}{9}(0.80)^9(0.20)^1 = 0.268.$$

▶ EXAMPLE 13.2

What is the probability that at least 9 out of 10 gears are bad if $p' = 0.80$?
    Remember that "at least 9" means 9 or more, which means that either 9 or 10 are bad (since $n = 10$). In other words, we are asking for the sum of the probabilities of exactly 9 and exactly 10 bad out of 10:

$$b(9; 10, 0.80) + b(10; 10, 0.80) = 0.268 + 0.107 = 0.375.$$

▶ EXAMPLE 13.3

What is the probability that at most 8 of these gears are bad if $p' = 0.80$?

$$B(8; 10, 0.8) = \sum_{d=0}^{8} b(d; 10, 0.80)$$

since "at most 8" means anything from $d = 0$ to 8. But rather than evaluate each of the 9 (0 to 8) probabilities required for this sum, it is easier to find the probability that at least 9 gears are bad and subtract the result from 1, the total probability. We thus obtain $(1 - 0.375) = 0.625$.

If n is large, the calculation of the binomial distribution can become quite tedious. Fortunately, good approximations are available for binomial probability calculations when the sample size is very large. One such approximation for small p' values is given by the Poisson probability distribution discussed later in this chapter.

Control Limits for the p Chart

In the preceding section it was shown that for random sampling under the assumptions of:

1. Part is either defective or nondefective,
2. Sampling with replacement, and
3. Independent trials,

the occurrence of the number of defectives d in a sample of size n is explained probabilistically by the binomial law:

$$P(d \text{ defectives in a sample of } n \text{ items}) = P(d)$$

$$P(d) = \frac{n!}{d! \, (n - d)!}(p')^{d}(1 - p')^{n-d}. \tag{13.6}$$

In quality control work, it has been found useful to use the fraction defective p for charting purposes. Since the sample fraction defective p is simply the number of defectives d divided by the sample size n, the mean and standard deviation for p may be obtained using the theorems on mathematical expectation in Chapter 3. The mean for p is

$$\mu_p = E\left(\frac{d}{n}\right)$$

$$= \frac{1}{n}E(d)$$

$$= \frac{1}{n}np'$$

$$= p'. \tag{13.7}$$

The variance of p is

$$\sigma_p^2 = \text{Var}\left(\frac{d}{n}\right)$$

$$= \frac{1}{n^2}\text{Var}(d)$$

$$= \frac{1}{n^2}np'(1 - p')$$

$$= \frac{p'(1 - p')}{n}.$$

The standard deviation of p is therefore

$$\sigma_p = \sqrt{\frac{p'(1 - p')}{n}}. \tag{13.8}$$

Hence, if p' were known, the centerline of the p chart would be at $\mu_p = p'$ and the control limits would be set at

$$p' \pm 3\sigma_p \tag{13.9}$$

or

$$p' \pm 3\sqrt{\frac{p'(1 - p')}{n}}. \tag{13.10}$$

Generally, p' is not known, so we must estimate p' from the sample data obtained. Given k rational samples of size n, the true fraction defective p' may be estimated by

$$\overline{p} = \sum_{i=1}^{k}\frac{p_i}{k}, \tag{13.11}$$

where $p_i = d_i/n$ for the ith sample (if the sample size varies, then the fraction defective should be calculated as $p_i = d_i/n_i$). A more general form is given by

$$\overline{p} = \frac{\displaystyle\sum_{i=1}^{k} d_i}{\displaystyle\sum_{i=1}^{k} n_i}. \tag{13.12}$$

The second formula is more general, since it is valid whether or not the sample size is the same for all samples. The first formula should be used only if the sample size n is the same for all k samples.

Therefore, given $\bar{p}$, the control limits for the p chart are then given by

$$\text{UCL}_p = \bar{p} + 3\sqrt{\frac{\bar{p}(1 - \bar{p})}{n}},$$

$$\text{LCL}_p = \bar{p} - 3\sqrt{\frac{\bar{p}(1 - \bar{p})}{n}}. \tag{13.13}$$

Thus only $\bar{p}$ has to be calculated for at least 25 samples of size n to set up a p chart. Sometimes the calculation for the lower control limit may yield a value less than zero. The binomial distribution is not symmetric and has a lower bound of $d = 0$. In this case a lower control limit of zero is used.

Return to the Injection Molding Example of p Chart Application

In this example we have been examining a process for injection molding of a portion of an automobile instrument panel. At one point in the process the partially completed panel is examined for the presence of four different types of molding defects. Recall that the data for 30 samples of size $n = 100$ were given in Table 13.1 and the fraction defective values were plotted in Fig. 13.4. Given the data in Table 13.1 and the results above based on the binomial distribution, the centerline and the control limits for the p chart for these 30 samples can be determined as shown below. Figure 13.8 shows the p chart for the fraction defective data of Table 13.1.

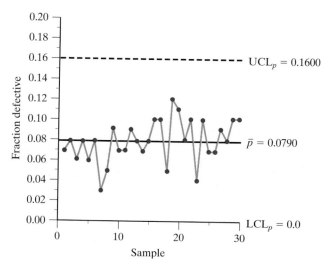

Figure 13.8 p chart for the fraction defective data of Table 13.1

$$\bar{p} = \frac{\text{sum of all defectives}}{\text{total number of units}}$$

$$= \frac{237}{3000}$$

$$= 0.0790.$$

For $\bar{p} = 0.0790$ and $n = 100$,

$$\hat{\sigma}_p = \sqrt{\frac{\bar{p}(1 - \bar{p})}{n}}$$

$$= \sqrt{\frac{(0.0790)(1 - 0.0790)}{100}}$$

$$= 0.0270,$$

$$\text{UCL} = \bar{p} + 3\hat{\sigma}_p$$

$$= 0.0790 + (3)(0.0270)$$

$$= 0.1600,$$

$$\text{LCL} = \bar{p} - 3\hat{\sigma}_p$$

$$= 0.0790 - (3)(0.0270)$$

$$= -0.002$$

$$= 0.0.$$

The interpretation of the p chart follows the same guidelines as other Shewhart charts in a general sense. That is, we look for the appearance of four basic types of behaviors in the data:

1. No points should exceed the control limits.
2. The points should be distributed randomly within the control limits (i.e., more points should be in the middle of the chart and fewer toward the control limits).
3. The points should show no evidence of trends or recurring cycles.
4. The points should look quite random with time (i.e., no patterns such as runs above or below the centerline are evident).

Since the p values do not follow the normal distribution, the tests based specifically on the three zones between the centerline and either of the control limits cannot be used (discussed in Chapter 6). However, the tests for runs and trends can be employed.

Examination of the p chart in Fig. 13.8 shows that no signals are present that would indicate the process is out of control. It would appear that a process subject only to common-cause variation is present. But while the process seems to be stable over the period studied, the fraction defective rate is quite high, on average at 8%; clearly this is unacceptable. And the chart indicates that the problem is a system problem.

An investigation was launched to determine the cause(s) of the high defective rate on the panels. The defect data were examined and it was determined that one of the defects, flash, was accounting for over 50% of all the defects. The process in question was examined and it was determined that certain molding process parameters—the pressure settings—needed some adjustment. The mold pressure settings were changed and the data collection process was continued.

Table 13.2 provides the fraction defective data for the 30 shifts that followed the adjustments on the mold pressure. These data are plotted as a continuation of the original chart (Fig. 13.8) using the same centerline and control limits and are shown in Fig. 13.9. It is clear from the general appearance of the chart that a significant change has occurred in the process. In examining Fig. 13.9 more carefully, we see that a run of 12 consecutive points below the centerline begins at sample 31, a run of 8 in a row below the centerline begins at sample 44, and a run of 8 in a row below the centerline begins at sample 53. In fact, only two of the second 30 sample fraction defective values fall at or above the centerline. It is clear that the process has changed—in particular, that the fraction defective rate has decreased considerably, perhaps as much as halved.

TABLE 13.2	Fraction Defective Data for 30 Shift Samples After the Pressure Adjustments Were Made for the Instrumental Panel Molding Example				
Sample	Number of Defectives	Fraction Defective	Sample	Number of Defectives	Fraction Defective
31	5	0.05	46	4	0.04
32	6	0.06	47	2	0.02
33	2	0.02	48	2	0.02
34	5	0.05	49	5	0.05
35	6	0.06	50	3	0.03
36	3	0.03	51	5	0.05
37	2	0.02	52	8	0.08
38	4	0.04	53	2	0.02
39	3	0.03	54	1	0.01
40	6	0.06	55	2	0.02
41	4	0.04	56	1	0.01
42	3	0.03	57	7	0.07
43	8	0.08	58	4	0.04
44	3	0.03	59	4	0.04
45	2	0.02	60	2	0.02
				114	

Figure 13.9   Continuation of the *p* chart of Fig. 13.4

The centerline and control limits for the *p* chart for samples 31 to 60 were recalculated using only the data from those samples. The calculations are given below, and the resulting *p* chart for these samples is shown in Fig. 13.10.

$$\bar{p} = \frac{\text{sum of all defectives}}{\text{total units}}$$

$$= \frac{114}{3000}$$

$$= 0.0380,$$

$$\hat{\sigma}_p = \left[\frac{0.038(1 - 0.038)}{100} \right]^{1/2}$$

$$= 0.0191,$$

$$\text{UCL}_p = 0.0380 + (3)(0.0191)$$

$$= 0.0953,$$

$$\text{LCL}_p = 0.0380 - (3)(0.0191)$$

$$= -0.0193$$

$$= 0.0.$$

From Fig. 13.10 two things are clear. First, the process is still running in a stable fashion after the machine adjustment. A careful examination of the data indicates that none of the tests/rules that apply here for out-of-control conditions are violated. Second, the fraction defective rate has been reduced dramatically,

Figure 13.10   *p* chart for samples 31 to 60 with recalculated control limits

to about one-half of its original level. A common-cause/system problem has been identified, adjustments have been made to the process, and the continued charting has demonstrated through the data that a significant improvement in the process has taken place. Figure 13.11 summarizes the charting of this process through the 60 samples and clearly shows the nature and magnitude of the change in the system.

13.3   Variable-Sample-Size Considerations for the *p* Chart

It is often the case that the sample size may vary from one time to another as data for the construction of a *p* chart are obtained. This may be the case if the data have been collected for other reasons or if a sample constitutes a day's production and production rates vary from day to day. Since the limits on a *p* chart depend on the sample size *n*, some adjustments must be made to ensure that proper interpretation of the chart is made.

There are several ways that the variable-sample-size problem can be handled. Some of the more common approaches are

1. Compute separate limits for each subgroup. This approach certainly leads to a correct set of limits for each sample but requires continual calculation of the control limits and a somewhat messy-looking control chart.

2. Determine an average sample size $\bar{n}$ and a single set of control limits based on $\bar{n}$. This method may be appropriate if the sample sizes do not vary greatly, perhaps no more than about 20%. However, if the actual *n* is less than $\bar{n}$, a

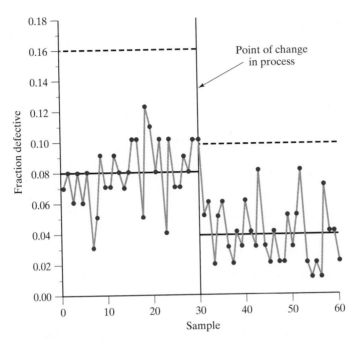

Figure 13.11   *p* chart for all of the samples analyzed for the instrument panel molding example

point above the control limit based on $\bar{n}$ may not be above its own true upper control limit. Conversely, if the actual n is greater than $\bar{n}$, a point may not be shown as out of control, when in fact it really is.

3. A third procedure for handling the variable-sample-size problem for the p chart is to express the fraction defective p in standard deviation units; that is, $Z = (p - \bar{p})/\hat{\sigma}_p$ becomes the value plotted on a control chart where the centerline is zero and the control limits are set at ± 3.0. This standardizes the plotted value even though n may be varying. Note that $Z = (p - \bar{p})/\hat{\sigma}_p$ is a familiar form; recall the standard normal distribution, $N(0, 1^2)$. For this method the continued calculation of the standardized variable is somewhat tedious, but the chart has a clean appearance, with constant limits always at ± 3.0 and a constant centerline at 0.0.

Variable-Sample-Size *p* Chart—Manufacture of a Package Tray

The following example deals with a process that manufactures package trays. The tray starts with a wood fiber/polymer composite material that is extruded into thin sheets. This material is then heated in a furnace and molded in a press that forms the shape of the tray. As part of the molding process the tray has a carpetlike (cloth) material applied by an adhesive to the surface, after which the formed and carpeted tray is trimmed to size. The process was launched on a pilot basis in the plant that ultimately was to manufacture this package tray. A semi-manual operation was developed to study the processing parameters for the furnace heating and the molding process to develop the operating conditions for the scaled-up process.

In an effort to evaluate the performance under operating conditions, the process was run semi-manually one shift per day. Depending on machine adjustments and other factors, somewhere between 200 and 300 trays could be made per shift. After some initial experiments were run to determine a first set of operating conditions, the process was run for several weeks, inspecting each tray for a series of defects, making minor process adjustments along the way, and carefully noting what the adjustments were and when they were made. After 30 shifts of data were available, a *p* chart was initiated and used to evaluate the influence of changes made in the process.

Seven different defects were noted in inspecting the trays. These were

- Cloth coverage on the hinge.
- Carpet coverage.
- Bleed-through (substrate showing through the carpet).
- Soiled carpet surface.
- Improper flex on hinge.
- Chips, scratches, abrasions on back surface.
- Wrinkles in the carpet.

Table 13.3 provides defective and fraction defective data for the first 30 samples collected. Remember, a sample is really an entire shift's production. For this example we construct and examine three different *p* charts, one with individual limits for each sample, one with constant limits based on an average

TABLE 13.3	Data for Package Tray Example					
Sample	*n*	*d*	*p*	LCL_p	UCL_p	*z*
1	238	11	0.046	0.026	0.131	−1.84
2	245	18	0.073	0.027	0.130	−0.28
3	270	17	0.063	0.029	0.127	−0.94
4	207	15	0.072	0.022	0.134	−0.31
5	251	11	0.044	0.027	0.129	−2.03
6	254	15	0.059	0.028	0.129	−1.14
7	236	19	0.081	0.026	0.131	0.13
8	245	20	0.082	0.027	0.130	0.19
9	246	35	0.142	0.027	0.130	3.74
10	269	14	0.052	0.029	0.127	−1.60

(Continued)

		TABLE 13.3	(*Continued*)			
Sample	*n*	*d*	*p*	*LCL$_p$*	*UCL$_p$*	*z*
11	223	7	0.031	0.024	0.132	−2.61
12	246	42	0.171	0.027	0.130	5.40
13	262	14	0.053	0.029	0.128	−1.50
14	258	15	0.058	0.028	0.128	−1.21
15	232	20	0.086	0.025	0.131	0.45
16	219	9	0.041	0.024	0.133	−2.05
17	263	23	0.087	0.029	0.128	0.55
18	244	11	0.045	0.027	0.130	−1.93
19	274	21	0.077	0.030	0.127	−0.10
20	245	37	0.151	0.027	0.130	4.24
21	233	16	0.069	0.026	0.131	−0.55
22	267	18	0.067	0.029	0.128	−0.66
23	254	20	0.079	0.028	0.129	0.03
24	264	16	0.061	0.029	0.128	−1.07
25	253	34	0.134	0.028	0.129	3.32
26	290	22	0.076	0.031	0.126	−0.15
27	231	9	0.039	0.025	0.131	−2.23
28	227	40	0.176	0.025	0.132	5.49
29	234	18	0.077	0.026	0.131	−0.08
30	253	15	0.059	0.028	0.129	−1.13

sample size, and one based on the standardization of the observed fraction defective value.

p Chart Based on Individual Limits.   The most obvious approach to the variable-sample-size problem for fraction defective data is simply to calculate separate limits for each sample based on its own sample size. The centerline of the chart will be the average fraction defective, $\bar{p}$. Calculations for the centerline and control limits for this chart are given below.

Centerline for all samples:

$$\Sigma n_i = 7433, \quad i = 1, 30,$$
$$\Sigma d_i = 582, \quad i = 1, 30,$$
$$\bar{p} = \frac{582}{7433}$$
$$= 0.0783.$$

Control limits for sample 1:

$$\text{UCL}_{p_1} \text{ and } \text{LCL}_{p_1}: \quad \bar{p} \pm 3\sqrt{\frac{\bar{p}(1 - \bar{p})}{n_1}}.$$

For sample 1, $n_1 = 238$,

$$0.0783 \pm 3 \left[\frac{0.0783(1 - 0.0783)}{238} \right]^{1/2}$$

$$0.0783 \pm 0.0527,$$

$$\mathrm{LCL}_{p_1} = 0.026, \quad \mathrm{UCL}_{p_1} = 0.131.$$

The control limits for each of the remaining 29 samples were calculated in a similar fashion and are given in Table 13.3. The associated p chart is shown in Fig. 13.12. From the figure we see that five of the sample p values fall above their respective upper control limits, indicating that at least for those shifts the true process fraction defective was higher than the centerline.

p Chart Based on Average-Sample-Size Control Limits.   When the sample size variation is not too great, average-sample-size control limits can be determined for the p chart. The advantage is that the chart looks "clean" because only a single set of control limits is present. The disadvantage is that we cannot be sure about the interpretation of the individual p values without checking them against their actual individual limits. This need be done only for those p values near the control limits, on either side.

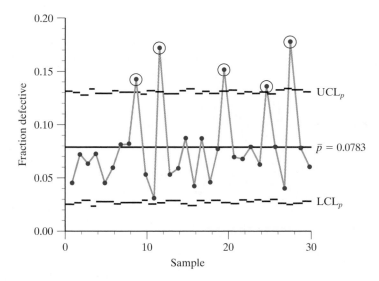

Figure 13.12   Variable-sample-size p chart with individual limits for each sample: package tray example

Calculations for the average sample size and the associated control limits are given below. The centerline of the chart remains the average fraction defective value previously calculated:

$$\Sigma n_i = 7433, \quad i = 1, 30,$$

$$\overline{n} = \frac{7433}{30} = 247.78,$$

$$\overline{n} \approx 248.$$

Upper and lower control limits:

$$0.0783 \pm 3\left[\frac{0.0783(1 - 0.0783)}{248}\right]^{1/2},$$

$$\text{UCL}_p = 0.1295, \quad \text{LCL}_p = 0.0271.$$

Figure 13.13 shows the associated p chart with a single set of control limits based on the average sample size. In this example, the individual sample sizes vary about the average by less than 20%, so it is likely that we would be safe in checking only those p values that are close to the limits. Samples 9 and 25 might qualify for further scrutiny. From Table 13.3 we see that sample 9 has a sample size of 246 and sample 25 has a sample size of 253. For sample 9, the sample size is very close to the average, so the interpretation of this point as being out of control, based on the average limit, is valid. In the case of sample 25, since its sample size is greater than the average, its own upper control limit

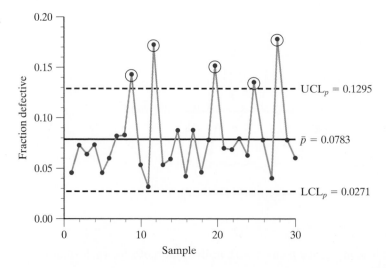

Figure 13.13   **Variable-sample-size p chart with average-sample-size limits: package tray example**

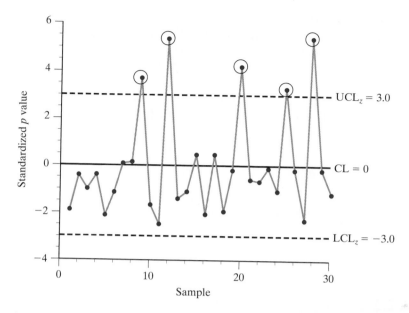

Figure 13.14   **Standardized variable-sample-size p chart: package tray example**

would be lower than the average limit, so again, the interpretation of this point as being out of control, based on the average limit, is valid.

p Chart Based on Standardized p Values.   A p chart for varying sample sizes that is very simple to interpret can be made by standardizing the individual p values. In this case the centerline of the chart is zero and the control limits are set at 3.0 and −3.0.

   The procedure here for variable sample size is to express the fraction defective p in standard deviation units [i.e., $Z = (p - \overline{p})/\hat{\sigma}_p$]. This means that the standard deviation must be calculated for each sample because σ_p depends on the sample size. Once this is done, the standardized value for each sample can be determined. These values are given in the last column of Table 13.3. The associated p chart is shown in Fig. 13.14. In the figure we see that the interpretation—that five samples have standardized p values above the upper control limit—is the same as that for Figs. 13.12 and 13.13. Although the calculations here are a bit more tedious, the common use of computers today for chart construction negates this apparent disadvantage of this chart.

13.4   Control Chart for Number of Defectives

Sometimes it is convenient to make a control chart simply by plotting the number of defectives d in the sample instead of the fraction defective p. This is particularly true if the sample size n is constant and the fraction defective

p is quite small. Such a chart, commonly referred to as an *np chart*, provides essentially the same information as the p chart. The basic difference between the p and np charts is a scaling of the vertical axis by a constant n. The np chart requires one less calculation, since $p = d/n$ for each sample need not be determined.

The centerline and control limits for the np chart are, of course, based on the binomial distribution. The average number of defectives expected in a sample of n is np' and the standard deviation for d is

$$\sigma_d = [np'(1 - p')]^{1/2}. \tag{13.14}$$

Hence the control limits for the np chart based on actual data are given by

$$n\bar{p} \pm 3\sqrt{n\bar{p}(1 - \bar{p})}. \tag{13.15}$$

▶ EXAMPLE 13.4   CONTROL CHART FOR NUMBER OF DEFECTIVES

To illustrate the relationship between the p and the np charts, the data of Table 13.1 will again be used here in the construction of an np chart. The p chart and the np chart for the same data should look the same, the only difference being the vertical scale.

Using the first 30 samples for the instrument panel assembly example, we find that the following calculations for the centerline and the upper and lower control limits lead to the np chart shown in Fig. 13.15.

Figure 13.15   Control chart for number of defectives using data of Table 13.1

$$\bar{p} = \frac{237}{3000}$$

$$= 0.079,$$

$$n\bar{p} = 100(0.079)$$

$$= 7.9.$$

Control limits:

$$n\bar{p} \pm 3\sqrt{n\bar{p}(1 - \bar{p})}$$

$$7.9 \pm 3\sqrt{7.9(1 - 0.079)},$$

$$\text{UCL}_{np} = 16.00,$$

$$\text{LCL}_{np} = -0.20, \quad \text{so a lower limit of 0.00 is used.}$$

Comparison of Figs. 13.8 and 13.15 shows the same pattern in the data, and the resulting analysis of this pattern for Fig. 13.15 is the same as for Fig. 13.8. Again, this should not be surprising, since the only difference in the two charts is the constant scaling of the vertical axis.

Use of the np chart has been very limited. One reason for not using np is that if the sample size varies, not only will the control limits vary but the centerline will vary from sample to sample as well. Such a chart is more difficult to interpret. On the other hand, the np chart is really plotting d, the number of defectives in the sample, and therefore one less calculation would be required. The problem, though, is that the number of defectives is a hard number to grasp and interpret without consideration of the sample size, hence the acceptance of the use of the fraction defective, p.

13.5   Control Chart for Number of Defects

Importance of Number of Defects per Sample

The p chart deals with the notion of a defective part or item, where "defective" means that the part has at least one nonconformity or disqualifying defect. It must be recognized, however, that the incidence of any one of several possible nonconformities would qualify a part for defective status. A part with 10 defects, any one of which makes it a defective, is on equal footing with a part with only one defect, in terms of being a defective.

Often, it is of interest to note every occurrence of every type of defect on a part and chart the number of defects per sample. A sample may only be one part, particularly if interest is focusing on final inspection of an assembled product, such as a washing machine. Inspection may focus on one type of defect, such as nonconforming rivets on an aircraft wing, or multiple defects, such as flash, splay, voids, and knit lines on an injection-molded part.

Considering an assembled product such as a washer, we find that the opportunity for the occurrence of a defect is quite large, and perhaps should be considered infinite. However, the probability of occurrence of a defect in any one spot arbitrarily chosen is probably very, very small. In this case the probability law that governs the incidence of defects is known as the *Poisson law* or *Poisson probability distribution*, where we denote c to be the number of defects per sample. It is important that the opportunity space for defects to occur be constant from sample to sample.

The Poisson distribution defines the probability of observing c defects in a sample, with c' as the expected mean rate of occurrence of defects per sample. To see more clearly how the Shewhart control chart model for the number of defects in a sample emerges, we now briefly examine this type of sampling situation probabilistically.

Poisson Probability Distribution

If the probability of an outcome, p', of a binomial random variable X is very small (or very large), n is large, and np' is moderately large but, say, less than 20, the binomial probability function of Eq. (13.6) can be approximated by a distribution called the Poisson distribution, with one parameter $\lambda = np'$, defined as follows:

$$P(X = x) = \frac{\lambda^x e^{-\lambda}}{x!}, \qquad x = 0, 1, 2, \dots, \tag{13.16}$$

where λ is the mean of X.

To illustrate the use of Poisson distribution as an approximation of a binomial distribution, we will consider an example. Suppose that some patients may suffer a bad reaction from taking a certain drug. If the probability of any such patient suffering a bad reaction is estimated to be 0.025, what is the probability that exactly 3 patients out of 200 who have taken the drug will have the bad reaction? If we assume that each of the 200 patients has the same probability $p' = 0.025$ of experiencing a bad reaction, the problem is basically a binomial random variable problem. Accordingly, $P(X = 3) = [200!/(3!\ 197!)][(0.025)^3(0.975)^{197}] = 0.1399995$. By Poisson approximation, we have $\lambda = np' = (200)(0.025) = 5$, and hence $P(X = 3) = 5^3 e^{-5}/3! = 0.1403739$, which is very close to the result from binomial calculation. Obviously, Eq. (13.16) is easier to work with. An acceptable rule of thumb is to use the Poisson distribution to approximate binomial probabilities if $n \geq 20$ and $p' \leq 0.05$. If $n \geq 100$, the approximation is excellent if $np' \leq 10$.

The Poisson probability distribution is a very important distribution in its own right. It is widely used in modeling various process and operational phenomena and many natural events, such as the weekly number of burned-out lights in a factory, the number of defects per bolt of a fabric, or the number of sunny days in a monsoon season. It is noted that a Poisson random variable is the number of occurrences of a certain event, which is often a rare event, in a clearly specified sample space such as a fixed time period or a fixed particular dimensional size (e.g., area, volume). The shape of a Poisson distribution is in general skewed to the right, as shown in Fig. 13.16.

Properties of Poisson Distribution.   The following are important properties of the Poisson distribution:

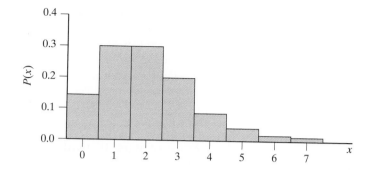

Figure 13.16   Histogram for Poisson distribution

1. The Poisson distribution has only one parameter, λ. If we look at Eq. (13.16), we find that we need to know only the mean, λ, to find the probability for any integer value of X from zero to infinity.

2. An interesting property of the Poisson distribution is that its mean is equal to its variance;
   a. Mean of $X = \lambda$.
   b. Variance of $X = \lambda$.

3. The mean of a Poisson variable is proportional to the size of the sampling unit. For example, if it is known that the average number of defects per square inch is 5, there are an average of 20 defects per 4 square inches and an average of 2.5 defects per $\frac{1}{2}$ square inch.

▶ EXAMPLE 13.5

Suppose that a telephone answering service at a company is designed to take 35 calls per minute. If it is known that the average number of calls is 20 per minute, what is the probability that in any given minute the service will be overloaded?

Solution

Here the expected number of calls per minute is 20 (i.e., $\lambda = 20$). What is the probability that X will exceed 35?

$$P(X > 35) = \sum_{x=0}^{\infty} P(X = x) - \sum_{x=0}^{35} P(X = x)$$

$$= 1 - \sum_{x=0}^{35} \frac{\lambda^x e^{-\lambda}}{x!}$$

$$= 1 - \sum_{x=0}^{35} \frac{20^x e^{-20}}{x!}$$

$$= 1 - 0.999 = 0.001.$$

> **EXAMPLE 13.6**

In another application of the Poisson distribution suppose it is known that in a certain enameling process the mean number of insulation breaks per yard is 0.07. What is the probability of finding exactly two such breaks in a piece of wire 16 yards long?

Solution

Here λ is the expected number of defects in a piece of wire 16 yards long, $\lambda = 16 \times 0.07 = 1.12$. Thus

$$P(X = x) = \frac{(1.12)^x}{x!} e^{-1.12},$$

and for $x = 2$,

$$P(2) = \frac{(1.12)^2}{2!} e^{-1.12} = 0.201.$$

Construction of c Charts from Sample Data

As we have just seen above, the number of defects, c, arises probabilistically according to the Poisson distribution. One important property of the Poisson distribution is that the mean and variance are the same value. Then, given c' (in general, we have used λ), the mean number of defects per sample, the 3 sigma limits for the c chart are given by

$$c' \pm 3\sqrt{c'}. \tag{13.17}$$

When c' is unknown, we must estimate it from the data. For a collection of k samples, each with an observed number of defects, c, the estimate of c' is

$$\bar{c} = \sum_{i=1}^{k} \frac{c_i}{k}, \tag{13.18}$$

where c_i is the number of defects in the ith sample. Hence control limits for the c chart may be established from the formula

$$\text{UCL}_c = \bar{c} + 3\sqrt{\bar{c}},$$
$$\text{LCL}_c = \bar{c} - 3\sqrt{\bar{c}}. \tag{13.19}$$

If the LCL_c is calculated to be less than zero, a value of zero is used.

Example of c Chart Construction

We consider next the process of the installation of front and rear bumpers on automobiles. The sample size here is the number of cars built on a given shift, since all autos were inspected for the bumper installation process. This number varied slightly from shift to shift, but only by about 2%, so a constant average sample size of $n = 560$ was used for the construction of the c chart. A total of 11 different types of defects were possible. An explanation of the defects follows.

1. PGM 180 degrees out: One of the mounting plates was placed with the wrong orientation, 180 degrees off.
2. Rear bumper boss missing: One of the screws on the bumper was not installed.
3. Rear bumper fit.
4. PGM missing: One of the mounting plates is not present.
5. Rear bumper damaged.
6. Wrong rear bumper.
7. Front bumper damaged.
8. Bumper no stocks: The bumper was not on the line at the point that it would typically be installed, and therefore was not placed on the car.
9. Front bumper loose.
10. Front bumper boss missing: One of the screws on the bumper was not installed.
11. Rear bumper loose.

   Charting was initiated after the defect data for 25 consecutive shifts were available. Table 13.4 provides the defect data for these 25 samples. A c chart for all defects was constructed. The calculations for the c chart centerline and upper and lower control limits are given below.

TABLE 13.4	Defect Data for First 25 Samples for the Bumper Assembly Example		
Sample	Number of Defects	Sample	Number of Defects
1	16	14	16
2	14	15	15
3	28	16	13
4	16	17	14
5	12	18	16
6	20	19	11
7	10	20	20
8	12	21	11
9	30	22	9
10	17	23	16
11	9	24	31
12	17	25	13
13	14		

Centerline:

$$\bar{c} = \frac{\text{total number of defects}}{\text{total number of samples}}$$

$$= \frac{400}{25}$$

$$= 16.$$

Control limits:

$$\text{UCL}_c, \text{LCL}_c = \bar{c} \pm 3\sqrt{\bar{c}}$$

$$= 16 \pm 3\sqrt{16},$$

$$\text{UCL}_c = 28,$$

$$\text{LCL}_c = 4.$$

Figure 13.17 shows the c chart for all defects for the 25 samples collected. By applying the same criteria for reading the statistical signals as were used for the p chart, it was determined that three separate instances indicating out-of-control conditions were evident in the data.

Sample	Signal
3	Point above upper control limit
9	Point above upper control limit
24	Point above upper control limit

In an effort to determine the possible reasons for the large numbers of defects associated with the three samples that have c values above the upper control limit, a Pareto diagram was constructed for all 11 defect types. This diagram is shown in Fig. 13.18.

Two of the more frequently occurring defects, front and rear bumper loose, were determined to be caused by essentially the same problem: a difficulty the assembly line workers were having in aligning the bumper with the mounting bracket under the car frame. A cartlike fixture was designed that would hold a bumper at the proper elevation, and the worker would then guide the bumper into place so that it could be tightened properly. After implementing this procedural change, the process charting was continued.

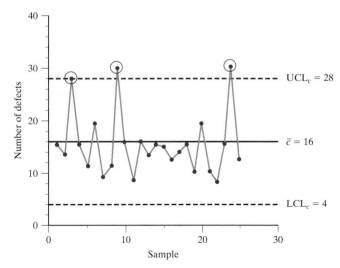

Figure 13.17   *c* chart for first 25 samples for the bumper assembly process

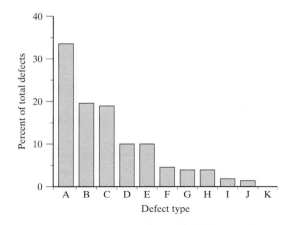

Defect types in descending order of importance are
A. Bumper no stocks: This means that the bumper was not on the line at the point at which it would typically be installed, and therefore was not placed on the car.
B. Front bumper loose.
C. Rear bumper fit.
D. Rear bumper loose.
E. PGM 180 deg out: One of the mounting plates was placed with the wrong orientation—180 degrees off.
F. Front bumper damaged.
G. Wrong rear bumper.
H. Rear bumper boss missing: One of the screws on the bumper was not placed.
I. Front bumper boss missing.
J. Rear bumper damaged.
K. PGM missing: One of the mounting plates is not present.

Figure 13.18   Pareto diagram for all defects for the data from samples 1 to 25 of the bumper assembly process

Table 13.5 gives the data for the next 12 shifts after the new bumper mounting fixture was put into use. Figure 13.19 shows the c chart continuation using these data and extending the centerline and control limits for the first 25 samples. It is clear from the chart that a change in process performance has occurred. At sample 32 a run of eight consecutive points below the centerline is detected. This run continues through the end of the chart—sample 37. This signal is indicating that the defect rate has been reduced. It would appear that the action taken at the process has led to a significant process improvement.

TABLE 13.5	Data for the Bumper Assembly Process After the Loose-Fits Problem Was Addressed
Sample	*Number of Defects*
26	12
27	9
28	11
29	12
30	13
31	12
32	15
33	8
34	7
35	8
36	11
37	9

Figure 13.19 c chart for all 37 samples of the bumper assembly process

13.6   Control Chart for Number of Defects per Unit

Although in c chart applications it is common to comprise a sample of only a single unit or item, the sample or subgroup may be comprised of several units. Further, the number of units per subgroup may vary, particularly if a subgroup is an amount of production for the shift or day, for example. In such cases the opportunity space for the occurrence of defects per subgroup changes from subgroup to subgroup, violating the "equal opportunity space" assumption on which the c chart is based. We need to create another statistic to overcome this problem. Such a statistic may be the average number of defects per unit or item. The symbol u is often used to denote the average number of defects per unit:

$$u = \frac{c}{n},$$

(13.20)

where c is the total number of defects per subgroup of n units. For k such subgroups gathered, the centerline on the u chart is

$$\bar{u} = \frac{\text{total defects in } k \text{ subgroups}}{\text{total number of items in } k \text{ subgroups}}$$

$$= \frac{\sum_{i=1}^{k} c_i}{\sum_{i=1}^{k} n_i}.$$

(13.21)

The control limits for the u chart are then given by

$$\bar{u} \pm 3\sqrt{\frac{\bar{u}}{n}}.$$

(13.22)

Example of the Use of the u Chart

The process by which moonroofs are installed in automobiles is subject to several types of critical defects:

- Wind noise.
- Water leaks.
- Binding during retraction.
- Squeaks and rattles.

Because of numerous customer complaints, the assembly plant was asked to collect some data on the occurrence of these concerns so that statistical charting and associated process study could take place. As a result of a shortage of labor

TABLE 13.6	Defect Data for Moonroof Installation Example				
Sample	Sample Size, n	Number of Defects per Sample, c	Average Number of Defects per Unit, u	LCL_u	UCL_u
1	16	23	1.44	0.49	2.25
2	20	30	1.50	0.59	2.16
3	26	35	1.35	0.68	2.06
4	8	12	1.50	0.13	2.61
5	22	29	1.32	0.62	2.12
6	29	35	1.21	0.72	2.02
7	31	50	1.61	0.74	2.00
8	13	15	1.15	0.40	2.35
9	28	36	1.29	0.71	2.04
10	23	38	1.65	0.64	2.10
11	19	24	1.26	0.57	2.18
12	23	32	1.39	0.64	2.10
13	14	24	1.71	0.43	2.31
14	29	34	1.17	0.72	0.72
15	27	38	1.41	0.70	2.05
16	15	25	1.67	0.46	2.28
17	22	26	1.18	0.62	2.12
18	22	24	1.09	0.62	2.12
19	14	22	1.57	0.43	2.31
20	16	17	1.06	0.49	2.25
21	22	33	1.50	0.62	2.12
22	16	21	1.31	0.49	2.25
23	14	18	1.29	0.43	2.31
24	5	9	1.80	0.00	2.94
25	13	18	1.38	0.40	2.35
26	19	26	1.37	0.57	2.18
27	10	12	1.20	0.26	2.48

time from one day to another and the time required to make the water test, the data collected varied in sample size, one sample per shift being collected.

Table 13.6 provides the defect data for 27 samples collected over a period of 9 days. The table includes the sample size for each shift, the total number of defects observed, and u, the average number of defects per unit. Upon using the data in Table 13.6, the centerline and control limits for the u chart to be constructed were determined as shown below. Since the sample size varies, the control limits will also vary from one sample to another.

$$\overline{u} = \frac{\sum_{i=1}^{k} c_i}{\sum_{i=1}^{k} n_i}$$

$$= \frac{706}{516}$$

$$= 1.37.$$

An example calculation for control limits follows.

$$\text{Control limits} = \bar{u} \pm 3\sqrt{\frac{\bar{u}}{n}}$$

$$= 1.37 \pm 3\sqrt{\frac{1.37}{n}}.$$

For sample 1, $n_1 = 16$, and therefore

$$\text{LCL}_{u_1} = 0.49,$$

$$\text{UCL}_{u_1} = 2.25.$$

Figure 13.20 shows the u chart for the first 27 samples for the moonroof installation example. Note that the control limits vary from sample to sample. Applying the same criteria for reading statistical signals that were used for the p and c charts, we find no reason to believe that the process is not operating in a state of statistical control. However, the defect rate is high, almost 1.4 defects per unit. It was therefore decided to investigate the cause(s) of the large number of defects.

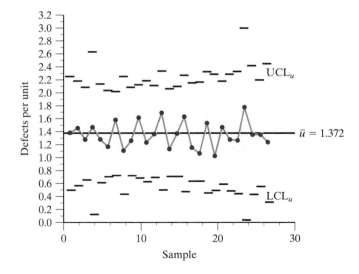

Figure 13.20 u chart for the first 25 samples for the moonroof installation example

An investigation of the specific nature of the moonroof defects revealed that the most prevalent defect was water leaks. A check of the warranty data confirmed that this seemed to be a serious problem. After some further study it was decided to try a new type of seal. Sampling was continued after use of the new seal was begun. The data in Table 13.7 show the results of the next seven shifts. These data were plotted as a continuation of the original u chart with the centerline extended and the control limits calculated using the original $\bar{u}$ value. The results are shown in Fig. 13.21.

TABLE 13.7	Data for the Moonroof Example After the New Seal Was Implemented				
Sample	Sample Size, n	Number of Defects per Sample, c	Average Number of Defects per Unit, u	LCL_u	UCL_u
28	10	8	0.80	0.26	2.48
29	14	14	1.00	0.43	2.31
30	11	8	0.73	0.31	2.43
31	29	14	0.48	0.72	2.02
32	19	7	0.37	0.57	2.18
33	19	12	0.63	0.57	2.18
34	45	25	0.56	0.85	1.90

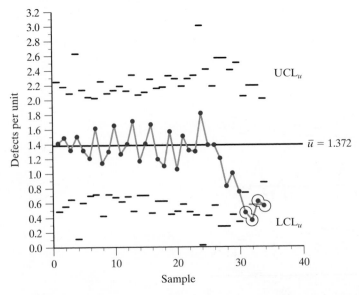

Figure 13.21   Continued u chart for the moonroof example after implementing the new seal

An evaluation of the u chart in Fig. 13.21 shows the following out-of-control signals to be evident:

Sample	Signal
31	Point below lower control limit
32	Point below lower control limit
33	Run below centerline
34	Point below lower control limit
34	Run below centerline

It is clear that the process has changed—in particular, that the average number of defects per unit has been significantly reduced. The u values are now in the range of 0.4 to 0.7, while the original centerline for the chart was almost 1.4.

■ EXERCISES

13.1 What types of situations call for quality characterization by attribute data instead of variable data? Give several examples.

13.2 Is it possible to use attribute data to describe quality characteristics that are inherently of a variable-measurement nature? Give an example of this type of situation and explain the possible advantages and disadvantages of doing so.

13.3 When using attribute data, why are operational definitions for defects important for proper sampling and charting?

13.4 What is meant by "using a product control mentality" in the development of operational definitions for defects for attribute control charting? What are the problems and shortcomings of this approach to the development of operational definitions?

13.5 Why might it be useful to record all occurrences of all defect types as opposed to simply denoting a part as defective as soon as the first instance of any one of many possible defects is found?

13.6 Given the sample of parts in the accompanying diagram having defect types A, B, C, and D, find:
(a) The number of defects.
(b) The number of defectives.
(c) The fraction defective.
(d) The average number of defects per unit.

Defect types: A, B, C, and D

13.7 After 100% inspection of a batch of 2000 parts, it was found that there were 73 defectives.

(a) How many different ways could a sample of $n = 10$ parts be selected from this batch to get 2 defectives and 8 nondefectives?

(b) Find the probability of the sum of all the outcomes in part (a).

(c) What is the probability of getting at most 3 defectives in a sample of size $n = 10$?

13.8 Over the last year, a manufacturer produced approximately 100,000,000 parts that were 100% inspected, and 300,000 defects on those parts were identified.

(a) Find the probability of getting 3 defects in a batch of $n = 1000$ parts.

(b) What is the probability of getting more than 3 defects in a sample of size $n = 1000$ parts?

13.9 Machined parts made from extruded aluminum bars are tumbled in a finishing operation to remove burrs. Samples of $n = 20$ parts are taken from each batch after it finishes the tumbling cycle and are checked for burrs. The fraction of parts with burrs is then plotted on a p chart. For some time, the process has been stable at an average fraction defective of 0.08. Occasionally, the person making the chart notices that a point falls on the lower control limit. These occurrences have been ignored because it is felt that such low p values are good and need not be worried about. His boss tells him that having points on the lower control limit means the process is out of control, and therefore he needs to be looking for special causes. Comment.

13.10 Discuss the pros and cons of using a u chart instead of c chart when:

(a) The sample sizes vary from one sample to the next.

(b) The sample size remains constant.

(c) The sample size is $n = 1$ item.

Give specific reasons in each case.

13.11 The metal body for a spark plug is made by a combination of cold extrusion and machining. The occurrence of surface cracking following

the extrusion process has been shown by Pareto diagrams to be responsible for producing virtually all of the defective parts. To identify opportunities for process improvement, it was decided to construct a p
chart for the process. During one shift, 25 samples of size $n = 100$ were
collected and the results shown in the table below were obtained.

Sample	Number of Defectives	Sample	Number of Defectives
1	1	14	4
2	6	15	3
3	3	16	3
4	1	17	2
5	3	18	4
6	2	19	4
7	0	20	4
8	2	21	2
9	4	22	5
10	1	23	2
11	5	24	6
12	4	25	3
13	5		

(a) Use these data to establish the centerline and control limits for a p
chart.
(b) Construct the p chart and comment on the state of the process. Sampling and charting was continued on the next shift, and the data for
the next 15 samples are as shown in the table below.

Sample	Number of Defectives	Sample	Number of Defectives
26	9	34	9
27	9	35	3
28	8	36	13
29	1	37	8
30	7	38	6
31	5	39	5
32	7	40	8
33	8		

(c) Plot the newly collected data on the chart constructed in part (b).
What can be said at this point regarding the state of the process?

13.12 Past examination of the process that produces automobile quarterpanels
has demonstrated that scratches are producing the majority of defective

panels. It is proposed to study the process with a c chart. The engineers involved with the process have kept track of the number of defects for samples of size $n = 100$ for the last 60 shifts, as shown in the table below.

Sample	Number of Defects	Sample	Number of Defects	Sample	Number of Defects
1	3	21	1	41	9
2	12	22	2	42	11
3	4	23	0	43	9
4	9	24	4	44	7
5	6	25	2	45	5
6	12	26	0	46	5
7	5	27	5	47	9
8	8	28	3	48	7
9	5	29	1	49	5
10	9	30	0	50	10
11	8	31	3	51	7
12	9	32	5	52	3
13	4	33	6	53	8
14	3	34	4	54	7
15	10	35	0	55	10
16	3	36	2	56	8
17	5	37	4	57	5
18	8	38	0	58	6
19	8	39	2	59	10
20	6	40	1	60	6

(a) Calculate the centerline and control limits for the c chart.
(b) Construct the c chart. Interpret the chart and comment on what may be occurring in this process.

13.13 In casting a certain type of flywheel, it was observed that many parts were defective and being scrapped because of hot tears. Hot tears are voids in the cast metal that are likely to occur at weak points in the shape, as shown. Sampling has indicated that 3.70% of the parts cast will be defective because of one or more hot tear defects.

Hot tears

(a) If a sample of size $n = 7$ is drawn, how many ways could two defectives occur? What is the probability of getting exactly two defectives?

(b) If a shipment of 50 parts is to be sent, what is the probability that none will be defective?

13.14 It is desired to study an injection molding process with a p chart. However, because the number of parts produced per hour varies, the individuals studying the process are uncertain as to how the data should be analyzed. The data shown in the table below summarize the number of parts produced and the number of defectives observed for 20 consecutive hours of operation for the process. Construct the three different variable-sample-size p charts and compare and contrast the information that each shows.

Sample	Sample Size	Number of Defectives	Sample	Sample Size	Number of Defectives
1	50	2	11	47	2
2	44	2	12	48	0
3	61	3	13	57	3
4	53	0	14	61	1
5	55	3	15	64	4
6	62	4	16	58	0
7	66	2	17	58	2
8	50	2	18	63	1
9	64	0	19	60	4
10	52	2	20	67	3

13.15 Suppose that a p chart is being used to monitor an electrostatic painting process. Parts are being painted in large batches. About 125 parts are hung by an operator on each rack, and 20 racks are painted per batch. Following the completion of the painting operation, a rack is sampled at random from the batch, and each part on the rack is examined. Samples from 30 consecutive batches were collected and the number of defective parts for each was noted; $\bar{p}$ was calculated to be 0.045. Since the sample size, n, varied between 115 and 129 with an average of $\bar{n} = 124$, a p chart with limits based on $\bar{n} = 124$ was constructed. One of the p values was 0.12 for a sample of size $n = 117$. Does this point initially show up on the p chart as an indication of an out-of-control condition? Is this actually an indication of a shift in the mean fraction defective?

13.16 Why should defects such as scratches on an injection molded part be classified as defects for the purpose of developing c charts for the molding process even if they are not on a surface that ultimately will be visible to the consumer? What is the major consequence of excluding such scratches from the data collection and charting process?

13.17 For each of the seven types of control charts presented in this chapter (p charts, variable-sample-size p charts, np charts, c charts, and u charts):

(a) Explain the situation for which each is most suited.

(b) Summarize the advantages and disadvantages associated with each.

13.18 In the production of a certain chemical liquid, a number of impurities can be present to degrade the quality of the resulting product. Four specific impurities are of interest. Fifty samples (1 liter each) were collected, and the number of suspended particles of each type of impurity were counted. The results of this data collection activity are shown in the accompanying table.

Sample	Impurity 1	Impurity 2	Impurity 3	Impurity 4
1	6	0	0	9
2	0	1	5	9
3	4	2	2	8
4	1	2	0	1
5	3	0	2	7
6	1	5	1	5
7	2	0	0	7
8	3	1	1	8
9	4	2	3	9
10	0	0	0	3
11	3	0	4	13
12	2	2	3	8
13	3	3	4	6
14	1	3	3	5
15	3	2	2	8
16	2	0	2	7
17	3	1	1	1
18	4	0	3	10
19	1	0	3	6
20	5	0	3	4
21	1	3	1	9
22	1	5	4	4
23	3	4	1	5
24	3	4	8	12
25	2	3	2	7
26	3	0	3	7
27	9	8	3	8
28	5	3	6	8
29	5	2	0	8
30	5	1	1	4
31	3	1	1	11
32	6	2	1	5
33	2	2	1	5
34	2	0	3	8
35	0	0	3	4

(Continued)

Sample	Impurity 1	Impurity 2	Impurity 3	Impurity 4
36	4	3	3	9
37	4	0	3	6
38	1	1	4	4
39	2	1	1	6
40	4	5	2	5
41	2	4	4	9
42	6	0	1	6
43	2	1	0	5
44	8	0	0	4
45	2	1	1	8
46	1	1	0	5
47	1	4	2	10
48	5	0	5	8
49	1	1	1	5
50	3	0	1	5

(a) Construct a c chart for this process from the data based on all defects, and a separate c chart for each defect. Comment on the appearance of each of the charts.

(b) Construct a Pareto chart for the defects. Which defect(s) should problem-solving efforts first focus on? Why? Consider in your answer the results from both the control charts and the Pareto diagram.

13.19 Five identical assembly lines are producing power supplies. The defective rate on one of the five lines is considerably higher than that on the other four, so an investigation into the reasons for this is begun. Samples of size $n = 50$ power supplies are gathered randomly throughout the day for each of two shifts for a total of 15 days. The results are shown in the first table.

Sample	Number of Defectives	Sample	Number of Defectives
1	6	16	4
2	4	17	2
3	5	18	5
4	3	19	4
5	1	20	10
6	8	21	6
7	2	22	2
8	4	23	4
9	3	24	4
10	3	25	5
11	7	26	4
12	2	27	6
13	6	28	2
14	2	29	3
15	4	30	1

(a) Construct a p chart from these data. Show all calculations.
(b) Comment on the appearance of the p chart and note any statistical signals that appear.
(c) Construct a scatter diagram of shift vs. the number of defectives. Assume that all odd-numbered samples were collected on the first shift, and all even-numbered samples were collected on the second shift. Comment.

One of the operators on the line suggests repairing the protective coverings on the sharp edges of the conveyor, thus eliminating the potential for scratches to occur at several locations. Sampling is continued after this improvement, with data as shown in the second table.

Sample	Number of Defectives	Sample	Number of Defectives
31	2	41	3
32	0	42	0
33	4	43	0
34	2	44	0
35	1	45	1
36	2	46	0
37	2	47	2
38	3	48	2
39	0	49	2
40	3	50	2

(d) Continue the p chart with the new data, using the same limits as before.
(e) Comment on the appearance of the chart now. What can be said about the change that was made?

13.20 The data shown in the accompanying table were collected for a process that involves the assembly of odometers that are then assembled into an instrument panel. The sample is collected during a 15-minute period of operation for a robotic assembly cell. Four such periods are monitored each shift.

Sample	Size	Number of Defectives	Sample	Size	Number of Defectives
1	93	3	11	83	5
2	87	3	12	86	5
3	61	4	13	92	1
4	94	4	14	80	2
5	79	5	15	89	3
6	87	3	16	78	5
7	90	5	17	84	6
8	88	6	18	82	4
9	95	4	19	90	1
10	79	4	20	87	2

(Continued)

Sample	Size	Number of Defectives	Sample	Size	Number of Defectives
21	95	4	26	91	2
22	86	2	27	88	4
23	77	1	28	78	6
24	83	4	29	93	5
25	83	4	30	85	2

(a) Make a p chart for these data using average limits. Show all calculations. Carefully interpret the chart.

(b) Make a p chart using individual limits for each sample. Show all calculations. Carefully interpret the chart.

(c) Make a standardized p chart. Again, show all calculations. Carefully interpret the chart.

(d) Compare the three charts and comment on any similarities or differences you see among the charts.

13.21 The data shown in the accompanying table were collected from a process making glass bottles for a carbonated beverage bottling company. The sample size was 500.

Sample	Number of Defectives	Sample	Number of Defectives
1	7	16	2
2	6	17	3
3	3	18	7
4	4	19	2
5	2	20	5
6	4	21	3
7	3	22	3
8	5	23	5
9	3	24	4
10	8	25	6
11	10	26	0
12	1	27	5
13	5	28	5
14	8	29	4
15	3	30	4

(a) Construct a p chart from these data. Show all calculations. Interpret the chart.

(b) Construct an np chart from these data. Again, show all calculations and interpret the chart.

(c) Compare the appearance of the two charts and comment on any similarities or differences between the charts.

13.22 An injection molding process is being monitored for the number of defective parts it is producing. Samples of 75 parts are collected on each shift, and the average number of defectives per sample is found to be rather high. See the data in the first table.

Sample	Number of Defectives	Sample	Number of Defectives
1	10	14	14
2	4	15	8
3	2	16	4
4	10	17	6
5	2	18	8
6	4	19	6
7	8	20	8
8	12	21	16
9	6	22	22
10	10	23	2
11	10	24	8
12	14	25	16
13	4		

(a) Construct a p chart from the data. Show all calculations.

(b) Analyze this chart and comment on any statistical signals.

It was suggested that perhaps the p chart was not the most appropriate chart to use for this process, that instead a c chart should be used. Again, data were collected from each shift for 25 shifts (sample size = 75) and are summarized in the second table.

Sample	Number of Defects	Sample	Number of Defects
1	10	14	9
2	8	15	13
3	12	16	9
4	10	17	12
5	12	18	11
6	13	19	16
7	14	20	17
8	25	21	25
9	10	22	19
10	7	23	14
11	12	24	27
12	14	25	15
13	10		

(c) Construct a c chart from these data. Show all calculations.

(d) Analyze this chart and comment on any statistical signals.

(e) What, if anything, does this chart tell you that the p chart did not? What is the next step that should be taken in analyzing this process? Explain your reasoning.

13.23 An injection molding process is making plastic gamepieces for games of checkers. There are two lines: one line makes the red pieces, and the other line makes the black pieces. Currently, the red line is producing a lot of defective parts for unknown reasons. Data collected from this line are shown in the accompanying table.

Sample	Size	Number of Defectives	Sample	Size	Number of Defectives
1	31	1	14	45	1
2	40	9	15	30	2
3	36	3	16	34	2
4	29	3	17	34	3
5	33	4	18	39	3
6	41	3	19	35	4
7	36	5	20	28	1
8	32	1	21	36	3
9	31	3	22	32	1
10	40	4	23	40	2
11	34	3	24	34	3
12	29	1	25	42	4
13	38	3			

(a) Construct a p chart from this data using individual limits. Show all calculations. Carefully interpret the chart.

(b) Construct an np chart from this data using individual limits. Show all calculations. Carefully interpret the chart.

(c) Compare the appearance of the two charts. Which is easier to interpret?

13.24 A process takes cut lengths of wire and bends them to form paper clips. Recently, a large number of paper clips have been malformed, resulting in a large quantity of scrap for this process. Data are collected on the number of defectives produced, and the results are shown in the accompanying table.

(a) Construct a p chart from these data using average limits. Show all calculations.

(b) Analyze this chart and comment on any statistical signals.

Sample	Size	Number of Defectives	Sample	Size	Number of Defectives
1	80	4	16	70	7
2	72	2	17	71	4
3	68	3	18	82	7
4	81	3	19	73	4
5	75	4	20	67	6
6	70	4	21	80	2
7	82	5	22	79	7
8	64	1	23	81	2
9	73	1	24	81	4
10	78	5	25	78	4
11	70	3	26	79	1
12	74	4	27	84	4
13	83	1	28	66	2
14	76	12	29	74	4
15	66	4	30	80	4

13.25 A process is cutting lengths of plastic tubing that will be used for fish tanks and aquariums. Data are collected on the number of defective pieces of tubing that are being produced to either verify or disprove the assumption that the process is in control with respect to the number of defectives being produced. A summary of the data collected is shown in the accompanying table.

(a) Construct a standardized p chart from these data. Show all calculations.

(b) Analyze this chart and comment on any statistical signals.

Sample	Size	Number of Defectives	Sample	Size	Number of Defectives
1	98	7	16	98	4
2	83	3	17	91	0
3	79	1	18	80	3
4	107	4	19	85	2
5	92	0	20	100	1
6	84	5	21	97	3
7	86	4	22	92	5
8	91	4	23	85	8
9	108	5	24	104	4
10	99	2	25	93	3
11	89	3	26	74	6
12	102	6	27	83	2
13	78	3	28	101	5
14	84	4	29	87	5
15	101	7	30	90	1

13.26 A process is making plastic water bottles for bicycles, and four defects can occur in the course of production: missing cap, scratches, flow lines, and excess flash. Data are collected on these defects in hopes that it will reveal opportunities for improvement. The results from the first 25 samples of size $n = 100$ are shown in the accompanying table.

Sample	Missing Cap	Scratches	Flow Lines	Excess Flash
1	6	6	4	7
2	3	4	1	8
3	8	5	0	8
4	3	3	1	5
5	9	3	3	6
6	3	1	3	5
7	4	3	2	3
8	5	2	6	10
9	4	4	7	6
10	6	5	3	6
11	4	1	4	7
12	3	6	3	8
13	2	3	2	11
14	3	0	3	5
15	5	3	3	7
16	5	0	1	6
17	5	1	3	7
18	4	3	4	5
19	6	3	3	9
20	2	2	3	5
21	4	1	2	10
22	6	2	4	12
23	5	1	6	12
24	2	1	2	8
25	2	1	7	5

(a) Construct a c chart from these data. Show all calculations.

(b) Analyze this chart and comment on any statistical signals.

(c) Construct individual c charts for each of the four defects. Comment on the similarities and differences among the four charts.

(d) Make a Pareto diagram of the four defects and comment on where problem-solving efforts should be concentrated, taking into consideration both the charting results and the Pareto diagram. Explain your reasoning.

13.27 From a previous study it was found that 99% of the defects that occur in an accounts payable process at a large manufacturer are missing purchase order numbers. Data are now being collected on the number of times this occurs each day in hopes that control charts made for this process will reveal opportunities for improvement. A summary of the data collected from 30 consecutive workdays is shown in the accompanying table.

Day	Missing POs	Day	Missing POs
1	12	16	4
2	6	17	6
3	8	18	4
4	16	19	14
5	14	20	8
6	6	21	8
7	12	22	10
8	8	23	10
9	16	24	2
10	4	25	12
11	12	26	12
12	10	27	8
13	12	28	4
14	6	29	4
15	10	30	8

(a) Construct a c chart from these data. Show all calculations.

(b) Analyze this chart and comment on any statistical signals.

13.28 The final inspection for a truck engine is used to check for visual defects. The three defects that occur on this product most frequently are paint runs, improper label alignment, and misstamped serial number. An investigation into the reason behind the high defect rate is begun, with samples collected every day for 25 days. The results are shown in the table below.

Sample	Number of Engines Examined	Paint Runs	Label Misaligned	Misstamped Serial No.
1	105	3	2	4
2	101	5	1	5
3	102	3	2	5
4	95	4	1	7
5	100	3	3	3
6	98	2	1	3
7	102	1	2	2
8	101	3	3	5
9	97	0	2	1
10	100	3	1	2
11	106	4	1	3
12	101	6	1	3
13	103	3	1	4
14	96	5	4	5
15	101	5	1	3

(Continued)

Sample	Number of Engines Examined	Paint Runs	Label Misaligned	Misstamped Serial No.
16	101	8	1	2
17	98	6	2	3
18	103	7	1	4
19	104	6	1	1
20	100	5	3	5
21	100	9	6	4
22	99	7	4	2
23	102	5	6	4
24	105	8	5	2
25	102	10	6	1

(a) Construct four u charts from these data. Show all calculations.

(b) Analyze these charts and comment on any statistical signals that appear.

(c) Make a Pareto diagram of the three defects and indicate which is occurring most frequently. Where should quality improvement efforts be focused?

13.29 A process that folds cardboard boxes that will eventually hold model airplane kits is experiencing problems with two frequently occurring defects: badly folded/mangled boxes and rips/tears in the cardboard. The data collected over the last 30 shifts are shown in the accompanying table.

Sample	Size	Bad Folds	Ripped Box	Sample	Size	Bad Folds	Ripped Box
1	52	3	1	16	47	2	3
2	50	6	3	17	58	10	2
3	47	0	2	18	49	3	1
4	50	3	3	19	47	0	3
5	50	2	2	20	52	6	2
6	52	5	3	21	50	2	2
7	51	3	1	22	50	3	4
8	54	5	0	23	53	6	4
9	48	2	1	24	51	3	5
10	57	9	2	25	52	4	3
11	55	6	1	26	51	4	5
12	50	4	1	27	53	5	7
13	50	3	4	28	50	3	6
14	52	6	6	29	46	2	5
15	50	5	1	30	49	3	3

(a) Construct three u charts from these data. Show all calculations.
(b) Analyze these charts and comment on any statistical signals.
(c) Construct a Pareto diagram of the defects and suggest where problem-solving efforts should concentrate. Explain your reasoning.

13.30 A supplier of plexiglass sheets is concerned that they contain an excessive number of scratches. It is decided to run a statistical study of the process to determine if it is in statistical control and to look for possible improvement opportunities. A defect is a scratch over 0.1 inch long. Each sample collected was actually 25 one-foot-square pieces of plexiglass. The study produced the data shown in the first table.

Sample	Number of Defects	Sample	Number of Defects
1	2	26	9
2	5	27	2
3	3	28	1
4	3	29	3
5	5	30	1
6	4	31	4
7	6	32	4
8	0	33	5
9	5	34	4
10	5	35	4
11	4	36	2
12	4	37	5
13	3	38	4
14	6	39	6
15	4	40	5
16	0	41	3
17	6	42	1
18	3	43	2
19	6	44	2
20	11	45	2
21	6	46	3
22	5	47	6
23	6	48	7
24	7	49	3
25	7	50	6

(a) Calculate the centerline and control limits for a c chart for these data. Plot the c chart and interpret it. Clearly label all out-of-control points and comment on the state of the process.
(b) Assume that you were able to find the source(s) of any problems identified by the signals on the chart and that appropriate corrective actions were taken at the process. Recalculate the control limits after removing these samples.

(c) Following these corrective actions, several additional samples were collected and are shown in the second table. Use the control limits calculated in part (b) as a basis to plot the new data. If you deem it appropriate, construct new control limits for the data and comment on the process.

Sample	Number of Defects	Sample	Number of Defects
51	3	76	3
52	2	77	1
53	4	78	3
54	1	79	2
55	0	80	1
56	1	81	2
57	1	82	0
58	0	83	1
59	3	84	3
60	0	85	4
61	2	86	2
62	1	87	1
63	3	88	2
64	0	89	0
65	3	90	3
66	3	91	0
67	2	92	2
68	0	93	0
69	1	94	1
70	2	95	2
71	3	96	1
72	1	97	1
73	4	98	3
74	5	99	1
75	3	100	0

13.31 A manufacturer produces a batch of 5000 gears and based on 100% inspection finds 300 defective gears. A sample of $n = 8$ gears is then randomly selected from the large batch.

(a) The sample of size $n = 8$ contains three defective gears. How many different ways could the sample be selected from the batch to obtain this result?

(b) Find the probability of obtaining the outcome in part (a).

(c) What is the probability of getting more than one defective in a sample of size $n = 8$?

13.32 As part of a study by the Internal Revenue Service (IRS), it is estimated that the rate of defects (e.g., missing Social Security numbers and math errors) on individual tax return filings is 0.15 defects per return.

(a) Find the probability of getting three defects in a sample of $n = 20$ returns.

(b) What is the probability of getting more than two defects in a sample of size $n = 20$ returns?

(c) Find the percent of returns that contain at least one defect.

13.33 Labels are applied to beverage bottles using a labeling machine. Some of the labels are found to be improperly aligned. To investigate this situation, 30 beverage bottle samples are collected and the number of misaligned labels is recorded. As is evident from the data in the accompanying table, the number of bottles in each sample varies. Construct and interpret a p chart with individual limits for the data. Construct and interpret a standardized p chart for the data. Compare and contrast the information that each chart shows.

Sample	Sample Size, n	Number of Defectives, d	Sample	Sample Size, n	Number of Defectives, d
1	29	1	16	31	1
2	18	2	17	40	2
3	21	7	18	43	3
4	32	1	19	38	1
5	48	4	20	24	2
6	47	1	21	30	3
7	29	5	22	34	5
8	41	1	23	26	3
9	27	3	24	22	4
10	27	5	25	27	3
11	31	6	26	37	4
12	39	6	27	39	1
13	36	1	28	45	1
14	27	1	29	40	1
15	27	1	30	48	2

13.34 A lithographic printer has a sheet-fed press that represents a capacity bottleneck. A study is undertaken to identify the frequency with which various types of problems (e.g., alignment errors, paper jams, electrical faults, ink problems, and operator mistakes) lead to production downtime on the press. Over the course of five weeks, data are collected every shift on the reason for unplanned press stoppages. These data are shown in the table below.

Shift	Alignment Error	Paper Jam	Electrical Fault	Ink Problem	Operator Mistake	Other
1	2	3	0	0	0	1
2	3	5	2	3	2	0
3	1	4	1	0	1	0
4	0	3	1	1	0	0

(Continued)

Shift	Alignment Error	Paper Jam	Electrical Fault	Ink Problem	Operator Mistake	Other
5	1	2	1	1	1	1
6	2	4	1	1	1	1
7	2	3	0	2	1	0
8	0	6	1	2	1	1
9	4	4	1	1	2	1
10	3	5	2	0	1	0
11	1	3	0	3	1	0
12	3	5	1	2	1	1
13	2	4	1	3	1	1
14	0	4	0	1	1	0
15	2	0	2	2	0	1
16	1	4	1	2	1	1
17	2	4	1	1	1	1
18	1	5	0	0	0	0
19	1	6	0	2	1	1
20	3	2	1	2	1	1
21	1	3	1	2	1	0
22	3	6	2	3	1	1
23	2	1	2	3	2	0
24	0	3	2	2	2	1
25	1	3	1	0	1	1

(a) Using the data in the table, construct a c chart based on the sum of all the defects per shift. Comment on the appearance of the chart.

(b) Construct a Pareto chart for the defects. Which defect(s) should problem-solving efforts first focus on? Why?

(c) For the purpose of focusing problem-solving efforts, what does it mean if a c chart is in control or out of control?

After the c chart was constructed, the production supervisor reminded those on the study team that the press was not operating continuously during all the shifts; the available production time was reduced for some of the shifts because of planned maintenance. Furthermore, in order to meet customer deadlines, several of the shifts lasted longer than eight hours. The actual available production time for each shift was as shown in the next table.

Shift	1	2	3	4	5	6	7	8	9	10
Available Hours	8	5	8	6	8	8	10	8	6	8
Shift	11	12	13	14	15	16	17	18	19	20
Available Hours	10	8	8	7	8	9	9	8	10	10
Shift	21	22	23	24	25					
Available Hours	8	6	8	8	8					

(d) Given that the opportunity for press stoppages differs from shift to shift, construct a u chart for the data. Comment on the appearance of the chart. What action should be taken?

13.35 An on-line grocery shopping and delivery service has had complaints from customers about order inaccuracies. To investigate this problem, data on the number of orders delivered and the number of those orders that are incorrect are recorded for a month for one of the delivery vans. As is evident from the data in the accompanying table, the number of deliveries varies from day to day. Construct and interpret a p chart with individual limits for the data. Construct and interpret a standardized p chart for the data. Compare and contrast the information that each chart shows. What action(s) should the on-line grocery undertake?

Sample	Number of Deliveries	Number of Incorrect Orders	Sample	Number of Deliveries	Number of Incorrect Orders
1	47	7	16	25	7
2	36	9	17	18	6
3	28	1	18	27	2
4	45	1	19	25	9
5	31	0	20	44	5
6	39	7	21	49	1
7	27	3	22	64	6
8	57	4	23	29	10
9	42	7	24	35	5
10	35	3	25	46	9
11	29	10	26	38	8
12	46	5	27	41	1
13	41	10	28	69	10
14	41	7	29	38	6
15	50	7	30	21	7

14

Attribute Control Chart Implementation for Process Improvement: Two Case Studies

14.1   Introduction

This chapter presents two case studies that document the use of statistical thinking and methods in the improvement of processes and systems. Both case studies deal with the characterization of quality on an attribute basis. The presentations of the case studies cover not only the statistical aspects but also some of the team-building and group dynamics aspects as well. The underlying purpose here is to develop a sense of the basic elements of the implementation methodology that must be invoked to realize full and lasting benefits from the use of statistical thinking and methods for process improvement. Although the situations described here are quite different from one another, it is important to recognize the elements of commonality in attacking and solving the problem.

14.2   Implementation of SPC Methods: A Case Study on "Press 120"—Visual Defects in a Molding Process

The following case study discussion[1] is grounded in fact and details the experiences of a plastics plant in applying statistical thinking and methods for quality and productivity improvement. Some liberties have been taken with the presentation to accommodate to space limitations and to emphasize certain key points.

---

[1]R. J. Deacon, "Reduction in Visual Defects in an Injection Molding Process," *Society of Plastics Engineers, 1983 National Technical Conference Proceedings*, p. 97.

Background of the Problem

A plastics plant, heavily involved in the injection molding of automobile components, was experiencing high and often erratic scrap rates on some products. To meet the shipping requirements it was necessary to run with a considerable amount of overtime and often to run backup dies on other presses to increase production quantity. Some troubleshooting activities were ongoing, and some improvement had been realized. As production orders increased during a time when automobile sales were increasing, the problem became more and more critical. During this period, the people in this plant were being introduced to the concepts and techniques of statistical process control, and management saw an excellent opportunity not only to attempt to solve a pressing problem using SPC, but also to give those in the plant some experience in implementing the techniques and some confidence in the power and importance of their use. The product involved in this particular case study is a radiator grille.

Formulating a Strategy for Implementation

The first step was to form a group of people who were involved in the process, including operators, setup persons, supervisors, molding engineers, quality people, area managers, and statistical methods facilitators. The group began meeting to try to understand the problem and get everyone's input. It was decided to begin to collect data from the process during production and to use statistical charting to study the process variations. This required that the group address such issues as: What quality characteristics should be monitored? How should these characteristics be measured/observed? How many samples should be taken? How often should they be taken? What other useful process information should be recorded? Other relevant questions were addressed as well. Because this was the first implementation experience in this plant, it was decided that the statistical methods facilitators would assume the primary data collection and analysis responsibilities. Ultimately, such activities would become a natural part of the system as the role of inspectors, supervisors, and others slowly changed under the adoption of the new way of doing business.

Prior to initiating the use of control charts, an orientation meeting was held for press operators and machine setup personnel to discuss the purpose of the study. The important role that these people would play in effecting improvement was discussed, and the management commitment to improvement was emphasized. Suggestions were sought, and as it turned out later, information obtained from press operators and setup persons was central to solving two major problems. The operators also attended an 8-hour presentation on basic chart concepts and philosophies so they could understand the importance of using the statistical approach to problem solving.

Operational Definitions

Perhaps the most difficult task in the entire process for the group was agreeing on what the defects were and how they were to be specifically defined. It was decided to focus on visual flaws such as scratches, black spots, and flow lines, but common agreement on what was important and what was not came slowly. The group realized the importance of writing down detailed operational definitions for defects so that all concerned were "playing by the same rules" and so that the most quantitative and sensitive measures of improvement opportunities could be developed and invoked. Initially, Table 14.1 was developed to serve as a basis of observation and measurement of quality. Later, the operational definitions were modified to provide for more sensitivity in finding the root causes of the problems. The first set of operational definitions, as seen in Table 14.1, were a bit too product control oriented. For example, scratches may be acceptable in a product control sense on the nonvisual areas of the grille, but the mechanism(s) putting the scratches on the grille may put them anywhere, depending on its random orientation at some point. Hence, all scratches should be observed and counted to increase the chart sensitivity.

Sampling Frequency and Data Collection

At the onset of the study, the occurrence of defects was rather high and process behavior was erratic. Because of this, it was decided to collect samples rather frequently so that the clues to the causes could be developed based on defect

TABLE 14.1	Operational Definitions for Defects
Defect	Criteria
Splay	Not acceptable on high surround or horizontal grille bars
Scratches/scuffs	Not acceptable on visual surround (use grease pencil test)
Flash	Not acceptable; must be repaired
Oil	Minimal amount; should be wiped clean so that residual amount will be removed by acid etch
Grease	Not acceptable if it cannot be wiped clean
Flow lines	Not acceptable in high surround (use grease pencil test); maximum $\frac{1}{4}$-inch flow line at base of surround (see sample); flow line acceptable in lower rail and between vertical ribs
Short shots	Front surface and visual ribs must be full; acceptable— only back rib behind paint step allowed short to $\frac{1}{8}$ inch maximum (see sample)
Burns	Unacceptable on visual surfaces; minimal amount acceptable on back rib surfaces
Sinks	Unacceptable on visual surface; acceptable on bottom base rail at outer vertical ribs (see sample)
Sprue on part	Part requires being scrapped
Material buildup	No residue allowed
Black spots	Unacceptable as groups on surface; acceptable in minimal number under surface
Defective trim	Gouges and chatter unacceptable; flash standing can be repaired

data and information collected on other process conditions. The group spent many hours discussing potential special and common causes of variation (defect occurrence). This was deemed important to understanding what process data (temperatures, pressures, and material conditions) ought to be collected to provide a basis for doing the detective work that would ultimately lead to the root causes of the defects. Among other things, a process condition sheet was placed at the machine each shift, and operators and setup persons were urged to jot down observations and/or thoughts they might have during the shift.

In the early stages of a study such as this, when process behavior seems erratic and the problems seem to be many, it is probably useful to collect samples quite frequently to reveal clearly the short-term and rapidly occurring changes in the process. For an attribute-based study such as this, sample sizes can be small if the occurrence of defects is great (an average of 2 to 3 defects per sample may be sufficient). As problems are found and solved, the process becomes more stable and defect rates fall, so less frequent sampling is in order, although sample size requirements will increase.

Fault Diagnosis Experience

After collecting a sufficient number of samples (25 is a bare minimum), a control chart for number of defects per item was constructed (sample size = 10). Figure 14.1 shows this initial u chart and indicates that a very erratic out-of-control process is present with a very high average defect rate ($\bar{u} = 4.5$ defects per item). To begin to shed some light on the problem, a Pareto chart for defects was also constructed and is shown in Fig. 14.2. As discussed in Chapter 6, a

Figure 14.1   **Initial control chart (u chart) for all defects from the Press 120 molding process**

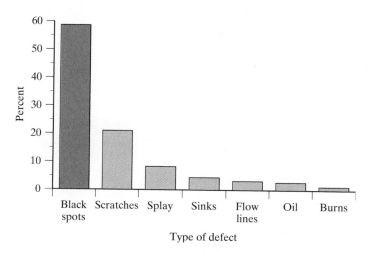

Figure 14.2   **Pareto diagram for defects on radiator grille**

Pareto chart is a bar graph that indicates the relative frequencies of occurrence of the various types of defects observed and is useful in identifying major problems, which may then be attacked in some precedence order. The Pareto diagram based on this initial u chart shows that black spots (degraded material at the surface of the part) constitute the vast majority of all defects at this point. For the time being, the group decided to focus its attention only on this problem.

In an effort to identify the root cause(s) of the black spot defect, the group used the "fishbone" or cause-and-effect diagram. Such a diagram, discussed in Chapter 6, is shown in Fig. 14.3; the major cause categories include the human element, methods, machines, and materials associated with the process. For each of these major potential cause categories, the group began to add increasing levels of detail to the diagram in an effort to reach the root cause level. For example, initial attention focused on the machine, in particular because the screw was felt to be worn out through extended processing of this raw material. It was reasoned that the screw had not been properly maintained and the material was adhering to the worn areas on the screw, which raised the temperature of the molten material to locally high levels at times, giving rise to degradation of the raw material. Particles of degraded material would then be carried along the screw and ultimately injected into the die cavity, where they would appear on the surface of the part. After considerable discussion, the group recommended that a new screw be ordered and installed.

After some delay, the new screw was installed in the machine. Statistical charting was continued all during the diagnostic period and after installation of the new screw as well. Figure 14.4 shows the u chart immediately before and after the screw was changed. Note that this chart is for black spot defects only and is based on a sample size of 5. Initially, black spots completely disappeared. However, a few days later the black spots returned with the same level

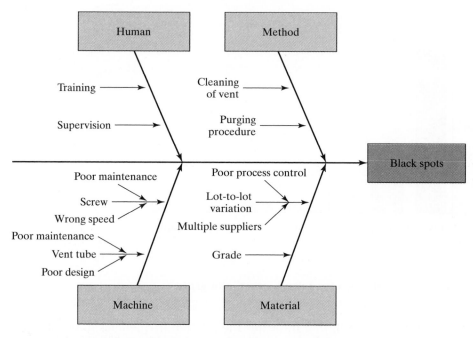

Figure 14.3   **Cause-and-effect diagram for black spot defects**

Figure 14.4   **Continuation of Press 120 u chart showing the effect of changing the screw**

of intensity as before the screw was changed. It was therefore evident that the root cause of black spots had not been properly identified. Because charting continued as it should, this fact was clearly borne out by the u chart. This points to the important role that statistical charting plays in showing clearly the short- and long-term effects (if any are present) of taking a certain remedial action. Clearly, changing the screw did have an impact on the problem of black spots, but this action did not lead to a permanent cure, so the root cause must still be active in the system. Some theorized that extensive cleaning of the inside of the barrel at the time the screw was changed may have led to the temporary reduction in black spots on the parts.

The group continued to meet to discuss possible solutions to the black spot problem. It was noted that the design of the gas vent tube on the barrel of the machine was somewhat different from that on most other machines and that operators complained of vent tube clogging and difficulty in subsequent cleaning. It was theorized that material accumulating in the vent tube port became overheated and then periodically broke free and continued down the barrel or was pushed back into the barrel during cleaning. A new vent tube port design was proposed to eliminate this problem. The maintenance department built the new vent system and installed it on the machine. Figure 14.5 shows the u chart after the new vent system was installed and indicates the immediate and lasting effect of this process change on the occurrence of black spots. Note that Fig. 14.5 is for black spot defects only.

With this major problem solved and a great deal of experience and confidence gained, the group enthusiastically continued to seek further improvement opportunities. As a result of the elimination of the black spot problem, the u chart was beginning to show signs of more stable process behavior, and the defect rate had dropped from 4.5 defects to slightly more than 1.0 defect per part on the average. The group prepared another Pareto diagram, which showed that scratches were now the most frequently occurring defect (see Fig. 14.6).

Figure 14.5   **Continuation of the Press 120 u chart showing the effect of the new vent tube port design**

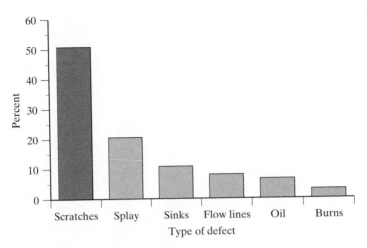

Figure 14.6   **Pareto diagram for defects on a radiator grille after the new vent tube system was installed**

Several theories for the occurrence of scratches were discussed, but the most popular was that offered by a press operator, who felt that many of the molded parts were being scratched by the metal lacings of the press conveyor belt. The logical solution would be to use a continuous (vulcanized joint) belt. However, this solution was not viewed favorably because its unit cost was approximately double that of the existing belt and the installation time was much longer, making it more difficult to remove the belt for cleaning and maintenance purposes. A considerable amount of resistance was therefore present and a strong effort was made by some to move away from this possible root cause and look for other possible causes of scratches.

One of the statistical methods facilitators in the plant was convinced that the belt lacings were causing many of the scratches. To test the belt-lacing theory, he applied a soft latex coating to the metal lacings. The u chart with $n = 8$ for scratches only, shown in Fig. 14.7, shows that scratches disappeared completely for some time. Eventually, the soft coating wore away and broke off with belt flexing, and scratches suddenly appeared again on the u chart (see Fig. 14.7). This emphasizes the important role that continual charting plays in verifying the effect that a given remedial action may or may not have on the process. With evidence from the u chart that the belt metal lacings were causing the problem, it was now possible to convince the area supervisor that action should be taken to install a vulcanized belt on the press. As evidenced by Fig. 14.8, this led to elimination of the problem of scratches.

It is important to point out that all presses in this plant had conveyor belts with metal lacings. Once this root cause was clearly demonstrated, the logical

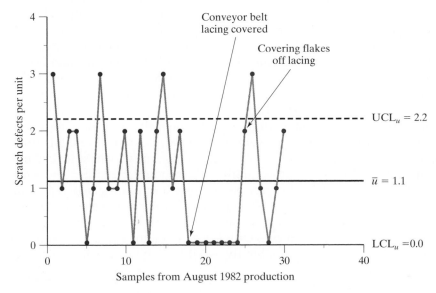

Figure 14.7   Control chart for scratch defects before and after latex coating was applied to the belt lacing

Figure 14.8   Control chart for scratch defects before and after the bonded belt was installed

action would be to replace the belts on all the machines. This emphasizes the fact that many problems revealed from charting on a single machine can be usefully extrapolated to other machines. Although such may not always be the case, given resources and priorities, it certainly helps in thinking through the problem of being faced with putting charts on perhaps hundreds of machines versus selectively charting a few key or representative processes.

With the scratch problem solved, the process defect rate had now fallen from 4.5 defects per part down to 0.17 defect per part. Scrap was noticeably reduced, overtime was eliminated, and running the backup die on another press was stopped. The economic implications here are obviously significant and need not be detailed. However, the group continued to work on improvements and began in earnest to attack the problem of splay. The splay phenomenon was found to be quite complicated, and as a result, statistical experiments were designed and conducted to develop a better understanding of the root cause of this process defect.

As an important footnote, it should be mentioned that over the period of about 3 months after the scratch problem was solved, the defect rate fell from 0.17 defect to 0.08 defect per part, for no apparent reason. The most logical explanation stems from the fact that as quality and productivity improvements were made, those involved in the process became more knowledgeable and concerned about the process. Press setup persons rarely needed to make adjustments, and management utilized the added free time at the press to perform more regular and thorough preventive maintenance on both the press and the mold. These factors undoubtedly would all give rise to improved process performance.

Postscript on This Case Study

The "Press 120" success story grew rapidly in fame, spreading across the company and upward to the highest levels of management. More than a year later, during a management review at this plant, the Division general manager recalled the "Press 120" story and asked to be taken to see the famous press! When he arrived in the area he immediately observed that a belt with metal lacings was installed on the press. When he asked why, the response was somewhat vague. Needless to say, the problem of holding the gains mentioned in Chapter 12 was again an issue here.

14.3   Improvement of an Accounts Payable Process

A significant inhibitor to the widespread use of statistical thinking and methods in "administrative processes" is the belief that these systems are unique and very different from manufacturing processes and that a unique and different approach and set of tools and methods are required to study them. This case study clearly demonstrates that the very same tools used in manufacturing process applications may be applied successfully to an administrative process. In this case study of an accounts payable system,[2] the following concepts and tools will be employed:

- Organization plan.
- Process flowcharts.
- Operational definitions for defects.

---

[2]R. J. Deacon, "Quality Improvement in an Accounts Payable System," Independent Study Project Report, Eastern Michigan University, College of Business, Ypsilanti, MI, 1986.

- p charts, c charts.
- Pareto diagram.
- Cause-and-effect diagrams.

As will become evident as this case study unfolds, the parallels between it and the injection molding study described previously are striking.

   This case study arose within a division of a major automotive manufacturer. This division was then comprised of five manufacturing plants and a central division office. Each plant has its own accounting department that handles its unique cost accounting and financial requirements, while the division accounting office handles accounting functions common to all plants in the division. One of these functions, accounts payable, is the focus of this case study. The accounts payable section in the operations accounting department processes approximately 12,000 invoices each month. The section employs 14 accountants, with each assigned to handle either the production (material that is part of the finished product) or nonproduction (spare parts) invoices for one plant (for the larger plants this task is divided among two people).

   During the summer of 1985, the accounts payable section was experiencing several undesirable and costly symptoms, all of which were a concern to management. Inefficiencies had developed that slowed the payables process to the extent that a backlog of over 6000 invoices had developed. This necessitated working the accountants overtime and hiring an outside agency to open incoming mail and sort the invoices by plant and type (production versus nonproduction). The cost of these inefficiencies was estimated to be approximately $15,000 per month or $180,000 per year.

Organizing for Improvement

Steering Group.   Experience has shown that many quality improvement efforts have failed because management did not establish a responsive environment that was suitable for action. In an attempt to get management support and their active participation in the activities necessary to resolve the problem, a committee consisting of the area's management (manager, section supervisor, and unit supervisor) and a few statistical methods specialists was formed to guide the project. Initially, the steering group described the accounts payable system in the form of a flowchart, and operationally defined the symptoms of poor system performance as well as the various terms and acronyms used in this accounting system. Later, this committee established a goal for the project, assigned others to working-level teams, and determined the role and responsibilities of each person involved with the project, to avoid duplication of effort.

Working-Level Teams.   Two diagnostic teams were utilized to provide more detailed analysis of the problem. A group consisting of the 14 accountants and their supervisor formed one group, which concentrated on analyzing and formulating theories on potential improvements to the accounts payable system. The second team consisted of the statistical methods people in the steering group, which provided services to the accountants' team such as data analysis, meeting facilitation, and training in statistical methods.

Project Goal.   The steering committee formulated the following goal as a focus for the efforts of all teams: "Identify and eliminate the faults in our system that cause inefficient processing of invoices. Eliminate the existing backlog of invoices, and institute actions as required to keep the system 'current' (invoices processed as close to, but not exceeding, 30 days) and under statistical control." The achievement of this difficult task would certainly not take place overnight and not without a lot of hard work, but the working-level teams agreed that it was a worthwhile project to undertake. These teams developed and utilized short-term action plans throughout the project to assure steady progress toward the goals.

Analysis of the Symptoms

Flowchart.   The use of a flowchart initially benefited the group through its visual illustration of the sequence of operations that should be followed by all persons involved with the accounts payable process, and subsequently the chart was used to formulate data collection strategies. Figure 14.9 shows the flowchart for this process. A review of the flowchart for the accounts payable

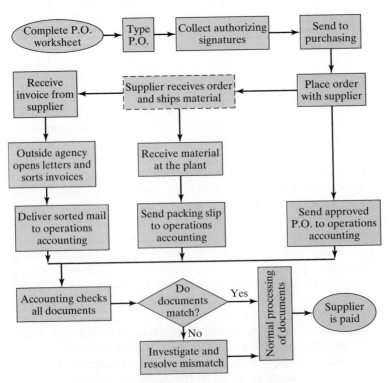

Figure 14.9   **Flowchart of accounts payable system**

process indicates that the purchase notification (order), packing slip, and invoice had to exactly match each other with respect to quantity, price, and so on, before the accountant could pay the supplier. If the documents did not match or there was missing or erroneous documentation, the invoice would go into the "backlog" until the discrepancies were resolved by the accountant (which involved time-consuming conversations with both the suppliers and the people at the plants).

Operational Definitions.   The accountants defined, in writing, the various types of symptoms encountered in the accounts payable process, as well as the acronyms commonly referred to in departmental dialect. This task served two purposes. First, the precise meaning of the words or phrases was clarified to people not directly involved with the process on a regular basis, and second, the working-level teams used these definitions in their efforts to identify the cause of the symptom under investigation. This actually improved the group's efficiency, because theories that did not relate to the symptoms were quickly discarded, which in effect allowed more time to discuss and evaluate theories that did. Some examples of the operational definitions for the defects are

- **No purchase order.** An instance when the packing slip and invoice are received at the accounts payable office before the purchase order.
- **Conversion issue.** Instances where a part is specified for shipment in one unit (by the pound) but purchased in another unit (by the yard).
- **Vendor number differs from log.** An instance where the wrong vendor number was logged in.
- **Logged-in nonproduction.** Situations where the production parts are confused with nonproduction parts.

A review of the operational definitions reveals that

1. They are worded in very simple terms to avoid misinterpretation.
2. They are worded in a way that avoids the description of their possible cause so as not to stifle the group's creativity in future brainstorming sessions, where possible theories on the causes are formulated and discussed.

Prioritizing the Critical Problem

After a review of some historical data that only enumerated the problem, the working-level teams were assigned the tasks of collecting data in a statistical format in order to clarify whether the problem was a chronic or a sporadic issue. This information was important to the steering group because it helped them assign specific tasks to the people in the best position to carry them out. This minimized the confusion, frustration, and cost of the problem-solving effort.

As is the case with manufacturing processes, sporadic problems suggest the presence of special causes of variation which are most efficiently resolved by the people directly connected with the operation (accountant, supplier, buyer), whereas the correction of problems chronically affecting the entire system in a stable manner requires management action on the system (usually through the efforts of a larger group). The team used the Shewhart control chart for the purpose of distinguishing between these two types of process variation. It was important that everyone involved in this quality improvement effort understood the concepts of variation and various tools used in statistical process control.

The goal of the initial data collection effort for this project was to assess the stability of the accounts payable process of each division location and to identify the most common symptoms of the problem. The *u* chart was used for this purpose because the accountant could encounter more than one problem trying to pay the invoice, and the steering group was interested in documenting the daily failure rate for all invoices processed (which meant dealing with uneven sample sizes). The data could also be used to construct a Pareto diagram in which the symptoms are displayed in descending order in terms of their frequency. Information for a *p* chart on the percentage of invoices not able to be processed was also collected, since the steering group was using this indicator to track the team's performance over time until the goal was achieved.

The accountants developed the data collection form that they would be using under the guidance of the statistical methods team, to ensure that the data would be rich in diagnostic content. From the statistical point of view, this involved making sure that "rational samples" were taken (separate data collection for each plant and process) to maximize the control chart's sensitivity in revealing variation due to special causes. From a team-building perspective, allowing the accountants to develop the data collection plan gave them some ownership in the project and set the stage for their support of the subsequent data analysis.

After two weeks of data collection, the *u* chart data were plotted via computer software and the *p* chart data were plotted by hand. The rationale for these actions is as follows:

- Even though the feedback loop with computerized charts is generally slower than that with manually maintained charts, the diagnostic capabilities of the software made its use worthwhile. Trial control limits could easily be developed (by removing samples due to special causes from the data set and/or separating the data into different time periods). Pareto analysis was performed on the symptoms automatically, and control charts for the major symptoms could easily be generated. Furthermore, once the group prioritized its effort on resolving the top concern, the original collection strategy would be discontinued in favor of another that provided more detailed information on that issue. This new data collection strategy will be discussed in more detail later.

- The steering group anticipated that in the future, each accountant would maintain a *p* chart at his or her desk to hold the gains. Charting the daily performance of their process by hand could be done very easily and the feedback loop was short, which would allow them to resolve the cause of any "special variation" on a timely basis.

An evaluation of the control charts revealed that the accountants were experiencing a very high (and unacceptable) failure rate in matching invoices with packing slips and purchase orders. With the exception of two plants, the control charts were stable, which indicated that the employees at the plants were doing the best they could within the constraints of the system. Figure 14.10 displays a typical, stable *u* chart for one of the plants. Since the process was stable, this indicated that the system needed modification (by management) to better suit it to the division's/plant's culture and method of operation. A review of the two charts that were out of control revealed that the cause of the special variation could be tracked down and resolved immediately by the accountant (without management or group assistance). Again, one can see that the interpretation and reaction to the charts in the administrative area is identical to that in the manufacturing area!

The Pareto diagram for the data collection on the process for nonproduction invoices is shown in Fig. 14.11. Upon its review, the steering group immediately agreed to prioritize working on identifying and correcting the cause of the "no purchase order" symptom. A very large return on investment could be realized by the group if this larger problem were corrected, whereas much smaller gains would be made by resolving the others.

Identifying the Cause of the Problem

The first step taken to identify the cause of the "no purchase order" symptom was to list the many possible factors that could explain why the purchase orders were so slow in moving through the system. The group utilized the cause-and-effect diagram to organize the theories. Instead of using the four major categories typically found in manufacturing cause-and-effect diagrams (human,

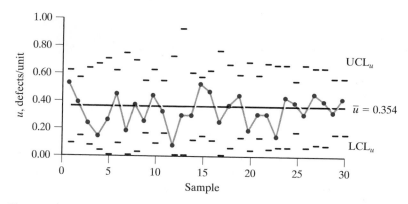

Figure 14.10   *u* chart for all defects from the accounts payable case study

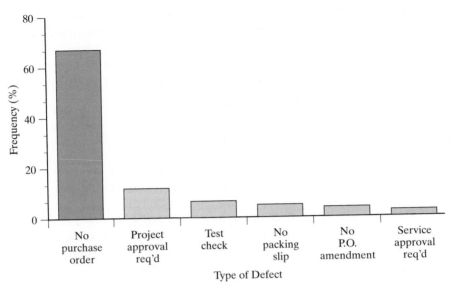

Figure 14.11   **Pareto chart for defects for nonproduction invoices**

methods, material, and machinery), the theories were listed among the four major activities involved with the administrative system (the *plant* placing the order, the *buyer* who quotes the order, the *supplier* receiving the order, and the *accountant* who pays the bill). Figure 14.12 is an incomplete reproduction of

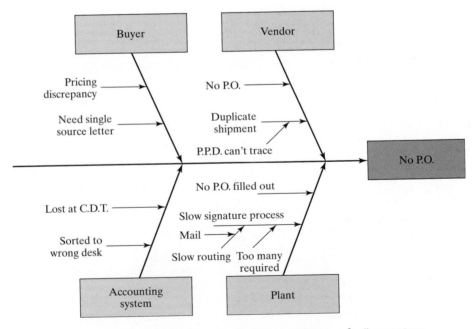

Figure 14.12   **Cause-and-effect diagram for "no purchase order" symptom**

the actual (and more detailed) cause-and-effect diagram. A review of this diagram reveals that the system is heavily dependent on the people at the plants, due to the number of possible factors that can go wrong, which slow the process.

The group utilized the information in the cause-and-effect diagram to develop their second data collection strategy. The goal was to test the theories and quantify their effect on the problem. The group chose to pursue this task by randomly sampling 10 purchase orders each day where 10 critical items were recorded for each order. The selection of the 10-piece sample was based on a trade-off between the cost of collecting the data (the accountants' time constraints) and the need to guarantee the integrity of the data (to make sure that the data provide accurate information on the process). After a month of data collection, the information was entered into a computer program and analyzed several different ways. At first, the data were characterized by calculating simple statistics on such key issues of the system as

- The average time required to obtain all of the signatures/approvals required.
- The average number of signatures on the purchase order.
- The average time needed to transfer the documents from the plant to the operations accounting office.
- The average elapsed time before the supplier's invoice reached the operations accounting office.
- The percentage of the problem contributed by the type of order and the department from which it was initiated.

Afterward, the group posed a few scenarios, such as "What takes place in situations where the material is ordered before the document is written and approved?" They had the computer sort and print out the data associated with such scenarios, and analyzed the data to find common elements that might explain the situation.

The analysis provided several interesting insights relative to the processing of a purchase order. It revealed that orders placed by the maintenance department (for the repair of broken equipment) quite often resulted in the "no purchase order" symptom at the accounts payable office. Analysis also showed that there was variation between the plants with respect to the number of signatures/approvals on the orders. This observation provided a major opportunity for improvement, since each additional signature lengthened the approval process one day! Each manager generally reviewed these documents at the end of his or her day, so the order would not be transferred to the next person until the following day. As far as additional data collection was concerned, the group decided to discontinue the collection of the 10-piece sample (since it had accomplished its goal of revealing possible causal factors), but continued to maintain the p chart to verify the effectiveness of the remedial actions on reducing the percentage of invoices containing errors.

Developing and Implementing Remedial Actions

Upon the conclusion of the data analysis, the group developed action plans to address each issue. The accountants had identified many excellent corrective actions that had proved to be effective in similar situations in their past working experiences. Their recommendations were quite heavily dependent on the amount of support and cooperation given by the people at the plants. Therefore, the initial step in implementing remedial actions was to hold a series of orientation meetings with key people throughout the division. The operations accounting manager met with the division and plant controllers, and later the steering group met with upper management at the division's largest plant to review the project and the data analysis.

It was important to bring out the fact that the control charts continued to show process stability because a natural reaction from the managers would be to accuse the people of violating the procedures set forth by the finance manual. This may be true, but the employees were just doing the best they could with a system that was not conducive to their job objectives. This was especially true with the maintenance people, who had the task of repairing equipment that was needed for manufacturing as quickly as possible. They sometimes could not spare time to fill out paperwork and get several approvals before placing the order, since they were often "under the gun" to get equipment up and running.

As a result of these sessions, the plant appointed its own small team to work with the staff accountants to implement the remedial actions. The plant team consisted of the cost accounting section supervisor, two people from the maintenance department who wrote most of the orders for their department, and the plant purchasing agent (buyer). The various working-level teams implemented many remedial actions, such as:

- Revising local purchase order procedures, such as reducing the number of signatures required and increasing the dollar limit of the local buyer's authority.
- Training the salaried workforce in the procedures for writing a purchase order.
- Publishing a list of account numbers not only to ensure that people used the proper account, but also to advise them of the manager responsible for approving the charges.
- Using laid-off hourly employees as couriers to transport documents between staff and plant accounting offices.

The accountants used the p chart to verify the effectiveness of these actions. Breakthroughs were apparent as the failure rate dropped closer to the steering group's goal.

Holding the Gains

As of early 1987 the group had not yet achieved its goal. The working-level teams at that point were investigating additional methods to either simplify or streamline the process. The performance trends for the three main plants in the division showed considerable improvement. One plant had been able to reduce its failure rate by over 60% (monthly average from 42% in May 1985 down to 15% in March 1986) due to the aforementioned actions. A slight degradation in the performance of this plant in the last half of 1986 was attributed to the increased number of people placing orders who were inexperienced in the purchasing system. This was a signal that more training was needed at the plant. Two other plants also experienced improvements in processing their purchase orders and invoices.

As they continued their never-ending pursuit of system improvements, the accountants continued to monitor the performance of the process via the p chart in order to hold the gains. Their task was to plot the daily reject rate, evaluate the chart's patterns, and resolve out-of-control conditions as soon as possible. Their supervisor's job was to reinforce good data collection behavior, provide any necessary coaching of the accountant, review the chart to identify long-term trends in either system improvement or degradation, and pursue further improvement actions.

To facilitate this process, the statistical methods group provided the accountants with some training on how to construct and maintain p charts. The sessions were customized to the needs of the group. In addition to teaching the charting, the quality and business philosophies of Deming and Juran were also reviewed. There was, however, a serious issue facing the statistical group. The concern was the amount of time needed to calculate the control limits associated with the variable sample sizes encountered by the accountants. Keeping in mind that the accountants were not (and probably did not care to be) statisticians, the statistical group provided each accountant with a table of upper and lower control limit values for the average ($\bar{p}$) failure rate of their process. As a result, they were easily able to note the limit values for the number of invoices they processed, make an assessment of process stability for the day, and react accordingly.

Conclusion

The approach utilized by the group on this administrative project was essentially the same methodology that is typically used to address manufacturing problems. Even though the management of the division had not achieved its goal on the accounts payable project at the time this case study was documented, it had reduced its direct costs by approximately $7500 per month, or $90,000 per year. It does appear that the approach utilized by the group was very efficient at identifying many of the factors causing the problems within the administrative system.

Product/Process Design and Improvement

15

Conceptual Framework for Planned Experimentation

15.1   Introduction

This chapter introduces some important notions and methods for the use of designed experiments. As such, it discusses five fundamental issues central to the successful use of design of experiments for quality design and improvement.

1. The *role* of design of experiments in quality design and control.
2. Several important *guiding principles* that drive sound design-of-experiments practice.
3. How to mitigate the *forces of noise* in the experimental environment.
4. The *contributions of Taguchi* to planned experimentation.
5. *Contrasting philosophies* in dealing with nuisance or noise factors.

The ability to successfully design, conduct, and interpret planned experiments for product and process design and improvement depends on how well these five issues are understood and implemented.

15.2   Role of Experimentation in Quality Design and Improvement

Earlier, considerable emphasis was placed on two important issues:

1. The need to focus on and continually study process performance with an eye toward identifying opportunities for improvement.
2. The need to push the quality issue farther and farther upstream into the engineering design arena.

In both situations, there is a need simultaneously to study the effects that those factors perceived to be important have on product or process performance when they are varied or changed purposefully.

SPC/DOE Relationship

We have seen that a crucial step in the "process" of process control is to diagnose or discover the fault that is causing the problem. Figure 15.1 shows the statistical process control "process" we discussed previously. Often, to get at the root cause of the problem we will need to experiment with the process, purposely changing certain factors with the hope of observing corresponding changes in the responses of the process.

Sometimes the problem is a system problem, as we saw in the gasket case study of Chapter 12. The process is in control, but the variation is too large, the defect rate is too high, and so on. This is signaling a fundamental problem that may not be revealed easily without a comprehensive study of process performance across a range of conditions, often governed by a large number of factors. Without an organized and systematic approach to experimentation, a costly and time-consuming "random walk" approach to looking for effects of change can lead to little or nothing in terms of enhanced knowledge of the process. The design-of-experiments methods discussed in the remainder of this book can help us to take this step in an efficient and reliable fashion.

From the results of the case study in Chapter 12 it is clear that the power of design of experiments can be greatly enhanced if the environment in which the experiments are conducted has been changed through variation reduction methods such as statistical process control. While statistical control of the process is not necessarily a prerequisite to drawing valid conclusions from the results of a designed experiment, it can greatly enhance the sensitivity of the design of the experiment in terms of its ability to detect the presence of real variable effects. A quieter process will allow the effects of small changes in the process parameters to be more readily observed.

Figure 15.1   Process of statistical process control

As we will see in this chapter, there are certain fundamental design measures that we can take to ensure experimental validity. SPC, however, is an important measure that we must consider whenever possible to improve experiment sensitivity. Furthermore, in cases where statistical control of a process has been established, subsequent experimentation and the associated improvement actions taken are more likely to be realized in future operation of the process, since when the process is under statistical control, the future is more predictable!

Role of DOE in the Design Process

As discussed in Chapter 1, a serious shortcoming of past approaches to quality has been the inability to deal rationally with the quality issue early in the product and process development life cycle. We learned in the early 1980's, largely through the work of Taguchi and others in the United States practicing these methods, that parameter selection at the early stages of design can be enhanced by measuring quality by functional variation during use and by using design-of-experiments methods. In particular, the concept of robust design, advocated by Taguchi as part of his model for the design process, shown again here in Fig. 15.2, has proved to be an effective tool for product and process design and improvement.

As we discussed previously, there is an important distinction to be made between testing and experimentation. While both have their rightful place, one should not serve as an alternative for the other. Many have used experimentation—design of experiments—for parameter selection at the product and process design stage. Here the object is to experiment with various combinations of the

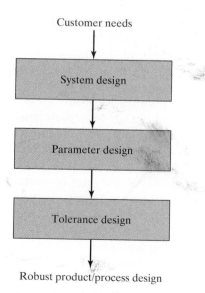

Figure 15.2   Taguchi's model for the design process

important design parameters for the purpose of identifying the particular combination(s) that optimize certain design criteria or performance measures. Prior to the 1980s, many placed a great deal of emphasis on testing—life testing—by subjecting many identical units to field conditions for the purpose of determining the life expectancy of performance. When a design fails, changes are made and the product is retested. The question is: How are those changes identified? The answer is: through a deliberate experimental approach or through more ad hoc procedures (e.g., "Let's try this now"). The life testing of product performance is important but is not a substitute for experimentation to determine what ought to be tested.

15.3   Some Important Issues in Planned Experimentation

Purpose and Nature of Planned Experimentation

Scientists and engineers are commonly involved in experimentation as a means to describe, predict, and control phenomena of interest. The collection of data is a fundamental activity toward the building and verification of mathematical models, whether such models are derived from first principles or are purely empirical in nature. Comparative experiments are an important means to discern differences in the behavior of processes, products, and other physical phenomena as various factors are altered in the environment.

Too often, data analysis, modeling, and inference are stressed at the expense of the activities that embrace the planning and execution of experiments. It is assumed that valid and meaningful data are available either from passive observation of the process or from purposeful experiments and that statistical methods embrace the analysis of such data. It is, however, the planning or design stage leading toward the collection of data that is critical and needs to receive more attention, and it is here that a statistical approach to the design of experiments is so important. If experiments have been designed and conducted properly, the analysis is usually straightforward and often quite simple.

The purpose of most experimental work is to discover the direction(s) of change that may lead to improvements in both the quality and the productivity of a product or process. In the past, there has been a tendency to conduct studies farther downstream, at the process, while the use of design of experiments has been less commonly embraced by the engineering community for product design purposes. This, of course, began to change in the 1980's with the emphasis placed by Taguchi and others on using design of experiments for product design. As recognition of the importance of pushing the quality issue farther and farther upstream into the engineering design process grew, the role of design of experiments in the earlier stages of the product development life cycle grew in proportion. Today, the concurrent or simultaneous engineering of products and their manufacturing processes is receiving widespread attention. Mathematical modeling, computer simulation, and the associated use of designed experiments are all playing a central role in this activity.

In this chapter we introduce some basic and important principles and methods for the design and conduct of experiments, thus laying the foundation for more detailed discussion of the methods, which follows in subsequent chapters. The recognition of these principles is crucial to the use of the tools and methods that follow.

In investigating the variation in performance of a given process, we focus on the identification of those factors that, when allowed to vary, cause performance to vary in some way. Some of these factors are qualitative in nature—categorical variables—while others are quantitative, possessing an inherent continuity of change. The situations we are about to examine may consider both qualitative and quantitative variables simultaneously. In fact, an important advantage of factorial designs that we will soon study is their ability to consider both types of variables within the same test plan.

But more important, there is a growing need for the precise recognition of the relative roles that factors play in governing the nature of product and process performance. Some factors are external and environmental in nature, not directly controllable but nonetheless having an important impact on performance. Some factors have a strong influence on performance on average, while others tend to influence the level of variation in performance. The specific definitions of the various roles that factors play in influencing performance are provided later, and more important, the way in which each type of factor is dealt with is discussed in detail.

Notion of the Mathematical Model

A fundamental task in design of experiments is that of selecting the appropriate arrangement of test points within the space defined by the independent variables. Although many different considerations must come into play in selecting a test plan, none can be more fundamental than the notion of the mathematical model.

Suppose that we are interested in a system involving a mean response η that depends on the input variables $x_1, x_2, \ldots, x_k$. Then we could write that

$$\eta = f(x_1, x_2, \ldots, x_k; \theta_1, \theta_2, \ldots, \theta_p).$$

That is, the mean response η could be expressed as a mathematical function f with independent variables $x_1, x_2, \ldots, x_k$ and a set of parameters $\theta_1, \theta_2, \ldots, \theta_p$.

Sometimes we know enough about the phenomenon under study so that we can use theoretical considerations to identify the form of the function f. For example, a chemical reaction may be described by a differential equation that when solved produces a theoretical relationship between the dependent and independent variables. Even if the form of f is known, the values of the parameters $\theta_1, \theta_2, \ldots, \theta_p$ will generally not be. They will need to be estimated from data collected during operation of the process using designed experiments. The observed data y are therefore represented as

$$y = f(x_1, x_2, \ldots, x_k; \theta_1, \theta_2, \ldots, \theta_p) + \epsilon,$$

where ϵ is the experimental error of observation, assumed to be NID$(0, \sigma_y^2)$.

The problem is that in most situations, little is known about the underlying mechanisms of the process. In most physical situations, at the most, we can say that f represents a relatively smooth response function. But without knowledge of the physical mechanisms, how can we proceed to explore the response surface with the ultimate goal, perhaps, of finding the values for $x_1, x_2, \ldots, x_k$ that optimize the mean response η?

When our knowledge is limited, we must rely on empirical models that serve to approximate the true but unknown model, describing relationships through the data—for example,

$$\eta = b_0 + b_1 x_1 + b_2 x_2,$$
$$\eta = b_0 + b_1 x_1 + b_2 x_2 + b_{12} x_1 x_2,$$
$$\eta = b_0 + b_1 x_1 + b_2 x_2 + b_{12} x_1 x_2 + b_{11} x_1^2 + b_{22} x_2^2.$$

It is important to plan experiments in such a way that several specific empirical forms can be contemplated and the best can be determined through the data once the experiment has been completed.

As an example, let us suppose that f can be represented, for a certain situation, by a simple and flexible graduating polynomial, say,

$$\eta = b_0 + b_1 x_1 + b_2 x_2 + b_{12} x_1 x_2 + b_{11} x_1^2 + b_{22} x_2^2.$$

If we can fit this function to data obtained from an experiment, we can use this equation to more clearly visualize and hence explore the complete nature of the response surface (e.g., perform optimization and sensitivity analysis). Figure 15.3 depicts this situation graphically in terms of both the shape of the true surface and its approximation based on the empirical model.

Whether explicitly recognized as such or not, most experimental studies are aimed either directly or indirectly at discovering the relationship between some performance response and a set of candidate variables that influence that response. In most studies the experimenter begins with a tentative hypothesis concerning the plausible model form(s) to be initially entertained. The experimenter must then select an experimental design that has the ability to produce data that will be capable of

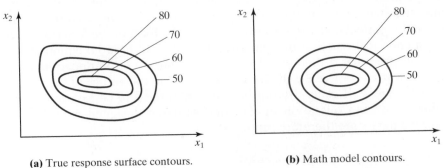

(a) True response surface contours.   (b) Math model contours.

Figure 15.3   **Math model approximation (right) to a true response surface (left)**

1. Fitting the model(s) proposed (i.e., estimate the model parameters).

2. Placing the model in jeopardy in the sense that inadequacies in the model can be detected through analysis.

The second consideration above is of particular importance, to ensure that through a series of iterations the most appropriate model can be determined while others may be proved less plausible through the data. If, for example, a quadratic relationship between temperature and reaction time in a chemical process is suspected, an experiment that studies the process at only two temperatures will be inadequate to reveal this possibility. An experiment with three levels of temperature would, however, allow this possibility to be considered. On the other hand, an experiment using five levels of temperature would be unnecessary and inefficient if the true relationship were quadratic. Figure 15.4 illustrates these situations:

- In Fig. 15.4(a), the relationship is actually curvilinear, but this will never be detected by the data of the experiment because the model chosen is a straight-line model, and, only two levels of the factor are tested.

- In Fig. 15.4(b), a poor model (straight line) has been hypothesized, but model checking will reveal this and help propose a better model because data are available to reveal the weakness of the model—the center point.

- In Fig. 15.4(c), if the relationship is known to be a straight line, many levels of temperature in the experiment are unnecessary, since only two points would be required to fit the line. The number of levels of the factors

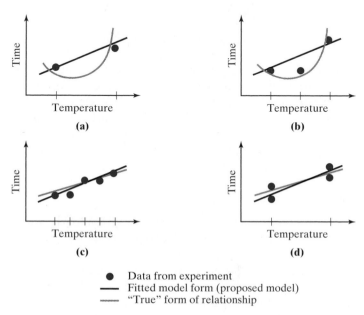

●   Data from experiment
—   Fitted model form (proposed model)
—   "True" form of relationship

Figure 15.4   Possible temperature–time relationships

in the experiment should bear a definite relationship to the order of the model proposed to describe the process through the data.

- In Fig. 15.4(d), if we know a priori that the relationship is a straight line, the best "test" plan would be to study only two, more extreme levels of temperature and use additional tests for replication—that is, to observe the amount of experimental error associated with the response y.

In short, the experimental design should be responsive to the mathematical model being proposed but have the ability to subject the model being considered to hazard, so that if it truly is inappropriate, that can be revealed through the data.

Sequential and Iterative Experimentation

There is always the temptation to carefully design one big experiment that will consider all aspects of the problem at hand. This is dangerous, for several reasons:

1. If erroneous decisions and hypotheses about the state of affairs are made, considerable time and experimental resources may be wasted and the end product may provide little useful information or direction in terms of what to do next.

2. If knowledge of the underlying situation is limited a priori, many factors may be suspected as being important, requiring a very large experiment in terms of number of tests. Ultimately, only a small subset of variables will be found to be of major significance.

3. In the early stages of experimentation, knowledge of the specific ranges of variables that ought to be studied is not always available. Furthermore, the metrics to be employed to define the variables, responses, or even what responses to observe may not always be clear in the early stages of experimental work.

4. One large experiment will necessarily cause the testing period to be protracted, making it more difficult to control the forces of nuisance variation.

For these reasons it is much more desirable to conduct an experimental program through a series of smaller, often interconnected experiments. Such provides for the opportunity to modify hypotheses about the state of affairs concerning the situation at hand, quickly discard variables shown to be unimportant, change the region of study of some or all of the variables, and/or define and include other measures of process performance. Experimental designs that can be combined sequentially are very useful in this regard.

Revelation of the Effects of Change

Often, the variables of importance that govern a process are not clearly known a priori. It is desirable to be able to study several variables together but independently observe the effect of a change in each of the variables. Furthermore, it may be important to know if a variable effect varies with the conditions of the process (i.e., when other variables take on varying levels). When such information is sought, the design—the arrangement of the tests—becomes very important. Consider the following example.

Figure 15.5   One-variable-at-a-time experiments

Suppose in a given situation that five variables seem to be important. One approach to experimentation would be to hold four constant and run tests varying only one variable. The results might look something like Fig. 15.5. This process would then be repeated for each of the other variables. There are several problems with this one-variable-at-a-time approach. One major problem is that most variables do not influence the response of interest independent of the others. The following example illustrates this fact.

EXAMPLE 15.1

Two factors, temperature and pressure, are thought to affect chemical reaction time. Two experimenters, Mr. X and Mr. Y, each run a series of tests to study the specific ways in which temperature and pressure affect reaction time.

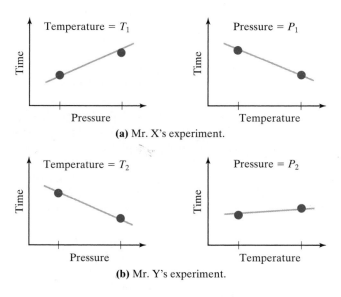

Figure 15.6   One-variable-at-a-time example experiment

Using the one-variable-at-a-time approach, we find that the results might be entirely different, due to different selections of the fixed levels for the variables that are not currently being tested. Figure 15.6 illustrates such possible outcomes as a result of different settings of fixed variables between Mr. X and Mr. Y. It is clear from Fig. 15.6 that the effect of temperature depends on the level of pressure and vice versa. We will later refer to such behavior as signaling the presence of variable interactions.

In summary, one-variable-at-a-time experimentation suffers from significant problems.

- It generally gives rise to a large number of tests—in particular, too many levels of each factor.
- It fails to recognize variable interdependencies, which results in a highly conditional or narrow interpretation.

We will soon see that such experimentation creates other problems:

- It promotes a systematic test sequence where it may be easy to have unknown forces changing over the experiment—biasing the results—so validity is questionable.
- It does not lend itself well to experimental considerations such as blocking, that is, the isolation of known effects of external variables that are not being studied.

In the example that follows we discuss further the problem of test arrangement to reveal important variable effects and offer some suggestions on how to deal with this problem.

Comparative Analysis Between Two Sandmills

A vinyl plant employs vertical sandmills for the dispersion of pigments. The pigment dispersion process produces colorants for the manufacture of vinyl film. As part of a quality improvement project, the plant was considering the acquisition of a new type of sandmill, a horizontal sandmill. The manufacturer of the new machine claimed that the horizontal sandmill would improve both productivity and quality. It was decided to perform some tests on the claimed merits of the horizontal sandmill in comparison with the vertical sandmill. Arrangements were made for some trial runs on the horizontal sandmill.

The measures used to judge the relative performance of the two sandmills were fineness of the grind and color strength. The grind fineness is measured on a Hegman block, which can be read as Hegman values on a scale from 0 (coarse) to 10 (fine). In this example we are concerned only with the grind fineness in terms of Hegman values.

At first glance it seemed attractive simply to run a few trials on the horizontal sandmill and compare the results with the available production records of batch

Hegman values of a representative vertical sandmill in the plant. It was quickly brought out that certain pigments were much more difficult to grind than others, depending on their hardness, or equivalently, the color of the pigment. For instance, red is soft and blue is hard. Also, it was conjectured that relative machine performance could be similar for some pigments but much different for others. Which pigment should be chosen for the trial runs on the horizontal sandmill? Or should more than one type of pigment be tested? If so, how many and what kind?

Another issue raised concerned the degree of pigment loading. The horizontal sandmill manufacturer maintained that his machine should be run with higher-percent-pigment mixes to realize the claimed merits in quality and productivity. Therefore, to run the comparative tests, at least two pigment loading levels must be chosen.

In addition to the two concerns above, there were many other factors to be considered before the trial runs could begin. One of these was to determine the settings of some key parameters, such as temperature and flow rate of the sandmill for the trial runs.

Finally, the idea of using historical Hegman values of the vertical sandmill for comparative analysis must be abandoned to avoid both large variability and any bias due to possible operational and environmental changes over time. Instead, the tests should be run side by side on vertical and horizontal sandmills to provide a more sensitive and valid comparison between the two.

It was ultimately decided for each particular machine setting that two types of pigment (red and blue) and two different pigment loadings (10% and 15%) should be tested on each sandmill. Now the questions that remain are: How many tests should be run, and how should the tests be arranged in the independent variable space? Consider the two possible arrangements illustrated in Fig. 15.7.

Arrangement 1 is somewhat haphazard, with test conditions randomly distributed in the pigment type, pigment loading space. Therefore, it is not possible to observe the effect of changing pigment loading only, for example, since two tests do not exist for which pigment loading changes but pigment type and machine type remain fixed. When we contrast the outcomes (Hegman values) of any two trials, several things are changing simultaneously. It is difficult to sort out exactly what is affecting change in the outcome and what is not.

Arrangement 2 in Fig. 15.7, however, provides for the opportunity to learn much about the relationships among the three variables and the Hegman value. By selecting the same combinations of pigment type and pigment loading for each sandmill, direct contrasts in Hegman values (shown by the dashed lines) between the two sandmills can be analyzed. Some specific features of arrangement 2 follow.

1. The effect of changing any of the three variables alone can be observed.

2. The possible joint effects of two or more variables on Hegman value can be observed. For example, if the Hegman value is influenced by the type of sandmill only at a certain level of pigment loading, or is influenced differently at different levels of pigment loading, this joint effect can readily be observed with arrangement 2 but not with arrangement 1.

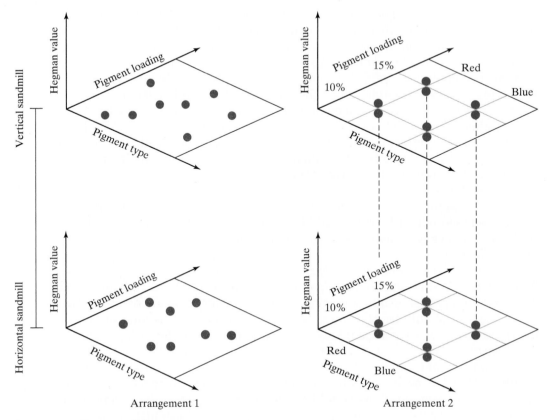

Figure 15.7   Two possible test arrangements

3. The total number of tests is generally much smaller using arrangement 2 than arrangement 1.

4. Because fewer levels are chosen in arrangement 2 for each of the variables to be tested, it is more feasible to replicate all the test conditions, as shown in Fig. 15.7. Replicated responses provide a database for the estimation of experimental error.

However, as shown, arrangement 1 is capable of estimating some curvilinear relationship of the variables on the response, if one exists. If there should be a need to fit a second- or higher-order model for a given situation, arrangement 2 may be extended or augmented to have three or more levels for some or all of the variables under study. Arrangement 2 is generally referred to as a factorial design. For the sandmill study this means to test all possible combinations of the three variables at two levels each, amounting to eight unique tests. Geometrically, the eight unique test conditions could be represented as a cube, as shown in Fig. 15.8. For example, the front lower-lefthand corner represents a test for which the conditions are:

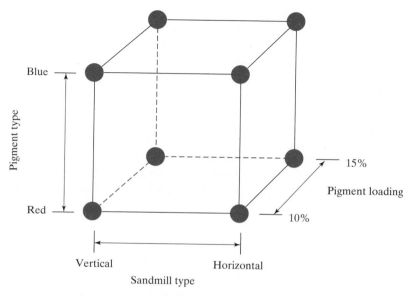

Figure 15.8   Geometric representation of the sandmill experiment

> *Sandmill type:* vertical.
> *Pigment type:* red.
> *Pigment loading:* 10%.

The information obtainable from this type of test plan is quite comprehensive. In particular, we can assess independently:

- The average individual effect of each variable on the response.
- The joint, or interdependency, effect of any two variables.
- The joint, or interdependency, effect of all three variables.

Such information would constitute a much more relevant assessment of relative sandmill performance than would a simple comparative analysis. The results of such an assessment might not only be useful for making the decision "to buy or not to buy" but could lead to the use of different loadings for different pigments on a sandmill to improve quality and productivity.

15.4   Dealing with Noise in the Experimental Environment

The experimental study of any phenomenon is made difficult by the presence of noise, sometimes called experimental error. Many factors not under study are varying during the experiment. Although the variation of any one such factor may produce only a very small change in the performance measure under study, the fact that many such forces are at work simultaneously produces a noticeable and sometimes sizable level of system noise. When the system is stable, these are the forces of common-cause variation. Sometimes we refer to such variation as *white noise*, to emphasize its random nature. Such variation may cloud or mask the effect of changes in the factors under study in an experiment.

Some noise sources in the system may be of a more spurious nature. Such sporadic disturbances, often of sizable magnitude, can further contribute to clouding the results of experiments and if occurring in a somewhat systematic fashion may even bias the results. Structured or systematic variation is sometimes referred to as *colored noise*. How can we better understand and perhaps mitigate the forces of noise in the experimental environment? There are several things we ought to consider at the planning stage to address noise in the environment. These will now be described.

General Strategies for Dealing with the Forces of Noise

Statistical Control/Stability Analysis

If the phenomenon under study is already a feasible and ongoing process, the pursuit of improvement opportunities through experimentation can be enhanced considerably by employing the techniques of statistical process control. In this way, spurious or sporadic sources of variation can be identified and through specific action removed. Achievement of a stable process will greatly contribute to our ability to observe more readily the effects of purposeful process change. Continued study in this fashion will further help us to observe the persistence of changes that might be introduced.

Once a process is stabilized, continued attack on the common-cause system will lead to a progressively "quieter and quieter" process, further enhancing our ability to observe the forces of purposeful process change through experimentation. As a process becomes "quieter" (i.e., the common-cause variation is reduced), it enhances the sensitivity of control charts in detecting the occurrences of certain special causes whose effects are relatively small.

Experimental Design Strategies

We must recognize that in many situations, the notion of a stable, ongoing process has little meaning. In the early stages of product or process design, prototype or pilot-plant testing, or in comparing performance of existing and proposed new methods, the luxury of having a stable, consistent process is not present. It is perhaps for this reason (among others) that the body of knowledge we know as experimental design was cultivated in the first place. Under such situations we must strive for four things:

1. Attempt to identify major sources of variation and take action to ensure that their presence is absent from the comparisons we make within an experiment. The technique of blocking is useful for this purpose.

2. Counteract the forces of unknown systematic variation over the period of the experiment by randomization of the tests so that such variation is uniformly and randomly distributed across the trials conducted.

3. Include replication in the experimental test plan. Multiple tests at the same conditions will provide comparisons that directly measure the amount of variation/experimental error. A priori knowledge of the approximate level of variation expected will enable us to make some judgment about the amount of replication required to discern the presence of real differences in performance of a certain amount.

4. Include confirmatory testing as part of the experimental strategy. The ultimate value of designed experimentation lies with the "persistence of effects" when improvement opportunities are revealed through the experiment. Hence it will be important that additional trials are run under specific conditions determined from the analysis to verify the improvement opportunities revealed from the experiment.

Two of these issues, blocking and randomization, are very fundamental issues that address the problems associated with the sensitivity and validity in making comparisons and drawing inferences on those comparisons within an experiment. We now briefly discuss these two important issues.

Blocking and Randomization in Experimental Designs

Blocking Known Sources of Variation

The example that follows illustrates the roles that the techniques of blocking and randomization play in improving the design of experiments. Nork Tool Company claims that their new cutting tools, called Nork-V tools, provide a longer life than the Cutgo-T tools for similar jobs. To check whether or not Nork's claim is valid, it is planned to machine nine 1048 steel bars using nine Nork tools and nine Cutgo-T tools, all on a single machine with the same machine setting. The tools and workpiece materials are selected as shown in Fig. 15.9.

The bar-to-bar material properties variation (mainly hardness) is known to have considerable effect on the tool life. We can see this effect in the data of Table 15.1 by looking up and down each of the columns under each tool type. The tool life values vary tremendously, even for the same tool. If we were to calculate an estimate of the experimental error using the tool life values within each column (tool), we would find these estimates inflated by the difference that the material—the bar—makes on the tool life. This large variation estimate would make it difficult to see a true difference in the two tools, if such really exists. Therefore, each bar is cut into halves, with one machined using a Nork tool and the other a Cutgo-T tool. This arrangement will block out any bar-to-bar material differences by allowing for the observation of the difference in tool life within the same bar. In such paired comparisons a much larger portion of the difference in tool life can be attributed to actual differences in tool performance, if such exists. Now, a measure of the difference in tool life between

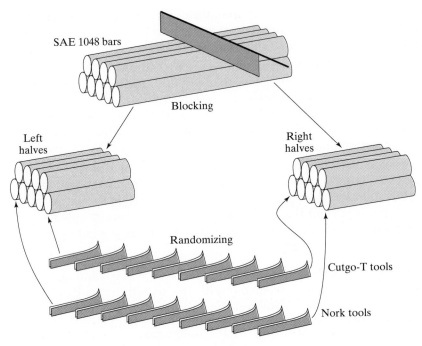

Figure 15.9   **Blocking and randomization**

the two tools can be determined within a bar, where a pair of tests are more similar to each other in terms of bar material properties.

When known sources of extraneous and unwanted variation can be identified, we can design the experiment in such a way as to eliminate their influence and provide a more sensitive test of significance for the variables under study. The technique we use is referred to as blocking. Blocking is an important concept in the design of experiments, and it is a method to eliminate the forces of

TABLE 15.1	Test Results: Tool Life (minutes)[*]		
Test	y_c *(Cutgo-T)*	y_n *(Nork-V)*	$d = y_n - y_c$
1 (Bar 1)	18 L	20 R	+2
2 (Bar 2)	16 R	14 L	−2
3 (Bar 3)	10 L	10 R	0
4 (Bar 4)	17 R	20 L	+3
5 (Bar 5)	17 L	19 R	+2
6 (Bar 6)	20 R	20 L	0
7 (Bar 7)	16 R	15 L	−1
8 (Bar 8)	28 L	28 R	0
9 (Bar 9)	24 L	29 R	+5

[*] L and R refer to the left half and right half of a bar, respectively.

extraneous variation from desensitizing the analysis when certain conditions of the experiment cannot be controlled or held constant. However, if the extraneous variation persists in the actual operating environment downstream, not just in the experimental environment, an alternative measure may be taken to include it purposefully in the designed experiment. One such measure, to be discussed later in the chapter, is robust product and process design.

Randomization for Unknown Sources of Variation

To further reduce possible bias due to material differences between the left half and the right half of each bar, the selection of either half of the bar for each test is made at random. Such random pairings would distribute evenly any possible material difference between the two halves within a bar. According to such a design the 18 test results are given in Table 15.1. The analysis can now proceed to check for a true tool life difference, estimated by the average difference across all bars, $\bar{d}$. By comparing this average difference against the variation in the differences, the bar-to-bar variation is eliminated from the analysis.

In this case, although the average difference from the test results shows a longer life by 1 minute for the Nork-V tools, it turns out that the statistical evidence is not strong enough to conclude that a true difference exists. This could be due to the relatively large common-cause variation in the machining process that masks any real difference in the tool life. Sometimes, however, there might be other systematic or assignable factors that were not blocked or randomized. In running these tool life tests, it was learned later that two tests were performed each day with the Nork-V tool tested in the morning and the Cutgo-T tool in the afternoon. The warm-up effect of the lathe tends to give a longer tool life for the afternoon tests. Therefore, a better design of the tests should consider randomizing the tools in time in addition to the random selection between left and right halves of the bar.

Often, we tend to assume that the sample observations are independently drawn from some normal distribution(s) with a constant mean and a constant variance (i.e., a process in statistical control). Unfortunately, unexpected changes often take place in a somewhat orderly fashion over a number of runs. As a result, the observations may be correlated in time, and hence the treatment of them as independent can cause misleading conclusions. Since such systematic external sources of variation are not easily blocked out, we must employ the concept of randomization to evenly distribute them between the two variables under consideration.

In general, there are two kinds of correlations among successive runs, positive correlation and negative correlation. As an example, suppose that an inexperienced operator was assigned to run the tool life tests of Table 15.1 and he ran the first nine tests with the Nork-V tools and the remainder with Cutgo-T tools. The Cutgo-T tools may show a higher average tool life because of his learning effect, which gradually gives a longer life of any tools. Such a systematic trend would produce positively correlated results. On the other hand, the warm-up effect of the lathe produces systematically higher and lower tool lives, alternating between A.M. and P.M. This oscillating behavior would give negatively correlated results. A plan should be carefully made to determine exactly how the tests are to be randomized to deal with each kind of possible correlation.

To deal with possible positive correlations or linear trends, randomizing adjacent runs within pairs is recommended because it avoids the large differences over the long trend. An example of such a randomization is shown as arrangement 1 in Table 15.2 for the tool life experiments. For situations with possible negative correlations or oscillating trends, running each pair of two random halves both at a high and a low level of the oscillating cycle, such as both in the A.M. or both in the P.M., will avoid direct comparison between adjacent runs, such as arrangement 2 in Table 15.2.

Another Example: Need for Randomization

As another example, suppose that a very simple sandmill experiment is conducted over two successive shifts, eight tests being conducted on each shift. The purpose is to compare the performance of a vertical and a horizontal sandmill in the dispersion of paint pigment to a prescribed level of fineness. Suppose further that the eight tests for the vertical machine are done on the day shift, followed by the eight tests for the horizontal machine on the evening shift. The two operators may, without us realizing it, operate the process somewhat differently, and as a result influence the outcomes. The resulting data could look something like the ones shown in Fig. 15.10.

The previous results seem to favor the horizontal sandmill (i.e., the fineness values seem consistently higher). This may have nothing to do with the machine performance itself but only be a reflection of a systematic difference in operator performance. How are we to deal with such systematic change when its presence and character are unknown to us?

TABLE 15.2	Specific Randomization Schemes for Positive and Negative Correlation Nuisance Variable Situations[*]			
	Arrangement 1		Arrangement 2	
Bar	*Cutgo-T*	*Nork-V*	*Cutgo-T*	*Nork-V*
1	La	Rb	LA	RA
2	Ld	Rc	RP	LP
3	Rf	Le	RP	LP
4	Lh	Rg	LA	RA
5	Lj	Ri	RA	LA
6	Rk	Ll	LP	RP
7	Rm	Ln	RP	LP
8	Rp	Lo	RA	LA
9	Lr	Rq	LP	RP

[*]L, left half; R, right half; A, A.M., P, P.M.; a to q are the time order of runs.

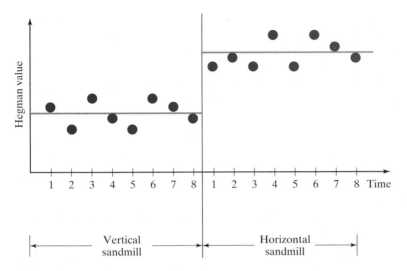

Figure 15.10   Shift-to-shift differences

Sir Ronald Fisher[1] has pointed out that in the face of systematic change of the testing environment, the fundamental precaution we must take to guarantee validity of the test of significance is randomization. If the 16 trials were randomly assigned to the two shifts, the statistical comparison that follows would be a valid one. The variation induced by any systematically changing unknown factor would certainly adversely affect the sensitivity of the test, but since it is randomly distributed across all trials, it would not bias the results. Since such systematic change could occur totally unknown to us, it will always be wise to randomize the experimental trials.

Another Viewpoint on Blocking and Randomization

The techniques of blocking and randomization are both aimed at improving the validity and sensitivity of comparative tests. There are, however, at least two concerns about the use of these techniques. One concerns blocking. If the factor(s) that are blocked out in our design of the tests are part of the system variation, we would not know what their effects might be in routine operation of the system. Therefore, for such cases, not only should we not block them out, we should purposely study their effects in addition to the tool difference on tool life. The second concern deals primarily with randomization. If there are several systematic effects, such as within-bar material differences, machine warm-up, learning curve, and so on, to be randomized to improve the sensitivity in data analysis, their joint effects due to some dependency

---

[1] R. A. Fisher, *The Design of Experiments*, 8th ed., Hafner, New York, 1966.

among them should be explicitly evaluated for better system design. In summary, when the environment surrounding the comparative tests involves too many variables that need to be blocked or randomized, one might consider them as additional factors to be experimented with using multifactor design strategies. In Chapters 16 to 21 we present some of these designs.

Summary

From the comments so far in this chapter we see that a number of characteristics or properties are desirable in sound experimental design test plans.

1. The ability to deal simultaneously with qualitative (categorical) and quantitative factors/variables.

2. The existence of the mathematical model concept as a guide to designing the experiment—that is, explicit recognition of the relationship between the proposed mathematical model and the experimental design that responds to that model.

3. Awareness of the notion of model building and the sequential and iterative nature of experimentation. In this regard, the ability to employ experiments that serve as building blocks, one upon the other, is important.

4. The ability of an experimental design to show the effects of changing factors individually (all other factors held constant) but at the same time doing this at several different combinations of the other factors so that factor interdependencies can be properly revealed. It is important to make comparisons that clearly show the effects of the several factors and to facilitate model identification—that is, determine just exactly which factors are important and how they are important.

5. The ability of the experimental design to show the forces of experimental error/system variability directly, so that comparisons of performance can be made in light of the magnitude of system noise. Replication of tests provides us with the ability to do this.

6. The ability of the experimental design to mitigate the forces of nuisance variation, explicitly identifiable or not, through the techniques of blocking and randomization.

In addition to these considerations, it is of course useful if the experimental design is

- Easy to set up and carry out.
- Simple to analyze and interpret.
- Simple to communicate or explain to others.

One important class of experimental designs that possesses many of these desirable features and is particularly useful in the formative stages of an investigation is the class of two-level factorial designs, such as the ones shown in Fig. 15.11. These experimental designs and derivatives of these designs,

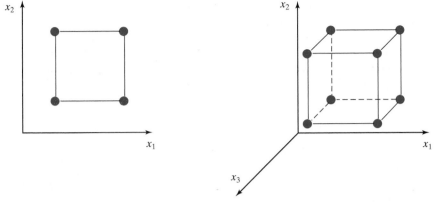

(a) Two variables; four tests required.   **(b)** Three variables; eight tests required.

Figure 15.11   **Two-level factorial designs**

two-level fractional factorial designs, are discussed in considerable detail through-out the balance of the book.

15.5   Role of Planned Experimentation in Taguchi Methods

Taguchi's Contribution to Quality Design

In the last two decades of the 20th century, many U.S. companies began to exa-mine new ways to improve their competitive position significantly in the long term. Companies sought to develop an overarching quality philosophy that could be implemented through policy and operating procedures to ensure that customer needs were met competitively. One thing that became clearly evident was the need to push the quality issue farther and farther upstream so that it became an integral part of every aspect of the product life cycle (e.g., market-ing, product design and engineering, process engineering, production).

Since the 1980's a great deal of attention has been given to the development of concepts and methods for product and process design and engineering. It is now well recognized that it is through engineering design that we have the greatest opportunity to influence the ultimate delivery of products that meet customer needs and expectations. In particular, the approach and techniques of Taguchi, commonly referred to as "Taguchi methods," have been widely adopted. These methods and the general issue of parameter design are the sub-ject of many books and papers.[2]

The Taguchi approach to quality design has a number of significant strengths that ought to be exploited. In particular, Taguchi has placed a great

---

[2]The reader is directed to entries in the References at the end of the book for Taguchi, Phadke, Kackar, Sullivan, and Clausing.

deal of emphasis on the importance of minimizing variation as a means to improve quality and to bring the mean of the process to the design-mandated target. The ideas of designing products whose performance is insensitive to environmental conditions, and of making this happen at the design stage through the use of design of experiments, have been cornerstones of the Taguchi methodology for quality engineering. The strengths of Taguchi's approach lie in the following areas:

1. Emphasis on quality during the engineering design process.
2. Recognition of the relative roles of factors in influencing the performance of the product/process.
3. Emphasis on variation reduction in quality engineering: use of the loss function and signal-to-noise ratio.
4. Robust design: the parameter design concept.

Engineering Design Process

As discussed in Chapter 1, Taguchi methods view the design process as evolving in three distinct phases or steps: system design, parameter design, and tolerance design (refer to Fig. 15.2). It is perhaps this broad umbrella placed over his concepts and methods for quality design and improvement that makes this approach so readily embraceable by the engineering community. As we have said earlier, Taguchi methods consider engineering design the central issue and statistical methods as just one of several tools to accomplish his objectives in engineering.

System design is the initial phase of any engineering design problem wherein knowledge from specialized fields (chemistry, electronics, etc.) is applied to develop the basic design alternative. This design alternative is then refined in the phases of parameter design and tolerance design. *Parameter design* is that phase of the design process in which the "best" nominal values for the important design parameters are selected. Taguchi defines the best values for the nominals as those values that minimize the transmitted variability resulting from the noise factors. This is why we sometimes refer to parameter design as robust design. *Tolerance design* is the final phase of Taguchi's design process model, in which the tolerances on the nominal values of critical design factors determined during parameter design are analyzed. In tolerance design the evaluation of tolerances is considered to be an economic issue and the loss function model is used as a basis to strike the economic trade-off between increased manufacturing cost and associated reduction in quality loss due to functional variation.

Recognition of the Relative Roles of Factors in Influencing the Performance of the Product/Process

Taguchi suggests a sound engineering interpretation of the varying roles that the important system factors/variables play in influencing the product/process performance. He emphasizes the importance of evaluating quality performance under field conditions, as part of the design process, and the fact that functional variation in performance is influenced by noise factors that vary in the environment in which the product/process is functioning. For example, suppose that one decided to run a simple experiment to study the effects that four factors—carburetor design, engine displacement, tire pressure, and driving speed—may have, either singly or in concert, on the performance of an automobile as measured by fuel economy in miles per gallon. A simple two-level factorial design could be employed with the variable levels as defined below.

	Type A	Type B
Carburetor design	Type A	Type B
Engine displacement (liters)	2.8	3.4
Tire pressure (psi)	22	28
Driving speed (mph)	45	55

A total of $2^4 = 16$ tests would be required and the purpose of the experiment might be to determine the "best" levels for each of the four factors (i.e., the levels that lead to the best fuel economy).

A more thoughtful consideration of the experiment above might lead one to believe that the notion of "best" levels does not have the same meaning for all four of the variables in question. In terms of the engineering design process, the selection of a "best" level for tire pressure and driving speed has very little meaning. In terms of product performance in the field, these two variables cannot be controlled to fixed levels. As a matter of fact, these two factors are likely to vary in a somewhat random fashion over time as the product performs in the field and the field environment (temperature, road terrain, etc.) varies. As tire pressure and driving speed change over time, their variation will be transmitted through the design and be felt in the output response of the product. As such, these two factors fit the definition of outer noise factors quite well.

The fact that some factors or variables influencing product performance in a substantial way cannot easily be controlled does not necessarily mean that we would not include them in an experiment aimed at understanding the nature of their influence and the influence of other, more readily controlled, factors. It does mean, however, that the structure of the experimental design might need to be adjusted from that defined by the more classical approach to the design of experiments.

Transfer Function Model: Classification of Variables

Figure 15.12 is a block diagram representation of a transfer function model for system performance. The transfer function describes the relationship between the inputs, defined by control factors and the outputs of the product/process, in the presence of noise factors varying in a generally uncontrollable fashion in the environment in which the product/process is functioning. When the state of knowledge permits, the transfer function might be derived from first principles and may be in the form of an explicit mathematical relationship. Often, however, such advanced knowledge is not available and hence we must rely on techniques such as design of experiments and empirical model building to develop the transfer function.

Figure 15.12   Transfer function model for product/process performance

Taguchi's contribution to the foregoing transfer function model is a more rigorous and unique classification of the nature of the factors that can influence product/process performance. In formulating the engineering design problem, Taguchi has found it appropriate to classify the factors that can influence product/process performance into four categories.

1. **Signal Factors.** These are the factors that may be adjusted by the user/operator to attain the target performance. The speed adjustment on a blender is a signal factor.
2. **Control Factors.** These are the product/process design parameters whose values are to be determined during the design process. The purpose of the design activity is to select the best levels of the control factors according to an appropriate design criterion.
3. **Noise Factors.** The noise factors, as defined previously, are the uncontrollable factors and the factors that although controllable in principle, are controlled only at considerable effort/expense and are therefore considered as noise factors for the purposes of design.
4. **Scaling/Leveling Factors.** These are special cases of control factors. They are factors that may easily be adjusted to achieve a desired functional relationship between a signal factor and the output response. For example, the gearing ratio in a steering mechanism can easily be adjusted to achieve a desired relationship between the turning radius and the steering angle.

The block diagram shown in Fig. 15.13 was adapted from Phadke[3] and provides a representation of a product or process in terms of the factors that govern its performance. The expression $f(x, m, n, r)$ is used to represent the relationship of the signal (m), control (x), noise (n), and scaling/leveling (r) factors to the response (y) of the product/process. Signal factors are those variables adjusted to attain the target/nominal performance. The control factors are those variables under the control of the designer. The selection of the nominal values for the control factors is the primary role of parameter design. Noise factors describe variables that are difficult or impossible to control but whose variation is transmitted through the design—as described by the transfer function—to the output.

---

[3]M. S. Phadke, *Quality Engineering Using Robust Design*, Prentice Hall, Englewood Cliffs, NJ, 1989.

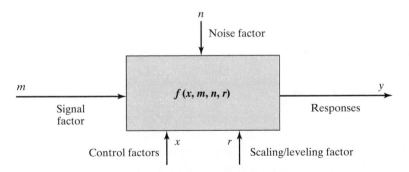

Figure 15.13   **Block diagram of a product or process**

The Concept of "Degree of Control"

Often the matter of control of a given factor is an economic one. Factors that appear to be capable of being treated as control factors might be treated as noise factors for the purpose of gaining economic advantage. This notion is particularly appealing in a simultaneous/concurrent engineering framework when product design and processing considerations are being dealt with jointly. The design-for-manufacturability concept can be served in a quantitative fashion by seeking levels for the control factors that minimize the transmitted variability to product performance of conditions present during processing—the manufacturing process.

Sometimes the identification of a factor as a noise factor is clear from the outset of the study. Only as product and process designs mature may it be possible to understand the economic ramifications of the control of certain factors. At other times, this is revealed only as the study progresses and data from experiments come to light. The opportunity to consider a factor as not needing to be actively controlled, or controlled at a lower level, may come to light only through the identification of nonlinear relationships that may exist between or among two or more variables.

The classification of factors in accordance with the roles they play in influencing the response/performance of the system is not meant to imply that the interpretation as to whether a certain factor is a control factor or a noise factor may not vary from one situation to another. The issue of the control of a certain factor is generally one of economics, and therefore any factor might be thought of in terms of its *degree of control* rather than as controllable or not controllable. Consider the following example, which illustrates this idea.

▶ **EXAMPLE 15.2**   DEGREE OF CONTROL

Suppose that we are studying an injection molding process because of a part shrinkage problem. A brainstorming session might lead to the following list of factors as having a potential influence on part shrinkage:

Cycle time	Mold temperature
Percent regrind vs. virgin raw material	Cavity surface finish
Gate configuration	Raw material moisture content
Holding pressure	Raw material melt flow index
Screw speed	Mold cooling water temperature

Although any or all of the factors above can influence part shrinkage, they are not all controlled or manipulated to the same degree or with the same ease economically. Some may be controlled at the engineering design process (e.g., the gate configuration or the cavity surface finish of the mold). Others are controllable at the molding machine (e.g., screw speed or cycle time). Still others are properties of the incoming raw material and are therefore under the control of the supplier (e.g., melt flow index). Finally, certain factors may best be categorized as defining part of the environment in which the process must function (e.g., mold cooling water temperature, percent regrind vs. virgin raw material).

It becomes readily apparent that it depends on who you are and what your point of view or objective is when it comes to determining whether a certain factor should be considered as a control or a noise factor. In dealing with an incoming raw material property, for example, one could either go back to the supplier and require a tightening of the specification to reduce the transmitted variability to the process output (part shrinkage), or one could treat that incoming material property as a noise factor and attempt to mitigate its influence on the output through adroit manipulation of other factors that are more easily controlled. In the illustration above, mold cooling water temperature and the virgin/regrind material mix are certainly factors that can be controlled, but if a part/die/process combination can be developed that is robust to variation in these factors, that would be economically attractive.

In a later chapter we examine in detail a case study dealing with rattles and closing effort for glove box assemblies in an instrument panel. In this case study, several factors were examined as to their influence on closing effort, including latch striker angle, panel foam thickness, and striker positioning during assembly. The first of these is controlled through the design of the die that stamps out the striker. The second is controlled through the foam molding process. The third is controlled by the worker at the assembly process. Clearly, if there were a problem with glove box closing effort, which might manifest itself through warranty complaints, one might consider treating the striker position variable as a noise factor, since it is more difficult/expensive to control, and attempt to overcome the problems this factor is creating by the informed manipulation of other factors, such as striker angle, that are considerably easier to change/control.

Signal-to-Noise Ratio as a Measure of Product/Process Performance

Taguchi methods emphasize the importance of variation reduction and the relationship of variation reduction to the meaning of quality. Taguchi provides a definition of quality that can successfully be made operational upstream at the point of engineering design as a design criterion. The concept of signal-to-noise ratio as a means to evaluate system performance is an important contribution to a better understanding of the importance of variation reduction. Furthermore, the loss function concept introduced in Chapter 2 and discussed in more detail in Chapters 10 and 11 is used to assess the specific economic consequences of variation.

In Chapter 1 we treated the concept of signal-to-noise ratio in a general way to examine quality design and improvement strategy in a global fashion. We will look at how Taguchi defines this measure as a more specific response in design of experiments.

In assessing the results of experiments, Taguchi suggests the use of statistics known as *signal-to-noise ratios* (S/N). Instead of calculating an average response value and a variation in response value, Taguchi suggests that the signal-to-noise ratio be calculated at each design point. The combinations of the design variables that maximize the signal-to-noise ratio are then generally selected for further consideration as product or process parameter settings.

Maximizing the S/N ratios is presumed to be equivalent to minimizing the average/expected loss associated with striving to be on target with smallest variation. Three cases are of practical significance.

Case 1: The S/N ratio that Taguchi defines for "smaller is better" is

$$S/N = -10 \log_{10}(\text{mean-squared response})$$

$$= -10 \log_{10} \frac{\sum y_i^2}{n}.$$

For this case, minimizing the average loss is equivalent to minimizing $\sum y_i^2/n$. Minimizing $\sum y_i^2/n$ is equivalent to maximizing $-10 \log_{10}(\sum y_i^2/n)$. Therefore, the S/N ratio is consistent with the loss function in this case.

Case 2: The S/N ratio that Taguchi defines for "larger is better" is

$$S/N = -10 \log_{10}(\text{mean square of the reciprocal response})$$

$$= -10 \log_{10} \left[\sum \frac{1/y_i^2}{n} \right].$$

For this case, minimizing the average loss is equivalent to minimizing $\Sigma(1/y_i^2)/n$. Minimizing $\Sigma(1/y_i^2)/n$ is equivalent to maximizing $-10 \log_{10}[\Sigma(1/y_i^2)/n]$. Therefore, the S/N ratio is consistent with the loss function in this case also.

Case 3: One S/N ratio that Taguchi defines for "nominal/target (m) is best" is

$$S/N = 10 \log_{10} \frac{\text{mean-squared response}}{\text{variance}}$$

$$= 10 \log_{10} \frac{\bar{y}^2}{s_y^2},$$

where

$$\bar{y} = \frac{\Sigma y_i}{n} \quad \text{and} \quad s_y^2 = \frac{\Sigma(y_i - \bar{y})^2}{n - 1}.$$

For this case, minimizing the average loss is equivalent to minimizing $E[(Y - m)^2]$. However, minimizing $\Sigma(y_i - m)^2/n$ is not necessarily equivalent to maximizing $10 \log_{10}[(\bar{y}^2/s_y^2)]$. Therefore, for this case the S/N ratio suggested by Taguchi is not necessarily consistent with the loss function.[4]

Parameter Design: The Robust Design Concept

Taguchi refers to parameter design as that phase of the design process in which the best nominal values for the important design parameters are selected. Taguchi defines the "best" values as those values that minimize the transmitted variability resulting from the noise factors. This is why we sometimes refer to parameter design as robust design.

Simple Example of Robust Design

Let us consider the following hypothetical example. Suppose that the setup of a certain injection molding machine is determined by the settings of five machine parameters and that these factors are likely to have some effect on part shrinkage. The five factors are

x_1:   nozzle temperature.
x_2:   holding pressure.
x_3:   holding time.
x_4:   mold temperature.
x_5:   screw speed.

In addition to the five control factors above, three other factors are known to have some effect on part shrinkage:

n_1:   mold cooling water temperature.
n_2:   percent regrind vs. virgin raw material.
n_3:   ambient temperature at the machine.

The three factors above are somewhat difficult to control and therefore we would like to view them as noise variables. In the interest of running the process "on target with smallest variation," interest will naturally focus on the selection

---

[4]For more on this issue the reader is directed to R. G. Hunter, J. W. Sutherland, and R. E. DeVor, "A Methodology for Robust Design Using Models for the Mean, Variance, and Loss," *ASME Symposium on Quality and Performance: Design, Evaluation, and Improvements, ASME Winter Annual Meeting*, December 1989, PED-Vol. 42, pp. 25–42.

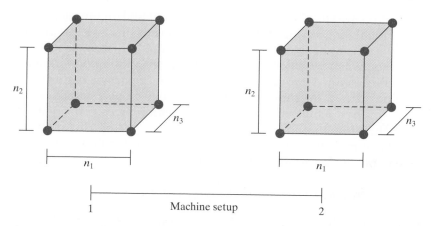

Machine setup

Figure 15.14   **Robust design experiment for the injection molding example**

of the five control variables that minimize part shrinkage and provide for the most consistently performing process in light of the presence of the three noise factors.

For the sake of simplicity in demonstrating the robust design concept, we will consider that the five control variables are collapsed into just one variable, the machine setup. Further, we will compare process performance across only two levels of the machine setup variable, machine setup 1 vs. machine setup 2. It is proposed to run an experiment in which the conditions of the three noise variables are purposely varied for both of the machine setups. A two-level factorial design is proposed as a suitable matrix to define the variations of the noise variables. A graphical representation of the design of the experiment is shown in Fig. 15.14. From the figure it is seen that a total of 16 trials are required for the experiment proposed. The complete design matrix for this experiment is given in Table 15.3. In the table the values of shrinkage are in units of 0.001 inch.

To analyze the data from a robust design point of view, we need to determine measures of both the average performance and the consistency of performance of the process at each of the two machine setup conditions. Calculating the average

TABLE 15.3	Robust Design Experiment for the Molding Example							
	Machine Setup 1				Machine Setup 2			
Test	n_1	n_2	n_3	Shrinkage (y)	n_1	n_2	n_3	Shrinkage (y)
1	−	−	−	10	−	−	−	7
2	+	−	−	10	+	−	−	10
3	−	+	−	12	−	+	−	15
4	+	+	−	11	+	+	−	11
5	−	−	+	10	−	−	+	11
6	+	−	+	12	+	−	+	8
7	−	+	+	11	−	+	+	12
8	+	+	+	12	+	+	+	14

response at each of the two machine setup conditions, we find an identical value of $\bar{y} = 11.0$ (10^{-3} inch) for each machine setup. This means that as far as average performance is concerned, the two machine setups cannot be distinguished from each other. However, when we calculate the standard deviation in performance, we find that machine setup 1 provides for much more consistent performance than does machine setup 2 over the variation in the outer noise variables. For machine setup 1 the standard deviation in performance is equal to $s_y = 0.93 \ 10^{-3}$ inch, while that for machine setup 2 is $s_y = 2.73 \ 10^{-3}$ inch. We would therefore favor machine setup condition 1 because it is more robust, that is, less sensitive to the noise variation in the process.

Inner and Outer Array Experimental Design Structures

The experimental design structure of Fig. 15.14 is sometimes referred to as an inner array/outer array structure. The inner array in that particular example considers only one factor at two levels, the control factor machine setup, and therefore contains only two test conditions. In Fig. 15.14 the outer array is a two-level factorial in the three noise factors and therefore is comprised of a total of eight test conditions. This design structure suggests that for each trial in the inner array we conduct the complete outer array of experiments. In this way we can see the variation in the response at each point in the inner array that results from the variation in the noise factors purposely created by the varying conditions of the outer array.

   The inner array/outer array structure as an experimental design structure responds to the classification of variables often employed with Taguchi methods. A simple example of the inner array/outer array experimental design structure is considered in Fig. 15.15. In this situation we have two control

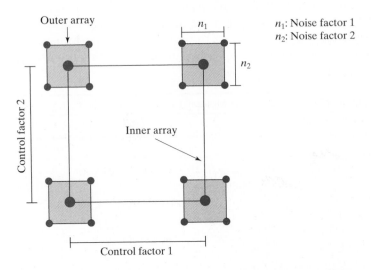

Figure 15.15   **Example of an inner/outer array design structure**

factors, each at two levels, and two noise factors, each also at two levels. The inner array would be defined by a 2^2 factorial, as would the outer array.

To examine the robustness of the product/process over the range of the two control factors as the two noise factors vary, we would conduct the outer array experiment for each of the four test conditions of the inner array. In this way we can see the variation in the response at each point in the inner array that results from the variation in the noise factors purposely created by the varying conditions of the outer array. Therefore, a total of 16 tests would need to be conducted. The results of such experiments may then be analyzed by calculating the signal-to-noise ratio at each of the points in the inner array.

When the number of control and/or design factors becomes large and it is still desirable to examine each factor at only two levels, Taguchi methods suggest the use of *orthogonal arrays* for either an inner array or an outer array, or both. In the context of the work herein, these would be derived from the useful class of two-level fractional factorial designs, to be discussed later.

The inner/outer array design structure seems to have a great deal of appeal when one breaks factors down as being either strictly control factors or strictly noise factors. There are at least three disadvantages, however, of using only the inner/outer array structure and analysis advocated by Taguchi methods.

1. The use of this structure requires that the distinction between control and noise factors be made prior to conduct of the experiment. This is somewhat contrary to the notion of degree of control and constrains the problem unnecessarily.

2. The use of this structure generally requires us to run a rather large number of experiments, since two separate designs are developed and then combined. As a result, there is a tendency to use very low resolution fractional factorials (to be discussed in Chapter 20).

3. The use of this structure makes it impossible to determine explicitly, and hence interpret, the interactions (nonlinearities) that may arise between control factors and noise factors. Yet it is these interdependencies that may be exploited to gain economic advantage.

The comments above suggest that it may be not only possible but also desirable to combine control and noise factors in a single factorial design scheme. Figure 15.16 provides an example of a situation in which a control factor that is a design feature related to an automobile engine block cylinder bore has a strong interdependency/interaction effect with a factor that arises through processing of the cylinder, namely, the bore offset. Although bore offset (the relationship between the bore axis and the boring process axis) may be controlled through the casting process and fixturing of the block at the boring station, the degree to which it needs to be controlled is the key issue. To control it to a point where it would not be a potential factor influencing final bore cylindricity would generally be considered cost prohibitive, particularly if an alternative design action is possible.

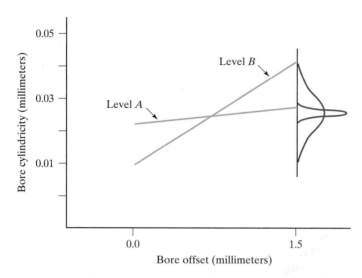

Figure 15.16   Two-factor interaction between a product design/control factor and a processing noise factor

Figure 15.16 shows that there is a strong two-factor interaction effect between bore offset and a certain design feature of the bore. When the design feature takes on level B, the changes in the amount of bore offset seem to have a strong bearing on the bore quality, in a variability sense. On the other hand, when the design feature takes on level A, the influence of bore offset on bore quality variation is greatly reduced. While level A and level B provide about the same bore cylindricity on average, the variation in cylindricity is greatly affected by this design feature. Hence, thinking of bore offset as a noise variable, it would appear that by selecting the design feature nominal to be at level A, a more robust, less variable product/process design is attained.

15.6   Contrasting Philosophies for Dealing with Nuisance or Noise Factors

Dealing with noise factors in the conduct of experimental designs has been the subject of considerable study for some time. From the standpoint of enhancing comparative experiments, two basic strategies emerged from the early work of Fisher: the techniques of blocking and randomization. For those nuisance/noise factors that are difficult to fix completely across the entire experiment, experimental subunits can be defined within which the nuisance factor can be controlled to a fixed level, and treatments are compared within the subunit, thereby blocking the impact of the noise factor in the response. The objective is to create a more sensitive comparison within the "block." At the same time, the experiment should give rise to variation from block to block that is representative of that encountered in the real world (i.e., outside the experiment). With the use of sound design-of-experiments practice, it is not only possible but generally useful to estimate directly the effects of the blocked factors. Such knowledge can ultimately lead to actions directed toward improved product design and/or process operation.

For those nuisance/noise variables that cannot be either fixed or blocked within the experiment (they may not even be identifiable), the technique of randomization ensures that their influence on the response is randomly distributed across the treatments, although the variation due to these factors does appear in the response. This measure guarantees the approximate validity of the statistical tests used to draw inferences on the data. Furthermore, the use of replication as a design strategy is also aimed at improving the sensitivity of the comparison of treatment means.

It is important to note that the experimental strategies above, as effective as they may be, are just that: design-of-experiments strategies, aimed at mitigating the forces of noise that could otherwise bias and/or desensitize the statistically based comparisons made through the experiment. The emphasis is clearly placed on enhancing the power of the statistical methods used to make inferences about the results of an experiment.

The strategic approach that Taguchi has taken in dealing with the forces of noise factors, as they may influence the performance response of interest, contains certain elements of the preceding discussion but at the same time assumes a fundamentally different posture. First, Taguchi is concerned about the influence of the nuisance/noise factors as they operate in the environment in which the product/process is functioning. In particular, his concern is focused on the extent to which this noise variation is transmitted through the product/process and hence influences the variation in the response from a field-performance perspective. Hence the primary emphasis is on the functional environment in contrast to the experimental environment. The issue is whether the noise factor(s) are viewed as a nuisance to the experiment or as a nuisance to performance in the field. In fact, one might consider the question: Under what circumstances would a factor be a nuisance to the experiment but not a nuisance to field performance?

Second, Taguchi's use of designed experiments in evaluation of the impact of noise factors on field performance means that regardless of whether the noise factors are explicitly recognized (and therefore part of the design structure) or not, their influence on the response is realized in the experiment, and hence the concept of robust/parameter design is preserved for both cases. The ability to do this in an unbiased manner, of course, depends on the use of good experimental design practice—randomization, for example. In a sense, the inner/outer array experimental design structure suggested by Taguchi may be thought of as a type of controlled replication where the influence of nuisance/noise factors, when varied over prescribed levels, can be observed and measured in the output performance.

But the purpose of this controlled replication is much different than that mentioned earlier in the context of the more traditional approach to experimental design. Taguchi's framework may be used to reveal interactions between the design/control factors and the noise factors that signal the presence of non-linearities to be exploited to obtain a more consistently functioning (robust) product/process. In the design strategy proposed herein, judgment is reserved on what is a control variable and what is a noise variable (the degree-of-control notion) until the effects are estimated and their economic consequences can be considered more fully. Such a strategy adds flexibility to the entire process and enables the interactions between the control factors and the noise factors to be explicitly determined.

At the same time it is clear that Taguchi's use of inner/outer array experimental design structures does contain the element of blocking, since the structure, although hierarchical in nature, is orthogonal, and hence the variation induced by purposely varying the noise factors equally influences the treatment responses for the control variables in an average sense. It is rather the variation transmitted to the performance response as the noise factors vary that may be different for different levels of the control factors. All of this, of course, needs to be viewed more carefully in light of the confounding of variable effects evident in most highly fractionated designs.

■ EXERCISES

15.1 Suppose that you are asked to develop an experimental plan to evaluate possible differences in performance between two comparably priced subcompact cars. Propose an experiment(s) that may be used to determine if there is a detectable difference in the performance of the two cars. You are to develop a proposal for the experiments you will run. Discuss the important things you must take into consideration when you design comparative experiments for such situations.

15.2 Two high-performance automobiles (type A and type B) are being studied. Seven cars of type A and five cars of type B are randomly selected and tested. The driver(s) are asked to drive the automobiles for 5 minutes around a 2-mile oval test track at 60 miles per hour. The amount of fuel consumed (liters) for each trial is given in the table below.

Car	1	2	3	4	5	6	7
Type A	1.65	1.52	1.62	1.73	1.78	1.68	1.76
Type B	1.60	1.42	1.70	1.38	1.53		

(a) What kind of an analysis might you propose for the data?
(b) What would you like to know about the process by which the data were collected?

15.3 In an effort to improve inventory control procedures, it has been suggested that marking items (applying an inventory code number) with a new "gun" may be faster than the presently used "stamp" method. To examine the time reduction claimed by the gun method, five different items of varying sizes, shapes, and inventory number placement locations were chosen, and the number of items that could be marked in 5 minutes were recorded using both methods. The data are shown in the table below.

Item	Gun Method	Stamp Method
1	106	99
2	49	50
3	147	132
4	88	77
5	46	44

(a) Discuss what important consideration was obviously taken into account as an experimental design strategy during the design of this experiment.

(b) State the null and alternative hypotheses to be tested in this situation.

15.4 Explain why a process that is not in statistical control might not be a good candidate for experimentation. Suggest an alternative course of action that would be more effective in studying the process and explain your reasoning.

15.5 Explain the fundamental difference between experimentation and what is commonly referred to in the engineering world as testing. Why do you think experimentation is more useful than testing to the product design process?

15.6 Returning to Exercise 15.2, suppose it is learned that the tests were run under the following conditions: (1) All the type A cars were tested on Tuesday, and (2) all the type B cars were tested on Wednesday. Additionally, only two drivers were used for the tests, with one driver being used on Tuesday and the other on Wednesday. Express your concerns about the experiment.

15.7 A set of experiments has been planned to test the color intensity of batches of latex paint made on a certain machine. It has been decided to run five tests, one per day for an entire week, each time increasing the amount of dye added by 1%. Explain why this is a poor testing scheme and propose an alternative plan that will yield more useful data.

15.8 Randomization and blocking are two experimental design strategies used to deal with sources of variation in the experimental environment. Explain the difference between these two strategies.

15.9 An experiment is being planned to test an injection molding process that makes snap-together toys that are put in boxes of children's cereal. Part quality, as measured by the frequency of various types of surface defects, is to be observed. Several variables that seem to be important are (1) machine, (2) raw material batch, (3) temperature of molten plastic, (4) injection pressure, (5) injection rate, (6) cooling time, and (7) size of the part being made. In designing a traditional experiment, which of these variables might be considered as variables to be blocked in the experiment? How would this be done, in principle? What would be the motivation for blocking these variables? If a Taguchi-like experiment were being planned, how would these variables be handled? Discuss the differences in these two approaches to handling the influence of these factors on the results of the experiments.

15.10 Assume that each of the following equations represents a certain relationship between the response of a process and the process variables. Sketch each function and suggest an appropriate experimental design (minimum number of levels of each factor) that would be required to verify this relationship.

(a) $\eta = 2x_1 + 5x_2 + 14$.

(b) $\eta = x_1^2 - 5$.

(c) $\eta = x_1 + 3x_2^2$.

(d) $\eta = 3x_1 + 10$.

15.11 Explain what is meant by the statement that one of the strengths of Taguchi's approach to quality is that "its center of gravity lies in the engineering design process."

15.12 For the process of cooking a pot of chili, define each of the following as a signal, a control, or noise factor:

(a) The list of ingredients.

(b) The amount of chili powder added "according to taste."

(c) The length of time the chili simmers.

(d) The consistency of the stove burner's temperature.

(e) The ability of the pot to distribute heat evenly.

(f) The brand of kidney beans used.

(g) The type of meat used (e.g., ground round, ground sirloin).

15.13 From your personal experience/knowledge, give two examples each to illustrate the three cases that Taguchi considers when maximizing the S/N ratio (i.e., "smaller is better," "larger is better," and "target/nominal is best").

15.14 Provide the geometrical representation of an inner/outer array, where the inner array is a 2^2 full factorial experiment and the outer array is a 2^3 full factorial experiment. How many tests are required for this experimental design?

15.15 In an experiment designed to test the performance of the punching process in a job shop that makes a variety of products out of sheet metal, consider the following variables:

(a) Operator.

(b) Sheet metal thickness.

(c) Number of parts made per length of run.

(d) Shift (first or second).

(e) Ambient temperature.

(f) Size of the part.

(g) Die/punch tool material.

(h) Press used (within range of acceptable tonnage).

Performance is measured as the fraction defective produced over the length of the run. For each of these variables, indicate whether or not you would choose to invoke randomization and/or blocking to deal with any or all of these variables as you design your experiment. Explain your decision in each case.

15.16   In considering the roles of various factors in influencing the response of a given process, discuss what is meant by "degree of control"; specifically, how is this concept invoked in the design and/or interpretation of an experiment?

15.17   In using the inner array/outer array structure for a designed experiment, how are the data generally analyzed according to the methods of Taguchi?

15.18   Suppose that an experiment is to be designed to study the coating used on the bottom surface of skis used for common recreational purposes by a broad class of users. Two different coatings are to be tested. It is desired to design an experiment and perform an associated statistical test to determine if there is a difference between the mean wear observed after a substantial period of use for the two coatings. The experiment designed should be as sensitive as possible. Describe the experiment to be performed.

15.19   Under Taguchi's concept of robust design, how might the experiment described in Exercise 15.18 be viewed? Explain and provide an alternative design of experiment.

15.20   Discuss the importance of statistical process control (SPC) in terms of its relationship to the design of experiments (DOE). How can SPC favorably affect DOE? How can DOE be used in aiding the work of SPC?

15.21   Explain why planned factorial experiments are preferred over one-variable-at-a-time approaches to experimentation.

15.22   Describe the purposes of each step in Taguchi's three-stage model of the design process.

15.23   Explain why experiments should be conducted in an iterative and sequential manner instead of conducting one large comprehensive experiment.

15.24   A chemical engineer is studying a new, fairly complicated process. The goal is to optimize the chemical yield for the process. Explain why the first experimental design that the engineer should conduct will consider the processing time and temperature at only two levels each.

15.25   A bank is studying the factors that influence the amount of time a customer must spend in the queue before being waited on by a teller. Some of the factors identified include (1) number of tellers on duty, (2) teller skill level, (3) time of day, (4) number of transactions per customer, and (5) transaction complexity. Design an experiment to study the system described. Indicate how you plan to deal with each variable.

16

Design and Analysis of Simple Comparative Experiments

16.1   Introduction

In Chapter 15, some desirable characteristics associated with experimental designs were presented, including:

- The ability to deal with nuisance variation factors, whether explicitly identified or not, through the techniques of blocking and randomization.
- The ability to observe the effects of changing factors individually and to reveal factor interdependencies.
- The ability to observe the forces of experimental error/system variability directly so that comparisons can be made in light of the magnitude of system noise.
- The ability to explicitly recognize the relationship between a proposed mathematical model and the experimental design that responds to that model.

In this chapter, we will keep these characteristics in mind as we explore several useful experiments that range from making comparisons between two or more levels (or treatments) of a single variable to making comparisons among multiple levels of multiple variables.

16.2   Comparing Two Alternatives Using an External Reference Distribution—Car Mileage Example

Over the next three sections, we will follow an example that will be used to introduce several experimental designs and associated analysis procedures.

These designs are focused on comparing the results of two treatments or two levels of a single variable (e.g., two types of an engine, two materials for a medical implant, two instruments for measuring the chemical composition of substance, and two temperatures for processing a certain chemical). Sometimes a variable in question will experience only certain fixed levels, e.g., two types of engines, while other times a variable in question will have "continuous" or what is often referred to as random levels, e.g., processing temperature. In either case, the designs and associated analyses will be the same, although inferences/predictions based on the results may be different. In some cases, particularly in the following chapters, variables with fixed and random levels will be mixed in the same experiment.

An automobile manufacturer has developed a new engine (B) that it hopes is superior to the engine that it has been producing for the last several years (engine A). If the car company establishes convincing evidence that the new engine B is superior to the existing engine A, then it will introduce the new engine into its vehicle.

The car company wishes to conduct some experimental trials in order to assess the performance of engine B relative to engine A. Of course, the performance of the engine can be judged based on a number of different criteria: emissions, fuel economy, etc. As a first step, it is desired to determine if new engine B is superior to existing engine A in terms of fuel economy (miles per gallon). For each fuel economy trial, an engine is selected and fitted into a test vehicle. The driver then maneuvers the vehicle on a 20-mile test route, and the fuel consumed on the route is measured in order to calculate the fuel economy in miles per gallon.

Initial Approach—Use of an External Reference Distribution

Over the last few years, a large number of trials have been performed with engine A to measure the fuel economy. These trials have provided a wealth of information on the fuel economy of engine A under a wide range of conditions. This information has been compiled over the years and is shown in Fig. 16.1 as a dot diagram for the historical data on engine A's performance in terms of mpg.

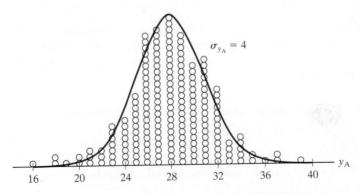

Figure 16.1   Engine A reference distribution for fuel economy, mpg

Let us first examine the appearance of the historical data, previously referred to as the reference distribution, shown in Fig. 16.1. We see that on the average, the fuel economy based on the engine A trials is $\mu_{y_A} = 28$ mpg. The distribution appears to be normal and has a standard deviation of $\sigma_{y_A} = 4$ mpg. In this case, we are assuming that the mean and standard deviation are known essentially with certainty, since the calculated average and standard deviation are based on a considerable amount of data. In considering the form of the distribution, we might ask ourselves, "Why does it have a normal shape?" To begin to answer this question, note that many factors collectively influence the fuel economy measurement obtained from each trial:

- Small manufacturing variations among engines of type A that are tested.
- Slight differences in the vehicle setup for each test.
- Route-to-route differences.
- Tire pressure differences from trial to trial.
- Environmental conditions (e.g., rain and wind).
- Driver-to-driver differences (e.g., speed differences).
- Fuel differences from trial to trial.

Factors such as these may influence the fuel economy by only a small amount individually (and independently), but collectively they produce a pattern of variability, which could be considerable over a long period of time, approximately characterized by the normal distribution (due to the central limit theorem).

Given the availability of historical data on fuel economy for engine A, one attractive experimental approach seems to be the following: Run several trials with engine B, then compare the results to the reference distribution of fuel economy results for engine A. Since no engine A trials will be part of the experiment in this case—the engine B trials will be judged based on information external to the experiment—we use the term the *external reference distribution* for the situation portrayed in Fig. 16.1. As we make our comparison/judgment we seek to answer the statistical question: Are the observed results for engine B substantially different from those historically observed for engine A? Or, conversely: Are the results for engine B consistent with what we have seen in the past for engine A? We will answer this question by simply comparing the sample statistic from the experimental trials on engine B to the sampling distribution for the statistic based on the historical data compiled for engine A.

The Experiment

Four ($n = 4$) fuel economy trials are performed with engine B, producing the following results: 26, 32, 35, and 31 mpg. Based on these 4 trials, the sample mean, $\bar{y}_B$, is calculated to be 31. Figure 16.2 shows the sampling distribution for a sample mean based on $n = 4$ trials, centered at the mean of the historical data for engine A, $\mu_{y_A} = \mu_{y_B} = 28$. We note that the sampling distribution is normal and has

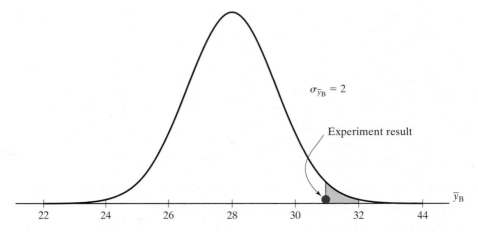

Figure 16.2   Distribution of the engine B sample mean

a standard deviation of $\sigma_{\bar{y}_B} = \sigma_{y_A}/\sqrt{n} = 4/2 = 2$. Comparing the engine B test statistic $\bar{y}_B = 31$ against this sampling distribution, we can calculate the probability of getting such an experimental result, or one more extreme. We can then compare that probability to the risk we are willing to take to say that engine B's performance is different from that of engine A even if it really is not. For this illustration, we will set that risk at $\alpha = 0.02$.

To evaluate the experiment result, we may determine how far it lies from the mean in standard deviation units. In terms of the standard normal variable Z, we obtain

$$z_{calc} = \frac{31 - 28}{4/\sqrt{4}} = \frac{3}{2} = 1.5.$$

To make our statistical decision on whether or not the performance of engine B differs from that of engine A, we must compare the probability of getting a Z value of 1.5 or greater to our risk level, in this case $\alpha/2 = 0.01$. Since

$$\text{Prob}(\bar{y}_B \geq 31) = \text{Prob}(Z \geq z_{calc}) = \text{Prob}(Z \geq 1.5) = 0.067,$$

we conclude that the value we have obtained for $\bar{y}_B$ is not an unlikely occurrence, given the historical data for engine A; that is, we cannot see any difference in the performance of engine B, in terms of mpg, from that which we historically have seen for engine A.

Assessing the Experiment

Let us consider the experiment and analysis that was performed—were there any problems? In fact, there was nothing wrong with our statistical analysis, but the *design* of the comparative test was poor. In particular, the character of the historical data may be suspect. The data used to create the reference distribution for engine A were collected under a variety of conditions over a long period of time. In fact, we listed seven likely sources of variation, and more certainly exist. These sources introduced a considerable amount of variability— and inflated the variance of the external reference distribution. As a result, the comparative test we performed was not a very *sensitive* comparison. For the experimental design we used here, an experimental result was compared with an external reference distribution based on historical data. To control conditions and reduce variability, a better comparison may have been possible by comparing the two engines side-by-side within the same experiment. This can be achieved by running trials for each engine under conditions as similar as possible. In our discussion of the sandmill experiment in Chapter 15, we reached a similar conclusion. When trials for both engines are run under similar conditions within the same experiment, we shall see that a sampling distribution for the comparative test statistic can be constructed from the data within the experiment itself to form an *internal reference distribution*. The experimental data can then be judged using this internal reference distribution.

16.3   Comparing Two Alternatives via an Internal Reference Distribution

Based on the discussion at the end of the previous section, the car company decided to design and conduct a second experiment with the hope that a more sensitive and fair comparison between engine A and engine B performance in terms of mpg could be developed. One engine of each type (A and B) was selected at random. The engines were installed in two different vehicles but of identical make and model. Several fuel economy trials were conducted for both engines using the same driver for all the trials. For each trial, one after another, the driver randomly selected the route from a group of several possible routes and also selected which particular engine to test on a given trial at random. The tests were run over a two-day period and there was time each day to run eight trials. The results of these sixteen trials are shown in Table 16.1.

TABLE 16.1	Results from the Second Engine Experiment (mpg)

y_A: 23, 30, 29, 24, 25, 28, 34, 27, 31
y_B: 26, 31, 30, 34, 35, 29, 33

	Engine A	Engine B
Sample Size	9	7
Sample Mean	$\bar{y}_A = 27.889$	$\bar{y}_B = 31.143$

Statistical Test of Hypothesis

Let us recall the discussion of the "statistical test of hypothesis" of Chapter 4. Such a test was said to consist of the following steps:

1. State the null and alternative hypotheses clearly. Define the test statistic used to analyze the situation.
2. Determine the significance level (risk level α) at which the test will be made.
3. Collect the data and calculate the test statistic.
4. Define the reference distribution for the test statistic.
5. Compare statistically the test statistic and its reference distribution, under H_0.
6. Assess the risk.

We will now proceed following these six simple steps.

1. State the null and alternative hypotheses clearly. Define the test statistic used to analyze the situation.

It is desired to determine if the true mean for engine B is different from the true mean of engine A. We are not concerned with whether B is better or worse than A, only different. Therefore, the null and alternative hypotheses are defined as

$$H_0: \mu_{y_B} - \mu_{y_A} = 0, \quad \text{to be tested against} \quad H_1: \mu_{y_B} - \mu_{y_A} \neq 0.$$

Since we are comparing means, it seems appropriate to use the respective averages from the data to form the test statistic, namely, $\bar{y}_B - \bar{y}_A$.

2. Determine the significance level (risk factor α) at which the test will be made.

As we did in analyzing the results of the first experimental design, we select a risk value: $\alpha = 0.02$ or $\alpha/2 = 0.01$.

3. Collect the data and calculate the test statistic.

From the data in Table 16.1,

$$\bar{y}_B - \bar{y}_A = 31.143 - 27.889 = 3.254.$$

4. Define the reference distribution for the test statistic.

We need to understand what the distribution of $\bar{y}_B - \bar{y}_A$ looks like. In particular, we need to determine the mean and variance of the distribution of $\bar{y}_B - \bar{y}_A$, and we need to determine the shape of the distribution. We state here without proof that, if $\bar{y}_A$ and $\bar{y}_B$ are independent, then

$$E[\bar{y}_B - \bar{y}_A] = \mu_{\bar{y}_B} - \mu_{\bar{y}_A} \qquad (16.1)$$

and

$$\text{Var}[\bar{y}_B - \bar{y}_A] = \text{Var}(\bar{y}_B) + \text{Var}(\bar{y}_A) = \sigma_{\bar{y}_B}^2 + \sigma_{\bar{y}_A}^2 = \frac{\sigma_{y_B}^2}{n_B} + \frac{\sigma_{y_A}^2}{n_A}. \quad (16.2)$$

Further, since we assumed that the performance of the two engines may vary so far as their means are concerned but not their levels of variability, then $\sigma_{y_A}^2 = \sigma_{y_B}^2 = \sigma_y^2$, and

$$\sigma_{(\bar{y}_B - \bar{y}_A)}^2 = \sigma_y^2 \left(\frac{1}{n_B} + \frac{1}{n_A} \right). \quad (16.3)$$

Furthermore, as far as the shape of $\bar{y}_B - \bar{y}_A$ distribution is concerned, if y_A and y_B are normal, then so is $\bar{y}_B - \bar{y}_A$. At the very least, $\bar{y}_B - \bar{y}_A$ is approximately normal regardless of the distribution of y_A and y_B. This is so since $\bar{y}_B - \bar{y}_A$ is a linear combination of 16 observations/responses (with an assumed known common variance),

$$\bar{y}_B - \bar{y}_A = \frac{y_{B_1} + y_{B_2} + \cdots + y_{B_6} + y_{B_7}}{7} - \frac{y_{A_1} + y_{A_2} + \cdots + y_{A_8} + y_{A_9}}{9}.$$

Therefore, we can say that the distribution of $(\bar{y}_B - \bar{y}_A)$ is approximately normally distributed with mean $\mu_{y_B} - \mu_{y_A}$ and standard deviation $\sigma_{(\bar{y}_B - \bar{y}_A)}$. The test statistic becomes:

$$z_{\text{calc}} = \frac{(\bar{y}_B - \bar{y}_A) - (\mu_{y_B} - \mu_{y_A})}{\sigma_{y(\bar{y}_B - \bar{y}_A)}} = \frac{(\bar{y}_B - \bar{y}_A) - (\mu_{y_B} - \mu_{y_A})}{\sigma \sqrt{1/n_A + 1/n_B}}. \quad (16.4)$$

In this case, however, σ_y^2 is not known and must be estimated from the data of the experiment. Two estimates are available for the unknown variance σ_y^2, namely, $s_{y_A}^2$ and $s_{y_B}^2$. Furthermore, these two estimates of the unknown common variance may be "pooled" to provide a better (more precise) estimate of the variance that contains all the information in each of the individual estimates,

$$s_{\text{pooled}}^2 = s_p^2 = \frac{\nu_A s_{y_A}^2 + \nu_B s_{y_B}^2}{\nu_A + \nu_B}, \quad (16.5)$$

where ν_A and ν_B are the degrees of freedom associated with each variance estimate ($\nu_A = n_A - 1$ and $\nu_B = n_B - 1$).

For the data of Table 16.1, the sample variances for engine A ($s_{y_A}^2$) and engine B ($s_{y_B}^2$) are 12.611 and 9.810, respectively. Thus, the pooled variance is

$$s_p^2 = \frac{8(12.611) + 6(9.810)}{8 + 6} = 11.410.$$

We now can substitute s_p^2 for σ_y^2 in Eq. (16.4) above, but in so doing, we can no longer consider the reference distribution as normal. Recalling our discussion in Chapter 4, we know that the reference/sampling distribution becomes Student's t-distribution; that is, the quantity

$$t_{\text{calc}} = \frac{(\bar{y}_B - \bar{y}_A) - (\mu_{y_B} - \mu_{y_A})}{s_{(\bar{y}_B - \bar{y}_A)}} \tag{16.6}$$

is distributed according to a t-distribution with $\nu = \nu_A + \nu_B = 14$ degrees of freedom. If we use s_p^2 as an estimate of σ_y^2, then our estimate of $\sigma_{(\bar{y}_B - \bar{y}_A)}^2$ becomes

$$s_{(\bar{y}_B - \bar{y}_A)}^2 = s_p^2 \left(\frac{1}{n_B} + \frac{1}{n_A} \right). \tag{16.7}$$

For the data in Table 16.1, we obtain:

$$s_{(\bar{y}_B - \bar{y}_A)}^2 = 11.410 \left(\frac{1}{7} + \frac{1}{9} \right) = 2.898$$

and

$$s_{(\bar{y}_B - \bar{y}_A)} = \sqrt{2.898} = 1.702,$$

where it may be noted that $s_{(\bar{y}_B - \bar{y}_A)}^2$ has $\nu_A + \nu_B = \nu_{\text{pooled}} = 8 + 6 = 14$ degrees of freedom. In determining the degrees of freedom associated with a t-distribution, we are always correct if we ask: "How many degrees of freedom did we have available for our estimate of the variance?" The answer here is $(n_A - 1) + (n_B - 1) = 14$.

For the data of Table 16.1 our calculated value for t becomes

$$t_{\text{calc}} = \frac{3.254 - 0}{1.702} = 1.912.$$

5. Compare statistically the test statistic and its reference distribution, under H_0.

We now need to compare our calculated value for t_{calc} to the appropriate t-distribution with $\nu = 14$ degrees of freedom. This situation is depicted in Fig. 16.3. To determine how extreme $t_{\text{calc}} = 1.912$ is, we need to determine $\text{Prob}(t \geq t_{\text{calc}})$— this is the shaded area in the figure. The probability of a t value with 14 degrees of freedom being larger than t_{calc} is about 0.038.

6. Assess the risk.

The probability associated with our calculated t value is 0.038, which is larger than the specified risk level of $\alpha/2 = 0.01$. Alternatively, we may compare our

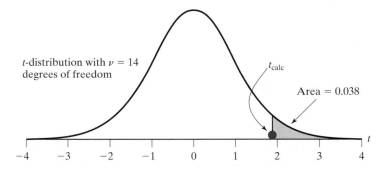

Figure 16.3   *t*-distribution and calculated *t* value

calculated *t* value to the critical *t* value corresponding to our risk level. Since $t_{calc} = 1.912$ is less than the critical *t* value, $t_{\nu,1-\alpha/2} = t_{14,0.99} = 2.62$ (from Table A.3 in the Appendix), we have no evidence to reject the null hypothesis, at our defined risk level of $\alpha = 0.02$. It appears as if the observed difference $(\bar{y}_B - \bar{y}_A)$ could reasonably be obtained simply due to the inherent variation in the experiment. We have no evidence of any real difference in performance between the two engines. With this in mind, since engine B has not been demonstrated to be superior to engine A, we therefore conclude that there is no reason to adopt engine B.

Confidence-Interval Approach to Testing the Difference in the Two Means

For the above engine problem, we may also develop a confidence interval for the difference between the two engine means: $\mu_{y_B} - \mu_{y_A}$. For this case, where the variance must be estimated, the $100(1 - \alpha)\%$ confidence interval for the difference between the true means of the two independent samples is given by

$$(\bar{y}_B - \bar{y}_A) \pm t_{\nu,1-\alpha/2} s_{(\bar{y}_B-\bar{y}_A)}, \tag{16.8}$$

where ν is the degrees of freedom associated with the variance estimate $s^2_{(\bar{y}_B-\bar{y}_A)}$.

For $\alpha = 0.02$, $\nu = 14$, and $t_{14,0.99} = 2.62$ the 98% confidence interval may be calculated as

$$(\bar{y}_B - \bar{y}_A) \pm t_{14,0.99} s_{(\bar{y}_B-\bar{y}_A)} = 3.254 \pm 2.62(1.702) = 3.254 \pm 4.459,$$

which provides the interval: $(-1.21, 7.71)$.

Since the confidence interval includes the possibility that $\mu_{y_B} - \mu_{y_A} = 0$, "0" represents a plausible value for the true mean difference. In fact, any hypothesis for $\mu_{\bar{y}_B} - \mu_{\bar{y}_A}$ that lies on the interval above cannot be rejected with 98% confidence. So, as was noted using the hypothesis testing approach, we cannot refute the notion that the two engines provide a comparable performance.

Thus, based on the experiment that was performed, there does not appear to be any need to introduce the new engine type (B).

Assessing the Experiment

There are several positive aspects of the second experiment performed by the car company relative to the first experiment of Section 16.2. The first experiment relied on an external reference distribution—formed from data collected over a long period of time and under a variety of conditions. In general, such historical data are often suspect, since little is known about the conditions under which they were collected and they generally have a large spread (level of variability). The second experiment constructed an internal reference distribution using the data from the experiment itself. In general, such an internal reference distribution is preferred, since the data are collected under much more tightly controlled conditions and will be subject to fewer sources of variability.

In considering how the second experiment was performed, there are a few important issues of note. Several actions were taken to reduce the level of variability in the experimental environment: only one engine of each type, only one driver, cars of identical make and model, and all the tests performed over a two-day period. These actions, focused on fixing factors that varied during the experiment described in the previous section, serve to reduce the variability of the internal reference distribution. This leads to a more sensitive experiment and comparison. Of course, while fixing these variables does reduce the level of variability, it may constrain our ability to apply the experiment results. For example, if the results of the experiment showed that engine B was indeed superior to engine A, might this be true for only the driver considered in the experiment, or the car make and model chosen for the test?

At the start of this section, it was noted that for each trial in the testing sequence, the engine type was randomly selected. This procedure serves to avoid any confusing of the engine type with extraneous factors in the experimental environment. Failure to take this "randomization" precaution, discussed in the last chapter, jeopardizes the *validity* of the experiment. For example, if the engine A trials were run on one day and the engine B trials on the next, it would not be possible to tell if a large value obtained for $\bar{y}_B - \bar{y}_A$ was due to a true engine difference or to some day-to-day effect (e.g., sunny one day and rainy the next), which could significantly influence the mpg response. In the absence of randomization, the experimental design could lead to erroneous results. The first experiment was a bad design—it had poor sensitivity—but it was not an invalid design. It is worth noting that in the experiment described in this section, the route used for each trial was also randomly selected; this safeguard avoids a pitfall like running all the engine A trials on a hilly route and all the engine B trials on a flat route. Randomization ensures the validity of the experiment.

One might question the advisability of considering a number of different routes in the experiment. Certainly, the inclusion of a variety of different routes will allow differences in the routes to produce a wider range of mpg results for the trials and therefore inflate the variability of the reference distribution and decrease the sensitivity of the comparison. Fixing the route might be considered as an option, but which specific route should we choose for all the trials? Furthermore, any real difference between the engines may be dependent on the route (e.g., engine A performs better on a hilly terrain), and we would definitely like to know if this is the case. This presents a challenge—we would like to include route-to-route differences in the experiment, but doing so seems to reduce the sensitivity of the comparison. One method by which we can respond to this challenge is described in the next section.

16.4   Comparing Two Alternatives Using a Blocked Design

The experimental structure and associated analysis described in the previous section is also referred to as an "independent *t*-test," since it considers the data for each treatment as an independent group. We will not dwell on this term, since it seems to place attention on the analysis rather than the experiment design. While the experiment was conducted in a manner that produced a valid test of the two engines—namely, it employed the important design tool called *randomization*—some questions were raised concerning the sensitivity of the experiment. In particular, the route used on any given trial was selected at random from a pool of many candidate routes. These routes varied in terms of road conditions, which could influence the mileage obtained. Furthermore, the tests were performed over a two-day period—again each trial selected at random— but the change in weather from one day to the next was also a source of variation that could influence the mileage obtained on a given trial. While these sources of nuisance variation were randomly distributed across the experiment, making it a valid test, they may be producing a pattern of variation large enough to mask or obscure any real difference that might exist between the two engine types, as far as mileage is concerned.

In an effort to overcome some of the possible problems with the previous design of experiment, let us consider yet another experiment that compares engine A and engine B in terms of fuel economy. For this experiment, as many potential noise factors as possible were fixed, with a few exceptions. Again, since only eight mileage trials could be performed each day, two days were required to perform the entire experiment—recall that the weather might be different across the two days of testing and this might influence the mileage for a given trial. Given the discussion at the end of the previous section, it was again decided to consider a wide variety of different routes in the experiment, with the variety representative of what the engines would be exposed to during field use. But, this time eight specific routes were chosen in advance from among those in the pool of routes. Each engine type was tested on each route, with the day of the test (Monday or Tuesday) being selected randomly. If one engine type was tested on a given route on Monday, then the other engine type was also tested on that route on Monday, and vice versa. Such "pairs" of trials were conducted consecutively, with engine type chosen in a random order in time

TABLE 16.2	Results from Third Engine Experiment	
Route	Engine A	Engine B
1	29 (T)	32 (T)
2	32 (T)	39 (T)
3	27 (T)	30 (T)
4	26 (M)	30 (M)
5	29 (M)	30 (M)
6	32 (M)	31 (M)
7	24 (T)	29 (T)
8	25 (M)	27 (M)

within the pair. The experimental data are shown in Table 16.2 (the day on which the trial was performed is shown in parentheses).

Independent t-test Analysis

Let us quickly analyze the data using the independent t-test approach of the previous section. The experiment has been suitably randomized to avoid any influence of systematic factors in the experimental environment, such as day of the test and order each day, so this analysis should produce a valid comparison of the two engines. Using the data in Table 16.2, we obtain

$$\bar{y}_A = 28, \quad \bar{y}_B = 31, \quad \bar{y}_B - \bar{y}_A = 3,$$

$$s_{y_A}^2 = 9.143, \quad \nu_A = 7, \quad s_{y_B}^2 = 12.571, \quad \nu_B = 7,$$

$$s_p^2 = \frac{s_{y_A}^2 \nu_A + s_{y_B}^2 \nu_B}{\nu_A + \nu_B} = \frac{9.143(7) + 12.571(7)}{7 + 7} = 10.857,$$

which serves as an estimate of the common population variance with $\nu_p = 14$ degrees of freedom. Note that this estimate of the variance is similar to that obtained from the experimental data in the previous section. Then, we may estimate the variance of the difference in the averages as

$$s_{(\bar{y}_B - \bar{y}_A)}^2 = s_p^2 \left(\frac{1}{n_A} + \frac{1}{n_B} \right) = 10.857 \left(\frac{1}{8} + \frac{1}{8} \right) = 2.714 \quad \text{and}$$

$$s_{(\bar{y}_B - \bar{y}_A)} = \sqrt{2.714} = 1.647.$$

Therefore, the calculated t value, under the null hypothesis that the means are equal, is given by

$$t_{calc} = \frac{(\bar{y}_B - \bar{y}_A) - \mu_{(\bar{y}_B - \bar{y}_A)}}{s_{(\bar{y}_B - \bar{y}_A)}} = \frac{3 - 0}{1.647} = 1.82.$$

For $\nu_p = 14$ degrees of freedom, the Prob($t \geq t_{\text{calc}}$) = 0.045, and since this value is larger than our selected risk value of $\alpha/2 = 0.01$, we cannot reject the null hypothesis that the true means for the two engines are equal. Note that this conclusion is consistent with the results from the preceding section. But have we used the best analysis technique here? Recall that we purposely *designed the experiment* by "pairing up" each engine type within a given route and on a given day. Looking back at Table 16.2, we see that each route forms an experiment subunit, the variation within which is less than that across all such subunits (refer to Fig. 16.4). In particular, each engine type can be compared within each route and on the same day. In short, we have made use of the important design technique of "blocking," described first in Chapter 15, in the *design* of the experiment, but we have not made use of this technique in the analysis of the data.

Blocked or "Paired" Data Analysis

Recall that in our experiment we deliberately paired up engines A and B for each route. This action allows us to compare engine types within a route as opposed to comparing the engines across all the routes. To compare mileages for each route, we may calculate $d_i = y_{Bi} - y_{Ai}$ for the ith route ($i = 1, \ldots, 8$); this quantity "blocks" out the variability due to route-to-route differences. Since each pair was tested on the same day, in consecutive trials, we have also blocked out day-to-day variation and minimized any effect of short-term variation within a day. Since we randomly decided whether a given pair should be run on Monday or Tuesday, we have distributed this variation factor randomly across the experiment, so validity is ensured as far as this factor is concerned. The engine differences, d_i values, are provided in Table 16.3, and are analyzed below to hopefully produce a more sensitive comparison of the two engines.

Hypothesis Testing Approach

$H_0: \mu_d = 0$, $H_1: \mu_d \neq 0$, $\alpha = 0.02$, $\alpha/2 = 0.01$.

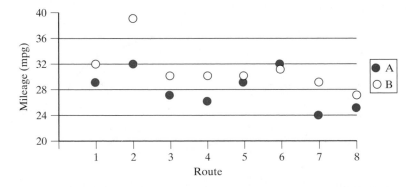

Figure 16.4   **Mileages for engines A and B for the eight routes**

TABLE 16.3	Engine A and Engine B Differences for Each Route							
	Route							
	1	2	3	4	5	6	7	8
Engine A	29	32	27	26	29	32	24	25
Engine B	32	39	30	30	30	31	29	27
$d_i = y_{Bi} - y_{Ai}$	3	7	3	4	1	−1	5	2

We will use $\bar{d}$ to test the hypothesis, where $\bar{d}$ is given by

$$\bar{d} = \sum_{i=1}^{n} d_i/n$$

$$= \frac{24}{8} = 3. \tag{16.9}$$

The reference distribution for $\bar{d}$ is shown in Figure 16.5.
We may then estimate σ_d^2 and $\sigma_{\bar{d}}^2$ using the relations

$$s_d^2 = \frac{\sum_{i=1}^{n}(d_i - \bar{d})^2}{n - 1} \quad \text{and} \quad s_{\bar{d}}^2 = s_d^2/n. \tag{16.10}$$

Applying Eq. (16.10) to the data of Table 16.3, we obtain

$$s_d^2 = 6.0, \quad s_d = 2.45, \quad \nu_d = 7, \quad \text{and} \quad s_{\bar{d}} = \frac{s_d}{\sqrt{n}} = \frac{2.45}{\sqrt{8}} = 0.866.$$

Comparing $s_{\bar{d}}$ with $s_{(\bar{y}_B - \bar{y}_A)}$, we can see how $s_{(\bar{y}_B - \bar{y}_A)} = 1.647$ was inflated by route-to-route and day-to-day differences. The appropriate t-statistic to be determined, under H_0, is

$$t_{\text{calc}} = \frac{(\bar{d}) - \mu_{\bar{d}}}{s_{\bar{d}}}. \tag{16.11}$$

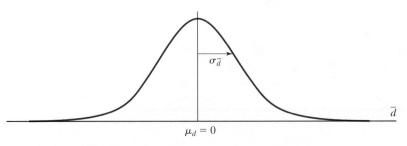

Figure 16.5   Distribution of $\bar{d}$

For our data, under H_0, this becomes

$$t_{\text{calc}} = \frac{(\bar{d}) - \mu_{\bar{d}}}{s_{\bar{d}}} = \frac{3 - 0}{0.866} = 3.46.$$

Now we need to compare our calculated value of t_{calc} to the appropriate t-distribution with $\nu = 7$ degrees of freedom. This situation is depicted in Figure 16.6. We may then calculate the probability of obtaining a t value more extreme than t_{calc}: $\text{Prob}(t \geq t_{\text{calc}}) = 0.0053$—this area is shown in Figure 16.6. Note that this probability is less than our risk level of $\alpha/2 = 0.01$. So, we reject H_0 and conclude that there is strong evidence to suggest that a real difference exists between the two engines so far as their mileage performance is concerned.

The lesson to be learned here is twofold. First, to create an experiment that produces a more sensitive comparison of the difference between two treatments, the design technique of "blocking" may be employed by creating experimental subunits or "blocks." Within each block the conditions for comparison of the two treatments are much less variable than they are across the blocks. Second, blocking is a design tool that requires an associated change in the analysis procedure to capitalize on the blocked design. The unblocked analysis that was first conducted—the independent t test—produced a valid analysis but did not take advantage of the blocking incorporated in the design of the experiment. It is also important to understand that since blocking is a design strategy, employed before the trials are run, we cannot now go back to the data in Table 16.1 and somehow "pair up" the data so that a blocked analysis based on the paired t test can be conducted. Blocking is a design strategy!

Confidence-Interval Approach

For completeness, we may also apply the confidence-interval approach to analyze the data—the d_i's—of the blocked experiment. In general, the $100(1 - \alpha)\%$ confidence interval for μ_d is given by

$$\bar{d} \pm t_{\nu, 1 - \alpha/2} s_{\bar{d}}, \tag{16.12}$$

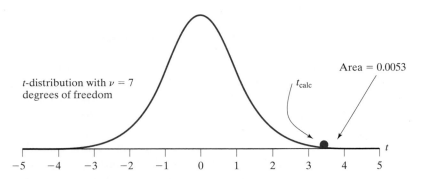

Figure 16.6   *t*-Distribution and calculated *t* value

and for our data this simplifies to

$$\overline{d} \pm t_{7,0.99}s_{\overline{d}} = 3 \pm 2.998(0.866).$$

Therefore, the 98% confidence interval for the true mean difference is (0.404, 5.596).

Since zero does not lie on the confidence interval, we reject H_0 and conclude that engine B produces a higher mpg value. While the evidence suggests that $\mu_{y_B} \neq \mu_{y_A}$, we note that since "0" is relatively close to the lower limit of the confidence interval, our conclusion might be challenged as not being strong enough evidence. Further experimentation might be contemplated with the objective of improving the strength of the evidence. Some prefer the use of the confidence-interval approach for hypothesis testing precisely because of this numerical scale for the interval.

Randomization Distribution

Assuming that the experiment is completely randomized and that there is no difference between the true means for engines A and B (i.e., H_0 is true), the label assigned to an engine (A or B) within a route is arbitrary. If we were to switch the labels on the responses for a given route, the sign on the value for d would switch. The observed sequence of eight d values is one of $2^8 = 256$ possible sets of d values that could be obtained by switching labels. Each set of these values produces a different $\overline{d}$. The complete set of $\overline{d}$ values is given by

$$\overline{d} = \frac{\sum\limits_{i=1}^{n} \pm d_i}{n},$$

which for our data becomes

$$\overline{d} = \frac{\pm 3 \pm 7 \pm 3 \pm 4 \pm 1 \pm 1 \pm 5 \pm 2}{8}.$$

If all the $\overline{d}$ values provided by this equation are compiled, they form what is termed the *randomization distribution*. The randomization distribution for our data is shown in Fig. 16.7.

We may refer our calculated value of $\overline{d} = 3$ to the randomization distribution. We observe that the value we have calculated for $\overline{d}$ is as large as all but one other potential $\overline{d}$ value. Significance levels for the randomization distribution closely match those of the t-distribution with appropriate degrees of freedom. In our case, only 1 out of 256 $\overline{d}$ values is larger than ours— this corresponds to a significance level of $1/256 = 0.0039$. Note that this significance level is fairly close to the probability value obtained using the t-distribution (0.0053). One benefit of using the randomization distribution is that it requires no assumptions about the underlying distribution. Of course,

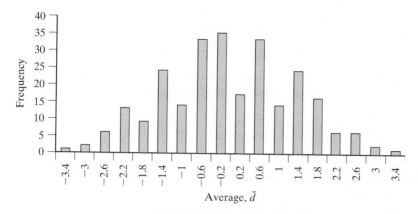

Figure 16.7   **Randomization distribution for blocked treatment comparison**

we can only avail ourselves of the benefits of the randomization distribution if the experiment is properly designed with the inclusion of randomization in the first place. Therefore, provided we randomize the experiment, we may use the randomization distribution without making any assumptions about the parent population.

Summary

Over the last three sections, three important concepts have been explored:

- *Internal vs. External Reference Distribution.* We have seen that we should generally avoid using historical data to form an external reference distribution against which our experiment results may be compared. Historical data are often acquired under a wide range of conditions, which serves to inflate the variance of the external reference distribution. It is generally preferable to employ the data within the experiment itself to form an internal reference distribution.

- *Blocking.* Identify important variation factors ahead of time, and design the experiment so that comparisons may be made between "treatments" (engine A vs. engine B) within each block or experimental subunit.

- *Randomization.* It is important for two reasons:

  - Experimental validity. We do not want any known or unknown factors confounded with the treatments (for example, all engine A's on Monday and all engine B's on Tuesday). Randomization was achieved in our paired or blocked experiment by simply tossing a coin to determine the day that a given route was tested and the order of the test, within each route/block.

  - Applicability of randomization distribution. When the experiment is randomized, a randomization distribution may be constructed that requires no assumptions regarding the underlying distributions of the response of interest. The significance levels associated with the randomization distribution closely match the probabilities obtained from appropriate statistical distributions.

As we move forward, it is worth summarizing several items that should guide our actions as we design our experiments. These "rules to live by" are

- At the outset of designing an experiment, identify the "important" factors that influence the response of interest.
- Fix as many nuisance variables as possible.
- Block those factors that can be controlled but not fixed, or those factors that we do not wish to fix.
- Randomize the experiment to handle unknown factors or factors that are not blocked or fixed. This ensures the validity of the experiment as well as the approximate validity of the statistical tests.

In short, "block the factors that you should and can, and randomize to handle the factors that are not blocked."

16.5   Comparison of k Treatments Using Analysis of Variance—One-Way ANOVA

In our previous discussions, we considered two levels for the variable under study (i.e., engine type A vs. engine type B). We will now examine a technique that can be used for examining multiple (two or more) levels for the variable being studied. This technique is referred to as *analysis of variance*, or ANOVA. It is widely used for decomposing the total variation in a set of data into sources of variation, each of which is considered important if it contributes a significant amount of variation to the experiment results—more than would be expected given the level/amount of random/chance variation. Before proceeding to our discussion of ANOVA, we must first introduce some basic concepts.

Sampling Distributions Related to the Sample Variance, s_y^2

Analysis of variance involves working with and comparing variation sources that are characterized by sample variances that we will call *mean squares*. With this in mind, we need to understand the sampling distribution of the sample variance.

Sampling Distribution of the Sample Variance, s_y^2

Up to this point, we have concerned ourselves with evaluating hypotheses concerning the true mean of a distribution. For example, the true mean foundry workers' wage rate is $16, the true mean difference between two engine mileages is 0, and so on. Hypotheses may also be made concerning the true variance of a distribution. The procedure to test a hypothesis concerning the variance is exactly the same as that described previously for the mean.

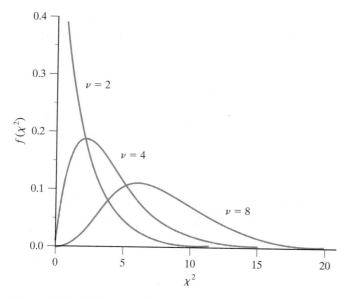

Figure 16.8 Chi-square distributions for ν = 2, 4, and 8

Let us assume that a set of individual measurements is normally distributed, and that the variance of these individuals, σ_y^2, is known or hypothesized to be a specific value. If a sample of size n is drawn from this distribution and is used to calculate the sample variance, s_y^2, the quantity

$$\frac{(n-1)s_y^2}{\sigma_y^2} \tag{16.13}$$

is distributed according to the *chi-square* (χ^2) *distribution*. The chi-square distribution defines the distribution of sample variances and is only a function of the degrees of freedom, $\nu = n - 1$, used to calculate the sample variance. The probability density functions for several chi-square distributions (different values for ν) are illustrated in Fig. 16.8. Table A.5 provides a listing of chi-square values beyond which the probability of the random variable χ^2 is α for various degrees of freedom.

▶ EXAMPLE 16.1

A manufacturer of hip implants claims that the hardness variability from implant to implant is $\sigma_y^2 = 0.01$ Vickers². The hardness of the implants is known to be normally distributed, and for a sample of size $n = 5$ implants, the sample variance is calculated to be $s_y^2 = 0.02$. Evaluate the company's claim at $\alpha = 0.05$.

Solution. Using Eq. (16.13), the quantity $\chi_{calc}^2 = \dfrac{(5-1)0.02}{0.01} = 8$ will follow a

χ^2-distribution with $\nu = n - 1 = 4$ degrees of freedom. From Table A.5, $\chi^2_{4,0.975} = 11.1$, and since χ^2_{calc} is less than this value, we cannot refute the manufacturer's claim.

Distribution of the Ratio of Two Sample Variances

Let s_1^2 and s_2^2 be the sample variances of two independent samples of size n_1 and n_2, taken from two normal populations having variances of σ_1^2 and σ_2^2, respectively. We know that (s_1^2/σ_1^2) is (χ_1^2/ν_1)-distributed (where $\nu_1 = n_1 - 1$) and that (s_2^2/σ_2^2) is (χ_2^2/ν_2)-distributed (where $\nu_2 = n_2 - 1$). The ratio $\left(\dfrac{\chi_1^2/\nu_1}{\chi_2^2/\nu_2}\right)$ is said to follow an F-distribution with ν_1 and ν_2 degrees of freedom (corresponding to the numerator and denominator, respectively). Therefore,

$$\frac{s_1^2/\sigma_1^2}{s_2^2/\sigma_2^2} \sim F_{\nu_1,\nu_2} \quad \text{or} \quad \frac{s_1^2}{s_2^2} \sim \frac{\sigma_1^2}{\sigma_2^2} F_{\nu_1,\nu_2}. \tag{16.14}$$

For the case where $\sigma_1^2 = \sigma_2^2$, Eq. (16.14) simplifies to

$$\frac{s_1^2}{s_2^2} \sim F_{\nu_1,\nu_2}. \tag{16.15}$$

The F-distribution is generally a skewed distribution and is defined by the two parameters ν_1 and ν_2, which are the degrees of freedom associated with s_1^2 and s_2^2, respectively.

Figure 16.9 shows some typical F-distributions where $F_{\nu_1,\nu_2,1-\alpha}$ is the critical value of F such that the probability is α that F is equal to or greater than $F_{\nu_1,\nu_2,1-\alpha}$. Table A.6 gives the critical F values for various combinations of ν_1 and ν_2 and several values for α. In using these tables, note that ν_1 always refers to the degrees of freedom associated with sample variance in the numerator of the F-ratio.

Water Hardness Example

A water company has four wells, and each well serves one of the four neighborhoods in a certain city. The water company believes that the water hardness (in terms of calcium carbonate level in mg/L) is the same for each well/neighborhood. To check this belief, the company randomly takes 16 water samples from homes across the four neighborhoods. Of interest is whether the water hardness differs among the neighborhoods. The water samples were taken in random order and once they were collected, the hardness tests were also performed in random order. Table 16.4 summarizes the water hardness responses for the experiment. Table 16.4 also displays the average for each treatment (or neighborhood) as well as the grand average. In general, if the response for the jth trial within the ith treatment is y_{ij}, then the ith treatment average is given by

$$\bar{y}_i = \sum_{j=1}^{n_i} y_{ij}/n_i, \tag{16.16}$$

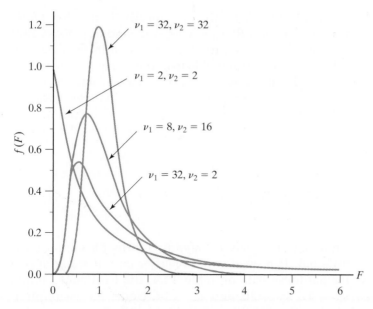

Figure 16.9 *F*-distributions for $\nu_1, \nu_2 = (32, 32), (2, 2), (8, 16),$ and $(32, 2)$

TABLE 16.4	Water Hardness Responses (CaCO$_3$ in mg/L) for Four Neighborhoods			
	Neighborhood			
	A	B	C	D
	123	116	138	123
	129	122	129	126
	134	122	135	122
	126		131	117
			127	
Treatment average	128	120	132	122
Grand average	126.25			

where n_i is the number of trials for the ith treatment. The grand average may be calculated via

$$\bar{\bar{y}} = \sum_{i=1}^{k}\sum_{j=1}^{n_i} y_{ij}/N, \qquad (16.17)$$

where k is the number of treatments and $N = \sum_{i=1}^{k} n_i$. For our water hardness example, $k = 4$, $n_i = 4, 3, 5, 4$ for $i = 1, \ldots, 4$, respectively, and $N = 16$. The

grand average and treatment averages are as specified in the table:
$\bar{\bar{y}} = 126.25, \bar{y}_1 = \bar{y}_A = 128, \bar{y}_2 = \bar{y}_B = 120, \bar{y}_3 = \bar{y}_C = 132$, and $\bar{y}_4 = \bar{y}_D = 122$.

ANOVA is a technique that decomposes the total variation in the data into component variations, each of which is attributable to a source, such as treatment means and experimental error. The mean squared variation associated with each source may then be compared against the mean squared variation due to experimental error to determine its significance using the appropriate F-distribution.

For the ith experimental treatment, the mean squared error, or the sample variance of treatment i, is calculated by:

$$s_i^2 = \frac{\sum_{j=1}^{n_i} (y_{ij} - \bar{y}_i)^2}{\nu_i}, \tag{16.18}$$

where $\nu_i = n_i - 1$ is the number of degrees of freedom for the ith sample variance. This is really no different than what we did earlier for our study of two treatments. For the water hardness data of Table 16.4, the treatment sample variances are $s_{y_A}^2 = 22$, $s_{y_B}^2 = 12$, $s_{y_C}^2 = 20$, and $s_{y_D}^2 = 14$. Under the assumption that all the treatments have an equal/common variance, σ_y^2, then the sample treatment variances may be pooled to provide an estimate of the unknown variance as follows:

$$s_{pooled}^2 = \frac{\sum_{i=1}^{k} \nu_i s_{y_i}^2}{\sum_{i=1}^{k} \nu_i}. \tag{16.19}$$

For our data, the pooled sample variance becomes

$$s_{pooled}^2 = \frac{\nu_A s_{y_A}^2 + \nu_B s_{y_B}^2 + \nu_C s_{y_C}^2 + \nu_D s_{y_D}^2}{\nu_A + \nu_B + \nu_C + \nu_D}$$
$$= \frac{3(22) + 2(12) + 4(20) + 3(14)}{3 + 2 + 4 + 3} = \frac{212}{12} = 17.667.$$

This estimate of the variance is based on the within-treatment variation of the hardness responses. It is, therefore, called the within-treatment mean square and denoted by MS(within) or s_{wt}^2. Another way of defining this mean square, consistent with the ANOVA method, is

$$s_{wt}^2 = \frac{\text{within-treatment sum of squares}}{\text{within-treatment degrees of freedom}} = \frac{SS_{wt}}{\nu_{wt}}$$

$$= \frac{\sum_{i=1}^{k}\sum_{j=1}^{n_i}(y_{ij} - \bar{y}_i)^2}{N - k} = s_{pooled}^2 = 17.667. \tag{16.20}$$

Now, we are interested in determining if at least one of the neighborhoods has a water hardness that is truly different from that of the other neighborhoods. This suggests that we should propose the following null hypothesis, $H_0: \mu_{y_A} = \mu_{y_B} = \mu_{y_C} = \mu_{y_D}$, to be tested against an alternative hypothesis (H_1) that at least one of the mean hardnesses differs from the others. Now, even if the hypothesis is true, the observed sample treatment averages of the experiment are likely to differ from one another due to experimental variation. The question is, "Is the observed difference between a given treatment average and the grand average, appropriately scaled by the number of observations used to determine the treatment average, greater than what would be expected if only the experimental variation is at work?"

Under our null hypothesis that all the treatment means are equal, another estimate of the error variance may be calculated by summing the squared differences of the treatment averages from the grand average, and since $\sigma_{\bar{y}}^2 = n\sigma_{\bar{y}}^2$, the following relation is obtained:

$$\hat{\sigma}_{\bar{y}}^2 = \frac{\sum\limits_{i=1}^{k} n_i(\bar{y}_i - \bar{\bar{y}})^2}{k - 1}. \tag{16.21}$$

where n_i is the number of trials for the ith treatment.

Of course, this estimate of the error variance will be inflated if our null hypothesis is not true. It may also be noted that this estimate is based on variation between treatments. For this reason, the above estimate is termed the between-treatments mean square, MS(between) or s_{bt}^2. Another way of defining this mean square, consistent with the ANOVA method, is

$$s_{bt}^2 = \frac{\text{between-treatments sum of squares}}{\text{between-treatments degrees of freedom}} = \frac{SS_{bt}}{\nu_{bt}} = \frac{\sum\limits_{i=1}^{k} n_i(\bar{y}_i - \bar{\bar{y}})^2}{k - 1}. \tag{16.22}$$

For our water hardness data, the between-treatments mean square becomes

$$s_{bt}^2 = \frac{SS_{bt}}{\nu_{bt}}$$

$$= \frac{4(128 - 126.25)^2 + 3(120 - 126.25)^2 + 5(132 - 126.25)^2 + 4(122 - 126.25)^2}{4 - 1},$$

$$s_{bt}^2 = \frac{367}{3} = 122.3333.$$

The within-treatment mean square value (s_{wt}^2) represents our "best" estimate of the experimental error variance, since it does not rely on the assumptions associated with H_0. For this reason, we often refer to s_{wt}^2 as the pure error mean square or $s_{\text{pure error}}^2$, since it is based on replicates within a given treatment.

The analysis that follows is exactly justified if the hardness data are random samples from four normal populations, and is approximately justified based on the randomization distribution idea presented previously.

Under the null hypothesis that there is no difference among the treatment means, we have established two estimates, s_{wt}^2 and s_{bt}^2, of the same experimental error variance, σ_y^2. When the observed value for s_{bt}^2 is much larger than s_{wt}^2, the added variation in s_{bt}^2 can be attributed to real differences that exist among the treatment means. This is the rationale behind the use of the F-distribution for our significance tests in ANOVA. Thus, under H_0, the ratio

$$F_{\text{calc}} = s_{bt}^2 / s_{wt}^2 \tag{16.23}$$

will follow an F-distribution with ν_{bt} and ν_{wt} degrees of freedom. For our data $F_{\text{calc}} = 122.3333/17.6667 = 6.92$. We recognize that the value for F_{calc} will be inflated if our assumptions relating to the null hypothesis are not satisfied. To assess how extreme F_{calc} is, we can compare $\text{Prob}(F_{\nu_1,\nu_2} \geq F_{\text{calc}}) = \text{Prob}(F_{3,12} \geq 6.92) = 0.0059$ to our critical risk level of α as shown in Fig. 16.10. For $\alpha = 0.05$, the probability associated with the calculated F is far less than the critical risk level. Alternatively, we can use Table A.6 and compare F_{calc} with the critical F value at a specified significance level, α, from the table. For example, $F_{3,12,0.99} = 5.95$, and our F_{calc} is greater than this critical F at $\alpha = 0.01$. Therefore, we reject the null hypothesis that the mean water hardness is the same for the four neighborhoods.

Variation Decomposition

As noted at the outset, the ANOVA procedure can decompose the variation within the data into a number of sources, viz.,

$$y_{ij} = \bar{\bar{y}} + (\bar{y}_i - \bar{\bar{y}}) + (y_{ij} - \bar{y}_i), \tag{16.24}$$

Figure 16.10 F_{calc} compared with appropriate F-distribution with ($\nu_1 = 3$, $\nu_2 = 12$)

where it may be noted that the left- and righthand sides of Eq. (16.24) are equal. It may also be observed that the righthand side of the equation includes terms associated with the grand average, the deviation of the ith treatment average from the grand average, and the deviation of the jth hardness response of the ith treatment from its average. If we square both sides of Eq. (16.24) and sum over all trials and treatments, we obtain

$$\sum_{i=1}^{k}\sum_{j=1}^{n_i} y_{ij}^2 = \sum_{i=1}^{k}\sum_{j=1}^{n_i} \{\bar{\bar{y}}^2 + 2\bar{\bar{y}}(\bar{y}_i - \bar{\bar{y}}) + 2\bar{\bar{y}}(y_{ij} - \bar{y}_i) + (\bar{y}_i - \bar{\bar{y}})^2$$

$$+ 2(\bar{y}_i - \bar{\bar{y}})(y_{ij} - \bar{y}_i) + (y_{ij} - \bar{y}_i)^2\}. \qquad (16.25)$$

Since

$$\sum_{i=1}^{k}\sum_{j=1}^{n_i} 2\bar{\bar{y}}(\bar{y}_i - \bar{\bar{y}}) = 2\bar{\bar{y}}\left[\sum_{i=1}^{k}\sum_{j=1}^{n_i}\bar{y}_i - \sum_{i=1}^{k}\sum_{j=1}^{n_i}\bar{\bar{y}}\right]$$

$$= 2\bar{\bar{y}}\left[\sum_{i=1}^{k} n_i\bar{y}_i - N\bar{\bar{y}}\right] = 2\bar{\bar{y}}[N\bar{\bar{y}} - N\bar{\bar{y}}] = 0,$$

$$\sum_{i=1}^{k}\sum_{j=1}^{n_i} 2\bar{\bar{y}}(y_{ij} - \bar{y}_i) = 2\bar{\bar{y}}\left[\sum_{i=1}^{k}\sum_{j=1}^{n_i}y_{ij} - \sum_{i=1}^{k}\sum_{j=1}^{n_i}\bar{y}_i\right]$$

$$= 2\bar{\bar{y}}\left[N\bar{\bar{y}} - \sum_{i=1}^{k} n_i\bar{y}_i\right] = 2\bar{\bar{y}}[N\bar{\bar{y}} - N\bar{\bar{y}}] = 0,$$

and

$$\sum_{i=1}^{k}\sum_{j=1}^{n_i} 2(\bar{y}_i - \bar{\bar{y}})(y_{ij} - \bar{y}_i) = 2\left[\sum_{i=1}^{k}\sum_{j=1}^{n_i}(\bar{y}_i y_{ij} - \bar{y}_i^2) - \sum_{i=1}^{k}\sum_{j=1}^{n_i}(\bar{\bar{y}}y_{ij} - \bar{\bar{y}}\bar{y}_i)\right]$$

$$= 2\left[\sum_{i=1}^{k}(n_i\bar{y}_i^2 - n_i\bar{y}_i^2) - \bar{\bar{y}}\sum_{i=1}^{k}(n_i\bar{y}_i - n_i\bar{y}_i)\right] = 0,$$

Eq. (16.25) simplifies to

$$\sum_{i=1}^{k}\sum_{j=1}^{n_i} y_{ij}^2 = \sum_{i=1}^{k}\sum_{j=1}^{n_i} \{\bar{\bar{y}}^2 + (\bar{y}_i - \bar{\bar{y}})^2 + (y_{ij} - \bar{y}_i)^2\} \qquad (16.26)$$

or

$$\sum_{i=1}^{k}\sum_{j=1}^{n_i} y_{ij}^2 = N\bar{\bar{y}}^2 + \sum_{i=1}^{k} n_i(\bar{y}_i - \bar{\bar{y}})^2 + \sum_{i=1}^{k}\sum_{j=1}^{n_i}(y_{ij} - \bar{y}_i)^2. \qquad (16.27)$$

We note that the second and third terms on the righthand side of Eq. (16.27) are the numerators of the variance estimates we used in Eqs. (16.20) and

(16.22)—based on the variation within treatment trials/replicates and between treatment averages, respectively. The total sum of squares of Eq. (16.27) may be decomposed into three components, each of which is a sum of squares (SS), and is often expressed as

$$SS_{total} = SS_{mean} + SS_{between\ treatments} + SS_{within\ treatment}. \qquad (16.28)$$

This variation decomposition may be succinctly captured in an ANOVA table such as Table 16.5. We note in this table that the sum of squares and degrees of freedom columns are additive (e.g., the sum of squares associated with the mean, between treatments, and within treatment add up to produce the total sum of squares).

For our water hardness data, the ANOVA table becomes:

Source of Variation	Sum of Squares	Degrees of Freedom	Mean Square	F_{calc}
Mean	255,025	1	255,025	14,432.65
Between treatments	367	3	122.33	6.92
Within treatment	212	12	17.67	
Total	255,604	16		

In interpreting the ANOVA table, we must compare the calculated F values to the appropriate critical F values from Table A.6. In considering $F_{calc} = 6.92$ associated with "between treatments," we recall our previous discussion where we noted that $F_{3,12,0.99} = 5.95$. Since F_{calc} is larger than this value, we conclude that the between-treatments source of variation is significant. For the "mean"

TABLE 16.5	ANOVA Table for Comparison of k Treatments

Source of Variation	Sum of Squares	Degrees of Freedom	Mean Square	F_{calc}
Mean	$N\bar{\bar{y}}^2$	1	$N\bar{\bar{y}}^2$	$\dfrac{MS(mean)}{MS(within)}$
Between treatments	$\sum_{i=1}^{k} n_i(\bar{y}_i - \bar{\bar{y}})^2$	$k - 1$	$\dfrac{\sum_{i=1}^{k} n_i(\bar{y}_i - \bar{\bar{y}})^2}{k - 1}$	$\dfrac{MS(between)}{MS(within)}$
Within treatment	$\sum_{i=1}^{k}\sum_{j=1}^{n_j}(y_{ij} - \bar{y}_i)^2$	$N - k$	$\dfrac{\sum_{i=1}^{k}\sum_{j=1}^{n_j}(y_{ij} - \bar{y}_i)^2}{N - k}$	
Total	$\sum_{i=1}^{k}\sum_{j=1}^{n_j} y_{ij}^2$	N		

source of variation $F_{\text{calc}} = 14{,}432.65$, which is much larger than the critical value of $F_{1,12,0.99} = 9.33$—indicating that the mean is different from zero.

Assessing the Treatment Averages Using the t-Distribution

Let us see if we can gain some additional insight into what the analysis of variance is telling us. To this point we have learned that the null hypothesis is not true, i.e., the neighborhoods do not have identical mean water hardness. It is not clear whether this suggests that one of the neighborhood means is different, all of the neighborhoods have different means, or something else. If σ_y was known and the number of trials for each treatment were the same, then under the null hypothesis the treatment averages could be thought of as outcomes from a common normal (reference) distribution with standard deviation $\sigma_y/\sqrt{n}$. We can imagine this situation graphically by constructing the normal reference distribution and sliding it along the horizontal axis; if the null hypothesis cannot be rejected, there should be some location at which the treatment averages could reasonably be explained as random samples from the same reference distribution.

For our water hardness example, σ_y is not known and the sample sizes are not equal. For such a situation, a normal distribution may not be used in our graphical approach; rather, we may compare our treatment averages to the appropriate t-distribution scaled by $\left(s_y/\sqrt{n_{\text{avg}}}\right)$, where n_{avg} is the average number of trials for the treatments. Since our best estimate of the error variance is $s_y^2 = \text{MS(within)}$, the value for s_y is $\sqrt{17.67} = 4.203$. Figure 16.11 displays the treatment averages on a number line, and a t-distribution with $\nu = 12$ degrees of freedom, scaled by $\left(s_y/\sqrt{4}\right)$, is shown above the line. This distribution may be thought of as an approximate reference distribution for the treatment averages.[1]

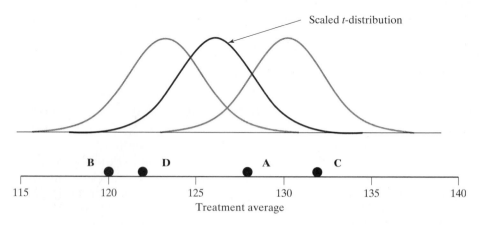

Figure 16.11   Treatment averages plotted relative to scaled t-distribution with 12 degrees of freedom

[1]G. E. P. Box, W. G. Hunter, and J. S. Hunter, *Statistics for Experimenters*, John Wiley, New York, 1978.

We may construct the scaled reference t-distribution using Table A.4 in the Appendix. The table provides ordinates of the t-distribution for different values of t for various degrees of freedom ν. For our water hardness example, $\nu = 12$ and the scale factor $(s_y/\sqrt{n_{avg}})$ is 2.102. Using this information in concert with Table A.4 we obtain:

value of t	0.0	0.5	1.0	1.5	2.0	2.5	3.0
$t \times 2.102$	0.0	1.051	2.102	3.153	4.204	5.255	6.306
t ordinate	0.391	0.342	0.232	0.128	0.060	0.026	0.010

To plot the reference distribution, we select an arbitrary origin (say the grand average, 126.25) and then plot the ordinates associated with 126.25, 126.25 ± 1.051, 126.25 ± 2.102, etc., and place a smooth curve through the points.

In examining Figure 16.11, it appears that we cannot slide the reference distribution to any axis location where all the treatment averages could be typical random outcomes from the same reference distribution. The figure seems to suggest that the averages for neighborhoods B and D differ dramatically from the average for neighborhood C. The average for neighborhood A falls in between the average for neighborhood C and the averages for the other neighborhoods. This finding may provide guidance in the planning of subsequent experiments.

Model for the Response

In Chapter 15 we emphasized that the design of the experiment should respond to the model proposed to describe the data. In this case, the underlying model is

$$\eta_i = \mu + \tau_i \qquad (16.29)$$

where μ is the intercept and τ_i is the effect associated with the ith treatment. Equation (16.29) specifies that the mean for the ith treatment (η_i) is equal to a constant plus a treatment effect. When the experimental error (ϵ) is considered, the response, y, takes the form:

$$y_{ij} = \eta_i + \epsilon_{ij}$$
$$= \mu + \tau_i + \epsilon_{ij}. \qquad (16.30)$$

It is typically assumed that ϵ is a normally and independently distributed variable with zero mean and fixed variance, σ_y^2. Based on the proposed model form, an experiment was performed that was capable of extracting information on μ and each τ_i. The responses for the experiment were shown in Table 16.4. In Eq. (16.30), the parameter values μ and τ_i ($i = 1, \ldots, k$) are unknown but can be estimated from the data. Based on Eq. (16.24) the parameter estimates are

$$\hat{\mu} = \bar{\bar{y}} \qquad (16.31)$$

and

$$\hat{\tau}_i = \bar{y}_i - \bar{\bar{y}}. \tag{16.32}$$

For the water hardness data, these parameter estimates are $\hat{\mu} = \bar{\bar{y}} = 126.25$ and $\hat{\tau}_i = \bar{y}_i - \bar{\bar{y}} = [1.75, -6.25, 5.75, -4.25]$. Based on these estimates, we may establish the following predictive equation for the response:

$$\hat{y}_i = \hat{\mu} + \hat{\tau}_i. \tag{16.33}$$

Note that we have included the treatment effect in our model for the predicted response because our ANOVA calculations revealed that there was a difference in the water hardness among neighborhoods. For our data, the predicted responses for each treatment, $\hat{y}_i$, are then: 128, 120, 132, and 122.

Diagnostic Checking of the Model

We have now established a model for the predicted water hardness for each neighborhood. As noted in Chapter 15, we should always critically examine such models to search for inadequacies and opportunities for further model improvement. Most often, this examination involves the careful scrutiny of the model residuals, i.e., the variation in the data not described by the model. These model residuals may be calculated as

$$e_{ij} = y_{ij} - \hat{y}_i, \tag{16.34}$$

where e_{ij} is the model residual for the jth trial within the ith treatment. If the model contains all the necessary terms to adequately predict the response, y, the model residuals should contain no structured variation at all; that is, they should behave in the following way:

- They should be centered about a mean of zero.
- They should tend to be normally distributed.
- They should have a scatter that does not vary as a function of the predicted response ($\hat{y}$) or any of the individual treatments.
- They should not be correlated with any variable related to the experiment, e.g., time. They should be truly random errors.

In short, the model residuals should have the same properties as we assumed for the errors (ϵ). The residuals for the water hardness data are provided in Table 16.6. We may now perform a series of checks on the residuals, e_{ij}. These should include:

- a normal probability plot of the residuals,
- a plot of the residuals vs. time/run order of the trials,
- a plot of the residuals vs. the predicted response, and

TABLE 16.6	Model Residuals for Water Hardness Response		
	Neighborhood		
A	B	C	D
−5	−4	6	1
1	2	−3	4
6	2	3	0
−2		−1	−5
		−5	

- plots of the residuals vs. the independent variables of the experiment (in this case, the treatment type).

Figures 16.12–16.15 show several of these plots. From these plots, we can say that the model residuals appear to behave in a random manner. They do not seem to show any trends with respect to the predicted response or the treatment. The normal probability plot suggests that the model residuals are normally distributed with the mean at zero.

16.6   Comparison of Multiple Treatments Using a Blocked Design—Two-Way ANOVA

The following example has been adapted from a study reported in the *Journal of Food Science.*[2] Through the example, we will illustrate how the design principle of blocking can be applied to the comparison of multiple treatments.

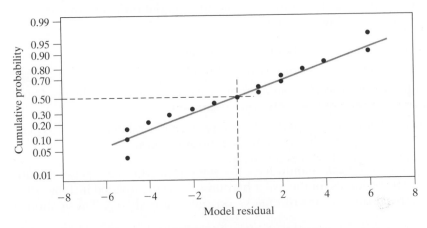

Figure 16.12   Normal probability plot of model residuals

---

[2]R. J. Matulis, F. K. McKeith, J. W. Sutherland, and M. S. Brewer, "Sensory Characteristics of Frankfurters as Affected by Fat, Salt, and pH," *J. of Food Science*, Vol. 60, No. 1, 1995, pp. 42–47.

Figure 16.13   Model residuals vs. run order

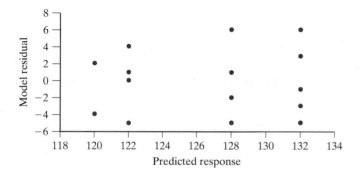

Figure 16.14   Model residuals vs. the predicted response

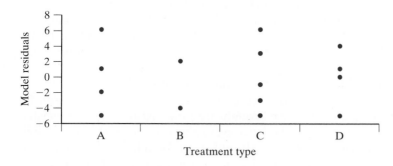

Figure 16.15   Model residuals vs. treatment type

Hot Dog Example

New regulations have relaxed requirements on salt, protein, fat, and filler contents in hot dogs. A company is seeking to develop a "healthy hot dog" that is low in fat and salt. Of course, reducing the levels of fat and salt may have negative effects on such hot dog performance responses as flavor, nutritional value (ability to retain vitamins), mouth feel, and texture. The texture is assessed by trained panelist tasters, who evaluate a cooked hot dog sample on a 0.0–14.0 scale (0: too soft/mushy, 14: too hard/chewy). The company has developed four recipes that are of interest for study, and the response to be considered in the analysis below is the texture. Three panelists are available for the texture testing trials. In planning for the experiment there are a number of questions to be answered, including, "Is there a texture difference among the four hot dog formulations/recipes?" and "Are the panelists consistent with one another?"

It is expected that both the recipe and the panelist may have an effect on the texture response (y). Furthermore, there is some feeling that the panelists may respond differently to the different recipes (i.e., there may be a recipe–panelist interaction). With this in mind, the mean response, η, could be postulated to relate to the recipes and panelists with the following form:

$$\eta_{ij} = \mu + \tau_j + \rho_i + \kappa_{ij}$$

where μ is the intercept, τ_j is the effect of the jth recipe on the mean response, ρ_i is the effect of the ith panelist on the mean response, and κ_{ij} is the interaction effect associated with the jth recipe and ith panelist. When the experimental error is considered, the response, y, for the kth replicate associated with the ith panelist $\times$ jth recipe combination may be represented by the model:

$$y_{ijk} = \eta_{ij} + \epsilon_{ijk}$$
$$= \mu + \tau_j + \rho_i + \kappa_{ij} + \epsilon_{ijk}. \qquad (16.35)$$

Based on the proposed model form, it is clear that the experimental design structure employed in the previous section will not work in this case. We now have two factors that need to be investigated: recipe and panelist. The recipe factor is the primary focus of the experiment, while the panelist factor is somewhat analogous to the route and day factors of the engine experiment in Section 16.3— it is a factor that varies and could obscure the effect of recipe differences that may exist. However, we certainly would like to know if a panelist effect exists, because it may make a difference in future panelist training efforts.

Based on this discussion, a total of nine hot dog samples were prepared for each recipe. The three panelists were brought into the testing room and seated side-by-side. Each panelist was asked to assess the texture of three hot dogs from each recipe; however, the overall order in which the 12 hot dogs (4 recipes $\times$ 3 trials/recipe) were tested was randomized. The ordering of the hot dogs was different from panelist to panelist, and the panelists did not know which hot dog recipe they were tasting on any trial. After assessing each hot dog, the panelists cleansed their palates with a small glass of apple juice. The responses from the experiment are provided in Table 16.7, with the run order shown in parentheses. Each panelist–recipe combination therefore constitutes an experimental subunit or block, the variability within which is unaffected by either panelist-to-panelist or recipe-to-recipe differences; i.e., the responses within each combination are replicates. Furthermore, in this design all the hot dog recipes are exposed to all the panelists, and vice versa.

TABLE 16.7	Hot Dog Surface Texture for 4 Recipes, 3 Panelists, and 3 Replications			

			Hot Dog Recipe		

Panelist	A	B	C	D	
1	7.6 (2) 6.5 (6) 7.2 (9)	11.4 (8) 7.6 (10) 9.5 (11)	6.3 (3) 7.9 (4) 6.8 (5)	2.7 (1) 3.1 (7) 1.7 (12)	$\bar{\bar{y}}_1 = 6.525$
2	7.2 (4) 13.6 (9) 10.7 (12)	12.9 (5) 12.4 (7) 10.7 (11)	11.1 (1) 6.9 (3) 9.0 (6)	3.3 (2) 1.9 (8) 2.3 (10)	$\bar{\bar{y}}_2 = 8.500$
3	7.0 (3) 10.2 (5) 8.3 (11)	10.2 (1) 9.0 (6) 8.7 (12)	6.8 (2) 9.2 (9) 11.0 (10)	3.7 (4) 2.4 (7) 3.2 (8)	$\bar{\bar{y}}_3 = 7.475$
	$\bar{\bar{y}}_A = 8.700$	$\bar{\bar{y}}_B = 10.267$	$\bar{\bar{y}}_C = 8.333$	$\bar{\bar{y}}_D = 2.700$	$\bar{\bar{\bar{y}}} = 7.500$

As is evident from Table 16.7, there are three trials for each combination of recipe and panelist—in other words, each combination has been repeated three times. The texture for the ith panelist, jth recipe, and kth replicate is y_{ijk}, where $i = 1, \ldots, n_p$, $j = 1, \ldots, n_r$, and $k = 1, \ldots, n_t$.

The grand average, $\bar{\bar{\bar{y}}}$, recipe averages, $\bar{\bar{y}}_j$, and panelist averages, $\bar{\bar{y}}_i$, shown in the table are calculated based on the following equations:

$$\bar{\bar{\bar{y}}} = \sum_{i=1}^{n_p}\sum_{j=1}^{n_r}\sum_{k=1}^{n_t} y_{ijk}/(n_t \cdot n_r \cdot n_p), \tag{16.36}$$

$$\bar{\bar{y}}_j = \sum_{i=1}^{n_p}\sum_{k=1}^{n_t} y_{ijk}/(n_t \cdot n_p), \tag{16.37}$$

and

$$\bar{\bar{y}}_i = \sum_{j=1}^{n_r}\sum_{k=1}^{n_t} y_{ijk}/(n_t \cdot n_r). \tag{16.38}$$

The averages for each cell, $\bar{y}_{ij}$, may be obtained with the following equation and are provided in Table 16.8.

$$\bar{y}_{ij} = \sum_{k=1}^{n_t} y_{ijk}/n_t. \tag{16.39}$$

Variance Decomposition

As suggested by the model form of Eq. (16.35), the variation within the data can be decomposed into a number of sources:

$$y_{ijk} = \bar{\bar{\bar{y}}} + (\bar{\bar{y}}_j - \bar{\bar{\bar{y}}}) + (\bar{\bar{y}}_i - \bar{\bar{\bar{y}}}) + (\bar{\bar{\bar{y}}} - \bar{\bar{y}}_i - \bar{\bar{y}}_j + \bar{y}_{ij}) + (y_{ijk} - \bar{y}_{ij}),$$

TABLE 16.8	Averages for Each Recipe and Panelist Combination			
	Hot Dog Recipe			
Panelist	A	B	C	D
1	7.1	9.5	7.0	2.5
2	10.5	12.0	9.0	2.5
3	8.5	9.3	9.0	3.1

where it may be noted that the left- and righthand sides of the equation are equal, i.e., the righthand side of the equation can be arithmetically reduced to y_{ijk}. It may also be observed that the righthand side of the equation includes a term associated with the grand average, the deviation of the treatment average from the grand average, the deviation of the block average from the grand average, an interaction term, and an error term. If we square both sides of the equation and sum over all replicates, treatments, and blocks, we obtain

$$\sum_{i=1}^{n_p}\sum_{j=1}^{n_r}\sum_{k=1}^{n_t} y_{ijk}^2 = N\bar{\bar{\bar{y}}}^2 + \sum_{i=1}^{n_p} n_t n_r (\bar{\bar{y}}_i - \bar{\bar{\bar{y}}})^2 + \sum_{j=1}^{n_r} n_t n_p (\bar{\bar{y}}_j - \bar{\bar{\bar{y}}})^2$$

$$+ \sum_{i=1}^{n_p}\sum_{j=1}^{n_r} n_t (\bar{\bar{\bar{y}}} - \bar{\bar{y}}_i - \bar{\bar{y}}_j + \bar{y}_{ij})^2 + \sum_{i=1}^{n_p}\sum_{j=1}^{n_r}\sum_{k=1}^{n_t} (y_{ijk} - \bar{y}_{ij})^2. \quad (16.40)$$

Equation (16.40) may be summarized as

$$SS_{total} = SS_{mean} + SS_{panelist} + SS_{recipe} + SS_{interaction} + SS_{error}. \quad (16.41)$$

These terms may be used to develop an ANOVA for this design. This ANOVA is given in Table 16.9, based on Eqs. (16.36)–(16.40).

For the hot dog example, the ANOVA table becomes:

Source of Variation	Sum of Squares	Degrees of Freedom	Mean Square	F_{calc}
Mean	2025.000	1	2025.0	816.53
Panelists	23.415	2	11.71	4.72
Recipes	295.460	3	98.49	39.71
Interaction	16.405	6	2.73	1.10
Pure Error	59.460	24	2.48	
Total	2419.740	36		

In interpreting the ANOVA table, we must compare the calculated F values to the appropriate critical F values from Table A.6 for a given α value ($\alpha = 0.01$). For the "Mean," $F_{calc} = 816.53$ may be compared to $F_{1,24,0.99} = 4.26$—since F_{calc} is

TABLE 16.9   ANOVA Table for Comparison of Multiple Treatments with Blocking

Source of Variation	Sum of Squares	Degrees of Freedom	Mean Square	F_{calc}
Mean	$N\bar{\bar{y}}^2$	1	$N\bar{\bar{y}}^2$	$\dfrac{\text{MS(Mean)}}{\text{MS(PE)}}$
Panelists (P)	$\sum_{i=1}^{n_p} n_t n_r (\bar{\bar{y}}_i - \bar{\bar{y}})^2$	$n_p - 1$	$\dfrac{\sum_{i=1}^{n_p} n_t n_r (\bar{\bar{y}}_i - \bar{\bar{y}})^2}{n_p - 1}$	$\dfrac{\text{MS(P)}}{\text{MS(PE)}}$
Recipes (R)	$\sum_{j=1}^{n_r} n_t n_p (\bar{\bar{y}}_j - \bar{\bar{y}})^2$	$n_r - 1$	$\dfrac{\sum_{j=1}^{n_r} n_t n_p (\bar{\bar{y}}_j - \bar{\bar{y}})^2}{n_r - 1}$	$\dfrac{\text{MS(R)}}{\text{MS(PE)}}$
Interaction (I)	$\sum_{i=1}^{n_p}\sum_{j=1}^{n_r} n_t (\bar{\bar{y}} - \bar{\bar{y}}_i - \bar{\bar{y}}_j + \bar{y}_{ij})^2$	$(n_p - 1)(n_r - 1)$	$\dfrac{\sum_{i=1}^{n_p}\sum_{j=1}^{n_r} n_t (\bar{\bar{y}} - \bar{\bar{y}}_i - \bar{\bar{y}}_j + \bar{y}_{ij})^2}{(n_p - 1)(n_r - 1)}$	$\dfrac{\text{MS(I)}}{\text{MS(PE)}}$
Pure Error (PE)	$\sum_{i=1}^{n_p}\sum_{j=1}^{n_r}\sum_{k=1}^{n_t} (y_{ijk} - \bar{y}_{ij})^2$	$n_p n_r (n_t - 1)$	$\dfrac{\sum_{i=1}^{n_p}\sum_{j=1}^{n_r}\sum_{k=1}^{n_t} (y_{ijk} - \bar{y}_{ij})^2}{n_p n_r (n_t - 1)}$	
Total	$\sum_{i=1}^{n_p}\sum_{j=1}^{n_r}\sum_{k=1}^{n_t} y_{ijk}^2$	$N = n_r n_p n_t$		

larger than this value, we conclude that the "Mean" source of variation is significant; we may reject the underlying null hypothesis that the mean is zero. For the "Panelists," $F_{calc} = 4.72$ may be compared to $F_{2,24,0.99} = 3.40$—since F_{calc} is larger than this value, we conclude that the "Panelist" source of variation is significant; we may reject the underlying null hypothesis that the panelist effect is zero. For the "Recipes," $F_{calc} = 39.71$ may be compared to $F_{3,24,0.99} = 3.01$—since F_{calc} is larger than this value, we conclude that the "Recipes" source of variation is significant; we may reject the underlying null hypothesis that the recipe effect is zero. For the "Interaction," $F_{calc} = 1.10$ may be compared to $F_{6,24,0.99} = 2.51$— since F_{calc} is smaller than this value, we have no evidence that the "Interaction" source of variation is significant, i.e., no evidence that the interaction effect is nonzero. One final point with respect to the ANOVA table is that the entry for the pure error mean square provides the best estimate of the error variance, σ_y^2.

Assessing Significant Effects Using the t-Distribution

As in the previous section, we may gain additional insight into what the ANOVA results are telling us by comparing our recipe and panelist averages to the appropriate t-distribution ($\nu = 24$) scaled by $(s_y/\sqrt{n})$. Since our best estimate of the error variance is MS(pure error) = 2.48, the value for s_y is 1.574. Figure 16.16

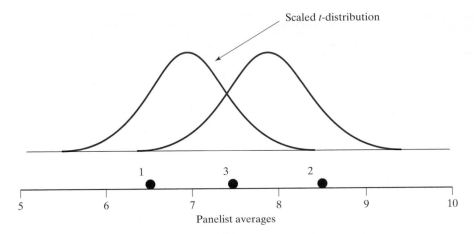

Figure 16.16   **Panelist averages plotted relative to scaled t-distribution with $\nu = 24$**

displays the panelist averages as well as the approximate reference distribution at different horizontal positions. The scaling factor in this case is $\left(1.574/\sqrt{12}\right)$, since 12 observations were used to calculate each panelist average. For the panelists, the figure appears to show that no single panelist is responsible for the ANOVA results—panelists 1 and 2 appear to be different from each other while panelist 3 cannot be distinguished from either 1 or 2. This finding may suggest that prior to future experiments, the panelists should receive additional training to improve their texture-tasting consistency with one another.

Figure 16.17 shows the recipe averages displayed against their approximate reference distribution—in this case, a t-distribution for $\nu = 24$ scaled by $\left(1.574/\sqrt{9}\right)$, since 9 trials were used to obtain each recipe average. The figure seems to indicate that B may be different from A and C, which are similar, and that D is substantially different from all the other recipes. Given that the texture

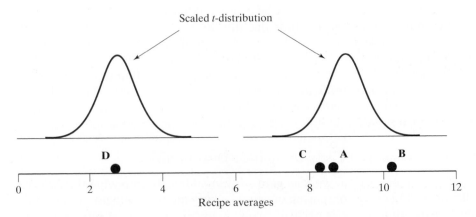

Figure 16.17   **Recipe averages plotted relative to scaled t-distribution with $\nu = 24$**

for recipe D is so low, it probably should not be considered further in assessing other measures of hot dog performance, unless some modification is made to improve its texture firmness.

Diagnostic Checking of the Model

At the outset of our investigation into the study of the hot dog texture problem, we proposed a model of the following form [shown previously as Eq. (16.35)]:

$$y_{ijk} = \eta_{ij} + \epsilon_{ijk}$$
$$= \mu + \tau_j + \rho_i + \kappa_{ij} + \epsilon_{ijk}.$$

The ANOVA has demonstrated that the κ_{ij} term (interaction between panel and recipe) is insignificant, so the model for the predicted response is

$$\hat{y}_{ij} = \hat{\mu} + \hat{\tau}_j + \hat{\rho}_i \qquad (16.42)$$

Based on the responses, the values for the estimates in Eq. (16.42) are $\hat{\mu} = 7.5$, $\hat{\tau}_j = [1.2, 2.767, 0.833, -4.8]$ for $j = 1, \dots, 4$ recipes, and $\hat{\rho}_i = [-0.975, 1.0, -0.025]$ for $i = 1, \dots, 3$ panelists. Equation (16.42) may now be used to make predictions for each panelist–recipe combination. These predictions are shown in Table 16.10.

We have now established a model for the predicted hot dog texture as a function of recipe and panelist. We must now investigate the model residuals using the guidelines established in the previous section to search for potential inadequacies. The model residuals can be calculated using

$$e_{ijk} = y_{ijk} - \hat{y}_{ij}, \qquad (16.43)$$

where e_{ijk} is the model residual for the jth recipe, ith panelist, and kth replicate. The model residuals are provided in Table 16.11. Several representative plots of the residuals are shown in Fig. 16.18–16.22.

Figure 16.18 displays the model residuals as a function of the predicted response; the residuals are centered about zero and appear to be randomly distributed. A normal probability plot of the residuals is shown in Fig. 16.19, and they appear to be well described by a straight line indicating that a single normal distribution can be used to describe them. A line describing the residuals

TABLE 16.10	Model Predictions for Each Recipe and Panelist Combination			
	Hot Dog Recipe			
Panelist	A	B	C	D
1	7.725	9.292	7.358	1.725
2	9.700	11.267	9.333	3.700
3	8.675	10.242	8.308	2.675

TABLE 16.11	Model Residuals for Hot Dog Texture Response			
	Hot Dog Recipe			
Panelist	A	B	C	D
	−0.125	2.108	−1.058	0.975
1	−1.225	−1.692	0.542	1.375
	−0.525	0.208	−0.558	−0.025
	−2.500	1.633	1.767	−0.400
2	3.900	1.133	−2.433	−1.800
	1.000	−0.567	−0.333	−1.400
	−1.675	−0.042	−1.508	1.025
3	1.525	−1.242	0.892	−0.275
	−0.375	−1.542	2.692	0.525

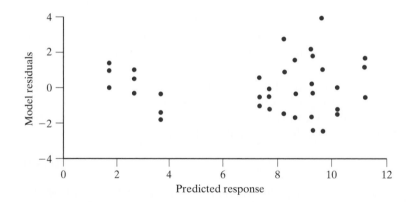

Figure 16.18   Model residuals vs. predicted response

Figure 16.19   Normal probability plot of the model residuals

passes near the point (0, 0.50)—again, this indicates that the residuals are centered about zero. Figures 16.20 and 16.21 illustrate the residuals vs. the run order and the recipe, respectively; no evidence of any trends with respect to either run order or the recipe are evident in the figures. Figure 16.22 is another plot that is often examined for diagnostic checking of the model. For situations where there is replication, i.e., multiple trials for each combination of variable levels, we may directly examine the replication error. For our experiment we have employed replication, three trials for each recipe–panelist combination; the difference between the trial responses (y_{ijk}) and the average of the replicates ($\bar{y}_{ij}$) may be calculated and studied as a function of the replicate average, $\bar{y}_{ij}$. Figure 16.22 shows no dependence of the replication error on the average. In considering all the residual plots, it appears that the model residuals behave randomly. There does not appear to be any indication from the residual plots that the assumptions associated with the residuals are unjustified—the diagnostic checking of the residuals has revealed no model inadequacies.

Figure 16.20   Model residuals vs. run order

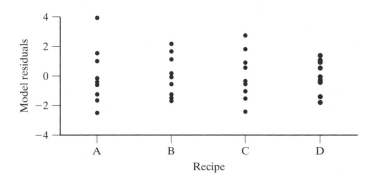

Figure 16.21   Model residuals vs. hot dog recipe

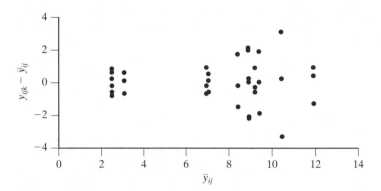

Figure 16.22   Difference between the response replicate and replicate average vs. the replicate average

■ CHAPTER SUMMARY

This chapter has covered a number of "simple comparative experiments" and in so doing has introduced some fundamental experiment design and analysis principles that will run through the remaining chapters in this book. Hence, it is important that these principles are thoroughly understood. Let us briefly review them now.

- Design experiments in such a way so as to make the desired comparisons among levels of a given factor **internal** to the experiment, so as to manage and control variation sources that could otherwise mask or obscure the comparisons.

- Use the **hypothesis testing formulation** initially introduced in Chapter 4 to guide the proper design of the experiments and the associated analysis of data that follows.

- Use the **design principle of randomization**
  - to distribute the forces of nuisance variation, usually coming from unidentifiable sources, across the design space so that they do not bias or otherwise influence in a prejudicial way the treatment effects being sought, thereby guaranteeing approximate validity of the experiment, and
  - so that the concept of the **randomization distribution** may be usefully employed in understanding the notion of separating real factor effects from the forces of experimental error/chance variation.

- Use the **design principle of blocking** to eliminate large and identifiable sources of nuisance variation, which cannot be controlled by fixing experimental conditions, from masking or obscuring the treatment effects that are being sought, thereby producing a more sensitive or discriminatory comparison of the treatments under study.

- Apply the **principles of analysis of variance (ANOVA)** as a powerful analysis method to decompose and assess the components of variation in the observed response of the trials of a designed experiment.

- Use the **mathematical model** as a way to guide the formulation of the experiment design and associated data analysis and as a useful tool for predictive purposes.

- Employ the analysis tool of **model checking via the analysis of the model residuals**, which is useful in revealing flaws in the experiment design and/or its conduct and in identifying model inadequacies that might be exploited in follow-on experimentation.

In the chapters that follow, these design and analysis principles will recur over and over again. In general, what we will see as we move to the next chapter is that we will begin to consider problems and associated design-of-experiment situations where the number of variables of interest will increase, in some cases considerably, while the number of levels (treatments) considered for each variable will decrease, in most cases being only two. As a result, the application of the above principles will become of increasing significance to avoid erroneous conclusions and associated inappropriate actions based on the results of experiments.

∎ EXERCISES

16.1 A manufacturer is considering the purchase of a new robotic welder to help fabricate steel enclosures for products. The size of the weld seam produced by the welder is of interest. Historically, the seam widths produced by traditional welding operations have had a mean of 0.15 inch and a standard deviation of 0.010 inch, and the widths are normally distributed. Five trials with the robotic welder produce an average seam width of 0.14 inch. Using a risk level of $\alpha = 0.05$, decide whether the seam width produced by the new welder differs from that obtained with the traditional method.

16.2 The air quality (for particulate matter less than 10 μm) within a manufacturing plant may be characterized with a daily average mass concentration level. The daily values for the last several years may be characterized with a normal distribution centered at a mean value of 4 mg/m^3 and a standard deviation of 1.0 mg/m^3. New dust collectors are installed throughout the plant and 10 days' worth of air quality measurements are collected, producing a sample mean of 3 mg/m^3. Using $\alpha = 0.01$, determine whether the new dust collector can be said to affect the particulate mass concentration level.

16.3 A biotech company has developed a new organic fertilizer that it believes is superior to an existing fertilizer. For many years, the existing fertilizer has been widely employed within one geographic region producing mean corn yields of 150 bushels/acre. The yields are known to be normally distributed with a standard deviation of 25 bushels/acre. Four field trials of the new fertilizer across the region produce a sample mean yield of 160 bushels/acre. Use $\alpha = 0.02$ to determine if the new fertilizer has a significant effect on the corn yield.

16.4 Evaluate the experimental approach followed in Exercise 16.3. Propose an improved experimental plan. Be sure to indicate why your new plan is superior to the original approach.

16.5 Two different alloys are formed through powder metallurgy methods. The only difference between the two powder mixes is that alloy A contains a slightly larger amount of manganese than alloy B. Metallurgists want to know if there is a significant difference in the hardness between the two alloys. Eight hardness-testing specimens were prepared from each powder mix, producing the hardness values shown in the accompanying table.

| Alloy A | 206 | 210 | 207 | 199 | 205 | 207 | 211 | 214 |
| Alloy B | 208 | 213 | 197 | 209 | 215 | 213 | 214 | 213 |

Assuming a risk level of $\alpha = 0.01$, perform a statistical test of hypothesis to determine whether the manganese content affects the hardness.

16.6 Using the data of Exercise 16.5, establish a 99% confidence interval for the true mean difference between the hardness of the two alloys. Based on this confidence interval, draw a conclusion about whether the manganese content influences the hardness.

16.7 The biotech firm of Exercise 16.3 conducts another experiment to examine potential differences between a new organic fertilizer and their traditional fertilizer product. They divide an 80-acre parcel of land into eight 10-acre fields. Four of the fields receive the new fertilizer and the other four the traditional fertilizer. The corn yield for each of the 10-acre regions is shown in the accompanying table. Perform a statistical test of hypothesis using a risk level of $\alpha = 0.05$ to determine whether the new fertilizer has a significant effect on the corn yield.

Old Fertilizer	New Fertilizer
157	158
139	169
164	146
142	181

16.8 Using the data of Exercise 16.7, establish a 95% confidence interval for the true mean difference between the two fertilizers. Based on this confidence interval, draw a conclusion about whether the new fertilizer is different than the old one.

16.9 Nork Industrial Supply has developed a new coating for their carbide milling cutters. They want to know if the new coating (coating B) will outperform the old coating (A) in terms of tool life. Twelve milling cutters are selected (seven with coating A and five with coating B) at random, and tool life tests are performed with each tool. The results in the table below (tool life in minutes) are obtained.

Coating A	55.4	60.4	38.2	61.1	45.0	45.9	56.3
Coating B	59.1	69.5	56.5	71.9	70.2		

Assuming $\alpha = 0.05$, perform a statistical test of hypothesis to determine whether the coating type has an effect on the tool life.

16.10 Using the data of Exercise 16.9, establish a 95% confidence interval for the true mean difference between the tool lives of the two coating types. Based on this confidence interval, draw a conclusion about whether the tool lives obtained with new coating are significantly different than those obtained with the old coating.

16.11 The vibration of two washing machine models (Superior and Ultimo) is under investigation. Several machines of each model type are selected at random and tested under various loading conditions. An accelerometer is used to measure the maximum level of lateral vibration (mm/sec^2) during each trial, and the results of the trials are shown in the table below.

Superior	386	426	381	363	422	475			
Ultimo	272	350	357	373	421	356	329	317	389

Perform a statistical test of hypothesis using a risk level of $\alpha = 0.02$ to determine whether the vibration level differs significantly between the two models.

16.12 Using the data of Exercise 16.11, establish a 98% confidence interval for the true mean difference between the vibration levels of the two washing machine models. Based on this confidence interval, draw a conclusion about whether a significant difference exists between the vibration levels of the two models.

16.13 A new method has been proposed for changing the fixturing on 11 similar machines in a production facility. A trial using each method was performed on each of the machines; the two trials for each machine were performed consecutively and in a random order. The fixture change time (minutes) was recorded for each trial. The results are shown in the table below.

						Machine					
	1	2	3	4	5	6	7	8	9	10	11
Old Method	11.4	16.0	16.5	8.3	11.5	17.9	7.2	14.0	10.4	15.7	17.1
New Method	11.2	8.7	11.5	5.9	11.1	15.0	4.6	12.1	7.7	16.3	13.8

(a) Evaluate the results of the experiment, using the fact that blocking has been employed. You may assume an α value of 0.05. Draw appropriate conclusions.

(b) Evaluate the results of the experiment using an "independent t-test analysis." In other words, do not use the fact that blocking has been employed. Assume an α value of 0.05. Draw appropriate conclusions.

(c) Comment on the impact that blocking has on the experimental conclusions.

16.14 Using the blocked data of Exercise 16.13, establish a 95% confidence interval for the true mean difference between the fixturing times associated with the two methods. Based on this confidence interval, draw a conclusion about whether a significant difference exists between the two methods.

16.15 To determine the effectiveness of a series of industrial safety workshops, a year's worth of data before and after the workshops are considered. The data are available for twelve plants. The response selected for study is the average weekly loss of person-hours due to accidents, and these responses are shown in the accompanying table for all twelve plants.

						Plant						
	1	2	3	4	5	6	7	8	9	10	11	12
Before workshops	37.1	72.5	26.6	125	45.8	54.3	13.2	79.5	12.6	34.9	39.3	26.6
After workshops	28.0	59.3	24.7	120	46.2	43.6	15.4	75.1	18.3	29.7	35.0	24.8

Analyze the data, making whatever assumptions you deem appropriate, and draw conclusions. Comment on the experiment that was performed. What further experiments and data analysis would you recommend?

16.16 Using the blocked data of Exercise 16.15, establish a confidence interval for the true mean difference between the accident rate before and after the safety workshops. Based on this confidence interval, draw a conclusion about the effectiveness of the safety workshops.

16.17 As part of an effort to understand the effect, if any, of the type of measurement device on part measurements, a study was conducted. An inspector was asked to measure the diameter (in millimeters) of ten machined shafts with each device. The results are shown in the table below.

Shaft	Device 1	Device 2	Shaft	Device 1	Device 2
1	30.004	29.997	6	34.993	34.995
2	45.003	44.994	7	25.004	24.998
3	50.009	49.995	8	45.000	45.004
4	30.011	30.003	9	50.006	50.003
5	40.007	40.002	10	25.008	25.002

Evaluate the results of the experiment assuming an α value of 0.02. Draw appropriate conclusions. Comment on the experiment that was performed. What further experiments and data analysis would you recommend?

16.18 Using the data of Exercise 16.17, establish a 98% confidence interval for the true mean difference between the diameters of the two measurement devices. Based on this confidence interval, draw a conclusion about the difference between the two devices.

16.19 A company manufactures bottle rockets. The company has developed a new chemical formulation for the rockets. Develop and describe an experimental plan to investigate the difference, if any, between the old and the new chemical formulations for the rockets. Be sure to explain carefully how the experiment will be performed.

16.20 A golf ball manufacturer would like to understand whether its ball performs better than the ball produced by its chief competitor. The principal response of interest is the distance the golf ball travels when hit with a driver. Develop and describe an experimental plan to investigate the difference, if any, between the golf balls of the manufacturer and its competitor. Be sure to explain in detail how the experiment will be performed.

16.21 A company has established a new, less expensive, formulation for one of its popular beverage products. Develop and describe an experimental plan to evaluate the new beverage formulation relative to the existing formulation. Be sure to explain carefully how the experiment will be performed.

16.22 A manufacturer believes that the variability of one of its processes is $\sigma_y^2 = 0.0005$ millimeter^2. A sample is taken from the process (sample size $n = 6$) and the sample variance is calculated to be $s_y^2 = 0.0020$. Assuming that the measurements follow a normal distribution, evaluate the manufacturer's claim assuming an α value of 0.02.

16.23 The part weight is the quality characteristic of interest for a die-casting process, and the process weights are known to be normally distributed. The process variability is believed to be $\sigma_y^2 = 25$ grams^2. A sample of eight parts is collected from the process, and the sample variance is calculated to be 42.86 grams^2. What is the probability of obtaining a sample variance this large or larger?

16.24 One of the key measures of performance of a golf ball is its durability. A testing laboratory compares the durability of four different brands of golf balls (A, B, C, and D). A number of balls of each type are tested. A durability trial involves placing a ball into the testing machine, which applies a number of hits to the ball. The number of hits taken until the outer covering of the ball cracks/splits is recorded for each ball. The balls are tested in random order; the data from the experiment are shown in the table below.

A	B	C	D
12	93	27	76
38	70	60	99
51	76	18	65
56	53	50	90
43	76	46	76
33	58	57	
	70	63	
	51		
	58		

(a) Perform an appropriate analysis of variance (ANOVA) for the data shown in the table. Construct an ANOVA table and determine whether there is a difference in the durability among the four brands of balls (assume a significance level of $\alpha = 0.05$).

(b) Using the method described in the chapter, assess the treatment means using the t-distribution.

(c) Find the model for the response, and then calculate and interpret the model residuals.

(d) Summarize what has been learned from the experiment.

16.25 The performance of three different headphone models is under study. The response is the amount of noise reduction (in dB). The test engineer randomly selects several headphones of each type, and in a random order tests each headphone. The results of the tests are shown in the table below.

Model A	Model B	Model C
43	32	32
32	27	46
35	35	40
38	40	47
	35	45
	37	

(a) Perform an appropriate analysis of variance (ANOVA) for the data shown in the table. Construct an ANOVA table and determine whether there is a difference in the noise reduction provided by the three headphone models (assume a significance level of $\alpha = 0.05$).

(b) Using the method described in the chapter, assess the treatment means using the t-distribution.

(c) Find the model for the response, and then calculate and interpret the model residuals.

(d) Summarize what has been learned from the experiment. What further experiments and data analysis would you recommend?

16.26 A research facility is evaluating five different processes of extracting starch from corn flour. The percent yield of starch is the response of interest. Several trials were performed with each process (the trials were performed in a random order) producing the results shown in the table below.

Process 1	Process 2	Process 3	Process 4	Process 5
75.9	77.7	63.5	61.0	76.3
72.3	74.9	60.4	64.0	75.3
64.7	78.3	65.3	55.9	71.5
68.7	77.3	62.6		81.0
73.4		59.3		
77.7				

(a) Perform an appropriate analysis of variance (ANOVA) for the data shown in the table. Construct an ANOVA table and determine whether there is a difference in the starch yield achievable with the processes (assume a significance level of $\alpha = 0.01$).

(b) Using the method described in the chapter, assess the treatment means using the t-distribution.

(c) Find the model for the response, and then calculate and interpret the model residuals.

(d) Summarize what has been learned from the experiment.

16.27 The purchasing department for a large company wants to understand whether the time required to perform a given task differs among the three methods used by employees to complete the task. An employee was selected at random and asked to perform the task multiple times with each method. The time to complete the task (seconds) was recorded, and the results of the trials are shown in the table below.

Method A	Method B	Method C
533	410	549
427	439	601
338	411	585
484	377	509
437	332	544
471		577
		641

(a) Perform an appropriate analysis of variance (ANOVA) for the data shown in the table. Construct an ANOVA table and determine whether there is a difference in the time associated with the three methods (assume a significance level of $\alpha = 0.05$).

(b) Using the technique described in the chapter, assess the treatment means using the t-distribution.

(c) Find the model for the response, and then calculate and interpret the model residuals.

(d) Summarize what has been learned from the experiment.

16.28 A textile manufacturer wishes to understand the effect, if any, of four different dyes on the strength of a particular type of cloth. An experiment was planned to investigate the effect. Because there may be differences among the strengths of cloth manufactured at different points in time, it was decided to use a blocked design for the experiment. Five large sections of cloth were selected for the experiment, with each section of cloth considered as a block. In preparing for the experiment, eight small pieces were taken from random locations on each section of cloth, and each dye type was applied to two of the small pieces selected at random. Tensile tests were then performed on each small piece of cloth, producing the tensile strengths (MPa) shown in the accompanying table.

	Dye A	Dye B	Dye C	Dye D
Section 1	163	143	120	170
	123	164	155	138
Section 2	161	157	100	151
	127	145	124	143
Section 3	81	130	126	156
	129	109	116	143
Section 4	145	163	124	177
	97	166	121	173
Section 5	147	177	141	162
	149	137	146	159

(a) Perform an appropriate analysis of variance (ANOVA) for the data shown in the table. Construct an ANOVA table and determine whether there is a significant difference between the dye types. Is there a significant difference between sections? (Assume a significance level of $\alpha = 0.05$.) Is the interaction significant?

(b) Using the method described in the chapter, assess the treatment and block means using the t-distribution.

(c) Calculate and interpret the model residuals.

(d) Summarize what has been learned from the experiment.

16.29 An engineer suspects that the surface finish of a machined part is influenced by the feed and the depth of cut used to produce the part. She decides to perform an experiment, for which she selects three feeds and four depths of cut. Three trials are performed for each feed–depth-of-cut combination. The data in the accompanying table are the surface finish values (microinches) obtained from the experiment.

	Depth of Cut (inch)			
Feed (in/rev)	0.02	0.04	0.06	0.08
0.005	24	21	21	26
	28	22	31	25
	31	18	22	25
0.01	118	89	114	118
	91	110	79	94
	91	113	83	111
0.015	245	201	239	227
	259	249	250	239
	247	243	227	218

(a) Perform an appropriate analysis of variance (ANOVA) for the data shown in the table. Construct an ANOVA table and determine whether the feed and depth of cut influence the surface finish.

(Assume a significance level of $\alpha = 0.05$.) Is the interaction effect significant?

(b) Using the method described in the chapter, assess the feed and depth of cut means using the t-distribution.

(c) Calculate and interpret the model residuals.

(d) If the magnitude of the residuals appears to be proportional to the response, apply a log transformation to the data in the table and repeat parts (a) and (c).

(e) Summarize what has been learned from the experiment.

16.30 Consider the results of the replicated blocked design shown in the table below. The design considers five treatments (A–E) and three blocks (I, II, and III).

| | Treatment | | | | |
Block	A	B	C	D	E
I	43.9, 33.4, 37.0	19.5, 22.3, 22.0	28.0, 31.9, 33.4	23.6, 20.6, 25.2	25.0, 24.4, 22.4
II	36.8, 35.1, 41.6	24.8, 19.1, 25.0	30.4, 37.5, 35.5	20.5, 21.5, 26.6	22.8, 22.5, 26.6
III	31.8, 45.2, 38.0	18.2, 24.9, 18.9	39.0, 27.2, 35.2	20.8, 22.9, 30.7	25.0, 18.3, 29.2

(a) Perform an appropriate analysis of variance (ANOVA) for the data shown in the table. Construct an ANOVA table and determine whether there is a significant difference among treatments. Is there a significant difference among blocks? (Assume a significance level of $\alpha = 0.01$.) Is the interaction effect significant?

(b) Using the method described in the chapter, assess the treatment and block means using the t-distribution.

(c) Calculate and interpret the model residuals.

(d) Summarize what has been learned from the experiment.

16.31 You have been hired as a consultant by a local restaurant to help them develop the best omelet in town. It is decided to adopt an experimental approach with respect to determining the best omelet recipe and cooking procedure. Develop an experimental plan for the proposed investigation. Which factors will be fixed? Identify the variables that you will study and the levels that will be investigated. What responses are of interest and how will you measure them? Describe the planned experiment (the complete conditions for all experimental trials) in detail. How does the plan reflect blocking and randomization considerations? How do you plan to analyze the data obtained from the experiment?

16.32 The biotech firm of Exercise 16.7 conducts yet another experiment to examine potential differences between a new organic fertilizer and a traditional fertilizer product. A large parcel of land is broken down into regions, and two trials with each fertilizer are used on plots of land within each region. The corn yield obtained for each of the plots is shown in the accompanying table.

Region	Old Fertilizer	New Fertilizer
1	147	160
	151	162
2	156	161
	151	151
3	165	159
	166	138
4	158	132
	149	159
5	139	164
	131	164
6	146	168
	118	169
7	161	158
	164	147
8	143	174
	147	157

(a) Perform an appropriate analysis of variance (ANOVA) for the data shown in the table. Construct an ANOVA table and determine whether there is a significant difference among the fertilizers. Is there a significant difference among regions? (Assume a significance level of $\alpha = 0.05$.) Is the interaction effect significant?

(b) Using the method described in the chapter, assess the treatment and block means using the t-distribution.

(c) Calculate and interpret the model residuals.

(d) Summarize what has been learned from the experiment.

17

Design and Interpretation of 2^k Factorial Experiments

17.1 Introduction

In Chapters 15 and 16 we discussed a number of considerations that collectively constitute good practice in the design of experiments. These include:

1. The ability to deal simultaneously with both categorical (fixed level) and quantitative (random level) factors/variables.

2. The ability of an experimental design to observe the effects of changing factors individually (all other factors held constant) and also at different combinations of the other factors so that factor interdependencies can be properly revealed.

3. The ability of the experimental design to deal with the forces of nuisance variation, explicitly identifiable or not, through the techniques of blocking and randomization.

4. The ability of the experimental design to observe the forces of experimental error/system variability directly so that comparisons of performance can be made in light of the magnitude of system noise. Replication of tests provides us with the ability to do this.

5. Explicit recognition of the relationship between the proposed mathematical model and the experimental design that responds to that model.

6. Awareness of the notion of model building and the sequential and iterative nature of experimentation. In this regard, the ability to employ experiments that serve as building blocks, one upon the other, is important.

7. Recognition of the concepts of Taguchi, including parameter/robust design in the use of designed experiments for product/process design and improvement.

In addition, it would be useful if the experimental design were

- Easy to set up and carry out.
- Simple to analyze and interpret.
- Simple to communicate or explain to others.

We also suggested in Chapter 15 that a potentially useful class of experimental designs that exhibits many of these desirable properties is the class of two-level factorial designs. Over the next five chapters we study these designs and their sister designs, two-level fractional factorials, in considerable detail. We introduce the two-level factorial design through the use of a case study.

17.2   Process Study Using a Two-Level Factorial Design: Chemical Calibration for the Foam Molding Process[1]

In the manufacture of a new design of an automobile instrument panel that would be "soft to the touch," it was necessary to make some changes in certain raw materials. One such change was to reformulate the substrate foam padding material. This reformulation caused some processing difficulties, mainly due to excessive foam voids. Foam voids are air pockets under the vinyl skin that are not completely filled with foam. With the help of c charts monitoring the frequency and location of foam voids, the study team made a number of improvements that reduced the average defect rate from 8.6 voids per sample of five parts to 0.46 void per sample of five parts over a period of several months. A summary of this study is shown in Fig. 17.1.

As can be seen from the foam molding study history of Fig. 17.1, the major improvement actions involved the maintenance of material-flow systems. These were primarily local faults that were detected and eliminated through application of the principles and methods of SPC. Although a significant reduction in the average rate of foam voids was achieved, the team of engineers wanted to learn more about the fundamental relationship between some of the variables of the foam molding process and the occurrence of foam voids. This called for experimental investigation of the entire processing system. Since the system was now running "in control" with relatively low common-cause variability, it was considered an appropriate time to begin an experimental program.

In developing a strategy to experiment with the molding process, the team of engineers quickly realized that the experiments must be carried out in a series of at least three stages. They found out that two of the potential control variables of the foam molding process were themselves output responses from the foam calibration stand. Furthermore, four input variables of the foam calibration stand were responses associated with two chemicals at the chemical calibration stand upstream. A three-stage experimentation program was finally developed to study the entire foam molding system, with each stage building

---

[1]R. J. Deacon and P. M. Belaire, "Quality Improvement: Design of Experiments Methodology," *IMPRO 88 Conference Proceedings*, Juran Institute, Chicago, 1988.

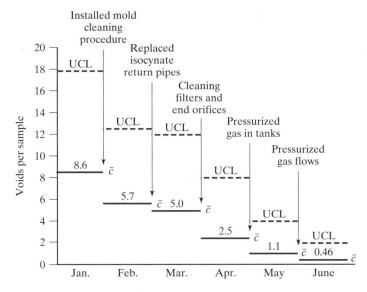

Figure 17.1   Foam molding improvement summary

information for the next stage. In the following sections we present an experimental study of the calibration of isocynate—in particular, the impingement pressure of the isocynate, one of the inputs to the foam calibration process. Figure 17.2 shows the three-stage process under investigation.

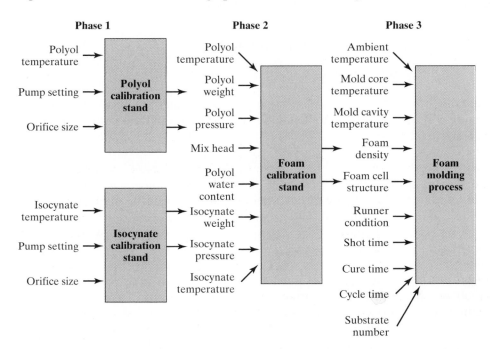

Figure 17.2   Foam molding process experimental design flow diagram

TABLE 17.1	Variable Levels for the Isocynate Calibration Experiment		
Variable	*Unit*	*Low Level*	*High Level*
Orifice size, O	mm	1.30	1.50
Pump setting, P		4.00	4.50
Isocynate temperature, T	°C	22	30

The experimental design formulated by the team for this first stage was a two-level, three-variable factorial design, simply designated as a 2^3 factorial design. The objective of this experiment was to determine the possible effects of three control variables on the isocynate weight and impingement pressure. The three variables were orifice size (diameter) of the mix head, pump setting for the isocynate tanks, and isocynate temperature. Two levels were chosen for each variable. The high and low levels of the three variables are listed in Table 17.1. These particular levels were centered around the current operating conditions and represented a range that consumed what was believed to be about 30% of the feasible operating range.

To adopt a notation that will be the same for all two-level factorial designs, the variables are coded such that the high level will be denoted by +1 and the low level by −1. By so doing, regardless of the physical conditions represented by the two levels, the basic design of any two-level factorial becomes a simple tabulation of a systematic arrangement of plus and minus "ones."

17.3   Construction of the 2^3 Factorial Design

Standard Order for the Test Conditions

A complete 2^3 factorial design contains $2^3 = 8$ unique test conditions. One systematic way of writing down the eight test conditions in their coded forms is to proceed as follows:

1. For variable 1 (x_1), write down a column of the values −1, +1, −1, +1, −1, +1, −1, +1. The signs alternate each time.
2. For variable 2 (x_2), write down a column of the values −1, −1, +1, +1, −1, −1, +1, +1. The signs alternate in pairs.
3. For variable 3 (x_3), write down a column of the values −1, −1, −1, −1, +1, +1, +1, +1. The signs alternate in groups of four.

By writing down these three columns next to one another as shown in Table 17.2, we obtain the desired 2^3 factorial design matrix, which consists of the eight distinct sets of test conditions, in both coded and uncoded forms. The eight sets of test conditions are given by the eight rows, corresponding to test numbers 1 to 8. These constitute the recipes for running the tests. For instance, the actual calibration of the isocynate for test number 1 (denoted by −1, −1, −1)

TABLE 17.2	Coded and Uncoded Test Conditions in Standard Order					
	Coded Test Conditions			*Actual Test Conditions*		
Test	x_1	x_2	x_3	*O* (mm)	*P*	*T* (°C)
1	−1	−1	−1	1.30	4.0	22
2	+1	−1	−1	1.50	4.0	22
3	−1	+1	−1	1.30	4.5	22
4	+1	+1	−1	1.50	4.5	22
5	−1	−1	+1	1.30	4.0	30
6	+1	−1	+1	1.50	4.0	30
7	−1	+1	+1	1.30	4.5	30
8	+1	+1	+1	1.50	4.5	30

was to use an orifice of 1.30 millimeters in diameter with a pump set at 4.0 and temperature at 22°C. Table 17.2 is often referred to as the *design matrix*; it defines the tests associated with the experimental design.

In general, there are 2^k unique sets of test conditions for a 2^k factorial design (i.e., all possible combinations of k factors at two levels each). We can obtain the 2^k sets of coded test conditions by writing down the columns as follows:

1. For x_1, write down a column of −1, +1, −1, +1, −1, +1, −1, +1,.... The signs alternate each time (i.e., $2^0 = 1$, one alternation every time). The column length is 2^k.

2. For x_2, write down a column of −1, −1, +1, +1, −1, −1, +1, +1,.... The signs alternate in pairs (i.e., $2^1 = 2$, alternate in pairs).

3. For x_3, write down a column of −1, −1, −1, −1, +1, +1, +1, +1,.... The signs alternate in groups of four (i.e., $2^2 = 4$, alternate in groups of four).

4. For x_4, the signs alternate in groups of eight ($2^3 = 8$).

5. Proceed in a similar way for $x_5, x_6, \ldots, x_k$. For x_k, write down a 2^{k-1} number of −1's, followed by a 2^{k-1} number of +1's.

This method will yield all of the 2^k distinct sets of coded test conditions. We refer to this particular ordering of the tests as *standard order*. Such an arrangement of test conditions guarantees that all columns in the matrix are orthogonal to each other so that independent estimates of the factor effects may be obtained.

Geometric Representation of the 2^3 Factorial Design

If we consider the three variables as three mutually perpendicular coordinate axes x_1, x_2, and x_3, the 2^3 factorial design can be represented geometrically as a cube as shown in Fig. 17.3. The numbers encircled at the eight corner points of the cube represent the corresponding test numbers in standard order. The eight actual test conditions are given in brackets. This geometric configuration will be

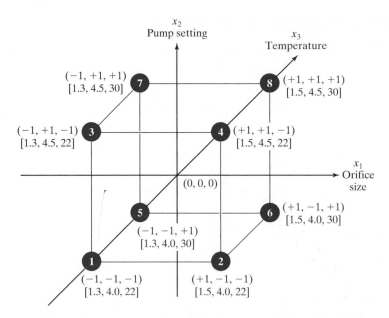

Figure 17.3   **Geometric representation of the 2^3 factorial design**

useful later when we calculate the main and interaction effects of the variables, and when we interpret the meaning of main effects and two-factor and three-factor interaction effects. The geometrical representation is also useful from the standpoint of communicating the design intent to others who are less informed.

Test Conduct: Random Order

In performing experiments, randomization of the test order should be exercised whenever possible. This is important, since the standard order discussed previously is a very systematic ordering of the tests in terms of the pattern of the levels of the variables from test to test. Imagine that the 2^3 experiment was to be performed in 2 days and it was determined that two trials could be run in the morning and two in the afternoon. Suppose that due to changing ambient conditions from the morning to the afternoon, afternoon tests tend to produce uniformly higher isocynate impingement pressures. Since variable 2 (in standard order) has all (-1) level tests in the morning and all $(+1)$ level tests in the afternoon, as shown in Table 17.3, the effects of ambient condition and variable 2 would be hopelessly confused or confounded. To avoid or minimize such confounding problems, the test order should be randomized.

17.4   Calculation and Interpretation of Main Effects

After the tests in the design matrix were performed in a random order, the eight responses of impingement pressure in pounds per square inch (psi) were

TABLE 17.3	Correlation of Standard Test Ordering with Ambient Conditions			
Test	x_1	x_2	x_3	Ambient Condition
1	−1	−1	−1	Morning
2	+1	−1	−1	Morning
3	−1	+1	−1	Afternoon
4	+1	+1	−1	Afternoon
5	−1	−1	+1	Morning
6	+1	−1	+1	Morning
7	−1	+1	+1	Afternoon
8	+1	+1	+1	Afternoon

obtained. These responses will be used to calculate the main and interaction effects on impingement pressure for the three variables under study. These responses are shown in the righthand column of Table 17.4. Note that the column test order in the table shows the order in which the tests were actually run.

Given the results in Table 17.4, how could we evaluate the individual and joint influences of orifice size, pump setting, and isocynate temperature on impingement pressure of the isocynate? To help us see what these variable effects are—what they mean and how they may be calculated—we employ the geometrical representation of the experimental design. The cube in Fig. 17.4 depicts the 2^3 design geometrically. The response (impingement pressure in psi) for each test is given at the appropriate corner of the cube.

Calculation of the Main Effect of Orifice Size

We might first ask: How is the isocynate impingement pressure influenced by a change in any one of the three variables while the other two are being held constant? Suppose that we consider this question for orifice size. A glance at the cube in Fig. 17.4 indicates that there are four comparisons or contrasts of test results that indicate how impingement pressure changes when orifice size is changed from low (small) to high (large), with pump setting and isocynate temperature each being held at a constant level.

TABLE 17.4	Test Results for Isocynate Calibration Experiment				
Test	x_1	x_2	x_3	Test Order	Response, y (psi)
1	−1	−1	−1	6	1550
2	+1	−1	−1	8	1925
3	−1	+1	−1	1	2150
4	+1	+1	−1	2	2350
5	−1	−1	+1	5	1525
6	+1	−1	+1	3	1800
7	−1	+1	+1	4	2175
8	+1	+1	+1	7	2200

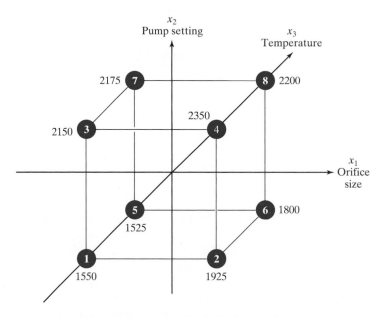

Figure 17.4   Geometrical representation of the test results

The following pairs of tests may be compared to see the effect of orifice size:

- y_2 and y_1.
- y_4 and y_3.
- y_6 and y_5.
- y_8 and y_7.

The differences in the results within each of the four pairs of tests just defined reflect the effect of orifice size alone on impingement pressure. The differences in units of psi are

Individual Contrasts

Tests 1 and 2:	$y_2 - y_1 = 1925 - 1550 = 375$ psi.
Tests 3 and 4:	$y_4 - y_3 = 2350 - 2150 = 200$ psi.
Tests 5 and 6:	$y_6 - y_5 = 1800 - 1525 = 275$ psi.
Tests 7 and 8:	$y_8 - y_7 = 2200 - 2175 = 25$ psi.

The average effect of orifice size, designated by E_1, is defined to be the average of the four differences or contrasts above. Note that this average effect is also commonly referred to as a *main effect* or *location effect*. That is,

$$E_1 = \left(\tfrac{1}{4}\right)[(y_2 - y_1) + (y_4 - y_3) + (y_6 - y_5) + (y_8 - y_7)]$$
$$= \left(\tfrac{1}{4}\right)[(375) + (200) + (275) + (25)]$$
$$= 218.75 \text{ psi.}$$

Geometrical Representation of the Main Effect of Orifice Size

Geometrically, the average (main) effect of orifice size, E_1, is the difference between the average test result on plane II (high level of orifice size) and the average test result on plane I (low level of orifice size), as shown in Fig. 17.5.

This can be seen by rearranging the previously defined average effect equation for E_1 as

$$E_1 = \frac{y_2 + y_4 + y_6 + y_8}{4} - \frac{y_1 + y_3 + y_5 + y_7}{4}.$$

It is noted that:

- All tests on plane II have one thing in common—a high level of orifice size.
- All tests on plane I have one thing in common—a low level of orifice size.

The main effect of orifice size tells us that on the average, over the ranges of the variables studied in this investigation, the effect of changing the orifice size from its low level to its high level is to increase the impingement pressure by 218.75 psi. Notice, however, that the individual differences (375, 200, 275, and 25 psi) are actually quite different. The average effect, therefore, must be interpreted with considerable caution, since this effect is not particularly consistent over the four unique combinations of pump setting and isocynate temperature (x_2 and x_3), as shown in Fig. 17.6.

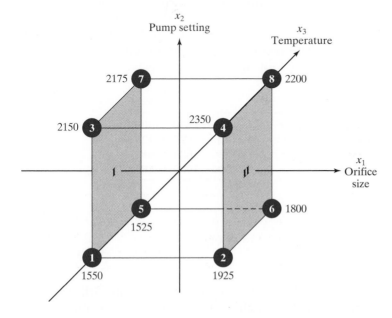

Figure 17.5   Geometric representation of the main effect of orifice size

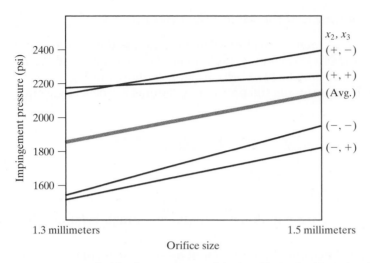

Figure 17.6   Individual contrasts and main effect of orifice size (heavy line)

Main Effect of Pump Setting, E_2

Returning to Table 17.4, consider tests 1 and 3. For these two tests, the orifice size (x_1) and isocynate pressure (x_3) are the same, but for test 1 the pump setting (x_2) is at the low level, and for test 3, the pump setting is at the high level. The difference in the two test results can be attributed to the effect of pump setting alone. Similarly, for tests 2 and 4, 5 and 7, and 6 and 8, the orifice size and isocynate temperature are the same within each pair, but the pump setting is at different levels.

Again, referring to Fig. 17.4, compare or contrast the four pairs of tests to see the effect of pump setting. The average effect of pump setting, E_2, can be obtained by taking the average of the four individual contrasts:

Individual Contrasts

Tests 1 and 3:	$y_3 - y_1 = 2150 - 1550 = 600$ psi.
Tests 2 and 4:	$y_4 - y_2 = 2350 - 1925 = 425$ psi.
Tests 5 and 7:	$y_7 - y_5 = 2175 - 1525 = 650$ psi.
Tests 6 and 8:	$y_8 - y_6 = 2200 - 1800 = 400$ psi.

The average (main) effect of x_2 is, therefore,

$$E_2 = \left(\tfrac{1}{4}\right)(600 + 425 + 650 + 400)$$
$$= 518.75 \text{ psi.}$$

Geometrically, the average effect of a pump setting is the difference between the average result on plane IV (high level of pump setting) and the average result on plane III (low level of pump setting) as shown in Fig. 17.7. The

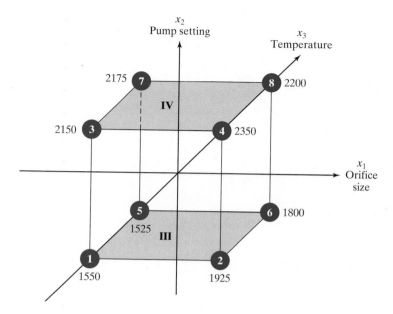

Figure 17.7   Main effect of pump setting

average effect of pump setting is 518.75 psi, which tells us that, on the average, an increase in pump setting from its low level to high level gives rise to an increase in impingement pressure of 518.75 psi.

Main Effect of Isocynate Temperature, E_3

Similarly, the average effect of isocynate temperature, E_3, is the corresponding comparison between plane VI (high level of isocynate temperature) and plane V (low level of isocynate temperature), as indicated in Fig. 17.8. This average or main effect is calculated as follows:

$$E_3 = \left(\tfrac{1}{4}\right)[(y_5 - y_1) + (y_6 - y_2) + (y_7 - y_3) + (y_8 - y_4)]$$
$$= \left(\tfrac{1}{4}\right)[(-25) + (-125) + (25) + (-150)]$$
$$= -68.75 \text{ psi.}$$

Interpretation of the Main Effects

We have defined the average or main effect of a variable as the amount of change we witness in the response, on the average, when only that variable changes from its low to high level. It is an average effect because within a 2^k experimental design there are 2^{k-1} individual comparisons or contrasts that measure how much the response changes when any one variable changes. The average of all such contrasts across the entire experimental design is called the

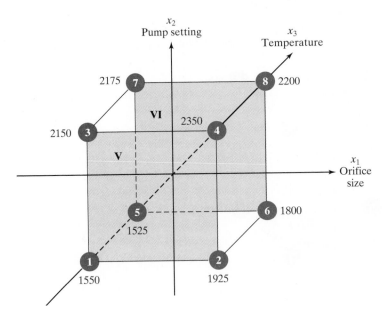

Figure 17.8   Main effect of isocynate temperature

average effect, main effect, or *location effect* of that factor. Both the sign and magnitude of an average (main) effect mean something:

- The sign tells us the direction of the effect—that is, whether the response increases or decreases.
- The magnitude tells us the strength of the effect.

Summary: Main Effects of the Isocynate Calibration Experiment

The relative importance of the three main effects of the control variables on impingement pressure is shown graphically in Fig. 17.9. The figure seems to show that pump setting has the largest average effect among the three factors. But an average effect is just that—an average of individual contrasts taken over all possible combinations of levels of the other factors. If the individual contrasts comprising this average are quite similar in magnitude and in sign, the effect of that particular variable tends to be independent of the levels of other variables. On the other hand, if the individual contrasts comprising this average are quite different in magnitude and/or sign, the effect of that particular variable tends to be dependent on the levels of one or more of the other variables. In this case, the average (main) effects may not mean much, and we must look at what happens to the response of interest when two or more variables change simultaneously.

17.5   Calculation and Interpretation of Interaction Effects

In calculating the main effect of pump setting, we saw that the amount of change in impingement pressure with change in pump setting seemed to depend on the

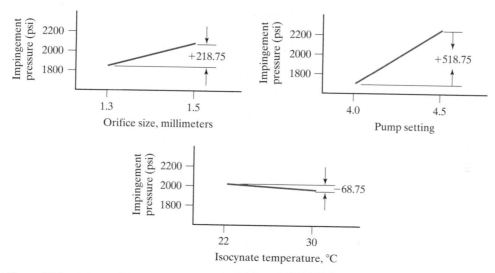

Figure 17.9   Main effects of orifice size, pump setting, and isocynate temperature

particular levels of orifice size and isocynate temperature. Table 17.5 shows this. What is the particular nature of the dependency of the effect of pump setting on orifice size and isocynate temperature? Looking more closely at the table, we see that:

1. The average effect of pump setting is very different, depending on the level of orifice size:

   (a) *Orifice size = 1.30 millimeters* (averaged over both levels of isocynate temperature).

   $$E_{\text{pump}} = \frac{600 + 650}{2} = 625 \text{ psi.}$$

   (b) *Orifice size = 1.50 millimeters* (averaged over both levels of isocynate temperature).

   $$E_{\text{pump}} = \frac{425 + 400}{2} = 412.50 \text{ psi.}$$

TABLE 17.5	Interdependency Among Variable Effects	
Orifice Size (mm)	Isocynate Temperature (°C)	Effect of Pump Setting (psi)
1.30	22	600
1.50	22	425
1.30	30	650
1.50	30	400

2. The average effect of pump setting is about the same, regardless of the level of isocynate temperature:

   (a) *Isocynate temperature = 22 degrees Celsius* (averaged over both levels of orifice size).

$$E_{\text{pump}} = \frac{600 + 425}{2} = 512.5 \text{ psi.}$$

   (b) *Isocynate temperature = 30 degrees Celsius* (averaged over both levels of orifice size).

$$E_{\text{pump}} = \frac{650 + 400}{2} = 525 \text{ psi.}$$

In short, the effect of pump setting on impingement pressure seems to be

   • Indifferent to the magnitude of isocynate temperature.
   • Quite dependent on the magnitude of orifice size.

   The degree of dependency of the pump setting effect on the particular levels of orifice size and isocynate temperature may be appreciated via the *two-way diagrams* in Fig. 17.10. Figure 17.10(a) indicates that the change in impingement pressure with a change in pump setting is the same, regardless of the level of isocynate temperature. There appears, therefore, to be no dependency or interaction between isocynate temperature and pump setting when averaged over orifice size. Again, however, we must interpret this interaction effect with caution since it, too, is an average effect, averaged over the high and low levels of orifice size. Figure 17.10(b) suggests that the change in impingement pressure with a change in pump setting will depend on the level of orifice size. The average impingement pressure, as pump setting is raised from

(a) Effect of pump setting for varying isocynate temperatures.

(b) Effect of pump setting for varying orifice sizes.

Figure 17.10   Two-way diagrams of interaction effects

low to high, increases by 412.5 psi with a larger orifice size, and 625 psi with a smaller orifice size. Since the pump effect depends on the size of the orifice, there is a certain interaction between orifice size and pump setting.

Still another way to see or appreciate the nature of the interaction that may exist between two factors is to construct another type of *two-way diagram*. Referring to the geometrical representation of the 2^3 factorial in Fig. 17.11(a), we can examine the interaction between orifice size and pump setting by compressing

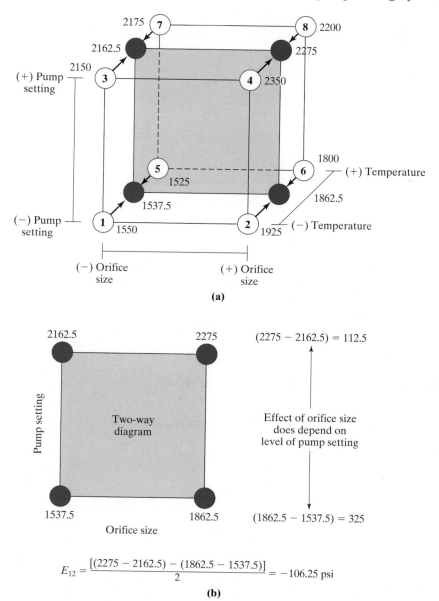

Figure 17.11   Two-factor interaction: orifice size and pump setting

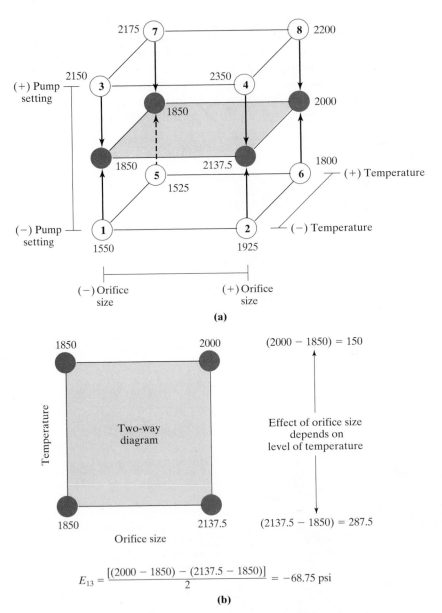

Figure 17.12   Two-factor interaction: temperature and orifice size

the cube along the isocynate temperature axis. By compressing the cube we mean that we average the response values for a given orifice size and pump setting combination [say $(-)$, $(-)$] across the high and low levels of isocynate temperature. The result is that the cube becomes a square, as shown in Fig. 17.11(b). Similarly, the orifice–temperature and pump–temperature interactions can be represented by their respective squares, called two-way diagrams. Figures 17.12 and 17.13 illustrate these two interactions.

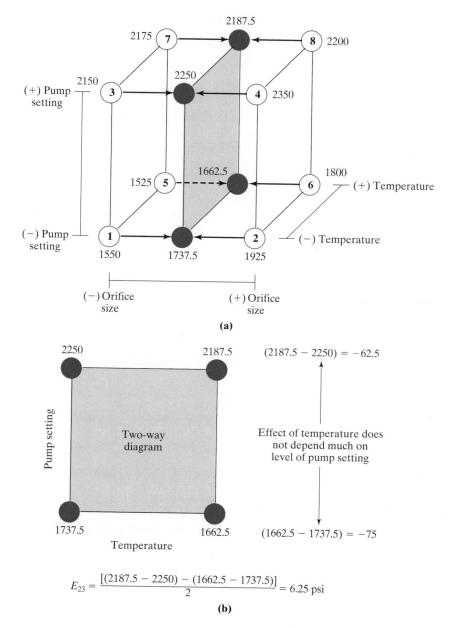

Figure 17.13   Two-factor interaction: pump setting and temperature

To characterize the orifice size–pump setting interaction numerically, let us examine Fig. 17.11(b). As is evident, the effect of orifice size is 112.5 psi when the pump setting is at its high level, while the effect of orifice size is 325 psi when the pump setting is at its low level. Earlier, we said that the average of these two effects was the average effect of the orifice size:

$$E_1 = \frac{112.5 + 325.0}{2}$$

$$= 218.75 \text{ psi}.$$

The difference between these two effects (112.5 and 325.0 psi) divided by 2 measures the dependency of the effect of x_1 on the level of x_2—that is, the interaction effect between the orifice size and the pump setting. Thus the orifice size–pump setting interaction (the "12" interaction) is

$$E_{12} = \frac{(2275 - 2162.5) - (1862.5 - 1537.5)}{2}$$

$$= \frac{112.5 - 325}{2}$$

$$= -106.25.$$

In a like manner, the two-factor interactions orifice size–isocynate temperature and pump setting–isocynate temperature may also be calculated (refer to Figs. 17.12 and 17.13) as

$$E_{13} = [(\text{effect of orifice size at high level for temperature})$$
$$- (\text{effect of orifice size at low level for temperature})]/2$$

$$= \frac{(2000 - 1850) - (2137.5 - 1850)}{2}$$

$$= \frac{150 - 287.5}{2}$$

$$= -68.75 \text{ psi},$$

$$E_{23} = [(\text{effect of temperature at high level for pump setting})$$
$$- (\text{effect of temperature at low level for pump setting})]/2$$

$$= \frac{(2187.5 - 2250) - (1662.5 - 1737.5)}{2}$$

$$= \frac{-62.5 - (-75.0)}{2}$$

$$= 6.25 \text{ psi}.$$

Three-Factor Interaction: Orifice Size–Pump Setting–Isocynate Temperature

In our previous discussion of two-factor interactions we "compressed the cube" (i.e., averaged over the variable not under examination). Suppose that we look at the two-factor interaction between orifice size and pump setting locally; that is, we

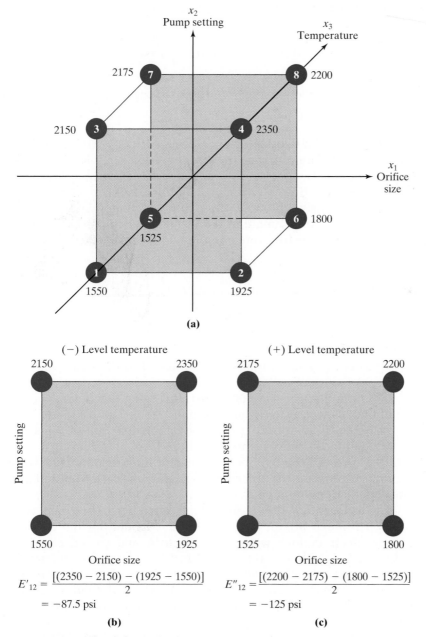

Figure 17.14 Three-factor interaction

do not compress the cube along the isocynate temperature axis. Figure 17.14(b) and (c) shows these two local two-factor interactions, one at the low level of temperature and the other at the high level of temperature. These two squares are each a two-way diagram that can be used to calculate two-factor interactions of orifice size and pump setting E'_{12} and E''_{12}, as shown in Fig. 17.14.

Again, taking the average difference of the two contrasts for orifice size, at high and low pump settings, we have

$$E'_{12} = \tfrac{1}{2}[(2350 - 2150) - (1925 - 1515)]$$
$$= -87.5 \text{ psi},$$

$$E''_{12} = \tfrac{1}{2}[(2200 - 2175) - (1800 - 1525)]$$
$$= -125.0 \text{ psi}.$$

Note that the nature of the two-factor interaction between orifice size and pump setting is not too different across the two levels of isocynate temperature. This means that the two-factor interaction between orifice size and pump setting does not depend much on isocynate temperature, or the three-factor interaction of orifice size, pump setting, and isocynate temperature is not likely to be important. An estimate of the three-factor interaction can be obtained by finding the average difference in the two local two-factor interactions, E'_{12} and E''_{12}:

$$E_{123} = \frac{E''_{12} - E'_{12}}{2}$$
$$= \frac{(-125) - (-87.5)}{2} = -18.75 \text{ psi},$$

which is relatively small in magnitude when compared to the other variable effects.

Previously, we found that $E_{12} = -106.25$ psi. But again this is an average. It is the average of the 12 interaction when 3 is high ($+$) and the 12 interaction when 3 is low ($-$). If these two local two-factor interactions are of the same sign and comparable in magnitude, regardless of the level of 3, then E_{123} may not be important. If, however, the two local two-factor interaction effects were quite different in magnitude and/or different in sign, a three-factor (and higher order, in general) interaction might be at work.

Again, the three-factor interaction is an average difference in two local two-factor interactions. Although this procedure can be extended for the calculation of interaction effects of four factors, five factors, and so on, clearly a simpler method is needed. Although the geometrical representation and graphical analysis of the experimental design has been quite useful in our understanding of the design and the calculation of effects, we will need a simpler and more general method for effect calculations. One such general method is presented in Section 17.6.

Through the use of a geometrical representation of this 2^3 factorial experiment for the isocynate calibration study, we have been able to see how we can calculate variable effects (both main and interaction) and, hopefully, how we might interpret them. But proper interpretation of the existence of real effects cannot be made without any knowledge of the experimental error. Some methods for the analysis of experimental results and statistical interpretation of the effects are discussed in Chapter 18.

17.6   Generalized Method for the Calculation of Effects

The purpose of going through the average effects and interactions in a rather detailed and descriptive manner, using the geometrical representation of the design, is to provide a basic understanding and better appreciation of the meaning of these effects. However, the usefulness of this approach for calculation purposes is somewhat limited, since extension to more than three variables is cumbersome. A simplified calculation procedure, which is easily extended for analyzing two-level factorial designs in any number of variables, is therefore needed and is now described.

Let us refer again to the design matrix for the chemical calibration example (Table 17.4).

Test	x_1	x_2	x_3	y (psi)
1	−1	−1	−1	1550
2	+1	−1	−1	1925
3	−1	+1	−1	2150
4	+1	+1	−1	2350
5	−1	−1	+1	1525
6	+1	−1	+1	1800
7	−1	+1	+1	2175
8	+1	+1	+1	2200

We now notice that if we were to multiply the column of ±1 values associated with x_1 by the column of data y and then divide the sum by $N/2 = 8/2 = 4$, we would have the main effect of orifice size (N = total number of tests, $2^3 = 8$):

x_1		y
−1	×	1550
+1	×	1925
−1	×	2150
+1	×	2350
−1	×	1525
+1	×	1800
−1	×	2175
+1	×	2200
Sum	=	875
Sum/4	=	875/4
E_1	=	218.75

This result should not be too surprising; it is simply a rearrangement of the individual results that form the four contrasts that were previously averaged.

A similar application of the signs of the x_2 and x_3 columns to the data followed by a summing operation and, finally, dividing by 4, produces the main effects of pump setting and isocynate temperature, respectively.

x_2		y	x_3		y
−1	×	1550	−1	×	1550
−1	×	1925	−1	×	1925
+1	×	2150	−1	×	2150
+1	×	2350	−1	×	2350
−1	×	1525	+1	×	1525
−1	×	1800	+1	×	1800
+1	×	2175	+1	×	2175
+1	×	2200	+1	×	2200
Sum	=	2075	Sum	=	−275

$$E_2 = \frac{2075}{4} = 518.75 \text{ psi}, \quad E_3 = \frac{-275}{4} = -68.75 \text{ psi}.$$

Again, this result is not a surprise to us, since in the geometrical representation, the average effect was simply an average of four contrasts, each contrast being the difference between a test result at a high level (+1) for that factor and one at a low level (−1) for that factor.

Although the simplified and generalized method for calculating the average (main) effects via the design matrix is somewhat obvious, it may not be immediately obvious how to extend this to the calculation of interaction effects. To see how we might handle the calculation of an interaction effect, we can examine the mathematical model associated with a 2^2 factorial:

$$y = b_0 + b_1 x_1 + b_2 x_2 + b_{12} x_1 x_2 + \epsilon.$$

The coefficients b_1 and b_2 correspond to the average or main effects of variables 1 and 2, respectively, while the coefficient b_{12} corresponds to the interaction effect between variables 1 and 2.

Given y, b_0, b_1, b_2, and particular values for x_1 and x_2, the key to obtaining a solution for b_{12} is to form the cross product $x_1 x_2$. Similarly, the associated mathematical model for a 2^3 factorial design is

$$y = b_0 + b_1 x_1 + b_2 x_2 + b_3 x_3 + b_{12} x_1 x_2 + b_{13} x_1 x_3$$
$$+ b_{23} x_2 x_3 + b_{123} x_1 x_2 x_3 + \epsilon.$$

The estimates of the b's are actually exactly one-half the value of the E's because the coefficient b measures the incremental change in y per *one* unit of change in the corresponding x (e.g., from 0 to 1), while the effect E measures the incremental change in y for a *two*-unit change in x (from −1 to +1).

To obtain the interaction effects (E_{12}, E_{13}, E_{23}, and E_{123}) we may form the cross-product columns $x_1 x_2$, $x_1 x_3$, $x_2 x_3$, and $x_1 x_2 x_3$, respectively, as shown in Table 17.6. This matrix will be referred to as the *calculation matrix*. The cross-product columns of ± signs are simply the inner products of the individual columns [e.g., $x_1 x_2 = (x_1)(x_2)$].

TABLE 17.6		Calculation Matrix for 2^3 Design							
		Main Effects			**Interactions**				
Test	I	x_1	x_2	x_3	x_1x_2	x_1x_3	x_2x_3	$x_1x_2x_3$	y (psi)
1	+	−1	−1	−1	+1	+1	+1	−1	1550
2	+	+1	−1	−1	−1	−1	+1	+1	1925
3	+	−1	+1	−1	−1	+1	−1	+1	2150
4	+	+1	+1	−1	+1	−1	−1	−1	2350
5	+	−1	−1	+1	+1	−1	−1	+1	1525
6	+	+1	−1	+1	−1	+1	−1	−1	1800
7	+	−1	+1	+1	−1	−1	+1	−1	2175
8	+	+1	+1	+1	+1	+1	+1	+1	2200

Note that Table 17.6 also contains a column of all + signs, denoted I. This column of + signs, when multiplied by the y column, summed, and divided by 8, provides an estimate of b_0, the mean response.

The two- and three-variable interactions can now easily be calculated. For instance, to calculate the interaction between orifice size and pump setting, E_{12}, we multiply the column x_1x_2 by the column y, sum algebraically, and then divide the sum by $2^3/2 = 4$.

x_1x_2		y		
(+1)	×	(1550)		+1550
(−1)	×	(1925)		−1925
(−1)	×	(2150)		−2150
(+1)	×	(2350)	=	+2350
(+1)	×	(1525)		+1525
(−1)	×	(1800)		−1800
(−1)	×	(2175)		−2175
(+1)	×	(2200)		+2200
		Sum	=	−425

Dividing the sum by 4, we obtain the answer that was determined previously, namely $E_{12} = -106.25$ psi, for this two-variable interaction. Similarly, all other interaction effects can be calculated the same way using appropriate interaction columns.

The foregoing algebraic method using a calculation matrix is a general method, because it can easily be extended for the calculation of all the effects for two-level factorial designs of any number of variables, k. Since the method is systematic, it lends itself readily to the computer.

17.7   Mathematical Model of the Response

Recall that the objective of the isocynate calibration experiment was to learn about how isocynate impingement pressure can be controlled to prepare for the experiments to study the foam calibration stand, and ultimately, the foam molding machine to discover the root causes of foam voids. Since the impingement pressure is one of the control variables for foam calibration, the team of engineers wanted to be able to calibrate with precision the isocynate to produce the impingement pressure at some appropriate levels for foam experiments. In other words, they hoped to know, from the isocynate experiment, how to set some or all of the three factors—orifice size, pump setting, and temperature—to produce impingement pressure at some target value. We will now evaluate the results of the isocynate experiment to see if a solution could be developed to meet the intended objective, by way of a fitted mathematical model.

First, this 2^3 factorial design was used with the following mathematical model in mind:

$$y = b_0 + b_1 x_1 + b_2 x_2 + b_3 x_3 + b_{12} x_1 x_2 + b_{13} x_1 x_3$$
$$+ b_{23} x_2 x_3 + b_{123} x_1 x_2 x_3 + \epsilon,$$

where values for the estimates of all the b's, except b_0, are the corresponding $\frac{1}{2}E$'s, and $\hat{b}_0$ is an average of all eight observed responses:

$$\hat{b}_0 = \left(\tfrac{1}{8}\right)(y_1 + y_2 + y_3 + y_4 + y_5 + y_6 + y_7 + y_8),$$

which could also be thought of as the response value at the center of the cube. Finally, the ϵ represents the random noise or experimental error associated with the response y, and it has an expected value of zero. Additionally, let us assume that the model adequately represents the functional relationship between y and the x's.

From the results of the isocynate experiment, our estimate of the mean impingement pressure is given by

$$\hat{b}_0 = \left(\tfrac{1}{8}\right)(1550 + 1925 + 2150 + 2350 + 1525 + 1800 + 2175 + 2200)$$
$$= 1959.375 \text{ psi}$$

and the other coefficients are estimated by

$$\hat{b}_1 = \frac{E_1}{2} = \frac{218.75}{2} = 109.375,$$

$$\hat{b}_2 = \frac{E_2}{2} = \frac{518.75}{2} = 259.375,$$

$$\hat{b}_3 = \frac{E_3}{2} = \frac{-68.75}{2} = -34.375,$$

$$\hat{b}_{12} = \frac{E_{12}}{2} = \frac{-106.25}{2} = -53.125,$$

$$\hat{b}_{13} = \frac{E_{13}}{2} = \frac{-68.75}{2} = -34.375,$$

$$\hat{b}_{23} = \frac{E_{23}}{2} = \frac{6.25}{2} = 3.125,$$

$$\hat{b}_{123} = \frac{E_{123}}{2} = \frac{-18.75}{2} = -9.375.$$

Note that the seven $\hat{b}$'s, or equivalently the E's, are quite different from one another in both magnitude and sign. This shows at least that not all effects are equally important. In fact, some of the smaller effects, positive or negative, could be just sample reflections of the experimental error, while their true effects are zero. The importance of an estimated effect should be judged by comparing it with the experimental error if it is known or can be estimated. We will address this issue in detail in Chapter 18.

Building the Math Model: Significance of Variable Effects

A closer examination of the calculated effects seems to show that some of them are quite large (259.375, 109.375, −53.125) while others are fairly small (−9.375, 3.125). The large sample effects may be important in affecting impingement pressure, while the smaller effects may result only from the forces of error/random variation and therefore could be ignored. Rigorous procedures for assessing the relative importance of the variable effects in light of random variation are presented in Chapter 18. To show how the estimated effects may be used through the mathematical model, we will tentatively assume that E_3, E_{13}, E_{23}, and E_{123} are small enough to be considered as completely explainable by experimental error. Equivalently, we assume that b_3, b_{13}, b_{23}, and b_{123} are zero and their associated terms can be dropped from the postulated model. Tentatively, our fitted model could be expressed with some rounding off as

$$\hat{y} = 1959 + 109x_1 + 259x_2 - 53x_1x_2, \tag{17.1}$$

where $\hat{y}$ is the predicted impingement pressure. This fitted model should be subjected to further analysis and confirmatory tests to check its adequacy as a prediction equation of the impingement pressure for any values of x_1 and x_2 within their high and low levels. Complete procedures for model building are discussed in Chapter 19.

Use of the Fitted Model

Assuming that the fitted model is judged to be adequate, we can use it to represent or approximate the true response surface of the impingement pressure in the space of orifice size and pump setting as shown in Fig. 17.15. Note that the surface shown with heavy lines is not a flat plane because of the interaction term, E_{12}, in the fitted model.

Once this estimated response surface model has been properly validated, it can be used for the selection of the levels or settings of x_1 and x_2 for any given response target (y_0). Conversely, given any combination of x_1 and x_2, the surface can be used to predict the impingement pressure, $\hat{y}$. In either case,

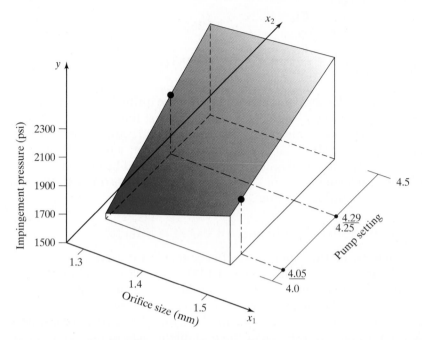

Figure 17.15   **Predicted response surface of isocynate impingement pressure as a function of orifice diameter and pump setting**

some confirmatory tests should be run to check the agreement between the test results and the predicted values from the response surface. In the following discussions, we assume that the model is valid within the domain of the experiment.

Since x_3, the isocynate temperature, was found to have no significant effect on impingement pressure, we can take x_3 out of our design matrix and rewrite a design matrix of x_1 and x_2 alone, with each unique test condition replicated as shown in Table 17.7. At least three useful observations can be made from the resulting math model and its geometrical interpretation in Fig. 17.15.

Observation 1. An increase in either the orifice size or the pump setting, or both, increases the average impingement pressure. It is, however, the pump setting that has the most influence on the impingement pressure (two and a half

TABLE 17.7	Design Matrix for x_1 and x_2						
Test	x_1	x_2	x_1x_2	Replicated Responses		Average Response	Replicate Difference
1	−1	−1	+1	1550	1525	1537.5	−25
2	+1	−1	−1	1925	1800	1862.5	+125
3	−1	+1	−1	2150	2175	2162.5	−25
4	+1	+1	+1	2350	2200	2275.0	+150

times greater than the effect of orifice size). This could also be seen from the relative values of $\hat{b}_1$ and $\hat{b}_2$ in the fitted model (109 versus 259). These average effects of x_1 and x_2 are also called location effects because they change the level or location of the response.

Observation 2. Because of the presence of the interaction effect, $\hat{b}_{12}$, the slope of the response surface in Fig. 17.15 along either the x_1 or x_2 axes is changing as a function of either x_2 or x_1, respectively. For example, for $x_1 = -1$ (1.3 mm orifice size), the pump-setting effect on pressure is given by

$$\hat{y} = 1959 + 109(-1) + 259x_2 - 53(-1)x_2$$
$$\hat{y} = 1850 + 312x_2.$$

For $x_1 = +1$ (1.5 mm orifice size), however, the pump-setting effect on pressure is given by

$$\hat{y} = 1959 + 109(+1) + 259 - 53(+1)x_2$$
$$\hat{y} = 2068 + 206x_2.$$

Note that the increase in pressure with increase in pump setting is 50% greater for the small orifice size than for the large orifice size.

Observation 3. Suppose that for further experimentation on the foam molding process it was desired to set the impingement pressure at 1900 psi. What values of orifice size and pump setting could/should be used? The model above offers an answer to this question. If the small (-1 level) orifice size is used, the pump setting should be

$$1900 = 1850 + 312x_2$$
$$x_2 = 0.16 \quad \text{(a pump setting of 4.29 in uncoded units).}$$

On the other hand, if the large orifice size ($+1$ level) is used, the pump setting should be

$$1900 = 2068 + 206x_2$$
$$x_2 = -0.81 \quad \text{(a pump setting of 4.05 in uncoded units).}$$

The two points are shown on the response surface in Fig. 17.15.

Finding Pump-Setting Levels for the Next Experiment. For the next experiment on the foam molding process, suppose that it is desired to set the high and low levels of the impingement pressure at 2050 and 1600 psi, respectively. Further suppose that an orifice setting of 1.3 millimeter (-1) is used. The pump settings can be determined using the response model. For the impingement pressure $\hat{y}$ to be equal to an average of 2050 psi, the fitted model [Eq. (17.1)] may be written as

$$2050 = 1959 + (109)(-1) + (259)(x_2) - (53)(-1)(x_2)$$

or

$$x_2 = \frac{200}{306} = 0.654,$$

which should be decoded to the actual units:

$$\text{pump setting} = (4.25) + (0.654)(4.50 - 4.25) = 4.4135.$$

Similarly for $\hat{y} = 1600$, using Eq. (17.1), $x_2 = -250/306 = -0.817$, which is converted to the actual units of pump setting $= 4.04575$.

The foregoing solution for the pump settings may be determined graphically using the predicted response surface as illustrated in Fig. 17.16. The two levels of the selected impingement pressure are indicated by the [+] and [−] on the vertical axis in Fig. 17.16. By projecting these two points first to the surface and then down to the orifice-size–pump-setting plane, the desired levels of the orifice size and the pump setting can be read from the x_1 and x_2 axes. If somewhat larger orifice sizes are preferred for other considerations, the response surface of Fig. 17.16 may be used as an aid in choosing a particular combination of orifice size and pump setting for each impingement pressure selected.

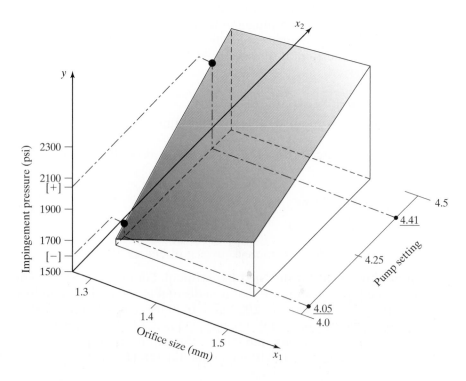

Figure 17.16   Use of response surface to select levels of orifice size and pump setting

17.8   Least-Squares Approach to Model Fitting

To this point in this chapter, we have focused on the development of models using the results from two-level factorial designs. Now we will examine a more general approach for determining the model coefficient estimates from data.

Least-Squares Method

Let us return to the millbase dispersion process previously discussed in Chapter 9. To review, a paint plant uses white millbase in the preparation of the topcoat. As part of an investigation into the process for mixing the resin and pigment, it became necessary to establish a relationship between the pigment concentration and the unit weight of the resulting slurry mixture. To explore this relationship, six slurries were prepared with varying pigment concentrations, and the resulting unit weight was recorded as shown in Table 17.8.

These data points, i.e., (x_i, y_i) for $i = 1$ to 6, were then plotted as shown in Fig. 17.17. Upon examination of the data in the figure, it was suggested that the

TABLE 17.8	Unit Weights of Slurry Mixtures for Different Pigment Concentrations	
Pigment Concentration (%)	Unit Weight (pounds/gallon)
10	13.1
12	13.8
14	13.9
16	13.9
18	14.5
20	14.4

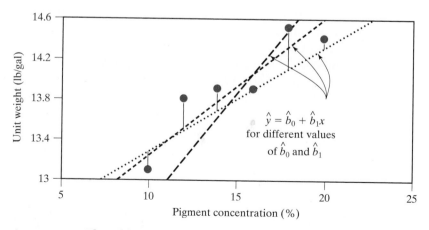

Figure 17.17   Plot of (x_i, y_i) data and possible fitted lines to the data

relationship between the pigment concentration (independent variable) and weight (dependent variable or response) is of the form:

$$y = b_0 + b_1 x + \epsilon \tag{17.2}$$

where b_0 and b_1 are the parameters associated with the intercept and slope, respectively, and ϵ is the experimental error, assumed to be normally and independently distributed with mean of zero and variance, σ_y^2. The data from our tests can be used to estimate the unknown parameters b_0 and b_1, or in other words, fit the model to the data. The fitted model becomes:

$$\hat{y} = \hat{b}_0 + \hat{b}_1 x, \tag{17.3}$$

where $\hat{b}_0$ and $\hat{b}_1$ are the fitted values for the intercept and slope, respectively, and $\hat{y}$ is the predicted response. One can envision that for a variety of different intercepts ($\hat{b}_0$) and slopes ($\hat{b}_1$), the values obtained for $\hat{y}$ may match the actual responses (y) well, or not so well, as shown in Fig. 17.17. Clearly, we are interested in selecting values for the slope and intercept that provide a good match between the model predicted and actual responses.

While there are several estimation criteria that may be employed to find the "best" values for the model parameters $\hat{b}_0$ and $\hat{b}_1$ from the data, the one that seems to have found the most favor is the method of least squares. With least squares, our objective is to minimize the sum of the squared deviations of the responses (y_i's) from their vertical projections onto the fitted line ($\hat{y}_i$'s). Figure 17.17 shows a set of these deviations for one of the lines; for each value of x_i the deviation is displayed as a vertical line segment. Mathematically the sum of the squared deviations may be expressed as

$$SS(\hat{b}_0, \hat{b}_1) = \sum_n (y_i - \hat{y}_i)^2 = \sum_n (y_i - \hat{b}_0 - \hat{b}_1 x_i)^2, \tag{17.4}$$

where the number of data points/tests, n, is 6 for our data. In Eq. (17.4), it may be noted that the value obtained for the sum of squares, $SS(\hat{b}_0, \hat{b}_1)$, depends on the values for $\hat{b}_0$ and $\hat{b}_1$ at which SS is evaluated. Since the least-squares method is focused on finding values for $\hat{b}_0$ and $\hat{b}_1$ that minimize SS, the above equation may be differentiated with respect to $\hat{b}_0$ and $\hat{b}_1$. The resulting two equations may then be set equal to zero. When this is done, we obtain the following, which we will refer to as the normal equations:

$$\sum_n \hat{b}_0 + \hat{b}_1 \sum_n x_i = \sum_n y_i, \tag{17.5}$$

$$\hat{b}_0 \sum_n x_i + \hat{b}_1 \sum_n x_i^2 = \sum_n x_i y_i. \tag{17.6}$$

In considering the normal equations of Eq. (17.5) and (17.6), we note that we have two independent equations and two unknowns, $\hat{b}_0$ and $\hat{b}_1$. As a result, we may solve these two normal equations simultaneously to obtain the least-squares fit. For our millbase dispersion data, we have

$$\sum_{i=1}^{6} x_i = 90, \quad \sum_{i=1}^{6} y_i = 83.6, \quad \sum_{i=1}^{6} x_i^2 = 1420, \quad \sum_{i=1}^{6} x_i y_i = 1262.6.$$

Thus, Eqs. (17.5) and (17.6) become

$$6\hat{b}_0 + 90\hat{b}_1 = 83.6,$$
$$90\hat{b}_0 + 1420\hat{b}_1 = 1262.6.$$

Solving simultaneously, we find that $\hat{b}_0 = 12.09$ and $\hat{b}_1 = 0.1229$. Thus, the least-squares fit to the data is

$$\hat{y} = 12.09 + 0.1229x. \tag{17.7}$$

This model for the predicted response is shown in Fig. 17.18. The figure also illustrates the model residuals, the deviations between the actual and the predicted responses. Of course, with the model established, the model residuals should be examined and interpreted to search for model inadequacies.

Least Squares—Matrix Approach

When working with a simple linear model for the response such as that of Eq. (17.2) it was relatively easy to take the derivative of the sum-of-squares function and find the resulting two normal equations. For more complex model forms, we often employ matrix methods to undertake least-squares model fitting. Suppose, for example, that the following second-order model is

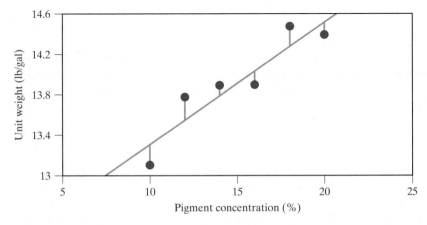

Figure 17.18   Millbase dispersion data and associated model for the predicted response based on the least-squares fit

proposed to describe a response as a function of two independent variables, x_1 and x_2:

$$y = b_0 + b_1x_1 + b_2x_2 + b_{12}x_1x_2 + b_{11}x_1^2 + b_{22}x_2^2 + \epsilon, \tag{17.8}$$

where it is noted that $p = 6$ parameters are included in the second-order model in this case. If the response is governed by Eq. (17.8), the associated equation for the predicted response is

$$\hat{y} = \hat{b}_0 + \hat{b}_1x_1 + \hat{b}_2x_2 + \hat{b}_{12}x_1x_2 + \hat{b}_{11}x_1^2 + \hat{b}_{22}x_2^2. \tag{17.9}$$

The sum-of-squares function is then:

$$\begin{aligned} SS &= \sum_n (y_i - \hat{y}_i)^2 \\ &= \sum_n (y_i - \hat{b}_0 - \hat{b}_1x_1 - \hat{b}_2x_2 \\ &\quad - \hat{b}_{12}x_1x_2 - \hat{b}_{11}x_1^2 - \hat{b}_{22}x_2^2)^2. \end{aligned} \tag{17.10}$$

Applying the least-squares method discussed above to Eq. (17.10), we must take the partial derivative of the function with respect to each $\hat{b}$. We then obtain a set of six independent equations that involve the six unknowns: $\hat{b}_0$, $\hat{b}_1$, $\hat{b}_2$, $\hat{b}_{12}$, $\hat{b}_{11}$, and $\hat{b}_{22}$. The unknowns can be solved for in these equations using various mathematical procedures; however, the solution becomes ever more cumbersome as the number of parameters to be estimated/number of equations increases. For this reason, we directly employ a matrix approach to the least squares problem.

Paralleling our previous least-squares development, let us assume that a set of response values can be modeled with the relationship:

$$\mathbf{y} = \mathbf{Xb} + \epsilon, \tag{17.11}$$

where $\mathbf{y}$ is the $(n \times 1)$ vector of responses where the ith response in the vector corresponds to the ith test, $\mathbf{X}$ is the $(n \times p)$ design matrix where the ith row contains the independent variable settings for the ith test, $\mathbf{b}$ is the $(p \times 1)$ vector of parameter values where the jth value in the vector corresponds to the jth parameter, and ϵ is the $(n \times 1)$ vector of experimental errors. The fitted model is then:

$$\hat{\mathbf{y}} = \mathbf{X}\hat{\mathbf{b}}, \tag{17.12}$$

where $\hat{\mathbf{y}}$ is the $(n \times 1)$ vector of predicted responses and $\hat{\mathbf{b}}$ is the $(p \times 1)$ vector of parameter estimates. The vector of deviations is given by $(\mathbf{y} - \hat{\mathbf{y}})$, and the least-squares problem is focused on finding the set of parameter estimates that minimize the sum of squared deviations, expressed mathematically by Eq. (17.13).

$$SS(\hat{\mathbf{b}}) = \left\| \mathbf{y} - \hat{\mathbf{y}} \right\|^2 = (\mathbf{y} - \mathbf{X}\hat{\mathbf{b}})^T(\mathbf{y} - \mathbf{X}\hat{\mathbf{b}})$$

$$= (\mathbf{y}^T\mathbf{y}) - (\hat{\mathbf{b}}^T\mathbf{X}^T\mathbf{y}) - (\mathbf{y}^T\mathbf{X}\hat{\mathbf{b}}) + (\hat{\mathbf{b}}^T\mathbf{X}^T\mathbf{X}\hat{\mathbf{b}})$$

$$= (\mathbf{y}^T\mathbf{y}) - 2(\hat{\mathbf{b}}^T\mathbf{X}^T\mathbf{y}) + (\hat{\mathbf{b}}^T\mathbf{X}^T\mathbf{X}\hat{\mathbf{b}}). \qquad (17.13)$$

Again, since the least-squares method is focused on finding values for $\hat{\mathbf{b}}$ that minimize the sum-of-squares function, we may differentiate the above equation with respect to $\hat{\mathbf{b}}$ and set the result equal to zero, viz.,

$$0 = -2(\mathbf{X}^T\mathbf{y}) + 2(\mathbf{X}^T\mathbf{X}\hat{\mathbf{b}})$$
$$\mathbf{X}^T\mathbf{X}\hat{\mathbf{b}} = \mathbf{X}^T\mathbf{y}.$$

Upon simplification, the solution to the least-squares problem using the matrix approach is

$$\hat{\mathbf{b}} = (\mathbf{X}^T\mathbf{X})^{-1}(\mathbf{X}^T\mathbf{y}). \qquad (17.14)$$

The following example illustrates how the matrix least-squares method may be employed for model fitting.

Grind Fineness Example

An ink maker is concerned about the size of the pigment particles within each batch. The pigment is broken down into small particles using a milling process, and the fineness of the grind (the response) is characterized by particle size (μm). As part of an effort to improve the process and achieve a smaller grind fineness, it is desired to understand the relationship between two of the variables associated with the process, media size and flow rate, and the fineness response (particle size, μm). Table 17.9 shows a set of tests that were performed to develop this relationship.

TABLE 17.9	Mill Tests and Associated Grind Fineness Responses				
	Actual Variable Levels		Coded Variable Levels		
Test	Media Size (μm)	Flow Rate (gal/hr)	x_1	x_2	Grind Fineness— Particle Size (μm)
1	10	100	-1	-1	33
2	20	100	1	-1	39
3	10	200	-1	1	48
4	10	200	1	1	52
5	7.93	125	$-\sqrt{2}$	0	42
6	22.07	125	$+\sqrt{2}$	0	48
7	15	89.65	0	$-\sqrt{2}$	31
8	15	160.36	0	$+\sqrt{2}$	53
9	15	125	0	0	38
10	15	125	0	0	40

The test conditions shown in Table 17.9 collectively form what is termed a central composite design (CCD), an experimental design that will be explained further in Chapter 23. It is apparent that CCD consists of 10 tests performed under a variety of variable-level combinations. As is evident, unlike the class of 2^k factorial designs, the central composite design considers more than two levels for each variable under study. The CCD is an experiment design that is ideally suited for responding to a proposed second-order model form such as that depicted in Eq. (17.8). Based on the proposed model of Eq. (17.8), we can use the data in the table to develop a fitted model.

Following the matrix least-squares approach, our principal task is to form the appropriate vectors and the design matrix in order to utilize Eq. (17.14). To do this, one can envision that the predicted response for the ith test is given by:

$$\hat{y}_i = \hat{b}_0 + \hat{b}_1 x_{1i} + \hat{b}_2 x_{2i} + \hat{b}_{12} x_{1i} x_{2i} + \hat{b}_{11} x_{1i}^2 + \hat{b}_{22} x_{2i}^2, \qquad (17.15)$$

where it is noted that the coded variables (x_1 and x_2) are being used in this case. The subscripts on the x's in Eq. (17.15) refer to the variable number and the test number, respectively. The $(p \times 1)$ vector associated with the least squares parameter estimates is

$$\hat{\mathbf{b}} = \begin{bmatrix} \hat{b}_0 \\ \hat{b}_1 \\ \hat{b}_2 \\ \hat{b}_{12} \\ \hat{b}_{11} \\ \hat{b}_{22} \end{bmatrix}.$$

Given the form of the vector for the parameter estimates, it is apparent that the corresponding ith row in the design matrix must take the following form:

$$\begin{bmatrix} \vdots \\ 1 & x_{1i} & x_{2i} & x_{1i}x_{2i} & x_{1i}^2 & x_{2i}^2 \\ \vdots \end{bmatrix}.$$

Therefore, the design matrix and response vector are

$$\mathbf{X} = \begin{bmatrix} +1 & -1 & -1 & +1 & +1 & +1 \\ +1 & +1 & -1 & -1 & +1 & +1 \\ +1 & -1 & +1 & -1 & +1 & +1 \\ +1 & +1 & +1 & +1 & +1 & +1 \\ +1 & -\sqrt{2} & 0 & 0 & 2 & 0 \\ +1 & +\sqrt{2} & 0 & 0 & 2 & 0 \\ +1 & 0 & -\sqrt{2} & 0 & 0 & 2 \\ +1 & 0 & +\sqrt{2} & 0 & 0 & 2 \\ +1 & 0 & 0 & 0 & 0 & 0 \\ +1 & 0 & 0 & 0 & 0 & 0 \end{bmatrix}, \quad \mathbf{y} = \begin{bmatrix} 33 \\ 39 \\ 48 \\ 52 \\ 42 \\ 48 \\ 31 \\ 53 \\ 38 \\ 40 \end{bmatrix}.$$

The least-squares solution is given by Eq. (17.14), so we need to determine $(\mathbf{X}^T\mathbf{X})^{-1}$ and $(\mathbf{X}^T\mathbf{y})$.

$$(\mathbf{X}^T\mathbf{X}) = \begin{bmatrix} 10 & 0 & 0 & 0 & 8 & 8 \\ 0 & 8 & 0 & 0 & 0 & 0 \\ 0 & 0 & 8 & 0 & 0 & 0 \\ 0 & 0 & 0 & 4 & 0 & 0 \\ 8 & 0 & 0 & 0 & 12 & 4 \\ 8 & 0 & 0 & 0 & 4 & 12 \end{bmatrix},$$

$$(\mathbf{X}^T\mathbf{X})^{-1} = \begin{bmatrix} 0.5 & 0 & 0 & 0 & -0.25 & -0.25 \\ 0 & 0.125 & 0 & 0 & 0 & 0 \\ 0 & 0 & 0.125 & 0 & 0 & 0 \\ 0 & 0 & 0 & 0.125 & 0 & 0 \\ -0.25 & 0 & 0 & 0 & 0.21875 & 0.09375 \\ -0.25 & 0 & 0 & 0 & 0.09375 & 0.21875 \end{bmatrix},$$

$$(\mathbf{X}^T\mathbf{Y}) = \begin{bmatrix} 424.000 \\ 18.485 \\ 59.113 \\ -2.000 \\ 352.000 \\ 340.000 \end{bmatrix}.$$

So, the least-squares solution for the parameter estimates of the grind fineness example becomes:

$$\hat{\mathbf{b}} = \begin{bmatrix} \hat{b}_0 \\ \hat{b}_1 \\ \hat{b}_2 \\ \hat{b}_{12} \\ \hat{b}_{11} \\ \hat{b}_{22} \end{bmatrix} = \begin{bmatrix} 39.000 \\ 2.311 \\ 7.389 \\ -0.500 \\ 2.875 \\ 1.375 \end{bmatrix}.$$

Thus, the fitted model based on the method of least squares can be written as

$$\hat{y} = 39.0 + 2.3x_1 + 7.4x_2 - 0.5x_1x_2 + 2.9x_1^2 + 1.4x_2^2.$$

The surface associated with this equation for the predicted response (response surface) is shown in Fig. 17.19, where it may be noted that the actual units have been shown on the figure rather than the coded units. Figure 17.20 displays the equation using a different graphical display, a contour plot.

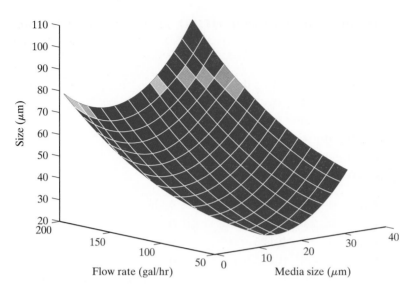

Figure 17.19   **Predicted grind fineness response surface**

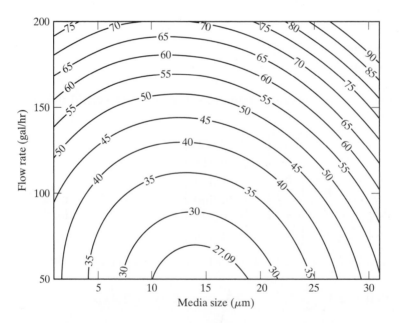

Figure 17.20   **Contours of constant predicted grind fineness (μm)**

With a model for the predicted response now established, we should now examine the model residuals to search for model shortcomings. Assuming the model is judged to be adequate, either representation of the predicted response may be used to identify promising conditions that provide improved grind fineness. Based on the figures, a flow rate of 50 gal/hr and a media size of 15 μm should produce a grind fineness (particle size) of approximately 27 μm. Of course, additional tests at these proposed settings must be performed to confirm the predictive ability of the model.

Relationship Between the E's and the $\hat{b}$'s for Two-Level Designs

We have stated earlier in this chapter that the calculated main and interaction effects for a two-level design are directly related to the least squares estimates of the parameters in the model for the predicted response ($\hat{b} = E/2$). We now demonstrate this through an illustration.

Suppose it is desired to establish a linear model based on the results of a 2^2 experiment in two variables x_1 and x_2. We know for a 2^2 design that the calculation matrix is

I	x_1	x_2	x_{12}	Response, y
1	-1	-1	1	y_1
1	1	-1	-1	y_2
1	-1	1	-1	y_3
1	1	1	1	y_4

and the effect estimates are

$$E_0 = \bar{y} = \tfrac{1}{4}(y_1 + y_2 + y_3 + y_4), \quad E_1 = \tfrac{1}{2}[(y_2 - y_1) + (y_4 - y_3)],$$
$$E_2 = \tfrac{1}{2}[(y_3 - y_1) + (y_4 - y_2)], \quad E_{12} = \tfrac{1}{2}[(y_4 - y_3) - (y_2 - y_1)].$$

In our earlier discussions, we said these effect estimates would give us the least-squares estimates of the parameters in the model for the predicted response, $\hat{y} = \hat{b}_0 + \hat{b}_1 x_1 + \hat{b}_2 x_2 + \hat{b}_{12} x_1 x_2$, by

$$\hat{b}_0 = \bar{y} = E_0, \quad \hat{b}_1 = \frac{E_1}{2}, \quad \hat{b}_2 = \frac{E_2}{2}, \quad \text{and} \quad \hat{b}_{12} = \frac{E_{12}}{2}.$$

To demonstrate that the above is true, we write the design matrix

$$\mathbf{X} = \begin{bmatrix} +1 & -1 & -1 & +1 \\ +1 & +1 & -1 & -1 \\ +1 & -1 & +1 & -1 \\ +1 & +1 & +1 & +1 \end{bmatrix}$$

and the response vector

$$\mathbf{y} = \begin{bmatrix} y_1 \\ y_2 \\ y_3 \\ y_4 \end{bmatrix}.$$

Since the least-squares matrix solution is $\hat{\mathbf{b}} = (\mathbf{X}^T\mathbf{X})^{-1}\mathbf{X}^T\mathbf{y}$, we need to determine $(\mathbf{X}^T\mathbf{X})^{-1}$ and $(\mathbf{X}^T\mathbf{y})$.

$$\mathbf{X}^T\mathbf{X} = \begin{bmatrix} +1 & +1 & +1 & +1 \\ -1 & +1 & -1 & +1 \\ -1 & -1 & +1 & +1 \\ +1 & -1 & -1 & +1 \end{bmatrix} \begin{bmatrix} +1 & -1 & -1 & +1 \\ +1 & +1 & -1 & -1 \\ +1 & -1 & +1 & -1 \\ +1 & +1 & +1 & +1 \end{bmatrix} = \begin{bmatrix} 4 & 0 & 0 & 0 \\ 0 & 4 & 0 & 0 \\ 0 & 0 & 4 & 0 \\ 0 & 0 & 0 & 4 \end{bmatrix},$$

producing

$$(\mathbf{X}^T\mathbf{X})^{-1} = \begin{bmatrix} \frac{1}{4} & 0 & 0 & 0 \\ 0 & \frac{1}{4} & 0 & 0 \\ 0 & 0 & \frac{1}{4} & 0 \\ 0 & 0 & 0 & \frac{1}{4} \end{bmatrix}$$

and

$$\mathbf{X}^T\mathbf{y} = \begin{bmatrix} +y_1 + y_2 + y_3 + y_4 \\ -y_1 + y_2 - y_3 + y_4 \\ -y_1 - y_2 + y_3 + y_4 \\ +y_1 - y_2 - y_3 + y_4 \end{bmatrix}.$$

Therefore, the least-squares estimates are

$$\hat{\mathbf{b}} = \begin{bmatrix} \hat{b}_0 \\ \hat{b}_1 \\ \hat{b}_2 \\ \hat{b}_{12} \end{bmatrix} = \begin{bmatrix} \frac{1}{4} & 0 & 0 & 0 \\ 0 & \frac{1}{4} & 0 & 0 \\ 0 & 0 & \frac{1}{4} & 0 \\ 0 & 0 & 0 & \frac{1}{4} \end{bmatrix} \begin{bmatrix} +y_1 + y_2 + y_3 + y_4 \\ -y_1 + y_2 - y_3 + y_4 \\ -y_1 - y_2 + y_3 + y_4 \\ +y_1 - y_2 - y_3 + y_4 \end{bmatrix}$$

$$= \begin{bmatrix} \frac{1}{4}(+y_1 + y_2 + y_3 + y_4) \\ \frac{1}{4}(-y_1 + y_2 - y_3 + y_4) \\ \frac{1}{4}(-y_1 - y_2 + y_3 + y_4) \\ \frac{1}{4}(+y_1 - y_2 - y_3 + y_4) \end{bmatrix} = \begin{bmatrix} \bar{y} \\ E_1/2 \\ E_2/2 \\ E_{12}/2 \end{bmatrix}$$

This completes the demonstration that the least-squares estimates, $\hat{b}$'s, of a first-order model can be easily calculated from the effect estimates E's of a 2^k experiment.

EXERCISES

17.1 A 2^3 factorial design was conducted to study the influence of temperature (x_1), pressure (x_2), and cycle time (x_3) on the occurrence of splay in an injection molding process. For each of the eight unique trials, 50 parts were made, and the response observed was the number of incidences of the occurrence of splay on the part surface across all 50 parts. The design matrix and test responses are given in the table below.

Test	1	2	3	y
1	−	−	−	12
2	+	−	−	15
3	−	+	−	24
4	+	+	−	17
5	−	−	+	24
6	+	−	+	16
7	−	+	+	24
8	+	+	+	28

(a) Draw the geometrical representation of this experiment and label each corner of the cube with the corresponding test conditions (e.g., $+, -, +$) as well as the test response.

(b) Calculate the main effects of all three variables.

(c) Calculate all the two- and three-factor interaction effects.

(d) Write down the appropriate mathematical model, assuming that all estimated effects are to be included in the model.

17.2 Suppose that it is proposed to run the tests outlined in Exercise 17.1 using the test order specified. Additionally, it is planned to run two tests per shift, and the operation is to be run for two shifts per day. If the eight tests are performed in this manner, what possible problem(s) could arise in the ultimate interpretation of the results? How can this situation be rectified?

17.3 A 2^2 factorial experiment was conducted to determine the effects that tool surface coating (x_1) and the application of cutting fluid (x_2) might have on the surface finish of a turned part. The low level for each of the two variables represented the absence of the technology in question (i.e., no surface coating, no cutting fluid). The high level represented the technology at some commercially recommended level. The response is the part surface finish in microinches (R_a value).

x_1	x_2	R_a
−	−	44
+	−	32
−	+	55
+	+	20

(a) Draw the geometrical representation of this experiment, labeling the corners with the corresponding test conditions (e.g., $+$, $-$).

(b) Calculate all the variable effects.

(c) Sketch a graphical representation of the two-factor interaction effect using a two-way diagram and explain the nature of the interaction effect in words.

17.4 A 2^3 factorial design was constructed to test the effect of three different variables on the alertness of students in an early morning class. The variables tested, the design matrix, and the responses are shown in the accompanying tables.

Variable	Low Level	High Level
1. Hours of sleep	4	8
2. Ounces of coffee	6	12
3. Number of donuts	1	2

Test	1	2	3	Alertness (%)
1	$-$	$-$	$-$	56
2	$+$	$-$	$-$	72
3	$-$	$+$	$-$	68
4	$+$	$+$	$-$	89
5	$-$	$-$	$+$	43
6	$+$	$-$	$+$	62
7	$-$	$+$	$+$	59
8	$+$	$+$	$+$	75

(a) Calculate estimates of all the main and interaction effects, using the algebraic method based on the calculation matrix for this design.

(b) Interpret the two-factor interaction effects geometrically by making two-way diagrams for all interaction effects. Briefly summarize in words the information that each graph yields.

(c) Explain the physical meaning of a three-factor interaction effect. Use any graphical summaries that you think are useful to your explanation.

17.5 Draw the geometrical representation of a 2^4 factorial experiment and label the diagram in appropriate places with the corresponding test conditions.

17.6 A 2^3 factorial design was used to study the possible effects of temperature, time, and pressure on the output of a chemical process. The data, including experimental results, are shown in the accompanying tables.

(a) Draw the geometrical representation of this experiment and label each corner of the cube with the corresponding test conditions (e.g., $+$, $-$, $+$) as well as the test responses.

Variable	Low Level	High Level
1. Temperature (°F)	80	100
2. Time (min)	5	7
3. Pressure (psi)	120	140

Test	1	2	3	Response
1	−	−	−	10
2	+	−	−	20
3	−	+	−	4
4	+	+	−	10
5	−	−	+	8
6	+	−	+	18
7	−	+	+	6
8	+	+	+	12

(b) Use the geometrical representation you created in part (a) to determine the value of the main effects and the two-factor interaction effects.

(c) Write down the calculation matrix and use it to obtain all main and interaction effect estimates. [Compare with the effect estimates obtained in part (b).]

16-7

17.7 A 2^3 factorial experiment was designed and conducted to determine the effects of packing material, distance shipped, and home shipments per load used on the average amount of damage sustained per load by a moving company. The variable levels used and the results of the experiment are shown in the first table.

Variable	Low Level	High Level
1. Packing material	Shredded paper	Foam "peanuts"
2. Distance traveled	Less than 500 miles	More than 1000 miles
3. Home shipments per load	One	Two or more

The response observed was the number of damaged items per 100 packed boxes, shown in the second table.

Test	1	2	3	Response
1	−	−	−	2.55
2	+	−	−	3.95
3	−	+	−	1.55
4	+	+	−	2.50
5	−	−	+	1.15
6	+	−	+	2.15
7	−	+	+	2.55
8	+	+	+	3.75

(a) Write out the calculation matrix and use it to calculate all the main and interaction effects.
(b) Interpret the interaction effects geometrically by making two-way diagrams for all interaction effects. Briefly summarize the information each graph yields.
(c) Write down the appropriate mathematical model in terms of the coded variables, assuming that all effects may be important.
(d) Determine the predicted value, $\hat{y}$, for each test condition.

16.8 **17.8** A 2^2 factorial design experiment was conducted, and the results are shown in the table below.

Test	1	2	Response
1	−	−	45.0
2	+	−	30.5
3	−	+	39.2
4	+	+	28.9

(a) Draw the geometrical representation of this experiment and label your diagram with the corresponding test conditions (e.g., +, −).
(b) Calculate all the main and interaction effects.
(c) Draw the two-way diagram for the interaction effect, and interpret this effect.

16.9 **17.9** A chemical engineer performed the experiments shown in the first table, randomizing the order of the tests within each week. The response of interest is the chemical yield, and the run order is given in parentheses.

Test	Temp.	Catalyst	pH	Week 1	Week 2
1	−	−	−	64.0 (5)	61.2 (10)
2	+	−	−	72.3 (8)	75.5 (11)
3	−	+	−	60.2 (1)	61.4 (13)
4	+	+	−	65.3 (3)	68.7 (16)
5	−	−	+	62.8 (7)	66.2 (9)
6	+	−	+	83.6 (4)	84.7 (14)
7	−	+	+	68.9 (6)	71.7 (12)
8	+	+	+	76.5 (2)	78.1 (15)

The variable levels are given in the second table.

Level	Temperature (°F)	Catalyst	pH
−	160	A	6
+	180	B	7

(a) Sketch the geometrical representation of the 2^3 design, labeling each corner with the corresponding test conditions.

(b) Calculate the average response for each test condition, and add these values to the drawing in part (a).

(c) Write out the calculation matrix and calculate all the effects.

(d) Interpret the three-factor interaction, using a pair of two-way diagrams associated with any one of the two-factor interactions.

17.10 A re-examination of the experiment performed in Exercise 17.9 revealed that all the tests performed in week 1 were performed by one operator, and the tests performed in week 2 were performed by another operator. Another interpretation of the experiment may therefore be that it is a 2^4 factorial, varying temperature, catalyst, pH, and operator.

(a) Write out the design matrix in standard order for the 2^4 design.

(b) Write out the calculation matrix for the 2^4 design.

(c) Calculate all the effects.

(d) Comment on the advisability of the manner in which the 2^4 factorial experiment was performed.

17.11 The results of a 2^3 factorial design are given in standard order in the table below.

Test	Response, y
1	48.8
2	91.6
3	40.8
4	60.4
5	48.8
6	89.2
7	40.4
8	60.0

Let x_1, x_2, and x_3 correspond to the coded variables.

(a) Write out the calculation matrix and use it to calculate estimates of all the main and interaction effects.

(b) Interpret the interaction effects geometrically by making two-way diagrams for all interaction effects. Briefly summarize the information each graph yields.

17.12 A 2^2 factorial experiment was conducted to test the effect of lighting and watchman on the average number of car burglaries per month from four different parking garages, with data as shown in the accompanying tables.

Variable	Low Level	High Level
1. Lighting	Poor	Good
2. Watchman	No	Yes

Test	1	2	y
1	−	−	2.80
2	+	−	1.00
3	−	+	2.40
4	+	+	0.75

(a) Draw the geometrical representation of this experiment and label your diagram with the corresponding test conditions (e.g., +, −).
(b) Calculate all the main and interaction effects.
(c) Interpret the results.

16.13   **17.13** A 2^4 factorial experiment was conducted, and the results are shown in the table below.

Test	1	2	3	4	y
1	−	−	−	−	34.62
2	+	−	−	−	41.35
3	−	+	−	−	23.08
4	+	+	−	−	18.27
5	−	−	+	−	26.92
6	+	−	+	−	33.65
7	−	+	+	−	31.73
8	+	+	+	−	21.15
9	−	−	−	+	30.77
10	+	−	−	+	37.50
11	−	+	−	+	24.04
12	+	+	−	+	23.08
13	−	−	+	+	16.35
14	+	−	+	+	13.46
15	−	+	+	+	25.00
16	+	+	+	+	19.23

(a) Sketch the geometrical representation of this design, and place the test results on this representation.
(b) Write out the calculation matrix and calculate all the main effects and interaction effects.

16.14   **17.14** Suppose that the investigator decided to ignore the presence of variable 4 in Exercise 17.13 and treat the design as a 2^3 factorial with replication.
(a) How does this change the estimates of the effects involving variables 1, 2, and 3?
(b) How does this change the actual interpretation of the results (i.e., how does the interpretation change between the 2^4 and 2^3 designs)?

17.15 A 2^3 factorial design was constructed to study the effects of lighting, ambient temperature, and years of service on the average number of injuries per month in three different factories. The results are shown in the accompanying tables.

Variable	Low Level	High Level
1. Lighting	Poor	Well-lit
2. Average ambient temperature (°F)	68	75
3. Years of service	Below 5	Above 5

Test	1	2	3	y
1	−	−	−	13.7
2	+	−	−	11.0
3	−	+	−	10.2
4	+	+	−	6.4
5	−	−	+	17.3
6	+	−	+	12.6
7	−	+	+	13.5
8	+	+	+	11.1

(a) Draw the geometrical representation of this experiment and label each corner of the cube with the corresponding test conditions (e.g., +, −, +) as well as the test response.

(b) Use the geometrical representation you created in part (a) to determine the value of the main effects and the two-factor interaction location effects.

(c) Write down the calculation matrix and use it to obtain all main and interaction effects. [Compare with the effects obtained in part (b).]

17.16 A 2^3 factorial experiment was designed to test the effect of three different variables on the ultimate tensile strength of a weld.[2] The three variables and the levels used are shown in the accompanying tables together with the test results. Note that this experiment was replicated.

Variable	Low Level	High Level
1. Ambient temperature (°F)	0	70
2. Wind velocity (mph)	0	20
3. Bar size (in.)	4/8	11/8

Test	1	2	3	y_1	y_2
1	−	−	−	84.0	91.0
2	+	−	−	90.6	84.0
3	−	+	−	69.6	86.0
4	+	+	−	76.0	98.0
5	−	−	+	77.7	80.5
6	+	−	+	99.7	95.5
7	−	+	+	82.7	74.5
8	+	+	+	93.7	81.7

---

[2] S. M. Wu, "Analysis of Rail Steel Bar Welds by Two-Level Factorial Design," *Welding Journal Research Supplement*, 1964, pp. 1795–1835.

(a) Sketch the geometrical representation of the 2^3 design, labeling each corner with the corresponding test conditions.

(b) Calculate the average response for each test condition, and add these values to the drawing in part (a).

(c) Write out the calculation matrix, and calculate all the effects.

17.17 An investigator performed a 2^3 factorial experiment, and he calculated the average of all eight tests and the main effect of variable 1 before leaving the office one evening. Upon returning to work the next day, the investigator was able to find only the six test results shown in the table below, and the numerical values for the average (25.00375) and E_1 (10.2675).

1	2	3	y
−	+	−	13.76
−	−	−	30.72
+	−	−	37.81
−	−	+	12.40
+	+	−	25.69
+	−	+	21.79

(a) Write down the test recipes for the two missing tests.

(b) Calculate the remaining main and interaction effects.

17.18 A 2^2 factorial design was run to study the effects of product model (x_1) and fixture type (x_2) on the productivity of an assembly process. The responses, parts assembled per hour, are shown below. The next day the same experiment was repeated with a completely different set of workers doing the assembly.

Day	Test	1	2	y
1	1	−	−	59.0
	2	+	−	76.3
	3	−	+	28.9
	4	+	+	51.4
2	5	−	−	20.3
	6	+	−	41.6
	7	−	+	48.3
	8	+	+	73.1

(a) Calculate E_1, E_2, and E_{12} for each day's experiment.

(b) Compare the results of the two experiments, and offer an explanation for possible differences observed.

(c) Calculate the main effect of worker set on assembly time.

(d) Calculate the two-factor interaction of fixture type by worker set.

(e) Comment on the way in which this experiment was performed. Does the manner in which the experiment was performed affect the interpretation of the results?

17.19 A 2^3 factorial design is constructed in an attempt to determine the effects of the amount of rest, the amount of preparation, and the exam room's temperature on the ability of a student to score well on an exam. The variable levels, design matrix, and responses are shown in the accompanying tables.

Variable	Low Level	High Level
1. Sleep (hr)	4	8
2. Preparation (hr)	2	6
3. Temperature (°F)	65	75

Test	1	2	3	Score
1	−	−	−	60
2	+	−	−	70
3	−	+	−	85
4	+	+	−	95
5	−	−	+	55
6	+	−	+	65
7	−	+	+	80
8	+	+	+	95

(a) Which factor has the largest average effect on exam score?
(b) Calculate and interpret the three two-factor interaction effects.
(c) For persons who have prepared very little for the exam, what action should be taken if they wish to score well? Explain.

17.20 While in the midst of making off with the results of a 2^3 factorial experiment, a disgruntled employee is interrupted and obtains only the seven test results shown in the table below.

1	2	3	y
+	−	−	157.0
−	−	−	161.3
+	+	+	141.1
−	+	−	152.1
+	+	−	160.9
−	−	+	131.6
−	+	+	139.2

(a) What is the recipe for the missing test?
(b) What assumption could reasonably be made to permit calculation of all main and interaction effects?
(c) Using the assumption of part (b), calculate all the main and interaction effects.

17.21 For each diagram shown, determine whether a two-factor interaction exists. Calculate the interaction effects in each case.

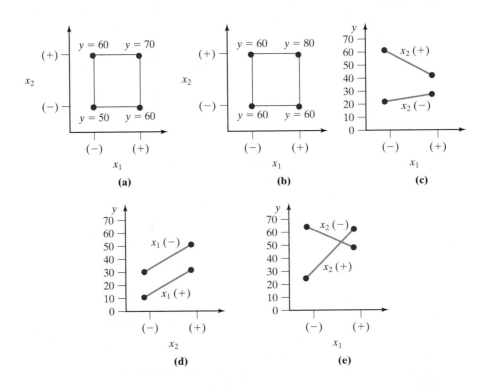

(a)   **(b)**   **(c)**

(d)   **(e)**

17.22 A 2^2 factorial design has produced the following model relating machined part surface finish to feed rate (x_1) and nose radius (x_2) of the tool $(x_1, x_2$ are in coded units of $\pm 1)$:

$$\hat{y} = 90 + 10x_1 - 15x_2 + 5x_1x_2.$$

(a) What is the average surface finish across the experiment?

(b) What are the values for E_1, E_2, and E_{12} as obtained from the experiment?

(c) Sketch the response surface in the x_1, x_2, y coordinate space.

(d) What value for the surface finish does the model predict when the feed rate and nose radius are at their high levels for the experiment?

(e) If a surface finish of 85 is desired, what value of feed rate (coded units) should be used if the nose radius takes on the low level used in the experiment?

17.23 Apply the responses $y = 5$, $y = 10$, and $y = 15$ to the graphical representation shown (a response may be repeated or not used at all) to satisfy the following four different situations:

(a) Positive main effects of x_1 and x_2 and interaction effect $= 0$.
(b) Positive main effect of x_1, main effect of $x_2 = 0$, and interaction effect $= 0$.
(c) Main effects of x_1 and $x_2 = 0$, interaction effect $\neq 0$.
(d) Positive main effect of x_1, negative main effect of x_2, and interaction effect $\neq 0$.

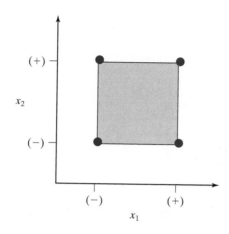

17.24 A 2^4 factorial experiment was conducted producing the following test results, in standard order: 10, 12, 12, 14, 8, 7, 9, 16, 8, 9, 17, 12, 10, 11, 6, 16.
  (a) Calculate the value of the two-factor interaction E_{34}.
  (b) Calculate the value of the three-factor interaction E_{124}.
  (c) Show the two-factor interaction E_{23} in graphical form (i.e., construct the two-way diagram to interpret this interaction).

17.25 The incidence of splay defects (measured in percent of part coverage) is being studied as a function of temperature and holding pressure for an injection molding process. The results of a 2^2 factorial experiment are depicted graphically as shown.

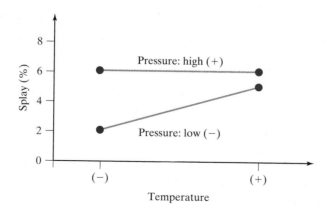

(a) Write down the design matrix, and indicate the numerical value of the four test results.

(b) Calculate E_1, E_2, and E_{12}.

(c) Interpret, in your own words, the two-factor interaction (i.e., what is the specific nature of the pressure–temperature dependency?).

17.26 Use the four responses (5, 10, 15, 20) of a 2^2 factorial design to produce the three different result scenarios defined below. Place the response values at the corners of the squares to indicate the experimental results that will yield the effects noted in each case. For each of the three scenarios, each response is used only once.

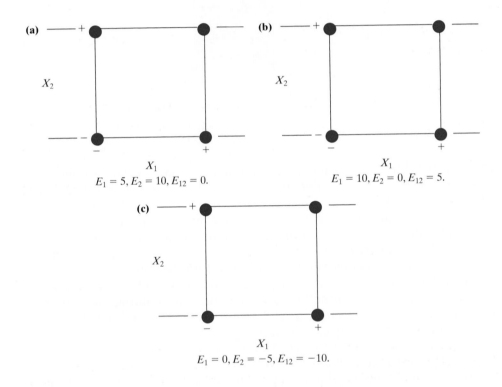

(a)

X_2

X_1

$E_1 = 5, E_2 = 10, E_{12} = 0.$

(b)

X_2

X_1

$E_1 = 10, E_2 = 0, E_{12} = 5.$

(c)

X_2

X_1

$E_1 = 0, E_2 = -5, E_{12} = -10.$

17.27 The results of a 2^3 factorial design are shown in the table below.

Test	1	2	3	y (ksi)
1	−	−	−	87.5
2	+	−	−	87.3
3	−	+	−	77.8
4	+	+	−	87.0

(Continued)

Test	1	2	3	y (ksi)
5	−	−	+	79.1
6	+	−	+	97.6
7	−	+	+	78.6
8	+	+	+	87.7

(a) Provide the geometrical representation for this experiment. Be sure to label properly all relevant entities.

(b) Using the geometric representation, estimate the main effect of variable 2.

(c) Using the geometric representation, estimate the two-factor interaction effect E_{13}.

17.28 A 2^3 factorial design yielded the following test results, given in standard order:

$$48.8, 91.6, 40.8, 60.4, 48.8, 89.2, 40.4, 60.0.$$

(a) Determine the estimate of the two-factor interaction effect E_{13}.

(b) Construct a two-way diagram for the interpretation of this interaction effect.

(c) What is this two-way diagram telling us about the nature of this interaction?

17.29 The model $\mathbf{y} = \mathbf{Xb} + \epsilon$ is to be fit by the method of least squares to the experimental results given by $\mathbf{y}^T = (16, 27, 21, 34)$, where $\mathbf{X}$ is:

$$\begin{bmatrix} +1 & -1 & -1 & +1 \\ +1 & +1 & -1 & -1 \\ +1 & -1 & +1 & -1 \\ +1 & +1 & +1 & +1 \end{bmatrix}$$

(a) Determine the estimates of the model parameters.

(b) Prepare a graph that displays the behavior of the fitted model.

17.30 The data in the accompanying table were collected using a two-level factorial design. The influence of two factors on the temperature reached in a certain chemical reaction is sought via the establishment of a math model for the temperature response. Values for x_1 and x_2 are the concentrations (in percent) of two of the reactants.

Test	x_1	x_2	Temperature (°C)
1	10	5	113
2	15	5	141
3	10	7	90
4	15	7	161

(a) Use the least-squares criterion to fit the most general math model consistent with the design of experiment. Show all of your work, clearly defining all relevant quantities.

(b) Provide the results in the form of the factor effects, E. Use a graphical representation to interpret the two-factor interaction effect.

(c) List the basic stages of the general model-building procedure we have discussed in the book. Briefly define and explain each stage.

18

Analysis of Two-Level Factorial Designs

18.1   Introduction

This chapter takes an in-depth look at the analysis of the results of two-level factorial designs. The emphasis is placed on ways to evaluate the level of noise (experimental error) inherent in the environment in which the tests were conducted, and how this noise influences the *effect estimates*. We would like to distinguish those effects that are truly important from those that are not, in light of the experimental error in the testing environment. It will be seen that experimental replication provides one method by which the important variable effects may be identified. For unreplicated experiments, two techniques are presented that may also be used to distinguish significant effects from those that have arisen solely due to the forces of error.

Chapter 17 presented a detailed account of the calculation and interpretation of variable effects for a chemical calibration problem. At one point we simply assumed certain main and interaction effects to be unimportant because their calculated magnitudes were relatively small. We know that if the entire experiment were repeated (another run for each of the eight unique test conditions), the results and hence the effect estimates would be somewhat different. The question is, how different? To answer this question in a somewhat general way, consider the following.

If we ran the experiment five times and for E_{pump} (the main effect of pump setting) obtained the results

$$533 \text{ psi}, \quad 496 \text{ psi}, \quad 524 \text{ psi}, \quad 509 \text{ psi}, \quad 477 \text{ psi},$$

our attitude would be to consider the effect of pump setting as important—real. It would be considered important because all five effect values are comparable and their average is quite different from zero relative to the variability among them. However, if we ran the experiment five times and for E_{pump} got the results

$$150 \text{ psi}, \quad -420 \text{ psi}, \quad 50 \text{ psi}, \quad 525 \text{ psi}, \quad -295 \text{ psi},$$

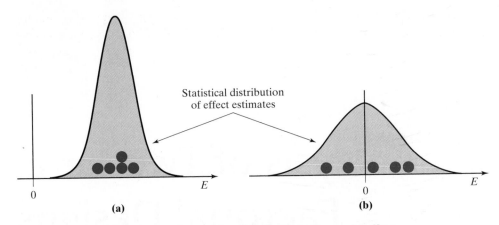

Figure 18.1   **Graphical depiction of the distribution of sample effects**

our attitude would be quite different concerning the importance of the pump effect, because these five values are widely different and average out close to zero. Figure 18.1 is a graphical depiction of these two hypothetical situations.

In the second case [Fig. 18.1(b)], some individual effect estimates are quite different from zero, but as a group their average is nearly zero. In some cases increasing pump setting increases impingement pressure, while in other cases it decreases pressure. On the other hand, for the distribution of sample effects that covers the effect estimates in Fig. 18.1(a), zero looks to be a quite implausible value to arise as a sample effect. That is, the sample effects, as a group, appear to be distinctly different from zero.

We must have some measure of the effect variability before we can make any judgment about the importance of an effect. Since the variations in effects are the result of experimental errors in the individual responses, it is best to have an estimate of the variance of the responses and then develop from it a variance estimate for the effects. This requires replicated runs of at least some of the test settings. Unfortunately, often we can run the experiment only once (without any replications), particularly when the design calls for a large number of tests. For these unreplicated designs, we will have to develop an estimate of the effect variance from the observed effects directly, usually with additional assumptions. In this chapter we discuss methods of judging the importance of effects for both replicated and unreplicated experiments.

18.2   Analysis of Replicated 2^k Experiments

To judge the relative importance of a variable effect based on our calculated main and interaction effects, we can make use of test replications to estimate the error we expect to see in an effect estimate. In particular, we will

1. Estimate the experimental error via the variance of individual observations.
2. Estimate the variance associated with the main or interaction effects, which are simply a linear combination of individual observations.

3. Construct the statistical distribution of effect estimates and examine the range of estimates that could reasonably arise, say, if the true effect were really zero. (Due to the central limit theorem, the distribution of effect estimates is generally assumed to be normal, as in the two illustrations of Fig. 18.1.)

Glove Box Door Alignment Experiment

To present the method of analysis for replicated two-level factorial experiments, we will use an automobile glove box door alignment study. An automobile assembly plant using an automated fixture to center a glove box door to the instrument panel was experimenting with ways to improve the door parallelism during assembly. The actual assembly operation at the time was producing an average deviation of 2 millimeters off-parallel, with some assemblies exceeding 4 millimeters. The purpose of the study was to find better ways to set up the fixtures to improve the door parallelism measure to within ±1.5 millimeters. At one phase of this study, the following four variables were examined over two levels each to investigate their possible effects on parallelism in millimeters.

Variable	Low (−)	High (+)
x_1: RH cowl fore/aft movement	Nominal	−5 mm
x_2: Center brace attachment sequence	Before	After
x_3: Plenum gasket	No	Yes
x_4: Evaporator case setup, fore/aft	Nominal	−5 mm

To examine the possible influence of these factors on parallelism, the engineers decided to use a 2^4 design with each test replicated once. The total of 32 tests were conducted in a random order. The results are given in Table 18.1.

   To obtain the effect estimates, we first augment the design matrix of Table 18.1, with columns for the two-, three-, and four-factor interactions. These new columns are formed from the design matrix by taking all possible products of the main variable columns two at a time, three at a time, and four at a time. The resulting matrix is referred to as the *calculation matrix* and also contains a column of all "+1" values for the average response. This additional column is generally denoted as the I or identity column. The calculation matrix should be a square matrix; that is, the number of columns/effects is equal to the number of rows/test conditions. The calculation matrix with experimental results of the 2^4 design used for the glove box door study is shown in Table 18.2. The last two columns in Table 18.2 are the average response, $\bar{y}_i = (y_{i1} + y_{i2})/2$, and the difference $d_i = (y_{i1} - y_{i2})$ for each of the $i = 1, 2, 3, \ldots, 16$ unique test conditions. The differences in the last column of Table 18.2 will be used later to estimate the experimental error variance for an individual test observation.

TABLE 18.1	2^4 Design and Results of the Glove Box Door Experiment (Test Order in Parentheses)

| | | | | | Parallelism (mm) | |
| | | | | | Run 1 | Run 2 |
Test	x_1	x_2	x_3	x_4	y_{i1}	y_{i2}
1	−	−	−	−	−1.44 (7)	−0.08 (28)
2	+	−	−	−	−1.79 (10)	−1.01 (24)
3	−	+	−	−	0.39 (14)	0.17 (32)
4	+	+	−	−	−0.50 (2)	−0.24 (21)
5	−	−	+	−	−0.20 (9)	0.17 (27)
6	+	−	+	−	−0.79 (6)	−0.64 (30)
7	−	+	+	−	1.22 (13)	0.28 (20)
8	+	+	+	−	0.21 (8)	0.28 (18)
9	−	−	−	+	−0.40 (1)	−0.65 (31)
10	+	−	−	+	−0.63 (15)	−1.19 (25)
11	−	+	−	+	0.47 (3)	0.44 (17)
12	+	+	−	+	−0.01 (5)	−0.03 (23)
13	−	−	+	+	1.29 (12)	0.64 (29)
14	+	−	+	+	−1.17 (4)	0.14 (19)
15	−	+	+	+	0.48 (16)	1.06 (22)
16	+	+	+	+	0.40 (11)	0.34 (26)

TABLE 18.2	Calculation Matrix of the Glove Box Door Experiment

Test	I	1	2	3	4	12	13	14	23	24	34	123	124	134	234	1234	$\bar{y}_i$	d_i
1	+	−	−	−	−	+	+	+	+	+	+	−	−	−	−	+	−0.76	−1.36
2	+	+	−	−	−	−	−	−	+	+	+	+	+	+	−	−	−1.40	−0.78
3	+	−	+	−	−	−	+	+	−	−	+	+	+	−	+	−	0.28	0.22
4	+	+	+	−	−	+	−	−	−	−	+	−	−	+	+	+	−0.37	−0.26
5	+	−	−	+	−	+	−	+	−	+	−	+	−	+	+	−	−0.02	−0.37
6	+	+	−	+	−	−	+	−	−	+	−	−	+	−	+	+	−0.72	−0.15
7	+	−	+	+	−	−	−	+	+	−	−	−	+	+	−	+	0.75	0.94
8	+	+	+	+	−	+	+	−	+	−	−	+	−	−	−	−	0.25	−0.07
9	+	−	−	−	+	+	+	+	−	+	−	−	+	+	+	−	−0.53	0.25
10	+	+	−	−	+	−	−	+	+	−	−	+	−	−	+	+	−0.91	0.56
11	+	−	+	−	+	−	+	−	+	−	+	−	+	−	−	+	0.46	0.03
12	+	+	+	−	+	+	−	+	−	+	−	−	−	+	−	−	−0.02	0.02
13	+	−	−	+	+	+	−	−	−	−	+	+	+	−	−	+	0.97	0.65
14	+	+	−	+	+	−	+	+	−	−	+	−	−	+	−	−	−0.52	−1.31
15	+	−	+	+	+	−	−	−	+	+	+	−	−	−	+	−	0.77	−0.58
16	+	+	+	+	+	+	+	+	+	+	+	+	+	+	+	+	0.37	0.06

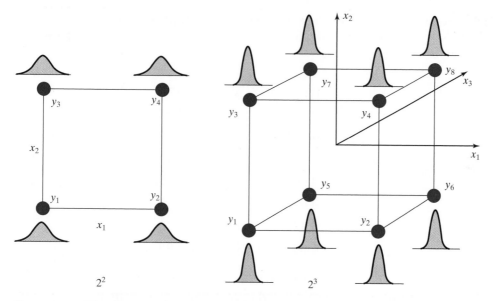

Figure 18.2   Distribution of responses on factorial designs

Estimation of Variance of an Individual Observation

Let us assume that the amount of error in an observation is the same for all observations (i.e., that the true variance is the same for all observations) and that the observations are independent. In Chapter 19 we check this assumption of equal variance through a statistical test. Figure 18.2 depicts this situation for 2^2 and 2^3 designs. When we calculate location effects (main and interaction effects on the mean response) we are really determining the extent to which the mean level of the response distribution is changing from one level of a variable to another. However, if the variance of the response distribution is dependent on one of the variables, we say that there exists a significant dispersion effect on the response. Dispersion effects or effects on variation of the response have received some attention in the technical literature.[1]

We will now describe a procedure for the estimation of the variance of an individual observation under the assumption of a common variance for all of the responses. For test condition 1 of the glove box door parallelism experiment, the two replicated observations are -1.44 and -0.08 millimeters. The sample variance for this test condition, designated as s_1^2, can be calculated as

$$s_1^2 = \frac{(y_{11} - \bar{y}_1)^2 + (y_{12} - \bar{y}_1)^2}{2 - 1}$$
$$= [-1.44 - (-0.76)]^2 + [-0.08 - (-0.76)]^2$$
$$= 0.9248.$$

[1] G. E. P. Box and R. D. Meyer, "An Analysis for Unreplicated Fractional Factorials" and "Dispersion Effects from Fractional Designs," *Technometrics*, Vol. 28, 1986, pp. 11–18 and 19–27.

In this example, we can calculate 16 independent sample variances $s_1^2, s_2^2, \ldots, s_{16}^2$, one for each unique test condition of the design. The 16 sample variances have been calculated to be

$$
\begin{aligned}
s_1^2 &= 0.92480, & s_9^2 &= 0.03125, \\
s_2^2 &= 0.30420, & s_{10}^2 &= 0.15680, \\
s_3^2 &= 0.02420, & s_{11}^2 &= 0.00045, \\
s_4^2 &= 0.03380, & s_{12}^2 &= 0.00020, \\
s_5^2 &= 0.06845, & s_{13}^2 &= 0.21125, \\
s_6^2 &= 0.01125, & s_{14}^2 &= 0.85805, \\
s_7^2 &= 0.44180, & s_{15}^2 &= 0.16820, \\
s_8^2 &= 0.00245, & s_{16}^2 &= 0.00180.
\end{aligned}
$$

Since we are assuming that there is a common variance for all 32 observations, each sample variance calculated is an independent estimate of that common variance, σ_y^2. A more precise, single estimate of this common variance may be obtained by combining, or pooling, the 16 independent estimates of the variance. The pooled sample variance, s_p^2, is, in general, given by

$$
s_p^2 = \frac{\nu_1 s_1^2 + \nu_2 s_2^2 + \cdots + \nu_m s_m^2}{\nu_1 + \nu_2 + \cdots + \nu_m} = \frac{\sum\limits_{i=1}^{m} \nu_i s_i^2}{\sum\limits_{i=1}^{m} \nu_i}, \tag{18.1}
$$

where ν_1, ν_2, and so on, are the degrees of freedom associated with the variance estimates s_1^2, s_2^2, and so on, and m is the number of independent estimates of the common variance σ_y^2 that are available. Determining the pooled variance simply amounts to calculating s_p^2 as a weighted average of the individual s_i^2, where the weights are the degrees of freedom associated with each estimate, s_i^2. When all tests are replicated n times each (i.e., $n_1 = n_2 = \cdots = n_m = n$), the formula for s_p^2 reduces to

$$
s_p^2 = \frac{s_1^2 + s_2^2 + \cdots + s_m^2}{m}. \tag{18.2}
$$

If the number of replicates is equal to two for all tests (i.e., $n = 2$), a convenient form for s_i^2 may be developed in terms of the difference $d_i = (y_{i1} - y_{i2})$:

$$
s_i^2 = \frac{\left(y_{i1} - \dfrac{y_{i1} + y_{i2}}{2}\right)^2 + \left(y_{i2} - \dfrac{y_{i1} + y_{i2}}{2}\right)^2}{2 - 1} = \left(\frac{d_i}{2}\right)^2 + \left(\frac{d_i}{2}\right)^2 = \frac{d_i^2}{2}. \tag{18.3}
$$

Using this relationship, we can express the pooled variance for this particular case as

$$s_p^2 = \sum_{i=1}^{m} \frac{d_i^2/2}{m}. \tag{18.4}$$

For the case at hand, therefore, using Eq. (18.2), the pooled variance is

$$s_p^2 = \frac{s_1^2 + s_2^2 + \cdots + s_{16}^2}{16} = \frac{0.9248 + \cdots + 0.0018}{16}$$

$$= 0.20242,$$

or alternatively, using Eq. (18.4) and the d_i values in Table 18.2, it is

$$s_p^2 = \frac{d_1^2/2 + \cdots + d_{16}^2/2}{16} = \frac{(-1.36)^2 + \cdots + (0.06)^2}{32}$$

$$= 0.20242.$$

Estimation of the Variances Associated with the Main Effects and Interaction Effects

Given the data of Table 18.2, we find that the main effect of variable 1, E_1, is

$$E_1 = \left(\tfrac{1}{8}\right)[(\bar{y}_2 - \bar{y}_1) + (\bar{y}_4 - \bar{y}_3) + \cdots + (\bar{y}_{14} - \bar{y}_{13}) + (\bar{y}_{16} - \bar{y}_{15})].$$

Since each $\bar{y}_i$ is an average of two observations y_{i1} and y_{i2}, E_1 can be written as

$$E_1 = \frac{\dfrac{y_{2,1} + y_{2,2}}{2} - \dfrac{y_{1,1} + y_{1,2}}{2} + \cdots + \dfrac{y_{16,1} + y_{16,2}}{2} - \dfrac{y_{15,1} + y_{15,2}}{2}}{8}$$

or

$$E_1 = \frac{y_{2,1} + y_{2,2} - y_{1,1} - y_{1,2} + \cdots + y_{16,1} + y_{16,2} - y_{15,1} - y_{15,2}}{16}.$$

Thus we see that an effect estimate is a linear combination of $(n \cdot m = 32)y_{ij}$ values.

By using the calculation matrix and the algebraic method of effect calculation, all the estimated effects for the glove box parallelism experiment were calculated, and are (in millimeters)

$$\text{average} = -0.087,$$
$$E_1 = -0.654, \quad E_3 = 0.638, \quad E_{12} = 0.147,$$
$$E_2 = 0.794, \quad E_4 = 0.322, \quad E_{13} = -0.117,$$

$$E_{14} = -0.031, \quad E_{34} = 0.009, \quad E_{134} = -0.138,$$
$$E_{23} = -0.191, \quad E_{123} = 0.172, \quad E_{234} = -0.104,$$
$$E_{24} = -0.154, \quad E_{124} = 0.101, \quad E_{1234} = 0.121.$$

Under the assumption that the observations y_{ij} are independent, each with an unknown but common variance σ_y^2, the variance of E_1, $\text{Var}(E_1)$, is

$$\text{Var}(E_1) = \text{Var}\left(\frac{y_{2,1} + y_{2,2} - y_{1,1} - y_{1,2} + \cdots + y_{16,1} + y_{16,2} - y_{15,1} - y_{15,2}}{16} \right)$$

$$= \left(\frac{1}{256} \right) [\text{Var}(y_{2,1}) + \text{Var}(y_{2,2}) + \cdots + \text{Var}(y_{15,1}) + \text{Var}(y_{15,2})]$$

$$= \left(\frac{1}{256} \right) (\sigma_y^2 + \sigma_y^2 + \cdots + \sigma_y^2 + \sigma_y^2)$$

$$= \frac{32}{256}\sigma_y^2 = \frac{4\sigma_y^2}{32} = \frac{\sigma_y^2}{8},$$

where the variance of an observation σ_y^2 is to be estimated by the sample (pooled) variance, s_p^2.

Since each effect is a linear combination of the same set of 32 responses, the variances of all other main effects and the interaction effects are the same as the variance of E_1:

$$\text{Var}(E_1) = \text{Var}(E_2) = \text{Var}(E_3) = \text{Var}(E_4) = \text{Var}(E_{12}) = \text{Var}(E_{13})$$
$$= \text{Var}(E_{14}) = \text{Var}(E_{23}) = \text{Var}(E_{24}) = \text{Var}(E_{34}) = \text{Var}(E_{123})$$
$$= \text{Var}(E_{124}) = \text{Var}(E_{134}) = \text{Var}(E_{234}) = \text{Var}(E_{1234})$$
$$= \frac{\sigma_y^2}{8}.$$

Substituting for σ_y^2 the pooled sample variance, s_p^2, we obtain the sample variance of the effects for this case as

$$s_{\text{effect}}^2 = \frac{s_p^2}{8}.$$

In general, an effect estimate is determined from the equation

$$E = \frac{2}{N}(\pm y_1 \pm y_2 \pm \cdots \pm y_N), \tag{18.5}$$

where N is the total number of tests conducted, including all replicates. Therefore, the variance of an effect estimate is given as

$$\text{Var}(E) = \text{Var}\left[\frac{2}{N}(\pm y_1 \pm y_2 \pm \cdots \pm y_N)\right] = \frac{4}{N^2}(N\sigma_y^2) = \frac{4\sigma_y^2}{N}. \quad (18.6)$$

Thus, in general, the sample variance of an effect is calculated by

$$s_{\text{effect}}^2 = \frac{4s_p^2}{N}.$$

For the glove box door experiment, $N = 32$ (16 unique test conditions replicated once) and $s_p^2 = 0.20242$. The estimated variance of an effect is therefore

$$s_{\text{effect}}^2 = \frac{4s_p^2}{32}$$
$$= \frac{s_p^2}{8}$$
$$= \frac{0.20242}{8}$$
$$= 0.0253.$$

The square root of s_{effect}^2 is sometimes called the standard error (s.e.) of an effect, which will be used for significance tests applied on the effects. For our example,

$$\text{s.e.} = s_{\text{effect}} = 0.159 \text{ mm}.$$

The variance of the average response is the familiar form:

$$\text{Var}(\text{average}) = \text{Var}\left(\frac{y_1 + y_2 + \cdots + y_N}{N}\right) = \frac{1}{N^2}(N\sigma_y^2) = \frac{\sigma_y^2}{N}. \quad (18.7)$$

Thus the sample variance of the average, s_{ave}^2, is s_p^2/N. For our example, the numerical value for s_{ave} is 0.0795, which is observed to be one-half of the standard error for an effect.

18.3   Determination of Statistically Significant Effects

Hypothesis Testing Formulation

Recall that based on the test results from the 2^4 glove box door alignment experiment, the following effect estimates were calculated:

$$E_1 = -0.654, \quad E_3 = 0.638, \quad E_{12} = 0.147,$$
$$E_2 = 0.794, \quad E_4 = 0.322, \quad E_{13} = -0.117,$$

$$E_{14} = -0.031, \qquad E_{34} = 0.009, \qquad E_{134} = -0.138,$$
$$E_{23} = -0.191, \qquad E_{123} = 0.172, \qquad E_{234} = -0.104,$$
$$E_{24} = -0.154, \qquad E_{124} = 0.101, \qquad E_{1234} = 0.121.$$

It is desired at this point to assess which of these estimates, if any, are important—that is, statistically significant.

In Chapter 4 the concept of hypothesis testing was developed in a general context and was subsequently used to formulate the Shewhart control chart model. The hypothesis testing idea may also be applied in the situation at hand to judge whether any of the calculated effect estimates are important.

In conducting a hypothesis test on the effect estimates above, we will assume that the true effect values are zero, that is, $\mu_{\text{effect}} = 0.0$. This assumption serves as our null hypothesis (H_0) and is depicted in Fig. 18.3. If a calculated effect estimate is not consistent with H_0, we will reject H_0 and accept the alternative hypothesis (H_1) that the true value for that effect is nonzero. Let us also assume that the effect estimates are normally distributed and that their variability is characterized by σ^2_{effect}. The assumption of normality is not unreasonable, since the estimates are a linear combination of N responses [refer to Eq. (18.5)], and such linear combinations are approximately normally distributed due to the central limit theorem.

As we saw in Chapter 4, to calculate the probability of obtaining a given value for E_i or one larger (or smaller), requires that we perform a transformation to the standard normal distribution,

$$Z = \frac{E_i - \mu_{\text{effect}}}{\sigma_{\text{effect}}}.$$

Generally, we do not know the true variance of an effect, σ^2_{effect}, but we may estimate it with s^2_{effect}. We might, therefore, consider performing the following transformation:

$$\frac{E_i - \mu_{\text{effect}}}{s_{\text{effect}}}.$$

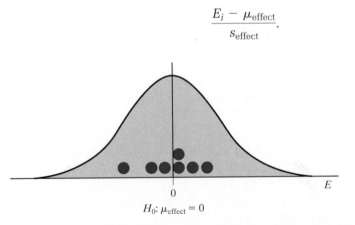

$H_0: \mu_{\text{effect}} = 0$

Figure 18.3 **Distribution of the effect estimates under the null hypothesis**

TABLE 18.3	Effect Estimates and Associated t Values for the Glove Box Door Alignment Study				
Effect	Effect Estimate	Associated t Value	Effect	Effect Estimate	Associated t Value
E_1	−0.654	−4.1132	E_{24}	−0.154	−0.9686
E_2	0.794	4.9937	E_{34}	0.009	0.0566
E_3	0.638	4.0126	E_{123}	0.172	1.0818
E_4	0.322	2.0252	E_{124}	0.101	0.6352
E_{12}	0.147	0.9245	E_{134}	−0.138	−0.8679
E_{13}	−0.117	−0.7358	E_{234}	−0.104	−0.6541
E_{14}	−0.031	−0.1950	E_{1234}	0.121	0.7610
E_{23}	−0.191	−1.2013			

The quantity given above no longer follows the standard normal distribution; rather, it follows the t-distribution that was introduced in Chapter 4.

For our hypothesis test on the estimated effects, we may calculate the t-statistic for each estimate, under H_0: $\mu_{effect} = 0$, using the transformation

$$t = \frac{E_i - \mu_{effect}}{s_{effect}}$$

$$= \frac{E_i - 0.0}{0.159}.$$

Note that the estimate of the true standard deviation of the effects used in this transformation is the standard error of the effects calculated previously, $s_{effect} = 0.159$ mm. Note further that this estimate of the standard deviation has $\nu = 16$ degrees of freedom (each unique test condition supplies an estimate of variance with 1 degree of freedom, and the pooled variance based on these 16 estimates then has 16 degrees of freedom). The degrees of freedom associated with a particular t statistic are always defined by the degrees of freedom associated with the variance estimate. The t values associated with the effect estimates are given in Table 18.3. For $E_1 = -0.654$,

$$t = \frac{-0.654 - 0}{0.159}$$

$$= -4.1132.$$

From Table A.3 it is seen that for a t-distribution with $\nu = 16$ degrees of freedom, values of t such that the probability of occurrence of that value or one more extreme (larger or smaller) is 0.025 are

$$t_{16,0.025} = -2.120,$$

$$t_{16,0.975} = 2.120,$$

where the subscripts on t represent the degrees of freedom and the area under the t-distribution to the left of that value, respectively. Thus any calculated t value with an absolute value greater than 2.120 is judged to be too large to have arisen under H_0. The risk that we will reject H_0 here, when it is true, is 0.05.

Upon examining Table 18.3, we see that the t values associated with E_1, E_2, and E_3 are greater in absolute value than $t_{16,0.975} = 2.120$; thus we must reject the null hypothesis for those effects. We conclude, therefore, that the true means of the effect values for E_1, E_2, and E_3, namely μ_{E_1}, μ_{E_2}, and μ_{E_3}, are nonzero. Figure 18.4 depicts the situation at hand. Note that we cannot infer from the results that the other effects are unimportant, only that we have no reason to doubt the null hypothesis.

Calculation of Confidence Intervals for the True Mean Effects

Through the hypothesis testing procedure described previously, the main effects of variables 1, 2, and 3 have been identified as important. Alternatively, we might say that the true mean effect values μ_{E_1}, μ_{E_2}, and μ_{E_3} are nonzero. The effect estimates obtained from the experiment, $E_1 = -0.654$, $E_2 = 0.794$, and $E_3 = 0.638$, are simply values that have been drawn from distributions centered at μ_{E_1}, μ_{E_2}, and μ_{E_3}, respectively. Although the hypothesis test above has led us to believe that these three μ_{E_i} are nonzero, one might ask more specifically: What value or values for μ_{E_i} are likely, given the data?

Unfortunately, we cannot determine exact values for the true mean effect values. We can, however, estimate them using the effect estimates. Specifically, $E_1 = -0.654$ is an estimate of μ_{E_1}, $E_2 = 0.794$ is an estimate of μ_{E_2}, and $E_3 = 0.638$ is an estimate of μ_{E_3}. We say that such effect estimates are point estimates of the true mean effect values. In short, the effect estimates are the best single values that we have for the true mean effects.

Although we recognize that the calculated effect value is the best estimate that we have for a true mean of an effect, we also recognize that the true mean of an effect may be larger or smaller than the estimate. We might therefore ask: How small (or large) could the true mean of an effect be? To answer this question, let us consider posing a succession of null hypotheses where μ_{effect} takes on values in the vicinity of E_i. The smallest and largest values for μ_{effect} for

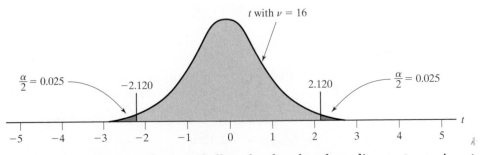

Figure 18.4 t test for significance of effects for glove box door alignment experiment

which the null hypothesis cannot be rejected (i.e., for which the observed effect estimate is a plausible realization) are given by

$$Z_{1-\alpha/2} = \frac{E_i - \mu_{low}}{\sigma_{effect}}, \quad Z_{\alpha/2} = \frac{E_i - \mu_{high}}{\sigma_{effect}}.$$

With a bit of algebra, we can rewrite the above as

$$\mu_{low} = E_i - Z_{1-\alpha/2}(\sigma_{effect}), \quad \mu_{high} = E_i - Z_{\alpha/2}(\sigma_{effect}),$$

or, simplifying,

$$\mu_{low}, \mu_{high} = E_i \pm Z_{1-\alpha/2}(\sigma_{effect}). \tag{18.8}$$

The situation depicted mathematically by Eq. (18.8) is illustrated graphically in Fig. 18.5. Equation (18.8) is termed an interval estimate for the true mean of an effect μ_{effect} and is centered at the point estimate, E_i. As we saw in our development of the hypothesis testing idea, we must be willing to accept some risk in the expression of our interval estimate. This risk arises from the fact that the calculated effect estimate, E_i, may not be a typical realization from the distribution of effect estimates centered at μ_{effect}. This risk is expressed in the present context by referring to the interval estimate as being a *confidence interval*. Thus Eq. (18.8) should be properly referred to as a $100(1 - \alpha)\%$ confidence interval for the true mean of an effect. Recall we first introduced this concept in Chapter 4.

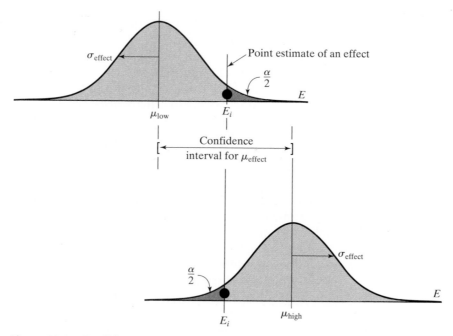

Figure 18.5   Confidence interval idea

In developing the confidence interval of Eq. (18.8) it may be observed that σ_{effect} was assumed to be known. As we saw previously, we may replace σ_{effect} with its estimate, s_{effect}, to produce the following form of the confidence interval:

$$\mu_{\text{low}}, \mu_{\text{high}} = E_i \pm t_{\nu, 1-\alpha/2} s_{\text{effect}}. \tag{18.9}$$

Note that the replacement of σ_{effect} with s_{effect} forces us to replace $Z_{1-\alpha/2}$ with $t_{\nu, 1-\alpha/2}$.

Confidence intervals may now be developed for all the true mean effects based on the calculated effect estimates $(E_1, E_2, \ldots, E_{1234})$ and the standard error of the effects $(s_{\text{effect}} = 0.159, \nu = 16)$. For a $100(1 - \alpha)\%$ confidence interval with $\alpha = 0.05$ (95% confidence interval), $t_{16, 0.975}$ is 2.120. The 95% confidence interval for μ_{effect} is, therefore,

$$E_i \pm t_{16, 0.975} s_{\text{effect}}$$
$$E_i \pm (2.120)(0.159)$$
$$E_i \pm 0.337.$$

As an example, the 95% confidence interval for the true mean effect of variable 1, the fore/aft movement of the righthand cowl, on parallelism (in millimeters) is

$$E_1 \pm 0.337$$
$$-0.654 \pm 0.337$$
$$-0.991, -0.317.$$

Thus we are 95% confident that the true mean effect of variable 1 lies on the interval from -0.991 to -0.317. Table 18.4 gives the 15 effect estimates together with the corresponding 95% confidence intervals for their true effect values.

A confidence interval may also be used to judge the statistical significance of variable effects. We will reject the null hypothesis, H_0: true mean effect $= 0$, if the confidence interval does not include "zero." In other words, we will conclude that an effect is statistically nonzero if the confidence-interval values are strictly either positive or negative. From Table 18.4 it would appear that only three of the estimated effects are important based on a 95% confidence level for their true mean effect values. The confidence intervals associated with significant variable effects, E_1, E_2, and E_3, are identified by an asterisk in Table 18.4. It should come as no surprise that the conclusions made as a result of hypothesis testing are exactly the same as those made based on this confidence-interval approach. The two approaches essentially supply identical information concerning the *significance* of variable effects.

Graphically, we can present these confidence intervals as shown in Fig. 18.6. From the figure we see that only the first three confidence intervals do not include zero, which was the hypothesized true mean for all the main and interaction effects under H_0.

In addition to the confidence intervals developed for each of the true mean effects above, one may also determine such an interval for the mean response.

TABLE 18.4	95% Confidence Intervals for True Mean Effects of the Glove Box Door Alignment Study Based on Replicated Experiment

Main Effects	95% Confidence Interval
RH cowl fore/aft (E_1)	-0.654 ± 0.337 mm[*]
Center brace (E_2)	0.794 ± 0.337 mm[*]
Plenum gasket (E_3)	0.638 ± 0.337 mm[*]
Evaporator case (E_4)	0.322 ± 0.337 mm

Two-Variable Interactions	95% Confidence Interval
RH cowl × center brace (E_{12})	0.147 ± 0.337 mm
RH cowl × plenum gasket (E_{13})	-0.117 ± 0.337 mm
RH cowl × evaporator case (E_{14})	-0.031 ± 0.337 mm
Center brace × plenum gasket (E_{23})	-0.191 ± 0.337 mm
Center brace × evaporator case (E_{24})	-0.154 ± 0.337 mm
Plenum gasket × evaporator case (E_{34})	0.009 ± 0.337 mm

Three-Variable Interaction	95% Confidence Interval
Cowl × brace × plenum (E_{123})	0.172 ± 0.337 mm
Cowl × brace × evaporator (E_{124})	0.101 ± 0.337 mm
Cowl × plenum × evaporator (E_{134})	-0.138 ± 0.337 mm
Brace × plenum × evaporator (E_{234})	-0.104 ± 0.337 mm

Four-Variable Interaction	95% Confidence Interval
Cowl × brace × plenum × evaporator (E_{1234})	0.121 ± 0.337 mm

[*]Confidence interval shows significant effect.

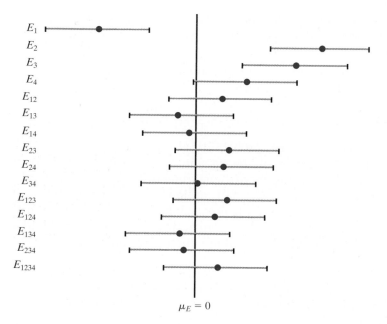

Figure 18.6   95% confidence intervals for the true mean effects (only the intervals based on E_1, E_2, and E_3 do not include zero)

Such an interval may be of special concern if it is desired to ascertain if the average response $(\hat{b}_0)$ is to be included in a math model such as that developed in Chapter 17. Therefore, we may wish to check to see if the calculated estimate for the average or mean response could have been a sample observation drawn from a distribution with a true mean equal to zero. Our calculated average parallelism for the 32 responses is -0.087 millimeter. The corresponding 95% confidence interval for the mean is then

$$\text{average} \pm t_{16,0.975}\frac{s_p}{\sqrt{32}}$$

$$-0.087 \pm 0.1685,$$

which shows that the true mean parallelism is not significantly different from zero. Hence if it were desired to write down a math model for parallelism, based on the results of the experiment, the $\hat{b}_0$ term would be left out.

18.4   Analysis of the Results of Unreplicated Two-Level Factorials

There are times when it is neither feasible nor desirable (due to perhaps very limited resources or due to a large number of variables) to include replication in a two-level factorial design. In such cases it is not possible to obtain an independent estimate of the experimental error for judging the relative importance of the main and interaction effects. There are, however, at least two methods that can be employed to aid in making such decisions:

1. The use of normal probability plots of the effect estimates to isolate the important effects from those that appear to be driven by the forces of experimental error alone.

2. The use of higher-order interaction effect estimates to estimate directly the variance of the effects for significance testing.

Normal Probability Plot to Analyze Unreplicated Designs

We saw in Chapter 3 that the normal probability plot is a graphical method to check whether or not a set of data could be considered as arising from a normal distribution. If a normal probability plot is constructed for data drawn from a single normal distribution, the data will fall randomly along/about a straight line. In addition, if all the data in a sample do come from the same normal distribution with the exception of a few data points, the normal probability plot will show these exceptional points to distribute themselves off a line that passes through the other points.

To use a normal probability plot to judge the importance of effects estimated from two-level factorial experiments, we make use of two important facts surrounding the situation at hand:

1. Since effect estimates are linear combinations of the individual response values, they tend to be normally distributed due to the central limit theorem.

2. Our purpose in making a test of significance for the effect estimates is to sort them into two piles: those that are insignificant (their true mean values are zero) and those that are significant (their true mean values are nonzero).

In addition, the variance of the effect values should be equal, an assumption we have made earlier. The first fact allows us to use normal probability plots as a means to evaluate the effect estimates, while the second, when taken with the first, means that all the insignificant effect estimates should fall on a straight line on the plot centered near zero.

We will illustrate this method using the data from the glove box door parallelism study. We will assume now that we have no way of estimating the variance of an individual response value.

Following the procedure outlined in Chapter 3, we need to first arrange the effect estimates in ascending order and estimate their corresponding cumulative probabilities, using

$$p_i = \frac{100(i - 1/2)}{2^4 - 1}, \quad i = 1, 2, \ldots, (2^4 - 1).$$

Table 18.5 contains all the information required to make the normal probability plot, and Fig. 18.7 gives the resulting normal plot. In the figure, a straight line has been drawn through a large number of the effect estimates, passing near the (0, 50%) point such that their scatter about this line is fairly random.

TABLE 18.5	Data Points for the Normal Probability Plot (Glove Box Door Parallelism Study)		
Order Number, i	Effect Estimates, E_i	Identity of Effects	Cumulative Probability, p_i (%)
1	−0.654	1	3.3
2	−0.191	23	10.0
3	−0.154	24	16.7
4	−0.138	134	23.3
5	−0.117	13	30.0
6	−0.104	234	36.7
7	−0.031	14	43.3
8	0.009	34	50.0
9	0.101	124	56.7
10	0.121	1234	63.3
11	0.147	12	70.0
12	0.172	123	76.7
13	0.322	4	83.3
14	0.638	3	90.0
15	0.794	2	96.7

Figure 18.7   Normal probability plot of the sample effects: glove box door parallelism experiment

It is seen that three effect estimates fall well off this line. We conclude, therefore, that effects 1, 2, and 3 (circled and marked in Fig. 18.7) are important; that is, they are not likely to be arising from a normal distribution with a mean of zero. This conclusion agrees with the one we reached by the method for replicated experiments in the preceding section. This example demonstrates the effectiveness of the use of normal probability plots for analyzing the results of unreplicated experiments.

To ensure an effective application of the normal probability plot for the identification of significant effects, it is recommended that it be used for two-level designs with at least four variables. With less than four variables (e.g., a 2^3 factorial design), if several of the true mean effects are nonzero, it becomes very difficult to decide which set of estimates has arisen from a normal distribution with mean zero.

Being a graphical method, the use of normal probability plots always involves a certain degree of subjective judgment. Sometimes, there may be a few points that are not easy to be ruled as off-line or on-line. If we follow the principle of parsimony in model building, we first only select the points that are obviously well off the line and conduct further math model checks (to be discussed later). If these checks indicate some model inadequacies or a general inability to confirm model predictions through additional testing, other effect estimates close to the line might be included in the model.

Use of Higher-Order Interaction Effects to Estimate Error and Assess the Importance of Variable Effects

The interpretation of the normal plot of estimated effects in Fig. 18.7 suggests that we might be able to obtain an estimate of the standard error of estimated effects from the higher-order interaction effects. Why is this so? The answer comes as we reexamine the normal plot. We see two things:

1. The majority of the insignificant effects that comprise the straight line about zero are interaction effect estimates, and in fact all the third- and higher-order interactions are randomly distributed about this line. This seems to suggest that at some point we might be able to assume that interactions of a certain order and above may be neglected—their true mean values are assumed to be zero.

2. When we collapse the effect estimates that fall on the straight line about zero to a "pile" in a dot diagram form, we can see the amount of scatter (the magnitude of the standard error) we expect to be associated with effect estimates. We actually see a data-based representation of the reference distribution of effects, as shown at the bottom of Fig. 18.7.

Based on the two observations above, let us assume that all third- and higher-order interaction effects are negligible. This means that we are declaring that the true mean effects of interactions 123, 124, 134, 234, and 1234 are zero. With this assumption we can estimate the variance of an effect directly from the formula

$$s_{\text{effect}}^2 = \sum_{\substack{\text{higher-order} \\ \text{interactions}}} \frac{(E_i - \mu_{E_i})^2}{\text{no. of higher-order interactions}}. \tag{18.10}$$

This is a sample variance formula where we can substitute 0 for the true mean effect μ_{E_i} that is assumed to be known. The number of higher-order interactions available in this case is $n = 5$, and we divide by $n = 5$ instead of $n - 1 = 4$, since the mean μ_{E_i} is assumed to be known.

Recall that the effect estimates for the third- and higher-order interactions were

$$E_{123} = +0.172, \quad E_{234} = -0.104, \quad E_{1234} = +0.121.$$
$$E_{124} = +0.101, \quad E_{134} = -0.138,$$

From Eq. (18.10) with $\mu_{E_i} = 0$ we obtain

$$s^2_{\text{effect}} = \frac{[(0.172 - 0)^2 + (0.101 - 0)^2 + (-0.104 - 0)^2 + (-0.138 - 0)^2 + (0.121 - 0)^2]}{5}$$

$$= 0.0168572.$$

Therefore, the estimated standard error of an effect is

$$s_{\text{effect}} = \text{s.e.} = 0.1298.$$

When we use this standard error to construct a 95% confidence interval with $t_{5,0.975} = 2.571$, we have

$$\text{effect estimate} \pm (2.571)(0.1298)$$
$$E_i \pm 0.334.$$

With this confidence interval, our conclusions about the importance of the effects are

- The main effects of variables 1, 2, and 3 are statistically significant.
- The remaining effects are not likely to be important.

Again, the same conclusion was reached previously by the other methods. It is interesting to note that the half-width of the 95% confidence interval calculated from the replicated responses was 0.337, which is about the same as the one above.

■ CHAPTER SUMMARY

In this chapter we have presented techniques for the assessment of important variable effects for both replicated and unreplicated two-level factorial designs. In identifying the important effects, our ultimate goal will be to develop a mathematical model for the process that can then be used for design and/or control purposes. In the next chapter we will discuss the building of such models.

The effects we have been studying up to now describe how variables and interactions between variables affect the centering or location of a distribution of responses. For this reason we term the types of effects described in this chapter as "location effects." In developing techniques to judge location effect importance, we have relied on the assumption that the variance of the observations is constant. Accordingly, we have assumed that none of the variables being studied in the experiment have any effect on the level of noise associated with the response. In the next chapter we check for the possible presence of "dispersion" or variability effects.

EXERCISES

18.1 The yield of a chemical process has been studied as a function of three factors, at the following experimental levels:

Variable	Low Level	High Level
1. Temperature (°C)	80	120
2. Pressure (psi)	50	70
3. Reaction time (min)	5	15

A 2^3 factorial design was performed. Each test, or unique combination of the three variables, was performed three times. The table below provides the results of the tests.

Test	Temperature	Pressure	Time	Trial 1	2	3	Average
1	80	50	5	61.43	58.58	57.07	59.03
2	120	50	5	75.62	77.57	75.75	76.31
3	80	70	5	27.51	34.03	25.07	28.87
4	120	70	5	51.37	48.49	54.37	51.41
5	80	50	15	24.80	20.69	15.41	20.30
6	120	50	15	43.58	44.31	36.99	41.63
7	80	70	15	45.20	49.53	50.29	48.34
8	120	70	15	70.51	74.00	74.68	73.07

(a) Calculate all the main and interaction effects.

(b) Calculate the sample pooled variance, s_p^2, based on the replication within each test run.

(c) Calculate the sample variance of an effect based on the sample pooled variance.

(d) Using a 95% confidence interval, find the statistically significant effects.

18.2 A 2^2 factorial design was conducted to study the effect of welding temperature (degrees Fahrenheit) and flux quantity (inches per minute) on the resulting weld strength. The accompanying table describes the conditions that were examined in the experiment and the associated measured weld strengths (note that each unique combination of temperature and flux quantity was tested three times). The table also shows the average and sample variance for the weld strength calculated for each unique combination of the temperature (variable 1) and flux quantity (variable 2).

Test	Temperature	Flux Quantity	Trial 1	Trial 2	Trial 3	Average	Sample Variance
1	300	10	3088	2724	3586	3132.67	187,257.33
2	500	10	799	1202	1079	1026.67	42,656.33
3	300	20	813	1124	1101	1012.67	30,032.33
4	500	20	2806	2946	3336	3029.33	75,433.33

(a) Calculate the effect estimates (i.e., the average, E_1, E_2, and E_{12}).

(b) Based on the replicated trials for each unique test combination, develop a 95% confidence interval to determine which effect estimates are significant.

(c) Prepare graphical aids that help to explain *all* significant effects (other than the average). Briefly explain/interpret the meaning of all significant effects.

18.3 An unreplicated 2^3 factorial design was conducted. Three variables (x_1, x_2, and x_3) were considered, with the following results:

$$\text{average} = 60.0, \qquad E_{12} = 3.0,$$
$$E_1 = -12.0, \qquad E_{13} = 23.0,$$
$$E_2 = -3.0, \qquad E_{23} = -2.0,$$
$$E_3 = -18.0, \qquad E_{123} = 1.0.$$

(a) Determine which effect estimates are significant by plotting them on normal probability paper. Show all your calculations.

(b) Write down what appears to be the most appropriate math model for the response.

18.4 A 2^3 factorial design was conducted. The design matrix and data are shown in the table below.

Test	1	2	3	y
1	−	−	−	12
2	+	−	−	15
3	−	+	−	24
4	+	+	−	17
5	−	−	+	24
6	+	−	+	16
7	−	+	+	24
8	+	+	+	28

(a) Calculate all the effects.

(b) Plot the effect estimates on normal probability paper. Based on the normal probability plot, indicate which effects seem to be significant.

18.5 A replicated 2^2 factorial experiment was performed, the results of which are shown in the accompanying table.

x_1	x_2	y_{i1}	y_{i2}
−	−	0.50	0.46
+	−	0.42	0.37
−	+	0.33	0.30
+	+	0.11	0.16

(a) Calculate the effect estimates (i.e., the average, E_1, E_2, and E_{12}).

(b) Based on the replicated trials for each unique test combination, develop a 90% confidence interval to determine which effect estimates are significant.

18.6 A local nightclub owner has performed a two-level factorial experiment to investigate the effects of five factors on the business (sales) at his club during a single evening (measured in hundreds of dollars).

Variable	Low Level	High Level
1. Price of beverage	$1.00	$2.00
2. Jukebox	Off	On
3. Bartender	Sam	Woody
4. Waitress	Diane	Carla
5. Mirror-ball	Off	On

The following effects have been calculated based on the 32 combinations of the five variables:

$$\text{average} = 13.00, \quad E_{14} = -0.50, \quad E_{123} = 0.24, \quad E_{245} = 0.37$$
$$E_1 = -4.30, \quad E_{15} = -0.20, \quad E_{124} = 0.57, \quad E_{345} = 0.82,$$
$$E_2 = 0.03, \quad E_{23} = 0.29, \quad E_{125} = 0.38, \quad E_{1234} = -0.60,$$
$$E_3 = -0.50, \quad E_{24} = 0.26, \quad E_{134} = -1.00, \quad E_{1235} = 0.13,$$
$$E_4 = 0.75, \quad E_{25} = -0.50, \quad E_{135} = -1.40, \quad E_{1245} = -0.40,$$
$$E_5 = 7.80, \quad E_{34} = -0.60, \quad E_{145} = 0.35, \quad E_{1345} = 0.22,$$
$$E_{12} = -0.40, \quad E_{35} = 0.01, \quad E_{234} = 0.00, \quad E_{2345} = -0.50,$$
$$E_{13} = -0.80, \quad E_{45} = 6.81, \quad E_{235} = -0.20, \quad E_{12345} = 0.48.$$

(a) Plot the effects on normal probability paper.

(b) Based on the normal probability plot, indicate which effects appear to be significant.

18.7 Referring to Exercise 18.6:

(a) Use the third- and higher-order interactions to calculate the sample variance of an effect.

(b) Develop 95% confidence intervals for each of the true mean effects. Which effects appear to be statistically significant?

(c) Are the results here consistent with those of Exercise 18.6? Explain.

18.8 Based on the significant effects determined in Exercise 18.6 or 18.7, what strategy would you take as the owner to maximize your revenues? Be specific. Provide graphical interpretations for all important main and interaction effects.

18.9 After completion of the analysis for the experiment of Exercise 18.6, it was found that the jukebox was on for all 32 tests. Reanalyze the data of this experiment, given this new information and the additional analysis capability it provides. Are the results here consistent with those of Exercise 18.6? Comment.

18.10 A 2^3 factorial design was conducted to study the effect of three variables on a response of interest. The table describes the conditions that were examined in the experiment and the associated response (note that each unique combination of the variables was tested four times). Also evident from an examination of the table is the fact that the average and sample variance for the response have been calculated for each unique combination of the variables.

			Trial					Sample
x_1	x_2	x_3	1	2	3	4	Average	Variance
−	−	−	13	28	12	31	21	98
−	+	−	9	18	30	23	20	78
+	+	+	38	33	53	32	39	94
+	+	−	57	55	39	45	49	72
−	−	+	52	34	49	41	44	66
−	+	+	30	42	43	33	37	42
+	−	−	41	53	60	62	54	90
+	−	+	38	41	50	27	39	90

The following effect estimates have been calculated:

$$\text{average} = 37.875, \quad E_{12} = 0.75,$$
$$E_1 = 14.75, \quad E_{13} = -16.25,$$
$$E_2 = -3.25, \quad E_{23} = ?,$$
$$E_3 = ? \quad E_{123} = 2.75.$$

(a) Calculate E_3 and E_{23}.

(b) Based on the replicated trials for each unique combination, calculate an estimate of the standard error of an effect. Using a 95% confidence interval, which effect estimates are statistically significant?

(c) Graphically explain the numerical value obtained for E_{13} by using a two-way diagram. Explain in your own words the meaning of the graphical display.

18.11 A fisherman has conducted a 2^5 factorial design to determine the effects of five factors on the number of fish he catches in a four-hour period. The variables studied and their levels for experiment are shown in the table below.

Variable	Low Level	High Level
1. Fishing location	Pier	Boat
2. Bait type	Worms	Night crawlers
3. Time of day	Day	Night
4. Weather	No rain	Rain
5. Hook size	Small	Large

The experimental trials were all performed at Jack's Fish Farm in northern Wisconsin, stocked with bluegills. The following table gives the results of the experiment.

Test	x_1	x_2	x_3	x_4	x_5	y
1	−	−	−	−	−	25
2	+	−	−	−	−	26
3	−	+	−	−	−	31
4	+	+	−	−	−	34
5	−	−	+	−	−	24
6	+	−	+	−	−	26
7	−	+	+	−	−	32
8	+	+	+	−	−	36
9	−	−	−	+	−	30
10	+	−	−	+	−	36
11	−	+	−	+	−	40
12	+	+	−	+	−	43
13	−	−	+	+	−	30
14	+	−	+	+	−	34
15	−	+	+	+	−	40
16	+	+	+	+	−	42
17	−	−	−	−	+	2
18	+	−	−	−	+	21
19	−	+	−	−	+	26
20	+	+	−	−	+	43
21	−	−	+	−	+	4
22	+	−	+	−	+	20
23	−	+	+	−	+	25
24	+	+	+	−	+	44
25	−	−	−	+	+	10
26	+	−	−	+	+	27
27	−	+	−	+	+	33
28	+	+	−	+	+	51
29	−	−	+	+	+	9
30	+	−	+	+	+	27
31	−	+	+	+	+	32
32	+	+	+	+	+	53

(a) Calculate all effect estimates.

(b) Assuming that the third- and higher-order interactions are negligible, determine an estimate for the variance of an effect and construct 95% confidence intervals based on the effect estimates.

(c) Based on part (b), which effects appear to be important?

(d) Construct two-way diagrams for all important two-factor interactions. Comment.

18.12 Using the data presented in Exercise 18.11:

(a) Calculate all effect estimates.

(b) Construct a normal probability plot of the effect estimates.

(c) Interpret the normal probability plot. Which effects appear to be significant?

(d) Construct two-way diagrams for all important two-factor interactions. Comment.

18.13 A 2^3 factorial design was used to study the possible effects of temperature, time, and pressure on the yield of a chemical process. The data, including experimental results, are shown in the accompanying tables.

Variable	Low Level	High Level
1. Temperature (°F)	80	100
2. Time (min)	5	7
3. Pressure (psi)	120	140

Test	1	2	3	Trial 1	Trial 2
				Response	
1	−	−	−	9	11
2	+	−	−	18	22
3	−	+	−	4	4
4	+	+	−	9	11
5	−	−	+	7	9
6	+	−	+	17	19
7	−	+	+	5	7
8	+	+	+	10	14

(a) Calculate all the effects.

(b) Using the replicated results, determine an estimate for the variance of the effects.

(c) Evaluate the statistical significance of each of the effects by calculating a t value for each effect estimate and comparing it to the critical t value at a 5% significance level.

(d) Based on your interpretation of the results, what action(s) might be taken to increase the yield?

18.14 Using the data presented in Exercise 18.13:
  (a) Calculate all the effect estimates.
  (b) Make a normal plot of the effect estimates, and interpret the plot.
  (c) Based on your interpretation of the results, what action(s) might be taken to increase the yield?

18.15 A 2^2 factorial experiment is conducted with each unique test condition being performed three times (giving 12 tests total). $100(1 - \alpha)\%$ confidence intervals are to be constructed for the true mean effects based on the effect estimates. How many degrees of freedom are associated with the Student's t-statistic value to be used in construction of the confidence intervals?

18.16 A 2^2 factorial design was conducted and each unique test condition was run four times. The sample variance of an effect estimate was calculated and found to be $s^2_{\text{effect}} = 9$. The calculated effect estimates were: average $= 7.5$, $E_1 = -6.0$, $E_2 = 8.0$, and $E_{12} = -5.5$. Using 95% confidence intervals, determine which of the effect estimates appear to be significant.

18.17 A 2^5 factorial experiment is conducted. It is assumed that the third- and higher-order interactions can be neglected, and these interactions are used to estimate the variance of an effect. How many degrees of freedom are associated with the t value that will be used to test the significance of the variable effects?

18.18 A 2^3 factorial design is to be conducted. The variance of an individual observation is $\sigma^2_y = 4$. Suppose that we want the length of a 95% confidence interval to be 1.8 or smaller. What is the minimum number of runs at each distinct test condition that will satisfy the foregoing requirement, assuming that each test will be run an equal number of times?

18.19 The results of a 2^3 factorial design are given in standard order in the table below.

Test	Response, y
1	48.8
2	91.6
3	40.8
4	60.4
5	48.8
6	89.2
7	40.4
8	60.0

Let x_1, x_2, and x_3 correspond to the coded variables.
  (a) Calculate the effects.
  (b) Using a normal probability plot, which effects appear to be significant?

18.20 An experiment was conducted to compare the strength of two brands of facial tissues, A and B. Each brand was tested under two different conditions, dry and damp.

Variable	Low	High
Brand	A	B
Condition	Dry	Damp

The experiment was repeated four times for each combination of brand and condition, and the results were:

		Brand	
		−	+
Condition	−	14,13 12,11	10,11 11,12
	+	4,4 3,5	5,5 6,6

(a) Calculate the variable effects using the calculation matrix method.
(b) Determine the statistical significance of the variable effects. Assume $\alpha = 0.05$ for your analysis.
(c) Interpret the results of the experiment.

18.21 A replicated 2^3 experiment is conducted with the following results.

X_1	X_2	X_3	y_1	y_2	s_i^2
−	−	−	54	58	8
+	−	−	73	71	2
−	+	−	65	71	18
+	+	−	90	88	2
−	−	+	44	42	2
+	−	+	62	62	0
−	+	+	63	55	32
+	+	+	74	76	2

(a) Estimate all the main and interaction effects for this experiment.
(b) Using the information from the test replication, determine 95% confidence intervals for the true effect values.
(c) Comment on the statistical significance of the estimated main and interaction effects.

18.22 A 2^4 factorial design produced the following effect estimates:

$$E_1 = -12.25, \quad E_{12} = -7.125, \quad E_{24} = 11.125, \quad E_{134} = 2.0,$$
$$E_2 = 17.625, \quad E_{13} = 8.75, \quad E_{34} = 1.5, \quad E_{234} = 2.375,$$
$$E_3 = -1.375, \quad E_{14} = -0.5, \quad E_{123} = -1.875, \quad E_{1234} = -1.125.$$
$$E_4 = 0.25, \quad E_{23} = 14.875, \quad E_{124} = 0.125,$$

Use a 95% confidence interval to determine the significance of the relevant variable effects. Clearly state any assumptions that you make to perform the analysis.

18.23 The results of a 2^3 factorial design are given in standard order in the table below.

Test Number	Response, y
1	49
2	92
3	41
4	y_4
5	49
6	90
7	40
8	60

Note that the value of the response for trial 4 was not obtained and is unknown to the analyst.

(a) Under the assumption that the interaction E_{123} is zero, find the value for y_4.

(b) Find estimates of the other effects using the value for y_4 calculated in part (a).

(c) Plot all of the effect estimates on normal probability paper.

(d) Comment on the normal probability plot. Is there any reason to suspect that the estimate of y_4 is not a good estimate? Explain your answer.

18.24 The effect estimates for a 2^3 factorial are given below:

$$E_1 = 8.0, \quad E_2 = -6.0, \quad E_3 = 0.05, \quad E_{12} = -2.0.$$
$$E_{13} = 0.0, \quad E_{23} = 2.0, \quad E_{123} = -0.02,$$

(a) Plot these effects on normal probability paper.

(b) Comment on the relative importance of the variable effects.

18.25 The data below were collected using a two-level factorial design. The influence of two factors on the temperature (°C) reached in a certain chemical reaction is sought via the establishment of a math model for the temperature response, y. The values for x_1 and x_2 are the concentrations in percent of two different reactants.

Test	x_1	x_2	y_1	y_2
1	10	5	113	115
2	15	5	141	138
3	10	7	90	94
4	15	7	160	164

(a) Determine the variable effects E_1, E_2, and E_{12}.
(b) Assess the statistical significance of these effect estimates via the confidence-interval approach. Assume $\alpha = 0.05$ for your analysis.

18.26 The following are the effect estimates from a two-level factorial design:

$$
\begin{aligned}
E_1 &= 8, & E_{12} &= -2, \\
E_2 &= -6, & E_{13} &= -0.5, \\
E_3 &= 0, & E_{23} &= 2. \\
E_{123} &= 0.5, &
\end{aligned}
$$

(a) Construct a piece of normal probability paper. Carefully show all of the steps. Be sure that you show exactly how you scaled the axes for the paper.
(b) Using the normal probability paper from (a), plot the effect estimates.
(c) Interpret the results.

19

Model Building for Design and Improvement Using Two-Level Factorial Designs

19.1   Introduction

In Chapter 17 we examined how a performance measure or response could be effectively studied through the use of two-level factorial designs. It was seen that the dependence of an experimental response on one or more variables, both singly and in concert with each other, could be characterized through the calculation of effects. In Chapter 18 we examined several techniques for separating the effects that are truly important from those that are not, in light of the experimental error in the testing environment. In developing these techniques, some assumptions about the character of the noise were made that need to be examined further to judge their appropriateness.

In this chapter we also consider two-level factorial designs, focusing on the development of empirical models for design, prediction, and control purposes. The development of such models requires that we check the appropriateness and adequacy of the prediction model. It is, at least in considerable part, for the building of such models for quality/productivity design and improvement that experiments are designed, conducted, and analyzed in the first place.

General Procedures for Model Building

It is useful to think of the model-building activity as an iterative procedure evolving in four distinct stages.

1. Postulation of a tentative general model form.
2. Design of an appropriate experiment.
3. Fitting the general model form to the data.
4. Diagnostic checking of the fitted model.

Postulation of a tentative model may be based on prior knowledge of the underlying behavior of the response over the ranges of the variables under study. When little is known in this regard and/or the variable level ranges are not large, low-order polynomial forms may be useful.

Given a tentative model form, an experiment must be designed that has the capacity to estimate the parameters/coefficients of the variable terms in the model. When low-order polynomial forms are proposed, two- and three-level factorials serve this purpose well.

Once the data are at hand, the model may be fit to the data. For two-level factorials and their associated polynomial forms, the effect estimates E_i are employed. These estimates, when suitably scaled, have been shown to be the least-squares estimates of the model coefficients and as a result have several desirable properties.

The fact that the initially postulated model is fit to the data is no guarantee that the model will provide good predictions of the process behavior. If the model postulation is misguided, the predictions will be poor even if the experiment is well designed and carried out. Remember, the experimental design is in response to the model postulated, good or bad. In diagnostically checking the model, two basic questions are addressed:

1. Are all the terms initially included in the model necessary?
2. Are the terms initially included in the model sufficient?

The first question may be dealt with, given the experimental results at hand. Dealing with the second may require additional experimentation.

19.2   Glove Box Door Alignment Study Revisited

Review of Previous Analysis

In Chapter 18 a study was presented that focused on improving the parallelism of a glove box door with respect to the instrument panel. Four variables were studied over two levels each to investigate their possible effects on parallelism. These variables and their low and high levels are given in the table below.

Variable	Low (−)	High (+)
x_1: RH cowl fore/aft movement	Nominal	−5 mm
x_2: Center brace attachment sequence	Before	After
x_3: Plenum gasket	No	Yes
x_4: Evaporator case setup, fore/aft	Nominal	−5 mm

A replicated 2^4 design was performed, and the parallelism was recorded for each of the 32 tests. The results of the experiment are given in Table 19.1.

Selection of this experimental design means that the tacit assumption has already been made that a model of the form of Eq. (19.1) may provide for an adequate representation of the phenomenon under study:

$$y = b_0 + b_1x_1 + b_2x_2 + b_3x_3 + b_4x_4 + b_{12}x_1x_2 + b_{13}x_1x_3 + b_{14}x_1x_4$$
$$+ b_{23}x_2x_3 + b_{24}x_2x_4 + b_{34}x_3x_4 + b_{123}x_1x_2x_3$$
$$+ b_{124}x_1x_2x_4 + b_{134}x_1x_3x_4 + b_{234}x_2x_3x_4$$
$$+ b_{1234}x_1x_2x_3x_4 + \epsilon, \tag{19.1}$$

where the ϵ are the deviations between the data y and the mean response η, represented here by the polynomial form. It is typically assumed that ϵ should behave as a normally and independently distributed variable with zero mean and fixed variance σ_y^2. Since the ϵ's may be thought of as an aggregation of many small random errors, it is not unreasonable to expect the above assumptions to be met if the model is adequate.

TABLE 19.1 2^4 Design and Results of the Glove Box Door Experiment

Test	x_1	x_2	x_3	x_4	Run 1 y_{i1}	Run 2 y_{i2}	$\bar{y}_i$	d_i	s_i^2
1	−	−	−	−	−1.44	−0.08	−0.760	−1.36	0.92480
2	+	−	−	−	−1.79	−1.01	−1.400	−0.78	0.30420
3	−	+	−	−	0.39	0.17	0.280	0.22	0.02420
4	+	+	−	−	−0.50	−0.24	−0.370	−0.26	0.03380
5	−	−	+	−	−0.20	0.17	−0.015	−0.37	0.06845
6	+	−	+	−	−0.79	−0.64	−0.715	−0.15	0.01125
7	−	+	+	−	1.22	0.28	0.750	0.94	0.44180
8	+	+	+	−	0.21	0.28	0.245	−0.07	0.00245
9	−	−	−	+	−0.40	−0.65	−0.525	0.25	0.03125
10	+	−	−	+	−0.63	−1.19	−0.910	−0.56	0.15680
11	−	+	−	+	0.47	0.44	0.455	0.03	0.00045
12	+	+	−	+	−0.01	−0.03	−0.020	0.02	0.00020
13	−	−	+	+	1.29	0.64	0.965	0.65	0.21125
14	+	−	+	+	−1.17	0.14	−0.515	−1.31	0.85805
15	−	+	+	+	0.48	1.06	0.770	−0.58	0.16820
16	+	+	+	+	0.40	0.34	0.370	0.06	0.00180

Based on the results in Table 19.1, the following effect estimates were calculated:

$$\text{average} = -0.087,$$

$$
\begin{aligned}
E_1 &= -0.654, & E_{13} &= -0.117, & E_{123} &= 0.172, \\
E_2 &= 0.794, & E_{14} &= -0.031, & E_{124} &= 0.101, \\
E_3 &= 0.638, & E_{23} &= -0.191, & E_{134} &= -0.138, \\
E_4 &= 0.322, & E_{24} &= -0.154, & E_{234} &= -0.104, \\
E_{12} &= 0.147, & E_{34} &= 0.009, & E_{1234} &= 0.121.
\end{aligned}
$$

To judge the relative importance of these effects, we may use the replicated responses from the experiment. As noted in Chapter 18, if we assume that the responses are all drawn from distributions with the same variance, the sample variances determined for each pair of replicated tests may be pooled to provide for a more precise estimate of this common variance,

$$s_p^2 = \frac{s_1^2 + s_2^2 + \cdots + s_{16}^2}{16} = \frac{0.9248 + \cdots + 0.0018}{16}$$

$$= 0.20243.$$

Recall that the variance of an effect for a two-level factorial design is given by

$$\sigma_{\text{effect}}^2 = \frac{4\sigma_y^2}{N}.$$

For this experiment the sample variance of an effect is

$$
\begin{aligned}
s_{\text{effect}}^2 &= \frac{4s_p^2}{32} \\
&= \frac{s_p^2}{8} \\
&= \frac{0.20243}{8} \\
&= 0.0253.
\end{aligned}
$$

A 95% confidence interval may be developed for the true mean value of each effect estimate,

$$E \pm t_{16,0.975}s_{\text{effect}} = E \pm (2.120)(0.159) = E \pm 0.337.$$

Table 19.2 summarizes the 95% confidence intervals based on each of the effect estimates, and also includes a confidence interval for the mean response. It may be seen in Table 19.2 that the confidence intervals for μ_{E_1}, μ_{E_2}, and μ_{E_3} do

TABLE 19.2	Confidence Intervals for True Mean Variable Effects of the Glove Box Door Alignment Study		
Effect	*95% Confidence Interval*	*Effect*	*95% Confidence Interval*
Average	-0.087 ± 0.169	E_{23}	-0.191 ± 0.337
E_1	**-0.654 ± 0.337**	E_{24}	-0.154 ± 0.337
E_2	**0.794 ± 0.337**	E_{34}	0.009 ± 0.337
E_3	**0.638 ± 0.337**	E_{123}	0.172 ± 0.337
E_4	0.322 ± 0.337	E_{124}	0.101 ± 0.337
E_{12}	0.147 ± 0.337	E_{134}	-0.138 ± 0.337
E_{13}	-0.117 ± 0.337	E_{234}	-0.104 ± 0.337
E_{14}	-0.031 ± 0.337	E_{1234}	0.121 ± 0.337

not include zero (bold type in the table). Therefore, it may be concluded that the true mean effects of variables 1, 2, and 3 differ significantly from zero.

Development of a Prediction Model Based on the Experiment Results

In our examination of two-level factorial designs we have seen that important variables and interactions can be distinguished from those that are not important using several techniques. One of the primary motivations in our study of two-level factorial designs was to mathematically describe the relationships between the experimental response and the variables manipulated during the experiment. Let us consider, therefore, how such a relationship or model can be developed.

As we have said earlier, in analyzing a 2^k factorial design we may express the experimental response using a model of the form of Eq. (19.1). It should be noted that for a 2^k factorial design there is a model coefficient associated with each of the $2^k - 1$ main and interaction effects, plus the mean response b_0. For a 2^k factorial design, estimates of the model coefficients may be obtained directly from the effect estimates, which are, in fact, the least-squares estimates. The coefficient estimates are simply the effect estimates divided by 2:

$$\hat{b} = \frac{E}{2}.$$

Thus a calculated effect, E_i, is an estimate of the true mean effect, and $\hat{b}$ is an estimate of the model coefficient, b.

A common question asked with respect to obtaining estimates of the model coefficients from the effect estimates is: "Why is the effect estimate divided by 2?" To understand the reason behind this, consider the graphical representation of the meaning of the main effect of variable 1, which is illustrated in Fig. 19.1. As can be seen from the figure, changing the level of the RH cowl movement from the nominal value (-1 level) to -5 millimeters ($+1$ level) produces a change in the parallelism of -0.6544 millimeters. In other words, a two-unit change in variable x_1 (from -1 to $+1$) produces a -0.6544 millimeter change in

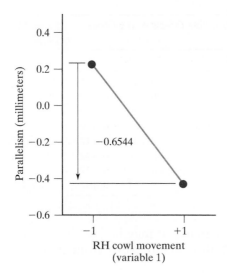

Figure 19.1   Graphical aid for the interpretation of the main effect of variable 1

the response. The dependency of the parallelism on x_1 may also be expressed via the slope of the line in Fig. 19.1. This slope, $\hat{b}_1$ is given by

$$\hat{b}_1 = \frac{-0.6544}{1 - (-1)} = -0.327$$

$$= \frac{E_1}{2}.$$

If estimates for all the model coefficients are inserted into Eq. (19.1), an equation is produced that may be used to predict the response as a function of the independent variables:

$$\begin{aligned}
\hat{y} = {} & \hat{b}_0 + \hat{b}_1 x_1 + \hat{b}_2 x_2 + \hat{b}_3 x_3 + \hat{b}_4 x_4 + \hat{b}_{12} x_1 x_2 + \hat{b}_{13} x_1 x_3 \\
& + \hat{b}_{14} x_1 x_4 + \hat{b}_{23} x_2 x_3 + \hat{b}_{24} x_2 x_4 + \hat{b}_{34} x_3 x_4 + \hat{b}_{123} x_1 x_2 x_3 \\
& + \hat{b}_{124} x_1 x_2 x_4 + \hat{b}_{134} x_1 x_3 x_4 + \hat{b}_{234} x_2 x_3 x_4 \\
& + \hat{b}_{1234} x_1 x_2 x_3 x_4,
\end{aligned} \tag{19.2}$$

where $\hat{y}$ is the predicted response. Examining Eq. (19.2), we see that there are terms associated with all the variables and their interactions. We know from Table 19.2, however, that only a few of the coefficients associated with these terms are important. If we drop the insignificant terms from the model, we obtain

$$\hat{y} = \hat{b}_1 x_1 + \hat{b}_2 x_2 + \hat{b}_3 x_3. \tag{19.3}$$

Recall that the mean response $\hat{b}$ could not be distinguished from zero and therefore does not appear in the model. If the calculated values for the $\hat{b}$'s are inserted into Eq. (19.3), the following equation is produced:

$$\hat{y} = -0.327x_1 + 0.397x_2 + 0.319x_3. \tag{19.4}$$

This equation provides us with a tool that can be used to study the response as a function of variables 1, 2, and 3 (RH cowl movement, attachment sequence, and plenum gasket). However, before we exercise this model for quality design and improvement purposes, we should examine the model more closely to determine if it is an adequate representation of the phenomenon under study. If it is not, some improvement may be possible given the existing data. However, it may be necessary to conduct more experiments to develop an adequate model.

19.3   Diagnostic Checking of the Fitted Model

Calculation of Predicted Responses and Model Residuals

In the case of our glove box door parallelism study, we identified the main effects of variables 1, 2, and 3 as the only important effects and therefore we obtained the model for the prediction of the door parallelism given by Eq. (19.4).

$$\hat{y} = -0.327x_1 + 0.397x_2 + 0.319x_3.$$

This equation may be used to calculate a predicted response, $\hat{y}_i$, for each of the test conditions. If this model contains all the necessary terms to adequately predict the response, y, the model residual errors (the differences between the data and the model predictions), $e_{ij} = (y_{ij} - \hat{y}_i)$, should contain no structured variation at all; that is, they should behave in the following way:

1. They should be centered about a mean of zero.
2. They should tend to be normally distributed.
3. They should have a scatter that does not vary as a function of the predicted response ($\hat{y}$'s).
4. They should not be correlated with the independent variables x, time order of the testing, or any other variables related to the experiment. They should be truly random errors.

In short, the model residuals (e's) should have the same properties as we assumed for the errors (ϵ's). If the model residuals do not demonstrate these properties, this may be an indication that the assumptions made with regard to the errors (ϵ's) do not hold—perhaps because the model η is inadequate.

To check to see if the residuals behave as above, the procedure outlined below should be followed:

- Use the model to predict the parallelism for each of the tests performed.
- Compare the model prediction with the observed data and calculate the residual errors, e_{ij}, for all i and j.
- Perform a series of tests and checks on the residuals to see if they do behave as they should if the predicting model is adequate.

At each of the 16 unique combinations of x_1, x_2, x_3, and x_4 in the experimental design, we can predict the response, $\hat{y}$, using Eq. (19.4). Since only x_1, x_2, and x_3 are important, the 16 test conditions will produce only eight unique predicted responses. For the first test (in standard order), the levels for x_1, x_2, and x_3 are all -1. Therefore, the predicted response is

$$\hat{y}_1 = -0.327(-1) + 0.397(-1) + 0.319(-1) = -0.389.$$

For test 1, there are two observed responses, $y_{11} = -1.44$ and $y_{12} = -0.08$. The model residual errors are then $e_{1j} = (y_{1j} - \hat{y}_1), j = 1, 2; e_{11} = -1.051$; and $e_{12} = 0.309$. Table 19.3 gives the predicted responses and residual errors associated with the prediction model over the entire experiment.

TABLE 19.3	Model Predictions and Residuals of the Parallelism Prediction Model (Glove Box Door Study)								
Test	x_1	x_2	x_3	x_4	Predicted Response, $\hat{y}_i$	Observed Response, y_{i1}	Run Order	Model Residual, e_{i1}	Observed Response, y_{i2}	Run Order	Model Residual, e_{i2}
---	---	---	---	---	---	---	---	---	---	---	---
1	$-$	$-$	$-$	$-$	-0.389	-1.440	(7)	-1.051	-0.080	(28)	0.309
2	$+$	$-$	$-$	$-$	-1.043	-1.790	(10)	-0.747	-1.010	(24)	0.033
3	$-$	$+$	$-$	$-$	0.405	0.390	(14)	-0.015	0.170	(32)	-0.235
4	$+$	$+$	$-$	$-$	-0.249	-0.500	(2)	-0.251	-0.240	(21)	0.009
5	$-$	$-$	$+$	$-$	0.249	-0.200	(9)	-0.449	0.170	(27)	-0.079
6	$+$	$-$	$+$	$-$	-0.405	-0.790	(6)	-0.385	-0.640	(30)	-0.235
7	$-$	$+$	$+$	$-$	1.043	1.220	(13)	0.177	0.280	(20)	-0.763
8	$+$	$+$	$+$	$-$	0.389	0.210	(8)	-0.179	0.280	(18)	-0.109
9	$-$	$-$	$-$	$+$	-0.389	-0.400	(1)	-0.011	-0.650	(31)	-0.261
10	$+$	$-$	$-$	$+$	-1.043	-0.630	(15)	0.413	-1.190	(25)	-0.147
11	$-$	$+$	$-$	$+$	0.405	0.470	(3)	0.065	0.440	(17)	0.035
12	$+$	$+$	$-$	$+$	-0.249	-0.010	(5)	0.239	-0.030	(23)	0.219
13	$-$	$-$	$+$	$+$	0.249	1.290	(12)	1.041	0.640	(29)	0.391
14	$+$	$-$	$+$	$+$	-0.405	-1.170	(4)	-0.765	0.140	(19)	0.545
15	$-$	$+$	$+$	$+$	1.043	0.480	(16)	-0.563	1.060	(22)	0.017
16	$+$	$+$	$+$	$+$	0.389	0.400	(11)	0.011	0.340	(26)	-0.049

Residual Analysis

We may now perform a series of checks on the residuals, e_{ij}. These should include:

- A normal probability plot of the residuals.
- A plot of the residuals vs. the time/run order of the tests.
- A plot of the residuals vs. the predicted response.
- Plots of the residuals vs. the independent variables.

Figure 19.2 shows a normal probability plot of the residuals. Most of the residuals seem to fall on a straight line on the normal probability paper, with the exception of one or two points, and are centered approximately at zero. These off-line points should be checked for possible causes, such as unusual events surrounding the environment of their respective tests, possible deviations from designed test conditions, and possible errors in the measurements and recording of these tests. On the whole, the plot looks as if the residual errors are quite normal.

Figures 19.3 through 19.9 give some additional plots of the residuals to further reveal possible inadequacies of the prediction model for the glove box door study. Even though $\hat{b}_4$ was not included in the model, a residual plot vs. x_4 should be made to reveal the possible presence of structure in the data with respect to x_4. The residuals seem to be random and to have a consistent amount of scatter, as none of the residual plots of Figs. 19.3 to 19.9 show any obvious relationship with the time order of running the tests or the levels of the independent variables. The residuals are centered about zero and there does

(Continued on page 707)

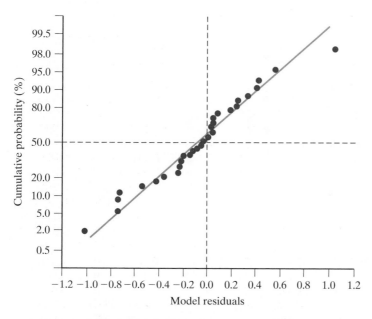

Figure 19.2   Normal probability plot of the residuals

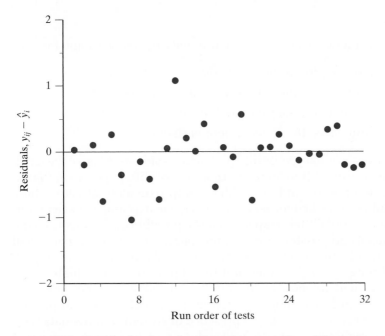

Figure 19.3   Residuals plotted against the run order of the tests

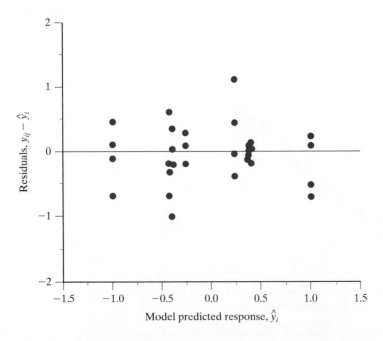

Figure 19.4   Residuals plotted against the predicted responses

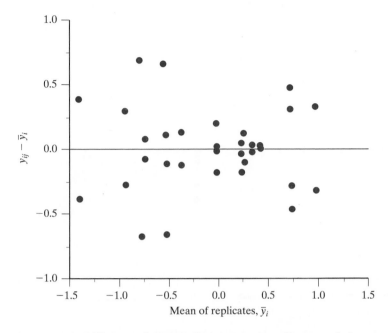

Figure 19.5   **Difference between the response replicate and the average of the replicates plotted against the average of the replicates**

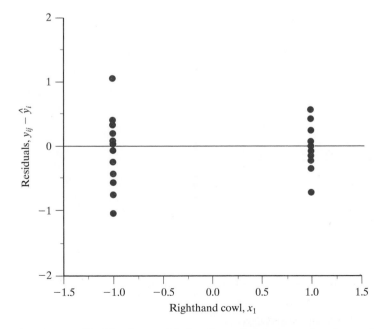

Figure 19.6   **Residuals vs. righthand cowl movement level: nominal (−1), −5 millimeters (+1)**

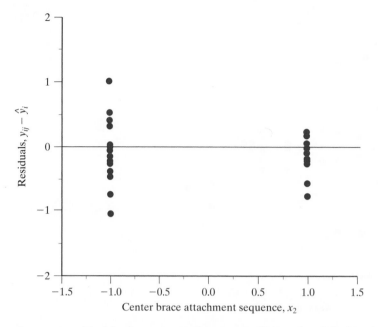

Figure 19.7   **Residuals vs. center brace attachment level: before (−1), after (+1)**

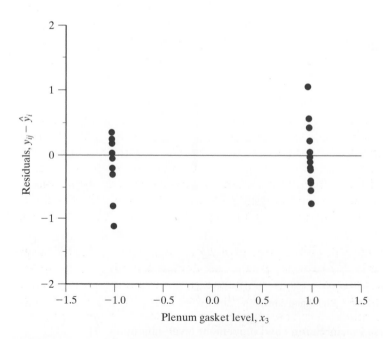

Figure 19.8   **Residuals vs. plenum gasket level: no (−1), yes (+1)**

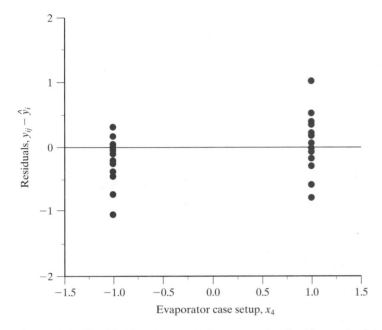

**Figure 19.9   Residuals vs. evaporator case setup level: nominal (−1),
−5 millimeters (+1)**

not appear to be any structure within the residuals. Figure 19.5 examines the
extent to which the magnitude of replication error may be related to the magnitude of the response. This check addresses the issue of constant variance,
which will be examined in more detail later in this chapter.

Some Nonrandom Patterns of Residual Plots

Residual plots are used primarily to check for model inadequacies. Sometimes
they may also reveal other concerns, such as unequal variance of the responses
or problems with the actual conduct or experimental environment of the tests.
To assist in the identification of residual plots with nonrandom patterns, several residual plots with common nonrandom patterns are shown in Fig. 19.10.
It is noted that all 10 nonrandom patterns of Fig. 19.10 have a zero mean, except pattern (b). A brief explanation of some typical causes of each nonrandom
pattern follows.

1. The residuals form a linear trend (upward or downward). If it is a trend
   over time, this might be due to some consistent and gradual buildup effect of certain external factors (which are not being studied), such as the
   moisture content of the raw material that was continuously decreasing.
   If the trend is a function of the predicted response $\hat{y}$, it would indicate
   that the model is inadequate.

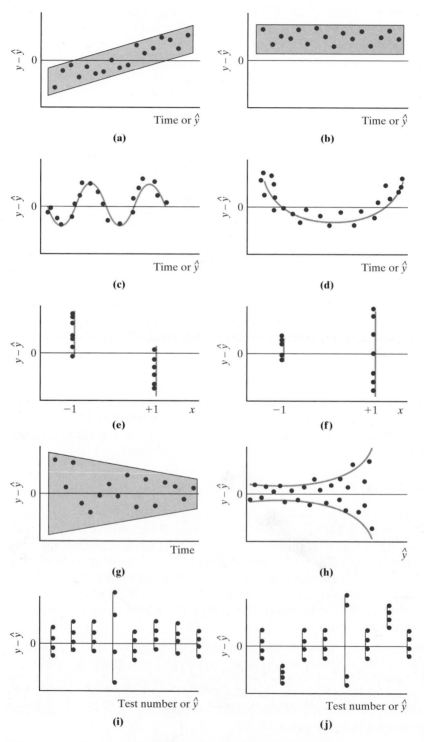

Figure 19.10   Some nonrandom patterns of residual plots

2. The model has a constant bias (positive or negative). Possible causes include error in the estimation of the mean and missing effects in the model. Also check the response data, particularly for one or two possible "outliers."

3. The residuals have a periodic trend. If it is seen from the time plot, it is probably caused by an unknown cyclic factor such as day–night–day. If it shows up on the $\hat{y}$ plot, it means that a higher-order model would be needed.

4. A quadratic trend (concave or convex). This may be revealing the presence of higher-order effects.

5. The model is missing a main or interaction effect associated with an x_i. If the control variable x_i is not in the model, consider possible inclusion of it. If x_i is one of the variables already included in the model, check for inclusion of some interaction effects involving x_i.

6. The model is not uniform in variability. This is a strong indication of a dispersion effect of x_i; that is, the variation level varies as a function of x_i.

7. Instability in the experiment process. This would indicate certain instability of the experimental environment for the tests that run earlier in the series. A learning effect of the experimenter or the transient state of certain new experimental equipment are examples of causes for this pattern. The use of data transformations such as a log transformation or adding a time-dependent term in the model could save the experiment in this case if the instability was well behaved.

8. There is an increasing response variability. The variability of the response is proportional to the magnitude of the response. The response data may have to be transformed to obtain a constant variability in the transformed responses. The use of log transformation often works well.

9. Outlier responses among the replicates. If the replicated responses are fairly uniform among all test conditions with the exception of one or two whose responses vary widely, check to verify the proper settings of these test conditions. Perhaps these test conditions are much more difficult to set up accurately each time.

10. Uneven residual variability. When several test conditions have widely different distributions for their respective residuals, a number of things might be at work. Possible causes include inaccurate measurement method, missing interaction effects in the model, and/or the presence of dispersion effects.

19.4 Checking the Assumption of Common Variance of the Responses

Several of the residual plots described previously either directly or indirectly focus on checking to see if the variability of the responses is constant. We have repeatedly relied on the assumption of constant variance throughout our analysis of two-level factorials. When an experiment is replicated as in the case study

TABLE 19.4	Variance Estimates from Glove Box Door Alignment Study		
Test	s_i^2	*Test*	s_i^2
1	0.92480	9	0.03125
2	0.30420	10	0.15680
3	0.02420	11	0.00045
4	0.03380	12	0.00020
5	0.06845	13	0.21125
6	0.01125	14	0.85805
7	0.44180	15	0.16820
8	0.00245	16	0.00180

above, we are afforded the opportunity to check the assumption of constant variance through a procedure that is a bit more rigorous than simply checking residual plots.

Recall from Table 19.1 that variance estimates for the response, door parallelism, were obtained for all 16 of the unique test conditions. These variance estimates are repeated in Table 19.4.

At this point, we might ask: "Is it reasonable to expect that all these sample variances are estimates of the same common variance?" In other words, "Is the variation among these sample variances greater than would be expected if only random error variation was at work?"

To answer this question, suppose that based on prior knowledge we believe that the parallelism response has a variance, σ_y^2, of 0.2. We may use the χ^2 distribution to judge whether any of the sample variances in Table 19.5 provides evidence that the true variance is not 0.2 for a given level of risk. Using the fact that the quantity $(n - 1)s_y^2/\sigma_y^2$ is χ^2 distributed, we may calculate χ^2 values for each of the 16 sample variances, as shown in Table 19.5. (Note that since two responses were used to obtain each of the sample variances, the value for n is 2.)

These calculated χ^2 values may be referred to the χ^2 distribution with $\nu = 1$ degree of freedom. Referring to Table A.5, it is seen that for 1 degree of freedom, χ^2 values of approximately 0.001 and 5.02 bracket 95% of all sampled

TABLE 19.5	Sample Variances and Calculated χ^2 Values				
Test	s_i^2	χ^2	*Test*	s_i^2	χ^2
1	0.92480	4.62400	9	0.03125	0.15625
2	0.30420	1.52100	10	0.15680	0.78400
3	0.02420	0.12100	11	0.00045	0.00225
4	0.03380	0.16900	12	0.00020	0.00100
5	0.06845	0.34225	13	0.21125	1.05625
6	0.01125	0.05625	14	0.85805	4.29025
7	0.44180	2.20900	15	0.16820	0.84100
8	0.00245	0.01225	16	0.00180	0.00900

χ^2 values. Since none of the calculated χ^2 values fall outside this range, we will not reject the hypothesis that the true common variance is 0.20 at a significance level of 0.05 (5%). Note that this significance level places 0.025 (2.5%) in each tail of the chi-square distribution.

Test for the Homogeneity of the Variance

The following widely used procedure, known as *Bartlett's test*,[1] may be used to test for equality of the variance (i.e., the null hypothesis below):

$$H_0: \ \sigma_1^2 = \sigma_2^2 = \cdots = \sigma_m^2 = \sigma_y^2,$$
$$H_1: \ \text{at least one } \sigma_i^2 \neq \sigma_j^2, \quad i \neq j.$$

The procedure computes a statistic whose sampling distribution is closely approximated by a chi-square distribution with $(m - 1)$ degrees of freedom. In this case $m = 2^k$ is the number of random samples of size n_i drawn from independent normal populations. For m sample variances calculated based on a total of N responses $(N = n_1 + n_2 + \cdots + n_m)$, the test statistic is given by

$$\chi_{\text{calc}}^2 = \frac{M}{c}, \tag{19.5}$$

where

$$M = (N - m) \ln s_p^2 - \sum_{i=1}^{m} (n_i - 1) \ln s_i^2,$$

$$c = 1 + \frac{1}{3(m - 1)} \left[\left(\sum_{i=1}^{m} \frac{1}{n_i - 1} \right) - \frac{1}{N - m} \right],$$

$$s_p^2 = \frac{\sum_{i=1}^{m} (n_i - 1) s_i^2}{N - m}.$$

The value for M will be large if the sample variances, s_i^2, differ greatly in magnitude, and will be zero if all the sample variances are exactly equal. We will reject H_0 if χ_{calc}^2 is too large; that is, we reject H_0 if

$$\chi_{\text{calc}}^2 > \chi_{m-1,\alpha}^2,$$

where $\chi_{m-1,\alpha}^2$ places α in the upper tail of the chi-square distribution with $m - 1$ degrees of freedom. Bartlett's test is sensitive to the normality assumption and should not be applied in cases where the distribution of the responses is not normal. Furthermore, it has been suggested that the test should be viewed with some caution if any of the n_i are less than 4.

---

[1]M. S. Bartlett, "The Use of Transformations," *Biometrics*, Vol. 3, 1947, pp. 39–52.

We now apply Bartlett's test to the sample variances obtained for the glove box door alignment study to test the common variance assumption. Using the sample variances of Table 19.5, the following may be calculated:

$$s_p^2 = \frac{(2-1)0.92480 + (2-1)0.30420 + \cdots + (2-1)0.00180}{32-16} = 0.20243,$$

$$
\begin{aligned}
M = {} & (32-16)\ln(0.20243) \\
& -[(2-1)\ln 0.92480 + (2-1)\ln 0.30420 + \cdots + (2-1)\ln 0.00180] \\
= {} & 28.1695,
\end{aligned}
$$

$$
\begin{aligned}
c = {} & 1 + \frac{1}{3(16-1)}\left(\frac{1}{2-1} + \frac{1}{2-1} + \cdots + \frac{1}{2-1} - \frac{1}{32-16}\right) \\
= {} & 1.3542,
\end{aligned}
$$

and the test statistic is

$$\chi_{\text{calc}}^2 = \frac{28.1695}{1.3542} = 20.8016.$$

Since $\chi_{15,0.05}^2 = 25.0$, we cannot reject the null hypothesis that the variances are equal. Note that since $n = 2$ for this example, this conclusion must be viewed with caution; however, the conclusion is consistent with the interpretations we made based on our study of the residual plots.

19.5   Using the Fitted Model for Quality Improvement

Recall that the objective of the glove box door alignment study was to find a fixturing method for the assembly of the door to the instrument panel so that the amount of off-parallelism would be within ± 1.5 millimeters. Therefore, we would like to find a solution that produces a zero off-parallelism on the average. To find such a solution, we will use the model that was developed previously for the parallelism response.

The fitted model we developed for the parallelism response is

$$\hat{y} = -0.327x_1 + 0.397x_2 + 0.319x_3.$$

All residual checks and modeling assumption checks have shown this model to be an adequate representation of the phenomenon under study. Some confirmatory tests were conducted and showed no reason for us to doubt the legitimacy of the model.

From the fitted model it is clear that the parallelism is a simple additive function of the linear effects of RH cowl movement (variable 1), brace attachment sequence (variable 2), and plenum gasket (variable 3). The levels for these variables that were used during the conduct of the experiments are summarized

Variable	Low Level (−1)	High Level (+1)
x_1: RH cowl movement	0 mm	−5 mm
x_2: Brace attachment sequence	Before	After
x_3: Plenum gasket	No	Yes

ent
give
$(x_2$
−3.

fron
dict
is ta

Con

It m
mor
of th
deve
the c
tion,

in the accompanying table. An examination of this table shows that while variable 1 is a continuous variable that can take on any value between −1 (0 millimeters) and +1(−5 millimeters), variables 2 and 3 are discrete variables that can only take on values of −1 or +1. The four combinations of the levels for variables 2 and 3 were substituted into Eq. (19.4) (repeated above) to produce the following simplified equations:

$$x_2 = -1, \quad x_3 = -1, \quad \hat{y}_1 = -0.716 - 0.327x_1. \tag{19.6}$$

$$x_2 = +1, \quad x_3 = -1, \quad \hat{y}_2 = 0.078 - 0.327x_1. \tag{19.7}$$

$$x_2 = -1, \quad x_3 = +1, \quad \hat{y}_3 = -0.078 - 0.327x_1. \tag{19.8}$$

$$x_2 = +1, \quad x_3 = +1, \quad \hat{y}_4 = 0.716 - 0.327x_1. \tag{19.9}$$

Recall that it is desired to select levels for the three important variables that produce a zero off-parallelism on the average, or equivalently, values for x_1, x_2, and x_3 that make $\hat{y} = 0$. It is clear that for both Eqs. (19.6) and (19.9), no value for x_1 on the interval from −1 to +1 is capable of satisfying this requirement. Potentially, values for x_1 outside this interval may give a zero off-parallelism on the average; however, it is not advisable to use the model to extrapolate outside the experimental range, at least without exhausting other alternatives first.

Prev
coul
exan
thick
0 mi
in th

move
tour
Eq. (
predi
lelisn
x_1 an

C
press

Equations (19.7) and (19.8) suggest that a value for variable 1 (between −1 and +1) does exist that produces zero off-parallelism on the average. The behavior of the predicted response as a function of x_1 is illustrated in Fig. 19.11 for these two equations. To find the value of x_1 that gives a predicted response of 0 for $x_2 = +1$ and $x_3 = -1$, Eq. (19.7) may be set equal to zero and x_1 solved for:

$$\hat{y} = 0.0 = 0.078 - 0.327x_1,$$

$$x_1 = \frac{-0.078}{-0.327}$$

$$= 0.239.$$

To find the value of x_1 that gives a predicted response of 0 for $x_2 = -1$ and $x_3 = +1$, Eq. (19.8) may be set equal to zero and x_1 solved for:

To de
equal
to the

$$\hat{y} = 0.0 = -0.078 - 0.327x_1,$$

$$x_1 = \frac{0.078}{-0.327}$$

$$= -0.239.$$

It remains to transform these coded (−1, +1 in the experimental design) variable values back to the original variable units.

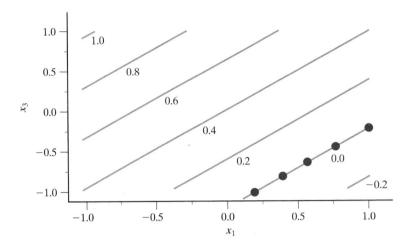

Figure 19.12   Contours of constant predicted parallelism

These points were used to construct the $\hat{y} = 0$ contour shown in Fig. 19.12. The table indicates that values of $x_1 = 1.00$ and $x_3 = -0.22$ produce a predicted response of 0.0 for the case where $x_2 = +1$ (center brace is attached after). The actual variable levels corresponding to the coded levels for variables 1 and 3 may also be found using Eq. (19.11):

$$\text{cowl movement} = \frac{0 + (-5)}{2} + 1\left(\frac{-5 - 0}{2}\right) = -5.00 \text{ mm},$$

$$\text{gasket thickness} = \frac{0 + 2}{2} + (-0.22)\left(\frac{2 - 0}{2}\right) = 0.78 \text{ mm}.$$

Thus a cowl movement of -5.00 millimeters and a gasket thickness of 0.78 millimeter produce a predicted off-parallelism value of zero on the average. Many other solutions are possible. It should be noted again, however, that any solution derived from the fitted model is only a prediction, and final validation of any prediction should be made through the results of confirmatory tests.

19.6   Another Case Study: Surface Finish of a Machined Part

In this chapter we have introduced and exercised a four-stage model building procedure using two-level factorial designs. Once a model has been developed and checked for potential inadequacies using this procedure, it may then be used for the purposes of quality and productivity design and improvement, as was seen in the example above. In this section we examine another case study concerned with finding variable settings that produce a desired surface finish for a machined part.

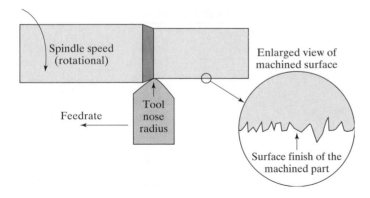

Figure 19.13   Variables considered in the surface finish case study

Background of Surface Finish Case Study

A manufacturer was concerned with the surface finish of a cylindrical part produced through a single-point turning machining operation. The specifications for the part called for a nominal surface finish of 100 microinches, and although the process producing the parts had been stabilized through the use of $\overline{X}$ and R charts, the average surface finish of the parts produced by the process was approximately 120 microinches. To identify how the process could be centered at the target value of 100 microinches, a study team was formed to examine the problem.

The study team identified three variables of interest: spindle speed, feedrate, and tool nose radius. These variables are defined in Fig. 19.13. The study team decided to perform a 2^3 factorial design to study the effects of the three variables on, and ultimately to develop a model for, the machined surface finish. In selecting the levels for these three variables for the experiment, the study team selected levels spaced about the current operating conditions for the process (spindle speed = 2500 revolutions per minute, feedrate = 0.008 inches per revolution, and nose radius = $\frac{1}{64}$ inch). The high and low levels selected for the three variables for the experiment are summarized in Table 19.6. The results of the 2^3 experiment are given in Table 19.7.

Calculation of Effect Estimates and Determination of Significant Effects

Given the experimental results of Table 19.7, the effect estimates were computed using the calculation matrix and the algebraic method of effect calculation. The

TABLE 19.6	Variable Levels for the Surface Finish Experiment	
Variable	Low (−1) Level	High (+1) Level
x_1: Spindle speed (rev/min)	2400	2600
x_2: Feedrate (in./rev)	0.005	0.010
x_3: Nose radius (in.)	1/64	1/32

TABLE 19.7	2^3 Design and Results of the Surface Finish Experiment				
Test	x_1	x_2	x_3	Surface Finish (μ in.)	Run Order
1	$-$	$-$	$-$	52	4
2	$+$	$-$	$-$	63	2
3	$-$	$+$	$-$	213	8
4	$+$	$+$	$-$	206	5
5	$-$	$-$	$+$	31	1
6	$+$	$-$	$+$	28	7
7	$-$	$+$	$+$	110	3
8	$+$	$+$	$+$	105	6

calculation matrix along with the response values is given in Table 19.8, and the computed effects are displayed in Table 19.9.

To identify those effect estimates that are distinguishable from the noise in the experimental environment, a normal probability plot may be constructed. The normal plot associated with the effects of Table 19.9 is shown in Fig. 19.14. In the figure a straight line has been drawn through four of the effect estimates that have values close to zero. This line passes near the (0, 50%) point, and the scatter of the four smaller effect estimates about this line appears fairly random. It is seen that the other three effect estimates fall well off the line. We therefore conclude that the main effects 2 and 3 and the interaction effect 23 are important.

TABLE 19.8	Calculation Matrix of the Surface Finish Experiment								
Test	I	1	2	3	12	13	23	123	y_i
1	$+$	$-$	$-$	$-$	$+$	$+$	$+$	$-$	52
2	$+$	$+$	$-$	$-$	$-$	$-$	$+$	$+$	63
3	$+$	$-$	$+$	$-$	$-$	$+$	$-$	$+$	213
4	$+$	$+$	$+$	$-$	$+$	$-$	$-$	$-$	206
5	$+$	$-$	$-$	$+$	$+$	$-$	$-$	$+$	31
6	$+$	$+$	$-$	$+$	$-$	$+$	$-$	$-$	28
7	$+$	$-$	$+$	$+$	$-$	$-$	$+$	$-$	110
8	$+$	$+$	$+$	$+$	$+$	$+$	$+$	$+$	105

TABLE 19.9	Effect Estimates for the Surface Finish Experiment (microinches)

$$\text{Average} = 101 \qquad E_{12} = -5$$
$$E_1 = -1 \qquad E_{13} = -3$$
$$E_2 = 115 \qquad E_{23} = -37$$
$$E_3 = -65 \qquad E_{123} = 4$$

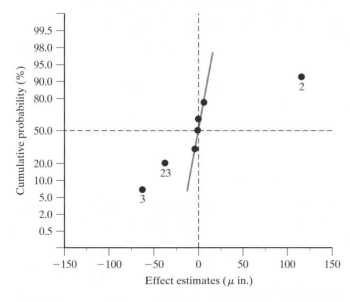

Figure 19.14   Normal probability plot of the effect estimates for the surface finish experiment

Model Development and Checking

For a 2^3 factorial design the response is assumed to be described by a model of the following form:

$$y = b_0 + b_1x_1 + b_2x_2 + b_3x_3 + b_{12}x_1x_2 + b_{13}x_1x_3 + b_{23}x_2x_3$$

$$+ b_{123}x_1x_2x_3 + \epsilon. \tag{19.14}$$

Since only the significant effects (model coefficients) need be included in the model, the fitted model for surface finish is given by

$$\hat{y} = \hat{b}_0 + \hat{b}_2x_2 + \hat{b}_3x_3 + \hat{b}_{23}x_2x_3, \tag{19.15}$$

and substituting the values for $\hat{b} = E/2$ and $\hat{b}_0 =$ average response, we obtain

$$\hat{y} = 101 + 57.5x_2 - 32.5x_3 - 18.5x_2x_3.$$

With a model for surface finish now developed, it must be checked for any potential inadequacies through study of the model residuals. Table 19.10 displays the measured and predicted responses along with the model residuals for each of the tests. The residuals of Table 19.10 should be randomly clustered about zero, and if plotted as a function of the test order, $\hat{y}$, x_1, x_2, and x_3 should show

TABLE 19.10			Predicted Response and Model Residuals for Surface Finish Experiment			
Test	x_1	x_2	x_3	y_i	$\hat{y}_i$	e_i
1	−	−	−	52	57.5	−5.5
2	+	−	−	63	57.5	5.5
3	−	+	−	213	209.5	3.5
4	+	+	−	206	209.5	−3.5
5	−	−	+	31	29.5	1.5
6	+	−	+	28	29.5	−1.5
7	−	+	+	110	107.5	2.5
8	+	+	+	105	107.5	−2.5

no trends/patterns, or in general, any structure whatsoever. Figures 19.15 to 19.20 display these residual plots. An examination of these plots shows no obvious structure remaining in the residuals, and thus we may conclude that our model, Eq. (19.15), has no apparent inadequacies.

Using the Model for Quality Improvement

With a prediction model now developed and checked for any potential inadequacies, the study team next turned its attention to using the model to find a solution to the problem at hand. In short, it was desired to find values for spindle speed, feedrate, and nose radius that center the turning process at a surface finish of 100 microinches, on the average. In terms of the fitted model, the team

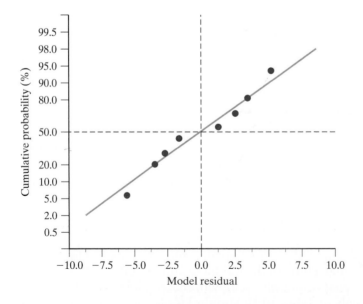

Figure 19.15   Normal probability plot of the model residuals

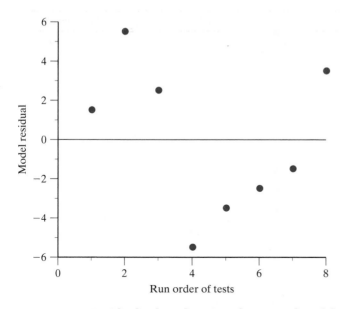

Figure 19.16   Residuals plotted against the run order of the tests

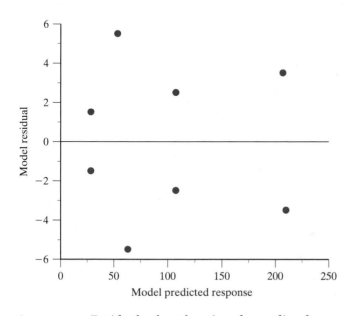

Figure 19.17   Residuals plotted against the predicted responses

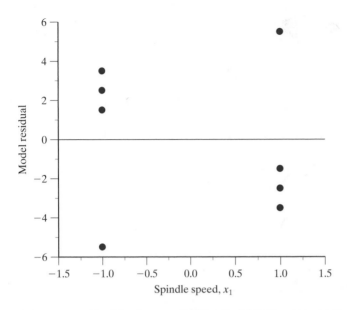

Figure 19.18   Residuals vs. spindle speed level: 2400 revolutions per minute (–1), 2600 revolutions per minute (+1)

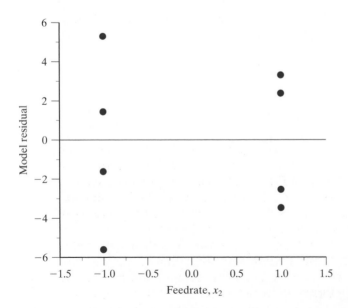

Figure 19.19   Residuals vs. feedrate level: 0.005 inch per revolution (–1), 0.010 inch per revolution (+1)

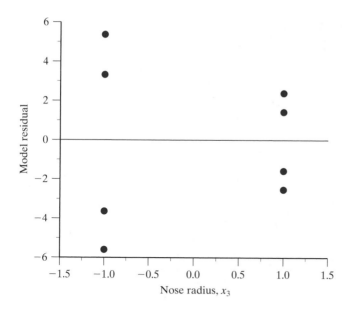

Figure 19.20   Residuals vs. nose radius level: $\frac{1}{64}$ inch (–1), $\frac{1}{32}$ inch (+1)

wanted to know values of x_1, x_2, and x_3 that produce a predicted surface finish of 100 microinches. Since the prediction model of Eq. (19.15) does not depend on the level of the spindle speed (x_1), values for x_2 and x_3 are sought that satisfy Eq. (19.15) when $\hat{y} = 100$.

Solving Eq. (19.15) for x_3 gives

$$x_3 = \frac{\hat{y} - 101 - 57.5x_2}{-32.5 - 18.5x_2}. \tag{19.16}$$

Using Eq. (19.16), we find that contours of constant predicted surface finish may be constructed as a function of x_2 and x_3. Such a contour plot was developed by the study team and is shown in Fig. 19.21. Table 19.11 shows some of the contour construction calculations for a contour value of 100 microinches.

An examination of Fig. 19.21 shows that a number of combinations of variables x_2 and x_3 produce a predicted surface finish of 100 microinches. Based on this contour plot, the study team decided to select the largest value for x_2 (feedrate), corresponding to an x_3 (nose radius) value of +1. The reasoning behind the selection of this particular x_2, x_3 combination was that in addition to its producing the predicted surface finish of 100 microinches, the larger value for x_2 would also lead to higher productivity. The same desire for higher productivity led to selection of the level for variable x_1 (i.e.,

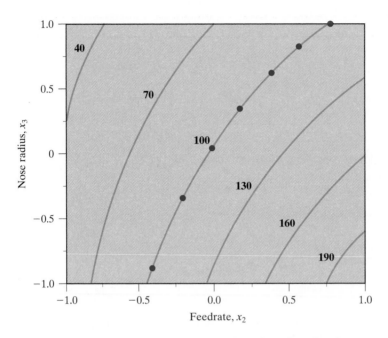

Figure 19.21   Contours of constant predicted surface finish in microinches

TABLE 19.11	Contour Calculations for $\hat{y} = 100$ microinches									
x_2 −1.0	−0.8	−0.6	−0.4	−0.2	0.0	0.2	0.4	0.6	0.8	1.0
x_3 −4.036	−2.542	−1.565	−0.876	−0.365	−0.031	0.345	0.601	0.814	0.994	1.147

$x_1 = +1$). In summary, therefore, the following levels were selected for x_1, x_2, and x_3:

$$x_1 = +1, \quad \text{spindle speed} = 2600 \text{ rev/min},$$
$$x_2 = +0.8, \quad \text{feedrate} = 0.0095 \text{ in./rev},$$
$$x_3 = +1, \quad \text{nose radius} = \tfrac{1}{32} \text{ in.}$$

When these conditions were applied to the turning process in confirmatory tests, the process was indeed recentered at approximately 100 microinches. Furthermore, the selected conditions provided for a very substantial increase in the productivity from the original operating conditions.

19.7   Use of Analysis of Variance for Two-Level Factorial Design Effect Evaluation

In Chapter 16 the Analysis-of-Variance (ANOVA) concept was introduced and used to decompose the total variation of a set of data into various sources. The

TABLE 19.12	Data for Replicated 2^2 Factorial Design						
Test	x_1	x_2	y_{i1}	y_{i2}	y_{i3}	y_{i4}	y_{i5}	$\bar{y}_i$
1	−1	−1	11	7	10	15	7	10
2	+1	−1	48	43	52	55	47	49
3	−1	+1	31	24	27	23	20	25
4	+1	+1	37	33	34	37	34	35

F-distribution was then used to judge whether the variation accounted for by a specific source was more than would arise simply due to the replication or pure error variation. We will now apply this ANOVA procedure to study models developed from two-level factorial designs.

Replicated 2^2 Factorial Design

Let us consider the replicated 2^2 factorial design presented in Table 19.12, from which we observe that there are $m = 2^k$ unique test conditions and $n = 5$ replicates for each condition.

For this factorial experiment, the following effects may be calculated:

$$\text{average} = 29.75, \quad E_1 = 24.50, \quad E_2 = 0.50, \quad E_{12} = -14.50.$$

Based on these calculated effect estimates, we may develop a model to predict the response. For now, let us make no decision as to which of the model coefficients are important and which are not. Let us initially consider a model with all possible terms. Under this condition, the prediction equation becomes

$$\hat{y} = \hat{b}_0 + \hat{b}_1 x_1 + \hat{b}_2 x_2 + \hat{b}_{12} x_1 x_2$$

or substituting for $\hat{b}$ the effect estimate, E, divided by 2,

$$\hat{y} = 29.75 + 12.25 x_1 + 0.25 x_2 + -7.25 x_1 x_2.$$

Recall that $\hat{b}_0$ is the sample mean or average response.

Let us now consider a decomposition of the variability in the data captured by the model into components due to each effect. As we know, the predicted response for the ith test may be expressed as

$$\hat{y}_i = \hat{b}_0 + \hat{b}_1 x_{1,i} + \hat{b}_2 x_{2,i} + \hat{b}_{12} x_{1,i} x_{2,i}.$$

Squaring both sides of this equation, and summing over all replicates, j, and tests, i, gives the following variability decomposition for the model, SS(model):

$$n \sum_{i=1}^{m} \hat{y}_i^2 = n \left(\sum_{i=1}^{m} \hat{b}_0^2 + \sum_{i=1}^{m} \hat{b}_1^2 x_{1,i}^2 + \sum_{i=1}^{m} \hat{b}_2^2 x_{2,i}^2 + \sum_{i=1}^{m} \hat{b}_{12}^2 x_{1,i}^2 x_{2,i}^2 \right). \quad (19.17)$$

It is noted that for the case of a 2^2 design (in fact, any 2^k design), each term in the parentheses on the righthand side of Eq. (19.17) can be simplified to the form $m\hat{b}^2$. Further, since $\hat{b} = E/2$ for 2^k designs, then for the example at hand,

$$SS(\text{model}) = (nm)(\text{average})^2 + \frac{(nm)E_1^2}{4} + \frac{(nm)E_2^2}{4} + \frac{(nm)E_{12}^2}{4}$$

$$= SS(\text{mean}) + SS(E_1) + SS(E_2) + SS(E_{12}).$$

In general, for a 2^k design where each of the $m = 2^k$ test conditions is performed n times, the sum of squares due to each effect, E, is given by

$$SS(E) = \frac{(nm)E^2}{4}, \text{ with one degree of freedom.}$$

To evaluate the appropriateness of a postulated model given the estimated effects from an experiment, we must determine the significance of each effect in terms of its contribution to the total sum of squares of the data. Only those effects that account for a large amount of variation will be included in the final prediction (fitted) model; the remaining effects will be ignored because their variability is not any larger than would be expected just due to experimental error. To evaluate the magnitude of the variability described by a single effect, we will compare it to an estimate of the experimental error (response) variance.

To illustrate the use of an ANOVA table for significance tests of the estimated effects and the prediction model, let us consider the replicated 2^2 factorial experiment that had the following effect estimates:

$$\text{average} = 29.75, \quad E_1 = 24.50, \quad E_2 = 0.50, \quad E_{12} = -14.50.$$

Based on these effect estimates the ANOVA table of Table 19.13 was constructed. We will now refer to that portion of the total variation in the data not explained by the model as the residual (i.e., that which is "left over"). In general, the residual

TABLE 19.13	ANOVA Table for the Model That Includes All Terms for Replicated 2^2 Factorial Design			
Source of Variation	Sum of Squares	Degrees of Freedom	Mean Square	Calculated F Value
Mean	17,701.25	1	17,701.250	1,323.46
E_1	3,001.25	1	3,001.250	224.39
E_2	1.25	1	1.250	0.09
E_{12}	1,051.25	1	1,051.250	78.60
Residual				
Pure error	214.00	16	13.375	
Total	21,969.00	20		

variation is composed of two parts, that due to pure error (replication) and that due to what we will call *lack of fit*. For now, this residual is composed solely of the pure error (replication) variation. Each of the calculated F values in the ANOVA table may be compared to an F-distribution with 1 and 16 degrees of freedom at $\alpha = 0.05$. Since $F_{1,16,0.95} = 4.49$, it may be concluded that the average, E_1, and E_{12} are statistically significant, since their calculated F values exceed 4.49.

Significance Tests for Model Lack of Fit

The fact that the initially postulated model is fit to the data is no guarantee that the model will provide good predictions of the process behavior. If the model postulation is misguided, the predictions will be poor even if the experiment is well designed and carried out. Earlier, we said that in diagnostically checking the model, two basic questions are addressed:

1. Are all the terms initially included in the model necessary?
2. Are the terms initially included in the model sufficient?

We have seen several ways that we can address the first of these two questions, both in Chapter 18 and in Chapter 19 to this point. The analysis of the residual plots has also given us some insights into the second question. We now will examine the latter in a more rigorous fashion through the technique of Analysis of Variance.

From the calculated F values of Table 19.13, we saw that two effects (E_1 and E_{12}) plus the average response were statistically significant—that is, their true effects differed from zero. One effect, E_2, was judged not to be important. Let us assume that the term associated with E_2 was never present in the prediction model. Thus the fitted prediction model for the replicated 2^2 factorial is

$$\hat{y} = 29.75 + 12.25x_1 - 7.25x_1x_2.$$

The predicted responses for the four tests based on this model are listed in Table 19.14.

We have shown earlier that the sum of squares due to a model is equal to the total of the sums of squares associated with each of the effects in the model. For our example,

TABLE 19.14	Data for Replicated 2^2 Factorial Design with Predicted Responses Using the Fitted Model							
Test	x_1	x_2	y_{i1}	y_{i2}	y_{i3}	y_{i4}	y_{i5}	$\bar{y}_i$	$\hat{y}_i$
1	−1	−1	11	7	10	15	7	10	10.25
2	+1	−1	48	43	52	55	47	49	49.25
3	−1	+1	31	24	27	23	20	25	24.75
4	+1	+1	37	33	34	37	34	35	34.75

$$SS(model) = SS(mean) + SS(E_1) + SS(E_{12})$$
$$= 17{,}701.25 + 3001.25 + 1051.25 = 21{,}753.75.$$

Since the total sum of squares has been previously calculated to be

$$SS(total) = 21{,}969.00,$$

the SS(residual) can be calculated by subtraction, as can the degrees of freedom associated with the residual sum of squares,

$$SS(residual) = SS(total) - SS(model)$$
$$= 21{,}969.00 - 21{,}753.75$$
$$= 215.25,$$

$$DOF(residual) = DOF(total) - DOF(model)$$
$$= 20 - 3$$
$$= 17.$$

Alternatively, we could calculate the residual sum of squares as the sum of the squared deviations of the data from their model predictions,

$$SS(residual) = \sum_{i=1}^{m} \sum_{j=1}^{n} (y_{ij} - \hat{y}_i)^2$$
$$= (11 - 10.25)^2 + (7 - 10.25)^2 + \cdots + (37 - 34.75)^2$$
$$+ (34 - 34.75)^2$$
$$= 215.25.$$

Evaluating the residual sum of squares based on this equation, we find that the SS(residual) is, again, equal to 215.25.

We now note that the residual sum of squares is really composed of two parts, the pure error sum of squares due to replication and the remainder of all of the variation in the data that is (1) not accounted for by the model and (2) not a result of the replication in the data. This latter component we will refer to as the "lack-of-fit" sum of squares. Since the pure error sum of squares has already been determined to be 214.00, then, by subtraction,

$$SS(lack\ of\ fit) = SS(residual) - SS(pure\ error)$$
$$= 215.25 - 214.00$$
$$= 1.25.$$

All of the above can be laid out in an ANOVA table, as shown in Table 19.15.

TABLE 19.15	ANOVA for Checking the Model for the Replicated 2^2 Factorial Experiment			
Source of Variation	Sum of Squares	Degrees of Freedom	Mean Square	Calculated F Value
Model	21,753.75	3		
Mean	17,701.25	1	17,701.250	1,323.46
Model – (mean)	4,052.50	2	2,026.250	151.50
Residual	215.25	17		
LOF	1.25	1	1.250	0.09
Pure error	214.00	16	13.375	
Total	21,969.00	20		

The lack-of-fit variation will be indistinguishable from the pure error variation if the model is an adequate one. Therefore, a model will be considered inadequate (suffer from lack of fit) if the calculated F value for the ratio of the MS(LOF) to the MS(pure error) exceeds the tabulated value for F with the appropriate degrees of freedom,

$$F_{\text{LOF}} = \frac{\text{MS(LOF)}}{\text{MS(pure error)}}$$

$$= \frac{1.25}{13.375}$$

$$= 0.09.$$

Since the calculated F value for the lack of fit, $F_{\text{LOF}} = 0.09$, is less than $F_{1,16,0.95} = 4.49$, we have no evidence to suggest any significant lack of fit.

It should be noted that for this example, the lack-of-fit sum of squares is the sum of squares associated with the estimated main effect of variable 2, the term in the original model that was found to be insignificant,

$$\text{SS}(E_2) = \frac{(nm)E_2^2}{4}$$

$$= \frac{(4)(5)(0.5)^2}{4}$$

$$= 1.25.$$

Once either dropped from the model, or not included in the first place, such terms are part of the residual.

The purpose of checking the model for lack of fit is to reveal problems with the model-building process. In dealing with 2^k experiments that involve a larger number of variables, say four or more, we might simply assume that higher-order interactions can be neglected (third and higher, perhaps) from the outset and not even estimate these effects; that is, we may not include them in the calculation matrix. Subsequently, examination of the model lack of fit might reveal inadequacies that could be overcome if one or more of these interactions were added to the model. It may also be possible that the phenomenon under study may contain structure—say, a quadratic effect—that cannot be captured through the data by the associated model. Such will not be revealed through the lack-of-fit analysis. In this case, additional experimentation will be required to build an adequate model.

An Overall Measure of Model Performance

The fact that a model is significant (contains one or more important terms) and that the model does not suffer from lack of fit does not necessarily mean that the model is a good one. If the experimental environment is quite noisy or some important variables have been left out of the experiment, then it is possible that the portion of the variability in the data not explained by the model (the residual) could be large. To quantify this, we introduce a measure of the model's overall performance, a quantity called the *coefficient of determination*, denoted by R^2. This R^2 (which is not the square of a range) is the percentage of the total variability in the data that is accounted for by the model. The percent variation accounted for by the model is

$$R^2 = \frac{\text{SS(model)}}{\text{SS(total)}}. \tag{19.18}$$

For our replicated 2^2 factorial we obtain

$$R^2 = \frac{21,753.75}{21,969} = 0.9902 \text{ or } 99.02\%.$$

The R^2 value is given by the equation above; however, often it is more meaningful to adjust or correct the SS(model) and SS(total) for the mean. Hence we define an R^2_{cor} value given by

$$R^2_{\text{cor}} = \frac{\text{SS(model)} - \text{SS(mean)}}{\text{SS(total)} - \text{SS(mean)}}. \tag{19.19}$$

For our example, R^2_{cor} is, therefore,

$$R^2_{\text{cor}} = \frac{21,753.75 - 17,701.25}{21,969 - 17,701.25} = 0.9496 \text{ or } 94.96\%.$$

The interpretation of R^2_{cor} is that the variable terms in the fitted model (E_1 and E_{12}) explain about 95% of the variability in the responses about the average of 29.75.

19.8   Use of Analysis of Variance on the Glove Box Door Alignment Study

Let us return to the glove box door parallelism experiment and examine how the ANOVA procedure described in the preceding section may be applied to this problem. The designed experiment and responses are given again in Table 19.16. The calculated effect estimates for this experiment are also summarized below.

$$\text{average} = -0.087,$$

$E_1 = -0.654,$	$E_{13} = -0.117,$	$E_{123} = 0.172,$
$E_2 = 0.794,$	$E_{14} = -0.031,$	$E_{124} = 0.101,$
$E_3 = 0.638,$	$E_{23} = -0.191,$	$E_{134} = -0.138,$
$E_4 = 0.322,$	$E_{24} = -0.154,$	$E_{234} = -0.104,$
$E_{12} = 0.147,$	$E_{34} = 0.009,$	$E_{1234} = 0.121.$

As we saw in Section 19.7 for replicated 2^k designs, the formulas for calculating the various sum-of-squares values are

$$\text{SS(mean)} = nm\hat{b}_0^2,$$

$$\text{SS(effect)} = \frac{nmE^2}{4} \quad \text{for each of the}(m-1) \text{ effects,}$$

TABLE 19.16	2^4 Design and Results of the Glove Box Door Experiment						
					Run 1,	Run 2,	Average,
Test	x_1	x_2	x_3	x_4	y_{i1}	y_{i2}	$\bar{y}_i$
1	$-$	$-$	$-$	$-$	-1.44	-0.08	-0.760
2	$+$	$-$	$-$	$-$	-1.79	-1.01	-1.400
3	$-$	$+$	$-$	$-$	0.39	0.17	0.280
4	$+$	$+$	$-$	$-$	-0.50	-0.24	-0.370
5	$-$	$-$	$+$	$-$	-0.20	0.17	-0.015
6	$+$	$-$	$+$	$-$	-0.79	-0.64	-0.715
7	$-$	$+$	$+$	$-$	1.22	0.28	0.750
8	$+$	$+$	$+$	$-$	0.21	0.28	0.245
9	$-$	$-$	$-$	$+$	-0.40	-0.65	-0.525
10	$+$	$-$	$-$	$+$	-0.63	-1.19	-0.910
11	$-$	$+$	$-$	$+$	0.47	0.44	0.455
12	$+$	$+$	$-$	$+$	-0.01	-0.03	-0.020
13	$-$	$-$	$+$	$+$	1.29	0.64	0.965
14	$+$	$-$	$+$	$+$	-1.17	0.14	-0.515
15	$-$	$+$	$+$	$+$	0.48	1.06	0.770
16	$+$	$+$	$+$	$+$	0.40	0.34	0.370

$$SS(\text{pure error}) = m(n-1)s_p^2,$$

$$SS(\text{total}) = \sum_{i=1}^{m}\sum_{j=1}^{n}y_{ij}^2,$$

where n is the number of repeat trials (replications) for each unique test condition, and m is equal to 2^k. It is assumed here that each unique test condition is repeated an equal (n) number of times.

Alternatively, in this case the pure error sum of squares may be calculated by subtraction:

$$SS(\text{pure error}) = SS(\text{total}) - SS(\text{mean}) - [\text{all}(m-1)\ SS(E)\text{'s}].$$

ANOVA Table

Based on the equations above and the replicated 2^4 design given in Table 19.16, the ANOVA table in Table 19.17 was constructed. The calculated F values in the ANOVA table, under the null hypothesis that the true variable effects are zero, are distributed according to the F-distribution with 1 and 16 degrees of freedom. Since the calculated F values associated with E_1, E_2, and E_3 exceed the critical value for F at a 5% significance level ($F_{1,16,0.95} = 4.49$), we may conclude that their true mean effects are not zero. In other words, the variability accounted for by E_1, E_2, and E_3 is larger than what can be explained solely by the forces of experimental error. Note that this is the same conclusion that was

TABLE 19.17	ANOVA for the Effects of the Glove Box Alignment Study			
Source of Variation	Sum of Squares	Degrees of Freedom	Mean Square	Calculated F Value
Mean	0.24325	1	0.24325	1.202
E_1	3.42565	1	3.42565	16.922
E_2	5.04825	1	5.04825	24.938
E_3	3.25763	1	3.25763	16.092
E_4	0.82883	1	0.82883	4.094
E_{12}	0.17258	1	0.17258	0.853
E_{13}	0.10928	1	0.10928	0.540
E_{14}	0.00750	1	0.00750	0.037
E_{23}	0.29070	1	0.29070	1.436
E_{24}	0.19065	1	0.19065	0.942
E_{34}	0.00070	1	0.00070	0.003
E_{123}	0.23633	1	0.23633	1.167
E_{124}	0.08100	1	0.08100	0.400
E_{134}	0.15263	1	0.15263	0.754
E_{234}	0.08715	1	0.08715	0.431
E_{1234}	0.11640	1	0.11640	0.575
Pure error	3.23895	16	0.20243	
Total	17.48750	32		

reached based on using the confidence-interval approach. Note again that the mean response is not different from zero.

An alternative method to check the significance of the effects is to compare the MS(E)'s against a critical value of the mean square that can be obtained as the product of MS(pure error) and the critical F value. This approach requires fewer calculations, in that F values need not be calculated for each effect. Since the critical F value ($F_{1,16,0.95}$) is 4.49, the critical MS value is given by

$$\text{MS(critical)} = \text{MS(pure error)}F_{1,16,0.95} = (0.2024)(4.49) = 0.9088.$$

Comparing the MS(E)'s with MS(critical), only MS(E_1), MS(E_2), and MS(E_3) exceed the critical mean square. As before, we conclude that the main effects of variables 1, 2, and 3 are the only effects that are statistically different from zero.

Checking for Model Adequacy and Lack of Fit

We have concluded that three effects (E_1, E_2, and E_3) differ significantly from zero. Consequently, a model to predict the parallelism response need only contain terms associated with these effects. The model for the predicted response is therefore given by

$$\hat{y} = -0.327x_1 + 0.397x_2 + 0.319x_3. \tag{19.20}$$

The sum of squares accounted for by the model may be obtained by adding the sums of squares due to E_1, E_2, and E_3:

$$\text{SS(model)} = \text{SS}(E_1) + \text{SS}(E_2) + \text{SS}(E_3) = 3.42565 + 5.04825 + 3.25763$$
$$= 11.73153.$$

Based on this model, predictions may be made for each of the $2^4 = 16$ test conditions. Since the model residuals are the differences between the model predictions and measured responses, the residual sum of squares is given by (refer to Table 19.3 for the model predictions):

$$\text{SS(residual)} = \sum_{i=1}^{m}\sum_{j=1}^{n}(y_{ij} - \hat{y}_i)^2$$
$$= [-1.44 - (-0.389)]^2$$
$$+ [-0.080 - (-0.389)]^2 + \cdots + (0.400 - 0.389)^2$$
$$+ (0.340 - 0.389)^2$$
$$= 5.75597.$$

Alternatively, since the SS(model) + SS(residual) is equal to the total sum of squares [SS(total) = 17.4875], the SS(residual) could have been obtained by subtraction.

The residual sum of squares is the variability in the data not described by the model. Part of this residual variability is due to the pure error sum of squares, SS(pure error) = 3.23895, with the remainder equal to the lack-of-fit sum of squares, SS(LOF). The lack-of-fit sum of squares, SS(LOF), may be calculated by subtracting SS(pure error) from SS(residual) [i.e., SS(LOF) = 5.75597 − 3.23895 = 2.51702]. The SS(LOF) is in reality the total of the sums of squares of the 12 insignificant effects plus the sum of squares of the sample mean response. When we add up all these sums of squares from Table 19.17, we likewise get SS(LOF) = 2.51702 with 13 degrees of freedom.

We may now evaluate the performance of the fitted prediction model by checking to see if it has a significant amount of lack of fit. This is most conveniently done via the ANOVA table shown in Table 19.18. The critical value of F with 3 and 16 degrees of freedom at a 5% level for checking the overall model significance is 3.239, which is smaller than the F calculated for the model, shown in the last column of Table 19.18. This suggests that the fitted model is statistically significant in its explanation of the variation in the responses. To check for potential model lack of fit, we compare the calculated F for the LOF/pure error ratio against the critical F value with 13 and 16 degrees of freedom, at a 5% level. Since this critical F of 2.397 is greater than the calculated F of 0.956, we have no evidence to suggest any significant lack of fit.

We may also examine the performance of the fitted model by calculating the coefficient of determination, R^2,

$$R^2 = \frac{\text{SS(model)}}{\text{SS(total)}} = \frac{11.73153}{17.4875} = 0.6709 \text{ or } 67.09\%.$$

The interpretation of R^2 is that the fitted model explains about 67% of the variability in the responses. Approximately one-third of the response variability is not explained by the fitted model. This may be indicating that variables not considered in the original design (i.e., variables that are currently part of the noise structure) are important. Potentially, if these variables were added to the model, the model would describe more of the variability in the alignment response. Of course, it is also possible (and perhaps likely) that noise factors operating in the experimental environment are responsible for much of the unexplained variation.

TABLE 19.18	ANOVA for Checking the Model of Glove Box Door Parallelism			
Source of Variation	Sum of Squares	Degrees of Freedom	Mean Square	Calculated F Value
Model	11.73153	3	3.91051	19.318
Residual	5.75597	29		
Lack of fit	2.51702	13	0.19362	0.956
Pure error	3.23893	16	0.20243	
Total	17.48750	32		

The use of SPC prior to the experiment or techniques such as blocking during the experiment could alleviate this problem.

EXERCISES

19.1 Describe the four stages associated with the model-building process. For each stage discuss the specific task(s) and associated methods that may be used.

19.2 What two specific problems can arise with a model developed from any designed experiment? That is, given a fitted model, what specific types of inadequacies can this model have in terms of its ability to describe the phenomenon of interest?

19.3 A replicated 2^4 factorial design is performed, and a model is developed based on the experimental results. List the types of residual plots that should be prepared to check the model. In each case, describe the patterns in the residual plot that would lead you to believe that the model is adequate.

19.4 A 2^3 factorial design was conducted to study the influence of temperature (x_1), pressure (x_2), and cycle time (x_3) on the occurrence of splay in an injection molding process. For each of the eight unique trials, 50 parts were made, and the response observed was the number of incidences of the occurrence of splay on the part surface across all 50 parts. The design matrix and test responses are given in the table below.

Test	1	2	3	y
1	−	−	−	12
2	+	−	−	15
3	−	+	−	24
4	+	+	−	17
5	−	−	+	24
6	+	−	+	16
7	−	+	+	24
8	+	+	+	28

(a) Calculate all variable effects.

(b) Construct a normal probability plot for the effect estimates. Which effects appear to be statistically significant?

(c) Write down the appropriate mathematical model, including only those terms that appear to be significant.

(d) Calculate the model residuals.

(e) Prepare all the residual plots necessary to check the model diagnostically. Comment.

19.5 A 2^3 factorial design was used to study the possible effects of temperature, time, and pressure on the yield of a chemical process. The data, including experimental results, are shown in the accompanying tables.

Variable	Low Level	High Level
1. Temperature (°F)	80	100
2. Time (min)	5	7
3. Pressure (psi)	120	140

Test	1	2	3	y
1	−	−	−	10
2	+	−	−	20
3	−	+	−	4
4	+	+	−	10
5	−	−	+	8
6	+	−	+	18
7	−	+	+	6
8	+	+	+	12

(a) Calculate the effects and construct a normal probability plot for the effect estimates. Which of the effects appear to be statistically significant?

(b) Write down the appropriate mathematical model, including only those terms that appear to be significant. Calculate the model residuals.

(c) Prepare all the residual plots necessary to check the model diagnostically.

(d) Prepare a contour plot for predicted yield that displays temperature and time on the horizontal and vertical axes, respectively, for a pressure of 130 psi.

(e) Within the range of the variables studied, where is the yield a maximum? What is the predicted yield at this point?

19.6 Consider again the 2^3 factorial design of Exercise 18.13.

(a) Calculate the variable effect estimates.

(b) Construct an ANOVA table to identify which of the effect estimates are statistically significant.

(c) Develop a model for the response based on the results of part (b). Construct an ANOVA table to see if the developed model suffers from lack of fit.

19.7 A chemical engineer performed the experiments shown in the first table, randomizing the order of the tests within each week. The response of

Test	Temperature	Catalyst	pH	Week 1	Week 2	Week 3
1	−	−	−	60.4 (5)	62.1 (10)	63.4 (19)
2	+	−	−	64.1 (8)	79.4 (11)	74.0 (23)
3	−	+	−	59.6 (1)	61.2 (13)	57.5 (17)
4	+	+	−	66.7 (3)	67.3 (16)	68.9 (22)
5	−	−	+	63.3 (7)	66.0 (9)	65.3 (24)
6	+	−	+	91.2 (4)	77.4 (14)	84.9 (18)
7	−	+	+	68.1 (6)	71.3 (12)	68.6 (21)
8	+	+	+	75.3 (2)	77.1 (15)	76.1 (20)

interest is the chemical yield, and the run order is given in parentheses. The variable levels are given in the second table.

Level	Temperature (°F)	Catalyst	pH
−	170	A	6
+	180	B	7

(a) Calculate the effects.

(b) Calculate the sample pooled variance, s_p^2, based on the replication for each test condition, and use this estimate to determine the sample variance of an effect. Develop 95% confidence intervals for each effect estimate. Find the statistically significant effects.

(c) Write down the appropriate mathematical model, including only those terms found to be significant. Calculate the model residuals, and prepare all the residual plots necessary to check the model diagnostically.

(d) Prepare a contour plot for predicted yield that displays temperature and pH on the horizontal and vertical axes, respectively, for catalyst type B.

(e) Within the range of the variables studied, where is the yield a maximum? What is the predicted yield at this point?

19.8 Consider again the data presented in Exercise 19.7, and use Bartlett's test to check the homogeneity of the variance of the response. Prepare a graphical representation of the experimental design, and graphically depict the result of Bartlett's test.

19.9 Recall the injection molding study of Chapter 17.

Variable	Unit	Low Level	High Level
Orifice size, O	mm	1.30	1.50
Pump setting, P		4.00	4.50
Isocynate temperature, T	°C	22	30

Test	x_1	x_2	x_3	Test Order	y
1	−1	−1	−1	6	1550
2	+1	−1	−1	8	1925
3	−1	+1	−1	1	2150
4	+1	+1	−1	2	2350
5	−1	−1	+1	5	1525
6	+1	−1	+1	3	1800
7	−1	+1	+1	4	2175
8	+1	+1	+1	7	2200

The calculated effect estimates for the impingement pressure (psi) response were

$$\text{average} = 1959.375, \quad E_{12} = -106.25,$$
$$E_1 = 218.75, \quad E_{13} = -68.75,$$
$$E_2 = 518.75, \quad E_{23} = 6.25,$$
$$E_3 = -68.75, \quad E_{123} = -18.75.$$

(a) Plot these effect estimates on normal probability paper, and postulate a model for the impingement pressure.

(b) Construct a contour plot for the response surface over the range of variables tested.

(c) If it is desired to control the isocynate impingement pressure to values of 1750 and 1950 psi for the next phase of the experiment (refer to Fig. 17.2), identify appropriate settings for the orifice size, pump setting, and isocynate temperature.

19.10 Referring to Exercise 19.9, suppose that the model proposed to describe the impingement pressure response was

$$\hat{y} = 1959.375 + 259.375x_2.$$

(a) Calculate all the model residuals.

(b) Construct all appropriate residual plots.

(c) If the proposed model is found to be inadequate, propose a new model, and conduct the necessary model diagnosis procedures.

19.11 A 2^3 factorial experiment was designed to test the effect of three different variables on the ultimate tensile strength of a weld. The three variables and the levels used are shown along with the test results in the tables below.

Variable	Low Level	High Level
1. Ambient temperature (°F)	20	40
2. Wind velocity (mph)	0	20
3. Bar size (in.)	1/2	3/4

Test	1	2	3	y_1	y_2
1	−	−	−	88.6	90.2
2	+	−	−	92.6	83.4
3	−	+	−	70.3	74.9
4	+	+	−	82.6	99.2
5	−	−	+	78.7	80.1
6	+	−	+	99.7	95.5
7	−	+	+	80.7	76.3
8	+	+	+	93.7	91.7

Use Bartlett's test to check the homogeneity of the variance of the response. Prepare a graphical representation of the experimental design, and graphically depict the result of Bartlett's test.

19.12 Consider again the 2^3 factorial design of Exercise 19.11. Assume that the following model form is postulated to describe the data given:

$$y = b_0 + b_1 x_1 + b_2 x_2 + b_3 x_3 + \epsilon.$$

(a) Estimate the variable effects and use them to estimate the coefficients in the model above. Construct an ANOVA table to determine which of the model coefficients are significant.

(b) Construct an ANOVA table to see if the model developed suffers from lack of fit.

(c) If the model does suffer from lack of fit, postulate a new model and perform a complete analysis of the new model.

19.13 A 2^3 factorial design was performed. Each test, or unique combination of the three variables, was performed three times. The variable levels and results of the tests are shown in the tables below.

Variable	Low Level	High Level
1. Temperature (°C)	80	120
2. Pressure (psi)	50	70
3. Reaction time (min)	5	15

				Replicate			
Test	x_1	x_2	x_3	1	2	3	Average
1	80	50	5	61.43	58.58	57.07	59.03
2	120	50	5	75.62	77.57	75.75	76.31
3	80	70	5	27.51	34.03	25.07	28.87
4	120	70	5	51.37	48.49	54.37	51.41
5	80	50	15	24.80	20.69	15.41	20.30
6	120	50	15	43.58	44.31	36.99	41.63
7	80	70	15	45.20	49.53	50.29	48.34
8	120	70	15	70.51	74.00	74.68	73.07

Use Bartlett's test to check the homogeneity of the variance of the response. Prepare a graphical representation of the experimental design, and graphically depict the result of Bartlett's test.

19.14 An in-depth examination of the experiment described in Exercise 19.13 revealed that the eight tests (24 trials) were actually conducted over a two-week period. During the first week, tests 1, 4, 6, and 7 were performed, and during the second week tests 2, 3, 5, and 8 were performed. Concern then arose as to whether the experimental error variation was the same during weeks 1 and 2. Perform a statistical test to see if the pure error variation, as characterized by the replicated responses, is different from one week to another.

19.15 Consider again the 2^3 factorial design presented in Exercise 19.13.
  (a) Calculate the between-test and within-test (pure error) sums of squares for the data.
  (b) Using the results of part (a), calculate the associated mean squares and conduct a test to determine if the variation between tests is larger than one might expect, given the variation within tests.

19.16 Consider again the factorial design presented in Exercise 19.13.
  (a) Calculate the variable effect estimates.
  (b) Construct an ANOVA table to identify which effects appear to be important.
  (c) Develop a model for the response based on the results of part (b). Construct an ANOVA table to see if the model developed suffers from lack of fit.

19.17 This problem is adapted from Example 9.2 in Montgomery.[2] A factorial experiment was performed in a pilot plant to study the effects of four factors: temperature (x_1), pressure (x_2), concentration of reactant (x_3), and stirring rate (x_4) on the filtration rate of a product. A complete analysis of the experiment led to the following model, which predicts filtration rate as a function of these four factors:

$$\hat{y} = 70.06 + 10.82x_1 + 4.94x_3 + 7.32x_4 - 9.07x_1x_3 + 8.32x_1x_4.$$

The filtration rate is measured in units of percent of theoretical value, and the levels of the variables over which the experiment was conducted were as given in the table below.

Variable	Low Level (−)	High Level (+)
Temperature (°C)	140	180
Pressure (psi)	100	120
Concentration of reactant (%)	4	8
Stirring rate (rpm)	50	70

  (a) Construct a contour plot for the following stirring rates: 50, 60, and 70 rpm.
  (b) Make recommendations for the process operation so that the filtration rate is closest to the predicted theoretical value (i.e., the response is closest to 100%).

19.18 A 2^2 factorial design was run to test the relative importance of fixture type (x_1) and workplace layout (x_2) on the time to complete an assembly task. Each of the four test conditions was run four times. Each trial observed the total time to complete 20 consecutive assemblies. The data are given in the accompanying table.
  (a) Calculate all the variable effects. Use the four replicates for each test condition to estimate the variance of an effect. Develop 95% confidence intervals to determine which effect estimates are significant.

[2]D. C. Montgomery, *Design and Analysis of Experiments*, 3d ed., Wiley, New York, 1990.

	Workplace Layout			
	A		B	
Fixture Type 1	10.24	9.78	14.30	13.48
	9.96	10.23	12.10	11.98
Fixture Type 2	9.90	8.50	9.82	10.56
	8.86	8.54	9.42	9.50

    (b) Calculate the model residuals. Construct all appropriate residual plots. Comment on the appearance of the plots.

19.19 Consider again the replicated 2^2 factorial design of Exercise 19.18.
    (a) Construct an ANOVA table to see if the variation between the various fixture/layout combinations is statistically significant.
    (b) Calculate the variable effect estimates.
    (c) Construct an ANOVA table to identify which of the effect estimates are statistically significant.

19.20 Consider the following residual plots developed from the results of two-level factorial designs.

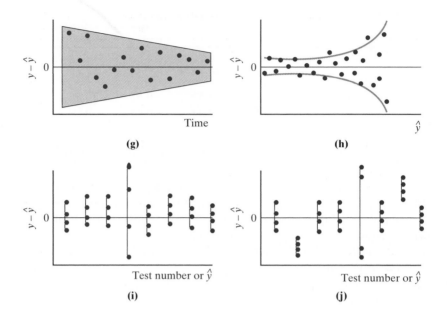

The residuals show the discrepancies between the model and the actual data. For each residual plot, describe in words what could be causing the residual patterns displayed. Be very specific. Where appropriate, present a model form that might describe the response of interest, and then propose an improved model.

19.21 A 2^4 factorial design was conducted to study the effect of four variables [side rake angle (degrees), cutting speed (surface feet per minute), feedrate (inches per revolution), and tool nose radius (inches)] on the finish of a machined surface. The table below describes the conditions for the 16 tests that were performed as well as the measured surface finish (R_a) value that was observed.

Test	Side Rake Angle (deg)	Cutting Speed (sf/min)	Feedrate (in./rev)	Nose Radius (in.)	Surface Finish (μ in.)
1	0	600	0.005	1/64	74
2	10	600	0.005	1/64	85
3	0	900	0.005	1/64	56
4	10	900	0.005	1/64	37
5	0	600	0.010	1/64	132
6	10	600	0.010	1/64	284
7	0	900	0.010	1/64	170
8	10	900	0.010	1/64	202
9	0	600	0.005	1/32	22
10	10	600	0.005	1/32	15
11	0	900	0.005	1/32	25
12	10	900	0.005	1/32	13

(Continued)

Test	Side Rake Angle (deg)	Cutting Speed (sf/min)	Feedrate (in./rev)	Nose Radius (in.)	Surface Finish (μ in.)
13	0	600	0.010	1/32	156
14	10	600	0.010	1/32	76
15	0	900	0.010	1/32	98
16	10	900	0.010	1/32	136

(a) Calculate the effects and construct a normal probability plot for the effect estimates. Which of the effects appear to be statistically significant?

(b) Write down the appropriate mathematical model, including only those terms that appear to be significant. Calculate the model residuals.

(c) Prepare all the residual plots necessary to check the model diagnostically.

(d) If the proposed model is found to be inadequate, propose a new model, and conduct the steps necessary to develop and check the new model.

19.22 A replicated 2^2 factorial experiment was performed, the results of which are shown in the table below.

x_1	x_2	y_{i1}	y_{i2}
−	−	0.50	0.46
+	−	0.42	0.37
−	+	0.33	0.30
+	+	0.11	0.16

(a) Calculate the effect estimates (i.e., the average, E_1, E_2, and E_{12}).

(b) Construct an ANOVA table to identify which of the effect estimates are significant.

19.23 A 2^2 factorial design was conducted to study the effect of welding temperature (°F) and flux quantity (in./min) on the resulting weld strength (psi). The table below describes the conditions that were examined in the experiment and the associated measured weld strengths (note that each unique combination of temperature and flux quantity was performed three times). The table also shows the average and sample variance for the weld strength calculated for each unique combination of the temperature (variable 1) and flux quantity (variable 2).

Test	x_1, Temp.	x_2, Flux Quantity	Trial 1	Trial 2	Trial 3	Average	Sample Variance
1	300	10	3088	2724	3586	3132.67	187,257.33
2	500	10	799	1202	1079	1026.67	42,656.33
3	300	20	813	1124	1101	1012.67	30,032.33
4	500	20	2806	2946	3336	3029.33	75,433.33

(a) It is assumed that a model of the form $y = b_0 + b_1x_1 + b_2x_2 + \epsilon$ may be used to describe the response. Estimate the coefficients in the postulated model using calculated variable effect estimates.

(b) Construct an ANOVA table to see if the model developed suffers from lack of fit.

(c) If the model examined in part (b) suffers from lack of fit, propose a new model, and construct an ANOVA table to check it for lack of fit.

19.24 A 2^4 factorial design was conducted to study the effect of four variables on a response of interest. The table below shows the conditions that were examined in the experiment and the associated response (note that each unique combination of the variables was performed twice).

x_1	x_2	x_3	x_4	Run 1	Run 2
−	−	−	−	13	28
−	+	−	−	9	18
+	+	+	−	38	33
+	+	−	−	57	55
−	−	+	−	52	34
−	+	+	−	30	42
+	−	−	−	41	53
+	−	+	−	38	41
−	−	−	+	12	31
−	+	−	+	30	23
+	+	+	+	53	32
+	+	−	+	39	45
−	−	+	+	49	41
−	+	+	+	43	33
+	−	−	+	60	62
+	−	+	+	50	27

(a) Construct an ANOVA table to see if the variation between the tests is statistically significant.

(b) Calculate the variable effect estimates.

(c) Construct an ANOVA table to identify which of the effect estimates are statistically significant.

(d) Develop a model for the response based on the results of part (b). Construct an ANOVA table to see if the model developed suffers from lack of fit.

19.25 A convenience store owner is trying to determine if the sales are different for four types of soda stocked in the cooler. The sales (in dollars) for each type for three randomly selected weeks are to be used for the analysis.

(a) Construct a simple ANOVA table to determine whether or not there exists a significant difference among the various soda types.

(b) Summarize and comment on the results.

Soda Brand	Regular	Diet
Caffeine-Free Cola	65	79
	68	71
	73	85
Cola-Caf	80	71
	74	83
	77	75

19.26 A dog food company is interested in how dogs respond to the taste of one of its new products. Two males and two females were selected from each of two different breeds of dogs for the experiment. To assess the taste of the new product, the number of ounces of the product consumed by each dog during one meal was recorded. The amount of the new product consumed by each dog is shown in the table below.

Dog Breed	Male	Female
Collie	15, 14	12, 14
Dachshund	7, 9	8, 6

(a) Construct an ANOVA table to see if the variation among the various breed/sex combinations is statistically significant.
(b) Calculate the variable effect estimates.
(c) Construct an ANOVA table to identify which of the effect estimates are statistically significant.

19.27 An experimenter is studying the number of defectives observed during a one-hour time period for four different inspectors and three different fixturing devices. The data were collected by randomly selecting two observations (one-hour inspection results) for each inspector/fixture combination from the scrap report record for a given week.

Fixture	A	B	C	D
I	7	5	2	6
	6	6	4	6
II	4	2	0	0
	4	4	2	2
III	5	4	4	3
	4	5	4	4

Inspector (column header spanning A, B, C, D)

The data displayed in the table may be described with a model of the following form:

$$y_{ijk} = \bar{\bar{\bar{y}}} + (\bar{\bar{y}}_i - \bar{\bar{\bar{y}}}) + (\bar{\bar{y}}_j - \bar{\bar{\bar{y}}}) + \text{interaction} + (y_{ijk} - \bar{y}_{ij}),$$

where y_{ijk} = response for the ith fixture, jth inspector, and kth replicate,

$\bar{\bar{\bar{y}}}$ = grand average over all 24 trials,

$\bar{\bar{y}}_i$ = average response for the ith fixture,

$\bar{\bar{y}}_j$ = average response for the jth inspector,

$\bar{y}_{ij}$ = average response for the ith fixture and jth inspector.

(a) Find the expression for the interaction term in the model.

(b) Derive expressions for the sum of squares due to the grand average, fixtures, inspectors, fixture/inspector interaction, and pure error.

(c) Construct an ANOVA table for the data. Perform the tests of significance and comment on the results.

19.28 A fisherman has conducted a 2^5 factorial design to determine the effects of five factors on the number of fish he catches in a four-hour period. The variables studied and their levels for experiment are shown in the table below.

Variable	Low Level	High Level
1. Fishing location	Pier	Boat
2. Bait type	Worms	Night crawlers
3. Time of day	Day	Night
4. Weather	No rain	Rain
5. Hook size	Small	Large

The experimental trials were all performed at Jack's Fish Farm in northern Wisconsin—stocked with bluegills. The following table gives the results of the experiment.

Test	x_1	x_2	x_3	x_4	x_5	y
1	−	−	−	−	−	25
2	+	−	−	−	−	26
3	−	+	−	−	−	31
4	+	+	−	−	−	34
5	−	−	+	−	−	24
6	+	−	+	−	−	26
7	−	+	+	−	−	32
8	+	+	+	−	−	36
9	−	−	−	+	−	30
10	+	−	−	+	−	36
11	−	+	−	+	−	40
12	+	+	−	+	−	43
13	−	−	+	+	−	30
14	+	−	+	+	−	34

(Continued)

Test	x_1	x_2	x_3	x_4	x_5	y
15	−	+	+	+	−	40
16	+	+	+	+	−	42
17	−	−	−	−	+	2
18	+	−	−	−	+	21
19	−	+	−	−	+	26
20	+	+	−	−	+	43
21	−	−	+	−	+	4
22	+	−	+	−	+	20
23	−	+	+	−	+	25
24	+	+	+	−	+	44
25	−	−	−	+	+	10
26	+	−	−	+	+	27
27	−	+	−	+	+	33
28	+	+	−	+	+	51
29	−	−	+	+	+	9
30	+	−	+	+	+	27
31	−	+	+	+	+	32
32	+	+	+	+	+	53

(a) Calculate all effect estimates.

(b) Calculate an estimate of the error variation using the sums of squares associated with three-factor and higher-order interactions.

(c) Construct an ANOVA table to identify which of the effect estimates are statistically significant.

19.29 The following factorial design was conducted with the results as given in the table.

Test No.	x_1	x_2	x_3	Test Order	y
1	−1	−1	−1	6	550
2	+1	−1	−1	8	925
3	−1	+1	−1	1	1150
4	+1	+1	−1	2	1350
5	−1	−1	+1	5	525
6	+1	−1	+1	3	800
7	−1	+1	+1	4	1175
8	+1	+1	+1	7	1450

Suppose that the model below was proposed to describe the response, y, based on analysis of the data:

$$\hat{y} = 990.625 + 140.625x_1.$$

(a) Using the proposed model, determine the model residuals.

(b) Construct residual plots vs. the independent variables x_1, x_2, and x_3. Comment on the appearance of the plots.

(c) Construct a residual plot based on the time order of the tests and comment.

(d) If the residual plots suggest that the proposed model does not adequately describe the response, propose a change in the model.

19.30 The following effects have been calculated based on the 32 combinations of five variables at two levels each in a 2^5 factorial design:

$$\bar{y} = 13.00,\ E_{13} = -0.80,\ E_{35} = \quad 0.01,\ E_{145} = \quad 0.35,\ E_{1245} = -0.40,$$
$$E_1 = -4.30,\ E_{14} = -0.50,\ E_{45} = \quad 6.81,\ E_{234} = \quad 0.00,\ E_{1345} = \quad 0.22,$$
$$E_2 = \quad 0.03,\ E_{15} = -0.20,\ E_{123} = \quad 0.24,\ E_{235} = -0.20,\ E_{2345} = -0.50,$$
$$E_3 = -0.50,\ E_{23} = \quad 0.29,\ E_{124} = \quad 0.57,\ E_{245} = \quad 0.37,\ E_{12345} = \quad 0.48.$$
$$E_4 = \quad 0.75,\ E_{24} = \quad 0.26,\ E_{125} = \quad 0.38,\ E_{345} = \quad 0.82,$$
$$E_5 = \quad 7.80,\ E_{25} = -0.50,\ E_{134} = -1.00,\ E_{1234} = -0.60,$$
$$E_{12} = -0.40,\ E_{34} = -0.60,\ E_{135} = -1.40,\ E_{1235} = \quad 0.13,$$

(a) Using the ANOVA method and the fact that higher-order effects can be used to estimate the error, determine which lower-order effects are significant. Be specific in stating the assumption(s) that you make.

(b) Write down the appropriate math model based on your results in (a).

19.31 A 2^3 factorial design has been conducted to determine the effect of three process factors on the output pressure of the process, in psi. The test results are given below in standard order.

$$(164,\ 328,\ 71,\ 234,\ 136,\ 350,\ 86,\ 222)$$

An analysis of the data led to the following fitted model:

$$\hat{y} = 198.875 + 84.625x_1 - 45.625x_2.$$

(a) Calculate the model residuals.

(b) Plot the residuals as a function of x_3 and comment.

DOE WORKSHOP #1: TURNING SIMULATION, TURN5DOE

Introduction

This workshop utilizes TURN5DOE, an MS Excel–based Visual Basic program, to simulate a turning operation that is being used to manufacture a bronze bushing. The goal of the workshop is to apply the general class of 2^k factorial designs to improve the centering of the two lathes that perform the turning operations in order to improve the process capability.

Learning Objective

The objective of this computer workshop is to provide students with practice in the planning, conduct, and interpretation of designed experiments. Specifically, the student is asked to determine, and then validate, the variable levels that will improve the process capability.

Background

In SPC Workshop #2, a turning process producing a bronze bushing was studied, with the finish of the outside surface serving as the quality characteristic of interest. After corrective actions were taken to eliminate special causes, the $\overline{X}$ and R charts for each machine showed good statistical control. Investigation of process capability for the individual machines revealed that two problems still existed:

1. Only one of the three machines was operating at the nominal surface roughness requirement of 70 microinches.

2. All three of the machines exhibited excessive variability in surface roughness values.

It was decided to attack the centering problem first, using statistical design-of-experiment techniques. Several individuals associated with the process developed a list of five potentially important factors that could influence the surface roughness. The feasible ranges for these factors are shown in the accompanying table.

Factor	Feasible Range for Experiments	Current Operating Condition
1. Use of cutting fluid	Either present or absent	No cutting fluid is being used
2. Machine used	Only the Nacirema and LeLathe lathes can be used for experimentation	Nacirema, LeLathe, Rex
3. Tool nose radius	0.0156″ to 0.0625″ in 0.0156″ increments	Fixed at 0.0312″
4. Feedrate	0.0070–0.0095 ipr	0.008″–0.0089″ ipr
5. Cutting speed	900–1150 sfpm	1000–1100 sfpm

Description of TURN5DOE

The TURN5DOE program has been developed to simulate running experiments with the five process variables. The program produces a surface roughness value as a test result for any user-specified combination of process settings for the five variables.

Prior to running TURN5DOE, an appropriate factorial design should be created. The decision as to the levels of the variables in the experiment is up to the experimenter. Then, TURN5DOE can be run to obtain the test results associated with the specific tests/runs of the design, and the data can be analyzed as appropriate. Once a model has been established for the response, plotting software can be used to characterize the response surface(s) and identify the levels for the processing variables that will center each machine. Ultimately, validation tests should be performed to confirm that the identified processing conditions do indeed result in centered machines.

On the Web

TURN5DOE is available for download on the Web at http://www.me.mtu.edu/quality_book/. Also available at the site is a more detailed workshop explanation and description of TURN5DOE usage. The website also has a suggested student assignment in connection with the TURN5DOE workshop.

20

Two-Level Fractional Factorial Designs

20.1   Rationale for, and Consequences of,
        Fractionation of Two-Level Factorials

In previous chapters we have studied a most useful class of experimental designs—two-level factorial designs. These designs constitute a powerful way to experiment with several factors simultaneously, revealing the extent and nature of variable interactions in a straightforward manner. These designs have great attraction because they are easy to set up, require relatively simple calculations to analyze the data, and have an appealing geometric representation that greatly enhances the interpretation of the results. Two-level factorial designs can deal simultaneously with variables whose levels are both continuous and discrete (i.e., quantitative or qualitative in nature).

Redundancy in Two-Level Factorials

Although the class of two-level factorial designs appears to be an efficient way to deal with several factors simultaneously, this efficiency seems to disappear as the number of variables to be studied grows. Since the two-level factorial requires the consideration of all possible combinations of k variables at two levels each, a 10-variable experiment would require $2^{10} = 1024$ tests. Such a test plan is simply prohibitive in size. Using the data from a 2^{10} factorial experiment the investigator could, in theory, estimate the following variable effects:

1	Mean response
10	Main effects
45	Two-factor interaction effects
120	Three-factor interaction effects
210	Four-factor interaction effects
252	Five-factor interaction effects

210	Six-factor interaction effects
120	Seven-factor interaction effects
45	Eight-factor interaction effects
10	Nine-factor interaction effects
1	Ten-factor interaction effect
1024	Mean + 1023 variable effects.

Although the volume of information from the experiment may seem impressive at first glance, one may wish to question the practical significance of the vast majority of it.

In dealing with phenomena involving continuous variables, it is not unreasonable to expect that the relationship which describes the influence of the variables on a response of interest constitutes a relatively smooth response surface. When dealing with qualitative (discrete) variables, usually the responses are similar at different levels of such variables. It should therefore not be too surprising to find that higher-order variable interactions are often very small—negligible—in magnitude. In fact, usually interaction effects involving three factors or more can simply be ignored without causing problems in interpretation of results.

Since the vast majority of the estimatable variable effects in a 2^{10} factorial are three-factor and above interactions (968 of the 1023 variable effects), it would appear that the more relevant information—main and two-factor interaction effects—could be obtained through an experiment requiring many fewer tests than the full factorial.

Experimentation with large numbers of variables generally arises out of uncertainty as to which variables have the dominant influence on the response of interest. In the early stages of an investigation the combined opinions of those involved with the situation under study can lead to an often long list of potentially important variables. In the end, however, only a subset of all of these variables will be proven to be important. Therefore, when considering 10 variables that together constitute 10 average/main effects and 45 two-factor interactions, perhaps as few as four to eight of these effects will really prove to be important.

The negligible magnitude of the many higher-order interactions, together with the fact that only a few variables will have significant influence on the response of interest, constitutes a tremendous amount of redundancy in two-level full factorials dealing with large numbers of variables; that is, many of the variable effects will be shown to be essentially zero. G. E. P. Box has referred to this phenomenon as the *sparsity of variable effects*.[1] If this is the case, one should be able to conduct fewer than the full factorial number of tests without loss of relevant information or confusion.

Consequences of Fractionating a Full Factorial

Suppose that an investigator wished to study three variables—ambient temperature, wind velocity, and bar size—to determine their possible effect on the

---

[1]G. E. P. Box and R. D. Meyer, "An Analysis for Unreplicated Fractional Factorials," *Technometrics*, Vol. 28, 1986, pp. 11–18.

ultimate strength of welded rail steel bars. A 2^3 full factorial was designed, providing for the calculation of three main effects, three two-factor interaction effects, and one three-factor interaction effect.

Suppose that the investigator then wished to consider a fourth variable, type of welding flux, but that the full factorial, $2^4 = 16$ tests, could not be considered. Rather, only eight tests could be run. Is it possible to conduct an experiment on four factors with only eight tests, and if so, what useful information can be obtained from the results?

Based on our assumption about three-factor and higher-order interaction effects being negligible, one could consider assigning the column of plus and minus signs associated with the 123 interaction to the fourth variable. This should be acceptable, since we believe that the 123 interaction is very small in magnitude. Therefore, we can use this column to define the levels of variable 4 for the eight tests, and also use it when we estimate the main effect of variable 4.

In using the 123 column to introduce a fourth variable into the experiment, the new design matrix for the eight tests to be conducted becomes:

		Design Matrix		
Test	1	2	3	4 = 123
1	−	−	−	−
2	+	−	−	+
3	−	+	−	+
4	+	+	−	−
5	−	−	+	+
6	+	−	+	−
7	−	+	+	−
8	+	+	+	+

Once the tests are conducted in accordance with the test recipes defined by the design matrix, the calculation matrix is determined to provide for the estimation of the interaction effects. Expanding the design matrix above, we obtain the following calculation matrix by forming all possible products of columns 1 through 4.

							Calculation Matrix									
Test	I	1	2	3	4	12	13	14	23	24	34	123	124	134	234	1234
1	+	−	−	−	−	+	+	+	+	+	+	−	−	−	−	+
2	+	+	−	−	+	−	−	+	+	−	−	+	−	−	+	+
3	+	−	+	−	+	−	+	−	−	+	−	+	−	+	−	+
4	+	+	+	−	−	+	−	−	−	−	+	−	−	+	+	+
5	+	−	−	+	+	+	−	−	−	−	+	+	+	−	−	+
6	+	+	−	+	−	−	+	−	−	+	−	+	+	−	+	+
7	+	−	+	+	−	−	−	+	+	−	−	+	+	+	−	+
8	+	+	+	+	+	+	+	+	+	+	+	+	+	+	+	+

Examination of the calculation matrix reveals that many of the columns are identical. Of the 16 columns, only eight are unique; each unique column appears twice. The following pairs of variable effects are represented in the calculation matrix by the same column of plus and minus signs:

1 and 234	12 and 34
2 and 134	13 and 24
3 and 124	23 and 14
4 and 123	average and 1234.

What does all this mean? When, for example, you multiply the 12 column by the data, sum, and divide by 4, do you get an estimate of the two-factor interaction 12? Or the two-factor interaction 34? Or both? The interactions 12 and 34 are said to be *confounded* or *confused*.

The interactions 12 and 34 are said to be *aliases* of the unique column of plus and minus signs defined by $(+--++--+)$. Use of this column for effect estimation produces a number (estimate) that is actually the sum of the two-factor interaction effects 12 and 34. Similarly, 1 and 234 are confounded effects, 2 and 134 are confounded effects, and so on. It seems that the *innocent* act of using the 123 column to introduce a fourth variable into a 2^3 full factorial scheme has created a lot of confounding among the variable effects.

The eight unique columns in the calculation matrix are used to obtain the linear combinations $l_0, l_1, \ldots, l_{123}$ of confounded effects when their signs are applied to the data, and the result is summed and then divided by 4 (divide by 8 for l_0):

l_0 estimates mean $+ \left(\frac{1}{2}\right)(1234)$,	l_{12} estimates $12 + 34$,
l_1 estimates $1 + 234$,	l_{13} estimates $13 + 24$,
l_2 estimates $2 + 134$,	l_{23} estimates $23 + 14$,
l_3 estimates $3 + 124$,	l_{123} estimates $4 + 123$.

Some of this confounding can be eliminated by invoking the assumption that third- and higher-order effects are negligible, leading to clear estimates of all main effects. But the six two-factor interactions are still hopelessly confounded. If we assume three- and four-factor interactions can be neglected, the experiment produces the following linear combinations:

l_0 estimates mean,	l_{12} estimates $12 + 34$,
l_1 estimates 1,	l_{13} estimates $13 + 24$,
l_2 estimates 2,	l_{23} estimates $23 + 14$,
l_3 estimates 3,	l_{123} estimates 4.

The four-variable, eight-test, two-level experiment discussed thus far is referred to as a two-level fractional factorial design, since it considers only a fraction of the tests defined by the full factorial. In this case we have created a one-half fraction design. It is commonly referred to as a 2^{4-1} fractional factorial design. It is a member of the general class of 2^{k-p} fractional factorial designs. For these designs

1. k variables are examined
2. in 2^{k-p} tests
3. requiring that the p of the variables be introduced into the full factorial in $k - p$ variables
4. by assigning them to interaction effects in the first $k - p$ variables.

▶ EXAMPLE 20.1 2^{4-1} FRACTIONAL FACTORIAL

1. 4 variables are studied
2. in $2^{4-1} = 8$ tests
3. $p = 1$ of the variables is introduced into a 2^3 full factorial
4. by assigning it to the interaction 123 (i.e., let 4 = 123).

Many other useful fractional factorials can be developed, some dealing with rather large numbers of variables in relatively few tests. The 2^{4-1} fractional factorial design just examined is one of the more *simple* fractional factorial designs, although the extensive confounding in this design seems a little *confusing*. It can get much worse. Therefore, we need a system to set up such designs easily and to determine quickly the precise nature/pattern of the confounding of the variable effects.

20.2   System to Define the Confounding Pattern of a Two-Level Fractional Factorial

Suppose that an investigator wishes to study the potential effects that five variables may have on the output of a certain process using some type of two-level factorial experiment. If all possible combinations of five variables at two levels each are to be considered, then $2^5 = 32$ tests must be conducted. His boss informs him that due to time and budget limitations he will only be able to run 8 tests, not 32. How might the investigator reconsider his original test plan so as to accommodate his boss's constraints and still gain some useful information about the five variables?

Base Design[2] and Introduction of Additional Variables

If only eight tests are to be conducted using a two-level scheme, only three variables can be examined in a full two-level factorial test plan. The eight unique test

---

[2]This element of the fractional factorial design construction procedure was introduced by the authors of this book some time ago in their classroom lectures, in order to help avoid the problem that often arises of not properly identifying the calculation matrix for a given fractional factorial.

settings/conditions and the effect calculation matrix for a 2^3 full factorial design are as follows:

Test	Average I	1	2	3	12	13	23	123	y
1	+	−	−	−	+	+	+	−	y_1
2	+	+	−	−	−	−	+	+	y_2
3	+	−	+	−	−	+	−	+	y_3
4	+	+	+	−	+	−	−	−	y_4
5	+	−	−	+	+	−	−	+	y_5
6	+	+	−	+	−	+	−	−	y_6
7	+	−	+	+	−	−	+	−	y_7
8	+	+	+	+	+	+	+	+	y_8

We recall that:

- The columns with headings 1, 2, and 3 constitute the definition of the eight unique test settings for the three variables denoted heretofore simply as 1, 2, and 3 [e.g., test 1 is conducted with all three variables at their low (−) levels].
- All eight columns of + and − signs constitute the calculation matrix and provide for the independent estimation of the mean (column I), main effects of variables 1, 2, and 3 (columns 1, 2, and 3), the two-factor interaction effects among the three variables (columns 12, 13, and 23), and the three-factor interaction among all three variables (column 123).

The calculation matrix is obtained directly from the design matrix (columns 1, 2, and 3) and is used to estimate the main effects and interaction effects for the variables under study. The columns with headings 12, 13, 23, and 123 are each obtained by multiplying together for each row (eight rows) the signs of the columns in the design matrix.

Now the dilemma of our investigator is that he would like to study five variables, not three. How might this be accomplished within the structure of the 2^3 factorial design presented above? To define test conditions for five variables he will need five columns of + and − signs. In the 2^3 design he has only three such columns; 1, 2, and 3. Where will he find the other two?

In designing fractional factorial experiments, we introduce additional variables into the base design by borrowing columns initially assigned to interaction effects in the base design variables. *The base design is the full factorial design associated with the number of tests we wish to run.* For the case under consideration:

1. Five variables will be studied using only eight tests.
2. Only three variables can be accommodated in a complete or full factorial scheme with eight tests.

3. Therefore, a 2^3 design is the base design.

4. Two variables must be further introduced in the 2^3 base design.

5. Columns 12, 13, 23, and 123 are available to introduce these two additional variables.

6. The new test plan will be called a 2^{5-2} fractional factorial design:
  - Two levels of each variable.
  - Five variables under study.
  - $2^{5-2} = 8$ tests to be run.
  - Two variables introduced into the 2^3 base design.

For the five-variable, eight-test fractional factorial under study, let us introduce variables 4 and 5 into the 2^3 base design by assigning them to the 12 and 13 columns, respectively.

Test	I	1	2	3	4 12	5 13	23	123	y
1	+	−	−	−	+	+	+	−	y_1
2	+	+	−	−	−	−	+	+	y_2
3	+	−	+	−	−	+	−	+	y_3
4	+	+	+	−	+	−	−	−	y_4
5	+	−	−	+	+	−	−	+	y_5
6	+	+	−	+	−	+	−	−	y_6
7	+	−	+	+	−	−	+	−	y_7
8	+	+	+	+	+	+	+	+	y_8

In the matrix of + and − signs, the first five columns (excluding I) represent the eight unique test conditions for the five variables [e.g., for test 1, fix variables 1, 2, and 3 at their low (−) levels and variables 4 and 5 at their high (+) levels]. These five columns constitute the design (recipe) matrix for this 2^{5-2} fractional factorial design.

All seven columns plus a column of all plus signs (the I column) constitute the calculation matrix for this 2^{5-2} design. This means that all average/main and interaction effects among the five variables will be contained (confounded) within eight linear combinations of the data. The question that remains is to determine exactly which effects are confounded with each other.

Design Generators and the Defining Relation

From now on, when we refer to a column heading (e.g., 1 or 23 or 123) we should imagine a column of + and − signs directly under it. Our 2^{5-2} fractional factorial design was generated by setting the 4-column equal to the 12-column and the 5-column equal to the 13-column. In the interest of convenience we will denote these as $4 = 12$ and $5 = 13$, where the $=$ sign really implies an identity between columns of $+$ and $-$ signs. For example,

$$
\begin{array}{ccc}
4 & = & 12 \\
+ & & + \\
- & & - \\
- & & - \\
+ & = & + \\
+ & & + \\
- & & - \\
- & & - \\
+ & & + \, .
\end{array}
$$

Now, if any column of $+$ and $-$ signs is multiplied by itself, a column of all $+$ signs is produced. We will denote such a column by the heading I. For example,

$$
\begin{array}{ccccc}
4 & \times & 4 & = & I \\
+ & & + & & + \\
- & & - & & + \\
- & & - & & + \\
+ & \times & + & = & + \\
+ & & + & & + \\
- & & - & & + \\
- & & - & & + \\
+ & & + & & + \, .
\end{array}
$$

This simple operation will prove to be very useful. Since the 2^{5-2} fractional factorial design was generated by setting

$$
\begin{aligned}
4 &= 12, \\
5 &= 13,
\end{aligned}
$$

and given the definition of I above, if we multiply both sides of the two "equations" above by 4 and 5, respectively, then

$$
4 \times 4 = 12 \times 4, \quad 5 \times 5 = 13 \times 5,
$$

which reduces to

$$
I = 124, \quad I = 135.
$$

These two identities are referred to as our design *generators*. While both the left- and righthand sides of the equation above represent columns of all $+$ signs, the righthand side retains the individual column headings that produced the column of all + signs by their product. That is,

$$I \text{ column} = 1 \text{ column} \times 2 \text{ column} \times 4 \text{ column}.$$

Now since both the 124 and 135 columns equal I, their product must also equal I:

$$(124) \times (135) = I,$$

or, rearranging numbers (columns),

$$(1)(1)2345 = I.$$

But $(1)(1) = I$, and any column multiplied by a column of plus signs (I) remains unchanged. Therefore, I is also equal to 2345:

$$I = 2345.$$

Hence we have

$$I = 124 = 135 = 2345,$$

an identity comprised of the design generators and their products in all possible combinations (in this case only one product). The identity

$$I = 124 = 135 = 2345$$

is referred to as the *defining relation* of this 2^{5-2} fractional factorial design, and through it we can reveal the complete aliasing/confounding structure of this fractional factorial design.

Revelation of the Complete Confounding Pattern

Returning to the original base design calculation matrix, we recall that we have seven independent columns of + and − signs, and an eighth column for I (a column of all + signs):

Test	I	1	2	3	12	13	23	123
1	+	−	−	−	+	+	+	−
2	+	+	−	−	−	−	+	+
3	+	−	+	−	−	+	−	+
4	+	+	+	−	+	−	−	−
5	+	−	−	+	+	−	−	+
6	+	+	−	+	−	+	−	−
7	+	−	+	+	−	−	+	−
8	+	+	+	+	+	+	+	+
	l_0	l_1	l_2	l_3	l_{12}	l_{13}	l_{23}	l_{123}

By letting $4 = 12$ and $5 = 13$, we have created many aliased effects. To find the aliases of the column headings above $(1, 2, 3, \ldots, 123)$, we multiply each by every term (including I) in the defining relation:

$$\text{defining relation:}\quad I = 124 = 135 = 2345.$$

For column heading 1:

$$(1)I = (1)124 = (1)135 = (1)2345.$$

Removing all I's [recall that $(1)(1) = I$], we have

$$1 = 24 = 35 = 12345.$$

That is, the aliases of 1 are 24, 35, and 12345. Therefore, when we multiply the 1 column by the y column, sum, and divide by 4, we obtain an estimate of the sum (linear combination) of 1, 24, 35, and 12345.[3] We conveniently denote this sum of confounded variable effects as l_1 (l for linear combination of the effects). That is,

$$l_1 \text{ estimates } 1 + 24 + 35 + 12345.$$

Similarly, moving to column headings 2, 3, and so on, we find that

$$
\begin{array}{rrrr}
2 = & 14 = & 1235 = & 345 \\
3 = & 1234 = & 15 = & 245 \\
12 = & 4 = & 235 = & 1345 \\
13 = & 234 = & 5 = & 1245 \\
23 = & 134 = & 125 = & 45 \\
123 = & 34 = & 25 = & 145.
\end{array}
$$

Hence

$$
\begin{array}{llll}
l_2 \text{ estimates} & 2 + & 14 + 1235 + 345 \\
l_3 \text{ estimates} & 3 + 1234 + & 15 + 245 \\
& \vdots & \\
l_{123} \text{ estimates} & 123 + & 34 + & 25 + 145.
\end{array}
$$

We have now defined the complete confounding pattern of this 2^{5-2} fractional factorial design and we know precisely what effect combinations we can obtain from the data.

---

[3]Actually, l_1 estimates the sum of the expected value of effects, E_1, E_{24}, E_{35}, and E_{12345}. That is, l_1 estimates the sum of the mean values of these effects. From this point on, when we use the notation l_1 estimates $1 + 24 + 35 + 12345$, it should be understood that 1, 24, 35, 12345 are mean values, while E_1, E_{24}, E_{35}, E_{12345} continue to be considered estimates of these mean values.

Summary

1. We want to study five variables in only eight tests in a two-level factorial scheme.

2. The base design is therefore a $2^3 = 8$ test two-level full factorial in the base design variables 1, 2, and 3.

3. To the base design we introduce variables 4 and 5 by assigning them to interaction columns in the base design—for example,

$$4 = 12,$$
$$5 = 13.$$

4. The design generators are therefore

$$I = 124,$$
$$I = 135,$$

and hence the design defining relation is

$$I = 124 = 135 = 2345.$$

5. The defining relation produces the confounding pattern or alias structure:

$$
\begin{array}{rrrr}
I = & 124 = & 135 = & 2345 \\
1 = & 24 = & 35 = & 12345 \\
2 = & 14 = & 1235 = & 345 \\
3 = & 1234 = & 15 = & 245 \\
12 = & 4 = & 235 = & 1345 \\
13 = & 234 = & 5 = & 1245 \\
23 = & 134 = & 125 = & 45 \\
123 = & 34 = & 25 = & 145.
\end{array}
$$

For the confounding structure shown above, it should be noted that each column heading (I, 1, 2, 3, . . . , 123) is confounded with three effects (e.g., I is confounded with 124, 135, and 2345). Thus, each of the eight rows in the confounding structure contains four effects. For a full factorial in five variables ($2^5 = 32$ tests) there are 31 column headings plus I in the calculation matrix. In particular, for a 2^5 full factorial calculation matrix, there are columns for

1	Mean
5	Main effects
10	Two-factor interaction effects
10	Three-factor interaction effects
5	Four-factor interaction effects
1	Five-factor interaction effect
32	Mean + 31 variable effects.

Examining the aliasing structure for the 2^{5-2} design, we observed that all 32 variable effects (including the mean, I) are accounted for (8 rows $\times$ 4 effects/row). It is a good idea to verify that this is the case.

6. The eight columns (including I) in the 2^{5-2} fractional factorial design produce the following linear combinations of effects which can be estimated:[4]

$$
\begin{aligned}
l_0 &\text{ estimates mean} + \tfrac{1}{2}(124 + 135 + 2345) \\
l_1 &\text{ estimates} \quad 1 + \quad 24 + \quad 35 + 12345 \\
l_2 &\text{ estimates} \quad 2 + \quad 14 + 1235 + \quad 345 \\
l_3 &\text{ estimates} \quad 3 + 1234 + \quad 15 + \quad 245 \\
l_{12} &\text{ estimates} \quad 12 + \quad 4 + 235 + 1345 \\
l_{13} &\text{ estimates} \quad 13 + \quad 234 + \quad 5 + 1245 \\
l_{23} &\text{ estimates} \quad 23 + \quad 134 + 125 + \quad 45 \\
l_{123} &\text{ estimates} \quad 123 + \quad 34 + \quad 25 + \quad 145.
\end{aligned}
$$

7. Often, we assume that the majority of the variability in the data can be explained by the presence of main effects and two-factor interaction effects among the variables. Under such an assumption, the linear combinations of effects that can be estimated from the 2^{5-2} fractional factorial under study are

$$
\begin{aligned}
l_0 &\text{ estimates mean} \\
l_1 &\text{ estimates} \quad 1 + 24 + 35 \\
l_2 &\text{ estimates} \quad 2 + 14 \\
l_3 &\text{ estimates} \quad 3 + 15 \\
l_{12} &\text{ estimates } 12 + \quad 4 \\
l_{13} &\text{ estimates } 13 + \quad 5 \\
l_{23} &\text{ estimates } 23 + 45 \\
l_{123} &\text{ estimates } 34 + 25.
\end{aligned}
$$

20.3   Procedure for Design Characterization: Another Example

Suppose that six variables are to be studied in a two-level experiment but only eight tests can be conducted. How is such an experiment set up, and what is the precise nature of the confounding among all variable effects?

STEP 1   Define the base design.

Since only eight tests are to be conducted, the base design for this fractional factorial is a 2^3 full factorial. The complete calculation matrix for the 2^3 base design is

Test	I	1	2	3	12	13	23	123	y
1	+	−	−	−	+	+	+	−	y_1
2	+	+	−	−	−	−	+	+	y_2
3	+	−	+	−	−	+	−	+	y_3
4	+	+	+	−	+	−	−	−	y_4
5	+	−	−	+	+	−	−	+	y_5
6	+	+	−	+	−	+	−	−	y_6
7	+	−	+	+	−	−	+	−	y_7
8	+	+	+	+	+	+	+	+	y_8
Divisor	8	4	4	4	4	4	4	4	

---

[4]Recall that the divisor for the I (mean) column is *eight*, not *four*.

STEP 2 Introduce additional variables.

The interaction columns 12, 13, 23, and 123 are available for the introduction of the variables 4, 5, and 6. In some cases the choice of which interactions to use is critical. We discuss this choice later. In this case we will use the interactions 12, 13, and 23 to establish the design generators.

$$4 = 12,$$
$$5 = 13,$$
$$6 = 23.$$

The design matrix for this 2^{6-3} fractional factorial design is therefore

Test	1	2	3	4	5	6
1	−	−	−	+	+	+
2	+	−	−	−	−	+
3	−	+	−	−	+	−
4	+	+	−	+	−	−
5	−	−	+	+	−	−
6	+	−	+	−	+	−
7	−	+	+	−	−	+
8	+	+	+	+	+	+

The eight tests may now be conducted in accordance with these test recipes.

STEP 3 Obtain the defining relation for the design.

The design was set up by introducing variables 4, 5, and 6 using the interactions 12, 13, and 23, respectively. Therefore, the design generators are

$$I = 124, \quad I = 135, \quad I = 236.$$

Remembering that

$$4 = 12$$
$$4 \times 4 = 12 \times 4,$$

but since $4 \times 4 = I$, a column of $(+)$ signs, we have $I = 124$.

Next, we obtain the defining relation. The defining relation consists of the generators plus all of their products taken two at a time, three at a time, ..., p at a time ($p = 3$ for this case). The defining relation is given by

$$I = 124, \quad I = 135, \quad I = 236 \quad \text{(the generators)}$$

plus two-at-a-time products:

$$(124)(135) = 2345,$$
$$(124)(236) = 1346,$$
$$(135)(236) = 1256,$$

plus the three-at-a-time product:

$$(124)(135)(236) = 456.$$

The complete defining relation is therefore

$$I = 124 = 135 = 236 = 2345 = 1346 = 1256 = 456.$$

STEP 4 Reveal the complete confounding structure of the design.

Once the defining relation has been obtained, the complete confounding pattern may be determined. Since the experiment involves eight tests, eight unique linear combinations of effects can be estimated. These are defined by the eight columns of the calculation matrix including I. Moving from left to right, these columns are identified as $l_0, l_1, l_2, \ldots, l_{123}$:

l_0	l_1	l_2	l_3	l_{12}	l_{13}	l_{23}	l_{123}
I	1	2	3	12	13	23	123.

That is, the use of each column for estimation produces an estimate of a certain linear combination of effects. To determine each set of confounded effects, we multiply, in turn, each column heading by every term in the defining relation, including I. For example, for the column denoted 1,

$$(1)I = (1)124 = (1)135 = (1)236 = (1)2345 = (1)1346 = (1)1256 = (1)456,$$

or

$$1 = 24 = 35 = 1236 = 12345 = 346 = 256 = 1456.$$

Assuming that third- and higher-order interactions can be neglected, the main effect of variable 1 is therefore confounded with the two-factor interactions 24 and 35. That is,

$$l_1 \text{ estimates } 1 + 24 + 35.$$

Summary: Relevant Confounding Structure

For this 2^{6-3} design, assuming third- and higher-order interactions can be neglected, we obtain

$$
\begin{aligned}
&l_0 \quad \text{estimates mean} \\
&l_1 \quad \text{estimates} \quad 1 + 24 + 35 \\
&l_2 \quad \text{estimates} \quad 2 + 14 + 36 \\
&l_3 \quad \text{estimates} \quad 3 + 15 + 26 \\
&l_{12} \quad \text{estimates} \ 12 + \ 4 \ + 56 \\
&l_{13} \quad \text{estimates} \ 13 + \ 5 \ + 46 \\
&l_{23} \quad \text{estimates} \ 23 + \ 6 \ + 45 \\
&l_{123} \ \text{estimates} \ 34 + 25 + 16.
\end{aligned}
$$

It is important to check to make sure that all six main effects and all 15 two-factor interactions have been accounted for. A quick check above shows that they have been. In general, there are $k!/2!(k-2)!$ two-factor interactions.

20.4   Concept of Resolution of Two-Level Fractional Factorial Designs

We have seen previously that the introduction of additional variables into two-level full factorials gives rise to confounding or aliasing of variable effects. It would be desirable to make this introduction in such a way as to confound low-order effects (main effects and two-factor interactions) not with each other but with higher-order interactions. Then, under the assumption that third- and higher-order interactions can be neglected, the low-order effects become, in a sense, unconfounded by this assumption.

Selecting the Preferred Generators

To illustrate, consider the study of five variables in just 16 tests (the full factorial would require $2^5 = 32$ tests). One additional variable—the fifth variable—must be introduced into a $2^4 = 16$ run base design. Any of the interactions in the first four variables could be used for this purpose:

$$
\begin{aligned}
&12, 13, 14, 23, 24, 34 \\
&123, 124, 134, 234 \\
&1234.
\end{aligned}
$$

If any one of the two-factor interactions are used, say, $5 = 12$, the design generator becomes

$$
I = 125,
$$

which is also the defining relation. Therefore, at least some of the main effects will be confounded with two-factor interactions:

$$1 = 25, \quad 2 = 15, \quad 5 = 12.$$

If any one of the three-factor interactions is used to introduce the fifth variable, the situation is greatly improved, at least for the estimation of main effects. For example, if we let $5 = 123$, then

$$I = 1235$$

is the generator and defining relation. So some main effects are confounded with, at worst, three-factor interactions, while two-factor interactions are confounded with each other—for example,

$$1 = 235, \quad 2 = 135, \quad 3 = 125, \quad 5 = 123, \quad 12 = 35,$$
$$13 = 25, \quad 23 = 15.$$

If the four-factor interaction in the first four variables is used to introduce the fifth variable (i.e., $5 = 1234$), an even more desirable result is obtained (the best under these circumstances). The generator and defining relation for this situation is

$$I = 12345.$$

Therefore,

$$1 = 2345, \quad 2 = 1345, \quad 3 = 1245, \quad 4 = 1235, \quad 5 = 1234,$$
$$12 = 345, \quad 13 = 245, \quad 14 = 235, \quad 15 = 234, \quad 23 = 145,$$
$$24 = 135, \quad 25 = 134, \quad 34 = 125, \quad 35 = 124, \quad 45 = 123.$$

In this last case:

1. All main effects are confounded with four-factor interactions.
2. All two-factor interactions are confounded with three-factor interactions.

The varying confounding structures produced by using different orders of variable interactions to introduce the fifth variable in the example above are described by the concept of the resolution of fractional factorial designs: "The *resolution* of a two-level fractional factorial design is defined to be equal to the number of letters (numbers) in the shortest length word (term) in the defining relation, excluding *I*." If the defining relation of a certain design is

$$I = 124 = 135 = 2345,$$

the design is of resolution three, denoted as resolution III, since the words "124" and "135" have three letters each. If the defining relation of a certain design is

$$I = 1235 = 2346 = 1456,$$

then the design is of resolution IV (1235, 2346, and 1456 each have four letters). The last design we examined above, which had the defining relation ($I = 12345$), is a resolution V design.

- If a design is of resolution III, this means that at least some main effects are confounded with two-factor interactions.
- If a design is of resolution IV, this means that at least some main effects are confounded with three-factor interactions, while at least some two-factor interactions are confounded with other two-factor interactions.
- If a design is of resolution V, this means that at least some main effects are confounded with four-factor interactions and some two-factor interactions are confounded with three-factor interactions.

It may be noted at this point that the number of words in the defining relation for a 2^{k-p} fractional factorial design is equal to 2^p. Thus for a 2^{6-3} fractional factorial ($k = 6$ and $p = 3$), there are $2^3 = 8$ words in the defining relation.

▶ EXAMPLE 20.2 DESIGN RESOLUTION/SELECTION OF GENERATORS

A 2^{6-2} fractional factorial design is set up by introducing variables 5 and 6 via

$$5 = 123, \quad 6 = 1234.$$

What is the resolution of this design? The design generators are

$$I = 1235, \quad I = 12346.$$

The defining relation is

$$I = 1235 = 12346 = 456.$$

Therefore, the design is of resolution III. What would the resolution be if the generators were

$$5 = 123, \quad 6 = 124?$$

The defining relation is

$$I = 1235 = 1246 = 3456.$$

Now, the design is of resolution IV. It is clear that the selection of the proper design generators is very important.

Summary: The Concept of Design Resolution

From the discussion of design resolution above, several observations can be made:

1. Higher-resolution designs seem more desirable, since they provide the opportunity for low-order effect estimates to be determined in an unconfounded state, assuming that higher-order interaction effects can be neglected.
2. There is a limit to the number of variables that can be considered in a fixed number of tests while maintaining a prespecified resolution requirement.
3. No more than $(n - 1)$ variables can be examined in n tests (n is a power of 2, e.g., 4, 8, 16, 32, ...) to maintain a design resolution of at least III. Such designs are commonly referred to as saturated designs. Examples are

$$2^{3-1}, \ 2^{7-4}, \ 2^{15-11}, \ 2^{31-26}.$$

For saturated designs all interactions in the base design variables are used to introduce additional variables.

20.5   Case Study Application: The Sandmill Experiment Revisited

Previously, we described an investigation conducted at an automotive plant that was concerned with a process used to manufacture vinyl film. At the beginning of the study, the plant employed vertical sandmills to disperse pigments for the production of colorants for the vinyl film. As has been noted, the plant was considering the acquisition of a new type of sandmill, a horizontal sandmill. The manufacturer claimed that the horizontal sandmill would improve both productivity and quality. It was decided to perform some tests to assess the superiority of the horizontal sandmill with respect to the vertical sandmill, since the sandmill replacement cost would be approximately $180,000. Arrangements were made with the manufacturer for some trial runs on the horizontal sandmill.

As the planning of the sandmill comparison experiment continued to evolve, questions were raised concerning the operating conditions of the machines. Past experience with the vertical sandmill had shown that grind fineness (measured using the Hegman scale from 0/coarse to 10/fine) was affected by both the temperature setting of the process and the flow rate. While targeted values had been established some time in the past for these variables (as a function of pigment type and pigment concentration), it was not at all clear which values to use for the horizontal machine. In addition, there was some concern about the "goodness" of the target values actually being used for the vertical machine. Therefore, it was decided to consider the two process parameters (temperature and flow rate) in addition to the machine type, pigment type, and pigment concentration in an experiment. Table 20.1 lists the variables and variable levels that were chosen for the experiment.

TABLE 20.1	Variables and Their Levels for the Sandmill Experiment		

		Level	
Variable	Name	Low (−)	High (+)
1	Sandmill/machine type	Vertical	Horizontal
2	Pigment type	Blue	Red
3	Pigment concentration (%)	10	15
4	Processing temperature (°F)	140	160
5	Flow rate (gal/hr)	20	30

To consider this group of five factors in accordance with a two-level full factorial design scheme seemed prohibitive. A total of $2^5 = 32$ unique conditions would need to be tested even without any provision for replication. Given the time constraints and the expense of so many tests, this approach did not seem feasible. It was decided, therefore, to conduct a 2^{5-1} fractional factorial design (i.e., a one-half fraction design).

Design and Conduct of the Experiment

As stated, it was decided to use a 2^{5-1} fractional factorial design to study the five factors listed in Table 20.1. For a 2^{5-1} fractional factorial, the base design calculation matrix is based on a 2^4 full factorial design:

Test	I	1	2	3	4	12	13	14	23	24	34	123	124	134	234	1234
1	+	−	−	−	−	+	+	+	+	+	+	−	−	−	−	+
2	+	+	−	−	−	−	−	−	+	+	+	+	+	+	−	−
3	+	−	+	−	−	−	+	+	−	−	+	+	+	−	+	−
4	+	+	+	−	−	+	−	−	−	−	+	−	−	+	+	+
5	+	−	−	+	−	+	−	+	−	+	−	+	−	+	+	−
6	+	+	−	+	−	−	+	−	−	+	−	−	+	−	+	+
7	+	−	+	+	−	−	−	+	+	−	−	−	+	+	−	+
8	+	+	+	+	−	+	+	−	+	−	−	+	−	−	−	−
9	+	−	−	−	+	+	+	−	+	−	−	−	+	+	+	−
10	+	+	−	−	+	−	−	+	+	−	−	+	−	−	+	+
11	+	−	+	−	+	−	+	−	−	+	−	+	−	+	−	+
12	+	+	+	−	+	+	−	+	−	+	−	−	+	−	−	−
13	+	−	−	+	+	+	−	−	−	−	+	+	+	−	−	+
14	+	+	−	+	+	−	+	+	−	−	+	−	−	+	−	−
15	+	−	+	+	+	−	−	−	+	+	+	−	−	−	+	−
16	+	+	+	+	+	+	+	+	+	+	+	+	+	+	+	+

A fifth variable may be introduced into the structure by assigning it to one of the interaction columns. To obtain the highest design resolution, variable 5 was introduced using the 1234 interaction column. Using this assignment, the design matrix becomes

Test	1	2	3	4	5
1	−	−	−	−	+
2	+	−	−	−	−
3	−	+	−	−	−
4	+	+	−	−	+
5	−	−	+	−	−
6	+	−	+	−	+
7	−	+	+	−	+
8	+	+	+	−	−
9	−	−	−	+	−
10	+	−	−	+	+
11	−	+	−	+	+
12	+	+	−	+	−
13	−	−	+	+	+
14	+	−	+	+	−
15	−	+	+	+	−
16	+	+	+	+	+

If the +/− signs in the design matrix are replaced by the actual levels for the variables, we obtain the recipes for each test condition (Table 20.2). The test conditions in the design matrix were performed in a random order (using the indicated run order) and the Hegman values displayed in Table 20.2 were obtained.

TABLE 20.2   Test Condition Recipes

Test	Machine Type 1	Pigment Type 2	Pigment Conc. (%) 3	Processing Temp. (°F) 4	Flow Rate (gal/hr) 5	Hegman Value, y	Run Order
1	Vertical	Blue	10	140	30	6.25	5
2	Horizontal	Blue	10	140	20	6.25	11
3	Vertical	Red	10	140	20	7.75	8
4	Horizontal	Red	10	140	30	6.75	2
5	Vertical	Blue	15	140	20	6.25	13
6	Horizontal	Blue	15	140	30	5.25	10
7	Vertical	Red	15	140	30	6.75	4
8	Horizontal	Red	15	140	20	6.75	14
9	Vertical	Blue	10	160	20	7.00	6
10	Horizontal	Blue	10	160	30	5.25	12
11	Vertical	Red	10	160	30	8.00	16
12	Horizontal	Red	10	160	20	7.00	1
13	Vertical	Blue	15	160	30	5.50	15
14	Horizontal	Blue	15	160	20	5.25	7
15	Vertical	Red	15	160	20	7.75	3
16	Horizontal	Red	15	160	30	6.50	9

Confounding Structure and Linear Combinations of Effects

As has been noted, the 2^{5-1} fractional factorial design described above was obtained by introducing variable 5 through the 1234 interaction column. Hence the generator and defining relation for this design is

$$I = 12345.$$

As can be seen, this design is of resolution V, since the length of the smallest word in the defining relation is five (excluding I). Therefore, at least some of the main effects (in this case all) will be confounded with four-factor interactions, and at least some of the two-factor interactions (in this case all) will be confounded with three-factor interactions.

Given the defining relation, the patterns of confounded effects may be obtained by multiplying the column headings in the base design calculation matrix $(I, 1, 2, 3, 4, 12, 13, \ldots, 1234)$ by every term in the defining relation. For example, for the column denoted 1, we have

$$(1)I = (1)12345 \quad \text{or} \quad 1 = 2345.$$

As another example, for the column denoted 123, we have

$$(123)I = (123)12345 \quad \text{or} \quad 123 = 45.$$

The complete set of linear combinations of confounded effects is then

l_0 estimates mean $+ \frac{1}{2}(12345)$,		l_{23} estimates	$23 + 145$,
l_1 estimates $1 + 2345$,		l_{24} estimates	$24 + 135$,
l_2 estimates $2 + 1345$,		l_{34} estimates	$34 + 125$,
l_3 estimates $3 + 1245$,		l_{123} estimates $123 +$	45,
l_4 estimates $4 + 1235$,		l_{124} estimates $124 +$	35,
l_{12} estimates $12 +$	345,	l_{134} estimates $134 +$	25,
l_{13} estimates $13 +$	245,	l_{234} estimates $234 +$	15,
l_{14} estimates $14 +$	235,	l_{1234} estimates $1234 +$	5.

Under the assumption that three-factor interactions and higher are negligible, this set of linear combinations reduces to:

l_0 estimates mean,	l_{23} estimates 23,
l_1 estimates 1,	l_{24} estimates 24,
l_2 estimates 2,	l_{34} estimates 34,
l_3 estimates 3,	l_{123} estimates 45,
l_4 estimates 4,	l_{124} estimates 35,
l_{12} estimates 12,	l_{134} estimates 25,
l_{13} estimates 13,	l_{234} estimates 15,
l_{14} estimates 14,	l_{1234} estimates 5.

An examination of these linear combinations demonstrates the advantage of designs having resolution V or higher. Under the assumption that three-factor interactions and higher are negligible, such designs provide unaliased estimates of all main and two-factor interaction effects.

Estimation of Effects

The estimates (l_i's) associated with the linear combinations of confounded effects are obtained by multiplying each column of plus and minus signs in the base design calculation matrix by the column of responses (Hegman values), summing, and then dividing by 8 (for the average, divide by 16). As an example, consider the calculation of the estimates l_1:

Column 1		Column y	
−		6.25	
+		6.25	
−		7.75	
+		6.75	
−		6.25	
+		5.25	
−	×	6.75	$= (-6.25 + 6.25 - 7.75 + 6.75 - 6.25 + 5.25$
+		6.75	$- 6.75 + 6.75 - 7.00 + 5.25 - 8.00 + 7.00$
−		7.00	$- 5.50 + 5.25 - 7.75 + 6.50)/8$
+		5.25	$= -0.78125 = l_1$
−		8.00	
+		7.00	
−		5.50	
+		5.25	
−		7.75	
+		6.50	

The estimates are summarized as follows:

$$
\begin{aligned}
l_0 &= 6.51563 \rightarrow \text{estimates mean,} \\
l_1 &= -0.78125 \rightarrow \text{estimates 1,} \\
l_2 &= 1.28125 \rightarrow \text{estimates 2,} \\
l_3 &= -0.53125 \rightarrow \text{estimates 3,} \\
l_4 &= 0.03125 \rightarrow \text{estimates 4,} \\
l_{12} &= -0.03125 \rightarrow \text{estimates 12,} \\
l_{13} &= 0.15625 \rightarrow \text{estimates 13,} \\
l_{14} &= -0.28125 \rightarrow \text{estimates 14,} \\
l_{23} &= 0.09375 \rightarrow \text{estimates 23,} \\
l_{24} &= 0.28125 \rightarrow \text{estimates 24,} \\
l_{34} &= -0.03125 \rightarrow \text{estimates 34,}
\end{aligned}
$$

$$l_{123} = \quad 0.03125 \rightarrow \text{estimates } 45,$$
$$l_{124} = -0.03125 \rightarrow \text{estimates } 35,$$
$$l_{134} = \quad 0.15625 \rightarrow \text{estimates } 25,$$
$$l_{234} = \quad 0.09375 \rightarrow \text{estimates } 15,$$
$$l_{1234} = -0.46875 \rightarrow \text{estimates } 5.$$

Previously, in our analysis of two-level full factorial designs, we employed normal probability plots to judge the importance of variable effects. This same graphical technique may be employed usefully in some cases in analyzing two-level fractional factorial designs. Figure 20.1 displays a normal probability plot of the estimates (l_i's) given above.

Summary

An examination of the normal probability plot of Fig. 20.1 reveals that the following effects are important (under the assumption that three-factor and higher-order interactions are negligible):

1, 2, 3, 5, 14, 24.

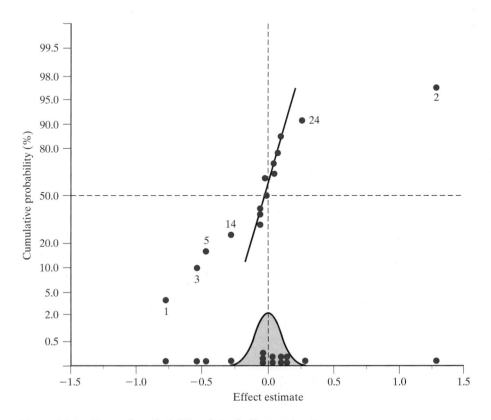

Figure 20.1 Normal probability plot of effect estimates

These main and two-factor interaction effects may be interpreted as follows:

1. **Main Effect of Machine Type (Variable 1).**   The vertical machine, on average, yields Hegman values 0.78125 higher than the horizontal machine. This effect was large relative to the other effects.
2. **Main Effect of Pigment Type (Variable 2).**   Blue pigment, on average, yields Hegman values 1.28125 lower than red pigment. Blue pigment is known to be generally more difficult to grind.
3. **Main Effect of Pigment Concentration (Variable 3).**   A 10% pigment concentration, on average, yields Hegman values 0.53125 higher than a concentration of 15%.
4. **Main Effect of Flow Rate (Variable 5).**   A 20 gal/hr flow rate, on average, gives Hegman values 0.46875 higher than a flow rate of 30 gal/hr.
5. **Two-Factor Interaction: Machine by Temperature (14 Interaction).**   This interaction (Fig. 20.2) indicates that the vertical machine performs somewhat better at higher temperatures, while the horizontal machine performs somewhat better at lower temperatures. However, the low-temperature horizontal mill results were poorer than either high- or low-temperature vertical mill results.
6. **Two-Factor Interaction: Pigment Type by Temperature (24 Interaction).**   This interaction (Fig. 20.3) indicates that for the blue pigment, better Hegman values are obtained at the low level of temperature, while for the red pigment, the high level of temperature is better. This result is important in that it begins to provide information of value with respect to the optimization of the process.

    Based on the results of this experiment, there appears to be no advantage in using the horizontal sandmill in place of the vertical sandmill. In fact, the

Figure 20.2   Machine-by-temperature interaction

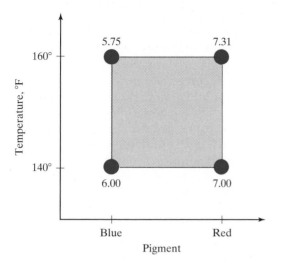

Figure 20.3   Pigment-type-by temperature interaction

vertical sandmill appears to be superior. Since a capital expenditure was being considered for the purchase of horizontal sandmills in the plant, this experiment effectively produced a savings of $180,000. Furthermore, it was learned that a lower pigment concentration actually produces higher/better Hegman values. This result is important because a reduction in the pigment concentration lowers the raw material costs. Finally, both processing variables, temperature and flow rate, greatly influence the grind fineness (the former through the temperature by pigment interaction). Further study of these two processing variables is needed from a process optimization standpoint.

20.6   Orthogonal Arrays and Two-Level Fractional Factorial Designs

The purpose of this section is to present some examples of the use of orthogonal arrays as recommended by Taguchi and to discuss the relationship between these designs and the more traditional treatment of fractional factorial designs. We do not attempt to contrast the overarching philosophical approaches encompassing these methods but, rather, attempt only to compare experimental design structures from a more mechanical point of view. However, it is impossible to separate the mechanics of the use of these experimental design structures completely from the frameworks in which they are implemented. We discuss only experimental designs that consider each variable at two levels. It should be noted that both in the context of Taguchi methods and in the more traditional design of experiments framework, designs that vary factors over more than two levels are commonly used.

The use of orthogonal arrays has been widespread for over 70 years. As pointed out by Box and Meyer,[5] Tippett[6] employed a 125th fraction of a 5^5 factorial for screening experiments on a cotton spinning machine. The theory of fractional factorial designs was first worked out by Finney[7] and Rao.[8] Other orthogonal arrays were introduced by Plackett and Burman.[9] Two- and three-level fractional factorial designs gained widespread attention and industrial application beginning in the 1950s with the publication of numerous papers on theory and applications. Papers such as "The 2^{k-p} Fractional Factorial Designs," Parts I and II, *Technometrics*, Vol. 3, p. 311, by G. E. P. Box and J. S. Hunter (1961) provided much useful guidance to the practitioner in the adroit use of these experimental design structures. Beginning around 1950, Taguchi and his colleagues began extensive use of this class of experimental designs. Since the early 1980s, the use of Taguchi methods and orthogonal arrays has been widespread in the United States.

Numerous examples of the use of orthogonal arrays can be found in the industrial experimental design literature of the Western world in the post–World War II era. In fact, these authors browsed through a 1950 publication of the Shell Oil Company that advocated the use of "magic squares" in industrial experimentation in the chemical processing industry.

Two-Level Fractional Factorial Design Structures: The Issue of Confounding

To many, the most significant property of the designs we are examining here is orthogonality—the ability to separate out the individual effects of several variables on a response of interest. The words "not mixed or confounded" have been used to describe the property of orthogonality in orthogonal arrays. Some caution might be exercised in the use of these descriptions, especially the words "not confounded," since it is not so much the design structure itself, but rather the assumptions used in its application, that justify the terminology. When anything less than a full factorial is under consideration, confounding or mixing of effects is present. Through thoughtful design, attractive confounding structures can be found, as we show in this chapter and in Chapter 21. Even so, confounding can be removed only by assumptions made about the physical system, not about the experimental design.

---

[5]G. E. P. Box and R. D. Meyer, "An Analysis for Unreplicated Fractional Factorials," *Technometrics*, Vol. 28, 1986, pp. 11–18.

[6]L. H. C. Tippett, *Applications of Statistical Methods to the Control of Quality in Industrial Production*, Manchester Statistical Society, Manchester, England, 1934.

[7]D. J. Finney, "The Fractional Replication of Factorial Arrangements," *Annals of Eugenics*, Vol. 12, 1945, pp. 291–301.

[8]C. R. Rao, "Factorial Experiments Derivable from Combinatorial Arrangements of Arrays," *Journal of the Royal Statistical Society, Ser. B*, Vol. 9, 1947, pp. 128–140.

[9]R. L. Plackett and J. P. Burman, "Design of Optimal Multifactorial Experiments," *Biometrika*, Vol. 23, 1946, pp. 305–325.

We have seen that an important feature of two-level factorial experiments is the potential for the estimation of variable interactions. A two-factor interaction measures the manner in which the effect of one factor changes/is influenced by the level of a second factor. When two-level factorials are fractionated, the ability to determine all of the interactions among a set of factors is lost. However, as we have seen, judicious fractionation leads to the ability to obtain knowledge on low-order interactions under the assumption that higher-order interactions are of negligible importance. We will soon see that orthogonal arrays are highly fractionated factorial designs. The information they produce is a function of two things: the nature of their confounding patterns and the assumptions made about the physical system to which they are applied.

It is well known that highly fractionated factorials suffer from the problem of having main effects of factors confounded/aliased with lower-order interactions effects, often with two-factor interactions. This was well documented earlier in this chapter and is discussed again in Chapter 21. Were it not for the *principle of redundancy*, which has been referred to as the *sparsity of variable effects*, and the "sequential assembly" (to be discussed in Chapter 21) of experimental designs, such highly fractionated designs might be of limited utility.

Analysis of the $L_8(2^7)$ Orthogonal Array Design Structure

Taguchi refers to the experimental design structure in Table 20.3 as an "$L_8(2^7)$ orthogonal array."[10] For the case illustrated, each of the seven columns is assigned to a different factor (A to G, 1 to 7). The two levels of each factor are designated 1 and 2.

We recognize this structure as representing the design matrix of a certain 2^{7-4} fractional factorial design. The alternative way that we are familiar with viewing this structure is shown in Table 20.4. The order of the columns, left to

TABLE 20.3	$L_8(2^7)$ Orthogonal Array							
				Factor				
	A	*B*	*C*	*D*	*E*	*F*	*G*	
Test	*1*	*2*	*3*	*4*	*5*	*6*	*7*	*Result*
1	1	1	1	1	1	1	1	y_1
2	1	1	1	2	2	2	2	y_2
3	1	2	2	1	1	2	2	y_3
4	1	2	2	2	2	1	1	y_4
5	2	1	2	1	2	1	2	y_5
6	2	1	2	2	1	2	1	y_6
7	2	2	1	1	2	2	1	y_7
8	2	2	1	2	1	1	2	y_8

---

[10]The next three examples in the chapter are based on experimental designs in G. Taguchi and Yu-In Wu, *Introduction to Off-Line Quality Control*, Central Japan Quality Control Association, Meieki Nakamura-Ku Magaya, Japan, 1979.

TABLE 20.4		2_{III}^{7-4} Fractional Factorial Design					
				−12	−13	−23	+123
Test	1	2	3	4	5	6	7
1	−	−	−	−	−	−	−
2	+	−	−	+	+	−	+
3	−	+	−	+	−	+	+
4	+	+	−	−	+	+	−
5	−	−	+	−	+	+	+
6	+	−	+	+	−	+	−
7	−	+	+	+	+	−	−
8	+	+	+	−	−	−	+
	D	B	A	F	E	C	G
	(4)	(2)	(1)	(6)	(5)	(3)	(7)

right, has been rearranged (note the cross-referenced column designations at the bottom), and we recognize this $L_8(2^7)$ orthogonal array as a member of the general family of 2^{7-4} fractional factorial designs. The concept of families of two-level fractional factorials will be discussed in Chapter 21. The particular member of this family shown above is defined by the generators

$$I = -124, \quad I = -135, \quad I = -236, \quad I = +1237.$$

It is noted that some of the generators carry a minus sign. The significance of this will be discussed in Chapter 21.

We have seen that when used to assign or consider seven variables, the $L_8(2^7)$ orthogonal array shown previously is equivalent to a particular 2_{III}^{7-4} fractional factorial design with generators: $I = -124$, $I = -135$, $I = -236$, and $I = 1237$. The "III" in the designation "2_{III}^{7-4}" refers to the resolution of the design, which is three, meaning that at least some (in this case, all) main effects are confounded or confused with two-factor interactions.

In the general context of the Taguchi methods, assignment of seven variables to the $L_8(2^7)$ orthogonal array and the subsequent interpretation of the experimental results assumes that variable interactions at all levels may be neglected unless specifically identified by a column in the array. In the context of the approach taken earlier in this chapter, the 2_{III}^{7-4} fractional factorial will be interpreted with more caution due to the explicit recognition of the potential importance of two-factor interaction effects, and the fact that they are confounded with the main effects of the seven variables under study.

Examination of the Use of an $L_{16}(2^{15})$ Orthogonal Array

The $L_{16}(2^{15})$ orthogonal array of Table 20.5 appears in *Introduction to Off-Line Quality Control* by Taguchi and Wu as Exercise 2 at the end of Chapter 6.

TABLE 20.5															$L_{16}(2^{15})$ Orthogonal Array
					A × B					B × D		A × D			
Test	F 1	A 2	e 3	B 4	e 5	B 6	E 7	C* 8	H 9	e 10	D 11	e 12	D 13	G 14	D 15	Response
1	1	1	1	1	1	1	1	1	1	1	1	1	1	1	1	
2	1	1	1	1	1	1	1	2	2	2	2	2	2	2	2	
3	1	1	1	2	2	2	2	1	1	1	1	2	2	2	2	
4	1	1	1	2	2	2	2	2	2	2	2	1	1	1	1	
5	1	2	2	1	1	2	2	1	1	2	2	1	1	2	2	
6	1	2	2	1	1	2	2	2	2	1	1	2	2	1	1	
7	1	2	2	2	2	1	1	1	1	2	2	2	2	1	1	
8	1	2	2	2	2	1	1	2	2	1	1	1	1	2	2	
9	2	1	2	1	2	1	2	1	2	1	2	1	2	1	2	
10	2	1	2	1	2	1	2	2	1	2	1	2	1	2	1	
11	2	1	2	2	1	2	1	1	2	1	2	2	1	2	1	
12	2	1	2	2	1	2	1	2	1	2	1	1	2	1	2	
13	2	2	1	1	2	2	1	1	2	2	1	1	2	2	1	
14	2	2	1	1	2	2	1	2	1	1	2	2	1	1	2	
15	2	2	1	2	1	1	2	1	2	2	1	2	1	1	2	
16	2	2	1	2	1	1	2	2	1	1	2	1	2	2	1	

Source: Adapted from G. Taguchi and Y. Wu, *Introduction to Off-Line Quality Control*, Central Japan Quality Control Association, Meieki Nakamura-Ku Magaya, Japan, 1979.
*Column 8 was labeled as variable D in Taguchi and Wu, but has been corrected here and labeled as variable C.

In this application of the $L_{16}(2^{15})$ orthogonal array, we see that the following assignments have been made:

- Eight columns have been assigned to variable main effects: A, B, C, D, E, F, G, H.

- Three columns have been assigned to two-factor interactions: A × B, B × D, A × D.

- Four columns, all designated as e, have been assigned to the estimation of error.

The $L_{16}(2^{15})$ orthogonal array in Table 20.5 can be examined through the eyes of the class of 2^{k-p} fractional factorial designs. Table 20.6 shows the design matrix for a 2^{8-4} fractional factorial design. This design matrix was constructed in the following way:

1. Columns 1 to 4 are the columns in the $L_{16}(2^{15})$ orthogonal array assigned to variables C, B, A, and F, respectively.

2. Columns 5 to 8 are the columns in the $L_{16}(2^{15})$ orthogonal array assigned to variables E, H, G, and D, respectively.

TABLE 20.6		2^{8-4} Fractional Factorial Design						
					234	− 14	123	− 1234
Test	1	2	3	4	5	6	7	8
1	−	−	−	−	−	−	−	−
2	+	−	−	−	−	+	+	+
3	−	+	−	−	+	−	+	+
4	+	+	−	−	+	+	−	−
5	−	−	+	−	+	−	+	+
6	+	−	+	−	+	+	−	−
7	−	+	+	−	−	−	−	−
8	+	+	+	−	−	+	+	+
9	−	−	−	+	+	+	−	+
10	+	−	−	+	+	−	+	−
11	−	+	−	+	−	+	+	−
12	+	+	−	+	−	−	−	+
13	−	−	+	+	−	+	+	−
14	+	−	+	+	−	−	−	+
15	−	+	+	+	+	+	−	+
16	+	+	+	+	+	−	+	−
	C	B	A	F	E	H	G	D

[Variable "names" given by Taguchi and Wu
in the $L_{16}(2^{15})$ orthogonal array]

It is interesting to note what we discover when we examine how columns 1 to 4 were used to define columns 5 to 8. Close examination of the $L_{16}(2^{15})$ orthogonal array of Table 20.5 (and Table 20.6) shows that the columns assigned to E, G, H, and D are the following products of the columns assigned to C, B, A, and F:

$$5 = 234, \qquad 6 = -14,$$
$$E = B \times A \times F, \qquad H = -(C \times F),$$
$$7 = 123, \qquad 8 = -1234,$$
$$G = C \times B \times A, \qquad D = -(C \times B \times A \times F).$$

Hence the design generators of this design are

$$I = 2345, \quad I = 1237, \quad I = -146, \quad I = -12348.$$

It follows, then, that the associated defining relation is

$$
\begin{aligned}
I = \quad & 2345 = -146 = 1237 = -12348, \\
= & -2356 = 1457 = -158 = -23467, \\
= & 2368 = -478 = 1678 = -23578, \\
= & 4568 = -567 = 12345678.
\end{aligned}
$$

The results here may be a little surprising. They show not only that the $L_{16}(2^{15})$ orthogonal array is a 2^{8-4} fractional factorial design, but that in particular, this one is a resolution III 2^{8-4} fractional factorial. Note the terms in the defining relation: −146, −258, −478, −567. This is surprising because we know that it is possible to design a 16-run two-level fractional factorial to examine eight variables that is resolution IV. In fact, Box, Hunter, and Hunter, in *Statistics for Experimenters*,[11] recommend the 2^{8-4} design with generators

$$5 = 234, \quad 6 = 134, \quad 7 = 123, \quad 8 = 124,$$
$$(I = 2345, \quad I = 1346, \quad I = 1237, \quad I = 1248).$$

This produces a resolution IV design. Recall that a resolution III design confounds at least some main effects with two-factor (and higher-order) interactions, while a resolution IV design confounds at least some main effects with three-factor (and higher-order) interactions (but not two-factor interactions).

Under the Taguchi methods approach to the design of experiments, the philosophy toward interaction effects is somewhat different. As a result, the bottom line is that unless expressly identified as present/real and assigned to the orthogonal array, interactions at *all* levels are assumed to be negligible. We must remember, however, that although different philosophies may be used to interpret the workings of the real world, those "workings" are unchanged by the assumption. Hence, it is probably always wise to select orthogonal arrays of the highest possible resolution so as to minimize the effect that the assumptions we do make about interaction effects has on the interpretation of the results.

Another Interesting Orthogonal Array Application

Another interesting example of the use of orthogonal arrays by Taguchi concerns an arc welding experiment performed by the National Railway Cooperation of Japan in 1959.[12] Nine variable main effects and four specific two-factor interactions were studied using the $L_{16}(2^{15})$ orthogonal array shown in Table 20.7.

TABLE 20.7	$L_{16}(2^{15})$ Orthogonal Array Example		
Factor	*Number of Levels*	*First Level*	*Second Level*
A: Kind of welding rods	2	A_1 = J100	A_2 = B17
B: Drying methods of rods	2	B_1 = no drying	B_2 = one-day drying
C: Welded materials	2	C_1 = SS41	C_2 = SB35
D: Thickness of welded material	2	D_1 = 8 mm	D_2 = 12 mm
E: Angle of welding device	2	E_1 = 70 degrees	E_2 = 60 degrees
F: Opening of welding device	2	F_1 = 1.5 mm	F_2 = 3 mm
G: Current	2	G_1 = 150 A	G_2 = 130 A
H: Welding methods	2	H_1 = weaving	H_2 = single
I: Preheating	2	I_1 = no preheating	I_2 = 150 degrees preheating

(continued)

---

[11]G. E. P. Box, W. G. Hunter, and J. S. Hunter, *Statistics for Experimenters*, Wiley, New York, 1978.
[12]See footnote 10.

Test	A	G	A × G	H	A × H	G × H	B	D	E	F	I	e	e	C	A × C
1	1	1	1	1	1	1	1	1	1	1	1	1	1	1	1
2	1	1	1	1	1	1	1	2	2	2	2	2	2	2	2
3	1	1	1	2	2	2	2	1	1	1	1	2	2	2	2
4	1	1	1	2	2	2	2	2	2	2	2	1	1	1	1
5	1	2	2	1	1	2	2	1	1	2	2	1	1	2	2
6	1	2	2	1	1	2	2	2	2	1	1	2	2	1	1
7	1	2	2	2	2	1	1	1	1	2	2	2	2	1	1
8	1	2	2	2	2	1	1	2	2	1	1	1	1	2	2
9	2	1	2	1	2	1	2	1	2	1	2	1	2	1	2
10	2	1	2	1	2	1	2	2	1	2	1	2	1	2	1
11	2	1	2	2	1	2	1	1	2	1	2	2	1	2	1
12	2	1	2	2	1	2	1	2	1	2	1	1	2	1	2
13	2	2	1	1	2	2	1	1	2	2	1	1	2	2	1
14	2	2	1	1	2	2	1	2	1	1	2	2	1	1	2
15	2	2	1	2	1	1	2	1	2	2	1	2	1	1	2
16	2	2	1	2	1	1	2	2	1	1	2	1	2	2	1

Nine columns are assigned to variables: $A, G, H, B, D, E, F, I, C$
Four columns are assigned to interactions: $A \times G, A \times H, G \times H, A \times C$
Two columns are assigned to error (e) estimation

This orthogonal array may be recognized as a 2^{9-5} fractional factorial as follows:

Test	D 1	H 2	G 3	A 4	B 5	E 6	F 7	I 8	C 9
1	−	−	−	−	−	−	−	−	−
2	+	−	−	−	−	+	+	+	+
3	−	+	−	−	+	−	−	−	+
4	+	+	−	−	+	+	+	+	−
5	−	−	+	−	+	−	+	+	+
6	+	−	+	−	+	+	−	−	−
7	−	+	+	−	−	−	+	+	−
8	+	+	+	−	−	+	−	−	+
9	−	−	−	+	+	+	−	+	+
10	+	−	−	+	+	−	+	−	−
11	−	+	−	+	−	+	−	+	−
12	+	+	−	+	−	−	+	−	+
13	−	−	+	+	−	+	+	−	−
14	+	−	+	+	−	−	−	+	+
15	−	+	+	+	+	+	+	−	+
16	+	+	+	+	+	−	−	+	−

This design is a 2_{III}^{9-5} fractional factorial with generators

$$I = 2345, \quad I = -146, \quad I = -137, \quad I = 1348, \quad I = -12349.$$

The defining relation for this design is

$$
\begin{aligned}
I = \ & 2345 = -146 = -137 = 1348 = -12349 \\
= \ & -12356 = 2567 = 12345678 = -156789 = -1345679 \\
= \ & -24568 = 135689 = 4569 = -12457 = -23578 = 145789 \\
= \ & 3579 = 1258 = -34589 = -159 = 3467 = 1678 = -2346789 \\
= \ & -12679 = -368 = 124689 = 2369 = -478 = 123789 = 2479 = -289.
\end{aligned}
$$

The linear combinations of aliased effects that can be estimated, assuming third- and higher-order interactions can be neglected, are as follows (in the regular calculation matrix order):

l_0 estimates I (mean),
l_1 estimates $1 - 46 - 37 - 59$,
l_2 estimates $2 - 89$,
l_3 estimates $3 - 17 - 68$,
l_4 estimates $4 - 16 - 78$,
l_5 estimates $12 + 58$ (error),
l_6 estimates $-7 + 13 + 48$,
l_7 estimates $-6 + 14 + 38$,

l_8 estimates $23 + 45 + 69$,
l_9 estimates $24 + 35 + 79$,
l_{10} estimates $34 + 25 + 18 + 67$,
l_{11} estimates $-27 - 49 - 56$,
l_{12} estimates $-26 - 39 - 57$ (error),
l_{13} estimates $8 - 36 - 47 - 29$,
l_{14} estimates $5 - 19$,
l_{15} estimates $-9 + 15 + 28$.

Now the $L_{16}(2^{15})$ or 2_{III}^{9-5} fractional factorial design above is of resolution III; that is, at least some (in this case, all) of the main effects are confounded with two-factor interactions. In fact, it is not possible to examine the effects of nine variables in a 16-run two-level fractional factorial and have a design resolution better than III. However, it is possible to find another resolution III 2^{9-5} design/orthogonal array that has potentially a more attractive alias structure.

As an alternative to the design above, Box, Hunter, and Hunter recommend a 2_{III}^{9-5} design with the following generators:

$$I = 1235, \quad I = 2346, \quad I = 1347, \quad I = 1248, \quad I = 12349.$$

The reason for this choice over the one above [the $L_{16}(2^{15})$] becomes apparent when we examine the complete alias structure of the design. For this 2_{III}^{9-5} design, the linear combinations of aliased effects are (again, assuming third- and higher-order interactions can be neglected)

l_0 estimates mean,
l_1 estimates $1 + 69$,
l_2 estimates $2 + 79$,
l_3 estimates $3 + 89$,

l_8 estimates $23 + 15 + 46 + 78$,
l_9 estimates $24 + 36 + 18 + 57$,
l_{10} estimates $34 + 26 + 17 + 58$,
l_{11} estimates $5 + 49$,

l_4 estimates $4 + 59$, l_{12} estimates $8 + 39$,

l_5 estimates $12 + 35 + 48 + 67$, l_{13} estimates $7 + 29$,

l_6 estimates $13 + 25 + 47 + 68$, l_{14} estimates $6 + 19$,

l_7 estimates $14 + 37 + 28 + 56$, l_{15} estimates $9 + 45 + 16 + 27 + 38$.

When we look closely at the above we notice something quite interesting. Although main effects are still confounded with two-factor interactions, eight of the nine main effects are confounded with only one two-factor interaction. Furthermore, in each case that two-factor interaction involves variable 9 (i.e., 69, 79, 89, 59, 49, 39, 29, 19). The one factor that is confounded with four two-factor interactions is variable 9.

Essentially, we might use the following design strategy: Among the nine factors to be studied select the one that is least likely to be important and/or to interact with other factors. Assign this factor as variable 9. If your thoughts about this factor's unimportance are correct, the design becomes effectively a resolution IV design (i.e., it is a 2_{IV}^{8-4} design with main effects "clear" of two-factor interaction aliases). Again, what we have witnessed in this example is the importance of design resolution and the judicious choice of design generators.

■ EXERCISES

20.1 Suppose that it is desired to study eight variables at two levels each in a full factorial design scheme.
  (a) How many tests are required?
  (b) How many effects can be estimated? How many two-factor interactions can be estimated? How many three-factor interactions? How many four-factor interactions and above?
  (c) Comment on the general advisability of running such an experiment.
  (d) If this experiment were performed, of what use might the higher-order interaction effect estimates be in the analysis?

20.2 It is desired to study the effect of five factors on the burr height in a metal stamping process. Only enough funds are available to conduct 16 tests; therefore, a 2^{5-1} fractional factorial design will be performed.
  (a) What generator should be used to get the highest-resolution fractional factorial? What is the defining relation?
  (b) Show the complete set of linear combinations of confounded effects that result from the generator that you chose.
  (c) How does this confounding structure simplify if third- and higher-order interactions are negligible?

20.3 Consider a five-variable, eight-run, two-level fractional factorial design.
  (a) Develop the confounding structure based on the following sets of generators:
      (i) $4 = 12, 5 = 13$.
      (ii) $4 = 12, 5 = 123$.
  (b) Does there appear to be any reason to choose one set of generators over another in part (a)? Indicate the reasons for your choice of generators clearly.

20.4 Consider a 2^{8-4} fractional factorial design.

    (a) Write down the defining relation for each of the following candidate designs:

        (i) $5 = 234$, $6 = 134$, $7 = 123$, $8 = 124$.

        (ii) $5 = 234$, $6 = 134$, $7 = 123$, $8 = 1234$.

    (b) Which of these designs is preferred? Why?

20.5 Consider a 2^{7-4} fractional factorial design with the generators $4 = 23$, $5 = 12$, $6 = 123$, and $7 = 13$.

    (a) Write down the defining relation for this design.

    (b) Write down the linear combinations of effects that may be estimated from such a design.

    (c) Under the assumption that three-factor interactions and higher are negligible, how do the linear combinations of effects that can be estimated simplify?

20.6 Consider a 2^{7-2} fractional factorial design with the generators $6 = 123$ and $7 = 345$.

    (a) Write down the defining relation for this design.

    (b) Write down the linear combinations of effects that may be estimated from such a design.

    (c) Under the assumption that three-factor interactions and higher are negligible, how do the linear combinations of effects simplify?

20.7 A 2^{6-1} fractional factorial design was conducted. The linear combinations of effects that were estimated based on the measured responses from the experiment are as follows.

$l_0 = 98.875$ estimates $I + 123456/2$, $l_{123} = -2.875$ estimates $123 + 456$,

$l_1 = 36.750$ estimates $1 + 23456$, $l_{124} = 4.625$ estimates $124 + 356$,

$l_2 = -5.375$ estimates $2 + 13456$, $l_{125} = -0.625$ estimates $125 + 346$,

$l_3 = 1.750$ estimates $3 + 12456$, $l_{134} = -0.500$ estimates $134 + 256$,

$l_4 = 0.750$ estimates $4 + 12356$, $l_{135} = 3.500$ estimates $135 + 246$,

$l_5 = 1.750$ estimates $5 + 12346$, $l_{145} = 3.500$ estimates $145 + 236$,

$l_{12} = -23.875$ estimates $12 + 3456$, $l_{234} = 2.875$ estimates $234 + 156$,

$l_{13} = 2.750$ estimates $13 + 2456$, $l_{235} = 2.125$ estimates $235 + 146$,

$l_{14} = -2.250$ estimates $14 + 2356$, $l_{245} = -2.625$ estimates $245 + 456$,

$l_{15} = -0.500$ estimates $15 + 2346$, $l_{345} = 2.000$ estimates $345 + 126$,

$l_{23} = 0.125$ estimates $23 + 1456$, $l_{1234} = 17.875$ estimates $1234 + 56$,

$l_{24} = -0.375$ estimates $24 + 1356$, $l_{1235} = -1.625$ estimates $1235 + 46$,

$l_{25} = 0.625$ estimates $25 + 1346$, $l_{1245} = -1.625$ estimates $1245 + 36$,

$l_{34} = -3.000$ estimates $34 + 1256$, $l_{1345} = -2.250$ estimates $1345 + 26$,

$l_{35} = 5.750$ estimates $35 + 1246$, $l_{2345} = -1.625$ estimates $2345 + 16$,

$l_{45} = -0.750$ estimates $45 + 1236$, $l_{12345} = -39.875$ estimates $12345 + 6$.

    (a) Write down the generator and the defining relation for this design.

    (b) Based on the assumption that third- and higher-order interactions are negligible, several of the linear combinations shown above may be used to estimate the standard error of an effect estimate. Under this assumption, develop 95% confidence intervals for the true mean effect values to determine which effect estimates/linear combinations of effects are significant. Comment on the results.

20.8 Consider a 2^{5-2} fractional factorial design with the generators $I = 124$ and $I = 135$.

(a) What is the defining relation for this design?

			Variable		
Test	1	2	3	4	5
1	−	−	−		
2	+	−	−		
3	−	+	−		
4	+	+	−		
5	−	−	+		
6	+	−	+		
7	−	+	+		
8	+	+	+		

(b) For this design, complete the recipe matrix shown in the table below.

(c) Write down the complete confounding pattern for this design; that is, what linear combinations of effects can be estimated from this design?

(d) What is the resolution of this design? What does the resolution tell us about our ability to interpret the results of such an experiment?

20.9 The table provides the design matrix for a certain two-level fractional factorial experimental design.

			Variable		
Test	1	2	3	4	5
1	−	−	−	+	−
2	+	−	−	−	+
3	−	+	−	−	+
4	+	+	−	+	−
5	−	−	+	+	+
6	+	−	+	−	−
7	−	+	+	−	−
8	+	+	+	+	+

(a) Write down the generators and the defining relation for this design.

(b) What is the resolution of this design?

(c) Write down all of the linear combinations of confounded effects that can be estimated from the results of this experiment. Assume that all interaction effects may be important.

(d) Suppose that it was decided to run only a four-variable experiment using eight tests (i.e., a 2^{4-1} fractional factorial design). What generator would you propose for this design? How would the design resolution differ, if at all, for this new 2^{4-1} design as compared with the original 2^{5-2} design?

20.10 Consider a 2^{7-3} fractional factorial design with the generators $I = 1235$, $I = 1246$, and $I = 1347$.

(a) Write down the defining relation for this design.

(b) What is the resolution of this design? What does this tell us?

(c) Write down all linear combinations of effects that can be estimated from this experiment. Assume that third- and higher-order interaction effects can be neglected.

20.11 Consider a 2^{8-4} fractional factorial design.

(a) How many variables does this design have?

(b) How many runs are involved in this design?

(c) How many levels are used for each of the variables?

(d) How many generators are there for this design?

(e) How many words are there in the defining relation (including *I*)?

20.12 Construct a 2^{6-2} fractional factorial design with as high a resolution as possible.

(a) Write down the design matrix and the calculation matrix for this design.

(b) What are the generators for your design?

(c) What is the defining relation for your design?

(d) What is confounded with the main effect of variable 3 in your design?

(e) What is confounded with the two-factor interaction 12 in your design?

20.13 It is proposed to conduct a 2^{8-3} fractional factorial design to study a certain process. After some discussion, it is decided that only 16 tests can be conducted, and so a 2^{8-4} design is proposed. Will the increased level of fractionation seriously affect the nature of the effect estimates, or the density of the confounding structure, that can be obtained from the experiment? Demonstrate your answer clearly.

20.14 A consulting firm engaged in road-building work is asked by one of its clients to carry out an experimental study to determine the effects of six variables on the physical properties of a certain kind of asphalt. Call these variables *A, B, C, D, E,* and *F*.

(a) If a two-level fractional factorial design of at least resolution IV is required for this situation, what is the smallest number of tests in which this can be accomplished?

(b) What would be the generator for the best 32-run design you would propose for this situation? What is the design resolution for this situation?

(c) If a seventh variable were to be added, what is the highest-resolution 32-run fractional factorial design that could be obtained? What generators would you propose for this design?

20.15 A welding experiment has been conducted to study the effect of five variables on the heat input in watts.[13] The variables and their settings are shown in the first table.

Variable	Low Level	High Level
x_1: Open-circuit voltage (V)	31	34
x_2: Slope	11	6
x_3: Electrode melt-off rate (in./min)	162	137
x_4: Electrode diameter (in.)	0.045	0.035
x_5: Electrode extension (in.)	0.375	0.625

The second table summarizes the levels of the five variables that were used for each of the tests (note that the tests are shown in the randomized order in which they were run), along with the measured heat input that was produced for that run.

Run	x_1	x_2	x_3	x_4	x_5	y
1	34	11	162	0.045	0.375	4141
2	31	6	162	0.045	0.375	3790
3	31	6	162	0.035	0.625	2319
4	34	6	137	0.045	0.375	3765
5	31	6	137	0.035	0.375	2485
6	34	6	162	0.045	0.625	4061
7	31	6	137	0.045	0.625	3507
8	34	11	137	0.045	0.625	3425
9	34	6	137	0.035	0.625	2450
10	34	11	137	0.035	0.375	2466
11	31	11	162	0.035	0.375	2580
12	31	11	162	0.045	0.625	3318
13	31	11	137	0.035	0.625	1925
14	34	11	162	0.035	0.625	2450
15	31	11	137	0.045	0.375	3431
16	34	6	162	0.035	0.375	3067

(a) The kind of experiment shown in the second table is a 2^{k-p} fractional factorial design. What are the values for k and p? Write out the complete calculation matrix for this design.

---

[13]D. A. J. Stegner, S. M. Wu, and N. R. Braton, "Prediction of Heat Input for Welding," *Welding Journal Research Supplement*, Vol. 1, March 1967.

(b) What is the generator for this fractional factorial? What is the defining relation?

(c) What effects are associated with each linear combination that may be calculated?

(d) What are the numerical values for each linear combination of effects?

(e) Under the assumption that three-factor and higher-order interactions are negligible, interpret the results of this experiment.

20.16 An experimenter wishes to study seven variables at two levels each and chooses to do so with a 32-run resolution IV design with the generators $I = 12346$ and $I = 12457$. A colleague argues that he can achieve the same resolution with a design that employs only 16 runs. Is the colleague correct? If so, write down generators and the defining relation for such a design. Write out all of the linear combinations of effects that can be estimated for each design. Comment on the relative merits of each design.

20.17 A 2^{4-1} fractional factorial design was conducted on a chemical process by assigning variable 4 to the 123 interaction column.

Variable	Low Level	High Level
1. Feedrate (liters/min)	5	20
2. Catalyst (%)	A	B
3. Temperature (°C)	200	220
4. Concentration (%)	5	7

The second table summarizes the eight tests that were run, including the levels of each of the four variables and the yield (% reacted) for each test.

Test	1	2	3	4	y (% reacted)
1	−	−	−	−	33
2	+	−	−	+	51
3	−	+	−	+	44
4	+	+	−	−	40
5	−	−	+	+	35
6	+	−	+	−	82
7	−	+	+	−	46
8	+	+	+	+	69

(a) Write down all of the linear combinations of effects that can be estimated from this experiment (assume that third- and higher-order interactions are negligible).

(b) Calculate numerical values for the effect estimates and determine which effects are significant using a normal probability plot of the effect estimates.

(c) Propose a tentative mathematical model for the response.

20.18 A newly hired engineer has been asked to analyze the experimental results of the fractional factorial design shown in the table below. It seems that this experiment was designed and performed by an employee who is now working for a competitor of the company. The response of interest is the number of particles per million of a certain contaminant. The response was available for all of the tests, but, as is evident, some of the test settings were not readable. The design matrix is given in the order in which the tests were performed.

Run	1	2	3	4	5	6	y
1	?	?	?	?	?	?	16
2	−	+	−	−	+	−	23
3	+	+	−	−	−	+	15
4	−	−	−	+	−	+	19
5	?	?	?	?	?	?	21
6	+	−	−	+	+	−	14
7	−	−	+	−	+	+	27
8	+	−	+	−	−	−	12

(a) What kind of a two-level design is this? Be as specific as you can in defining this experiment.

(b) Provide the missing information concerning the variable settings.

(c) Estimate whatever variable effects (or linear combination of effects) you can. Can you say anything about the results of this experiment?

20.19 Some tests were carried out on a newly designed automobile. Five variables were studied, as shown in the first table.

Variable	Low Level	High Level
1. Engine size (L)	4.0	4.5
2. Fuel octane	87	93
3. Tire pressure (psi)	22	28
4. Driving speed (mph)	45	55
5. Air conditioning	Off	On

The objective of the experiment was to determine the effects of these variables on the fuel economy. The experimental results are shown in the following table.

Test	x_1	x_2	x_3	x_4	x_5	y (mpg)
1	−1	−1	−1	−1	1	25
2	1	−1	−1	−1	−1	21
3	−1	1	−1	−1	−1	26
4	1	1	−1	−1	1	22
5	−1	−1	1	−1	−1	29
6	1	−1	1	−1	1	20

(Continued)

Test	x_1	x_2	x_3	x_4	x_5	y (mpg)
7	−1	1	1	−1	1	29
8	1	1	1	−1	−1	29
9	−1	−1	−1	1	−1	27
10	1	−1	−1	1	1	19
11	−1	1	−1	1	1	27
12	1	1	−1	1	−1	28
13	−1	−1	1	1	1	26
14	1	−1	1	1	−1	21
15	−1	1	1	1	−1	32
16	1	1	1	1	1	25

    (a) Completely define the linear combinations of effects that can be obtained from this experiment.

    (b) Analyze the data. Propose a tentative mathematical model and do a complete analysis of the model residuals. Revise the model, if necessary, and repeat the residual analysis.

20.20 A two-level fractional factorial design was run with 16 unique test conditions. One of the linear combinations of confounded effects that can be estimated from this particular design is 13 + 2345 + 1246 + 56. Determine the design generators. Show all work. What is the resolution of this design? What does *resolution* mean?

20.21 It is desired to study seven variables in a two-level factorial or fractional factorial scheme. Write down and discuss all the options available to the experimenter from the most highly fractionated design to the full factorial. Discuss the advantages and disadvantages of each design. Which design will be best if:

    (a) Only main effects are felt to be important?

    (b) Two or three of the variables are likely to have two-factor interactions with each other and/or some of the other factors?

    (c) Main effects and two-factor interactions may be important but the estimation of main effects is of particular importance?

20.22 Consider the following two 2^{9-4} fractional factorial designs:

        (i) $I = 23456, I = 13457, I = 12458, I = 12359.$
        (ii) $I = 1236, I = 1247, I = 1358, I = 3459.$

    (a) Determine the defining relation for each of these designs.

    (b) Which of these designs is preferred? Why?

    (c) If it is known that variables 6, 7, 8, and 9 cannot physically interact with each other or any of the other five variables, which design is preferred? Why?

20.23 Consider a 2^{9-3} fractional factorial design.

    (a) How many variables does the design consider?

    (b) How many tests are required, assuming no replication?

    (c) How many generators are there for this design?

    (d) How many words are there in the defining relation (including *I*)?

20.24 Is it possible to develop a resolution V two-level fractional factorial for six variables that requires only 16 tests? Explain your answer.

20.25 What is the maximum number of variables that could be examined in a 64-run two-level fractional factorial design and maintain at least a resolution V design? If it is desired to consider two additional variables in the scheme you just proposed, will the number of tests have to increase to maintain a resolution V design? How many tests are now required?

20.26 Consider the following defining relation for a two-level fractional factorial: $I = 1236 = 1247 = 23458 = 3467 = 1568 = 13578 = 25678$.
  (a) How many generators are there for this design?
  (b) What is the resolution of this design?
  (c) How many tests are required, assuming no replication?
  (d) If the number of tests were doubled, what would happen to the resolution of the design? (Assume the best-case situation.)
  (e) If one more variable is added to the design but the number of tests is held constant, will this affect the resolution of the design? What if two more variables are added?

20.27 Following are the generators for a 2^{9-5} resolution III design: $I = 1235$, $I = 2346$, $I = 1347$, $I = 1248$, $I = 12349$. This design has a somewhat interesting characteristic that could be particularly useful if a high premium is placed on keeping the number of tests in the experiment to a minimum. What is this characteristic? You will probably have to write down all the linear combinations of effects that can be estimated from this design to discover what is unique about this nine-variable, 16-run, two-level fractional factorial.

20.28 It is desired to study nine variables in a fractional factorial scheme requiring 128 tests or less. Write down and discuss all of the options available to the experimenter from the most highly fractionated to the least fractionated design (remember, no more than 128 tests). Discuss the advantages and disadvantages of each design.

20.29 Following are the generators for three 16-run fractional factorial designs:
  (i) $5 = 123, 6 = 234$.
  (ii) $5 = 123, 6 = 234, 7 = 134$.
  (iii) $5 = 123, 6 = 234, 7 = 134, 8 = 124$.
  (a) Determine the defining relation for each design. What is the resolution for each design?
  (b) Write out the linear combinations of effects that may be determined for each design, assuming that three-factor and higher-order interactions are negligible.
  (c) Explain what happens in this case to the confounding structure as the number of variables studied in 16 tests is increased from 6 to 7 to 8.

20.30 A 2^{6-2} fractional factorial design was performed to assess the effect of six variables on the cutting force produced during the machining of an aluminum alloy. The variables studied, and their levels for the experiment, are shown in the first table. The design matrix for the experiment and the response (cutting force, in pounds) are shown in the second table.

Variable	Low Level	High Level
1. Feedrate (in./rev)	0.005	0.010
2. Depth of cut (in.)	0.050	0.100
3. Cutting speed (sf/min)	2000	2500
4. Tool geometry	Positive	Negative
5. Machine	American	Monarch
6. Cutting fluid	Absent	Present

Test	1	2	3	4	5	6	y
1	−	−	−	−	−	−	21
2	+	−	−	−	+	−	59
3	−	+	−	−	+	+	57
4	+	+	−	−	−	+	122
5	−	−	+	−	+	+	29
6	+	−	+	−	−	+	52
7	−	+	+	−	−	−	45
8	+	+	+	−	+	−	110
9	−	−	−	+	−	+	31
10	+	−	−	+	+	+	70
11	−	+	−	+	+	−	65
12	+	+	−	+	−	−	119
13	−	−	+	+	+	−	35
14	+	−	+	+	−	−	61
15	−	+	+	+	−	+	57
16	+	+	+	+	+	+	121

(a) Completely define the linear combinations of effect that can be obtained from this experiment, assuming that three-factor and higher-order interactions are negligible.

(b) Calculate numerical values for the linear combinations of effects and determine which of these are significant, using a normal probability plot.

(c) Under the assumption that three-factor interactions and above are negligible, several of the linear combinations may be used to estimate the standard error of an effect estimate. Use this assumption to develop 95% confidence intervals for the true mean effect values. Which effects/linear combinations are significant? Comment on the results of the experiment.

20.31 An import/export company is working with a construction engineering firm to study the tensile strength of various concrete formulations. Six variables of interest have been identified, and are given in the first table. A replicated $L_8(2^7)$ orthogonal array was performed to study the effects of these factors, and the results of this experiment are summarized in the second table.

Variable	Low Level	High Level
1. Specimen size (ft^3)	2	4
2. Amount of water	Low	High
3. Curing time (hr)	24	48
4. Mixing technique	By hand	Machine
5. Aggregate size	Fine	Coarse
6. Cement concentration	Low	High

Test	1	2	3	4	5	e	6	Run 1	Run 2
1	1	1	1	2	2	2	1	2.3	2.2
2	2	1	1	1	1	2	2	3.5	3.3
3	1	2	1	1	2	1	2	3.0	2.9
4	2	2	1	2	1	1	1	2.1	1.9
5	1	1	2	2	1	1	2	3.5	3.6
6	2	1	2	1	2	1	1	2.6	2.6
7	1	2	2	1	1	2	1	2.9	2.9
8	2	2	2	2	2	2	2	3.9	4.0

(a) Using the assumptions typically made when analyzing orthogonal arrays, calculate estimates of all the variable effects.

(b) Construct an ANOVA table to identify which of the variable effects determined in part (a) are significant.

(c) Develop a model for the response based on the results of part (b). Construct an ANOVA table to see if the model developed suffers from lack of fit.

(d) If the model developed in part (c) demonstrates a lack of fit, reanalyze the data using the fractional factorial design techniques described in this chapter, and comment on what might be the problem here in terms of how the study was initially formulated and analyzed.

20.32 An $L_8(2^7)$ orthogonal array was conducted to study the effect of seven variables on the finish of a machined surface (roughness average value, in microinches). The first table describes the levels for the seven variables that were studied in the experiment.

Variable	Low Level	High Level
1. Cutting speed (sfpm)	600	900
2. Feedrate (in./rev)	0.005	0.010
3. Nose radius (in.)	1/64	1/32
4. Depth of cut (in.)	0.050	0.100
5. Side rake (deg)	0	10
6. Lead angle (deg)	0	30
7. Back rake (deg)	0	10

The second table summarizes the conditions for the eight tests that were performed as well as the measured surface finish (R_a) value that was observed for each of the conditions.

Test	x_1	x_2	x_3	x_4	x_5	x_6	x_7	y
1	1	1	1	2	2	2	1	71
2	2	1	1	1	1	2	2	57
3	1	2	1	1	2	1	2	130
4	2	2	1	2	1	1	1	171
5	1	1	2	2	1	1	2	21
6	2	1	2	1	2	1	1	24
7	1	2	2	1	1	2	1	148
8	2	2	2	2	2	2	2	100

(a) What are the generators and the defining relation for this design?

(b) Using the assumptions typically made when analyzing orthogonal arrays, calculate estimates of all the variable effects.

(c) Construct an ANOVA table to identify which of the variable effects determined in part (a) are significant.

(d) Develop a model for the response based on the results of part (c). Construct an ANOVA table to see if the model developed suffers from lack of fit.

(e) Prepare all the residual plots necessary to check the model diagnostically.

(f) If the model developed in part (d) demonstrates a lack of fit or, based on the residual plots, is found to be inadequate, postulate a new model. Conduct all the steps necessary to develop and check the new model.

20.33 What is the maximum number of variables that can be examined using an $L_8(2^7)$ orthogonal array? What are the assumptions that generally govern the construction and interpretation of orthogonal arrays as promoted by Taguchi?

20.34 In the selection, implementation, and interpretation of the results of orthogonal arrays, as advocated by Taguchi, why is the concept of design resolution so important?

20.35 Suppose that an investigator decides to use an $L_{16}(2^{15})$ orthogonal array to study the effect of 18 factors on the yield of a chemical etching process. What is the most basic problem that arises in this situation? How would you propose to solve the problem? Propose at least two alternatives.

20.36 In a 2^{7-3} experiment with generators $5 = 123$, $6 = 234$, and $7 = 134$, it was determined that l_1, l_4, l_{123}, and l_{1234} are significant.

(a) What variable effects are confounded in each of the four significant linear combinations of the data?

(b) Under the normal assumptions made for the interpretation of such designs, what can you say about the importance of certain variable effects?

20.37 Suppose that we wish to study eight variables at two levels each under the constraint that a maximum of 64 tests can be conducted.

(a) Write down all possible options available, giving the best generators in each case and the resolution of each design.

(b) What design would be best if only main effects are truly important? Why?

(c) What design would be best if it is suspected that some significant two-factor interactions are present? Why?

20.38 A certain two-level fractional factorial design has the following as one of its linear combinations of effects that can be estimated:

$$l_2 \text{ estimates } 2 + 14 + 1235 + 345.$$

(a) What is the defining relationship?

(b) What are the generators for this design?

(c) What is the resolution of the design?

20.39 The generators for a 2^{6-3} fractional factorial design are $I = 124$, $I = 135$, $I = 236$. Write down the complete set of linear combinations of effects that can be estimated from this design, assuming third- and higher-order interactions are neglected.

20.40 An experimenter has proposed to study five variables in eight tests, using the following fractional factorial design.

Test	1	2	3	4	5
1	−	−	−	+	−
2	+	−	−	−	+
3	−	+	−	−	+
4	+	+	−	+	−
5	−	−	+	+	+
6	+	−	+	−	−
7	−	+	+	−	−
8	+	+	+	+	+

The experimenter notes that the generators for the design are $4 = 12$ and $5 = 123$; the defining relation is therefore $I = 124 = 1235 = 345$. The experimenter then reports that the linear combinations of confounded effects that can be estimated from the data are

l_0	estimates	Mean	+	(1/2) (124 +	1235 +	345)
l_1	estimates	1	+	24 +	235 +	1345
l_2	estimates	2	+	14 +	135 +	2345
l_3	estimates	3	+ 1234 +	125 +	45
l_4	estimates	4	+	12 +	35 +	12345
l_5	estimates	5	+	34 +	123 +	1245

The experimenter then proposes that l_0 through l_5 be determined and interpreted once the data are collected. Comment on this analysis plan. Be specific in describing the pros and cons of the design.

21

Sequential and Iterative Nature of Experimentation

21.1 Value of Sequential Experimentation: A Case Study

In the early stages of an investigation it is often the case that many variables seem to be of potential importance. It would not be uncommon for a project group or task force to draw up a list that could range from as few as 5 up to 15 or more variables. In the final analysis, perhaps only two or three of these variables may prove to be important. The problem is: Which ones are important? It therefore seems that the first task at hand is to conduct some experiments that will quickly reduce the number of variables under study to a few seemingly important ones that will then be the focus of further experimentation. For this screening task, two-level fractional factorial designs constitute a powerful and efficient tool.

Sometimes, situations are presented that require troubleshooting. Perhaps, problems have arisen with a particular manufacturing process, a procedure such as assembly of a component; or the design of a component may seem to be giving trouble in the field. The initial task is to identify quickly the few key factors that seem to govern the situation at hand. If investigators attempt to get their arms around the entire problem by designing one comprehensive experiment, the resource requirements will probably be extensive and the final results of the experiment may be inconclusive due to poor selection of variable levels and/or a poorly controlled experimental environment precipitated by a large experiment.

In an experimental investigation, it is usually wise to take a sequential approach, building up knowledge gradually through a series of related experiments. In this regard, two-level fractional factorial designs serve as useful building blocks.

To this end, an application of fractional factorial designs will be illustrated via an example that is adapted from a study described in the *Journal of Food Science.*[1]

Description of the Problem and Initial Experiment

Nondairy whipped topping is a fabricated food product that serves as a substitute for whipped cream dessert topping. It is generally formulated with sodium caseinate, vegetable fat, carbohydrates, and emulsifiers. Additives such as stabilizers, corn syrup solids, and salts are also widely incorporated into various formulations. Owing to the relatively high price of sodium caseinate, an investigation was conducted into using an aqueous soybean emulsion in its place. For an initial investigation to study the effects of the various ingredients and their concentrations on the resulting whipped topping, the seven variables and variable levels in Table 21.1 were selected.

In the evaluation of a whipped topping, the following characteristics are typically of interest: homogeneity of air cell distribution, stability, stiffness, stand-up, richness, sweetness, gumminess, greasy aftertaste, and texture mouthfeel. The overrun, or whippability, of a given formulation is given by the equation

$$\% \text{ overrun} = 100 \times \frac{W_1 - W_2}{W_2},$$

where W_1 is the weight of a given volume of liquid and W_2 is the weight of the same volume after whipping. Although each characteristic described above was studied during the original investigation, the analysis that follows focuses only on the overrun performance measure.

		Level (weight %)	
Variable	Ingredient	Low (−)	High (+)
1	Soybean emulsion: 9.3% soybean solids	1.67	5.00
2	Vegetable fat: hydrogenated coconut oil	10.00	20.00
3	Carbohydrates: corn syrup solids	0.00	5.00
4	Emulsifiers: mono- and di-glycerides	0.17	0.50
5	Primary stabilizer: hydroxypropyl methyl cellulose	0.00	0.50
6	Secondary stabilizer: microcrystalline cellulose	0.00	0.25
7	Salt: sodium chloride	0.00	0.10

TABLE 21.1   Variables Under Study Together with Their Low and High Settings*

*Amount of sucrose kept constant at 7%; water added to balance to 100%.

---

[1]E. T. S. Chow, L. S. Wei, R. E. DeVor, and M. P. Steinberg, "Application of Two-Level Fractional Factorial Designs in Development of a Soybean Whipped Topping," *Journal of Food Science*, Vol. 48, 1983, p. 230.

TABLE 21.2	Base Design Calculation Matrix							
Test	I	1	2	3	12	13	23	123
1	+	−	−	−	+	+	+	−
2	+	+	−	−	−	−	+	+
3	+	−	+	−	−	+	−	+
4	+	+	+	−	+	−	−	−
5	+	−	−	+	+	−	−	+
6	+	+	−	+	−	+	−	−
7	+	−	+	+	−	−	+	−
8	+	+	+	+	+	+	+	+

The design that was used for the initial experiment was a 2^{7-4} fractional factorial design. The *base design calculation matrix* for a 2^{7-4} design is shown in Table 21.2. Since four additional variables are to be introduced here, all of the interaction columns will be used and hence the design is a saturated fractional factorial. The recipe (design) matrix that was selected for this 2^{7-4} design is displayed in Table 21.3. Each trial was performed only once, in random order. Table 21.3 also displays the overrun associated with each of the experimental conditions.

Generators, Defining Relation, and Confounding Structure

The 2^{7-4} fractional factorial design illustrated in the design matrix of Table 21.3 was obtained by introducing variables 4, 5, 6, and 7 through the following interaction columns in the 2^3 base design calculation matrix:

$$4 = 12, \qquad 5 = 13, \qquad 6 = 23, \qquad 7 = 123.$$

Hence the generators for this design are

$$I = 124, \qquad I = 135, \qquad I = 236, \qquad I = 1237.$$

TABLE 21.3	Design (Recipe) Matrix							
Test	1	2	3	4	5	6	7	Overrun (%)
1	−	−	−	+	+	+	−	115
2	+	−	−	−	−	+	+	81
3	−	+	−	−	+	−	+	110
4	+	+	−	+	−	−	−	69
5	−	−	+	+	−	−	+	174
6	+	−	+	−	+	−	−	99
7	−	+	+	−	−	+	−	80
8	+	+	+	+	+	+	+	63

The defining relation for this design is obtained by adding to the generators all of their products taken two, three, and four at a time. The complete defining relation is

$$I = 124 = 135 = 236 = 1237$$
$$= 2345 = 1346 = 347 = 1256 = 257 = 167$$
$$= 456 = 1457 = 2467 = 3567$$
$$= 1234567.$$

This relationship completely defines the confounding structure of this design. This design is clearly of resolution III, since the length of the smallest word in the defining relation is 3 (excluding I). Therefore, at least some of the main effects are confounded with two-factor interactions.

Obtaining the Linear Combinations of Confounded Effects

Given the defining relation, the sets of confounded effects can be determined by multiplying the column headings in the base design calculation matrix by every term in the defining relation, including I. For example, for the column denoted 1, we have

$$(1)I = (1)124 \quad = (1)135 \ = (1)236 \ = (1)1237$$
$$= (1)2345 \quad = (1)1346 = (1)347 \ = (1)1256 = (1)257 = (1)167$$
$$= (1)456 \quad \ = (1)1457 = (1)2467 = (1)3567$$
$$= (1)1234567,$$

which simplifies to

$$1 = 24 \quad \quad = 35 \ \ = 1236 \ = 237$$
$$= 12345 \quad = 346 = 1347 \ = 256 \quad = 1257 = 67$$
$$= 1456 \quad = 457 = 12467 = 13567$$
$$= 234567.$$

Assuming that third- and higher-order interactions are negligible, we arrive at

$$1 = 24 = 35 = 67.$$

Hence the main effect of variable 1 is confounded with the two-factor interactions 24, 35, and 67.
As another example, for the column denoted 12, we have

$$(12)I = (12)124 \quad = (12)135 \ = (12)236 \ = (12)1237$$
$$= (12)2345 \quad = (12)1346 = (12)347 \ = (12)1256 = (12)257 = (12)167$$
$$= (12)456 \quad \ = (12)1457 = (12)2467 = (12)3567$$
$$= (12)1234567,$$

which simplifies to

$$12 = 4 \qquad = 235 \quad = 136 \quad = 37$$
$$= 1345 \quad = 2346 = 12347 = 56 \qquad = 157 = 267$$
$$= 12456 = 2457 = 1467 \quad = 123567$$
$$= 34567.$$

Assuming that third- and higher-order interactions are negligible, we arrive at

$$12 = 4 = 37 = 56.$$

Hence the main effect of variable 4 is confounded with the two-factor interactions 12, 37, and 56.

The complete set of linear combinations of confounded effects to be estimated are (assuming third- and higher-order interactions can be neglected)

$$l_0 \quad \text{estimates mean,}$$
$$l_1 \quad \text{estimates} \quad 1 + 24 + 35 + 67,$$
$$l_2 \quad \text{estimates} \quad 2 + 14 + 36 + 57,$$
$$l_3 \quad \text{estimates} \quad 3 + 15 + 26 + 47,$$
$$l_{12} \text{ estimates} \quad 12 + \ 4 + 37 + 56,$$
$$l_{13} \text{ estimates} \quad 13 + \ 5 + 27 + 46,$$
$$l_{23} \text{ estimates} \quad 23 + \ 6 + 17 + 45,$$
$$l_{123} \text{ estimates} \quad 34 + 25 + 16 + \ 7.$$

As noted previously, this is a resolution III design; at least some of the main effects are confounded with a two-factor interaction. In this case, each of the main effects is confounded with three two-factor interactions.

Estimation of Effects

The estimates for the linear combinations of confounded effects are obtained by multiplying each column of plus and minus signs in the calculation matrix by the overrun response column of Table 21.3, summing, and then dividing by 4 (for the sample mean, divide by 8). The estimates are shown in Table 21.4. Figure 21.1 is a normal probability plot of the effect estimates of Table 21.4. Figure 21.1 reveals that three of the linear combinations have estimates associated with them that appear to be large relative to the others; that is, estimates that are not likely to have true mean values of zero.

Interpretation of Results

In attempting to explain the experimental results obtained up to this point, we might conclude that the three linear combinations of confounded effects that were determined above to be large, are so because the main effects of variables

TABLE 21.4	Linear Combinations and Estimates for the 2_{III}^{7-4} Fractional Factorial Design for Overrun (%)

$$l_0 = 98.875 \text{ and estimates mean}$$
$$l_1 = -41.750 \text{ and estimates } 1 + 24 + 35 + 67$$
$$l_2 = -36.750 \text{ and estimates } 2 + 14 + 36 + 57$$
$$l_3 = 10.250 \text{ and estimates } 3 + 15 + 26 + 47$$
$$l_{12} = 12.750 \text{ and estimates } 12 + 4 + 37 + 56$$
$$l_{13} = -4.250 \text{ and estimates } 13 + 5 + 27 + 46$$
$$l_{23} = -28.250 \text{ and estimates } 23 + 6 + 17 + 45$$
$$l_{123} = 16.250 \text{ and estimates } 34 + 25 + 16 + 7$$

1, 2, and 6 (the concentrations of soybean emulsion, vegetable fat, and secondary stabilizer) are the important variables. But we should bear in mind that there are many possible interpretations of these results. Looking at the confounded structure, we could say that:

1. The main effects of variables 1, 2, and 6 alone are important.
2. The main effects of variables 1 and 6, as well as the interactions 14 and/or 36, which involve variables with large main effects, could be responsible for the observed results.

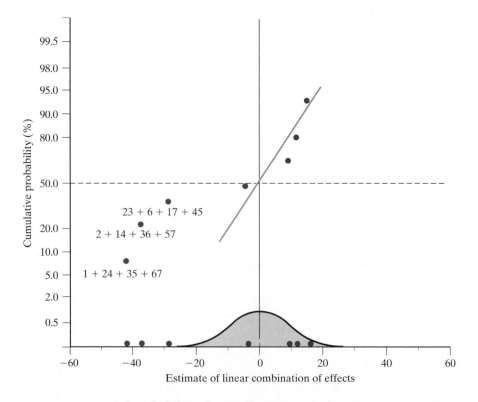

Figure 21.1   Normal probability plot of effect estimates based on tests 1 to 8

3. Similarly, we might conclude that l_{23} is large not because the main effect of variable 6 is large but rather because the two-factor interactions 17 and/or 23 are important.

4. A similar interpretation can be made in terms of variables 2 and 6 by hypothesizing that l_1 is large in magnitude only because the two-factor interactions 24 and/or 67 are large.

It is possible that one might wish to adopt interpretation 1 and run some additional tests, manipulating these factors to attempt to increase overrun. However, it is often the case that chemical processes of this nature are subject to nonlinear behavior, and hence a more conservative but revealing approach may be to run some additional tests to remove the ambiguity surrounding the results at this stage.

Mirror Image Design

If we wish to separate the main effects of the seven variables from the two-factor interactions, we can do this by using something called the *mirror image design*. This second 2^{7-4} fractional factorial is obtained by switching the signs (from $-$ to $+$, and vice versa) for **all** of the columns in the original *design (recipe) matrix* (Table 21.3). The mirror image design is shown in Table 21.5. Comparing the design matrix of Table 21.5 to the initial 2^{7-4} design (Table 21.3), we see that the signs for all seven columns (main variables) have been switched. These new eight tests were run and the results are given in the rightmost column.

Generators and Defining Relation for the Mirror Image Design

As we examine this new design matrix for the second 2^{7-4} fractional factorial design, we find that the generators of this new design are different from those of the first. For example, the signs in the column associated with the fourth variable are no longer the product of columns 1 and 2; rather, column 4 is the negative of this product (i.e., $4 = -12$). Columns for the fifth and sixth variables are of a similar nature, while for the seventh variable we have the relationship $7 = 123$, as before. Therefore, the generators for the mirror image design are

$$I = -124, \quad I = -135, \quad I = -236, \quad I = 1237.$$

TABLE 21.5	Design Matrix for the Mirror Image Design							
Test	1	2	3	4	5	6	7	Overrun (%)
9	+	+	+	−	−	−	+	84
10	−	+	+	+	+	−	−	69
11	+	−	+	+	−	+	−	56
12	−	−	+	−	+	+	+	161
13	+	+	−	−	+	+	−	56
14	−	+	−	+	−	+	+	40
15	+	−	−	+	+	−	+	92
16	−	−	−	−	−	−	−	208

Note that some sign changes have occurred, but the letters in the words of the generators have remained the same. The defining relation for the mirror image design is

$$I = -124 = -135 = -236 = 1237$$
$$= 2345 = 1346 = -347 = 1256 = -257 = -167$$
$$= -456 = 1457 = 2467 = 3567$$
$$= -1234567.$$

Again, while the effects in the linear combinations remain the same, some of the signs have changed. They have changed, however, in a way that will prove most useful.

When we use the defining relation to obtain the linear combinations of confounded effects, we find, for example, that (once again assuming that third- and higher-order interactions can be neglected)

$$l_1' \text{ estimates } 1 - 24 - 35 - 67.$$

Note that we write this linear combination with a prime (l_1') to distinguish it from l_1 from the first 2^{7-4} experiment. Note also the difference between l_1 and l_1':

$$l_1 \text{ estimates } 1 + 24 + 35 + 67,$$
$$l_1' \text{ estimates } 1 - 24 - 35 - 67.$$

Hence it appears that when the results of the first design are combined with those of the second design, some simplification will occur that will greatly reduce the ambiguity of the initial results.

Using the mirror image design, eight additional tests (test numbers 9 through 16) were performed accordingly. The results were shown in the Table 21.5. For these eight tests, we again calculate the estimate for each linear combination of confounded effects. These are compared in Table 21.6 with the results of the original eight tests.

TABLE 21.6 | Comparison of Original and Mirror Image Tests

Original Design	Mirror Image Design
l_0 = 98.875 and estimates mean	l_0' = 95.750 and estimates mean
l_1 = −41.750 and estimates 1 + 24 + 35 + 67	l_1' = −47.500 and estimates 1 − 24 − 35 − 67
l_2 = −36.750 and estimates 2 + 14 + 36 + 57	l_2' = −67.000 and estimates 2 − 14 − 36 − 57
l_3 = 10.250 and estimates 3 + 15 + 26 + 47	l_3' = −6.500 and estimates 3 − 15 − 26 − 47
l_{12} = 12.750 and estimates 12 + 4 + 37 + 56	l_{12}' = 63.000 and estimates 12 − 4 + 37 + 56
l_{13} = −4.250 and estimates 13 + 5 + 27 + 46	l_{13}' = 2.500 and estimates 13 − 5 + 27 + 46
l_{23} = −28.250 and estimates 23 + 6 + 17 + 45	l_{23}' = 35.000 and estimates 23 − 6 + 17 + 45
l_{123} = 16.250 and estimates 34 + 25 + 16 + 7	l_{123}' = −3.000 and estimates −34 − 25 − 16 + 7

Unconfounding the Main Effects

Based on the confounding structure of the original eight tests and that of the additional eight tests from the mirror image design and assuming that third- and higher-order interactions are negligible, we can break some of the main effects away from the two-factor interactions by adding l_i and l_i'. That is,

$$\frac{l_1 + l_1'}{2} = \left(\tfrac{1}{2}\right)[(-41.750) + (-47.500)]$$

$$\text{estimates } \left(\tfrac{1}{2}\right)[(1 + 24 + 35 + 67) + (1 - 24 - 35 - 67)]$$

$$= -44.625 \text{ estimates 1.}$$

Similarly,

$$\frac{l_2 + l_2'}{2} = -51.875 \text{ estimates 2,}$$

$$\frac{l_3 + l_3'}{2} = 1.875 \text{ estimates 3,}$$

$$\frac{l_{12} + l_{12}'}{2} = 37.875 \text{ estimates } 12 + 37 + 56,$$

$$\frac{l_{13} + l_{13}'}{2} = -0.875 \text{ estimates } 13 + 27 + 46,$$

$$\frac{l_{23} + l_{23}'}{2} = 3.375 \text{ estimates } 23 + 45 + 17,$$

$$\frac{l_{123} + l_{123}'}{2} = 6.625 \text{ estimates 7.}$$

We may also subtract l_i' from l_i to break some of the main effects away from the two-factor interactions. That is,

$$\frac{l_1 - l_1'}{2} = \left(\tfrac{1}{2}\right)[(-41.750) - (-47.500)]$$

$$\text{estimates } \left(\tfrac{1}{2}\right)[(1 + 24 + 35 + 67) - (1 - 24 - 35 - 67)]$$

$$= 2.875 \text{ estimates } 24 + 35 + 67,$$

$$\frac{l_2 - l_2'}{2} = 15.125 \text{ estimates } 14 + 36 + 57,$$

$$\frac{l_3 - l_3'}{2} = 8.375 \text{ estimates } 15 + 26 + 47,$$

$$\frac{l_{12} - l_{12}'}{2} = -25.125 \text{ estimates 4,}$$

$$\frac{l_{13} - l_{13}'}{2} = -3.375 \text{ estimates 5,}$$

$$\frac{l_{23} - l'_{23}}{2} = -31.625 \text{ estimates } 6,$$

$$\frac{l_{123} - l'_{123}}{2} = 9.625 \text{ estimates } 34 + 25 + 16.$$

An estimate of the mean based on the information from both the initial and the mirror image design may be obtained by averaging the linear combinations l_0 and l'_0. This calculation is shown below, where the complete forms for l_0 and l'_0 have been displayed; i.e., no assumption about the negligibility of higher-order effects has been made:

$$l_0 = 98.875 \text{ estimates } I + \left(\tfrac{1}{2}\right)(124 + 135 + 236 + 1237 + 2345 + 1346$$
$$+ 347 + 1256 + 257 + 167 + 456 + 1457 + 2467 + 3567$$
$$+ 1234567),$$

$$l'_0 = 95.570 \text{ estimates } I - \left(\tfrac{1}{2}\right)(124 + 135 + 236 - 1237 - 2345 - 1346$$
$$+ 347 - 1256 + 257 + 167 + 456 - 1457 - 2467 - 3567$$
$$+ 1234567),$$

$$\frac{l_0 + l'_0}{2} = 97.313 \text{ estimates } I + \left(\tfrac{1}{2}\right)(1237 + 2345 + 1346 + 1256 + 1457$$
$$+ 2467 + 3567).$$

Under the assumption that third- and higher-order interactions are negligible, this reduces to

$$\frac{l_0 + l'_0}{2} = 97.313 \text{ estimates mean.}$$

By taking the difference between l_0 and l'_0, the following result is obtained:

$$l_0 - l'_0 = 3.125 \text{ estimates } 124 + 135 + 236 + 347 + 257 + 167 + 456$$
$$+ 1234567.$$

If the assumption that third- and higher-order interactions are negligible is then applied to this result, we find that the difference $(l_0 - l'_0)$ has no effects associated with it. Thus the numerical value, 3.125, may be assumed to be due to the experimental error in the system. In general, if several of the linear combinations (or averages of linear combinations) have no effects associated with them, they may be assumed to be due to experimental error, and used to develop a confidence interval to assess the statistical significance of linear combinations.

Nature of the Two Combined 2_{III}^{7-4} Fractional Factorial Designs

When the mirror image design is combined with the original design, all of the main effects are uncoupled from the sets of three two-factor interactions they were confounded with—for example,

$$
\begin{aligned}
\textit{First design:} & \quad l_1 \text{ estimates } 1 + (24 + 35 + 67), \\
\textit{Mirror image:} & \quad l_1' \text{ estimates } 1 - (24 + 35 + 67), \\
\textit{Combined design:} & \quad \frac{l_1 + l_1'}{2} \text{ estimates } 1, \\
& \quad \frac{l_1 - l_1'}{2} \text{ estimates } 24 + 35 + 67.
\end{aligned}
$$

The combined design includes all 16 tests in the initial and mirror image designs. In the combined design, two-factor interactions are confounded with one another and the main effects are confounded with three-factor interactions. The combined (2^{7-3}) design is therefore of resolution IV. Earlier, it was seen that the average of l_0 and l_0' confounded the mean, I, with a set of four-factor interactions and, under the assumption of three-factor interactions and higher being negligible, produced an unaliased estimate of the mean. The four-factor interactions confounded with the mean were

$$I = 1237 = 2345 = 1346 = 1256 = 1457 = 2467 = 3567.$$

This expression specifies the defining relation for the combined design. Such a defining relation for a 2_{IV}^{7-3} design is produced with the generators

$$I = 1237, \qquad I = 2345, \qquad I = 1346.$$

Thus for the combined design, variables 5, 6, and 7 are introduced through the interactions 234, 134, and 123 in a 2^{7-3} (2^4 base design) calculation matrix.

Interpretation of Results

Let us now look at the relative sizes of the main effects and sets of confounded interactions to see what physical interpretation of the data is possible at this stage. The estimates of the main effects and sets of three two-factor interactions are summarized in Table 21.7 and have been used to construct the normal probability plot displayed in Fig. 21.2. An examination of Table 21.7 and Fig. 21.2 shows five values that are much larger (in absolute value) than the others. Increased concentrations of variables 1 (soybean emulsion), 2 (vegetable fat), 4 (emulsifiers), and 6 (secondary stabilizer) all appear to reduce the overrun significantly. The linear combination $(12 + 37 + 56)$ also appears to be large.

Although the linear combination of effects 12, 37, and 56 appears to be important, we do not know which interaction (or combination of interactions) is

TABLE 21.7	Results of the Combined Designs

Estimate of 1	=	−44.625
Estimate of 2	=	−51.875
Estimate of 3	=	1.875
Estimate of 12 + 37 + 56	=	37.875
Estimate of 13 + 27 + 46	=	−0.875
Estimate of 23 + 45 + 17	=	3.375
Estimate of 7	=	6.625
Estimate of error	=	3.125
Estimate of 24 + 35 + 67	=	2.875
Estimate of 14 + 36 + 57	=	15.125
Estimate of 15 + 26 + 47	=	8.375
Estimate of 4	=	−25.125
Estimate of 5	=	−3.375
Estimate of 6	=	−31.625
Estimate of 34 + 25 + 16	=	9.625

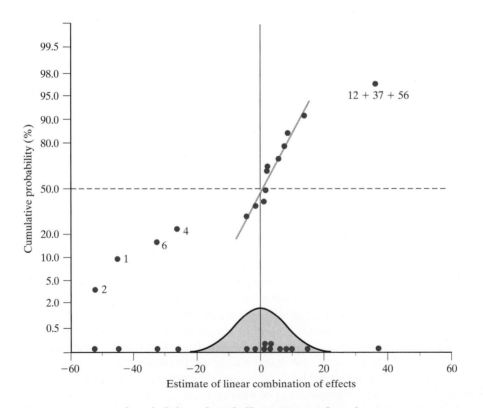

Figure 21.2   Normal probability plot of effect estimates based on tests 1 to 16

responsible. Since the largest effects in Table 21.7 are associated with variables 1 and 2, it is quite possible that the 12 interaction is the important one. Furthermore, since variables 3, 5, and 7 do not appear to have main effects that exert appreciable influence on the overrun, one interpretation forwarded might be that the interactions 37 and 56 do not influence the overrun either. One tentative conclusion that may therefore be drawn based on the results of the 16 tests performed is that variables 1, 2, 4, and 6 and the interaction 12 are important with respect to the overrun.

Two-Way Diagram Analysis

Based on the assumptions that we have made above about the possible importance of variables 1, 2, 4, 6 and interaction 12, a two-way diagram may be constructed to interpret the effects of variables 1 and 2. This diagram is shown in Fig. 21.3. It is interesting to note the following from the two-way diagram.

1. Using a high level of vegetable fat (20%) and changing the concentration of soybean emulsion from 5% to 1.67% increases the overrun by 6.75% (68% to 74.75%). This increase is negligible for all practical purposes.
2. Using a low level of vegetable fat (10%) and changing the concentration of soybean emulsion from 5% to 1.67% increases the overrun by 82.5% (82% to 164.5%).

Clearly, the two-factor interaction between soybean emulsion and vegetable fat appears to be very important.

Summary

The effect of changing the concentration of the emulsifiers (variable 4) and secondary stabilizer (variable 6) is fairly clear. Reducing the levels for these

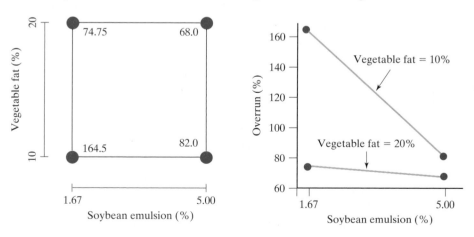

Figure 21.3   **Two-way diagram for variables 1 and 2**

variables appears to increase the amount of overrun. The concentrations of carbohydrates (3), primary stabilizer (5), and salt (7) appear to have no real effect on the response. Reducing the concentration for the soybean emulsion has little effect on the overrun when the vegetable fat concentration is at its high level. The key to a desirable (large) overrun response is the discovery that low concentrations for both the soybean emulsion and vegetable fat must be used simultaneously. "Hence main effects can be misleading without proper consideration of interaction effects."

This example illustrates how fractional factorial designs can be used for screening purposes. Only a relatively small number of tests were used to study the effects of many variables. If statistically designed experiments had not been used, chances are good that many more tests would have been required before similar conclusions could be reached. This example further reinforces the concept of *sequential assembly*—that is, combining the results of similarly structured experimental designs to increase our knowledge and reduce the ambiguity that may surround the results of a single experiment.

21.2   Sequential Assembly of Fractional Factorials

We have emphasized previously the importance of running experiments in a sequential manner, gradually improving on our knowledge base. This notion is important to ensure that a lack of knowledge early in an investigation will not lead to wasting experimental resources (e.g., inappropriate variables or variable levels chosen in the context of one large experiment). To avoid such problems, it is wise to run a series of smaller but related experiments. Ambiguities that may arise in an early experiment can be removed by making additional information available from a follow-up experiment. It is possible to identify related experimental designs which, when combined, produce a clearer picture of the relative importance of the factors under study. This was evidenced in the preceding section through the use of the mirror image design.

In the soybean whipped-topping example a 2_{III}^{7-4} fractional factorial design was run initially and the results were somewhat ambiguous. In particular, three of the linear combinations of effects seemed large, but the high degree of confounding among low-order effects (main effects and two-factor interactions) made several interpretations possible. Specifically, it was found from the first experiment that the seemingly large linear combinations of effects were

$$l_1 \text{ estimates } 1 + 24 + 35 + 67 = -41.750,$$
$$l_2 \text{ estimates } 2 + 14 + 36 + 57 = -36.750,$$
$$l_{23} \text{ estimates } 23 + 6 + 17 + 45 = -28.250.$$

The question is: Which effects are making these linear combinations large? One interpretation is that the three main effects (1, 2, and 6) are important. Other interpretations are possible, of course.

Alternative Experimental Strategies: The Notion of Families of Fractional Factorial Designs

Since the initial 2^{7-4} fractional factorial design results for the soybean whipped-topping example were so ambiguous, we performed the mirror image design to clarify things. One might ask: Why not perform the 2_{IV}^{7-3} 16-run experiment in the first place? This question is an important one and gets to the very heart of the sequential experimentation issue.

The 2^{7-4} design initially run, with generators

$$I = 124, \quad I = 135, \quad I = 236, \quad I = 1237,$$

is a $\frac{1}{16}$ fraction of the full 2^7 factorial (8 tests versus 128 tests). Among the 128 combinations of the seven variables, each over two levels, we selected a certain 8 tests based on the generators above. We should think of the 2^7 design as being comprised of 16 one-sixteenth fractions, of which the first design was one of the fractions and the mirror image design was another. This still leaves 14 other $\frac{1}{16}$ fractions that could have been chosen for the second design.

The two $\frac{1}{16}$ fractions of the 2^7 factorial that we ran were related through the following generators:

Original 2^{7-4}: $I =$ 124, $I =$ 135, $I =$ 236, $I = 1237$,
Mirror image: $I = -124$, $I = -135$, $I = -236$, $I = 1237$.

The terms remained the same; only the signs changed. The remaining 14 members of this *family* of fractional factorials are defined by the other possible sign combinations:

$$I = \pm124, \quad I = \pm135, \quad I = \pm236, \quad I = \pm1237.$$

The generalized generators above define a *family* of fractional factorials. The design with all + signs on the generators is called the *principal fraction*. The remaining, in this case 15, designs are the *alternate fractions*. The key point here is the following: Depending on the interpretation of the results of the principal fraction, we may choose any one of several other alternate fractions to achieve a particular result when the two fractions are combined.

As an alternative to the mirror image design, suppose that the larger linear combinations from the first experiment,

$$l_1 \text{ estimates } \quad 1 + 24 + 67 = -41.750,$$
$$l_2 \text{ estimates } \quad 2 + 14 + 36 = -36.750,$$
$$l_{23} \text{ estimates } 23 + \quad 6 + 17 = -28.250,$$

were interpreted somewhat differently. Suppose that our knowledge of the process led us to suspect a priori that variable 1 was a very important variable, and now the results seem to bear this out. In particular, we feel that the linear

TABLE 21.8	Principal Fraction Design Matrix							
Test	1	2	3	4	5	6	7	Overrun (%)
1	−	−	−	+	+	+	−	115
2	+	−	−	−	−	+	+	81
3	−	+	−	−	+	−	+	110
4	+	+	−	+	−	−	−	69
5	−	−	+	+	−	−	+	174
6	+	−	+	−	+	−	−	99
7	−	+	+	−	−	+	−	80
8	+	+	+	+	+	+	+	63

Generators: $I = 124$, $I = 135$, $I = 236$, $I = 1237$

combinations above are large primarily because of the importance of the main effect of variable 1 and its two-factor interactions 14 and 17.

This interpretation could be clarified if we could run an additional 2^{7-4} fractional factorial that when combined with the principal fraction produced cleared or unconfounded estimates of the main effect of 1 and all of its two-factor interactions. Such a fraction exists; it is one of the 15 *alternate fractions* of the family we are working with. It is generated by folding (switching the signs of) only the variable 1 column in the original design (recipe) matrix.

For the soybean whipped-topping experiment, this interpretation was in fact made, and the alternative fraction obtained by folding variable 1 was also performed. The principal and alternative fraction recipe matrices are shown in Tables 21.8 and 21.9 along with the overruns that were obtained for the tests.

The implications of folding the 1 column only can be seen clearly through an examination of the defining relation for the alternative fraction. The defining relation for this new second design is

$$
\begin{aligned}
I &= -124 & = -135 & = 236 = -1237 \\
&= 2345 & = -1346 = \ &347 = -1256 = 257 = -167 \\
&= 456 & = -1457 = 2467 = \ &3567 \\
&= -1234567.
\end{aligned}
$$

TABLE 21.9	Alternative Fraction Design Matrix							
Test	1	2	3	4	5	6	7	Overrun (%)
17	+	−	−	+	+	+	−	66
18	−	−	−	−	−	+	+	171
19	+	+	−	−	+	−	+	147
20	−	+	−	+	−	−	−	122
21	+	−	+	+	−	−	+	51
22	−	−	+	−	+	−	−	148
23	+	+	+	−	−	+	−	49
24	−	+	+	+	+	+	+	14

Generators: $I = -124$, $I = -135$, $I = 236$, $I = -1237$

The linear combinations of effects obtained using this defining relation are

$$
\begin{aligned}
l_0'' \text{ estimates} \quad & \text{mean,} \\
l_1'' \text{ estimates} \quad & 1 - 24 - 35 - 67, \\
l_2'' \text{ estimates} \quad & 2 - 14 + 36 + 57, \\
l_3'' \text{ estimates} \quad & 3 - 15 + 26 + 47, \\
l_{12}'' \text{ estimates} \quad & 12 - 4 - 37 - 56, \\
l_{13}'' \text{ estimates} \quad & 13 - 5 - 27 - 46, \\
l_{23}'' \text{ estimates} \quad & 23 + 6 - 17 + 45, \\
l_{123}'' \text{ estimates} \quad & -34 - 25 + 16 - 7.
\end{aligned}
$$

The l_i'' differ from the original l_i only in that in the new set of linear combinations, the signs of terms not including a 1 differ from the signs of terms that do have a 1 (i.e., 1, 12, 13, 14, 15, 16, 17). Therefore, if the l_i and l_i'' are added/subtracted, we obtain clear estimates of the main effect of variable 1 and all of its two-factor interactions, under the assumption that third- and higher-order interactions can be neglected.

Using the alternative fraction design, tests 17 through 24 were used to calculate the estimate for each linear combination of confounded effects. These are compared in Table 21.10 with the linear combinations based on the original eight tests. When the confounding structures and effect estimates of the original eight tests are combined with those of the eight tests from the alternative fraction, the following structure is obtained:

$$
\frac{l_0 + l_0''}{2} = 97.438 \text{ estimates mean,}
$$

$$
\frac{l_1 + l_1''}{2} = -38.625 \text{ estimates 1,}
$$

$$
\frac{l_2 + l_2''}{2} = -31.375 \text{ estimates } 2 + 36 + 57,
$$

$$
\frac{l_3 + l_3''}{2} = -25.375 \text{ estimates } 3 + 26 + 47,
$$

$$
\frac{l_{12} + l_{12}''}{2} = 39.125 \text{ estimates 12,}
$$

$$
\frac{l_{13} + l_{13}''}{2} = 0.125 \text{ estimates 13,}
$$

$$
\frac{l_{23} + l_{23}''}{2} = -35.125 \text{ estimates } 23 + 6 + 45,
$$

$$
\frac{l_{123} + l_{123}''}{2} = 8.375 \text{ estimates 16,}
$$

$$
l_0 - l_0'' = 2.875 \text{ estimates error,}
$$

$$\frac{l_1 - l_1''}{2} = -3.125 \text{ estimates } 24 + 35 + 67,$$

$$\frac{l_2 - l_2''}{2} = -5.375 \text{ estimates } 14,$$

$$\frac{l_3 - l_3''}{2} = 35.625 \text{ estimates } 15,$$

$$\frac{l_{12} - l_{12}''}{2} = -26.375 \text{ estimates } 4 + 37 + 56,$$

$$\frac{l_{13} - l_{13}''}{2} = -4.375 \text{ estimates } 5 + 27 + 46,$$

$$\frac{l_{23} - l_{23}''}{2} = 6.875 \text{ estimates } 17,$$

$$\frac{l_{123} - l_{123}''}{2} = 7.875 \text{ estimates } 34 + 25 + 7.$$

Interpretation of Results

Let us now look at results from the combined designs to see what physical interpretation of the data is possible based on the principal and alternative fractions. The linear combinations of effects and associated estimates are summarized below. Figure 21.4 displays a normal probability plot of the estimates.

Estimate of $1 = -38.625$,

Estimate of $2 + 36 + 57 = -31.375$,

Estimate of $3 + 26 + 47 = -25.375$,

Estimate of $12 = 39.125$,

Estimate of $13 = 0.125$,

Estimate of $23 + 6 + 45 = -35.125$,

TABLE 21.10	Comparison of Test Results		
	Principal Fraction Design		*Alternative Fraction Design*
	Tests 1–8		*Tests 17–24*
$l_0 = 98.875$ estimates mean		$l_0'' = 96.000$ estimates mean	
$l_1 = -41.750$ estimates $1 + 24 + 35 + 67$		$l_1'' = -35.500$ estimates $1 - 24 - 35 - 67$	
$l_2 = -36.750$ estimates $2 + 14 + 36 + 57$		$l_2'' = -26.000$ estimates $2 - 14 + 36 + 57$	
$l_3 = 10.250$ estimates $3 + 15 + 26 + 47$		$l_3'' = -61.000$ estimates $3 - 15 + 26 + 47$	
$l_{12} = 12.750$ estimates $12 + 4 + 37 + 56$		$l_{12}'' = 65.500$ estimates $12 - 4 - 37 - 56$	
$l_{13} = -4.250$ estimates $13 + 5 + 27 + 46$		$l_{13}'' = 4.500$ estimates $13 - 5 - 27 - 46$	
$l_{23} = -28.250$ estimates $23 + 6 + 17 + 45$		$l_{23}'' = -42.000$ estimates $23 + 6 - 17 + 45$	
$l_{123} = 16.250$ estimates $34 + 25 + 16 + 7$		$l_{123}'' = 0.500$ estimates $-34 - 25 + 16 - 7$	

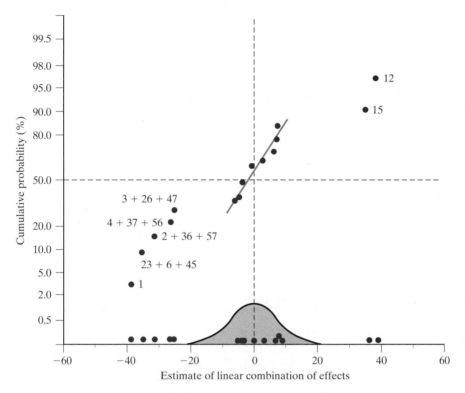

Figure 21.4   Normal probability plot of effect estimates based on tests 1 to 8 and 17 to 24

Estimate of $16 = 8.375$,

Estimate of error $= 2.875$,

Estimate of $24 + 35 + 67 = -3.125$,

Estimate of $14 = -5.375$,

Estimate of $15 = 35.625$,

Estimate of $4 + 37 + 56 = -26.375$,

Estimate of $5 + 27 + 46 = -4.375$,

Estimate of $17 = 6.875$,

Estimate of $34 + 25 + 7 = 7.875$.

In examining the effects, our a priori suspicion concerning the importance of variable 1 seems to be borne out, as the main effect of variable 1 as well as the two-factor interactions 12 and 15 appear to be important.

The estimates of the sets of confounded effects $(2 + 36 + 57)$, $(3 + 26 + 47)$, $(23 + 6 + 45)$, and $(4 + 37 + 56)$ are also relatively large. We do not know, however, which terms (or combination of terms) within the linear combinations are responsible. One conclusion might be that the four linear combinations are large because of the main effects of variables 2, 3, 6, and 4 embedded within them. However, we do not have much confidence in such a conclusion based solely on the results of these two fractions and may therefore more conservatively conclude that variable 1 and interactions 12 and 15 are important with respect to the overrun. But in this case, we previously obtained more reliable information on the importance of main effects 2, 4, and 6 (principal fraction + mirror image). That information, when combined with what we have learned above, leads us now to believe that 1, 2, 4, 6, 12, and 15 are important effects.

Summary

We have just witnessed an example of a more general strategy that might be employed in running successive experimental designs by choosing judiciously among the members of a given family of fractional factorials. In this case we have seen that by folding a single column in the design (recipe) matrix we create an alternate fraction, which when combined with the principal fraction provides clear estimates of the main effect of the variable that column represents and all of its two-factor interactions, when third- and higher-order interactions are neglected.

Clearly, given the principal fraction, we have several alternatives among the remaining alternative fractions, each providing a different set of information/ effect estimates from the combined design. Table 21.11 provides some relevant information concerning the family of fractional factorials we have been examining. We have seen that running the principal fraction followed by the mirror

TABLE 21.11	2_{III}^{7-4} Family of Fractional Factorials				
Fraction		Generators			Combined with Principal Fraction Gives Estimates of:*
Principal	$I = +124$	$I = +135$	$I = +236$	$I = +1237$	—
A_1	$I = -124$	$I = +135$	$I = +236$	$I = +1237$	4, 14, 24, 34, 45, 46, 47
A_2	$I = +124$	$I = -135$	$I = +236$	$I = +1237$	5, 15, 25, 35, 45, 56, 57
A_3	$I = -124$	$I = -135$	$I = +236$	$I = +1237$	
A_4	$I = +124$	$I = +135$	$I = -236$	$I = +1237$	6, 16, 26, 36, 46, 56, 67
A_5	$I = -124$	$I = +135$	$I = -236$	$I = +1237$	
A_6	$I = +124$	$I = -135$	$I = -236$	$I = +1237$	
A_7	$I = -124$	$I = -135$	$I = -236$	$I = +1237$	All main effects
A_8	$I = +124$	$I = +135$	$I = +236$	$I = -1237$	7, 17, 27, 37, 47, 57, 67
A_9	$I = -124$	$I = +135$	$I = +236$	$I = -1237$	
A_{10}	$I = +124$	$I = -135$	$I = +236$	$I = -1237$	
A_{11}	$I = -124$	$I = -135$	$I = +236$	$I = -1237$	1, 12, 13, 14, 15, 16, 17
A_{12}	$I = +124$	$I = +135$	$I = -236$	$I = -1237$	
A_{13}	$I = -124$	$I = +135$	$I = -236$	$I = -1237$	2, 12, 23, 24, 25, 26, 27
A_{14}	$I = +124$	$I = -135$	$I = -236$	$I = -1237$	3, 13, 23, 34, 35, 36, 37
A_{15}	$I = -124$	$I = -135$	$I = -236$	$I = -1237$	

*Assuming that third- and higher-order interactions are negligible.

image design (A_7) produced unconfounded estimates of all main effects and seven sets of two-factor interactions confounded in groups of three. If the estimate associated with one of these sets of two-factor interactions showed up to be large, we were forced to use our judgment to select the important two-factor interaction from a group of three.

When we combined the principal fraction with a fraction (A_{11}) to unconfound the main effect of 1 as well as all of its interactions (12, 13, 14, 15, 16, and 17), we saw that the remaining main effects were each confounded with a pair of two-factor interactions. If the estimate associated with one of these main effects plus a pair of two-factor interactions showed up to be large, it was difficult to conclude much about what term(s) were driving the estimate.

If, instead of following up the principal fraction with a mirror image design or a design to clear the main and interaction effects of a single variable, we were to run one of the designs A_3, A_5, A_6, A_9, A_{10}, A_{12}, or A_{15}, the structure shown in Table 21.12 would be obtained (assuming three-factor and higher-order interactions are negligible). The structure of the combined designs illustrated in Table 21.12 clearly is not appealing in that there are no unconfounded main effects or two-factor interactions. However, the distribution of the main and interaction effects is much more uniform across the structure. Thus if an effect estimate is large, the investigator is required only to ponder the significance of effects from two terms instead of three. In addition, two error estimates are available for this combined design as compared with one. Finally, suppose that based on a priori knowledge we know that variables 6 and 7 are not likely to interact with any other variables. Under these conditions, alternate fraction A_3 (with associated estimates l_i^*) will provide unconfounded estimates of variables 1, 2, 3, 4, 5, and 7 and interactions 12, 13, 14, and 15. Clearly, this alternate fraction does offer some advantages over the other designs described in detail previously.

TABLE 21.12	Principal Fraction Combined with Fraction A_3
$(l_0 + l_0^*)/2$	estimates mean
$(l_1 + l_1^*)/2$	estimates 1 + 67
$(l_2 + l_2^*)/2$	estimates 2 + 36
$(l_3 + l_3^*)/2$	estimates 3 + 26
$(l_{12} + l_{12}^*)/2$	estimates 12 + 37
$(l_{13} + l_{13}^*)/2$	estimates 13 + 27
$(l_{23} + l_{23}^*)/2$	estimates 23 + 6 + 17 + 45
$(l_{123} + l_{123}^*)/2$	estimates 16 + 7
$l_0 - l_0^*$	estimates error
$(l_1 - l_1^*)/2$	estimates 24 + 35
$(l_2 - l_2^*)/2$	estimates 14 + 57
$(l_3 - l_3^*)/2$	estimates 15 + 47
$(l_{12} - l_{12}^*)/2$	estimates 4 + 56
$(l_{13} - l_{13}^*)/2$	estimates 5 + 46
$(l_{23} - l_{23}^*)/2$	estimates error
$(l_{123} - l_{123}^*)/2$	estimates 34 + 25

The discussion above reinforces the value of running experiments sequentially and assembling them together to provide information deemed desirable based on the results of initial experiments. A key point here is that these related experiments can provide several different alternatives for a second experiment, depending on the results and inferences drawn from the first experiment.

21.3   Soybean Whipped-Topping Example: Summary and Interpretation

Two separate analyses have been performed that build on the results obtained from an initial principal fraction. In the first analysis, the principal fraction was followed up by a mirror image design (design A_7). These two designs, when combined (Table 21.13), provided unconfounded estimates of all main effects (under the assumption that three-factor and higher-order interactions are negligible). In the second analysis, the principal fraction was followed up by a design that folded over variable 1 (design A_{11}). These two designs, when combined (Table 21.14), provided unconfounded estimates of the effect of variable 1 and all its interactions.

A study of Tables 21.13 and 21.14 and the normal probability plots of Figs. 21.2 and 21.3 suggests that the following effects/linear combinations are important:

Principal Fraction + *Mirror Image Design* (A_7)		*Principal Fraction* + *Alternative Fraction* (A_{11})	
1 estimated as	−44.625	1 estimated as	−38.625
2 estimated as	−51.875	2 + 36 + 57 estimated as	−31.375
12 + 37 + 56 estimated as	37.875	12 estimated as	39.125
4 estimated as	−25.125	4 + 37 + 56 estimated as	−26.375
6 estimated as	−31.625	23 + 6 + 45 estimated as	−35.125
		3 + 26 + 47 estimated as	−25.375
		15 estimated as	35.625

TABLE 21.13   Results of Principal Fraction + Mirror Image Design (Design A_7)

$$
\begin{aligned}
(l_0 + l_0')/2 &= 97.313 && \text{estimates mean} \\
(l_1 + l_1')/2 &= -44.625 && \text{estimates } 1 \\
(l_2 + l_2')/2 &= -51.875 && \text{estimates } 2 \\
(l_3 + l_3')/2 &= 1.875 && \text{estimates } 3 \\
(l_{12} + l_{12}')/2 &= 37.875 && \text{estimates } 12 + 37 + 56 \\
(l_{13} + l_{13}')/2 &= -0.875 && \text{estimates } 13 + 27 + 46 \\
(l_{23} + l_{23}')/2 &= 3.375 && \text{estimates } 23 + 45 + 17 \\
(l_{123} + l_{123}')/2 &= 6.625 && \text{estimates } 7 \\
l_0 - l_0' &= 3.125 && \text{estimates error} \\
(l_1 - l_1')/2 &= 2.875 && \text{estimates } 24 + 35 + 67 \\
(l_2 - l_2')/2 &= 15.125 && \text{estimates } 14 + 36 + 57 \\
(l_3 - l_3')/2 &= 8.375 && \text{estimates } 15 + 26 + 47 \\
(l_{12} - l_{12}')/2 &= -25.125 && \text{estimates } 4 \\
(l_{13} - l_{13}')/2 &= -3.375 && \text{estimates } 5 \\
(l_{23} - l_{23}')/2 &= -31.625 && \text{estimates } 6 \\
(l_{123} - l_{123}')/2 &= 9.625 && \text{estimates } 34 + 25 + 16
\end{aligned}
$$

TABLE 21.14	Results of Principal Fraction + Alternative Fraction (Design A_{11})

$$(l_0 + l_0'')/2 = \quad 97.438 \qquad \text{estimates average}$$
$$(l_1 + l_1'')/2 = -38.625 \qquad \text{estimates } 1$$
$$(l_2 + l_2'')/2 = -31.375 \qquad \text{estimates } 2 + 36 + 57$$
$$(l_3 + l_3'')/2 = -25.375 \qquad \text{estimates } 3 + 26 + 47$$
$$(l_{12} + l_{12}'')/2 = \quad 39.125 \qquad \text{estimates } 12$$
$$(l_{13} + l_{13}'')/2 = \quad 0.125 \qquad \text{estimates } 13$$
$$(l_{23} + l_{23}'')/2 = -35.125 \qquad \text{estimates } 23 + 6 + 45$$
$$(l_{123} + l_{123}'')/2 = \quad 8.375 \qquad \text{estimates } 16$$
$$l_0 - l_0'' = \quad 2.875 \qquad \text{estimates error}$$
$$(l_1 - l_1'')/2 = -3.125 \qquad \text{estimates } 24 + 35 + 67$$
$$(l_2 - l_2'')/2 = -5.375 \qquad \text{estimates } 14$$
$$(l_3 - l_3'')/2 = \quad 35.625 \qquad \text{estimates } 15$$
$$(l_{12} - l_{12}'')/2 = -26.375 \qquad \text{estimates } 4 + 37 + 56$$
$$(l_{13} - l_{13}'')/2 = -4.375 \qquad \text{estimates } 5 + 27 + 46$$
$$(l_{23} - l_{23}'')/2 = \quad 6.875 \qquad \text{estimates } 17$$
$$(l_{123} - l_{123}'')/2 = \quad 7.875 \qquad \text{estimates } 34 + 25 + 7$$

In examining the information shown above, the following observations may be made:

1. Both combined designs produce an unconfounded estimate of the main effect of variable 1. These estimates (-44.625 and -38.625) appear to be fairly close in magnitude, reinforcing the importance of this effect.

2. Both combined designs have a linear combination that involves the interaction 12. The estimates associated with these combinations (37.875 and 39.125) are quite close in magnitude. This reinforces the importance of this interaction effect.

3. Both combined designs have a linear combination that includes the main effect of variable 4. The estimates associated with these combinations (-25.125 and -26.375) are very close in magnitude.

4. Both combined designs have a linear combination that involves the main effect of variable 6. The estimates associated with these combinations (-31.625 and -35.125) have a comparable magnitude.

5. Based on the combined design (principal fraction + A_{11}), the interaction effect 15 is important (35.625).

6. Although both combined designs have a linear combination that involves the main effect of variable 2, these estimates (-51.875 and -31.375) are not close at all in magnitude. In fact, the difference between these estimates $[-31.375 - (-51.875) = 20.5]$ is itself quite large in magnitude. This difference, 20.5, is an estimate of the linear combination $36 + 57$. Thus this linear combination also appears to be important.

As indicated in observation 6, we may integrate the information from both combined designs to further clear the confounding pattern.

$$\frac{l_{12} + l'_{12}}{2} - \frac{l_{12} + l''_{12}}{2} = -1.250 \text{ estimates } 37 + 56$$

$$\frac{l_{13} + l'_{13}}{2} - \frac{l_{13} + l''_{13}}{2} = -1.000 \text{ estimates } 27 + 46$$

$$\frac{l_{23} + l'_{23}}{2} - \frac{l_{23} + l''_{23}}{2} = -3.500 \text{ estimates } 23 + 45$$

$$\frac{l_2 - l'_2}{2} - \frac{l_2 - l''_2}{2} = 20.500 \text{ estimates } 36 + 57$$

$$\frac{l_3 - l'_3}{2} - \frac{l_3 - l''_3}{2} = -27.250 \text{ estimates } 26 + 47$$

$$\frac{l_{123} - l'_{123}}{2} - \frac{l_{123} - l''_{123}}{2} = 1.250 \text{ estimates } 34 + 25.$$

In examining the information shown above, it appears that the linear combinations $(36 + 57)$ and $(26 + 47)$ are important. In summary, when all three 2_{III}^{7-4} designs are examined together (combined), the following effects (and linear combinations of effects) are identified as important:

$$1, 2, 4, 6, 12, 15, (36 + 57), (26 + 47).$$

Although the linear combinations $(36 + 57)$ and $(26 + 47)$ are large, we do not know which interactions are responsible. If we consider the linear combination $(36 + 57)$, the main effect of neither variable 5 nor variable 7 has been identified as significant; therefore, it is unlikely that their interaction, 57, is important either. Thus one tentative conclusion is that this linear combination is important because of the two-factor interaction 36. If we consider the linear combination $(26 + 47)$, the main effects of both variables 2 and 6 have been identified as significant; therefore, it is reasonable to suspect that their interaction, 26, is also important. Thus one tentative conclusion is that this linear combination is important because of the two-factor interaction 26.

In summary, based on the results of the 24 tests performed, we have evidence that the variables 1, 2, 4, and 6, the interactions 12 and 15, and the linear combinations $(26 + 47)$ and $(36 + 57)$ are important with respect to overrun. A procedure is available[2] for unconfounding the remaining ambiguities associated with the $(26 + 47)$ and $(36 + 57)$ linear combinations of effects by performing a few additional tests. Alternatively, we might use the aforementioned logic and tentatively conclude that interactions 26 and 36 are important, and

---

[2]G. E. P. Box, W. G. Hunter, and J. S. Hunter, *Statistics for Experimenters*, Wiley, New York, 1978.

therefore postulate the following model to describe the overrun response:

$$y = b_0 + b_1x_1 + b_2x_2 + b_4x_4 + b_6x_6 + b_{12}x_1x_2 + b_{15}x_1x_5 + b_{26}x_2x_6$$
$$+ b_{36}x_3x_6 + \epsilon.$$

This model may be fit to the data for all 24 tests to develop contour plots. The model and contour plots may then be used to propose some confirmatory tests that ultimately will verify the ability of the model to adequately predict the response.

■ EXERCISES

21.1 A 2^5 factorial is to be run to study the effects of five factors on the burr height in a metal stamping process. After five tests have been performed (in a random order), the testing program is suspended, because it is realized that not enough funds are available for all 32 tests. It is then proposed to conduct a new design that requires fewer total tests (16 instead of 32)—that is, conduct a 2^{5-1} fractional factorial. At worst, how many of the original five tests can be saved? Clearly explain your answer.

21.2 Consider a 2^{5-2} fractional factorial design with the generators $I = -124$ and $I = 135$.
   (a) What is the defining relation for this design? Display the linear combinations of effects that can be estimated from this experiment. Assume that third- and higher-order interaction effects can be neglected. Write out the design (recipe) matrix for this design in standard order.
   (b) The results of the experiment in part (a) were somewhat ambiguous; therefore, a second fractional factorial design, the mirror image design, is to be performed. Display the design matrix for the mirror image design. What are the generators for this second design? Display the linear combinations of effects that can be estimated from the mirror image design, assuming that third- and higher-order interaction effects can be neglected.
   (c) When these two designs are combined, what effects, or linear combinations of effects, can be estimated? Once again, assume that third- and higher-order interactions are negligible.

21.3 Consider a 2^{5-2} fractional factorial design with the generators $I = 134$ and $I = -235$.
   (a) What is the defining relation for this design? Display the linear combinations of effects that can be estimated from this experiment. Assume that third- and higher-order interaction effects can be neglected. Write out the design (recipe) matrix for this design in standard order.
   (b) The results of the experiment in part (a) indicated that variable 3 and its interactions were likely to be important. A second fractional factorial design is to be performed that will unconfound the main effect of

variable 3 and all of its two-factor interactions. Display the design matrix for this second design. What are the generators for this second design? Display the linear combinations of effects that can be estimated, assuming that third- and higher-order interaction effects can be neglected.

(c) When these two designs are combined, what effects, or linear combinations of effects, can be estimated? Once again, assume that third- and higher-order interactions are negligible.

21.4 It is desired to perform a 2^{6-3} fractional factorial design with the generators $4 = -12$, $5 = -23$, and $6 = 123$.

(a) What is the defining relation for this design? Display the linear combinations of effects that can be estimated from this experiment. Assume that third- and higher-order interaction effects can be neglected. Write out the design (recipe) matrix for this design in standard order.

(b) The results of the experiment in part (a) were somewhat ambiguous; therefore, a second fractional factorial design, the mirror image design, is to be performed. Display the design matrix for the mirror image design. What are the generators for this second design? Display the linear combinations of effects that can be estimated from the mirror image design, assuming that third- and higher-order interaction effects can be neglected.

(c) When these two designs are combined, what effects, or linear combinations of effects, can be estimated? Once again, assume that third- and higher-order interactions are negligible.

21.5 It is desired to perform a 2^{6-3} fractional factorial design with the generators $4 = 13$, $5 = -12$, and $6 = -23$.

(a) What is the defining relation for this design? Display the linear combinations of effects that can be estimated from this experiment. Assume that third- and higher-order interaction effects can be neglected. Write out the design (recipe) matrix for this design in standard order.

(b) The results of the experiment in part (a) indicated that variable 5 and its interactions were likely to be important. A second fractional factorial design is to be performed that will unconfound the main effect of variable 5 and all of its two-factor interactions. Display the design matrix for this second design. What are the generators for this second design? Display the linear combinations of effects that can be estimated, assuming that third- and higher-order interaction effects can be neglected.

(c) When these two designs are combined, what effects, or linear combinations of effects, can be estimated? Once again, assume that third- and higher-order interactions are negligible.

21.6 The 2^{5-2} fractional factorial design shown in the accompanying table has been performed.

(a) What are the generators and defining relation for this design?

Test	1	2	3	4	5	y
1	−	−	−	−	+	76
2	+	−	−	+	−	45
3	−	+	−	+	+	33
4	+	+	−	−	−	48
5	−	−	+	−	−	28
6	+	−	+	+	+	41
7	−	+	+	+	−	75
8	+	+	+	−	+	54

(b) Display the complete set of linear combinations of effects that can be obtained from this experiment, assuming that three-factor and higher-order interactions are negligible. Calculate estimates for each of the linear combinations of effects. Prepare a normal probability plot for the estimates. Which linear combinations of effects appear to be statistically significant?

(c) It has been proposed to follow up the first design with a mirror image design. Write down the recipe matrix for this second design. What are the generators and defining relation for the mirror image design?

(d) Clearly explain why it might be inadvisable to conduct the proposed mirror image design, given the results of the first fractional factorial design. Propose a better design to follow up the first 2^{5-2} design.

21.7 Consider a 2^{5-2} fractional factorial design with the principal fraction defined by the generators $4 = 12$ and $5 = 13$.

(a) Write down the generators for the entire family of 2^{5-2} fractional factorials associated with this principal fraction.

(b) Identify the member of the family that when combined with the principal fraction unconfounds the main effects (assuming third- and higher-order interactions can be neglected).

(c) Identify the member of the family that when combined with the principal fraction unconfounds the main effect of variable 1 and all its two-factor interactions. Repeat for the main effect of variable 2 and its two-factor interactions, and for the main effect of variable 5 and its two-factor interactions. Assume that third- and higher-order interactions can be neglected.

21.8 The 2^{3-1} fractional factorial design shown in the table was run, and the results of the analysis were somewhat ambiguous.

Test	1	2	3
1	−	−	−
2	+	−	+
3	−	+	+
4	+	+	−

It was decided to run the mirror image design and combine its results with the design in the table.

(a) Write down the design matrix for the mirror image design.

(b) What is the resulting confounding pattern when the two designs are combined (i.e., what effects or confounded effects can be estimated when the results of the two designs are combined)?

21.9 Consider a 2^{7-3} fractional factorial design with the generators $I = -1235$, $I = 1246$, and $I = -1347$.

(a) Write down the defining relation for this design. What is the resolution of this design?

(b) Write down all linear combinations of effects that can be estimated from this experiment. Assume that third- and higher-order interaction effects can be neglected.

(c) Propose a second, follow-up design that will clear up some of the confounding associated with the first design. Assuming that three-factor interactions and higher can be neglected, which effects and linear combinations of effects can be estimated from the two designs?

21.10 An experimenter wishes to study eight variables in 16 tests using the highest-resolution design possible. Physical restrictions of the experimental testing process prohibit variables 1, 3, 4, and 6 from taking on their "+" levels simultaneously. Similarly, variables 1, 2, 3, and 5 cannot take on their "−" levels simultaneously.

(a) Find generators for a 2^{8-4} resolution IV design that will satisfy the restrictions above. What is the defining relation? Write out the design matrix.

(b) Under the assumption that three-factor and higher-order interactions are negligible, display the linear combinations of confounded effects.

(c) Explain why it would be inadvisable to follow up this 2^{8-4} design with the mirror image design. Propose a follow-up design that might be more appropriate.

21.11 A two-level fractional factorial design was run with 16 test conditions. Given that one of the linear combinations of confounded effects that can be estimated from this particular design is $123 - 245 + 6 - 13456$, determine the design generators. Show all work and explain how you arrived at your answer.

21.12 [*Source*: Advanced experimental design course at University of Wisconsin–Madison, 1969—Professor Irwin Guttman.] Ozzie Cadenza, owner and manager of Ozzie's Bar and Grill, recently decided to study the factors that influence the amount of business done at his bar. At first, he did not know which factors were important and which were not, but he drew up a list of six variables, which he decided to investigate by means of a fractional factorial experiment.

1. The amount of lighting in the bar.

2. The presence of free potato chips and chip dip at the bar.

3. The volume of the jukebox.

4. The presence of Ozzie's favorite customer, a young lady by the name of Rapunzel Freeny. Miss Freeny is a real "life of the party," continually chatting with the customers, passing around the potato chips, and so on, all of which make Ozzie feel that she has a real effect on the amount of bar business.

5. The presence of a band of roving Gypsies who have formed a musical group called the Roving Gypsy Band, and who have been hired by Ozzie to play a limited engagement at the bar.

6. The effect of a particular bartender who happens to be on duty. There were originally three bartenders: Tom, Dick, and Harry. Ozzie fired Harry, however, so that each of the factors for the experiment would have only two levels.

Low and high levels were selected for each of the variables as follows:

Variable	Low Level	High Level
1	Lights dim	Lights bright
2	No chips at bar	Chips at the bar
3	Jukebox playing softly	Jukebox blaring loudly
4	Miss Freeny stays home	Miss Freeny at the bar
5	No Gypsy Band	Gypsy Band playing
6	Tom is the bartender	Dick is the bartender

Ozzie decided to perform one "run" of the experiment every Friday night during the happy hour (4:30–6:30 p.m.). He thought that he should try a fractional factorial with as few runs as possible, since he was never quite sure when the band of Gypsies would pack up and leave town. He finally decided to use a member of the family of the 2^{6-3} designs with generators of $I = \pm 124$, $I = \pm 135$, and $I = \pm 236$. He had wanted to find a resolution III design, for which the jukebox would never be blaring away while the Gypsies were playing, but he found this requirement impossible.

(a) Why is the statement above true? Carefully explain.

Ozzie did insist, however, that no run of the experiment could simultaneously have variables 1, 3, and 5 at their high levels. This restriction was made necessary by the annoying tendency of all light fuses to blow whenever the Gypsies played their electric zither at the same time the lights were turned up brightly and the jukebox was playing at full volume. Note that this restriction makes it impossible for the principal fraction generators of the specified family to be used.

(b) What members of the given family of generators does the specified restriction above allow?

Ozzie finally settled on a 2^{6-3} fractional factorial design, the recipe matrix for which is given in the accompanying table. The response (income in dollars) associated with each run is also given.

1	2	3	4	5	6	y
+	+	+	+	−	+	3150
+	+	−	+	+	−	2050
+	−	+	−	−	−	2050
+	−	−	−	+	+	1550
−	−	−	+	−	+	2650
−	+	+	−	+	+	1250
−	+	−	−	−	−	1350
−	−	+	+	+	−	1950

(c) What are the generators and defining relation for this design?

(d) Assuming that third- and higher-order interactions are negligible, write down the estimates obtained from this experiment and tell what they estimate (e.g., "1 + 24 − 35 estimated to be 400").

Being partial to Miss Freeny, and encouraged by the results of the first fraction, Ozzie chose a second fraction that would give unaliased estimates of her and each of her two-factor interactions. The results of this second fraction in standard order are 1350, 1650, 2850, 1750, 2050, 1950, 2950, and 1450.

(e) What estimates are obtained by combining the results of both fractions?

(f) Offer a brief conjecture that might explain the presence and direction of the interactions involving Miss Freeny.

21.13 A 2^{7-4} fractional factorial design was conducted on a chemical process to assess the effect of the seven factors on the chemical yield (percent reacted).

Variable	Low Level	High Level
1. Temperature (°C)	150	200
2. Pressure	Low	High
3. Chemical A concentration (%)	3	5
4. Chemical B concentration (%)	2	8
5. Catalyst type	Brand X	Brand Y
6. Reaction time (min)	2	5
7. Flow rate	Low	High

The second table summarizes the results of the 2^{7-4} design. Included in the table are the levels for each of the seven variables and the yield (percent reacted) for each test. Note that the tests are displayed in the order in which they were performed.

(a) What are the generators and defining relation for this 2^{7-4} design? Display the complete set of linear combinations of effects that can be obtained from this experiment, assuming that three-factor and higher-order interactions are negligible. Calculate numerical values for each of the linear combinations of effects. Prepare a normal probability plot for the estimates. Which linear combinations of effects appear to be statistically significant?

Run	1	2	3	4	5	6	7	y
1	−	+	+	−	−	+	−	66.1
2	−	+	−	+	−	−	+	59.6
3	+	−	+	+	−	−	−	62.3
4	+	+	−	−	+	−	−	67.1
5	−	−	+	−	+	−	+	21.1
6	+	+	+	+	+	+	+	57.8
7	−	−	−	+	+	+	−	59.7
8	+	−	−	−	−	+	+	22.5

(b) Propose a follow-up design that you believe should be performed next. Clearly explain the logic behind your decision. When the results of the first design are combined with those of your proposed second design, what effects or linear combinations of effects can be estimated?

21.14 A chemical process has been studied through the use of a fractional factorial design to judge the effects of the following seven factors on the chemical yield (percent reacted).

Variable	Low Level	High Level
1. Temperature (°C)	150	200
2. Pressure	Low	High
3. Chemical A concentration (%)	3	5
4. Chemical B concentration (%)	2	8
5. Catalyst type	Brand X	Brand Y
6. Reaction time (min)	2	5
7. Flow rate	Low	High

Initially, a 2^{7-4} fractional factorial design was performed that utilized the following generators: $4 = 12$, $5 = 13$, $6 = 23$, and $7 = 123$. Based on the first design, the following linear combinations of effects (assuming that three-factor and higher-order interactions are negligible) and associated estimates were calculated:

$$l_0 = 51.9625 \text{ estimates mean,}$$
$$l_1 = 1.2250 \text{ estimates } 1 + 24 + 35 + 67,$$
$$l_2 = 21.3750 \text{ estimates } 2 + 14 + 36 + 57,$$
$$l_3 = -0.4250 \text{ estimates } 3 + 15 + 26 + 47,$$
$$l_{12} = -1.4250 \text{ estimates } 12 + 4 + 37 + 56,$$
$$l_{13} = 15.4750 \text{ estimates } 13 + 5 + 27 + 46,$$
$$l_{23} = -1.1750 \text{ estimates } 23 + 6 + 17 + 45,$$
$$l_{123} = -24.3750 \text{ estimates } 34 + 25 + 16 + 7.$$

To unconfound the main effects from the two-factor interactions, the mirror image design was performed, the results of which are shown in the table below (tests are displayed in the order in which they were performed).

Run	1	2	3	4	5	6	7	y
9	+	−	+	+	−	+	−	48.6
10	−	−	+	−	+	+	+	39.6
11	+	−	−	+	+	−	+	35.3
12	−	+	+	+	+	−	−	82.0
13	−	−	−	−	−	−	−	48.2
14	+	+	−	−	+	+	−	80.5
15	+	+	+	−	−	−	+	41.9
16	−	+	−	+	−	+	+	42.9

(a) Display the linear combinations of effects (assuming that three-factor and higher-order interactions are negligible) and associated estimates for the mirror image design.

(b) Combine the results of the two designs and prepare a normal probability plot to identify those effects or linear combinations of effects that appear to be statistically significant.

(c) Comment on the findings of the two experiments. Specifically, which variables are important, and how may the response be maximized?

21.15 Consider a 2^{6-3} fractional factorial design with the principal fraction defined by the generators $4 = 12$, $5 = 23$, and $6 = 123$.

(a) Write down the entire family of 2^{6-3} fractional factorials associated with this principal fraction.

(b) Identify the member of the family that when combined with the principal fraction unconfounds the main effects. Assume that third- and higher-order interactions can be neglected.

(c) Identify the member of the family that when combined with the principal fraction unconfounds the main effect of variable 1 and all its two-factor interactions. Repeat for the main effect of variable 3 and its two-factor interactions, and for the main effect of variable 6 and its two-factor interactions. Assume that third- and higher-order interactions can be neglected.

21.16 After two 2^{5-2} fractional factorials are conducted and analyzed, the following model is developed:

$$\hat{y} = 17.1 - 9.4x_2 + 7.7x_4 - 12.3x_5 + 8.5(x_2x_4 \text{ and/or } x_3x_5)$$
$$- 5.9(x_3x_4 \text{ and/or } x_2x_5).$$

Why have the interaction terms in the model been expressed in the manner indicated? Clearly explain (and demonstrate) your answer.

21.17 Consider a 2^{9-5} fractional factorial design with the generators $I = 1235$, $I = 1246, I = 1347, I = 2348$, and $I = 12349$.

(a) What is the defining relation for this design? Display the linear combinations of effects that can be estimated from this experiment. Assume that third- and higher-order interaction effects can be neglected. Write out the design (recipe) matrix for this design in standard order.

(b) The results of the experiment in part (a) were somewhat ambiguous; therefore, it is proposed to perform a second fractional factorial design, the mirror image design. Display the design matrix for the mirror image design. What are the generators for this second design? Display the linear combinations of effects that can be estimated from the mirror image design, assuming that third- and higher-order interaction effects can be neglected. When the principal fraction and mirror image designs are combined, what effects, or linear combinations of effects, can be estimated? Once again, assume that third- and higher-order interactions are negligible.

(c) Instead of performing the second design described above (the mirror image design), it is proposed to follow up the principal fraction with a 2^{9-5} design having the generators $I = -1235, I = -1246, I = -1347$, $I = -2348$, and $I = 12349$. This alternative fraction was serendipitously identified by T. S. Babin and J. W. Sutherland (while they were graduate students at the University of Illinois at Urbana-Champaign). Display the linear combinations of effects that can be estimated from the Babin–Sutherland design, assuming that third- and higher-order interaction effects can be neglected. When the principal fraction and Babin–Sutherland designs are combined, what effects, or linear combinations of effects, can be estimated? Once again, assume that third- and higher-order interactions are negligible.

(d) Considering the results of parts (b) and (c), which design should be performed to follow up the principal fraction? Clearly explain your answer.

21.18 A 2^{5-2} fractional factorial design is performed, utilizing the generators $I = 1234$ and $I = 135$. After analyzing the results, the investigator decided to run another 2^{5-2} design with the generators $I = -1234$ and $I = -135$.

(a) Write down the defining relations and linear combinations of confounded effects (assuming three-factor and higher-order interactions are negligible) for each of the designs above.

(b) Combine the two designs and indicate what effects or combinations of effects can be estimated from the two designs.

(c) What might have led the investigator to choose the specific second fraction that she did?

21.19 Consider a 2^{6-3} fractional factorial design with generators $4 = 12$, $5 = 13, 6 = 23$. Consider further that additional tests 9 through 16 will be conducted using the mirror image design, based on the first 8 tests.

(a) Write down the design matrix for the mirror image design. Base your answer on a standard-order representation of the design matrix for tests 1 through 8.

(b) Write down the linear combinations of effects l_{12} and l'_{12}. Demonstrate how these two companion experiments can be used to reduce the ambiguity in the results if each experiment is examined separately.

DOE WORKSHOP #2: TURNING SIMULATION, TURN9DOE

Introduction

This workshop utilizes TURN9DOE, an MS Excel–based Visual Basic program, to simulate a turning operation that is being used to manufacture a bronze bushing. The goal of the workshop is to apply the general class of 2^{k-p} fractional factorial designs to improve the centering of two lathes that perform the turning operations in order to improve the process capability.

Learning Objective

The objective of this computer workshop is to provide students with practice in the planning, conduct, and interpretation of designed experiments. It is expected that the student will need to combine the results of fractional factorial designs to unconfound variable effects. Finally, based on the experimental analysis, the student is asked to determine the variable levels that will improve the process capability and then confirm this with validation tests.

Background

In SPC Workshop #2, a turning process producing a bronze bushing was studied, with the finish of the outside surface serving as the quality characteristic of interest. After corrective actions were taken to eliminate special causes, the $\overline{X}$ and R charts for each machine showed good statistical control. Investigation of process capability for the individual machines revealed that two problems still existed:

1. Only one of the three machines was operating at the nominal surface roughness requirement of 70 microinches.

2. All three of the machines exhibited excessive variability in surface roughness values.

It was decided to attack the centering problem first, using statistical design-of-experiment techniques. Several individuals associated with the process developed a list of nine potentially important factors that could influence the mean surface roughness. The feasible ranges for these factors are shown in the accompanying table.

Factor	Feasible Range for Experiments	Current Operating Condition
1. Tool material	Two grades (A and B) of cemented carbide tools were considered applicable	Grade A is being used
2. Use of cutting fluid	Either present or absent	No cutting fluid is being used
3. Machine used	Only the Nacirema and LeLathe lathes can be used for experimentation	Nacirema, LeLathe, Rex
4. Tool nose radius	0.0156″ to 0.0625″ in 0.0156″ increments	Fixed at 0.0312″
5. Tool back rake angle	0 degrees to 30 degrees	Varies between 5 and 10 degrees
6. Tool side cutting edge angle	5 degrees to 30 degrees	Fixed at 10 degrees
7. Depth of cut for final pass	0.005″ to 0.030″	Varies between 0.005″ and 0.015″
8. Feedrate	0.0070–0.0095 ipr	0.008″–0.0089″ ipr
9. Cutting speed	900–1150 sfpm	1000–1100 sfpm

Description of TURN9DOE

The TURN9DOE program has been developed to simulate running experiments with the nine process variables. The program produces a surface roughness value as a test result for any user-specified combination of process settings for the nine variables.

Prior to running TURN9DOE, an appropriate fractional factorial design should be created. The decisions as to the specific choice of a 2^{k-p} design and the levels of the variables in the experiment are up to the experimenter. Then TURN9DOE can be run to obtain the test results associated with the design, and the data can be analyzed as appropriate. Multiple designs may need to be performed to unconfound the variable effects associated with a highly fractionated design. Once a model has been established for the response, plotting software can be used to characterize the response surface(s) and identify the levels for the processing variables that will center each machine. Ultimately, validation tests should be performed to confirm that the identified processing conditions do indeed result in centered machines.

On the Web

TURN9DOE is available for download on the Web at http://www.me.mtu.edu/quality-book/. Also available at the site is a more detailed workshop explanation and description of TURN9DOE usage. The website also has a suggested student assignment in connection with the TURN9DOE workshop.

22

Robust Design Case Studies

22.1   Taguchi's Approach to Robust Design

Taguchi views the design process as evolving in three distinct phases or steps: system design, parameter design, and tolerance design. It is perhaps this broad umbrella that he places over his concepts and methods for quality design and improvement that makes his approach so readily embraceable by the engineering community. As we said earlier, Taguchi considers engineering design as the central issue and statistical methods as just one of several tools to accomplish his objectives in engineering.

System design is the initial phase of any engineering design problem wherein knowledge from specialized fields (chemistry, electronics, etc.) is applied to develop the basic design alternative. This design alternative is then refined in the phases of parameter design and tolerance design. *Parameter design* is that phase of the design process in which the best nominal values for the important design parameters are selected. Taguchi defines the "best" values for the nominals as those values that *minimize the transmitted variability resulting from the noise factors.* This is why we sometimes refer to parameter design as *robust design. Tolerance design* is the final phase of Taguchi's design process model, in which the tolerances on the nominal values on some of the critical design parameters are selected. In tolerance design the setting of specifications is considered to be an economic issue and the loss function model is used as a basis to determine the allowable tolerance on the nominal values.

22.2   Overview of Approaches to Robust Design

In recent years a number of different approaches have been proposed as possible ways to implement the robust design concept of Taguchi. These vary from purely analytical approaches, to computer simulation using product/process mathematical models and/or Monte Carlo methods, to the use of physical experimentation.

In most of these approaches the use of experimental design strategies including two-level and multilevel factorial and fractional factorial designs has been extensive. Taguchi refers to these experimental design structures as *orthogonal arrays* and makes use of them for both inner and outer arrays. In Chapter 20 we discussed Taguchi's use of orthogonal arrays and their relationship to the class of two-level fractional factorial designs.

In addition to the different approaches to "generating" data on product/process performance, the issue of the specific measures of performance to use in the facilitation of parameter design needs to be considered. Taguchi and his colleagues make extensive use of the signal-to-noise ratio as a measure of performance. Others evaluate the mean and the variance or standard deviation of performance separately.

In this chapter we examine four case studies that exemplify the different ways in which the concept of parameter design can be implemented to develop robust product and process designs. These are presented under three basic approaches:

1. Experimental approaches to parameter design.
2. Analytical approaches to parameter design.
3. Computer simulation approaches to parameter design.

Experimental Approaches to Parameter Design

Two of the case studies illustrate how the parameter design method for the determination of a robust product/process can be implemented through physical experimentation. These are:

1. The explicit use of an inner array/outer array experimental design structure for robust design in the calibration and assembly of a fuel gage.
2. The exploitation of system nonlinearities in the form of variable interactions to effect a robust design in the study of glove box closing effort in an automobile assembly plant.

Analytical Approaches to Parameter Design

As Taguchi's methods for engineering design are studied, applied, and refined, it is important to keep a broad perspective on his contributions and the wide variety of ways in which they can be utilized. Perhaps the single overarching concept that drives the philosophies of Taguchi is the importance placed on the consistency of performance of products and processes—that is, the importance of variation reduction in achieving quality products and processes. What this essentially means in a broad sense is that as we formulate designs, either on paper or in hard materials, and we evaluate their performance, either analytically or experimentally, we should place as much or more emphasis on variation in performance as we do on performance on average. There are many ways to invoke this overarching concept.

Analytical methods proposed to date for the implementation of the parameter design concept find their roots in the foundations of classical optimization theory and statistical modeling. While much of the knowledge that we use in any given design situation is derived from experimentation, an ever-expanding baseline of knowledge is being derived from first principles/physical laws and the utilization of sophisticated modeling methods such as finite element modeling. As closed-form solutions become available, either directly or through approximations based on data, it becomes possible to evaluate them analytically or numerically using classical optimization methods. The performance function should be examined in terms of both its expected value and its variance function. Searching the variance function to find those values for the design/control variables that minimize the variance in performance then amounts to invoking the parameter design method. One of the case studies that follow uses this approach to examine automobile door closing effort.

Computer Simulation Approaches to Robust Design

With the strong efforts of recent years in the area of mathematical modeling of processes and systems, the opportunity for in-depth study of performance through computer simulation methods has grown considerably. In addition, the drive to push product and process design analysis farther and farther upstream has been considerable. Physical experimentation downstream is being replaced by model-based experimentation earlier in the product development life cycle. In this chapter we examine one such case study on the simultaneous engineering of automobile engine components and their associated manufacturing processes.

22.3   Case Study 1: Fuel Gage Calibration

The Problem.   The following case study was developed from the results of a larger experiment and is presented at this time to demonstrate how relatively simple experiments such as two-level factorial designs can be employed to study and improve the performance of an assembly process using the concept of robust design. Customer complaints concerning the difference between the fuel gage reading and the actual amount of fuel in the tank led to an investigation of the process of gage calibration. Seven variables were tentatively identified as being of possible importance to the position (percent deflection or fallback) of the indicator in the fuel gage.

Initial Design of the Experiment.   A 2^{7-3} fractional factorial design experiment was constructed and carried out. We examine a portion of the results, which involve four of the seven variables that were determined to be the most important.

Variable	Low Level	High Level
1. Spring tension	Loose	Tight
2. Method of pointer location	Visual	Other (standard)
3. Bimetal hook twist	Vendor A	Vendor B
4. Bimetal bracket bend	90 degrees	93 degrees

Variables 1 and 2 describe adjustments that can be made during the calibration process on the assembly line. Variables 3 and 4 describe the condition (considered undesirable) of certain purchased parts that are assembled into the gage and may affect the calibration process. The 2^{7-3} fractional factorial design was collapsed into a 2^4 full factorial design to study the effects of these variables, as they are purposely changed, on the response in question.

The design matrix (Table 22.1) shows the settings for the 16 tests conducted in terms of the four variables defined above. The tests are listed in standard order, although they were actually run in a randomized test sequence. The main effects of each of the four factors, as well as the two-, three-, and four-factor interactions, were estimated from the data and are as follows:

$$
\begin{aligned}
E_1 &= -2.50, & {}^* E_{24} &= 0.50, \\
{}^* E_2 &= 0.00, & E_{34} &= -1.25, \\
E_3 &= -1.75, & & \\
E_4 &= 3.00, & {}^* E_{123} &= 0.75, \\
& & E_{124} &= -2.50, \\
{}^* E_{12} &= 0.50, & {}^* E_{134} &= 0.75, \\
E_{13} &= 1.75, & {}^* E_{234} &= 0.25, \\
E_{14} &= -1.00, & & \\
E_{23} &= 2.25, & E_{1234} &= 2.25.
\end{aligned}
$$

The results of the larger experiment, in fact, showed that these four variables tended to exhibit the stronger effect estimates, both main effects and interactions. A normal probability plot of the estimated effects could be interpreted as

TABLE 22.1   Design Matrix

Test	x_1	x_2	x_3	x_4	y (%)
1	Loose	Visual	Vendor A	90°	14
2	Tight	Visual	Vendor A	90°	7
3	Loose	Other	Vendor A	90°	7
4	Tight	Other	Vendor A	90°	9
5	Loose	Visual	Vendor B	90°	9
6	Tight	Visual	Vendor B	90°	7
7	Loose	Other	Vendor B	90°	9
8	Tight	Other	Vendor B	90°	10
9	Loose	Visual	Vendor A	93°	15
10	Tight	Visual	Vendor A	93°	14
11	Loose	Other	Vendor A	93°	18
12	Tight	Other	Vendor A	93°	7
13	Loose	Visual	Vendor B	93°	10
14	Tight	Visual	Vendor B	93°	8
15	Loose	Other	Vendor B	93°	12
16	Tight	Other	Vendor B	93°	12

suggesting that only the "starred" effect estimates above might be considered insignificant. This interpretation leaves the investigator somewhat perplexed as to what exactly to do next. Many interactions seem important; it is not at all clear how best to adjust/alter the process.

It would appear that the main effects of variables 1 and 4 might be two of the more important effects.

$E_1 = -2.50$ implies that on the average going to tighter spring tension reduces fallback (deflection),

$E_4 = +3.00$ implies that larger bimetal bracket bend (93°) increases deflection, which is bad.

The latter inference suggests that we might have to go to the bimetal bracket supplier and request closer adherence to the 90° bend, which is the desired nominal value. Essentially we would require the supplier to tighten the tolerance on this angle. Rather than taking this more costly approach, it might be useful to consider the variation in this part dimension as a noise variable and try to determine how the transmission of the variation in the bimetal bracket bend could somehow be reduced through the manipulation of control factors during the calibration process—that is, develop a more robust calibration process.

Alternate Experimental Design and Analysis.   A more thoughtful examination of the four variables under study seems to indicate that they fall into two basic categories:

Variable 1: spring tension,

Variable 2: method of pointer location.

These two variables may be easily adjusted at the discretion of the operator/setup person.

Variable 3: bimetal hook twist,

Variable 4: bimetal bracket bend.

These two variables describe certain aspects of the condition of two purchased parts. Although we have specified certain requirements to the vendors, we do not have very strong control over these factors—the condition of the parts coming in—without tightening the specification requirements, which is a costly action. We might therefore consider these two variables to be *noise variables* and perhaps use the robust design/parameter design concept to see how adjustments in variables 1 and 2 could be made to reduce the transmitted variability due to variables 3 and 4.

Now, instead of thinking in terms of a 16-run experimental design that considers four variables at two levels each, we will think in terms of two 2^2 factorial experiments, an "inner" design and an "outer" design.

- The *inner design* is a 2^2 factorial in the two controllable variables, spring tension and method of pointer location.

- The *outer design* is a 2^2 factorial in the two noise variables, bimetal hook twist and bimetal bracket bend.

The outer design is conducted at each of the four unique variable settings of the inner design. A graphical representation of this experimental design is shown in Fig. 22.1. The numbers in the corners of each of the four outer design "squares" represent the fallback results for each of the 16 unique trials.

Referring to the data in Fig. 22.1, we now determine $\bar{y}$ and s_y at each of the four test combinations of the *inner design*.

TEST 1 $x_1(-), x_2(-)$
$\qquad \bar{y} = 12.0, \quad s_y = 2.94.$

TEST 2 $x_1(+), x_2(-)$
$\qquad \bar{y} = 9.0, \quad s_y = 3.37.$

TEST 3 $x_1(-), x_2(+)$
$\qquad \bar{y} = 11.5, \quad s_y = 4.8.$

TEST 4 $x_1(+), x_2(+)$
$\qquad \bar{y} = 9.5, \quad s_y = 2.08.$

Analyzing the responses above (average fallback and standard deviation of fallback) separately, we can determine the settings/values for the control variables

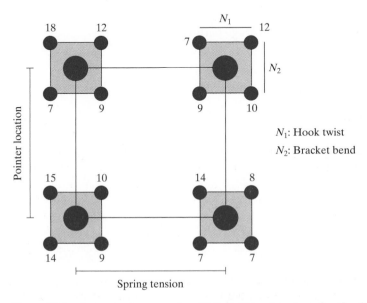

Figure 22.1   Inner/outer experimental design structure for the fuel gage study

that (1) minimize average fallback and (2) give rise to minimum variation in fall-back (i.e., the most robust process design).

Figures 22.2 and 22.3 provide the results of the analyses. The interpretation of the results is as follows:

- *In terms of average response*, percent deflection is reduced by increasing spring tension. This reduction is about the same for both methods of pointer location. On the average, pointer location method has little effect on percent deflection.

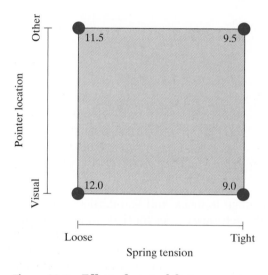

Figure 22.2   **Effect of control factors on average performance**

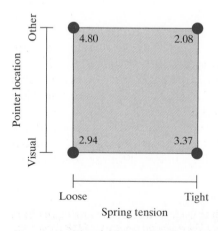

Figure 22.3   **Effect of control factors on variation in performance**

- *In terms of standard deviation*, it is best to use the other (standard) method of pointer location along with tighter spring tension. In the case of the variation response, the interaction of the two control variables seems important.

In this case study, it was decided to adjust the spring tension to a tighter level. Such did lead to a reduction in calibration error. A more in-depth study of the method of pointer location was also initiated.

If we were to look at the results of Figs. 22.2 and 22.3 from a loss function point of view, we could calculate the loss resulting from both deviation from the target on average and variation about the average:

$$\text{loss} = k[s_y^2 + (\bar{y} - m)^2],$$

where here $m = 0$ is the target for zero deflection/fallback. For the four cases characterized by the inner array/outer array experiment, the loss is smallest for the combination of tight spring tension and visual method of pointer location. Although less variation results from the noise factors when tight spring tension is used with the standard method of pointer location, the average for this setting is farther from the target. It should be pointed out, however, that the difference in loss for these two conditions is not great.

22.4   Case Study 2: Glove Box Closing Effort Analysis

Quality Concern.   This case study deals with a classic automobile assembly problem, that of glove box fits.[1] As a result of individual component design and the subassembly design configuration as well as manufacturing variations, quality concerns often arise. These problems manifest themselves as glove box assemblies which fit too loosely, so that a rattle condition is present, or too tightly, so that the closing effort required is excessive or the door simply cannot be closed. The issue in question here arose at the upper management level, as high warranty costs brought this quality concern to the forefront.

Figure 22.4 illustrates the situation under study. Curve (a) represents the statistical distribution for closing effort in units of inch-pounds centered at the target specified by the design process. The spread of this distribution reflects the state of affairs at the outset of this study. Because of excessive variation about the target, the rattle condition develops too frequently. In an effort to solve this problem, an action was taken at the assembly process level that led to an overall increase in closing effort caused by shifting the mean of the distribution as shown by curve (b) in Fig. 22.4. Unfortunately, due to the excessive

---

[1]P. M. Belaire and R. J. Deacon, "The Strategic Approach to Quality Improvement Using Design of Experiments Concepts," *Symposium on Quality: Design, Planning, and Control, ASME Winter Annual Meeting*, December 13–18, 1987, PED Vol. 27, pp. 157–167.

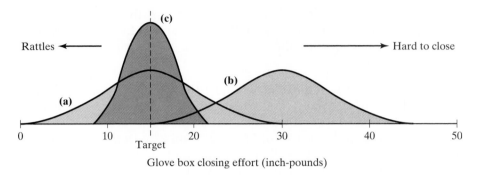

Glove box closing effort (inch-pounds)

Figure 22.4   Statistical representation of closing effort problem

variation in closing effort, the problem now became one of experiencing too many assemblies with intolerably high closing effort. Further adjustment of the mean in an effort to strike a balance between the two contrasting quality problems produced only the simultaneous occurrence of both.

It is clear from the discussion that the basic issue here is not simply the proper centering of a process but rather one of being "on target with smallest variation." It is unlikely that a satisfactory solution will be found without effecting a considerable reduction in process variation, as shown by curve (c) in Fig. 22.4. This can be accomplished in several fundamental ways. Study of the assembly process itself may reveal faults that when rectified could lead to improved consistency. Improving the process consistency for vendor-supplied parts may also yield improvement. Design changes in the form of requiring the manufacturing of components to tighter specifications may also lead to improved glove box fits. It is even possible (perhaps likely) that changes in component and/or subassembly designs may affect significant quality improvement.

Design of Experiment.   In an effort to better understand the glove box closing effort problem, it was decided to plan and conduct a statistically designed experiment to reveal the effects that several thought-to-be important factors might have on closing effort. Table 22.2 lists the five factors under study together with the low and high levels for each that were used in a two-level fractional factorial experiment.

TABLE 22.2	Design and Process Factors and Associated Levels for Experiment			
Variable Name	*Lower Level*	*Higher Level*		*Measure/Unit*
1. Panel foam	Low	High		Thickness
2. Bumper durometer	Soft	Hard		Hardness
3. Ramp angle	Present	Design change		Degrees
4. Striker position	Fore	Aft		Location
5. Striker angle	Present	Design change		Degrees

Figure 22.5   Glove box assembly

Figure 22.5 shows a sketch of the glove box assembly and illustrates the physical meaning of the five variables under study. Panel foam thickness refers to the foam-cushioned vinyl covering on the dash panel substrate. Bumper durometer is the hardness of rubber bumpers, which cushion impact and hold the door assembly firmly in place. Striker angle and ramp angle refer to two design configurations relating to the striker and the bracket it is mounted on. Striker position (fore and aft) is set by the operator during the assembly process. Three of the five factors relate to characteristics of purchased parts and can be altered through design changes (bumper hardness, ramp angle, and striker angle). Foam thickness is a function of the foam molding process. Striker position is a function of the actual assembly process and is operator controlled.

Table 22.3 provides the design matrix and test results for the 2^{5-1} fractional factorial design employed in this study. Under the assumption (which was used in this study) that third- and higher-order interactions may be neglected, the design used provides clear estimates of all the main effects and two-factor interactions among the five variables.

A normal probability plot of the effect estimates that are given in Table 22.4 led to the conjecture that variables 3, 4, and 5 have large main effects, while interactions E_{12}, E_{34}, E_{35}, and E_{45} may also be important. It is, however, the physical interpretation of these apparently important effects that is critical to determining a permanent and cost-effective quality improvement.

TABLE 22.3		Design Matrix and Test Results					
	(1)	(2)	(3)	(4)	(5)		
			Ramp	Striker	Striker	Response	Run
Test	Foam	Bumper	Angle	Position	Angle	(in.-lb)	Number
1	Low	Soft	Present	Fore	Change	150.0	8
2	High	Soft	Present	Fore	Present	150.0	2
3	Low	Hard	Present	Fore	Present	150.0	10
4	High	Hard	Present	Fore	Change	150.0	13
5	Low	Soft	Change	Fore	Present	17.3	15
6	High	Soft	Change	Fore	Change	24.7	11
7	Low	Hard	Change	Fore	Change	25.7	9
8	High	Hard	Change	Fore	Present	21.7	3
9	Low	Soft	Present	Aft	Present	22.7	5
10	High	Soft	Present	Aft	Change	150.0	7
11	Low	Hard	Present	Aft	Change	150.0	14
12	High	Hard	Present	Aft	Present	18.7	4
13	Low	Soft	Change	Aft	Change	19.7	16
14	High	Soft	Change	Aft	Present	8.0	12
15	Low	Hard	Change	Aft	Present	21.0	6
16	High	Hard	Change	Aft	Change	45.0	1

TABLE 22.4	Effect Estimates for the Glove Box Door Closing Effort Study

$$\text{Average} = 70.2813 \qquad E_{23} = 5.9625$$
$$E_1 = 1.4625 \qquad E_{24} = 3.6125$$
$$E_2 = 4.9625 \qquad E_{34} = 32.8625$$
$$E_3 = -94.7875 \qquad E_{45} = 36.3625$$
$$E_4 = -31.7875 \qquad E_{35} = -26.4375$$
$$E_{12} = -29.2875 \qquad E_{25} = 1.6125$$
$$E_{13} = 2.4625 \qquad E_{15} = 4.6125$$
$$E_{14} = 0.6145 \qquad E_5 = 38.2125$$

Discussion of Results.   Figure 22.6 is a graphical representation of the variable main effects that appear to be most dominant. The suggestion is that the proposed design change for the ramp angle may reduce closing effort, while that for the striker angle may increase effort. The aft position for striker appears to produce lower closing effort values. There are two problems associated with the interpretation of the main effects in Fig. 22.6. First, they are average effects; that is, they are averaged over the low and high levels of the other four factors in each case. If two-factor interactions are important, as appears to be the case here, these main effects must be interpreted with great caution. Second, these effects are measures of how changes in factor levels influence the mean response. As such they do not have a direct impact on the amount of variability

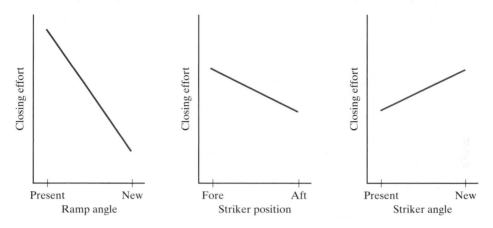

Figure 22.6   Graphical representation of important main effects

about the mean. However, as we will see shortly, a thoughtful interpretation of two of the important two-factor interactions will clearly reveal how improved consistency (reduced variation) about the target mean response can be achieved.

Figure 22.7 provides a graphical representation of two of the larger two-factor interaction effects. In both cases the variable striker position is involved. It was clear from the definition of the five factors under study that all are not controllable in either the same basic way (design versus process) or to the same degree. It may be useful to invoke Taguchi's concept of robust design/parameter design here and think of striker position as a noise variable and the ramp angle and striker angle as design/control variables.

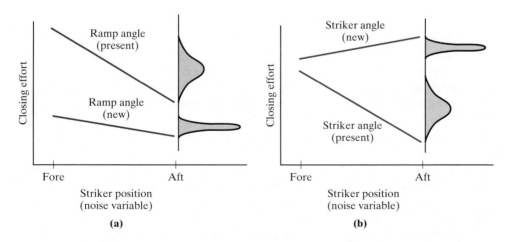

Figure 22.7   Graphical representation of two important two-factor interaction effects

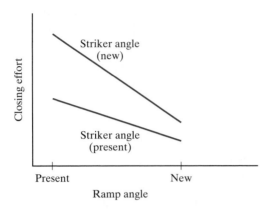

Figure 22.8   **Ramp angle–striker angle interaction effect**

It is clear from Fig. 22.7(a) that as the noise variable striker position varies fore to aft, the transmitted variation in closing effort is greatly reduced with a design modification to the ramp angle. One might propose that no action need be taken relative to the striker position if the ramp angle design change is made. Further, the mean response is reduced to levels very near the desired target. A similar situation is present in Fig. 22.7(b), the striker angle/striker position interaction. In this case, however, the mean closing effort is greatly increased, which is undesirable. Figure 22.8 shows that design changes in both ramp angle and striker angle may reduce closing effort on average.

The foregoing interpretations of some large two-factor interactions are important because they show how the discovery of such system nonlinearities can be used to develop more robust, less variable product designs in an economic fashion. Understanding Taguchi's concept of parameter design is important in using statistical design of experiments for quality improvement.

Improvement Action.   The results of the experimental design that relate to design changes for the striker angle and ramp angle were verified by building a number of glove box assemblies using the prescribed conditions and testing them for mean effort and variation in effort. Based on these verification tests, action was taken to implement the ramp angle change.

22.5   Case Study 3: Automobile Door Closing Effort

H. L. Oh[2] uses an explicit model for the mean square error of design performance in an automobile door closing effort problem to identify the values of

---

[2]H. L. Oh, "Variation Tolerant Design," *Symposium on Quality: Design, Planning, and Control, ASME Winter Annual Meeting*, December 13–18, 1987, PED Vol. 27, pp. 137–147.

the design parameters that provide the most consistent design configuration (minimum variance) while maintaining an acceptable level of average performance. In this case study the following logic/methodology is employed.

Problem Formulation. The first step in any design optimization problem is to formulate the design performance measure of interest. In this case study, car door fit is the issue—in particular, the sensitivity of the closing effort to variations in the door positioning as it is hung in place. Figure 22.9 shows a sketch of a car door and the locations of the three positioning points used to hang the door. The closing effort is a function of the compression of the weather strip around the door periphery. The closing effort can be expressed analytically as a function of the weather strip diametral stiffness, the weather strip diameter and length, and the nominal values of the door position points that define the door/frame gap. The basic design problem is then to find the nominal values for the door position points that minimize the variation in closing effort as a result of the variation in the door position points about their nominal values.

This problem is very similar to the glove box closing effort problem discussed previously in the sense that we are seeking to exploit the underlying nonlinearities in the system to improve performance consistency. This approach is an alternative to simply reducing the allowable tolerance on a more arbitrarily determined nominal value to reduce transmitted variation. Reducing the tolerance has been an altogether too common engineering design approach to improving consistency of performance, an approach that is far too costly and does not contribute substantially to the enhancement of competitive position.

Math Model Derivation. Under the assumption that the door gap is linearly proportional to the door position point locations, it is possible to derive the door closing effort (in stored energy) as a quadratic function of the position

Figure 22.9   Location points for positioning of door

point locations. Remember that the position point locations are the control factors in this problem. The quadratic function for closing effort is a function of the three hanging positions and has several unknown parameter values. Although the closing effort will clearly vary as a function of variation from car to car in weather strip thickness, car body size, and so on, only the variation in the door hanging position from car to car was examined in this case study, in the interest of simplicity.

Math Model Fitting.   The parameters of the quadratic function for closing effort were estimated by first calculating the door gap at 15 points around the door as a function of variations in the door hanging positions and then using the calculated gap values to determine the closing effort values. The variations in the hanging position control factors were introduced using a three-level factorial design, which required a total of 27 (3^3) evaluations of the gap and the stored energy at each of the 15 locations around the door. These data were then used to estimate the coefficients of the quadratic closing effort function.

Design Optimization.   The variance of the closing effort quadratic function may now be derived analytically, and this variance function may be subjected to classical optimization techniques. Such an optimization algorithm will seek the values of the three door position locations that minimize the variance of the closing effort. This optimization is subject to the constraint that the mean closing effort is less than or equal to a prespecified target value. The result is the determination of nominal values for the design/control factors, the door positioning locations, which minimize the closing effort variation due to variation in these locations. In this design optimization problem the control factors are the nominal values for the door position locations, and the noise factors are really the variations in the door position locations.

   Figure 22.10 shows a plot of the derived functional relationship between closing effort and the location of point 1 for positioning the door, for fixed values of location points 2 and 3. The graph shows clearly how the parameter design method exploits a basic system nonlinearity by seeking a new nominal value for the location of point 1, which reduces the transmitted variation to the closing effort that results from the variation in location of point 1 about its nominal value.

22.6   Case Study 4: Simultaneous Engineering of a Product and Its Manufacturing Process

Simultaneous Engineering Model

In this case study, the general class of products being studied is typified by castings with irregular shapes, thin-walled sections, and highly irregular and interrupted surfaces machined by processes such as face milling and

Figure 22.10   Closing efforts vs. upper hinge location

boring.[3] Such components may, for example, be found in automobile engines and power trains. Such parts are generally machined in high volumes on special-purpose machining systems such as flexible cells.

The purpose of the simultaneous engineering concept is to give increased consideration to manufacturing system design and the impact of manufacturing on the product design configuration at the earliest stages of product design. It is well recognized that lead times can be significantly reduced and a more globally optimal product and process combination obtained through interactive simultaneous design rather than a more sequentially based design strategy.

Simultaneous engineering is not simply a matter of applying current product and process design strategies at the same time. New models and methods must be forthcoming. Figure 22.11 provides a simultaneous engineering conceptual framework. The complete framework is composed of four models, three of which describe the product and the process, while the fourth represents the design process itself.

Product Design Model.   The product design model may take several forms. Its purpose is to reduce the design to a set of parameters that are related to relevant

[3]R. E. DeVor, S. G. Kapoor, R. Hayashida, G. Subramani, and T. Lindem, "A Methodology for the Simultaneous Engineering of Products and Manufacturing Processes," *Proc. 14th North American Manufacturing Research Conference*, 1986, pp. 535–542.

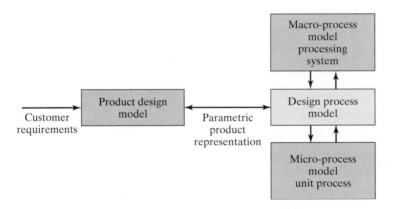

Figure 22.11   Conceptual framework for simultaneous engineering

performance measures. Simple performance measures might include size, weight, and/or shape. A finite element model may be used to define the values of the relevant structural dynamics of the product and/or characterize the stresses present in the product during field use.

Macro-Level Process Model.   The macro-level process model is a large-scale system representation of the processing system under design. The elements of the system (e.g., machines, conveyors, robotic manipulators, and gaging devices) could be modeled in a structure that allows for the simulation of the operation of the system. Simulation languages and models may be usefully employed to build up such a macro-level representation of a system consisting of a number of unit processes.

Micro-Level Process Model.   The micro-level process model provides a representation of a unit process (e.g., face milling, boring, reaming). Such a model is capable of simulating a cut or series of cuts with full consideration of the process geometry, tool geometry, part geometry, and the cutting mechanism itself. Tooling and fixturing are also part of the micro-level process model.

Design Process Model.   The design process model provides a framework and methodology to facilitate rapid convergence to a quality product and manufacturing process. Several design process models have been proposed. When merged with the classical methods of experimental design, the Taguchi design concepts provide a particularly useful model for simulation work.

Product and Process Models

To apply the Taguchi methods of parameter design through computer simulation requires performance models that can predict measures such as cutting forces and vibration/displacement magnitudes as a function of cutting conditions, cut geometry, and the cutting mechanism itself. A dynamic force model developed by Fu, DeVor, and Kapoor is used in this case study.[4] This model has three basic elements:

1. A static mechanistic model, which can predict instantaneous cutting forces based on the design parameters related to part, process, and cut geometries.
2. Finite element models of the part–fixture and cutter–spindle combinations.
3. A model that represents the feedback mechanism between the cutting process and the structural elements.

In the dynamic cutting force model, the cutting forces act as the source of excitation, and the structural elements of item 2 respond to this excitation through displacement in various directions. The cutting process is affected by the excited structural elements via a feedback loop.

Static Force Model.   The static force model may be used to predict the instantaneous X force, Y force, and Z force as a function of part geometry, tool geometry, cut geometry, the relative positions of cutter and workpiece (critical to certain highly irregular part designs), cutter offset, and runout (see Fig. 22.12).

Figure 22.12   Fixture/dynamometer/workpiece assembly

Product/Part Model.   The product/part model consists of the finite element models of the part–fixture combination and, cutter–spindle combination, and a methodology for determining the dominant modes of vibration and estimating the corresponding stiffness and damping of these modes. The part–fixture combination is shown in Fig. 22.12. The part–fixture was modeled with 1477 nodes and 821 elements, of which 128 were of the thin shell type and 693 were of the solid type. This modeling was done with the assumption that the machine table to which the fixture is bolted is perfectly rigid.

The cutter–spindle finite element model is based on the spindle design of a Kearney and Trecker horizontal milling machine on which the cutting tests for model calibration were performed. This model consists of 943 nodes and 706 solid elements. The first 12 dynamic modes of the part–fixture and the first three modes of the cutter–spindle were determined to be the most relevant. The information associated with these modes is shown in Table 22.5.

Experiment Design

The experimental design structure used in this simulation study was an inner/outer array. This structure is exemplified in Fig. 22.13, which illustrates the concept with four variables. In the figure, the inner design is a 2^2 factorial in two design variables: depth of cut and workpiece–fixture stiffness. The outer design, performed for each unique test combination in the inner design, is a 2^2 factorial in two noise variables: insert runout and workpiece surface error. The

TABLE 22.5	Dominant Frequencies and Stiffnesses for Part–Fixture and Cutter-Spindle Combinations	
Displacement Direction/Force Direction	Natural Frequency (hertz)	Stiffness (lb/in.)
XX	961.00	20,759,320.0
XY	984.00	8,964,032.0
XZ	1452.00	−5,662,946.0
YX	984.00	8,995,269.0
YY	285.00	24,641.3
YY	650.00	7,158.7
YY	961.00	24,802.7
YZ	1452.00	25,066.4
ZX	1452.00	−5,248,543.0
ZY	285.00	−2,506,015.0
ZZ	285.00	679,657.0
ZZ	417.00	800,714.7
ZZ	650.00	841,598.1
ZZ	961.00	813,324.8
ZZ	1452.00	429,528.0

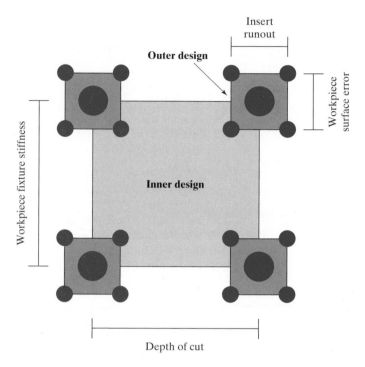

Figure 22.13   Inner design/outer design experimental design structure

idea of parameter design is to find the conditions of the design variables that minimize the functional variation in performance resulting from the presence of the noise variables.

Table 22.6 provides the variable partitioning for the combined inner and outer design simulation experiments conducted in this study. The high (+) and low (−) levels for each of these variables are also given in this table. Runout was considered as a noise variable and was induced by uniformly and progressively offsetting the inserts from no runout to a maximum single insert runout of 0.0015 inch. Surface error, when present, took the form of a cylindrically based "crown" on the workpiece, with a maximum height of 0.0035 inch at the center of the workpiece and of constant cross section along the direction of cutter feed motion. Table 22.6 also provides relevant fixed process conditions for the simulations.

For this study the inner design was a 2^{5-1} fractional factorial (generator: $I = 12345$) and the outer design was a 2^2 factorial. As a result, $16 \times 4 = 64$ separate tests—simulations—were required (i.e., the four simulations of the outer design were conducted for each of the 16 unique variable combinations of the inner design). Table 22.7 provides the combinations of the five variables for each of the 16 tests in the inner design.

TABLE 22.6	Conditions of Simulation Experiments

	Inner Design Control/Design Variables			Outer Design Noise Variables		
Variable	Low Level	High Level	Variable	Low Level	High Level	
1. Feedrate (in./tooth)	0.00225	0.00450	1. Runout (in.)	0	0.0015	
2. Depth of cut (in.)	0.020	0.040	2. Surface error (in.)	0	0.0035	
3. Number of inserts	12	18				
4. Cutter offset (in.)	0	0.900				
5. Workpiece fixture stiffness*	1	1.5				

Other Process Conditions	
Speed = 342 rev/min	Material: 390 cast aluminum alloy
Cutter diameter = 8 in.	Material density: 0.0968 lb/in.^{3}
Nose radius = 0.03 in.	Workpiece dimension: 13 in. × 6 in. × 2.5 in.
Lead angle = 20°	
Radial rake angle = 12°	
Axial rake angle = 17°	

*Low level of stiffness is the set of values obtained from the part model. High level equals 1.5 times the values for low level.

TABLE 22.7	Design Matrix for Inner Design Simulation Experiments

Test	Feedrate (in./tooth)	Depth of Cut (in.)	Number of Inserts	Cutter Offset (in.)	Stiffness (lb/in.)
1	0.00225	0.020	12	0.000	1.5
2	0.00450	0.020	12	0.000	1
3	0.00225	0.040	12	0.000	1
4	0.00450	0.040	12	0.000	1.5
5	0.00225	0.020	18	0.000	1
6	0.00450	0.020	18	0.000	1.5
7	0.00225	0.040	18	0.000	1.5
8	0.00450	0.040	18	0.000	1
9	0.00225	0.020	12	0.900	1
10	0.00450	0.020	12	0.900	1.5
11	0.00225	0.040	12	0.900	1.5
12	0.00450	0.040	12	0.900	1
13	0.00225	0.020	18	0.900	1.5
14	0.00450	0.020	18	0.900	1
15	0.00225	0.040	18	0.900	1
16	0.00450	0.040	18	0.900	1.5

The results of the dynamic force model cutting process simulations produce a great deal of information related to cutting forces, workpiece/cutter deflections, and dynamic process behavior. In this case study we examine as measures of product/process quality and productivity performance only three responses: maximum Y-direction cutting force (MAXFY), maximum Z-direction displacement (MAXDZ), and metal removal rate (MRR). These responses have been studied in the form of a single measure (weighted objective function F):

$$F_{ij} = W_1\left(\frac{MRR_{ij} - \overline{MRR}}{\hat{\sigma}_{MRR}}\right) - W_2\left(\frac{MAXFY_{ij} - \overline{MAXFY}}{\hat{\sigma}_{MAXFY}}\right)$$

$$- W_3\left(\frac{MAXDZ_{ij} - \overline{MAXDZ}}{\hat{\sigma}_{MAXDZ}}\right)$$

for $i = 1$, 16 tests in the inner design and for $j = 1$, 4 tests in the outer design, where W_1, W_2, W_3 are the response weights. Several sets of weights were examined. In this study the results are shown for $W_1 = 0.4$, $W_2 = 0.2$, and $W_3 = 0.4$. Each of the three responses was normalized using the average and standard deviation calculated over all 64 tests, in each case.

The results of the simulation experiments were examined in three basic ways in an effort to understand the manner in which the design and noise factors influence performance:

1. The objective function F was averaged over the $j = 1$, 4 conditions of the outer design for each of the $i = 1$, 16 conditions of the inner design, and these average objective function values were treated as responses to evaluate the inner design results.

2. The sample standard deviation of objective function values was determined over $j = 1$, 4 conditions of the outer design for each of the $i = 1$, 16 conditions of the inner design, and these standard deviations were treated as responses to evaluate the inner design results.

3. The inner design results in terms of variable effects on objective function values F were evaluated at each of the $j = 1$, 4 conditions of the outer design.

Discussion and Interpretation of Results

Effects of Inner Design Variables on the Average Objective Function Values. Under the assumption that third- and higher-order interactions can be neglected, the main and two-factor interaction effects among the five variables were estimated. The results indicated that all variables except feedrate have quite large (significant) main effects and that the interactions between depth of cut and number of inserts, and depth of cut and cutter offset, are also important. The smaller feedrate effect may be explained by the narrow range of feeds studied. In general, increasing stiffness and employing cutter offset tend to improve performance—increase F—while increasing depth of cut and the number of cutter inserts tends to reduce performance.

Of particular interest is the examination of the two important two-factor interaction effects. Figure 22.14 shows these interactions graphically. In the case of the depth of cut–number of inserts interaction [Fig. 22.14(a)], it appears that the deterioration of the objective function with an increase in depth of cut is greatly reduced when the number of inserts is at its lower level. However, for lower depths of cut, the use of 12 or 18 inserts causes very little change in the objective function.

Figure 22.14(b) shows that the use of cutter offset can greatly improve process performance and, in particular, can help counter the detrimental influence of increasing depth of cut. In this figure we see that the inclusion of cutter offset in the process causes a considerable improvement in the objective function when depth of cut is at its highest level. Additional experimentation (simulation) with the cutter offset variable showed two other positive results of using offset: (1) the positive offset effect was consistently observed for several other positions of the cutter along the feed direction on the workpiece, and (2) in several cases where a centered cutter gave rise to an unstable cutting condition, including offset in the process stabilized the cutting condition. The use of offset (Fig. 22.15) avoids the problem of having the point of maximum instantaneous chip load coincide with the most flexible point on the workpiece. The inclusion of surface error (in the form used in this case study) only makes matters worse for a centered cutter. It is, of course, true that these results are workpiece geometry dependent and may not in all cases be reproducible for other geometries. In fact, for this workpiece (see Fig. 22.15), offset to the other side is actually worse than a centered cutter due to the location of points of maximum flexibility.

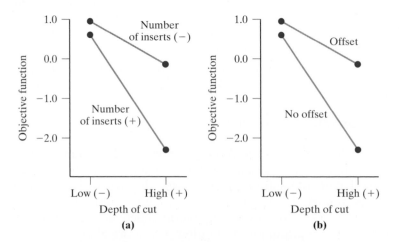

Figure 22.14   Graphical representation of depth of cut/number of inserts interaction effect and depth of cut/cutter offset interaction effect

Figure 22.15   Relative position of the cutter on the workpiece

Effects of Inner Design Variables on the Sample Standard Deviation of Objective Function Values.   An analysis of variable effects on the sample standard deviation of the objective function values for the 16 tests seems to indicate that the main effects of stiffness, offset, and depth of cut are large. Increases in depth of cut tend to magnify the effects of the noise factors, while increased stiffness and the use of offset tend to reduce the effects of the noise factors, precipitating a more robust process and product design. Figure 22.16 shows again how the presence of cutter offset can be a positive force, this time in terms of suppressing the variation in performance due to the noise factors. Again, offset of the cutter appears to counteract the deleterious effect of increasing the depth of cut.

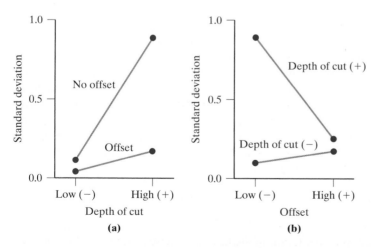

Figure 22.16   Graphical representation of the depth of cut/offset interaction effect

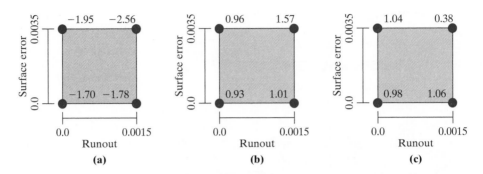

Figure 22.17 **Influence of noise factors on (a) depth-of-cut effect, (b) cutter offset effect, and (c) workpiece/fixture stiffness effect**

An analysis was also performed by examining the ratio of the average response to the sample standard deviation of the response for each condition of the inner design. This ratio may be interpreted as a signal-to-noise ratio. Analysis of the inner design variable effects on the signal-to-noise ratio shows that:

1. Increasing stiffness greatly increases the S/N ratio.
2. Increasing depth of cut greatly reduces the S/N ratio.
3. A large depth of cut–stiffness interaction exists, showing that an increase in stiffness can offset the deleterious depth of cut effect.
4. A large depth of cut–number of inserts interaction exists, showing that the positive influence of fewer inserts cannot be realized if the depth of cut is large.

Effect of the Inner Design Variables as a Function of the Levels of the Outer Design Variables. Figure 22.17 was constructed to show how the effects that the important design variables have on performance are influenced by the noise variables. In these figures the numbers at the corners of each "square" represent the main effect of a specified inner design variable at that particular combination of the noise variables. For example, Fig. 22.17(a) shows that the adverse effect of increasing depth of cut is worsened by the presence of higher levels of the noise variables. Conversely, Fig. 22.17(b) and (c) show that the positive effects of cutter offset and workpiece–fixture stiffness, respectively, are actually largest in the presence of higher levels of the noise variables. Such reinforces the suggestions discussed previously that these design factors may be important to the development of a robust product/process design.

Conclusions

This case study explores the use of computer simulation as a central tool in the simultaneous engineering of products and the processes used to manufacture them. Computer-based mathematical models for the part and for the machining process were used to investigate how changes in product and/or process design parameters could improve the quality and productivity of manufacture.

Taguchi's engineering design process model, in particular the method of parameter design, has been explored as a technique to understand the relationships among the design (product and process) parameters and the performance measures of interest. The method of robust design has been demonstrated by showing how manipulation of design parameters can mitigate the deleterious effects of the noise factors. In the specific context of this work it was seen that part stiffness and cutter offset are design parameters that can be usefully manipulated to reduce the effect of external/noise variables surrounding the process.

23

Modeling of Response Surfaces and Response Optimization

23.1 Introduction

Up to now our planned experiments have been aimed at determining **which** of a larger set of candidate variables are the most important ones in terms of their relative influence on the response of interest, and **how** specifically those few important variables influence the process response. Often once the important variables have been identified, we would like to determine the specific levels of these variables that will **maximize** (or **minimize**) the process response.

In the sandmill experiment discussed in Chapters 15 and 20, a sandmill was used to reduce the size of pigment particles and disperse them throughout the mixture. Two of the factors in the process, temperature, T, and flow rate, R, were found to have significant effects on the particle fineness, F, measured on a Hegman scale. Since both temperature and flow rate are controllable variables in the sandmill process, we would like to know what are the levels for these variables that will produce the maximum Hegman value, i.e., the finest dispersion.

Suppose there exists a relatively smooth response function that relates the response F to the control variables T and R. Figure 23.1 depicts such a hypothetical response surface within the feasible ranges of the two control variables. Mathematically, the function may be written as:

$$\text{fineness} = f(\text{flow rate, temperature}) + \epsilon.$$

We imagine that currently the process is operating at a flow rate R_1 and temperature T_1 and that the corresponding dispersion fineness is F_1 (denoted by

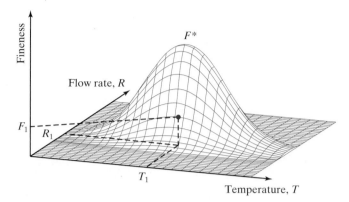

Figure 23.1   A hypothetical response surface

the blue dot on the response surface). We feel that we can achieve a much better fineness than the one we currently have—that we can increase the fineness by climbing to the top of the hill. The question is, "How do we do this?"

Now, if the precise form of the mathematical function f is known, then we can employ calculus or graphical methods to solve $F = f(R, T)$ for the optimal settings, R^* and T^*, which jointly maximize F. In practice, however, the function f is typically unknown to us! How then do we find our way to F^*, the maximum?

One way to search for the "top of the hill" is to search for the optimum value of one control variable at a time—in this case, running tests varying one factor while holding the other variable constant. We do this repeatedly, each time finding the variable level that maximizes the response by searching in one direction, then fixing the variable at that value, and then searching along the direction of the other variable.

While the one-variable-at-a-time approach can lead us in the proper direction, it generally requires a large number of tests to reach the maximum. Figure 23.2 uses a contour plot representation of a response surface to demonstrate the inefficient zig-zag pattern movement toward the optimum with the one-variable-at-a-time approach. Such a search method is not capable of revealing and taking advantage of any interaction effects that may exist between the control variables, R and T. Furthermore, this method gives only a maximum point solution without elucidation of the response surface in the vicinity of the optimum. Such could be important where process constraints need to be applied to the response or where multiple responses require striking some trade-off. Knowledge of the rate of change of the response as a function of each independent variable may also be important for process control purposes. As an improvement over one-variable-at-a-time searches, the technique of *response surface methodology* (RSM) will be presented in this chapter.

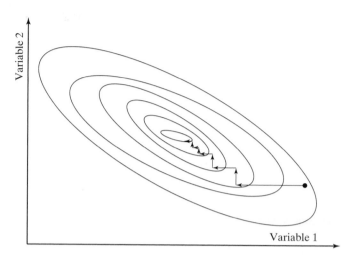

Figure 23.2   **One-variable-at-a-time search to reach the optimum**

23.2   Modeling the Response Surface

Suppose we are interested in a process response y that depends on a set of k input or control variables x_1 through x_k. Such a dependency relationship can be expressed as

$$y = \eta + \epsilon$$
$$= f(x_1, \ldots, x_k; \theta_1, \ldots, \theta_p) + \epsilon.$$

That is, the random variable y can be expressed by a mathematical function f of the independent variables x_1 through x_k and a set of parameters θ_1 through θ_p, which are coefficients of the function f, plus a random error, ϵ. The error, ϵ, in this case reflects the collective effects of all the other nuisance variables/factors that are at work in the experimental environment but not recognized or not controlled. The function f represents the true relationship, theoretically or mechanistically, between the response and the control variables.

The problem is that in most real-world situations, little is known about the underlying mechanisms of the process; i.e., f **is unknown**. How can we then proceed with any confidence to explore the unknown response surface with the ultimate goal of finding the values for x_1 through x_k, which jointly optimize the response y? Following the general approach adopted in statistical problem solving, we make certain assumptions about the problem at hand, proceed using the method that is appropriate under the assumed conditions, and finally check the results for any serious violations of initial assumptions. In the RSM technique, and more generally in the modeling of response surfaces, a basic assumption is that the true function f is not too complicated and can be approximated adequately by low-order polynomial functions. In other words, RSM will be successful if the response function is both relatively smooth and

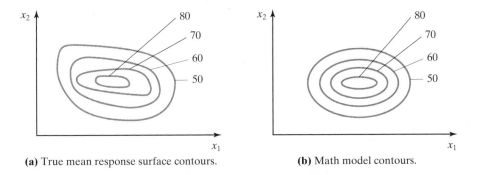

(a) True mean response surface contours.   **(b)** Math model contours.

Figure 23.3   Math model approximation to a true response surface

unimodal, at least within the domain of our exploration for the optimum. Such an approximating function, which has no physical meaning by itself, will better represent the true response surface when a statistical modeling procedure is followed using the data from properly designed experiments.

Approximating the True Response Function

Let the function f be approximated by a simple and flexible graduating polynomial, say,

$$y = b_0 + b_1 x_1 + b_2 x_2 + b_{12} x_1 x_2 + b_{11} x_1^2 + b_{22} x_2^2 + \epsilon.$$

Given the data from a designed experiment, we can estimate the b's (the model parameters) to obtain a fitted equation for calculating and analyzing the predicted response, viz.,

$$\hat{y} = \hat{b}_0 + \hat{b}_1 x_1 + \hat{b}_2 x_2 + \hat{b}_{12} x_1 x_2 + \hat{b}_{11} x_1^2 + \hat{b}_{22} x_2^2,$$

where the $\hat{b}$'s are the least-squares estimates of the b's. If $\hat{y}$ is found to be a statistically adequate fit to y, perhaps through residual analysis and confirmatory testing, then it can be used to determine the values of x_1 and x_2 that optimize the response. Additionally, the fitted equation $\hat{y}$ can be employed to more clearly visualize and hence explore and exploit more completely the nature of the response surface within the experimental region. As an illustration, the contour plots on the right of Fig. 23.3 were determined from a fitted quadratic equation for the true mean response represented by the corresponding contours on the left side.

23.3   Response Surface Methodology—A Brief Description

Response surface methodology (RSM) is a technique for experimental optimization that proceeds in two stages by:

1. Moving quickly and efficiently from some location on the response surface, often remote from the optimum, into the neighborhood of the optimum by employing simple two-level factorial experiments and associated planar models, and then,

2. Characterizing and then exploring and exploiting the neighborhood of the optimum region using multilevel experimental designs and associated higher-order models.

An important feature of RSM is that it attempts to characterize the unknown nature of a response surface through simple polynomial functions, first using linear models for determining the path of steepest ascent (descent) toward the optimum region, and then using quadratic or higher-order polynomials for characterizing the response surface in the vicinity of the optimum. RSM brings together simple experimental designs and least-squares modeling methods to provide an efficient experimental technique for process improvement. We will now briefly describe the RSM process using a simple example.

Two-Level Factorial Design and Linear Model

An RSM study begins by choosing a small, localized region in the experimental domain of interest, usually near an existing point of knowledge concerning the response, and performing a simple two-level factorial or fractional factorial experiment (2^k or 2^{k-p}) for the k control variables. If the region under study is not large and likely remote from the optimum, then a planar assumption for the response surface in this region is a good approximation. This is why a low-resolution 2^{k-p} design might be used, particularly if the number of variables under investigation is four or more. For example, let us consider the results of the 2^2 factorial design shown in Fig. 23.4. Perhaps this factorial design represents a set of conditions that surround the current operating conditions for the process. Based on the conditions in the figure, the following linear model can be established for the predicted response:

$$\hat{y} = 27.5 - 7.50x_1 + 12.5x_2, \tag{23.1}$$

where it is noted that the interaction term ($-2.5x_1x_2$) has not been included in the model. In general the linear model to be used for determining the path of steepest ascent will be of the form:

$$\hat{y} = \hat{b}_0 + \hat{b}_1x_1 + \cdots + \hat{b}_kx_k. \tag{23.2}$$

Figure 23.4 displays contours of constant predicted response based on the model.

Determining the Path of Steepest Ascent

It is evident from an examination of the contours in Fig. 23.4 that the response of interest may be improved by moving up and to the left—that is, by decreasing the value of x_1 while increasing the value of x_2. To determine the most

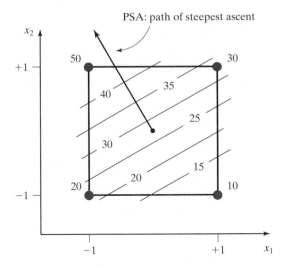

Figure 23.4   Factorial experiment, contours of constant predicted response, and path of steepest ascent

fruitful direction to move along, the path of *steepest ascent*, consider the contour that passes through the center point of the design ($x_1 = 0$, $x_2 = 0$), which is the locus of points with a predicted response of $\hat{y} = 27.5$. Given the predicted response function of Eq. (23.1), the relationship between x_1 and x_2 on the contour is

$$0 = -7.5x_1 + 12.5x_2 \ \text{ or } \ x_2 = \frac{-7.5x_1}{-12.5} \ \text{ or } \ x_2 = 0.60x_1. \qquad (23.3)$$

Equation (23.3) indicates that for every one-unit step (the distance in coded units from, say, 0 to +1) in the x_1 direction, we must take a 0.60-unit step in the x_2 direction in order to remain on the contour.

Given that the contour is defined by Eq. (23.3), to improve our response the quickest, we should move perpendicular to the contour. The equation of the line that is perpendicular to the contour that passes through the design center point is then given by

$$x_2 = \frac{-1}{0.60}x_1 = -1.667x_1. \qquad (23.4)$$

Equation (23.4) defines the *path of steepest ascent* (PSA). The PSA is a line that originates from the center of the experimental region and goes in the direction of increasing response perpendicular to the contour lines. The PSA given by Eq. (23.4) shows that if we were to take a −0.5-unit step in the x_1 direction, then we should take a (−0.5)(−1.667) = +0.833-unit step in the x_2 direction.

In general, the PSA may be determined directly from Eq. (23.2) by finding the vector that is associated with the gradient (direction of maximum improvement) of $\hat{y}$.

Moving on the Path of Steepest Ascent

Once the PSA has been determined, our next step will be to explore the response in this direction by running a series of experiments along the path. We will continue to take steps—run experiments—along the path so long as our response improves—in this case, increases. Of course, taking very small steps may require many tests to make progress, and taking steps that are too large may cause us to overshoot the high point of the response on the path. In order to take a measured and informed approach to this process, it is useful to set up a table, such as that shown below, to define a series of candidate tests along the PSA. Often it is most efficient to take the first step—run the first test—along the PSA just outside the 2^k(or 2^{k-p}) region.

Candidate Test	x_1	$x_2 = -1.667x_1$	Response
A	0	0	
B	−0.5	0.833	
C	−1.0	1.667	
D	−1.5	2.500	
E	−2.0	3.333	
F	−2.5	4.167	

In examining the table, it may be noted that candidate test A is the center point of the design. The next few candidate test points are obtained by incrementing the x_1 value by −0.5 unit. The choice for the magnitude of the step in the x_1 direction is somewhat subjective—we wish to avoid taking steps that are either too small or too large for the reasons indicated above. The values in the x_2 column are obtained using Eq. (23.4). Initially, the response column is empty—we will fill in specific values when we actually run the tests. Remember, the table only lists candidate test conditions—potential points on the PSA for which we may wish to determine, based on experiments, values for the response. When we move along the path of steepest ascent, we expect to see improvements in the response value. But while the model for the predicted response clearly indicates that this will be the case, the PSA is an extrapolation of the fitted model based on the experimental results in a relatively small region on the surface, and eventually it will "begin to depart" from the actual response surface. Accordingly, our experimental exploration of the PSA should proceed carefully, one step at a time, with experimental results used to verify the extrapolated expectations for the values of the process response. We will continue to move in the direction of the current PSA so long as the response continues to improve as verified by experimental results.

Of the PSA candidate tests listed in the table, the first point lying outside our factorial design region is test C. Suppose that we decide to begin our exploration of the PSA by running test C, and we secure a response of 56 (refer to Fig. 23.5). This response is higher than any response within the initial factorial design region, so we proceed to run the next candidate test, D. When we perform this test, we obtain a response of 63. Continuing on the PSA, test E is then run, producing a response of 71. When test F is run, the response drops to 65. So, our best response on the PSA is 71, obtained at test E. At this point, we could go back to the table and define additional candidate tests between points E and D or between points E and F. Alternatively, we may simply abandon the current PSA and determine a new PSA based on the results of a new factorial experiment, the center of which is at or near the "best" response value on the current PSA, viz., 71.

In setting up the levels of the variables for the next PSA determination it is often useful to make adjustments based on the linear coefficients from the previous analysis. If a coefficient of a variable is small, we might expand the range of the variable levels, while if it is large, we might reduce the range a bit. Sometimes a coefficient might be very small, indicating that either the variable is not very important or we are straddling a conditional optimum. Moving the center point for the next PSA design off the path and expanding the range for that variable might be a good strategy to test for these possibilities. If the linear coefficient then becomes large, the latter might have been the case, while if it remains small, the former is likely to be true.

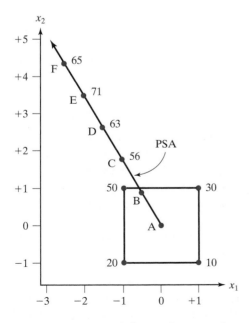

Figure 23.5   **Factorial experiment and tests on the path of steepest ascent**

Often, the process of performing factorial designs, defining the PSA, and then running tests along the PSA is repeated several times as we attempt to "climb" the response surface. As the summit of the surface is approached, it is likely that the actual surface will exhibit more curvature locally. One can check for this by either calculating the interaction term(s) for the variables or comparing the test results from a center point in the factorial design to those at the vertices of the square (cube, etc.) or both. As we repeat our process of performing PSA designs, we typically find that it becomes more and more difficult to take "extended walks" along the PSA without quickly seeing a change—in this case a drop in the response—from that predicted along the path by the model. Further analysis by moving along PSA's in such situations will not be fruitful. This again is a sign that we are nearing the neighborhood of the optimum. Once we believe that the neighborhood of the optimum has been reached, our experimental strategy will change.

Exploring the Neighborhood of the Optimum

When we believe that we have arrived at the neighborhood of the optimum (the response values are approaching those we had hoped to achieve and the surface is exhibiting some curvature locally), we will need to model the surface using a higher-order polynomial approximation. Generally, we will need to employ at least a quadratic model form, viz.,

$$ y = b_0 + \sum_{i=1}^{k} b_i x_i + \sum_{i=1}^{k-1} \sum_{j=i+1}^{k} b_{ij}\, x_i\, x_j + \sum_{i=1}^{k} b_{ii}\, x_i^2 + \epsilon. \qquad (23.5) $$

Such a full second-order polynomial has the ability to describe many different surface shapes. To capture the curvature in the response function, each control variable must take on at least three levels in the ensuing experimental design. Two experimental designs that are commonly used for this purpose are illustrated in Figs. 23.6 and 23.7 for cases with two and three independent variables. Each of these two designs has certain advantages or useful properties. In general, the

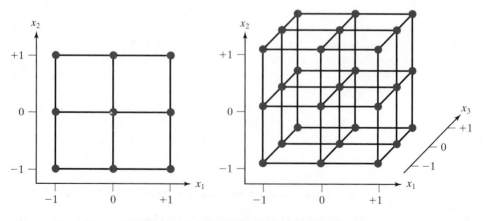

Figure 23.6   **Three-level factorial (3^k) designs for two and three variables**

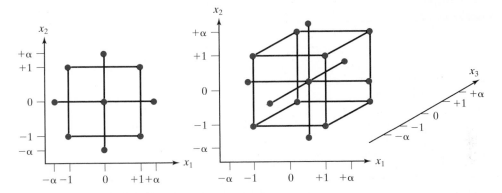

Figure 23.7   Central composite designs (CCD) for two and three variables

central composite design (CCD) structure is more efficient when there are more than two control/independent variables.

For three-level factorial designs the number of tests in the design is 3^k, where k is the number of variables being examined in the experiment. Therefore, the number of tests required for the design increases rapidly as k increases beyond three or four.

Variables	2	3	4	5
Tests	9	27	81	243

While 3^k designs can be fractionated to produce a more efficient design structure, the central composite design (CCD) is often preferred. The CCD consists of: i) points/tests based on a 2^k (or 2^{k-p}) base design, ii) two axial or "star" points/tests for each variable, and iii) center points. For the CCD, as the number of variables increases beyond four, using a 2^{k-p} base design may be wise. Such reduces the number of tests required and is justified as long as the design resolution is V or more and a quadratic approximation of the response surface is deemed to be adequate. Table 23.1 presents the CCD design parameters for up to $k = 6$ independent variables. The values for the axial points are based on the coded form for the 2^k (or 2^{k-p}) design level. In the table, the number of center points, n_c, is shown as a variable. In general, the number of center points may be selected to ensure certain properties of the resulting design,[1,2] but as a general guideline, the design should contain at least two or three center points to provide some replication for the estimation of the experimental error independent of the model.

---

[1]G. E. P. Box and N. R. Draper, *Empirical-Model Building and Response Surfaces*, John Wiley & Sons, New York, 1987.
[2]N. R. Draper, "Center Points in Second-Order Response Surface Designs," *Technometrics*, Vol. 24/2, 1982, pp. 127–133.

TABLE 23.1	Central Composite Designs					
			Number of Variables, k			
	2	*3*	*4*	*5*	*5*	*6*
Number of tests for 2^k (or 2^{k-p}) base design, N	4	8	16	32	(16)	(32)
Number of axial points $= 2k$	4	6	8	10	10	12
Distance of axial point from center point, $\alpha = [N]^{1/4}$	1.414	1.682	2.000	2.378	2.000	2.378
Number of center points, n_c	n_c	n_c	n_c	n_c	n_c	n_c
Total number of tests	$8 + n_c$	$14 + n_c$	$24 + n_c$	$42 + n_c$	$26 + n_c$	$44 + n_c$

Note: Here, $\alpha = [N]^{1/4}$ ensures the property of "rotatability," i.e., equal model predictive precision for points equidistant from the center of the design in any direction.

Once a 3^k factorial or CCD experiment has been conducted in the near-optimal region on the response surface and the generalized least-squares method has been used to estimate the coefficients in the model, e.g., establish a quadratic equation for the predicted response, the fitted model may then be used to identify the optimum. Plots of the response surface may be usefully employed to find the optimum, or calculus can be used as well. Once the optimal variable settings have been determined, validation tests should be performed to confirm the predictions of the response based on the model.

Summary

In this section, the procedures—experimental designs and associated math models—of the technique of response surface methodology have been described. Figure 23.8 shows that for the above example, the process of performing a factorial design, defining the PSA, and then running tests along the PSA was repeated twice. Following the second set of PSA trials, a 3^2 factorial design was performed, producing the predicted response contours shown in the figure. These contours (the fitted model) may be used to identify variable levels that predict the optimum response, a prediction that should be confirmed through validation tests.

23.4   RSM Example—A Pulp Whitening Process

We will now use RSM to study a process employed to whiten wood pulp that is subsequently used in a paper-making process. Previous experiments have revealed that two variables, processing temperature and bleach concentration, have a significant effect on the pulp whiteness. Currently, the process is running at a temperature T of 170°F and a concentration level C of 16%. For these conditions, the pulp whiteness response, y, is 20%. It is believed that by proper manipulation of the levels for temperature and concentration, the response can be increased to at least 90%. To find optimal levels for temperature and concentration, it was decided that the technique of RSM should be employed.

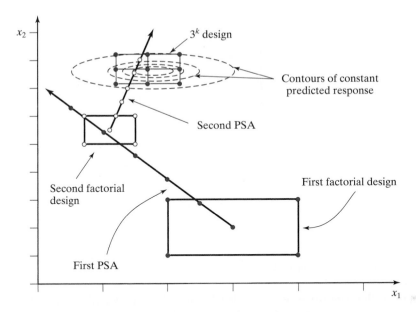

Figure 23.8   Exploring the response surface using RSM

Engineers have defined the experimental region that is feasible for the whitening process by specifying the experimental range for each of the two process variables as shown below.

Variable	Low Level	High Level
Concentration (%)	0	20
Temperature (°F)	110	180

First Design—2^2 Full Factorial

To begin the process of identifying the optimal response for the pulp whitening process, a 2^2 factorial experiment was conducted according to the test conditions given in Table 23.2. Figure 23.9 graphically displays the design and test results as well as the feasible region for the two variables under study.

TABLE 23.2   Conditions for Initial Factorial Design and Whiteness Response

	Actual Variable Units		Coded Variable Units		
Test	Concentration, C	Temperature, T	x_1	x_2	Whiteness %, y
1	14	160	−1	−1	32
2	18	160	+1	−1	30
3	14	180	−1	+1	19
4	18	180	+1	+1	13

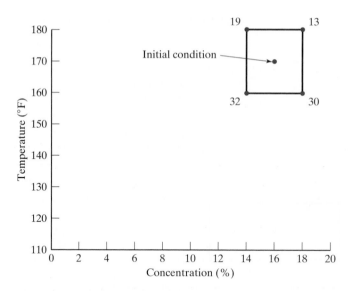

Figure 23.9 **Experimental conditions and results from the first experiment and process operation region**

Path of Steepest Ascent—First Design

Based on the test results of our first experiment (Table 23.2), we obtain the following effect estimates and associated model coefficients:

$$\bar{y} = \tfrac{1}{4}[32 + 30 + 19 + 13] = 23.5, \quad \hat{b}_0 = 23.5,$$

$$E_1 = \tfrac{1}{2}[-32 + 30 - 19 + 13] = -4, \quad \hat{b}_1 = -2.0,$$

$$E_2 = \tfrac{1}{2}[-32 - 30 + 19 + 13] = -15, \quad \hat{b}_2 = -7.5.$$

Therefore, the model for the predicted response based on this initial design is

$$\hat{y} = 23.5 - 2.0x_1 - 7.5x_2, \tag{23.6}$$

where it is noted that the model has been expressed in coded variable units. Based on the model, the PSA is

$$x_2 = (7.5/2.0)x_1 = 3.75x_1 \quad \text{or} \quad x_1 = (2.0/7.5)x_2 = 0.267x_2. \tag{23.7}$$

We note that Eq. (23.7) displays the PSA in two forms; x_2 as a function of x_1, and x_1 as a function of x_2. The first form uses x_1 as the base variable, and the second form uses x_2. We must choose one among the k variables as the base variable, and then calculate the step sizes for the other $k - 1$ variables to ensure that our steps exactly follow the PSA. In considering the two forms for the PSA, for our case, we decide to use the second form—the version that uses x_2

as the base variable, since this form will result in a smaller step size for the other variable.

The next issue in setting up a PSA table is the determination of a step size for the base variable. Here, experience in or knowledge of the process under study will be most helpful in making a step size determination. Too large a step size may prove to be misleading, since we could easily step over the maximum along the path and move away from the optimum. Too small a step leads to slow movement along the path and may result in the consumption of excessive experimental resources. In general, however, relatively large steps may be taken initially when the two-level design is remote from the neighborhood of the maximum/minimum response. As exploration progresses from a given design region to the next, smaller steps should be taken to avoid possible "overshoot."

Returning to our example, the step change of the base variable, x_2, was set at -0.5. We may now construct a table similar to that in the previous section to determine a series of tests along the PSA. As is evident from an examination of Table 23.3, several additional columns have been added. Notably, columns for the levels of the actual variables have been introduced—these columns will be helpful in running the experimental trials, and they may be determined from the following relationships between the actual and coded variables associated with the first factorial design:

$$C = 16 + (2)x_1, \tag{23.8}$$

$$T = 170 + (10)x_2. \tag{23.9}$$

Another column for "Test Number" has been added, so that we may keep track of the total number of experimental trials we have performed.

In Table 23.3, a number of candidate tests have been defined along the PSA based on the proportional steps for x_1 and x_2 that we calculated above. We are now ready to "walk/run tests along the PSA" in search for higher values of the

TABLE 23.3	Path of Steepest Ascent Table for First Design					
Candidate Test	x_1 = $0.267x_2$	x_2	$C(\%)$ = $16 + 2x_1$	$T(°F)$ = $170 + 10x_2$	Test Number	Whiteness %, y
A	0.0	0.0	16.00	170.0		
B	−0.133	−0.5	15.73	165.0		
C	−0.267	−1.0	15.47	160.0	5	29
D	−0.401	−1.5	15.20	155.0	6	35
E	−0.534	−2.0	14.93	150.0	7	39
F	−0.668	−2.5	14.67	145.0	8	43
G	−0.801	−3.0	14.40	140.0	9	47
H	−0.935	−3.5	14.13	135.0	10	43
I	−1.068	−4.0	13.87	130.0	11	40

whiteness response. Remember, every step we take going forward is a step that lowers the temperature by 5°F and the bleach concentration by 0.267%. For this example, trials were performed for the PSA candidate tests lying on or outside the factorial design region, with the resulting responses recorded in Table 23.3 and shown in Fig. 23.10.

Interpretation of Results

As can be seen from Table 23.3, the first five steps/tests along the PSA produced successively higher whiteness values. In the process, we have found a setting that produces a whiteness value more than twice that under the current operating condition. Beyond test #9, however, it appears that further steps along the PSA may not be productive, as the response drops on test #10. To make sure that such a decrease in the response was not a product of experimental variation, test #11 was then conducted. This test produced a much lower value for the response. Hence, one can be reasonably confident that further exploration along this PSA will not be fruitful. It is now time to abandon the current PSA and start searching along a new PSA by running a new 2^2 design centered at the conditions of test #9.

Second Design—2^2 Full Factorial

For the determination of a second PSA, we need to run another 2^2 experiment, fit a linear model, and repeat the procedure we followed for the first design. We believe we are still quite a distance from the expected optimum point. Given the relatively large effect of temperature in the first design, it was decided to reduce

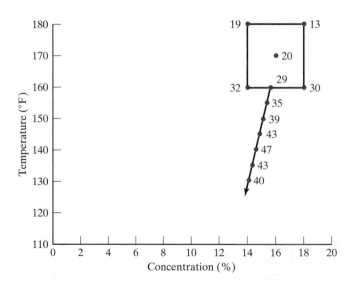

Figure 23.10   **Results of the first path of steepest ascent**

TABLE 23.4	Conditions for Second Factorial Design and Whiteness Response				
	Actual Variable Units		Coded Variable Units		
Test	Concentration, C	Temperature, T	x_1	x_2	Whiteness %, y
12	12	135	−1	−1	52
13	16	135	+1	−1	37
14	12	145	−1	+1	53
15	16	145	+1	+1	41

the range of the temperature variable levels by one-half. The range of the concentration variable levels was kept the same. This seems a reasonable approach, since the step and associated gain in the x_2 (temperature) direction was large while the proportional step in x_1 (concentration) was smaller. The conditions for the second factorial design are given in Table 23.4 along with the associated test results.

For the second design, the relationships between the actual and coded variables are

$$C = 14 + (2)x_1, \tag{23.10}$$

$$T = 140 + (5)x_2. \tag{23.11}$$

Path of Steepest Ascent—Second Design

Using the data in Table 23.4, the parameter estimates for the planar model approximating the true response surface in this region were determined:

$$\bar{y} = \tfrac{1}{4}[52 + 37 + 53 + 41] = 45.75, \quad \hat{b}_0 = 45.75,$$

$$E_1 = \tfrac{1}{2}[-52 + 37 - 53 + 41] = -13.5, \quad \hat{b}_1 = -6.75,$$

$$E_2 = \tfrac{1}{2}[-52 - 37 + 53 + 41] = 2.5, \quad \hat{b}_2 = 1.25.$$

Using the above parameter estimates, the linear model for the predicted response and the equation for the new PSA are

$$\hat{y} = 45.75 - 6.75x_1 + 1.25x_2,$$
$$x_2 = (-1.25/6.75)x_1 = -0.185x_1.$$

Examining the model for the predicted response, we note that now the x_1 coefficient is larger than the x_2 coefficient and the sign of the x_2 coefficient has changed from that of the first design. The new PSA will now lead us back in the direction of higher temperatures (x_2). Table 23.5 shows the candidate tests

TABLE 23.5	Path of Steepest Ascent Table for Second Design				
Candidate Test	x_1	x_2 = −0.185x_1	$C(\%)$ = 14 + 2x_1	$T(°F)$ = 140 + 5x_2	Test Number	Whiteness %, y
A	0.0	0.000	14.0	140.0		
B	−0.5	0.093	13.0	140.5		
C	−1.0	0.185	12.0	140.9	16	53
D	−1.5	0.278	11.0	141.4	17	56
E	−2.0	0.370	10.0	141.8	18	62
F	−2.5	0.463	9.0	142.3	19	67
G	−3.0	0.555	8.0	142.8	20	70
H	−3.5	0.648	7.0	143.2	21	70
I	−4.0	0.740	6.0	143.7	22	71
J	−4.5	0.833	5.0	144.2	23	72
K	−5.0	0.925	4.0	144.6	24	68
L	−5.5	1.018	3.0	145.1	25	66

for the second PSA. Note that now x_1 has been chosen as the base variable with a step size of −0.5, and the new PSA has smaller, but positive, increases in temperature (x_2) while continuing to reduce the concentration (x_1), but now in larger steps. Ten tests (16–25) were performed one at a time along the second PSA. Figure 23.11 graphically displays the results of these tests along the second PSA. Note that the nature of the process will dictate how exactly the tests should be done—one at a time or in batches—and how many tests can reasonably be conducted at one time. Both the cost of experiments and the time required for data processing/reduction could be important factors in making this decision.

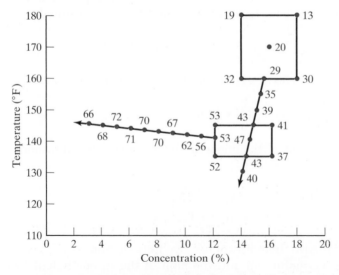

Figure 23.11   RSM investigation of pulp whitening process including second design and associated path of steepest ascent

TABLE 23.6	Conditions for Third Factorial Design and Whiteness Response				
	Actual Variable Units		*Coded Variable Units*		
Test	*Concentration, C*	*Temperature, T*	x_1	x_2	*Whiteness %, y*
26	4	140	-1	-1	66
27	6	140	$+1$	-1	62
28	4	150	-1	$+1$	81
29	6	150	$+1$	$+1$	82

Interpretation of Results

Movement along the second PSA study provided another big gain in the pulp whiteness, with response values now in the low 70%'s. This, however, is still not close to what we hope to achieve in our search for the maximum response. As the results of tests #24 and #25 indicate, this PSA has served its purpose and a new PSA should be pursued.

Third Design—2^2 Full Factorial

A third 2^2 factorial design was performed around the conditions of test #23. This time, since the concentration effect was large, the range for this variable was reduced by one-half. Table 23.6 shows the test conditions and the responses for the third factorial design.

For this design, the relationships between the actual and coded variables are

$$C = 5 + x_1, \tag{23.12}$$

$$T = 145 + (5)x_2. \tag{23.13}$$

Path of Steepest Ascent—Third Design

The effect estimates and associated linear model coefficients for the third PSA experimental design are

$$\bar{y} = \tfrac{1}{4}[66 + 62 + 81 + 82] = 72.75, \quad \hat{b}_0 = 72.75,$$

$$E_1 = \tfrac{1}{2}[-66 + 62 - 81 + 82] = -1.5, \quad \hat{b}_1 = -0.75,$$

$$E_2 = \tfrac{1}{2}[-66 - 62 + 81 + 82] = 17.5, \quad \hat{b}_2 = 8.75.$$

Using the above parameter estimates, the linear model for the third design and the new equation for the PSA (shown in Fig. 23.12) are

$$\hat{y} = 72.75 - 0.75x_1 + 8.75x_2,$$
$$x_1 = (-0.75/8.75)x_2 = -0.086x_2.$$

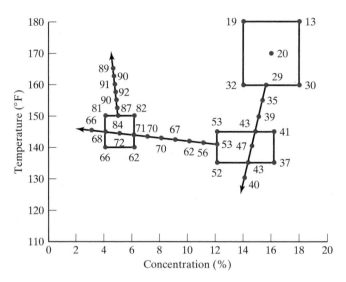

Figure 23.12   Third design with path of steepest ascent

Interpretation of Results

From Table 23.7 and Fig. 23.12, it is seen that the third PSA has led us into the vicinity of the optimum. The whiteness value is now greater than 90%, more than what we had hoped for at the beginning of the study. We also note that as we move from the center of the third design along the PSA the response is changing in a somewhat nonlinear way, indicating that the surface has curvature in this region. Our highest response along the PSA is for test #33. We might consider terminating the study at this point and running some confirmatory tests at the settings of test #33. If the analysis of these confirmatory tests indicates a whiteness reading of about 92 on the average, the study might be concluded with a recommendation to use the variable levels associated with

TABLE 23.7	Path of Steepest Ascent Table for Third Design				
Candidate	x_1		$C(\%)$	$T(°F)$	Test	Whiteness %,
Test	$= -0.086x_2$	x_2	$= 5 + x_1$	$= 145 + 5x_2$	Number	y
A	0.0	0.0	5.00	145.0		
B	−0.043	0.5	4.95	147.5		
C	−0.086	1.0	4.91	150.0	30	84
D	−0.129	1.5	4.87	152.5	31	87
E	−0.172	2.0	4.83	155.0	32	90
F	−0.215	2.5	4.78	157.5	33	92
G	−0.258	3.0	4.74	160.0	34	91
H	−0.301	3.5	4.70	162.5	35	90
I	−0.344	4.0	4.65	165.0	36	89

test #33 as a new operating point for the process. However, there is an opportunity to obtain even higher responses by exploring the neighborhood of the optimum region. Furthermore, knowledge of the precise nature of the response surface in this region could be useful for process control purposes. This will require a change in our strategy, now employing a higher-order design to capture the behavior of the response surface in this region.

Central Composite Design—Second-Order Model

In our investigation of the pulp whitening process we believe that we have now arrived at the neighborhood of the optimum response. We must now plan and conduct a higher-order design to respond to a proposed quadratic model for the response in this region, viz.,

$$y = b_0 + b_1x_1 + b_2x_2 + b_{12}x_1x_2 + b_{11}x_1^2 + b_{22}x_2^2 + \epsilon.$$

Based upon this model form, it was decided to perform a central composite design near the conditions of test #33. Table 23.8 shows the variable levels for this CCD in both coded form and the actual variable values. Table 23.9 displays the complete design matrix and the test results for this CCD. For this design, the relationships between the actual and coded variables are

$$C = 4.75 + x_1, \tag{23.14}$$

$$T = 157.5 + (5)x_2. \tag{23.15}$$

TABLE 23.8   Variable Levels for the CCD

	Level of the Coded Variable				
	$-\alpha = -\sqrt{2}$	-1	0	$+1$	$+\alpha = +\sqrt{2}$
Concentration, C (%)	3.34	3.75	4.75	5.75	6.16
Temperature, T (°F)	150.4	152.5	157.5	162.5	164.6

TABLE 23.9   Design Matrix for CCD and Test Results

Test Number	x_1	x_2	$C(\%) = 4.75 + x_1$	$T(°F) = 157.5 + 5x_2$	Whiteness %, y
37	-1	-1	3.75	152.5	84
38	1	-1	5.75	152.5	85
39	-1	1	3.75	162.5	94
40	1	1	5.75	162.5	82
41	$-\sqrt{2}$	0	3.34	157.5	91
42	$+\sqrt{2}$	0	6.16	157.5	81
43	0	$-\sqrt{2}$	4.75	150.4	83
44	0	$+\sqrt{2}$	4.75	164.6	89
45	0	0	4.75	157.5	92
46	0	0	4.75	157.5	93

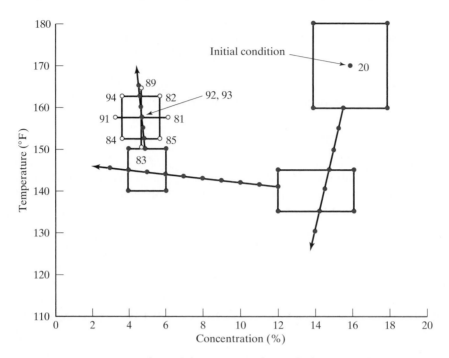

Figure 23.13   **Tests conducted for RSM study, including CCD for the near-optimum region**

The test conditions for the CCD (open dots) and the associated responses are shown in Fig. 23.13.

Fitting the Model

Using the results of the CCD, we may now determine the model coefficients using the method of least squares. The step-by-step calculations to obtain the $\hat{b}$'s are as follows:

$$
\mathbf{X} = \begin{bmatrix}
+1 & -1 & -1 & +1 & +1 & +1 \\
+1 & +1 & -1 & -1 & +1 & +1 \\
+1 & -1 & +1 & -1 & +1 & +1 \\
+1 & +1 & +1 & +1 & +1 & +1 \\
+1 & -\sqrt{2} & 0 & 0 & 2 & 0 \\
+1 & +\sqrt{2} & 0 & 0 & 2 & 0 \\
+1 & 0 & -\sqrt{2} & 0 & 0 & 2 \\
+1 & 0 & +\sqrt{2} & 0 & 0 & 2 \\
+1 & 0 & 0 & 0 & 0 & 0 \\
+1 & 0 & 0 & 0 & 0 & 0
\end{bmatrix}, \quad
\mathbf{y} = \begin{bmatrix}
84 \\
85 \\
94 \\
82 \\
91 \\
81 \\
83 \\
89 \\
92 \\
93
\end{bmatrix}
$$

$$(\mathbf{X}^T\mathbf{X}) = \begin{bmatrix} 10 & 0 & 0 & 0 & 8 & 8 \\ 0 & 8 & 0 & 0 & 0 & 0 \\ 0 & 0 & 8 & 0 & 0 & 0 \\ 0 & 0 & 0 & 4 & 0 & 0 \\ 8 & 0 & 0 & 0 & 12 & 4 \\ 8 & 0 & 0 & 0 & 4 & 12 \end{bmatrix}$$

$$(\mathbf{X}^T\mathbf{X})^{-1} = \begin{bmatrix} 0.5 & 0 & 0 & 0 & -0.25 & -0.25 \\ 0 & 0.125 & 0 & 0 & 0 & 0 \\ 0 & 0 & 0.125 & 0 & 0 & 0 \\ 0 & 0 & 0 & 0.125 & 0 & 0 \\ -0.25 & 0 & 0 & 0 & 0.21875 & 0.09375 \\ -0.25 & 0 & 0 & 0 & 0.09375 & 0.21875 \end{bmatrix}$$

$$(\mathbf{X}^T\mathbf{y}) = \begin{bmatrix} 874.000 \\ -25.142 \\ 15.485 \\ -13.000 \\ 688.993 \\ 688.993 \end{bmatrix}.$$

Therefore, the least-squares estimates of the full second-order model are

$$\hat{\mathbf{b}} = \begin{bmatrix} \hat{b}_0 \\ \hat{b}_1 \\ \hat{b}_2 \\ \hat{b}_{12} \\ \hat{b}_{11} \\ \hat{b}_{22} \end{bmatrix} = (\mathbf{X}^T\mathbf{X})^{-1}(\mathbf{X}^T\mathbf{y}) = \begin{bmatrix} 92.5000 \\ -3.1427 \\ 1.9357 \\ -3.2500 \\ -3.1875 \\ -3.1875 \end{bmatrix}.$$

Thus, the second-order fitted model is given by

$$\hat{y} = 92.5 - 3.14x_1 + 1.94x_2 - 3.25x_1x_2 - 3.19x_1^2 - 3.19x_2^2.$$

Using the Second-Order Model

After fitting the second-order model, we can determine the optimum values of concentration and temperature by taking derivatives of the predicted response function with respect to both x_1 and x_2. After setting the resulting equations equal to zero and simplifying, we obtain:

$$-6.38x_1 - 3.25x_2 = 3.14,$$
$$-3.25x_1 - 6.38x_2 = -1.94.$$

Solving these equations simultaneously produces the following predicted optimum settings for the two variables in their coded units:

$$x_1^* = -0.874,$$
$$x_2^* = 0.749.$$

Using Eqs. (23.14) and (23.15), which relate the actual and coded variable units, the values for the concentration, C (%), and temperature, T (°F), at the predicted optimum condition are 3.88% and 161.2°F, respectively. The predicted whitening response at this point is 94.60%.

Several tests were then performed at the predicted optimum point for the concentration and temperature. The whitening responses obtained for these validation tests were as follows: 95%, 97%, 96%, 95%, and 98%. The average for these responses is 96.2%, which is in fact slightly higher than the predicted response. The validation tests demonstrate that the proposed new levels for the concentration and temperature are vastly superior to the operating conditions that existed at the start of the RSM study.

Figure 23.14 shows the second-order approximation of the response surface contours overlaid on the actual response surface contours that were used to generate the test results for this example. Several things are worth noting in the figure:

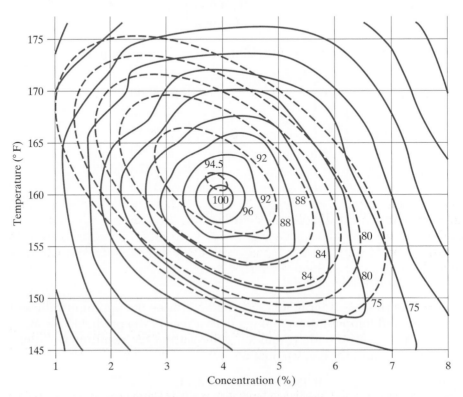

Figure 23.14   **Actual and predicted contours for the pulp whitening example**

- The predicted location of the optimum value for the response is actually quite close to the true optimum value.

- In the vicinity of the optimum the predicted response contours are close in appearance (magnitude and shape) to the actual response contours.

- In regions remote from the optimum, the quadratic approximation is poor— but this is unimportant.

Of course, in an actual case study, the true behavior of the response surface will be unknown to us—but this example serves to reassure us that models established based on an RSM study can be effectively used to optimize the process in situations like the one depicted, where the true response behavior is relatively smooth and unimodal in the operating region of interest.

Summary

This section has demonstrated that the technique of response surface methodology is a blend of structured quantitative tools and methods and sound judgment. Over the lifetime of an RSM study, the analyst will be required to interpret the results of experiments at each stage of the process and make certain logical judgments concerning future actions, e.g., the design of the next experiment. Analyzing a factorial design, fitting a linear model, and calculating a path of steepest ascent are all mechanics that can readily be mastered. But using RSM also involves certain more subjective, but informed, decisions to be made by the analyst all along the way, viz.,

- Which variables should be used?
- Can a variable be eliminated or dropped out?
- How should the levels of the variables be set?
- What size steps should be taken along the PSA?
- When should the PSA be redetermined?
- When should PSA analysis be abandoned?

Determining the answers to these questions is not always straightforward. The answers vary with each unique situation. The case study described in detail in the next section shows how a group of individuals faced with an RSM study responded to some of these questions and therefore should provide the reader with some useful guidance on RSM decision making.

23.5   Soldering Process Optimization—A Case Study

The following case study was adapted from an actual investigation conducted at Motorola in the early 1990s.[3] It describes the complete investigation of a

---

[3]T. S. Babin, "Modeling and Optimization in Electronic Assembly Operations," *Proceedings of the Second Industrial Engineering Research Conference*, May 1993, pp. 143–149.

process—beginning with an initial screening experiment and progressing through the optimization of the process.

The Problem

For the purpose of reducing cycle time and improving quality of an electronic assembly line for portable pager manufacture, a study was conducted to develop a method by which predictions of assembly cycle time and performance quality can be made at the earliest possible stages of process design. Figure 23.15 schematically shows the entire assembly line, where the flow of products through various stages is in the counterclockwise direction. The top- and bottom-side assembly operations, typically referred to as the front end, consist of attachment and soldering of components to the printed wiring board (PWB). The bottom section of the line is referred to as the back end, where the PWB is configured and assembled into the housing. There are 33 individual workstations associated with processing operations, including those for:

- Preparing a PWB for component insertion and/or placement;
- Insertion and/or placement of chip resistors, chip capacitors, integrated circuits, etc.;
- Assembly of mechanical components such as switches, displays, keypads, housings, etc.;
- Attachment of piece parts with soldering or fastening operations; and
- Electrical tuning (frequency adjustments, etc.) and board testing.

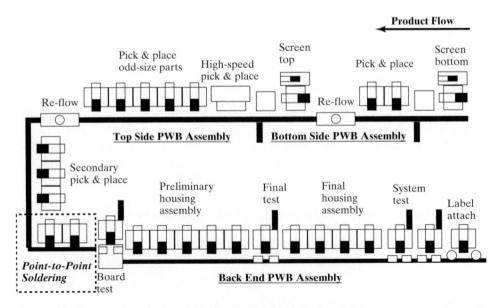

Figure 23.15   An electronic printed wiring board assembly line

Each automated tuning and testing station provides data on the failures or defects that have occurred at that station. Based on such data, the two point-to-point soldering operations in Fig. 23.15 were quickly identified as being the "quality bottleneck" of the process. The process engineers then conducted a brainstorming session to develop a list of possible contributing factors that might result in performance failures in soldering. An initial list of eight potentially important process variables was developed:

- Primary solder amount
- Secondary solder amount
- Temperature of iron
- Solder type
- Distance between joints
- Time between joints
- Heating time
- Preheating time.

It was then decided to conduct an initial experiment to determine which of these variables might be most important in terms of improving quality. The response selected for the experiment was the solder joint quality, which is actually a composite score of three quality concerns: joint coverage, joint surface roughness, and fillet angle. Referring to the sketches of Fig. 23.16, these three responses were more precisely described as:

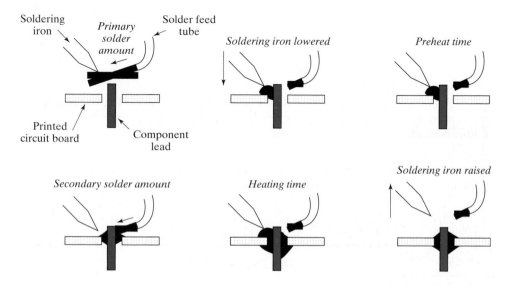

Figure 23.16 Schematic illustration of the point-to-point soldering operation

- Joint coverage—percentage of joint covered with solder (on a 0–200 scale);
- Solder finish—surface finish of the solder joint (on a 0–200 scale);
- Fillet angle—angle between the solder fillet and the circuit board (on a 0–200 scale).

The composite score was then found by summing the values associated with the individual responses, and then dividing the sum by 20, viz.,

$$composite\ score = (coverage + finish + angle)/20.$$

For the initial operating conditions of the process, composite score responses typically varied between 12 and 22. The engineers hoped to improve the score to 30 by finding optimum settings for the most dominant control variables.

First Design– 2^{8-4} Fractional Factorial

A 2^{8-4} fractional factorial experiment was conducted and then analyzed in order to

- Identify the key process variables that have an influence on performance,
- Obtain a model for quality score prediction based on the key factors, and
- Use the fitted model to develop process control measures for performance improvement.

The levels for each of the eight variables for the 2^{8-4} design are given in Table 23.10.

The 2^{8-4} fractional factorial design chosen for the experiment was of resolution IV. The engineers introduced variables x_5, x_6, x_7, and x_8 into the design using the following generators:

$$-5 = 123, \quad +6 = 124, \quad -7 = 134, \quad \text{and} \quad 8 = 234.$$

TABLE 23.10	Variables and Levels for the 2^{8-4} Soldering Process Experiment		
Variable		*Low Level*	*High Level*
1. Primary solder amount (mm)	x_1	2	3
2. Secondary solder amount (mm)	x_2	1	2
3. Iron temperature (°C)	x_3	170	190
4. Solder type	x_4	A	B
5. Distance between joints (mm)	x_5	3	5
6. Time between joints (sec)	x_6	0.5	1.2
7. Heating time (sec)	x_7	0.8	1.8
8. Preheating time (sec)	x_8	1.0	2.2

| TABLE 23.11 | Design Matrix and Responses for the 2^{8-4} Soldering Process Experiment |

Test	x_1	x_2	x_3	x_4	x_5	x_6	x_7	x_8	Response, y
1	−1	−1	−1	−1	1	−1	1	−1	15.00
2	1	−1	−1	−1	−1	1	−1	−1	12.75
3	−1	1	−1	−1	−1	1	1	1	18.85
4	1	1	−1	−1	1	−1	−1	1	16.70
5	−1	−1	1	−1	−1	−1	−1	1	18.65
6	1	−1	1	−1	1	1	1	1	16.10
7	−1	1	1	−1	1	1	−1	−1	22.75
8	1	1	1	−1	−1	−1	1	−1	20.05
9	−1	−1	−1	1	1	1	−1	1	16.20
10	1	−1	−1	1	−1	−1	1	1	13.05
11	−1	1	−1	1	−1	−1	−1	−1	19.70
12	1	1	−1	1	1	1	1	−1	17.40
13	−1	−1	1	1	−1	1	1	−1	15.65
14	1	−1	1	1	1	−1	−1	−1	12.80
15	−1	1	1	1	1	−1	1	1	19.35
16	1	1	1	1	−1	1	−1	1	17.00

Based on these generators, the experimental conditions for the (16) trials were determined and the trials were performed, in a randomized run order, using the variable levels shown in Table 23.11. The table also displays the result obtained for each test.

Based on the design generators, the defining relationship for the 2^{8-4} design was found to be

$$I = -1235 = +1246 = -1347 = +2348$$
$$= -3456 = +2457 = -1458 = -2367 = +1368 = -1278$$
$$= +1567 = -2568 = +3578 = -4678$$
$$= +12345678.$$

Linear Combinations of Confounded Effects

Assuming that three-factor and higher-order interaction effects can be neglected, the linear combinations of confounded effects and their corresponding estimates were then determined:

$$l_0 = 17.00 \rightarrow \text{estimates mean}$$
$$l_1 = -2.54 \rightarrow \text{estimates 1}$$
$$l_2 = 3.95 \rightarrow \text{estimates 2}$$
$$l_3 = 1.58 \rightarrow \text{estimates 3}$$
$$l_4 = -1.21 \rightarrow \text{estimates 4}$$
$$l_{12} = 0.16 \rightarrow \text{estimates } 12 - 35 + 46 - 78$$

$$l_{13} = -0.80 \rightarrow \text{estimates } 13 - 25 - 47 + 68$$

$$l_{14} = -0.13 \rightarrow \text{estimates } 14 + 26 - 37 - 58$$

$$l_{23} = 0.04 \rightarrow \text{estimates } 23 - 15 + 48 - 67$$

$$l_{24} = -0.01 \rightarrow \text{estimates } 24 + 16 + 38 + 57$$

$$l_{34} = -1.98 \rightarrow \text{estimates } 34 - 17 + 28 - 56$$

$$l_{123} = -0.08 \rightarrow \text{estimates } -5$$

$$l_{124} = 0.18 \rightarrow \text{estimates } 6$$

$$l_{134} = 0.14 \rightarrow \text{estimates } -7$$

$$l_{234} = -0.03 \rightarrow \text{estimates } 8$$

$$l_{1234} = -0.01 \rightarrow \text{estimates } -45 + 36 - 27 + 18$$

The estimates associated with the linear combinations of confounded effects shown above were then used to construct a normal probability plot as shown in Fig. 23.17. As can be seen from the figure, the linear combinations l_1, l_2, l_3, l_4, and l_{34} are clearly significant—they all lie well off the line associated with the effect estimates that appear to be arising from a normal distribution centered at zero. The linear combination l_{34} estimates the sum of the two-factor interactions 34, 17, 28, and 56. Based both on the fact that l_3 and l_4 were found to be significant and on prior knowledge of the process, the study team felt that the linear combination l_{34} was probably large because of the 34 interaction, i.e., the interaction between iron temperature and solder type. Thus, the following model for the predicted response was postulated:

$$\hat{y} = 17.00 - 1.27x_1 + 1.98x_2 + 0.79x_3 - 0.60x_4 - 0.99x_3x_4. \qquad (23.16)$$

Since the two-factor interaction 34 (iron temperature–solder type) was judged to be the important interaction within the linear combination of confounded effects associated with l_{34}, it was decided to prepare a two-way diagram to gain a better understanding of its influence on the process. As can be seen from Figure 23.18, the changes in the iron temperature have little effect on

Figure 23.17   **Normal probability plot of effect estimates for the 2^{8-4} design**

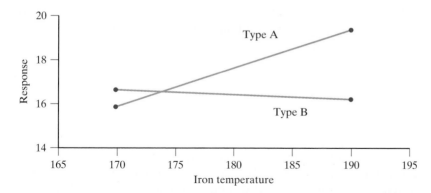

Figure 23.18   Two-way diagram for iron temperature–solder type interaction

the response for solder type B. On the other hand, with solder type A, a change in the iron temperature from the low to high level produces a substantial increase in the response. Since it is desired to obtain high values for the response, the use of solder type B seems to offer little promise. A large value for the response does appear to be possible if solder type A (low level for variable 4) is employed, assuming that a high level for iron temperature is utilized. With this in mind, it was decided to focus exclusively on the low level (-1) for variable 4—solder type A—in subsequent testing.

The results from the 2^{8-4} experiment also seemed to indicate that for the levels of the variables examined, four of the variables—distance between joints, time between joints, preheating time, and heating time—did not appear to influence the response. This finding suggested that levels for these variables may be selected based on other factors/considerations not related to the quality of the solder joint (e.g., production cycle time, energy minimization, and cost); therefore these variables were not considered in the balance of the RSM investigation.

Path of Steepest Ascent—First Design

Based on the foregoing discussion, particularly the decision to consider only the low level for variable 4 (i.e., use solder type A), the relation for the predicted response simplified to

$$\hat{y} = 17.61 - 1.27x_1 + 1.98x_2 + 1.78x_3. \qquad (23.17)$$

Based on Table 23.10 the relations between the actual and the coded variables are

$$P = 2.5 + (0.5)x_1,$$
$$S = 1.5 + (0.5)x_2,$$
$$T = 180 + (10.0)x_3,$$

where P, S, and T are the actual levels for the primary solder amount (mm), secondary solder amount (mm), and iron temperature (°C), respectively.

Based on the model for the predicted response, the vector that defines the PSA was found to be:

$$-1.27\vec{i} + 1.98\vec{j} + 1.78\vec{k}$$

or

$$-0.64\vec{i} + 1.00\vec{j} + 0.90\vec{k}, \tag{23.18}$$

where $\vec{i}$, $\vec{j}$, and $\vec{k}$ are the unit vectors associated with the x_1, x_2, and x_3 directions respectively. Equation (23.18) indicates that for a one-unit step in the x_2 direction, x_1 must change by -0.64 unit, and x_3 must increase by 0.90 unit. Based on the vector of Eq. (23.18), PSA candidate tests were defined as shown in Table 23.12. The tests along the calculated PSA were run one at a time, and the responses obtained are shown in Table 23.12.

Interpretation of Results

The process engineers conducted a total of nine tests along the PSA. Beginning with test #17, successive tests achieved higher responses until test #24, when the response decreased. An additional test, #25, was performed to confirm that the drop in the response on test #24 was not due to experimental error. The study team concluded that a new PSA should be established in the vicinity of test #23.

Second Design—2^3 Full Factorial

From the experience of the first PSA exploration (refer to Table 23.12) higher quality scores were obtained with smaller amounts of primary solder, larger

TABLE 23.12					Path of Steepest Ascent Table for First Design—Soldering Process Experiment			
Candidate Test	Test Number	x_1 $= -0.64x_2$	x_2	x_3 $= 0.90x_2$	$P(mm)$ $= 2.5 + 0.5x_1$	$S(mm)$ $= 1.5 + 0.5x_2$	$T(°C)$ $= 180 + 10x_3$	Response, y
A		0.0	0.0	0.0	2.50	1.50	180.0	
B		−0.32	0.5	0.45	2.34	1.75	184.5	
C	17	−0.64	1.0	0.90	2.18	2.00	189.0	22.10
D	18	−0.96	1.5	1.35	2.02	2.25	193.5	23.15
E	19	−1.28	2.0	1.80	1.86	2.50	198.0	23.65
F	20	−1.60	2.5	2.25	1.70	2.75	202.5	24.15
G	21	−1.92	3.0	2.70	1.54	3.00	207.0	24.90
H	22	−2.24	3.5	3.15	1.38	3.25	211.5	25.70
I	23	−2.56	4.0	3.60	1.22	3.50	216.0	26.35
J	24	−2.88	4.5	4.05	1.05	3.75	220.5	26.10
K	25	−3.20	5.0	4.50	0.89	4.00	225.0	25.05

amounts of secondary solder, and higher iron temperatures. Unlike the first design where eight variables were considered, only three variables now needed to be addressed in second design, and it was decided to run an unreplicated 2^3 design. Since the first PSA demonstrated the highest observed quality score near test #23, the 2^3 experiment was centered about the conditions of test #23 and employed the variable levels shown in Table 23.13. The design matrix for this experiment is given in Table 23.14. These eight tests were performed and the associated quality scores were determined. These results are also shown in Table 23.14.

Path of Steepest Ascent—Second Design

Based on the test results of Table 23.14, the following effects and associated linear model parameter estimates were calculated:

$$\bar{y} = \tfrac{1}{4}[24.95 + 24.80 + 28.25 + 28.30 + 23.40 + 23.50 + 25.25 + 25.05]$$

$$= 25.43, \quad \hat{b}_0 = 25.43,$$

$$E_1 = \tfrac{1}{2}[-24.95 + 24.80 - 28.25 + 28.30 - 23.40 + 23.50 - 25.25 + 25.05]$$

$$= -0.025, \quad \hat{b}_1 = -0.0125,$$

$$E_2 = \tfrac{1}{2}[-24.95 - 24.80 + 28.25 + 28.30 - 23.40 - 23.50 + 25.25 + 25.05]$$

$$= 2.525, \quad \hat{b}_2 = 1.2625,$$

$$E_3 = \tfrac{1}{2}[-24.95 - 24.80 - 28.25 - 28.30 + 23.40 + 23.50 + 25.25 + 25.05]$$

$$= -2.3, \quad \hat{b}_3 = -1.15.$$

TABLE 23.13	Variables and Levels for the Second Soldering Process Experiment		
Variable		Low Level	High Level
Primary solder amount (mm)	x_1	1	2
Secondary solder amount (mm)	x_2	3	4
Iron temperature (°C)	x_3	205	225

TABLE 23.14	Design Matrix and Responses for the 2^3 Soldering Process Experiment		
Test	x_1	x_2	x_3	Response, y
26	−1	−1	−1	24.95
27	1	−1	−1	24.80
28	−1	1	−1	28.25
29	1	1	−1	28.30
30	−1	−1	1	23.40
31	1	−1	1	23.50
32	−1	1	1	25.15
33	1	1	1	25.05

The following linear model was then used for determination of the new PSA:

$$\hat{y} = 25.43 - 0.01x_1 + 1.26x_2 - 1.15x_3. \tag{23.19}$$

The vector that defines the PSA was found to be

$$-0.01\vec{i} + 1.26\vec{j} - 1.15\vec{k}$$

or

$$-0.0099\vec{i} + 1.00\vec{j} - 0.91\vec{k}. \tag{23.20}$$

Based on Table 23.13 the relations between the actual and the coded variables are

$$P = 1.5 + (0.5)x_1,$$
$$S = 3.5 + (0.5)x_2,$$
$$T = 215 + (10.0)x_2.$$

Table 23.15 displays the candidate PSA tests associated with the second design along with the responses for the tests that were conducted.

Interpretation of Results

As can be seen from Table 23.15, the test results for the second PSA demonstrate the highest observed quality score at test #36. The response for this test, $y = 29.3$, is in the range of what was hoped for at the outset of the experimental investigation. An additional step beyond this point to test #37 resulted in a large drop in the response, indicating a considerable amount of curvature in the response surface in this area, so that further exploration along the PSA would not be advisable. Since the second PSA exploration has apparently moved very close to the vicinity of the maximum, it was decided to abandon the PSA strategy and move to the characterization of the behavior of the response surface in the near-optimum region.

TABLE 23.15		PSA Table for Second Design—Soldering Process Experiment					
Candidate Test	Test Number	x_1 $= -0.0099x_2$	x_2	x_3 $= -0.91x_2$	$P(mm)$ $= 1.5 + 0.5x_1$	$S(mm)$ $= 3.5 + 0.5x_2$	$T(°C)$ $= 215 + 10x_3$	Response, y
A		0.0	0.0	0.0	1.500	3.50	215.0	
B		−0.00495	0.5	0.455	1.497	3.75	210.5	
C	34	−0.00990	1.0	0.910	1.495	4.00	206.0	28.20
D	35	−0.01485	1.5	1.365	1.493	4.25	201.5	28.85
E	36	−0.01980	2.0	1.820	1.491	4.50	197.0	29.30
F	37	−0.02475	2.5	2.275	1.489	4.75	192.5	27.35

Third Design—3^2 Full Factorial

In preparation for the design of an experiment to establish a second-order model for the neighborhood of the optimum, a number of decisions had to be made. First, since the previous experiment had revealed a very small coefficient for variable x_1 relative to the coefficients/slopes associated with the other variables, it was decided to fix this variable at the level used in test #36. Thus, the characterization of the near-optimum region focused on the effects of variables x_2 and x_3 only. Second, the study team debated the type of higher-order design to perform. Owing to its convenience in terms of experimental setup and the communication of its nature and purpose to others, a 3^2 full factorial design was selected. Finally, the ranges for the variables under study were defined and are listed in Table 23.16.

Based on Table 23.16, the relationships between the actual and coded variable units are

$$S = 4.5 + (0.5)x_2 \quad \text{or} \quad x_2 = 2(S - 4.5) \tag{23.21}$$

and

$$T = 195 + (15)x_3 \quad \text{or} \quad x_3 = (T - 195)/15. \tag{23.22}$$

Table 23.17 displays the nine tests associated with the 3^2 full factorial design that was used to gather information on the behavior of the response (quality score) in the neighborhood of the optimum.

TABLE 23.16   Variable Levels for the 3^2 Full Factorial Soldering Process Experiment

	Level of the Coded Variable		
	−1	0	+1
Secondary solder amount (mm), x_2	4.0	4.5	5.0
Iron temperature (°C), x_3	180	195	210

TABLE 23.17   Design Matrix and Responses for the 3^2 Soldering Process Experiment

Test	x_2	x_3	$S(mm)$ = 4.5 + 0.5x_2	$T(°C)$ = 195 + 15x_3	Test Result, y
38	−1	−1	4.0	180	25.65
39	−1	0	4.0	195	27.90
40	−1	1	4.0	210	28.05
41	0	−1	4.5	180	25.30
42	0	0	4.5	195	29.20
43	0	1	4.5	210	28.35
44	1	−1	5.0	180	23.85
45	1	0	5.0	195	25.75
46	1	1	5.0	210	24.90

Fitting the Model

Using the results of the 3^2 factorial design shown in Table 23.17, the study team used the method of least squares to obtain estimates for the model parameters in the following quadratic equation used to characterize the response:

$$y = b_0 + b_2x_2 + b_3x_3 + b_{23}x_2x_3 + b_{22}x_2^2 + b_{33}x_3^2 + \epsilon. \qquad (23.23)$$

The step-by-step calculations to obtain the $\hat{b}$'s are as follows:

$$\mathbf{X} = \begin{bmatrix} +1 & -1 & -1 & +1 & 1 & 1 \\ +1 & -1 & 0 & 0 & 1 & 0 \\ +1 & -1 & 1 & -1 & 1 & 1 \\ +1 & 0 & -1 & 0 & 0 & 1 \\ +1 & 0 & 0 & 0 & 0 & 0 \\ +1 & 0 & +1 & 0 & 0 & 1 \\ +1 & +1 & -1 & -1 & 1 & 1 \\ +1 & +1 & 0 & 0 & 1 & 0 \\ +1 & +1 & +1 & +1 & 1 & 1 \end{bmatrix}, \quad \mathbf{y} = \begin{bmatrix} 25.65 \\ 27.90 \\ 28.05 \\ 25.30 \\ 29.20 \\ 28.35 \\ 23.85 \\ 25.75 \\ 24.90 \end{bmatrix}$$

$$(\mathbf{X}^T\mathbf{X}) = \begin{bmatrix} 9 & 0 & 0 & 0 & 6 & 6 \\ 0 & 6 & 0 & 0 & 0 & 0 \\ 0 & 0 & 6 & 0 & 0 & 0 \\ 0 & 0 & 0 & 4 & 0 & 0 \\ 6 & 0 & 0 & 0 & 6 & 4 \\ 6 & 0 & 0 & 0 & 4 & 6 \end{bmatrix}$$

$$(\mathbf{X}^T\mathbf{X})^{-1} = \begin{bmatrix} 0.55 & 0 & 0 & 0 & -0.33 & -0.33 \\ 0 & 0.167 & 0 & 0 & 0 & 0 \\ 0 & 0 & 0.167 & 0 & 0 & 0 \\ 0 & 0 & 0 & 0.25 & 0 & 0 \\ -0.33 & 0 & 0 & 0 & 0.5 & 0 \\ -0.33 & 0 & 0 & 0 & 0 & 0.5 \end{bmatrix}$$

$$(\mathbf{X}^T\mathbf{y}) = \begin{bmatrix} 238.95 \\ -7.10 \\ 6.50 \\ -1.35 \\ 156.10 \\ 156.10 \end{bmatrix}.$$

Therefore, the following least-squares estimates were obtained:

$$\hat{\mathbf{b}} = \begin{bmatrix} \hat{b}_0 \\ \hat{b}_2 \\ \hat{b}_3 \\ \hat{b}_{23} \\ \hat{b}_{22} \\ \hat{b}_{33} \end{bmatrix} = (\mathbf{X}^T\mathbf{X})^{-1}(\mathbf{X}^T\mathbf{y}) = \begin{bmatrix} 28.68 \\ -1.18 \\ 1.08 \\ -0.34 \\ -1.60 \\ -1.60 \end{bmatrix}.$$

The model for the predicted response is

$$\hat{y} = 28.68 - 1.18x_2 + 1.08x_3 - 0.34x_2x_3 - 1.60x_2^2 - 1.60x_3^2. \tag{23.24}$$

It may be noted that Eq. (23.24) can also be expressed in terms of the actual variable levels for secondary solder amount (S) and iron temperature (T) by replacing x_2 and x_3 with the expressions given in Eqs. (23.21) and (23.22).

Using the Model

The model for the predicted response given in Eq. (23.24) was found to be an adequate fit to the quality score responses based on an analysis of the model residuals. With this fitted model, the study team employed two commonly used methods for locating the true optimum as well as the optimal settings for the input/control variables. One method is to locate approximately the maximum response on a contour plot such as the one shown in Fig. 23.19. At the center of the contour plot, we see that the estimated maximum quality score is 29.17. When we project down from this center to the secondary solder amount axis, we read $S = 4.3$ mm, and when we project horizontally to the iron temperature axis, we read $T = 200°C$. These would then serve as estimates for the optimal settings of the control variables that should produce a quality score of about 29.

The other method of analysis is to determine the values for the input/control variables mathematically—i.e., using calculus. Again, based on the fitted second-order model given by Eq. (23.24) the optimum levels for the control variables are found by taking derivatives of the equation with respect to x_2 and x_3 and setting the results equal to zero, viz.,

$$-3.20x_2 - 0.34x_3 = 1.18,$$
$$-0.34x_2 - 3.20x_3 = -1.08.$$

When these equations are solved simultaneously, the following result is obtained:

$$x_2^* = -0.41,$$
$$x_3^* = 0.38.$$

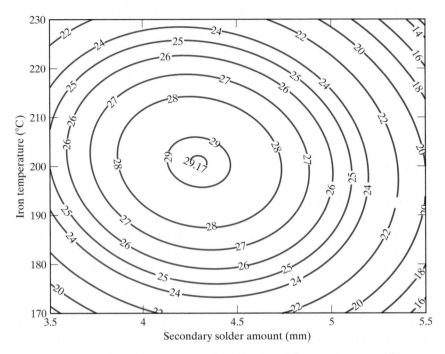

Figure 23.19 **Predicted contours in the vicinity of the maximum quality score**

Based on the relations between the actual and the coded variable units, the process engineers determined the following optimum values for secondary solder amount (S) and iron temperature (T): $S = 4.29$ mm and $T = 200.7°C$, respectively. The predicted response at this point is 29.17.

Based on the identified optimum levels for the secondary solder amount and the iron temperature, eight validation tests were performed by the study team. The resulting quality scores were: 29.10, 29.40, 29.25, 28.90, 29.60, 29.15, 29.45, and 29.05. The average for these responses is 29.24. Based on the results of these validation tests, the recommended optimal settings for the secondary solder amount and iron temperature were implemented.

Some important observations may be made in connection with the implementation of the optimal operating conditions for the soldering process. One concern at the outset of the study was that it would be necessary to increase the heating time—to produce better surface finish and coverage of solder joints, which in turn would increase greatly the process cycle time. The case study revealed that the quality of the soldering process could be improved without inflating the cycle time for the process. In fact, a quality score of 30, which was the goal of this study, was achieved through the manipulation of only three variables without altering the cycle time, with the other five variables thought initially to be important being fixed at levels that satisfied other process concerns.

Summary

An investigation of a process has been described where the goal was to identify variable settings that would significantly improve the quality of the solder joints produced by the process. The case study began with a fractional factorial design to identify a few important process variables from a larger set of factors, and continued through the optimization of the process. The case study demonstrates the following points:

- Screening experiments (fractional factorial designs) may be effectively used to identify important variables that merit further study.
- Focused experiments (two-level full factorial designs) may be employed to provide process understanding and to provide guidance on directions for improvement of the process performance.
- Response surface experiments can be used to establish predictive models in the vicinity of the optimum and may be employed to (1) identify variable settings that optimize the process, and (2) characterize the behavior of the response surface in the near-optimum region.

■ EXERCISES

23.1 Consider the results of the experimental design below, used in a process optimization study.

Test	x_1	x_2	y
1	20°F	160 psi	63
2	30°F	160 psi	79
3	20°F	180 psi	54
4	30°F	180 psi	68

Calculate the path of steepest ascent and define the initial six tests you would run along the path. Show all work.

23.2 At one point during an RSM procedure it was decided to run an experiment based on a central composite design. Briefly describe the features of this design.

23.3 The following model has been fit to the data obtained from a CCD:

$$\hat{y} = 20 - 1.50x_1^2 - 2.50x_2^2.$$

Prepare a contour plot for the predicted response surface, and comment on its form. For what values of x_1 and x_2 is the predicted response a maximum?

23.4 The following experiment was done to determine the path of steepest ascent in an RSM study:

Test	x_1	x_2	y
1	30	0.5	23
2	50	0.5	28
3	30	0.7	31
4	50	0.7	35

Choose a step size of 10 for x_1 and use that along with your other calculations to recommend the first two tests on the PSA. Show all of your work. Make a sketch of the situation, carefully showing and labeling all relevant entities.

23.5 When employing two-level fractional factorial designs for "path of steepest ascent" analysis in response surface methodology, discuss with brief explanations whether or not each of the following concerns is important or relevant.

(a) It is essential to use resolution IV or above designs so that main effects will not be confounded with two-factor interactions.

(b) Replication will be required, because the error cannot be estimated from the model.

(c) Resolution III designs provide an attractive alternative, given the assumptions normally made at this stage of the study.

(d) It is important that the generators of the design are not confounded with each other.

23.6 In the pulp whitening process example in the chapter, it was shown that after three PSA explorations the pulp whiteness had been improved from 20% to about 92%. It is noted that a total of 36 tests were run to arrive at the near-optimum neighborhood. Suppose we would like to find out how many tests it will take to search for the settings of C (bleach concentration) and T (process temperature) to produce a whiteness of 90 by the one-variable-at-a-time approach.

Of course, in actual studies, the true behavior of the response surface will be unknown to us—but for this exercise we have an actual (simulated) contour plot of the response surface, that is presented below. Therefore, one can read off the contours to obtain an approximate whiteness value for any pair of C and T values within their experimental ranges.

(a) Carry out a one-variable-at-a-time search for the optimal running conditions of C and T that will yield a maximum pulp whiteness. To do this, the initial search can start by holding C constant at its current setting while varying the T value five times, all within its range. Corresponding to each of the five pairs of C and T values, read the contour curves to find an approximate whiteness response y. The next search will hold T constant, where y is the

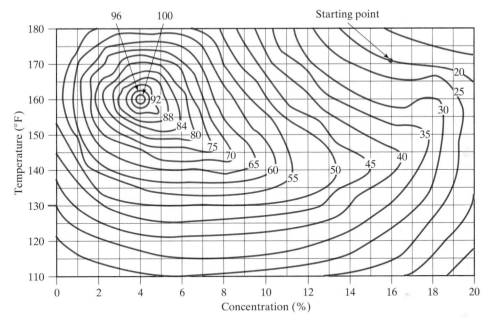

Figure E23.6   Actual Contours for the Pulp Whitening Example

largest among the five tests in the first search, and run five tests at five different values of C within its range. Repeat the process, with perhaps gradually reduced ranges, until the changes/improvements in y become negligible.

(b) Compare the maximum y obtained and the total number of tests performed by this method with those given by the PSA method in the chapter. Comment and discuss any advantages and disadvantages between these two methods of experimental optimization.

23.7 Refer to Exercise 21.14 of Chapter 21, where two designs were analyzed to study the effects of seven process variables on chemical yield of the process. The analysis showed that only the main effects of three variables—pressure, catalyst type, and flow rate—are important in determining the yield from the process.

(a) Use the eight available test results of the mirror image design with the variables renumbered as shown below, to develop a PSA and display it graphically in coded units. (Often it pays to review the experimental results carefully before marching into modeling.)

(b) With no opportunity for running any additional tests, you can only calculate a predicted response at each step on the PSA that you have established in part (a) above. Calculate an estimate for the % reaction

Test #	x_1 (pressure)	x_2 (catalyst type)	x_3 (flow rate)	y (% reaction)
1	−1	−1	−1	48.6
2	−1	+1	+1	39.6
3	−1	+1	+1	35.3
4	+1	+1	−1	82.0
5	−1	−1	−1	48.2
6	+1	+1	−1	80.5
7	+1	−1	+1	41.9
8	+1	−1	+1	42.9

on each of the first five steps you plan to explore outside the present experimental region. If two additional tests could actually be run to confirm two of the five predicted responses, which two would you select and why?

23.8 Some experimental results have shown that the following three variables may have significant effects on the weld strength produced by an inertia welding process:

x_1: Axial pressure during welding (in coded units),
x_2: Initial rotating speed of the flywheel (in coded units),
x_3: Total amount of inertia of the rotating mass (in coded units).

Now the engineers want to seek out the optimal levels of the three variables to maximize the weld strength y in ksi. To search for the maximum by RSM, the following planar model has been developed to approximate the surface for PSA exploration:

$$\hat{y} = 91.75 + 6.7x_1 + 5.2x_2 + 10.1x_3.$$

(a) Take three steps from the base along the path of steepest ascent using a step size for x_1 of 0.45 in coded units. Do this by completing the following table.

Path-of-Steepest-Ascent Table

Step	x_1	$x_2 = $ ___ x_1	$x_3 = $ ___ x_1	y
1				
2				
3				

(b) Suppose the actual test results of the weld strength at the three points on the PSA are: 103.8 ksi at the first step, 115.5 ksi at the second step, and 107.9 ksi at the third step. What would you recommend the engineers do next? Be specific and give explanations.

23.9 Suppose the engineers of the weld strength study of Exercise 23.8 have decided to abandon the PSA and to search for the optimum by a second-order model. In preparation for the development of a second-order model, a number of questions must be answered, including (1) Should a variable be dropped, such as x_3? (2) Which design should be chosen, e.g., a three-level design or a central composite design? And so on.

(a) Present your final recommendation with a design matrix of test settings.

(b) Provide a justification for each decision of the choices you made in part (a).

23.10 Using the contour plot of Exercise 23.6 for response approximations, conduct a series of PSA searches until you reach the near-optimum neighborhood. Suppose this time the response surface exploration starts at a concentration value of 15% and a temperature of 155° with a whiteness result of 37.

(a) Compare the responses obtained by this set of PSA searches with those presented in the chapter to see if there are significant differences between the two explorations with different starting points.

(b) Based on the comparisons obtained from part (a), comment on the robustness of the PSA method in search of better responses.

23.11 Second-order polynomial models may represent many widely different shapes. The success or effectiveness of the method of RSM in search of an optimum depends to a large degree on the true shape of the response surface that must be approximated during explorations by both PSA and near-optimum studies.

(a) Draw a contour plot for each of the following three second-order models:

    (i) $\hat{y} = 88 - 9x_1^2 - 2.1x_2^2$

    (ii) $\hat{y} = 88 - 9x_1^2 + 2.1x_2^2$

    (iii) $\hat{y} = 84.3 + 11x_1 + 4x_2 - 9.4x_1x_2 - 6.5x_1^2 - 0.43x_2^2$

(b) Examine each of the three contour plots in part (a) above for possible difficulties that would be encountered when employing RSM for experimental searching for the unknown optimum.

23.12 Whippability, the capacity to form stable foams with air, is an important functional property of proteins for applications in food products such as whipped toppings. Whipping properties are known to be affected by many processing and ingredient variables individually as well as possible interactions among them. A study[4] of the whipping properties of a full-fat soy protein product was undertaken employing the method of

---

[4]C. L. Lah, M. Cheryan, and R. E. DeVor, "A Response Surface Methodology Approach to the Optimization of Whipping Properties of an Ultrafiltered Soy Product," *Journal of Food Science*, Vol. 45, pp. 1720–1731, 1980.

RSM. Through analyses of experimental results of two-level fractional factorial designs of seven independent variables following the method of PSA, the experimenters advanced to the point where they neared the region of optimum "overrun," a measure of whippability. Furthermore, they found that three of the seven variables had become insignificant in this near-optimum region. Based on these analyses, a central composite design, similar to that shown in the table below, was chosen to search for the optimum overrun.

Factors	Symbol	Level (coded units)				
		−2	−1	0	1	2
Protein (%w/v)	x_1	1.0	2.5	4.0	5.5	7.0
Speed (rpm)	x_2	130	140	150	160	170
Time (min)	x_3	2	3	4	5	6
Sugar (g/300g)	x_4	0	4	8	12	16

Note: The other three variables were each set at a constant, desirable value.

In the study, 16 test results were already available from the final PSA step that employed a 2^{7-3} design. As a result, only 15 additional tests were needed to complete a 31-run CCD. The following table is adapted from the whippability study and lists 31 test conditions and the corresponding results for the overrun response.

Test No.	x_1	x_2	x_3	x_4	y (overrun, %)
1	−1	−1	−1	−1	458
2	+1	−1	+1	+1	323
3	−1	+1	+1	+1	355
4	+1	+1	−1	−1	353
5	−1	−1	+1	−1	404
6	+1	−1	−1	+1	360
7	−1	+1	−1	+1	376
8	+1	+1	+1	−1	314
9	−1	−1	−1	+1	404
10	+1	−1	+1	−1	361
11	−1	+1	+1	−1	361
12	+1	+1	−1	+1	388
13	−1	−1	+1	+1	377
14	+1	−1	−1	−1	393
15	−1	+1	−1	−1	427
16	+1	+1	+1	+1	312
17	−2	0	0	0	262

(Continued)

Test No.	x_1	x_2	x_3	x_4	y (overrun, %)
18	2	0	0	0	151
19	0	-2	0	0	441
29	0	2	0	0	348
21	0	0	-2	0	419
22	0	0	2	0	366
23	0	0	0	-2	514
24	0	0	0	2	442
25	0	0	0	0	548
26	0	0	0	0	532
27	0	0	0	0	526
28	0	0	0	0	541
29	0	0	0	0	518
30	0	0	0	0	537
31	0	0	0	0	516

Using the method of least squares, establish a second-order model for the overrun response.

23.13 In the whippability study described in Exercise 23.12 a second-order model for the overrun response was developed. Let us assume that the following model was obtained:

$$\hat{y} = 531.0 - 28.6x_1 - 20.9x_2 - 24.3x_3 - 12.0x_4 - 3.3x_1x_2$$
$$+ 11.9x_1x_3 + 9.4x_1x_4 - 4.8x_2x_3 + 3.3x_2x_4 + 1.1x_3x_4 - 80.0x_1^2$$
$$- 31.6x_2^2 - 31.5x_3^2 - 14.9x_4^2.$$

(a) Suppose the above model was found to adequately fit the data. It is desired to prepare two-dimensional contour plots as a function of x_1 and x_2 for several pairs of fixed values for x_3 and x_4. While many possible combinations of (x_3, x_4) may be held constant in preparing a contour plot within the (x_1, x_2) space let us consider only four pairs of constants for (x_3, x_4). Prepare a contour plot for each of the following settings of (x_3, x_4).

$$(0.0, 0.0) \quad (2.0, 2.0) \quad (-2.0, -2.0) \quad (-0.5, -0.5)$$

(b) Based on the contour plots prepared in part (a), find two alternative optimal settings for the four independent variables that are expected to produce the two highest overrun values.

(c) State some of the advantages of approximating the optimal settings from contour plots, if any, over the use of calculus to find the optimum mathematical solution.

23.14 To follow up your PSA analysis in Exercise 23.10, do the following.

(a) Fit a second-order model using a three-level design and center it around the highest test result on your last PSA.

(b) Plot contours of the estimated whiteness response using the fitted model of (a) above. Compare your findings of the optimum with the one given in the pulp whitening example in the chapter and comment on the robustness of RSM as a method of experimental optimization.

23.15 Consider a new project at its final development stage. You have been requested to propose a plan of experimentation on a new process to select settings for six key process variables that are thought to be important to the output response. The new process is supposed to be an improvement over an existing process with which most members of this project team have rich experience. The goal is to increase the output response value by at least 25%. As always, budget constraints as well as time pressure play a role, such that the total number of experiments to run to complete the study should not be more than about 60. Therefore, in your planning to allocate the 60 runs to each stage of this program of experiments, you need to weigh carefully the following:

(i) Screening experiments to identify/verify the really important variables with the hope that up to three of the initial six variables may be ignored for further experimentation.

(ii) PSA exploration that may require up to three iterations.

(iii) Tests to establish a second-order prediction equation that approximates the response surface in the neighborhood of the optimum.

(iv) Confirmation of the optimal settings of the process variables from (iii) above.

(a) Present your program of experiments in detail, including specific designs for each of the four stages. For each stage, state the basis for your selection of the design.

(b) In your presentation of your plan to the team, what are some of the most important instructions you should provide in explaining the execution of the experiments at each stage to ensure their validity and effectiveness?

(c) Suppose you are relatively new to the process and have limited knowledge of its technological background and operation. Prepare some questions that you may need the team to answer in order to complete your design of experiments at each stage.

(d) Suppose the team is very satisfied with the final results and decides to implement them. Propose some advice to the team as well as people who will be actually running this new process to ensure its continuing success.

23.16 The experiment on fuel economy of a new automobile was first presented in Exercise 20.19. Based on an analysis of the experimental results of a 2^{5-1} design, only three main effects were found to be significant, with the following model being found to fit adequately.

$$\hat{y} = 25.375 - 2.25x_1 + 1.875x_2 - 1.25x_3.$$

The three variables, renumbered for this problem, and their levels are shown in the following table.

		− 1 (Low Level)	+ 1 (High Level)
x_1	Engine size (liters)	4.0	4.5
x_2	Fuel octane no.	87	93
x_3	Air conditioning	Off	On

The original design matrix and all the test results are reproduced in a modified format below.

Test No.	x_1	x_2	x_3	y (mpg)
1	−1	−1	+1	25
2	+1	−1	−1	21
3	−1	+1	−1	26
4	+1	+1	+1	22
5	−1	−1	−1	29
6	+1	−1	+1	20
7	−1	+1	+1	29
8	+1	+1	−1	29
9	−1	−1	−1	27
10	+1	−1	+1	19
11	−1	+1	+1	27
12	+1	+1	−1	28
13	−1	−1	+1	26
14	+1	−1	−1	21
15	−1	+1	−1	32
16	+1	+1	+1	25

Note that this was a 16-test design for a study of five variables. Since two of the five variables were not found to have any significant effects on fuel consumption, we have equivalently a 2^3 design replicated once. It seems logical to analyze the eight average responses of replicates and fit a planar model accordingly.

(a) Fit a prediction model based to the eight average mpg values, compare it to the one obtained in Exercise 20.19, as shown above, and finally comment on any observed differences between these two prediction models.

(b) Someone has suggested that PSA explorations should be pursued in order to find a better way to operate this automobile for maximum mpg. Would you concur? Why or why not?

(c) Suppose it has been agreed to conduct a PSA study for better mpg values. How would you plan to conduct such a study?

(d) Suppose it is desired to find an approximate solution of the optimum mpg, using only the available test results. On the basis of your analysis, recommend a practically feasible way to operate this automobile for maximum mpg.

23.17 Prepare a one-page executive summary for the soldering optimization study of the chapter.

DOE WORKSHOP #3: RESPONSE SURFACE METHODOLOGY, RSMGAME

Introduction

This workshop utilizes RSMGAME, an MS Excel–based Visual Basic program, to simulate a chemical process where the yield is the response of interest. The goal of the workshop is to apply response surface methodology (RSM) to optimize the chemical process yield.

Learning Objective

The objective of this computer workshop is to provide students with practice in the application of response surface methodology.

Background

It is desired to find the optimal settings of temperature, catalyst concentration, and processing time that produce the maximum yield (in grams) for a chemical process and to describe the nature of the response surface in the near-optimum region using response surface methodology (RSM). The chemical process is subject to the following conditions:

1. Experience has led to the conclusion that the temperature must be at least 100°F in order to obtain an appreciable yield. At temperatures above 300°F, the reaction becomes highly unstable.

2. The catalyst employed is expensive and available only in limited quantities. It is therefore desirable to restrict the concentration used to less than 25%.

3. The production schedule requires that the processing time be less than 140 minutes per batch.

At present, the settings for the three variables are

Temperature:	180°F
Catalyst concentration:	15%
Time:	30 min

which gives rise to an average yield of 60 grams. It is believed that a yield of about 95 grams is achievable for the process.

Description of RSMGAME

The RSMGAME program has been developed to simulate the yield of a chemical process. The program produces a yield (in grams) as a test response for user-specified levels of the three process variables.

Prior to running RSMGAME, the student must plan an experiment near the current settings for the process operation. After responses are obtained from RSMGAME, the path of steepest ascent can be determined. Using this path and the results of subsequent experiments, based on the techniques of RSM, the near-optimal region for the response/yield should be determined. Ultimately, a model for the response surface in the near-optimum region should be established, and recommended settings for the process variables selected. Validation tests should also be performed to confirm that the recommended settings produce the desired yield levels.

On the Web

RSMGAME is available for download on the Web at http://www.me.mtu.edu/quality_book/. Also available at the site is a more detailed workshop explanation and description of RSMGAME usage. The website also has a suggested student assignment in connection with the RSMGAME workshop.

References and
Further Readings

Aaron, H. B., "Quality for Today and Tomorrow," seminar presented to Lafayette, Ind., ASQC Section, Mar. 17, 1988.

Adams, C. W., P. Gupta, and C. E. Wilson, Jr., *Six Sigma Development*, Butterworth-Heinemann, 2003.

The Asahi, Japanese Language Newspaper, Apr. 17, 1979. Reported by G. Taguchi during lectures at AT&T Bell Laboratories in 1980.

Asao, M., "A Simulation Game of Experimental Design," *Quality Engineering*, Vol. 4, No. 4, 1992, pp. 497–517.

Babin, T. S., "Modeling and Optimization in Electronic Assembly Operations," *Proceedings of the Second Industrial Engineering Research Conference*, May 1993, pp. 143–149.

Barker, T. B., *Quality by Experimental Design*, 3rd ed., CRC Press, Boca Raton, Fla., 2005.

Bartlett, M. S., "The Use of Transformations," *Biometrics*, Vol. 3, 1947, pp. 39–52.

Belaire, P. M., and Deacon, R. J., "The Strategic Approach to Quality Improvement Using Design of Experiments Concepts," *Symposium on Quality: Design, Planning, and Control, ASME Winter Annual Meeting*, December 13–18, 1987, PED Vol. 27, pp. 147–157.

Berger, P. D., and R. E. Maurer, *Experimental Design with Applications in Management, Engineering, and the Sciences*, Duxbury, Pacific Grove, Calif 2002.

Box, G. E. P., and N. R. Draper, *Empirical-Model Building and Response Surfaces*, John Wiley, New York, 1987.

Box, G. E. P., and N. R. Draper, *Evolutionary Operation: A Statistical Method for Process Improvement*, Wiley, New York, 1969.

Box, G. E. P., and J. S. Hunter, "Signal-to-Noise Ratios, Performance Criteria, and Transformations," *Technometrics*, Vol. 30, No. 1, 1988, pp. 1–40.

Box, G. E. P., and J. S. Hunter, "The 2^{k-p} Fractional Factorial Designs," Parts I and II, *Technometrics*, Vol. 3, 1961, pp. 331–351 and 449–458.

Box, G. E. P., W. G. Hunter, and J. S. Hunter, *Statistics for Experimenters*, Wiley, New York, 1978.

Box, G. E. P., and R. D. Meyer, "An Analysis for Unreplicated Fractional Factorials," *Technometrics*, Vol. 28, 1986, pp. 11–18.

Box, G. E. P., and R. D. Meyer, "Dispersion Effects from Fractional Designs," *Technometrics*, Vol. 28, 1986, pp. 19–27.

Box, G. E. P., and R. D. Meyer, "Finding the Active Factors in Fractionated Screening Experiments," *J. of Quality Technology*, Vol. 25, 1993, pp. 94–105.

Brady, J. E., and T. T. Allen, "Case Based Instruction of DOE and SPC," *The American Statistician*, Vol. 56, 2002, pp. 312–315.

Burr, I. W., *Statistical Quality Control Methods*, Marcel Dekker, New York, 1976.

Castillo, E. D., *Statistical Process Adjustment for Quality Control*, Wiley, New York, 2002.

Chan, L. K., S. W. Cheng, and F. A. Spiring, "A New Measure of Process Capability: C_{pm}," *Journal of Quality Technology*, Vol. 20, July 1988, pp. 162–175.

Chan, L. K., S. W. Cheng, and Z. M. Wang, "Applications of Loss Function and Tolerance Design," *ASQC Quality Congress Transactions*, 1991, pp. 539–546.

Chan, L. K., and P. H. Xiao, "Combined Robust Design," *Quality Engineering*, Vol. 8, No. 1, 1995, pp. 47–56.

Chandra, M. J., *Statistical Quality Control*, CRC Press, Boca Raton, Fla., 2001.

Charbonneau, H. C., and G. L. Webster, *Industrial Quality Control*, Prentice Hall, Englewood Cliffs, N.J., 1978.

Chen, A., and E. A. Elsayed, "Design and Performance Analysis of the Exponentially Weighted Moving Average Mean Estimate for Processes Subject to Random Step Changes," *Technometrics*, Vol. 44, 2002, pp. 379–389.

Chow, E. T. S., L. S. Wei, R. E. DeVor, and M. P. Steinberg, "Application of Two-Level Fractional Factorial Designs in Development of a Soybean Whipped Topping," *Journal of Food Science*, Vol. 48, 1983, p. 230.

Clausing, D. P., "Taguchi Methods Integrated into the Improved Total Development," *Proc. IEEE International Conference on Communications*, Philadelphia, June 1988, pp. 826–832.

Cobb, G. W., *Introduction to Design and Analysis of Experiments*, Springer-Verlag, New York, 1998.

"Continuing Process Control and Process Improvement," Corporate Quality Office, Ford Motor Company, Dearborn, Mich., Dec. 1987.

Conway, W., Videotape of presentation to Ford Motor Company executives, *Ford Motor Company, Radio, TV, and Film Dept.*, File No. 81V44-1, Dearborn, Mich., 1981.

Corbett, J., M. Dooner, J. Meleka, and C. Pym, *Design for Manufacture: Strategies, Principles and Techniques*, Addison-Wesley, Reading, Mass., 1991.

Cox, D. R., "The Use of Range in Sequential Analysis," *Journal of the Royal Statistical Society, Ser. B*, Vol. 11, 1949.

Czitron, V., "One-Factor-at-a-Time Versus Designed Experiments," *The American Statistician*, Vol. 53, 1999, pp. 126–131.

Daniel, C., *Applications of Statistics to Industrial Experimentation*, Wiley, New York, 1976.

Davies, O. L., ed., *The Design and Analysis of Industrial Experiments*, Hafner, Macmillan, 1954.

Davis, R. B., and W. H. Woodall, "Performance of the Control Chart Trend Rule Under Linear Shift," *Journal of Quality Technology*, Vol. 20, Oct. 1988, pp. 260–262.

Deacon, R. J., "Quality Improvement in an Accounts Payable System," Independent Study Project Report, Eastern Michigan University, College of Business, Ypsilanti, Mich., 1986.

Deacon, R. J., "Reduction in Visual Defects in an Injection Molding Process," *Society of Plastics Engineers, 1983 National Technical Conference Proceedings*, p. 97.

Deacon, R. J., and P. M. Belaire, "Quality Improvement: Design of Experiments Methodology," *IMPRO 88 Conference Proceedings*, Juran Institute, Chicago, 1988.

Dehnad, K., *Quality Control, Robust Design, and the Taguchi Method*, Wadsworth & Brooks/Cole, Advanced Book and Software, Pacific Grove, Calif., 1989.

Deming, W. E., "On the Distinction Between Enumerative and Analytical Surveys," *J. of American Statistical Association*, Vol. 48, 1953, pp. 244–255.

Deming, W. E., "On the Use of Judgement-Samples," *Reports of Statistical Application Research, JUSE*, Vol. 23, No. 1, March 1976, pp. 25–31.

Deming, W. E., *Out of the Crisis*, Massachusetts Institute of Technology, Center for Advanced Engineering Study, Cambridge, Mass., 1982.

Deming, W. E., *Quality, Productivity and Competitive Position*, Massachusetts Institute of Technology, Center for Advanced Engineering Study, Cambridge, Mass., 1982.

Deming, W. E., "The Logic of Evaluation," *Handbook of Evaluation Research*, Vol. 1, E. L. Struening and M. Guttentag, eds., Sage Publications, Beverly Hills, Calif., 1975, pp. 53–68.

Dessouky, M. I., S. G. Kapoor, and R. E. DeVor, "A Methodology for Integrated Quality Systems," *J. of Eng. for Ind., Trans. ASME*, Vol. 109, Feb. 1987, pp. 241–247.

DeVor, R. E., S. G. Kapoor, R. Hayashida, G. Subramani, and T. Lindem, "A Methodology for the Simultaneous Engineering of Products and Manufacturing Processes," *Proc. 14th North American Manufacturing Research Conference*, 1986, pp. 535–542.

Dodge, H. F., and H. G. Romig, *Sampling Inspection Tables: Single and Double Sampling*, 2nd ed., Wiley, New York, 1959.

Dooley, K. J., S. G. Kapoor, M. I. Dessouky, and R. E. DeVor, "An Integrated Quality Systems Approach to Quality and Productivity Improvement in Continuous Manufacturing Processes," *J. of Eng. for Ind., Trans. ASME*, Vol. 108, Nov. 1986, pp. 322–327.

Draper, N. R., "Center Points in Second-Order Response Surface Designs," *Technometrics*, Vol. 24, No. 2, 1982, pp. 127–133.

Draper, N. R., "Small Composite Designs," *Technometrics*, Vol. 27, 1985, pp. 187–194.

Draper, N. R., and D. K. J. Lin, "Small Response-Surface Designs," *Technometrics*, Vol. 32, 1990, pp. 187–194.

Draper, N. R., and H. Smith, *Applied Regression Analysis*, Wiley, New York, 1966.

Duncan, A. J., *Quality Control and Industrial Statistics*, 4th ed., Richard D. Irwin, Homewood, Ill., 1974.

Eckes, G., *The Six Sigma Revolution*, Wiley, New York, 2001.

Endres, W. J., J. W. Sutherland, and R. E. DeVor, "Quality Design Using a Computer-Based Dynamic Force Model for the Turning Process," *Proc. 6th International Conference on Computer-Aided Production Engineering*, Nov. 1990, pp. 29–42.

Feigenbaum, A. V., *Total Quality Control*, McGraw-Hill, New York, 1983.

Finney, D. J., "The Fractional Replication of Factorial Arrangements," *Annals of Eugenics*, Vol. 12, No. 4, 1945, pp. 291–301.

Fisher, R. A., *The Design of Experiments*, 8th ed., Hafner, New York, 1966.

"Ford Worldwide Quality System Standard Q-101," Corporate Quality Office, Ford Motor Company, Dearborn, Mich., 1990.

Forzan, C., B. R. Koko, F. Salvadov, W. Nie, and M. Rungtusanatham, "TQM Across Multiple Countries: Convergence Hypothesis Versus National Specificity Arguments," *J. of Operations Management*, Vol. 23, 2005, pp. 43–63.

Friedman, T. L., *The World Is Flat: A Brief History of the Twenty-first Century*, Farrar, Straus, and Giroux, New York, 2005.

Freund, R. A., "Definitions and Basic Quality Concepts," *Journal of Quality Technology*, Vol. 17, Jan. 1985, pp. 50–56.

Fu, H. J., R. E. DeVor, and S. G. Kapoor, "A Mechanistic Model for the Prediction of the Force System in Face Milling Operations," *J. Engr. for Ind., Trans. ASME*, Vol. 106, Feb. 1984, pp. 81–88.

Gitlow, H., and S. Gitlow, *The Deming Guide to Quality and Competitive Position*, Prentice Hall, Englewood Cliffs, N.J., 1987.

Gitlow, H., S. Gitlow, A. Oppenheim, and R. Oppenheim, *Tools and Methods for the Improvement of Quality*, Richard D. Irwin, Homewood, Ill., 1989.

Grant, E. L., and R. S. Leavenworth, *Statistical Quality Control*, 6th ed., McGraw-Hill, New York, 1988.

Haaland, P. D., *Experimental Design in Biotechnology*, Marcel Dekker, New York, 1989.

Hahn, G. J., "Statistical Assessment of a Process Change," *Journal of Quality Technology*, Vol. 14, Jan. 1982, pp. 1–9.

Hahn, G. J., N. Daganaksoy, and R. Hoerl, "The Evolution of Six Sigma," *Quality Engineering*, Vol. 12, 2000, pp. 317–326.

Harry, M., and R. Schroeder, *Six Sigma*, Doubleday, New York, 2000.

Hicks, C. R., *Fundamental Concepts in the Design of Experiments*, 3rd ed., Holt, Rinehart, & Winston, New York, 1982.

Hoerl, R., and R. Snee, *Statistical Thinking: Improving Business Performance*, Duxbury, Pacific Grove, Calif., 2001.

Hunter, J. S., "Statistical Design Applied to Product Design," *Journal of Quality Technology*, Vol. 17, No. 4, 1985, pp. 210–221.

Hunter, J. S., "The Exponentially Weighted Moving Average," *Journal of Quality Technology*, Vol. 18, Oct. 1986, pp. 203–209.

Hunter, R. G., J. W. Sutherland, and R. E. DeVor, "A Methodology for Robust Design Using Models for the Mean, Variance, and Loss," *ASME Symposium on Quality and Performance: Design, Evaluation, and Improvements, ASME Winter Annual Meeting*, December 1989, PED Vol. 42, pp. 25–42.

Ishikawa, K., *Guide to Quality Control*, Asian Productivity Organization, 2nd ed., Tokyo, 1982.

ISO (International Organization for Standardization), http://www.iso.org/iso/en/iso 9000-14000/understand/selection_use/selection_use.html (accessed December 2005).

Jeang, A., and E. Lue, "Robust Tolerance Design by Response Surface Methodology," *International Journal of Production Research*, Vol. 15, 1999, pp. 399–403.

Jessup, P. T., "The Value of Continuing Improvement " *Proc. IEEE International Communications Conference, ICC 85*, 1985.

Jiang, B. C., J. T. Black, D. W. H. Chen, and J. N. Hool, "Taguchi-Based Methodology for Determining/Optimizing Robust Process Capability," *IIE Transactions*, Vol. 23, 1991, pp. 169–184.

Juran, J. M., and F. M. Gryna, Jr., *Quality Planning and Analysis*, 2nd ed., McGraw-Hill, New York, 1980.

Juran, J. M., F. M. Gryna, Jr., and R. S. Bingham, Jr., *Quality Control Handbook*, 3rd ed., McGraw-Hill, New York, 1974.

Kackar, R. N., "Off-Line Quality Control, Parameter Design and the Taguchi Method," *Journal of Quality Technology*, Vol. 17, 1985, pp. 176–209.

Kackar, R. N., "Taguchi's Quality Philosophy: Analysis and Commentary," *Quality Progress*, Dec. 1986, pp. 21–29.

Kane, V. E., *Defect Prevention: Use of Simple Statistical Tools*, Marcel Dekker, ASQC Quality Press, Milwaukee, Wis., 1989.

Kane, V. E., "Process Capability Indices," *Journal of Quality Technology*, Vol. 18, Jan. 1986, pp. 41–52.

Keats, J. B., and N. F. Hubele, eds., *Statistical Process Control in Automated Manufacturing*, Marcel Dekker, New York, 1989.

Kenett, R., *Modern Industrial Statistics: Design and Control of Quality and Reliability*, Duxbury, Pacific Grove, Calif., 1998.

Kotz, S., and C. R. Lovelace, *Process Capability Indices in Theory and Practice*, Arnold, New York, 1998.

Kotz, S., and N. L. Johnson, *Process Capability Indices*, Chapman and Hall, 1993.

Krismann, C., *Quality Control: An Annotated Bibliography Through 1988*, Quality Resources, White Plains, N.Y., 1990.

Kushler, R. H., and P. Hurley, "Confidence Bounds for Capability Indices," *J. of Quality Technology*, Vol. 24, 1992, pp. 188–195.

Lah, C. L., M. Cheryan, and R. E. DeVor, "A Response Surface Methodology Approach to the Optimization of Whipping Properties of an Ultrafiltered Soy Product," *J. of Food Science*, Vol. 45, 1980, pp. 1720–1731.

Lamprecht, J. L., *Interpreting ISO9001:2000 with Statistical Methodology*, ASQ Quality Press, Milwaukee, Wis., 2001.

Lawson, J., and J. Erjavec, *Modern Statistics for Engineering and Quality Improvement*, Duxbury, Pacific Grove, Calif., 2001.

Lenth, R. V., "Quick and Easy Analysis of Unreplicated Factorials," *Technometrics*, Vol. 31, 1989, pp. 469–473.

Li, W., and C. F. J. Wu, "An Integrated Method of Parameter Design and Tolerance Design," *Quality Engineering*, Vol. 11, No. 3, 1999, pp. 417–425.

Lucas, J. M., and M. S. Saccucci, "Exponentially Weighted Moving Average Control Schemes: Properties and Enhancements," *Technometrics*, Vol. 32, 1990, pp. 1–12.

Matulis, R. J., F. K. McKeith, J. W. Sutherland, and M. S. Brewer, "Sensory Characteristics of Frankfurters as Affected by Fat, Salt, and pH," *J. of Food Science*, Vol. 60, No. 1, 1995, pp. 42–47.

Military Standard 105D, *Sampling Procedures and Tables for Inspection by Attributes*, U.S. Government Printing Office, Washington, D.C., 1963.

Miller, I., and J. E. Freund, *Probability and Statistics for Engineers*, 3rd ed., Prentice Hall, Englewood Cliffs, N.J., 1985.

Moen, R. D., T. W. Nolan, and L. P. Provost, *Improving Quality Through Planned Experimentation*, Wiley, New York, 1989.

Montgomery, D. C., *Design and Analysis of Experiments*, 5th ed., Wiley, New York, 2001.

Montgomery, D. C., *Introduction to Statistical Quality Control*, 4th ed., Wiley, New York, 2000.

Murdoch, J., *Control Charts*, Macmillan, New York, 1979.

Myers, R. H., "Response Surface Methodology: Current Status and Future Directions," *J. of Quality Technology*, Vol. 31, 1999, pp. 30–44.

Myers, R. H., and D. C. Montgomery, *Response Surface Methodology: Process and Product Optimization Using Designed Experiments*, 2nd ed., Wiley, New York, 2002.

Myers, R. H., G. G. Vining, A. Giovannitti-Jensen, and S. L. Myers, "Variance Dispersion of Second-Order Response Surface Designs," *J. of Quality Technology*, Vol. 24, 1992, pp. 1–11.

Nelson, L. S., "Calculation of New Limits for *X*bar, *R* Charts When Subgroup Size Is Changed," *Journal of Quality Technology*, Vol. 2, Apr. 1988, pp. 149–150.

Nelson, L. S., "The Calculation of Process Capability Indices," *J. of Quality Technology*, Vol. 31, 1999, pp. 249–250.

Nelson, L. S., "Control Charts for Individual Measurements," *Journal of Quality Technology*, Vol. 14, July 1982, pp. 172–173.

Nelson, L. S., "Control Charts: Rational Subgroups and Effective Applications," *Journal of Quality Technology*, Vol. 20, Jan. 1988, pp. 73–75.

Nelson, L. S., "An Early-Warning Test for Use with the Shewhart *p* Control Chart," *Journal of Quality Technology*, Vol. 15, Apr. 1983, pp. 68–71.

Nelson, L. S., "Interpreting Shewhart *X*bar Control Charts," *Journal of Quality Technology*, Vol. 17, Apr. 1985, pp. 114–116.

Nelson, L. S., "The Shewhart Control Chart: Tests for Special Causes," *Journal of Quality Technology*, Vol. 16, No. 4, 1984, pp. 237–239.

Newman, D., "The Distribution of the Range from a Normal Population, Expressed in Terms of an Independent Estimate of Standard Deviation," *Biometrika*, Vol. 31, 1939, pp. 20–30.

Ng, C. H., and K. E. Case, "Development and Evaluation of Control Charts Using Exponentially Weighted Moving Averages," *Journal of Quality Technology*, Vol. 21, No. 4, 1989, pp. 242–250.

NIST (National Institute of Standards and Technology), http://www.quality.nist.gov/ (accessed September 2005) .

Oakland, J. S., *Statistical Process Control*, William Heinemann, London, England, 1986.

Oh, H. L., "Variation Tolerant Design," *Symposium on Quality: Design, Planning, and Control, ASME Winter Annual Meeting*, Dec. 13–18, 1987, PED, Vol. 27, pp. 137–146.

Ott, E. R., *Process Quality Control: Trouble Shooting and Interpretation of Data*, McGraw-Hill, New York, 1975.

Ott, E. R., and E. G. Schilling, *Process Quality Control*, 2nd ed., McGraw-Hill, New York, 1990.

Pande, P., R. Neuman, and R. Cavanagh, *The Six Sigma Way*, McGraw-Hill, New York, 2000.

Parlar, M., and G. Wesolowsky, "Specification Limits, Capability Indices, and Process Centering in Assembly Manufacturing," *J. of Quality Technology*, Vol. 31, 1999, pp. 317–325.

Phadke, M. S., *Quality Engineering Using Robust Design*, Prentice Hall, Englewood Cliffs, N.J., 1989.

Phillips, M. D., and B. R. Cho, "An Empirical Approach to Designing Product Specifications: A Case Study," *Quality Engineering*, Vol. 11, No.1, 1998, pp. 91–100.

Pignatiello, J. J., and J. S. Ramberg, "Top Ten Triumphs and Tragedies of Genichi Taguchi," *Quality Engineering*, Vol. 4, 1997, pp. 211–225.

Plackett, R. L., and J. P. Burman, "Design of Optimal Multifactorial Experiments," *Biometrika*, Vol. 23, 1946, pp. 305–325.

"Planning for Quality," Corporate Quality Office, Ford Motor Company, Dearborn, Mich., Apr. 1990.

Proceedings of Supplier Symposia on Taguchi Methods, Apr. 1984, Nov. 1984, Oct. 1985–1989, Nov. 1990; American Supplier Institute, Inc., Dearborn, Mich.

Pugh, S., *Tool Design: Integrated Methods for Successful Product Engineering*, Addison-Wesley, Reading, Mass., 1991.

Quesenberry, C. P., *SPC Methods for Quality Improvement*, John Wiley and Sons, New York, 1997.

Quesenberry, C. P., "SPC Q Charts for a Binomial Parameter p: Short or Long Runs," *Journal of Quality Technology*, Vol. 23, 1991, pp. 239–246.

Quesenberry, C. P., "SPC Q Charts for Start-Up and Short or Long Runs," *Journal of Quality Technology*, Vol. 23, 1991, pp. 217–224.

Rahn, G. S., S. G. Kapoor, and R. E. DeVor, "Development of Operational Characteristics and Diagnostics for $\overline{X}$-Control Charts," *J. Engr. for Ind., Trans. ASME*, 1992.

Rao, C. R., "Factorial Experiments Derivable from Combinatorial Arrangements of Arrays," *Journal of the Royal Statistical Society, Ser. B*, Vol. 9, 1947, pp. 128–140.

Rekab, K., and S. Muzaffar, *Statistical Design of Experiments with Engineering Applications*, CRC Press, Boca Raton, Fla., 2005.

Roberts, S. W., "Control Charts Based on Geometric Moving Averages," *Technometrics*, Vol. 1, 1959, pp. 234–250.

Robinson, A., ed., *Continuous Improvement in Operations: A Systematic Approach to Waste Reduction*, Productivity Press, Cambridge, Mass., 1991.

Rodriguez, R. N., "Recent Development in Process Capability Analysis," *J. of Quality Technology*, Vol. 24, No. 4, 1992, pp. 167–187.

Roy, R. K., *Design of Experiments Using the Taguchi Approach: 16 Steps to Product and Process Improvement*, Wiley, New York, 2001.

Ryan, T. P., *Statistical Methods for Quality Improvement*, Wiley, New York, 1989.

Sadsworth, H. M., ed., *Taguchi Methods (Special Issue of Quality and Reliability Engineering)*, Wiley-Interscience, Chichester, England, 1988.

Scheffe, H., *The Analysis of Variance*, Wiley, New York, 1959.

Scherkenbach, W. W., *The Deming Route to Quality and Productivity: Road Maps and Roadblocks*, Mercury Press/Fairchild Publications, New York, 1987.

Scherkenbach, W., *Deming's Road to Continual Improvement*, SPC Press, Knoxville, Tenn., 1991.

Schey, J. A., *Introduction to Manufacturing Processes*, 2nd ed., McGraw-Hill, New York, 1987.

Senge, P. M., *The Fifth Discipline: The Art and Practice of the Learning Organization*, Doubleday, New York, 1990.

Shewhart, W. A., *Economic Control of Quality of Manufactured Products*, D. Van Nostrand, Princeton, N.J., 1931. Reprinted by the American Society for Quality Control, Milwaukee, Wis.

Shewhart, W. A., in *Statistical Method from the Viewpoint of Quality Control*, W. E. Deming, ed., The Graduate School, Department of Agriculture, Washington, D.C., 1939.

Smith, G. M., *Statistical Process Control and Quality Improvement*, 4th ed., Prentice Hall, Upper Saddle River, NJ, 2001.

Statistical Quality Control Handbook, Western Electric Company, Newark, N.J., 1956.

Stegner, D. A. J., S. M. Wu, and N. R. Braton, "Prediction of Heat Input for Welding," *Welding J. Res. Suppl.*, 1967.

"Student" (Gosset, W. S.), "The Probable Error of a Mean," *Biometrika*, Vol. 6, No. 1, 1908, pp. 1–25.

Sullivan, L. P., "Quality Function Deployment," *Quality Progress*, June 1986, pp. 39–50.

Sullivan, L. P., "Reducing Variability: A New Approach to Quality," *Quality Progress*, Vol. 17, No. 7, 1985, pp. 15–21.

Sullivan, L. P., "The Seven Stages of Company-Wide Quality Control," *Quality Progress*, May 1986.

Sutherland, J. W., R. G. Hunter, and R. E. DeVor, "A Quality Engineering-Based Approach to the Simultaneous Engineering of Products and Their Manufacturing Processes," *Proc. 5th International Conference on Computer-Aided Production Engineering*, Nov. 1989, pp. 1–12.

Sweet, A. L., "Control Charts Using Coupled Exponentially Weighted Moving Averages," *Transactions of the IIE*, Vol. 18, No. 1, 1986, pp. 26–33.

Taguchi, G., *Introduction to Quality Engineering*, Asian Productivity Organization, Tokyo, 1986. Distributed by the American Supplier Institute, Inc., Dearborn, Mich.

Taguchi, G., *On-Line Quality Control During Production*, Japanese Standards Association, Tokyo, 1981.

Taguchi, G., S. Chowdhury, and S. Taguchi, *Robust Engineering*, McGraw-Hill, New York, 1999.

Taguchi, G., A. Elsayed, and T. Hsiang, *Quality Engineering in Production Systems*, McGraw-Hill, New York, 1989.

Taguchi, G., and Y. Wu, *Introduction to Off-Line Quality Control*, Central Japan Quality Control Association, Meieki Nakamura-Ku Magaya, Japan, 1979.

Taylor, F. W., *Principles of Scientific Management*, Harper & Brothers, New York, 1911.

Thompson, J. R., and J. Koronacki, *Statistical Process Control*, 2nd ed., Chapman and Hall/CRC, Boca Raton, Fla 2001.

Tippett, L. H. C., *Applications of Statistical Methods to the Control of Quality in Industrial Production*, Manchester Statistical Society, Manchester, England, 1934.

Tool and Manufacturing Engineers Handbook, 4th ed., Vol. 1, Chapter 2, Society of Manufacturing Engineers, Dearborn, Mich, 1983.

Trietsch, D., *Statistical Quality Control: A Loss Minimization Approach*, World Scientific, Hackensack, NJ, 1999.

Unal, R., and E. B. Dean, *Design for Cost and Quality: The Robust Design Approach*, National Aeronautics & Space Administration, Washington, D.C., 1995.

Vance, L. C., "A Bibliography of Statistical Quality Control Chart Techniques, 1970–1980," *Journal of Quality Technology*, Vol. 15, Apr. 1983, pp. 59–62.

Vardeman, S., and J. A. Cornell, "A Partial Inventory of Statistical Literature on Quality and Productivity Through 1985," *Journal of Quality Technology*, Vol. 19, Apr. 1987, pp. 90–97.

Vuchkov, I. N., and L. N. Boyadjieva, *Quality Improvement with Design of Experiments*, Kluwer Academic Publishers, Dordrecht, The Netherlands, 2001.

Wadsworth, H. M., K. S. Stephens, and A. B. Godfrey, *Modern Methods for Quality Control and Improvement*, Wiley, New York, 1986.

Walpole, R. E., and R. H. Myers, *Probability and Statistics for Engineers and Scientists*, 3rd ed., Macmillan, New York, 1985.

Wheeler, D. J., *Understanding Industrial Experimentation*, Statistical Process Controls, Knoxville, Tenn., 1988.

Wheeler, D. L., T. M. Cavalier, and E. A. Lehtihet, "An Implicit Enumeration Approach to Probabilistic Tolerance Allocation under Conventional Tolerance Control," *International Journal of Production Research*, Vol. 37, 1999, p. 3773.

Wheeler, D. L., T. M. Cavalier, and E. A. Lehtihet, "An Implicit Enumeration Approach to Tolerance Allocation in Sequential Tolerance Control," *IIE Transactions*, Vol. 31, 1999, p. 75.

Woodall, W. H., "Conflicts Between Deming's Philosophy and the Economic Design of Control Charts," *Frontiers in Statistical Quality Control 3*, edited by H. J. Lenz, G. B. Wetherill, and P. T. Wilrich, Physica-Verlag, Heidelberg, Germany, 1987, pp. 242–248.

Woodall, W. W., "Controversies and Contradictions in Statistical Process Control (with Discussions)," *J. of Quality Technology*, Vol. 32, 2000, pp. 341–378.

Working, H., and E. G. Olds, *Manual for an Introduction to Statistical Methods of Quality Control in Industry: Outline of a Course of Lectures and Exercises*, Office of Production Research and Development, War Production Board, Washington, D.C., 1944.

Wortham, A. W., "The Use of Exponentially Smoothed Data in Continuous Process Control," *International Journal of Production Research*, Vol. 10, No. 4, 1972, pp. 393–400.

Wortham, A. W., and G. F. Heinrich, "Control Charts Using Exponential Smoothing Techniques," *Transactions of the AQC*, Vol. 26, 1972, pp. 451–458.

Wu, S. M., "Analysis of Rail Steel Bar Welds by Two-Level Factorial Designs," *Welding Journal Research Supplement*, 1964, pp. 179s–183s.

Appendix Tables

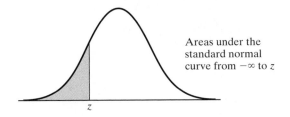

Areas under the standard normal curve from $-\infty$ to z

TABLE A.1	Areas Under the Normal Curve								
z	0.09	0.08	0.07	0.06	0.05	0.04	0.03	0.02	0.01	0.00
−4.00	0.00002	0.00002	0.00002	0.00003	0.00003	0.00003	0.00003	0.00003	0.00003	0.00003
−3.90	0.00003	0.00003	0.00004	0.00004	0.00004	0.00004	0.00004	0.00004	0.00005	0.00005
−3.80	0.00005	0.00005	0.00005	0.00006	0.00006	0.00006	0.00006	0.00007	0.00007	0.00007
−3.70	0.00008	0.00008	0.00008	0.00009	0.00009	0.00009	0.00010	0.00010	0.00010	0.00011
−3.60	0.00011	0.00012	0.00012	0.00013	0.00013	0.00014	0.00014	0.00015	0.00015	0.00016
−3.50	0.00017	0.00017	0.00018	0.00019	0.00019	0.00020	0.00021	0.00022	0.00022	0.00023
−3.40	0.00024	0.00025	0.00026	0.00027	0.00028	0.00029	0.00030	0.00031	0.00033	0.00034
−3.30	0.00035	0.00036	0.00038	0.00039	0.00040	0.00042	0.00043	0.00045	0.00047	0.00048
−3.20	0.00050	0.00052	0.00054	0.00056	0.00058	0.00060	0.00062	0.00064	0.00066	0.00069
−3.10	0.00071	0.00074	0.00076	0.00079	0.00082	0.00085	0.00087	0.00090	0.00094	0.00097
−3.00	0.00100	0.00104	0.00107	0.00111	0.00114	0.00118	0.00122	0.00126	0.00131	0.00135
−2.90	0.0014	0.0014	0.0015	0.0015	0.0016	0.0016	0.0017	0.0018	0.0018	0.0019
−2.80	0.0019	0.0020	0.0021	0.0021	0.0022	0.0023	0.0023	0.0024	0.0025	0.0026
−2.70	0.0026	0.0027	0.0028	0.0029	0.0030	0.0031	0.0032	0.0033	0.0034	0.0035
−2.60	0.0036	0.0037	0.0038	0.0039	0.0040	0.0041	0.0043	0.0044	0.0045	0.0047
−2.50	0.0048	0.0049	0.0051	0.0052	0.0054	0.0055	0.0057	0.0059	0.0060	0.0062
−2.40	0.0064	0.0066	0.0068	0.0069	0.0071	0.0073	0.0075	0.0078	0.0080	0.0082
−2.30	0.0084	0.0087	0.0089	0.0091	0.0094	0.0096	0.0099	0.0102	0.0104	0.0107
−2.20	0.0110	0.0113	0.0116	0.0119	0.0122	0.0125	0.0129	0.0132	0.0136	0.0139
−2.10	0.0143	0.0146	0.0150	0.0154	0.0158	0.0162	0.0166	0.0170	0.0174	0.0179
−2.00	0.0183	0.0188	0.0192	0.0197	0.0202	0.0207	0.0212	0.0217	0.0222	0.0228
−1.90	0.0233	0.0239	0.0244	0.0250	0.0256	0.0262	0.0268	0.0274	0.0281	0.0287
−1.80	0.0294	0.0301	0.0307	0.0314	0.0322	0.0329	0.0336	0.0344	0.0351	0.0359
−1.70	0.0367	0.0375	0.0384	0.0392	0.0401	0.0409	0.0418	0.0427	0.0436	0.0446
−1.60	0.0455	0.0465	0.0475	0.0485	0.0495	0.0505	0.0516	0.0526	0.0537	0.0548
−1.50	0.0559	0.0571	0.0582	0.0594	0.0606	0.0618	0.0630	0.0643	0.0655	0.0668
−1.40	0.0681	0.0694	0.0708	0.0721	0.0735	0.0749	0.0764	0.0778	0.0793	0.0808
−1.30	0.0823	0.0838	0.0853	0.0869	0.0885	0.0901	0.0918	0.0934	0.0951	0.0968
−1.20	0.0985	0.1003	0.1020	0.1038	0.1057	0.1075	0.1093	0.1112	0.1131	0.1151
−1.10	0.1170	0.1190	0.1210	0.1230	0.1251	0.1271	0.1292	0.1314	0.1335	0.1357
−1.00	0.1379	0.1401	0.1423	0.1446	0.1469	0.1492	0.1515	0.1539	0.1562	0.1587
−0.90	0.1611	0.1635	0.1660	0.1685	0.1711	0.1736	0.1762	0.1788	0.1814	0.1841
−0.80	0.1867	0.1894	0.1922	0.1949	0.1977	0.2005	0.2033	0.2061	0.2090	0.2119
−0.70	0.2148	0.2177	0.2207	0.2236	0.2266	0.2297	0.2327	0.2358	0.2389	0.2420
−0.60	0.2451	0.2483	0.2514	0.2546	0.2578	0.2611	0.2643	0.2676	0.2709	0.2743
−0.50	0.2776	0.2810	0.2843	0.2877	0.2912	0.2946	0.2981	0.3015	0.3050	0.3085
−0.40	0.3121	0.3156	0.3192	0.3228	0.3264	0.3300	0.3336	0.3372	0.3409	0.3446
−0.30	0.3483	0.3520	0.3557	0.3594	0.3632	0.3669	0.3707	0.3745	0.3783	0.3821
−0.20	0.3859	0.3897	0.3936	0.3974	0.4013	0.4052	0.4090	0.4129	0.4168	0.4207
−0.10	0.4247	0.4286	0.4325	0.4364	0.4404	0.4443	0.4483	0.4522	0.4562	0.4602
−0.00	0.4641	0.4681	0.4721	0.4761	0.4801	0.4840	0.4880	0.4920	0.4960	0.5000

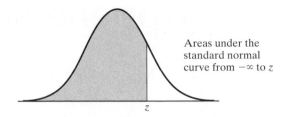

Areas under the
standard normal
curve from $-\infty$ to z

TABLE A.1	(Continued)								
z	0.00	0.01	0.02	0.03	0.04	0.05	0.06	0.07	0.08	0.09
0.00	0.5000	0.5040	0.5080	0.5120	0.5160	0.5199	0.5239	0.5279	0.5319	0.5359
0.10	0.5398	0.5438	0.5478	0.5517	0.5557	0.5596	0.5636	0.5675	0.5714	0.5753
0.20	0.5793	0.5832	0.5871	0.5910	0.5948	0.5987	0.6026	0.6064	0.6103	0.6141
0.30	0.6179	0.6217	0.6255	0.6293	0.6331	0.6368	0.6406	0.6443	0.6480	0.6517
0.40	0.6554	0.6591	0.6628	0.6664	0.6700	0.6736	0.6772	0.6808	0.6844	0.6879
0.50	0.6915	0.6950	0.6985	0.7019	0.7054	0.7088	0.7123	0.7157	0.7190	0.7224
0.60	0.7257	0.7291	0.7324	0.7357	0.7389	0.7422	0.7454	0.7486	0.7517	0.7549
0.70	0.7580	0.7611	0.7642	0.7673	0.7704	0.7734	0.7764	0.7794	0.7823	0.7852
0.80	0.7881	0.7910	0.7939	0.7967	0.7995	0.8023	0.8051	0.8079	0.8106	0.8133
0.90	0.8159	0.8186	0.8212	0.8238	0.8264	0.8289	0.8315	0.8340	0.8365	0.8389
1.00	0.8413	0.8438	0.8461	0.8485	0.8508	0.8531	0.8554	0.8577	0.8599	0.8621
1.10	0.8643	0.8665	0.8686	0.8708	0.8729	0.8749	0.8770	0.8790	0.8810	0.8830
1.20	0.8849	0.8869	0.8888	0.8907	0.8925	0.8944	0.8962	0.8980	0.8997	0.9015
1.30	0.9032	0.9049	0.9066	0.9082	0.9099	0.9115	0.9131	0.9147	0.9162	0.9177
1.40	0.9192	0.9207	0.9222	0.9236	0.9251	0.9265	0.9279	0.9292	0.9306	0.9319
1.50	0.9332	0.9345	0.9357	0.9370	0.9382	0.9394	0.9406	0.9418	0.9429	0.9441
1.60	0.9452	0.9463	0.9474	0.9484	0.9495	0.9505	0.9515	0.9525	0.9535	0.9545
1.70	0.9554	0.9564	0.9573	0.9582	0.9591	0.9599	0.9608	0.9616	0.9625	0.9633
1.80	0.9641	0.9649	0.9656	0.9664	0.9671	0.9678	0.9686	0.9693	0.9699	0.9706
1.90	0.9713	0.9719	0.9726	0.9732	0.9738	0.9744	0.9750	0.9756	0.9761	0.9767
2.00	0.9773	0.9778	0.9783	0.9788	0.9793	0.9798	0.9803	0.9808	0.9812	0.9817
2.10	0.9821	0.9826	0.9830	0.9834	0.9838	0.9842	0.9846	0.9850	0.9854	0.9857
2.20	0.9861	0.9864	0.9868	0.9871	0.9875	0.9878	0.9881	0.9884	0.9887	0.9890
2.30	0.9893	0.9896	0.9898	0.9901	0.9904	0.9906	0.9909	0.9911	0.9913	0.9916
2.40	0.9918	0.9920	0.9922	0.9925	0.9927	0.9929	0.9931	0.9932	0.9934	0.9936
2.50	0.9938	0.9940	0.9941	0.9943	0.9945	0.9946	0.9948	0.9949	0.9951	0.9952
2.60	0.9953	0.9955	0.9956	0.9957	0.9959	0.9960	0.9961	0.9962	0.9963	0.9964
2.70	0.9965	0.9966	0.9967	0.9968	0.9969	0.9970	0.9971	0.9972	0.9973	0.9974
2.80	0.9974	0.9975	0.9976	0.9977	0.9977	0.9978	0.9979	0.9979	0.9980	0.9981
2.90	0.9981	0.9982	0.9983	0.9983	0.9984	0.9984	0.9985	0.9985	0.9986	0.9986
3.00	0.99865	0.99869	0.99874	0.99878	0.99882	0.99886	0.99889	0.99893	0.99897	0.99900
3.10	0.99903	0.99907	0.99910	0.99913	0.99916	0.99918	0.99921	0.99924	0.99926	0.99929
3.20	0.99931	0.99934	0.99936	0.99938	0.99940	0.99942	0.99944	0.99946	0.99948	0.99950
3.30	0.99952	0.99953	0.99955	0.99957	0.99958	0.99960	0.99961	0.99962	0.99964	0.99965
3.40	0.99966	0.99968	0.99969	0.99970	0.99971	0.99972	0.99973	0.99974	0.99975	0.99976
3.50	0.99977	0.99978	0.99978	0.99979	0.99980	0.99981	0.99982	0.99982	0.99983	0.99984
3.60	0.99984	0.99985	0.99985	0.99986	0.99986	0.99987	0.99987	0.99988	0.99988	0.99989
3.70	0.99989	0.99990	0.99990	0.99990	0.99991	0.99991	0.99992	0.99992	0.99992	0.99993
3.80	0.99993	0.99993	0.99993	0.99994	0.99994	0.99994	0.99994	0.99995	0.99995	0.99995
3.90	0.99995	0.99995	0.99996	0.99996	0.99996	0.99996	0.99996	0.99996	0.99997	0.99997
4.00	0.99997	0.99997	0.99997	0.99997	0.99997	0.99997	0.99998	0.99998	0.99998	0.99998

TABLE A.2	Constants for Determining from $\bar{R}$ the 3 Sigma Control Limits for $\bar{X}$ and R Charts and for Estimating the Process Standard Deviation from $\bar{R}$			
Number of Observations in Subgroup/Sample	d_2	A_2	D_3	D_4
2	1.128	1.880	0	3.267
3	1.693	1.023	0	2.575
4	2.059	0.729	0	2.282
5	2.326	0.577	0	2.115
6	2.534	0.483	0	2.004
7	2.704	0.419	0.076	1.924
8	2.847	0.373	0.136	1.864
9	2.970	0.337	0.184	1.816
10	3.078	0.308	0.223	1.777
15	3.472	0.223	0.348	1.652
20	3.735	0.180	0.414	1.586

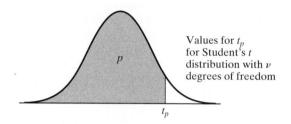

Values for t_p
for Student's t
distribution with ν
degrees of freedom

p

t_p

TABLE A.3		Values of t for Various Probability Levels, p									
ν	$t_{.60}$	$t_{.75}$	$t_{.85}$	$t_{.90}$	$t_{.95}$	$t_{.975}$	$t_{.99}$	$t_{.995}$	$t_{.9975}$	$t_{.999}$	$t_{.9995}$
1	0.325	1.000	1.963	3.078	6.314	12.71	31.82	63.66	127.3	318.3	636.6
2	0.289	0.816	1.386	1.886	2.920	4.303	6.965	9.925	14.09	22.33	31.60
3	0.277	0.765	1.250	1.638	2.353	3.182	4.541	5.841	7.453	10.21	12.92
4	0.271	0.741	1.190	1.533	2.132	2.776	3.747	4.604	5.598	7.173	8.610
5	0.267	0.727	1.156	1.476	2.015	2.571	3.365	4.032	4.773	5.893	6.869
6	0.265	0.718	1.134	1.440	1.943	2.447	3.143	3.707	4.317	5.208	5.959
7	0.263	0.711	1.119	1.415	1.895	2.365	2.998	3.499	4.029	4.785	5.408
8	0.262	0.706	1.108	1.397	1.860	2.306	2.896	3.355	3.833	4.501	5.041
9	0.261	0.703	1.100	1.383	1.833	2.262	2.821	3.250	3.690	4.297	4.781
10	0.260	0.700	1.093	1.372	1.812	2.228	2.764	3.169	3.581	4.144	4.587
11	0.260	0.697	1.088	1.363	1.796	2.201	2.718	3.106	3.497	4.025	4.437
12	0.259	0.695	1.083	1.356	1.782	2.179	2.681	3.055	3.428	3.930	4.318
13	0.259	0.694	1.079	1.350	1.771	2.160	2.650	3.012	3.372	3.852	4.221
14	0.258	0.692	1.076	1.345	1.761	2.145	2.624	2.977	3.326	3.787	4.140
15	0.258	0.691	1.074	1.341	1.753	2.131	2.602	2.947	3.286	3.733	4.073
16	0.258	0.690	1.071	1.337	1.746	2.120	2.583	2.921	3.252	3.686	4.015
17	0.257	0.689	1.069	1.333	1.740	2.110	2.567	2.898	3.222	3.646	3.965
18	0.257	0.688	1.067	1.330	1.734	2.101	2.552	2.878	3.197	3.610	3.922
19	0.257	0.688	1.066	1.328	1.729	2.093	2.539	2.861	3.174	3.579	3.883
20	0.257	0.687	1.064	1.325	1.725	2.086	2.528	2.845	3.153	3.552	3.850
21	0.257	0.686	1.063	1.323	1.721	2.080	2.518	2.831	3.135	3.527	3.819
22	0.256	0.686	1.061	1.321	1.717	2.074	2.508	2.819	3.119	3.505	3.792
23	0.256	0.685	1.060	1.319	1.714	2.069	2.500	2.807	3.104	3.485	3.768
24	0.256	0.685	1.059	1.318	1.711	2.064	2.492	2.797	3.091	3.467	3.745
25	0.256	0.684	1.058	1.316	1.708	2.060	2.485	2.787	3.078	3.450	3.725
26	0.256	0.684	1.058	1.315	1.706	2.056	2.479	2.779	3.067	3.435	3.707
27	0.256	0.684	1.057	1.314	1.703	2.052	2.473	2.771	3.057	3.421	3.690
28	0.256	0.683	1.056	1.313	1.701	2.048	2.467	2.763	3.047	3.408	3.674
29	0.256	0.683	1.055	1.311	1.699	2.045	2.462	2.756	3.038	3.396	3.659
30	0.256	0.683	1.055	1.310	1.697	2.042	2.457	2.750	3.030	3.385	3.646
40	0.255	0.681	1.050	1.303	1.684	2.021	2.423	2.704	2.971	3.307	3.551
50	0.255	0.679	1.047	1.299	1.676	2.009	2.403	2.678	2.937	3.261	3.496
60	0.254	0.679	1.045	1.296	1.671	2.000	2.390	2.660	2.915	3.232	3.460
70	0.254	0.678	1.044	1.294	1.667	1.994	2.381	2.648	2.899	3.211	3.435
80	0.254	0.678	1.043	1.292	1.664	1.990	2.374	2.639	2.887	3.195	3.416
90	0.254	0.677	1.042	1.291	1.662	1.987	2.368	2.632	2.878	3.183	3.402
100	0.254	0.677	1.042	1.290	1.660	1.984	2.364	2.626	2.871	3.174	3.390
110	0.254	0.677	1.041	1.289	1.659	1.982	2.361	2.621	2.865	3.166	3.381
120	0.254	0.677	1.041	1.289	1.658	1.980	2.358	2.617	2.860	3.160	3.373
∞	0.253	0.674	1.036	1.282	1.645	1.960	2.326	2.576	2.807	3.090	3.291

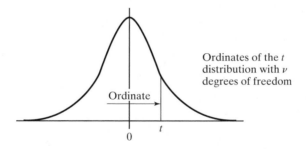

Ordinate

Ordinates of the t distribution with ν degrees of freedom

0 t

TABLE A.4	Ordinates of the t Distribution with ν Degrees of Freedom											
	t Value											
ν	0.00	0.25	0.50	0.75	1.00	1.25	1.50	1.75	2.00	2.25	2.50	2.75	3.00
1	0.318	0.300	0.255	0.204	0.159	0.124	0.098	0.078	0.064	0.053	0.044	0.037	0.032
2	0.354	0.338	0.296	0.244	0.192	0.149	0.114	0.088	0.068	0.053	0.042	0.034	0.027
3	0.368	0.353	0.313	0.261	0.207	0.159	0.120	0.090	0.068	0.051	0.039	0.030	0.023
4	0.375	0.361	0.322	0.270	0.215	0.164	0.123	0.091	0.066	0.049	0.036	0.026	0.020
5	0.380	0.366	0.328	0.276	0.220	0.168	0.125	0.091	0.065	0.047	0.033	0.024	0.017
6	0.383	0.369	0.332	0.280	0.223	0.170	0.126	0.090	0.064	0.045	0.031	0.022	0.015
7	0.385	0.372	0.335	0.283	0.226	0.172	0.126	0.090	0.063	0.044	0.030	0.021	0.014
8	0.387	0.373	0.337	0.285	0.228	0.173	0.127	0.090	0.062	0.043	0.029	0.019	0.013
9	0.388	0.375	0.338	0.287	0.229	0.174	0.127	0.090	0.062	0.042	0.028	0.018	0.012
10	0.389	0.376	0.340	0.288	0.230	0.175	0.127	0.090	0.061	0.041	0.027	0.018	0.011
11	0.390	0.377	0.341	0.289	0.231	0.176	0.128	0.089	0.061	0.040	0.026	0.017	0.011
12	0.391	0.378	0.342	0.290	0.232	0.176	0.128	0.089	0.060	0.040	0.026	0.016	0.010
13	0.391	0.378	0.342	0.291	0.233	0.177	0.128	0.089	0.060	0.039	0.025	0.016	0.010
14	0.392	0.379	0.343	0.292	0.234	0.177	0.128	0.089	0.060	0.039	0.025	0.015	0.009
15	0.392	0.380	0.344	0.292	0.234	0.178	0.128	0.089	0.059	0.038	0.024	0.015	0.009
16	0.393	0.380	0.344	0.293	0.235	0.178	0.128	0.089	0.059	0.038	0.024	0.015	0.009
17	0.393	0.380	0.345	0.293	0.235	0.178	0.128	0.089	0.059	0.038	0.023	0.014	0.009
18	0.393	0.381	0.345	0.294	0.235	0.178	0.129	0.088	0.058	0.037	0.023	0.014	0.008
19	0.394	0.381	0.345	0.294	0.236	0.179	0.129	0.088	0.058	0.037	0.023	0.014	0.008
20	0.394	0.381	0.346	0.294	0.236	0.179	0.129	0.088	0.058	0.037	0.023	0.014	0.008
22	0.394	0.382	0.346	0.295	0.237	0.179	0.129	0.088	0.058	0.036	0.022	0.013	0.008
24	0.395	0.382	0.347	0.296	0.237	0.179	0.129	0.088	0.057	0.036	0.022	0.013	0.007
26	0.395	0.383	0.347	0.296	0.237	0.180	0.129	0.088	0.057	0.036	0.022	0.013	0.007
28	0.395	0.383	0.348	0.296	0.238	0.180	0.129	0.088	0.057	0.036	0.021	0.012	0.007
30	0.396	0.383	0.348	0.297	0.238	0.180	0.129	0.088	0.057	0.035	0.021	0.012	0.007
35	0.396	0.384	0.348	0.297	0.239	0.180	0.129	0.088	0.056	0.035	0.021	0.012	0.006
40	0.396	0.384	0.349	0.298	0.239	0.181	0.129	0.087	0.056	0.034	0.020	0.011	0.006
45	0.397	0.384	0.349	0.298	0.239	0.181	0.129	0.087	0.056	0.034	0.020	0.011	0.006
50	0.397	0.385	0.350	0.298	0.240	0.181	0.129	0.087	0.056	0.034	0.020	0.011	0.006
∞	0.399	0.387	0.352	0.301	0.242	0.183	0.130	0.086	0.054	0.032	0.018	0.009	0.004

Values of χ_p^2
for the chi-square
distribution with ν
degrees of freedom

TABLE A.5	Values for χ^2 for Various Probability Levels														
ν	$\chi^2_{.001}$	$\chi^2_{.005}$	$\chi^2_{.010}$	$\chi^2_{.025}$	$\chi^2_{.050}$	$\chi^2_{.100}$	$\chi^2_{.250}$	$\chi^2_{.500}$	$\chi^2_{.750}$	$\chi^2_{.900}$	$\chi^2_{.950}$	$\chi^2_{.975}$	$\chi^2_{.990}$	$\chi^2_{.995}$	$\chi^2_{.999}$
1	0.0000	0.0000	0.0002	0.0010	0.0039	0.0158	0.102	0.455	1.32	2.71	3.84	5.02	6.63	7.88	10.8
2	0.0020	0.0100	0.0201	0.0506	0.103	0.211	0.575	1.39	2.77	4.61	5.99	7.38	9.21	10.6	13.8
3	0.0243	0.0717	0.115	0.216	0.352	0.584	1.21	2.37	4.11	6.25	7.81	9.35	11.3	12.8	16.3
4	0.0908	0.207	0.297	0.484	0.711	1.06	1.92	3.36	5.39	7.78	9.49	11.1	13.3	14.9	18.5
5	0.210	0.412	0.554	0.831	1.15	1.61	2.67	4.35	6.63	9.24	11.1	12.8	15.1	16.7	20.5
6	0.381	0.676	0.872	1.24	1.64	2.20	3.45	5.35	7.84	10.6	12.6	14.4	16.8	18.5	22.5
7	0.598	0.989	1.24	1.69	2.17	2.83	4.25	6.35	9.04	12.0	14.1	16.0	18.5	20.3	24.3
8	0.857	1.34	1.65	2.18	2.73	3.49	5.07	7.34	10.2	13.4	15.5	17.5	20.1	22.0	26.1
9	1.15	1.73	2.09	2.70	3.33	4.17	5.90	8.34	11.4	14.7	16.9	19.0	21.7	23.6	27.9
10	1.48	2.16	2.56	3.25	3.94	4.87	6.74	9.34	12.5	16.0	18.3	20.5	23.2	25.2	29.6
11	1.83	2.60	3.05	3.82	4.57	5.58	7.58	10.3	13.7	17.3	19.7	21.9	24.7	26.8	31.3
12	2.21	3.07	3.57	4.40	5.23	6.30	8.44	11.3	14.8	18.5	21.0	23.3	26.2	28.3	32.9
13	2.62	3.57	4.11	5.01	5.89	7.04	9.30	12.3	16.0	19.8	22.4	24.7	27.7	29.8	34.5
14	3.04	4.07	4.66	5.63	6.57	7.79	10.2	13.3	17.1	21.1	23.7	26.1	29.1	31.3	36.1
15	3.48	4.60	5.23	6.26	7.26	8.55	11.0	14.3	18.2	22.3	25.0	27.5	30.6	32.8	37.7
16	3.94	5.14	5.81	6.91	7.96	9.31	11.9	15.3	19.4	23.5	26.3	28.8	32.0	34.3	39.3
17	4.42	5.70	6.41	7.56	8.67	10.1	12.8	16.3	20.5	24.8	27.6	30.2	33.4	35.7	40.8
18	4.90	6.26	7.01	8.23	9.39	10.9	13.7	17.3	21.6	26.0	28.9	31.5	34.8	37.2	42.3
19	5.41	6.84	7.63	8.91	10.1	11.7	14.6	18.3	22.7	27.2	30.1	32.9	36.2	38.6	43.8
20	5.92	7.43	8.26	9.59	10.9	12.4	15.5	19.3	23.8	28.4	31.4	34.2	37.6	40.0	45.3
21	6.45	8.03	8.90	10.3	11.6	13.2	16.3	20.3	24.9	29.6	32.7	35.5	38.9	41.4	46.8
22	6.98	8.64	9.54	11.0	12.3	14.0	17.2	21.3	26.0	30.8	33.9	36.8	40.3	42.8	48.3
23	7.53	9.26	10.2	11.7	13.1	14.8	18.1	22.3	27.1	32.0	35.2	38.1	41.6	44.2	49.7
24	8.08	9.89	10.9	12.4	13.8	15.7	19.0	23.3	28.2	33.2	36.4	39.4	43.0	45.6	51.2
25	8.65	10.5	11.5	13.1	14.6	16.5	19.9	24.3	29.3	34.4	37.7	40.6	44.3	46.9	52.6
26	9.22	11.2	12.2	13.8	15.4	17.3	20.8	25.3	30.4	35.6	38.9	41.9	45.6	48.3	54.1
27	9.80	11.8	12.9	14.6	16.2	18.1	21.7	26.3	31.5	36.7	40.1	43.2	47.0	49.6	55.5
28	10.4	12.5	13.6	15.3	16.9	18.9	22.7	27.3	32.6	37.9	41.3	44.5	48.3	51.0	56.9
29	11.0	13.1	14.3	16.0	17.7	19.8	23.6	28.3	33.7	39.1	42.6	45.7	49.6	52.3	58.3
30	11.6	13.8	15.0	16.8	18.5	20.6	24.5	29.3	34.8	40.3	43.8	47.0	50.9	53.7	59.7
40	17.9	20.7	22.2	24.4	26.5	29.1	33.7	39.3	45.6	51.8	55.8	59.3	63.7	66.8	73.4
50	24.7	28.0	29.7	32.4	34.8	37.7	42.9	49.3	56.3	63.2	67.5	71.4	76.2	79.5	86.7
60	31.7	35.5	37.5	40.5	43.2	46.5	52.3	59.3	67.0	74.4	79.1	83.3	88.4	92.0	99.6
70	39.0	43.3	45.4	48.8	51.7	55.3	61.7	69.3	77.6	85.5	90.5	95.0	100	104	112
80	46.5	51.2	53.5	57.2	60.4	64.3	71.1	79.3	88.1	96.6	102	107	112	116	125
90	54.2	59.2	61.8	65.6	69.1	73.3	80.6	89.3	98.6	108	113	118	124	128	137
100	61.9	67.3	70.1	74.2	77.9	82.4	90.1	99.3	109	118	124	130	136	140	149
110	69.8	75.6	78.5	82.9	86.8	91.5	99.7	109	120	129	135	141	147	152	162
120	77.8	83.9	86.9	91.6	95.7	101	109	119	130	140	147	152	159	164	174

Values of F_p
for the F distribution
with ν_1 and ν_2
degrees of freedom

TABLE A.6 Critical Values of F for a Probability Level, p, of 0.75

ν_2 \ ν_1	1	2	3	4	5	6	7	8	9	10	12	15	20	24	30	40	60	120	∞
1	5.83	7.50	8.20	8.58	8.82	8.98	9.10	9.19	9.26	9.32	9.41	9.49	9.58	9.63	9.67	9.71	9.76	9.80	9.85
2	2.57	3.00	3.15	3.23	3.28	3.31	3.34	3.35	3.37	3.38	3.39	3.41	3.43	3.43	3.44	3.45	3.46	3.47	3.48
3	2.02	2.28	2.36	2.39	2.41	2.42	2.43	2.44	2.44	2.44	2.45	2.46	2.46	2.46	2.47	2.47	2.47	2.47	2.47
4	1.81	2.00	2.05	2.06	2.07	2.08	2.08	2.08	2.08	2.08	2.08	2.08	2.08	2.08	2.08	2.08	2.08	2.08	2.08
5	1.69	1.85	1.88	1.89	1.89	1.89	1.89	1.89	1.89	1.89	1.89	1.89	1.88	1.88	1.88	1.88	1.87	1.87	1.87
6	1.62	1.76	1.78	1.79	1.79	1.78	1.78	1.78	1.77	1.77	1.77	1.76	1.76	1.75	1.75	1.75	1.74	1.74	1.74
7	1.57	1.70	1.72	1.72	1.71	1.71	1.70	1.70	1.69	1.69	1.68	1.68	1.67	1.67	1.66	1.66	1.65	1.65	1.65
8	1.54	1.66	1.67	1.66	1.66	1.65	1.64	1.64	1.63	1.63	1.62	1.62	1.61	1.60	1.60	1.59	1.59	1.58	1.58
9	1.51	1.62	1.63	1.63	1.62	1.61	1.60	1.60	1.59	1.59	1.58	1.57	1.56	1.56	1.55	1.54	1.54	1.53	1.53
10	1.49	1.60	1.60	1.59	1.59	1.58	1.57	1.56	1.56	1.55	1.54	1.53	1.52	1.52	1.51	1.51	1.50	1.49	1.48
11	1.47	1.58	1.58	1.57	1.56	1.55	1.54	1.53	1.53	1.52	1.51	1.50	1.49	1.49	1.48	1.47	1.47	1.46	1.45
12	1.46	1.56	1.56	1.55	1.54	1.53	1.52	1.51	1.51	1.50	1.49	1.48	1.47	1.46	1.45	1.45	1.44	1.43	1.42
13	1.45	1.55	1.55	1.53	1.52	1.51	1.50	1.49	1.49	1.48	1.47	1.46	1.45	1.44	1.43	1.42	1.42	1.41	1.40
14	1.44	1.53	1.53	1.52	1.51	1.50	1.49	1.48	1.47	1.46	1.45	1.44	1.43	1.42	1.41	1.41	1.40	1.39	1.38
15	1.43	1.52	1.52	1.51	1.49	1.48	1.47	1.46	1.46	1.45	1.44	1.43	1.41	1.41	1.40	1.39	1.38	1.37	1.36
16	1.42	1.51	1.51	1.50	1.48	1.47	1.46	1.45	1.44	1.44	1.43	1.41	1.40	1.39	1.38	1.37	1.36	1.35	1.34
17	1.42	1.51	1.50	1.49	1.47	1.46	1.45	1.44	1.43	1.43	1.41	1.40	1.39	1.38	1.37	1.36	1.35	1.34	1.33
18	1.41	1.50	1.49	1.48	1.46	1.45	1.44	1.43	1.42	1.42	1.40	1.39	1.38	1.37	1.36	1.35	1.34	1.33	1.32
19	1.41	1.49	1.49	1.47	1.46	1.44	1.43	1.42	1.41	1.41	1.40	1.38	1.37	1.36	1.35	1.34	1.33	1.32	1.30
20	1.40	1.49	1.48	1.47	1.45	1.44	1.43	1.42	1.41	1.40	1.39	1.37	1.36	1.35	1.34	1.33	1.32	1.31	1.29
21	1.40	1.48	1.48	1.46	1.44	1.43	1.42	1.41	1.40	1.39	1.38	1.37	1.35	1.34	1.33	1.32	1.31	1.30	1.28
22	1.40	1.48	1.47	1.45	1.44	1.42	1.41	1.40	1.39	1.39	1.37	1.36	1.34	1.33	1.32	1.31	1.30	1.29	1.28
23	1.39	1.47	1.47	1.45	1.43	1.42	1.41	1.40	1.39	1.38	1.37	1.35	1.34	1.33	1.32	1.31	1.30	1.28	1.27
24	1.39	1.47	1.46	1.44	1.43	1.41	1.40	1.39	1.38	1.38	1.36	1.35	1.33	1.32	1.31	1.30	1.29	1.28	1.26
25	1.39	1.47	1.46	1.44	1.42	1.41	1.40	1.39	1.38	1.37	1.36	1.34	1.33	1.32	1.31	1.29	1.28	1.27	1.25
26	1.38	1.46	1.45	1.44	1.42	1.41	1.39	1.38	1.37	1.37	1.35	1.34	1.32	1.31	1.30	1.29	1.28	1.26	1.25
27	1.38	1.46	1.45	1.43	1.42	1.40	1.39	1.38	1.37	1.36	1.35	1.34	1.32	1.31	1.30	1.28	1.27	1.26	1.24
28	1.38	1.46	1.45	1.43	1.41	1.40	1.39	1.38	1.37	1.36	1.34	1.33	1.31	1.30	1.29	1.28	1.27	1.25	1.24
29	1.38	1.45	1.45	1.43	1.41	1.40	1.38	1.37	1.36	1.35	1.34	1.32	1.31	1.30	1.29	1.27	1.26	1.25	1.23
30	1.38	1.45	1.44	1.42	1.41	1.39	1.38	1.37	1.36	1.35	1.34	1.32	1.30	1.29	1.28	1.27	1.26	1.24	1.23
40	1.36	1.44	1.42	1.40	1.39	1.37	1.36	1.35	1.34	1.33	1.31	1.30	1.28	1.26	1.25	1.24	1.22	1.21	1.19
60	1.35	1.42	1.41	1.38	1.37	1.35	1.33	1.32	1.31	1.30	1.29	1.27	1.25	1.24	1.22	1.21	1.19	1.17	1.15
120	1.34	1.40	1.39	1.37	1.35	1.33	1.31	1.30	1.29	1.28	1.26	1.24	1.22	1.21	1.19	1.18	1.16	1.13	1.10
∞	1.32	1.39	1.37	1.35	1.33	1.31	1.29	1.28	1.27	1.25	1.24	1.22	1.19	1.18	1.16	1.14	1.12	1.08	1.00

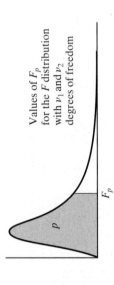

Values of F_p for the F distribution with ν_1 and ν_2 degrees of freedom

TABLE A.6. *(Continued)* Critical Values of F for a Probability Level, p, of 0.900

ν_1 / ν_2	1	2	3	4	5	6	7	8	9	10	12	15	20	24	30	40	60	120	∞
1	39.86	49.50	53.59	55.83	57.24	58.20	58.91	59.44	59.86	60.20	60.71	61.22	61.74	62.00	62.27	62.53	62.79	63.06	63.32
2	8.53	9.00	9.16	9.24	9.29	9.33	9.35	9.37	9.38	9.39	9.41	9.42	9.44	9.45	9.46	9.47	9.47	9.48	9.49
3	5.54	5.46	5.39	5.34	5.31	5.28	5.27	5.25	5.24	5.23	5.22	5.20	5.18	5.18	5.17	5.16	5.15	5.14	5.13
4	4.55	4.32	4.19	4.11	4.05	4.01	3.98	3.95	3.94	3.92	3.90	3.87	3.84	3.83	3.82	3.80	3.79	3.78	3.76
5	4.06	3.78	3.62	3.52	3.45	3.40	3.37	3.34	3.32	3.30	3.27	3.24	3.21	3.19	3.17	3.16	3.14	3.12	3.11
6	3.78	3.46	3.29	3.18	3.11	3.05	3.01	2.98	2.96	2.94	2.90	2.87	2.84	2.82	2.80	2.78	2.76	2.74	2.72
7	3.59	3.26	3.07	2.96	2.88	2.83	2.78	2.75	2.72	2.70	2.67	2.63	2.59	2.58	2.56	2.54	2.51	2.49	2.47
8	3.46	3.11	2.92	2.81	2.73	2.67	2.62	2.59	2.56	2.54	2.50	2.46	2.42	2.40	2.38	2.36	2.34	2.32	2.29
9	3.36	3.01	2.81	2.69	2.61	2.55	2.51	2.47	2.44	2.42	2.38	2.34	2.30	2.28	2.25	2.23	2.21	2.18	2.16
10	3.29	2.92	2.73	2.61	2.52	2.46	2.41	2.38	2.35	2.32	2.28	2.24	2.20	2.18	2.16	2.13	2.11	2.08	2.06
11	3.23	2.86	2.66	2.54	2.45	2.39	2.34	2.30	2.27	2.25	2.21	2.17	2.12	2.10	2.08	2.05	2.03	2.00	1.97
12	3.18	2.81	2.61	2.48	2.39	2.33	2.28	2.24	2.21	2.19	2.15	2.10	2.06	2.04	2.01	1.99	1.96	1.93	1.90
13	3.14	2.76	2.56	2.43	2.35	2.28	2.23	2.20	2.16	2.14	2.10	2.05	2.01	1.98	1.96	1.93	1.90	1.88	1.85
14	3.10	2.73	2.52	2.39	2.31	2.24	2.19	2.15	2.12	2.10	2.05	2.01	1.96	1.94	1.91	1.89	1.86	1.83	1.80
15	3.07	2.70	2.49	2.36	2.27	2.21	2.16	2.12	2.09	2.06	2.02	1.97	1.92	1.90	1.87	1.85	1.82	1.79	1.76
16	3.05	2.67	2.46	2.33	2.24	2.18	2.13	2.09	2.06	2.03	1.99	1.94	1.89	1.87	1.84	1.81	1.78	1.75	1.72
17	3.03	2.64	2.44	2.31	2.22	2.15	2.10	2.06	2.03	2.00	1.96	1.91	1.86	1.84	1.81	1.78	1.75	1.72	1.69
18	3.01	2.62	2.42	2.29	2.20	2.13	2.08	2.04	2.00	1.98	1.93	1.89	1.84	1.81	1.78	1.75	1.72	1.69	1.66
19	2.99	2.61	2.40	2.27	2.18	2.11	2.06	2.02	1.98	1.96	1.91	1.86	1.81	1.79	1.76	1.73	1.70	1.67	1.63
20	2.97	2.59	2.38	2.25	2.16	2.09	2.04	2.00	1.96	1.94	1.89	1.84	1.79	1.77	1.74	1.71	1.68	1.64	1.61
21	2.96	2.57	2.36	2.23	2.14	2.08	2.02	1.98	1.95	1.92	1.87	1.83	1.78	1.75	1.72	1.69	1.66	1.62	1.59
22	2.95	2.56	2.35	2.22	2.13	2.06	2.01	1.97	1.93	1.90	1.86	1.81	1.76	1.73	1.70	1.67	1.64	1.60	1.57
23	2.94	2.55	2.34	2.21	2.11	2.05	1.99	1.95	1.92	1.89	1.84	1.80	1.74	1.72	1.69	1.66	1.62	1.59	1.55
24	2.93	2.54	2.33	2.19	2.10	2.04	1.98	1.94	1.91	1.88	1.83	1.78	1.73	1.70	1.67	1.64	1.61	1.57	1.53
25	2.92	2.53	2.32	2.18	2.09	2.02	1.97	1.93	1.89	1.87	1.82	1.77	1.72	1.69	1.66	1.63	1.59	1.56	1.52
26	2.91	2.52	2.31	2.17	2.08	2.01	1.96	1.92	1.88	1.86	1.81	1.76	1.71	1.68	1.65	1.61	1.58	1.54	1.50
27	2.90	2.51	2.30	2.17	2.07	2.00	1.95	1.91	1.87	1.85	1.80	1.75	1.70	1.67	1.64	1.60	1.57	1.53	1.49
28	2.89	2.50	2.29	2.16	2.06	2.00	1.94	1.90	1.87	1.84	1.79	1.74	1.69	1.66	1.63	1.59	1.56	1.52	1.48
29	2.89	2.50	2.28	2.15	2.06	1.99	1.93	1.89	1.86	1.83	1.78	1.73	1.68	1.65	1.62	1.58	1.55	1.51	1.47
30	2.88	2.49	2.28	2.14	2.05	1.98	1.93	1.88	1.85	1.82	1.77	1.72	1.67	1.64	1.61	1.57	1.54	1.50	1.46
40	2.84	2.44	2.23	2.09	2.00	1.93	1.87	1.83	1.79	1.76	1.71	1.66	1.61	1.57	1.54	1.51	1.47	1.42	1.38
60	2.79	2.39	2.18	2.04	1.95	1.87	1.82	1.77	1.74	1.71	1.66	1.60	1.54	1.51	1.48	1.44	1.40	1.35	1.29
120	2.75	2.35	2.13	1.99	1.90	1.82	1.77	1.72	1.68	1.65	1.60	1.55	1.48	1.45	1.41	1.37	1.32	1.26	1.19
∞	2.71	2.30	2.08	1.95	1.85	1.77	1.72	1.67	1.63	1.60	1.55	1.49	1.42	1.38	1.34	1.30	1.24	1.17	1.00

Values of F_p for the F distribution with ν_1 and ν_2 degrees of freedom

p

F_p

TABLE A.6 (Continued) Critical Values of F for a Probability Level, p, of 0.950

ν_1 \ ν_2	1	2	3	4	5	6	7	8	9	10	12	15	20	24	30	40	60	120	∞
1	161.4	199.5	215.7	224.6	230.2	234.0	236.8	238.9	240.5	241.9	243.9	245.9	248.0	249.1	250.1	251.1	252.2	253.3	254.3
2	18.51	19.00	19.16	19.25	19.30	19.33	19.35	19.37	19.38	19.40	19.41	19.43	19.45	19.45	19.46	19.47	19.48	19.49	19.50
3	10.13	9.55	9.28	9.12	9.01	8.94	8.89	8.85	8.81	8.79	8.74	8.70	8.66	8.64	8.62	8.59	8.57	8.55	8.53
4	7.71	6.94	6.59	6.39	6.26	6.16	6.09	6.04	6.00	5.96	5.91	5.86	5.80	5.77	5.75	5.72	5.69	5.66	5.63
5	6.61	5.79	5.41	5.19	5.05	4.95	4.88	4.82	4.77	4.74	4.68	4.62	4.56	4.53	4.50	4.46	4.43	4.40	4.36
6	5.99	5.14	4.76	4.53	4.39	4.28	4.21	4.15	4.10	4.06	4.00	3.94	3.87	3.84	3.81	3.77	3.74	3.70	3.67
7	5.59	4.74	4.35	4.12	3.97	3.87	3.79	3.73	3.68	3.64	3.57	3.51	3.44	3.41	3.38	3.34	3.30	3.27	3.23
8	5.32	4.46	4.07	3.84	3.69	3.58	3.50	3.44	3.39	3.35	3.28	3.22	3.15	3.12	3.08	3.04	3.01	2.97	2.93
9	5.12	4.26	3.86	3.63	3.48	3.37	3.29	3.23	3.18	3.14	3.07	3.01	2.94	2.90	2.86	2.83	2.79	2.75	2.71
10	4.96	4.10	3.71	3.48	3.33	3.22	3.14	3.07	3.02	2.98	2.91	2.85	2.77	2.74	2.70	2.66	2.62	2.58	2.54
11	4.84	3.98	3.59	3.36	3.20	3.09	3.01	2.95	2.90	2.85	2.79	2.72	2.65	2.61	2.57	2.53	2.49	2.45	2.40
12	4.75	3.89	3.49	3.26	3.11	3.00	2.91	2.85	2.80	2.75	2.69	2.62	2.54	2.51	2.47	2.43	2.38	2.34	2.30
13	4.67	3.81	3.41	3.18	3.03	2.92	2.83	2.77	2.71	2.67	2.60	2.53	2.46	2.42	2.38	2.34	2.30	2.25	2.21
14	4.60	3.74	3.34	3.11	2.96	2.85	2.76	2.70	2.65	2.60	2.53	2.46	2.39	2.35	2.31	2.27	2.22	2.18	2.13
15	4.54	3.68	3.29	3.06	2.90	2.79	2.71	2.64	2.59	2.54	2.48	2.40	2.33	2.29	2.25	2.20	2.16	2.11	2.07
16	4.49	3.63	3.24	3.01	2.85	2.74	2.66	2.59	2.54	2.49	2.42	2.35	2.28	2.24	2.19	2.15	2.11	2.06	2.01
17	4.45	3.59	3.20	2.96	2.81	2.70	2.61	2.55	2.49	2.45	2.38	2.31	2.23	2.19	2.15	2.10	2.06	2.01	1.96
18	4.41	3.55	3.16	2.93	2.77	2.66	2.58	2.51	2.46	2.41	2.34	2.27	2.19	2.15	2.11	2.06	2.02	1.97	1.92
19	4.38	3.52	3.13	2.90	2.74	2.63	2.54	2.48	2.42	2.38	2.31	2.23	2.16	2.11	2.07	2.03	1.98	1.93	1.88
20	4.35	3.49	3.10	2.87	2.71	2.60	2.51	2.45	2.39	2.35	2.28	2.20	2.12	2.08	2.04	1.99	1.95	1.90	1.84
21	4.32	3.47	3.07	2.84	2.68	2.57	2.49	2.42	2.37	2.32	2.25	2.18	2.10	2.05	2.01	1.96	1.92	1.87	1.81
22	4.30	3.44	3.05	2.82	2.66	2.55	2.46	2.40	2.34	2.30	2.23	2.15	2.07	2.03	1.98	1.94	1.89	1.84	1.78
23	4.28	3.42	3.03	2.80	2.64	2.53	2.44	2.37	2.32	2.27	2.20	2.13	2.05	2.01	1.96	1.91	1.86	1.81	1.76
24	4.26	3.40	3.01	2.78	2.62	2.51	2.42	2.36	2.30	2.25	2.18	2.11	2.03	1.98	1.94	1.89	1.84	1.79	1.73
25	4.24	3.39	2.99	2.76	2.60	2.49	2.40	2.34	2.28	2.24	2.16	2.09	2.01	1.96	1.92	1.87	1.82	1.77	1.71
26	4.23	3.37	2.98	2.74	2.59	2.47	2.39	2.32	2.27	2.22	2.15	2.07	1.99	1.95	1.90	1.85	1.80	1.75	1.69
27	4.21	3.35	2.96	2.73	2.57	2.46	2.37	2.31	2.25	2.20	2.13	2.06	1.97	1.93	1.88	1.84	1.79	1.73	1.67
28	4.20	3.34	2.95	2.71	2.56	2.45	2.36	2.29	2.24	2.19	2.12	2.04	1.96	1.91	1.87	1.82	1.77	1.71	1.65
29	4.18	3.33	2.93	2.70	2.55	2.43	2.35	2.28	2.22	2.18	2.10	2.03	1.94	1.90	1.85	1.81	1.75	1.70	1.64
30	4.17	3.32	2.92	2.69	2.53	2.42	2.33	2.27	2.21	2.16	2.09	2.01	1.93	1.89	1.84	1.79	1.74	1.68	1.62
40	4.08	3.23	2.84	2.61	2.45	2.34	2.25	2.18	2.12	2.08	2.00	1.92	1.84	1.79	1.74	1.69	1.64	1.58	1.51
60	4.00	3.15	2.76	2.53	2.37	2.25	2.17	2.10	2.04	1.99	1.92	1.84	1.75	1.70	1.65	1.59	1.53	1.47	1.39
120	3.92	3.07	2.68	2.45	2.29	2.17	2.09	2.02	1.96	1.91	1.83	1.75	1.66	1.61	1.55	1.50	1.43	1.35	1.25
∞	3.84	3.00	2.60	2.37	2.21	2.10	2.01	1.94	1.88	1.83	1.75	1.67	1.57	1.52	1.46	1.39	1.32	1.22	1.00

Values of F_p
for the F distribution
with ν_1 and ν_2
degrees of freedom

TABLE A.6 (*Continued*) Critical Values of F for a Probability Level, p, of 0.950

ν_1 / ν_2	1	2	3	4	5	6	7	8	9	10	12	15	20	24	30	40	60	120	∞
1	4052	4999	5403	5625	5764	5859	5928	5982	6022	6056	6106	6157	6209	6235	6261	6287	6313	6339	6366
2	98.50	99.00	99.17	99.25	99.30	99.33	99.36	99.37	99.39	99.40	99.42	99.43	99.45	99.46	99.47	99.47	99.48	99.49	99.50
3	34.12	30.82	29.46	28.71	28.24	27.91	27.67	27.49	27.35	27.23	27.05	26.87	26.69	26.60	26.50	26.41	26.32	26.22	26.13
4	21.20	18.00	16.69	15.98	15.52	15.21	14.98	14.80	14.66	14.55	14.37	14.20	14.02	13.93	13.84	13.75	13.65	13.56	13.46
5	16.26	13.27	12.06	11.39	10.97	10.67	10.46	10.29	10.16	10.05	9.89	9.72	9.55	9.47	9.38	9.29	9.20	9.11	9.02
6	13.75	10.92	9.78	9.15	8.75	8.47	8.26	8.10	7.98	7.87	7.72	7.56	7.40	7.31	7.23	7.14	7.06	6.97	6.88
7	12.25	9.55	8.45	7.85	7.46	7.19	6.99	6.84	6.72	6.62	6.47	6.31	6.16	6.07	5.99	5.91	5.82	5.74	5.65
8	11.26	8.65	7.59	7.01	6.63	6.37	6.18	6.03	5.91	5.81	5.67	5.52	5.36	5.28	5.20	5.12	5.03	4.95	4.86
9	10.56	8.02	6.99	6.42	6.06	5.80	5.61	5.47	5.35	5.26	5.11	4.96	4.81	4.73	4.65	4.57	4.48	4.40	4.31
10	10.04	7.56	6.55	5.99	5.64	5.39	5.20	5.06	4.94	4.85	4.71	4.56	4.41	4.33	4.25	4.17	4.08	4.00	3.91
11	9.65	7.21	6.22	5.67	5.32	5.07	4.89	4.74	4.63	4.54	4.40	4.25	4.10	4.02	3.94	3.86	3.78	3.69	3.60
12	9.33	6.93	5.95	5.41	5.06	4.82	4.64	4.50	4.39	4.30	4.16	4.01	3.86	3.78	3.70	3.62	3.54	3.45	3.36
13	9.07	6.70	5.74	5.21	4.86	4.62	4.44	4.30	4.19	4.10	3.96	3.82	3.66	3.59	3.51	3.43	3.34	3.25	3.17
14	8.86	6.51	5.56	5.04	4.69	4.46	4.28	4.14	4.03	3.94	3.80	3.66	3.51	3.43	3.35	3.27	3.18	3.09	3.00
15	8.68	6.36	5.42	4.89	4.56	4.32	4.14	4.00	3.89	3.80	3.67	3.52	3.37	3.29	3.21	3.13	3.05	2.96	2.87
16	8.53	6.23	5.29	4.77	4.44	4.20	4.03	3.89	3.78	3.69	3.55	3.41	3.26	3.18	3.10	3.02	2.93	2.84	2.75
17	8.40	6.11	5.18	4.67	4.34	4.10	3.93	3.79	3.68	3.59	3.46	3.31	3.16	3.08	3.00	2.92	2.83	2.75	2.65
18	8.29	6.01	5.09	4.58	4.25	4.01	3.84	3.71	3.60	3.51	3.37	3.23	3.08	3.00	2.92	2.84	2.75	2.66	2.57
19	8.18	5.93	5.01	4.50	4.17	3.94	3.77	3.63	3.52	3.43	3.30	3.15	3.00	2.92	2.84	2.76	2.67	2.58	2.49
20	8.10	5.85	4.94	4.43	4.10	3.87	3.70	3.56	3.46	3.37	3.23	3.09	2.94	2.86	2.78	2.69	2.61	2.52	2.42
21	8.02	5.78	4.87	4.37	4.04	3.81	3.64	3.51	3.40	3.31	3.17	3.03	2.88	2.80	2.72	2.64	2.55	2.46	2.36
22	7.95	5.72	4.82	4.31	3.99	3.76	3.59	3.45	3.35	3.26	3.12	2.98	2.83	2.75	2.67	2.58	2.50	2.40	2.31
23	7.88	5.66	4.76	4.26	3.94	3.71	3.54	3.41	3.30	3.21	3.07	2.93	2.78	2.70	2.62	2.54	2.45	2.35	2.26
24	7.82	5.61	4.72	4.22	3.90	3.67	3.50	3.36	3.26	3.17	3.03	2.89	2.74	2.66	2.58	2.49	2.40	2.31	2.21
25	7.77	5.57	4.68	4.18	3.85	3.63	3.46	3.32	3.22	3.13	2.99	2.85	2.70	2.62	2.54	2.45	2.36	2.27	2.17
26	7.72	5.53	4.64	4.14	3.82	3.59	3.42	3.29	3.18	3.09	2.96	2.81	2.66	2.58	2.50	2.42	2.33	2.23	2.13
27	7.68	5.49	4.60	4.11	3.78	3.56	3.39	3.26	3.15	3.06	2.93	2.78	2.63	2.55	2.47	2.38	2.29	2.20	2.10
28	7.64	5.45	4.57	4.07	3.75	3.53	3.36	3.23	3.12	3.03	2.90	2.75	2.60	2.52	2.44	2.35	2.26	2.17	2.06
29	7.60	5.42	4.54	4.04	3.73	3.50	3.33	3.20	3.09	3.00	2.87	2.73	2.57	2.49	2.41	2.33	2.23	2.14	2.03
30	7.56	5.39	4.51	4.02	3.70	3.47	3.30	3.17	3.07	2.98	2.84	2.70	2.55	2.47	2.39	2.30	2.21	2.11	2.01
40	7.31	5.18	4.31	3.83	3.51	3.29	3.12	2.99	2.89	2.80	2.66	2.52	2.37	2.29	2.20	2.11	2.02	1.92	1.80
60	7.08	4.98	4.13	3.65	3.34	3.12	2.95	2.82	2.72	2.63	2.50	2.35	2.20	2.12	2.03	1.94	1.84	1.73	1.60
120	6.85	4.79	3.95	3.48	3.17	2.96	2.79	2.66	2.56	2.47	2.34	2.19	2.03	1.95	1.86	1.76	1.66	1.53	1.38
∞	6.63	4.61	3.78	3.32	3.02	2.80	2.64	2.51	2.41	2.32	2.18	2.04	1.88	1.79	1.70	1.59	1.47	1.32	1.00

Values of F_p
for the F distribution
with ν_1 and ν_2
degrees of freedom

p

F_p

| TABLE A.6 | (Continued) Critical Values of F for a Probability Level, p, of 0.999 |

ν_1 / ν_2	1	2	3	4	5	6	7	8	9	10	12	15	20	24	30	40	60	120	∞
1	4053*	5000*	5404*	5625*	5764*	5859*	5929*	5981*	6023*	6056*	6107*	6158*	6209*	6235*	6261*	6287*	6313*	6340*	6366*
2	998.5	999.0	999.2	999.2	999.3	999.3	999.4	999.4	999.4	999.4	999.4	999.4	999.4	999.5	999.5	999.5	999.5	999.5	999.5
3	167.0	148.5	141.1	137.1	134.6	132.8	131.6	130.6	129.9	129.2	128.3	127.4	126.4	125.9	125.4	125.0	124.5	124.0	123.5
4	74.14	61.25	56.18	53.44	51.71	50.53	49.66	49.00	48.47	48.05	47.41	46.76	46.10	45.77	45.43	45.09	44.75	44.40	44.05
5	47.18	37.12	33.20	31.09	29.75	28.84	28.16	27.64	27.24	26.92	26.42	25.91	25.39	25.14	24.87	24.60	24.33	24.06	23.79
6	35.51	27.00	23.70	21.92	20.81	20.03	19.46	19.03	18.69	18.41	17.99	17.56	17.12	16.89	16.67	16.44	16.21	15.99	15.75
7	29.25	21.69	18.77	17.19	16.21	15.52	15.02	14.63	14.33	14.08	13.71	13.32	12.93	12.73	12.53	12.33	12.12	11.91	11.70
8	25.42	18.49	15.83	14.39	13.49	12.86	12.40	12.04	11.77	11.54	11.19	10.84	10.48	10.30	10.11	9.92	9.73	9.53	9.33
9	22.86	16.39	13.90	12.56	11.71	11.13	10.70	10.37	10.11	9.89	9.57	9.24	8.90	8.72	8.55	8.37	8.19	8.00	7.81
10	21.04	14.91	12.55	11.28	10.48	9.92	9.52	9.20	8.96	8.75	8.45	8.13	7.80	7.64	7.47	7.30	7.12	6.94	6.76
11	19.69	13.81	11.56	10.35	9.58	9.05	8.66	8.35	8.12	7.92	7.63	7.32	7.01	6.85	6.68	6.52	6.35	6.17	6.00
12	18.64	12.97	10.80	9.63	8.89	8.38	8.00	7.71	7.48	7.29	7.00	6.71	6.40	6.25	6.09	5.93	5.76	5.59	5.42
13	17.81	12.31	10.21	9.07	8.35	7.86	7.49	7.21	6.98	6.80	6.52	6.23	5.93	5.78	5.63	5.47	5.30	5.14	4.97
14	17.14	11.78	9.73	8.62	7.92	7.43	7.08	6.80	6.58	6.40	6.13	5.85	5.56	5.41	5.25	5.10	4.94	4.77	4.60
15	16.59	11.34	9.34	8.25	7.57	7.09	6.74	6.47	6.26	6.08	5.81	5.54	5.25	5.10	4.95	4.80	4.64	4.47	4.31
16	16.12	10.97	9.00	7.94	7.27	6.81	6.46	6.19	5.98	5.81	5.55	5.27	4.99	4.85	4.70	4.54	4.39	4.23	4.06
17	15.72	10.66	8.73	7.68	7.02	6.56	6.22	5.96	5.75	5.58	5.32	5.05	4.78	4.63	4.48	4.33	4.18	4.02	3.85
18	15.38	10.39	8.49	7.46	6.81	6.35	6.02	5.76	5.56	5.39	5.13	4.87	4.59	4.45	4.30	4.15	4.00	3.84	3.67
19	15.08	10.16	8.28	7.26	6.62	6.18	5.85	5.59	5.39	5.22	4.97	4.70	4.43	4.29	4.14	3.99	3.84	3.68	3.51
20	14.82	9.95	8.10	7.10	6.46	6.02	5.69	5.44	5.24	5.08	4.82	4.56	4.29	4.15	4.00	3.86	3.70	3.54	3.38
21	14.59	9.77	7.94	6.95	6.32	5.88	5.56	5.31	5.11	4.95	4.70	4.44	4.17	4.03	3.88	3.74	3.58	3.42	3.26
22	14.38	9.61	7.80	6.81	6.19	5.76	5.44	5.19	4.99	4.83	4.58	4.33	4.06	3.92	3.78	3.63	3.48	3.32	3.15
23	14.19	9.47	7.67	6.69	6.08	5.65	5.33	5.09	4.89	4.73	4.48	4.23	3.96	3.82	3.68	3.53	3.38	3.22	3.05
24	14.03	9.34	7.55	6.59	5.98	5.55	5.23	4.99	4.80	4.64	4.39	4.14	3.87	3.74	3.59	3.45	3.29	3.14	2.97
25	13.88	9.22	7.45	6.49	5.88	5.46	5.15	4.91	4.71	4.56	4.31	4.06	3.79	3.66	3.52	3.37	3.22	3.06	2.89
26	13.74	9.12	7.36	6.41	5.80	5.38	5.07	4.83	4.64	4.48	4.24	3.99	3.72	3.59	3.44	3.30	3.15	2.99	2.82
27	13.61	9.02	7.27	6.33	5.73	5.31	5.00	4.76	4.57	4.41	4.17	3.92	3.66	3.52	3.38	3.23	3.08	2.92	2.75
28	13.50	8.93	7.19	6.25	5.66	5.24	4.93	4.69	4.50	4.35	4.11	3.86	3.60	3.46	3.32	3.18	3.02	2.86	2.69
29	13.39	8.85	7.12	6.19	5.59	5.18	4.87	4.64	4.45	4.29	4.05	3.80	3.54	3.41	3.27	3.12	2.97	2.81	2.64
30	13.29	8.77	7.05	6.12	5.53	5.12	4.82	4.58	4.39	4.24	4.00	3.75	3.49	3.36	3.22	3.07	2.92	2.76	2.59
40	12.61	8.25	6.60	5.70	5.13	4.73	4.44	4.21	4.02	3.87	3.64	3.40	3.15	3.01	2.87	2.73	2.57	2.41	2.23
60	11.97	7.76	6.17	5.31	4.76	4.37	4.09	3.87	3.69	3.54	3.32	3.08	2.83	2.69	2.55	2.41	2.25	2.08	1.89
120	11.38	7.32	5.79	4.95	4.42	4.04	3.77	3.55	3.38	3.24	3.02	2.78	2.53	2.40	2.26	2.11	1.95	1.76	1.54
∞	10.83	6.91	5.42	4.62	4.10	3.74	3.47	3.27	3.10	2.96	2.74	2.51	2.27	2.13	1.99	1.84	1.66	1.45	1.00

*Multiply entries by 100.

925

Index

Unblocked Comparison of Two Averages

$100(1 - \alpha)\%$ Confidence Interval for true mean difference $\mu_B - \mu_A$ (variances unknown):

$(\bar{y}_B - \bar{y}_A) \pm t_{\nu,1-\alpha/2} s_{\bar{y}_B - \bar{y}_A}$, where

$$s_{\bar{y}_B - \bar{y}_A}^2 = s_p^2 \left(\frac{1}{n_B} + \frac{1}{n_A} \right) \text{ and } s_p^2 = \frac{\nu_A s_{y_A}^2 + \nu_B s_{y_B}^2}{\nu_A + \nu_B}.$$

Blocked Comparison of Two Averages

$100(1 - \alpha)\%$ Confidence Interval for true mean difference $\mu_B - \mu_A$:

$\bar{d} \pm t_{\nu,1-\alpha/2} s_{\bar{d}}$, where

$d_i = y_{Bi} - y_{Ai}$

$$\bar{d} = \frac{\sum_{i=1}^{n} d_i}{n}, \quad s_d^2 = \frac{\sum_{i=1}^{n} (d_i - \bar{d})^2}{n - 1}, \quad \text{and } s_{\bar{d}}^2 = \frac{s_d^2}{n}$$

Residual Checking

- Normal probability plot of the residuals,
- A plot of the residuals versus time/run order of the trials,
- A plot of the residuals versus the predicted response,
- Plots of the residuals versus the independent variables of the experiment (e.g., treatment and block), and,
- A plot of the residuals versus any other variable in the experiment.

ANOVA for k-Averages Design

Source	Sum of Squares	DoF	Mean Square	F_{calc}
Mean	$N\bar{y}^2$	1	$N\bar{y}^2$	$\dfrac{MS(Mean)}{MS(Within)}$
Between Treatments	$\sum_{i=1}^{k} n_i (\bar{y}_i - \bar{\bar{y}})^2$	$k-1$	$\dfrac{\sum_{i=1}^{k} n_i (\bar{y}_i - \bar{\bar{y}})^2}{k-1}$	$\dfrac{MS(Between)}{MS(Within)}$
Within Treatment	$\sum_{i=1}^{k} \sum_{j=1}^{n_i} (y_{ij} - \bar{y}_i)^2$	$N - k$	$\dfrac{\sum_{i=1}^{k} \sum_{j=1}^{n_i} (y_{ij} - \bar{y}_i)^2}{N - k}$	
Total	$\sum_{i=1}^{k} \sum_{j=1}^{n_i} y_{ij}^2$	N		

ANOVA for Randomized Block Design

Source of Variation	Sum of Squares	Degrees of Freedom	Mean Square	F_{calc}
Mean	$N\bar{\bar{\bar{y}}}^2$	1	$N\bar{\bar{\bar{y}}}^2$	$\dfrac{MS(Mean)}{MS(PE)}$
Treatments – (T)	$\sum_{j=1}^{n_T} n_r n_B (\bar{\bar{y}}_j - \bar{\bar{\bar{y}}})^2$	$n_T - 1$	$\dfrac{\sum_{j=1}^{n_T} n_r n_B (\bar{\bar{y}}_j - \bar{\bar{\bar{y}}})^2}{n_T - 1}$	$\dfrac{MS(T)}{MS(PE)}$
Blocks – (B)	$\sum_{i=1}^{n_B} n_r n_T (\bar{\bar{y}}_i - \bar{\bar{\bar{y}}})^2$	$n_B - 1$	$\dfrac{\sum_{i=1}^{n_B} n_r n_T (\bar{\bar{y}}_i - \bar{\bar{\bar{y}}})^2}{n_B - 1}$	$\dfrac{MS(B)}{MS(PE)}$
Interaction – (I)	$\sum_{i=1}^{n_B} \sum_{j=1}^{n_T} n_r (\bar{\bar{\bar{y}}} - \bar{\bar{y}}_i - \bar{\bar{y}}_j + \bar{y}_{ij})^2$	$(n_B - 1)(n_T - 1)$	$\dfrac{\sum_{i=1}^{n_B} \sum_{j=1}^{n_T} n_r (\bar{\bar{\bar{y}}} - \bar{\bar{y}}_i - \bar{\bar{y}}_j + \bar{y}_{ij})^2}{(n_B - 1)(n_T - 1)}$	$\dfrac{MS(I)}{MS(PE)}$
Pure Error - (PE)	$\sum_{i=1}^{n_B} \sum_{j=1}^{n_T} \sum_{k=1}^{n_r} (y_{ijk} - \bar{y}_{ij})^2$	$n_B n_T (n_r - 1)$	$\dfrac{\sum_{i=1}^{n_B} \sum_{j=1}^{n_T} \sum_{k=1}^{n_r} (y_{ijk} - \bar{y}_{ij})^2}{n_B n_T (n_r - 1)}$	
Total	$\sum_{i=1}^{n_B} \sum_{j=1}^{n_T} \sum_{k=1}^{n_r} y_{ijk}^2$	$n_B n_T n_r$		